International Conference on Innovative Research (ICIR Euroinvent 2016)

IOP Conference Series: Materials Science and Engineering
Volume 133

Iasi, Romania
19 – 20 May 2016

Editors:

Andrei Victor Sandu
Mohd Mustafa al Bakri Abdullah
Petrica Vizureanu
Shayfull Zamree Abd Rahim
Ion Sandu

ISBN: 978-1-5108-2614-4
ISSN: 1757-8981

Printed from e-media with permission by:

Curran Associates, Inc.
57 Morehouse Lane
Red Hook, NY 12571

Some format issues inherent in the e-media version may also appear in this print version.

Printed by Curran Associates, Inc. (2016)

For permission requests, please contact the Institute of Physics
at the address below.

Institute of Physics
Dirac House, Temple Back
Bristol BS1 6BE UK

Phone: 44 1 17 929 7481
Fax: 44 1 17 920 0979

techtracking@iop.org

Additional copies of this publication are available from:

Curran Associates, Inc.
57 Morehouse Lane
Red Hook, NY 12571 USA
Phone: 845-758-0400
Fax: 845-758-2633
Email: curran@proceedings.com
Web: www.proceedings.com

Table of contents

Volume 133

International Conference on Innovative Research - ICIR Euroinvent 2016
19–20 May 2016, Iasi, Romania

Accepted papers received: 24 May 2016
Published online: 28 June 2016

Preface

Papers

Synthesis and Characterization of Materials

Procedures and Technologies for Materials Engineering

Materials & Life Science

IOP Conf. Series: Materials Science and Engineering **133** (2016) 011001 doi:10.1088/1757-899X/133/1/011001

Editorial

The volume contains selected peer-reviewed papers from the 2016 International Conference on Innovative Research, which was held in Iași, România from 19 to 20th of May 2016.

The ICIR Conference is organized under the auspices of EUROINVENT. Euroinvent is a joint event promoting creativity in a European context, by displaying the contributions of consecrated schools from higher education and academic research and also of individual inventors and researchers.

The EUROINVENT International Conference on Innovative Research (ICIR) brings together leading researchers, engineers and scientists who will present actual research results in the field of Materials Science and Engineering.

The conference aims to provide a high level international forum for researchers, engineers and scientists to present their new advances and research results in the field of materials science and engineering.

The volume covers all the aspects of materials science, from synthesis and characterization of materials to procedures and technologies for materials engineering, as well as materials application and their involvement in the life sciences.

All the papers have been reviewed by two expert referees in their relevant topic disciplines. The papers selected for the volume depended on their quality and relevancy to the conference.

The editors hope that this volume will provide the reader a broad overview of the latest advances in the field of materials science and engineering, and that they will be a valuable references source for further research.

The editors would like to express their sincere appreciations and thanks to all the committee members of the ICIR 2016 for their tremendous efforts. Thanks also to IOP Conference Series for producing this volume.

Finally, the editors would like to thank all the authors for their contribution to this valuable volume.

Andrei Victor SANDU

Mohd Mustafa al Bakri ABDULLAH

Petrică VIZUREANU

Shayfull Zamree Abd RAHIM

Ion SANDU

May, 2016

IOP Conf. Series: Materials Science and Engineering **133** (2016) 011001 doi:10.1088/1757-899X/133/1/011001

ICIR 2016 EUROINVENT
International Conference on Innovative Research

Organized by:

- **Romanian Inventors Forum**
- **Faculty of Materials Science and Engineering, The "Gheorghe Asachi" Technical University of Iasi, Romania**
- **Centre of Excellence Geopolymer and Green Technology CEGeoGTech), Universiti Malaysia Perlis (UniMAP)**
- **ARHEOINVEST Platform, Alexandru Ioan Cuza University of Iasi**
- **Malaysian Research & Innovation Society (MyRIS)**

With support of:

- **Ubudiyah University of Indonesia**
- **National Institute for Research and Development URBAN INCERC**
- **International Federation of Inventors' Associations - IFIA**
- **World Invention Intellectual Property Associations**

Scientific Advisory Board

Prof.Dr. Ion SANDU	*Alexandru Ioan Cuza University of Iasi, Romania*
Prof.Dr. Kamarudin HUSSIN	*Universiti of Malaysia Perlis (UniMAP)*
Prof.Dr.Eng. Dan CASCAVAL	*The "Gheorghe Asachi" Technical University of Iasi, Romania*
Prof.Dr. Christian CODDET	*Universite de Technologie Belfort-Montbeliard, France*
Prof.Dr.Eng. Neculai Eugen SEGHEDIN	*The "Gheorghe Asachi" Technical University of Iasi, Romania*
Prof.Dr.Eng. Catalin POPA	*Technical University of Cluj-Napoca, Romania*
Prof.Dr. Antonio CANDEIAS	*University of Evora, Portugal*
Prof.Dr. Vasile MEITA	*NIRD Urban Incerc*
Prof.Dr. Che Mohd Ruzaidi GHAZALI	*Universiti of Malaysia Perlis (UniMAP)*
Prof.Dr.Eng. Cornel SAMOILA	*Transilvania University of Brasov, Romania*
Prof.Dr. Mohd Mustafa Al Bakri ABDULLAH	*Malaysian Research & Innovation Society*
Prof.Dr.Eng. Cristian PREDESCU	*Politehnica University Bucharest, Romania*
Prof.Dr. Gabi DROCHIOIU	*Alexandru Ioan Cuza University of Iasi, Romania*

IOP Conf. Series: Materials Science and Engineering **133** (2016) 011001 doi:10.1088/1757-899X/133/1/011001

Conference Organization

Chairman

Prof.Dr.Eng. Petrica VIZUREANU
The "Gheorghe Asachi" Technical University of Iasi, Romania

Event Coordinator

Assist.Prof.Dr.Eng. Andrei Victor SANDU
Romanian Inventors Forum &
The "Gheorghe Asachi" Technical University of Iasi, Romania

Technical and Organizing Committee

Assoc.Prof.Dr.Eng. Dragos Cristian ACHITEI	*The "Gheorghe Asachi" Technical University of Iasi, Romania*
PhD student Madalina Simona BALTATU	*The "Gheorghe Asachi" Technical University of Iasi, Romania*
PhD Bogdan ISTRATE	*The "Gheorghe Asachi" Technical University of Iasi, Romania*
Assist.Prof.Dr.Eng. Mirabela Georgiana MINCIUNA	*The "Gheorghe Asachi" Technical University of Iasi, Romania*
PhD student Marius MUNTEANU	*Alexandru Ioan Cuza University of Iasi, Romania*
PhD student Vasile PELIN	*Alexandru Ioan Cuza University of Iasi, Romania*
Assoc.Prof.Dr.Eng. Manuela Cristina PERJU	*The "Gheorghe Asachi" Technical University of Iasi, Romania*
PhD student Daniel POTOLINCA	*Alexandru Ioan Cuza University of Iasi, Romania*
Assoc.Prof.Dr.Eng. Ioan Gabriel SANDU	*The "Gheorghe Asachi" Technical University of Iasi, Romania*
Assist.Prof.Eng. Catalin Andrei TUGUI	*The "Gheorghe Asachi" Technical University of Iasi, Romania*
Dr. Viorica VASILACHE	*Alexandru Ioan Cuza University of Iasi, Romania*

IOP Conf. Series: Materials Science and Engineering **133** (2016) 011002 doi:10.1088/1757-899X/133/1/011002

TOC ICIR EUROINVENT 2016

IOP Conf. Series: Materials Science and Engineering **133** (2016) 011002 doi:10.1088/1757-899X/133/1/011002

T2	20	Procedures and Technologies for Materials Engineering	Aizenshtein	Wetting and spontaneous infiltration: the case study of TaC/(Au, Al and Cu) compared to TiC/Cu
	21	Procedures and Technologies for Materials Engineering	Nuruddin	Geopolymer concrete for structural use: Recent findings and limitations
	22	Procedures and Technologies for Materials Engineering	Lute	Mixed Consolidation Solution for a Reinforced Concrete Structure
	23	Procedures and Technologies for Materials Engineering	Ahmad	Properties and Microstructural Characteristic of Kaolin Geopolymer Ceramics with Addition of Ultra High Molecular Weight Polyethylene
	24	Procedures and Technologies for Materials Engineering	Efzan	Effect of Fluxes on 60Sn-40Bi Solder Alloy on Copper Substrate
	25	Procedures and Technologies for Materials Engineering	Perju	ESD morphology deposition with WZr8 electrode on austenitic stainless steel support
	26	Procedures and Technologies for Materials Engineering	Zailan	Self-cleaning geopolymer concrete - A review
	27	Procedures and Technologies for Materials Engineering	Aisyah	Influence of different treatment condition on biopolymer yield production for coagulation-flocculation process
	28	Procedures and Technologies for Materials Engineering	Saad	Self-Cleaning Technology in Fabric: A Review
	29	Procedures and Technologies for Materials Engineering	Duceac	Morphological and physical - chemical issues of metal nanostructures used in medical field
	30	Procedures and Technologies for Materials Engineering	Domopoulou	Development of tailored ceramic microstructures using recycled marble processing residue as pore-former
	31	Procedures and Technologies for Materials Engineering	Baitanu	The aluminizing in powder technology of AISI 304 steel
T3	32	Materials Application	Tyagunov	Special features of structural changes of amorphizing nickel alloys in liquid and heterogeneous state
	33	Materials Application	Lute	Classic and New Materials Used for Structural Rehabilitation. Case Study
	34	Materials Application	Kadir	Physical and Mechanical Properties of Bodymill Sludge (BS) Incorporated Into Fired Clay Brick
	35	Materials Application	Buica	Leakage current measurement of protective equipment insulating materials used in electrical installations
	36	Materials Application	Kadir	Investigation on Leaching Behaviour of Fly Ash and Bottom Ash Replacement in Self-Compacting Concrete
	37	Materials Application	Cotfas	Investigation of the photovoltaic cell/ thermoelectric element hybrid system performance
	38	Materials Application	Ion	Effects of the restoration mortar on chalk stone buildings
	39	Materials Application	Kadir	Properties and Leachability of Self-Compacting Concrete Incorporated with Fly Ash and Bottom Ash

	40	Materials Application	Tudose	ZnO for photocatalytic air purification applications
	41	Materials Application	Kadir	The effect on slurry water as a fresh water replacement in concrete properties
	42	Materials Application	Wazien	Strength and Density of Geopolymer Mortar Cured at Ambient Temperature for Use as Repair Material
	43	Materials Application	Buzaianu	Some Analysis of Major Impact of Geothermal Fluid Components in Power Plant Equipment
	44	Materials Application	Hashim	Effect of Geopolymer filler in Glass Reinforced Epoxy (GRE) Pipe for Piping Application: Mechanical Properties
	45	Materials Application	Faris	Performance of Straight Steel Fibres Reinforced Alkali Activated Concrete
	46	Materials Application	Efzan	Microstructure and Mechanical Properties of Fly Ash Particulate Reinforced in LM6 for Energy Enhancement in Automotive Applications
	47	Materials Application	Csaki	Morphological characterisation of complex powder used for protective coatings for geothermal plant components
	48	Materials Application	Efzan	Mechanical and Physical Properties of In-Zn-Ga Lead-Free Solder Alloy for Low Energy Consumption
	49	Materials Application	Campian	Methods for elimination of dampness in Building walls
	50	Materials Application	Mahmad	Applications of Radio Frequency Identification (RFID) in Mining Industries
T4	51	Materials & Life Science	Suica-Bunghez	Antioxidant activity and phytochemical compounds of snake fruit (Salacca Zalacca)
	52	Materials & Life Science	Nuruddin	Evolution of geopolymer binders: a review
	53	Materials & Life Science	Zainab	Green bio-oil extraction for oil crops
	54	Materials & Life Science	Niculescu	Low cost method for manufacturing a data acquisition system with USB connectivity
	55	Materials & Life Science	Chiriac	Antimicrobial property of zinc based nanoparticles
	56	Materials & Life Science	Stoleriu	The effect of erosion and abrasion on surface properties of composite resin
	57	Materials & Life Science	Niculescu	Study of switching electric circuits with DC hybrid breaker, one stage
	58	Materials & Life Science	Hong	Sensory evaluation based fuzzy AHP approach for material selection in customized garment design and development process
	59	Materials & Life Science	Risteiu	Quality Engineering Tools Focused on Designing Remote Temperature Measurements for Inaccessible Locations by Using Light Components Parameterization of the Heated Materials
	60	Materials & Life Science	Munteanu	The implications of free 3D scanning in the conservation state assessment of old wood painted icon
	61	Materials & Life Science	Pelin	Colour change evaluation on UV radiation exposure for Păun-Repedea calcareous geomaterial
	62	Materials & Life Science	Azzeri	A concept design of three rudders-shaped like body in columns for low-drag USV
	63	Materials & Life Science	Soporan	Methodology for appreciation the manufacturing castings from perspective of circular economy

IOP Conf. Series: Materials Science and Engineering **133** (2016) 011003 doi:10.1088/1757-899X/133/1/011003

Peer review statement

All papers published in this volume of *IOP Conference Series: Materials Science and Engineering* have been peer reviewed through processes administered by the proceedings Editors. Reviews were conducted by expert referees to the professional and scientific standards expected of a proceedings journal published by IOP Publishing.

IOP Conf. Series: Materials Science and Engineering **133** (2016) 012001 doi:10.1088/1757-899X/133/1/012001

Production feature of soft magnetic amorphous alloys

A G Tyagunov[1], E E Baryshev[1] and K Yu Shmakova[1]

[1] Ural Federal University, Mira,19, Yekaterinburg, Russia

E-mail: e.e.barysev@urfu.ru

Abstract. Methods for making nanocrystalline alloys have been discussed. Temperature dependences of the surface tension (σ), electric resistivity (ρ), magnetic susceptibility (χ) and kinematic viscosity (ν) have been obtained. Comparison of the properties of amorphous ribbons obtained by the pilot and serial technologies has been conducted. Science-based technology of multi-component alloy smelting makes it possible to prepare equilibrium smelt, the structure of which has a significant effect on the properties of the amorphous ribbon before spinning and kinetics of its crystallization has been offered.

Introduction. Soft magnetic iron-nickel, iron-cobalt, iron-chromium and other high alloys have special physical and physicochemical characteristics. They are determined not only by an exact chemical composition, but also by factors of their production. Normal magnetic properties corresponding to the normative and technical documentation can be achieved by smelting in open furnaces. To obtain alloys with better and high magnetic properties vacuum units and smelting processes are used [1-4].

Distinctive features of soft magnetic alloys are their high magnetic permeability, low coercive force, ability to be treated by plastic deformation, and higher specific electrical resistivity.

Such indicators are achieved by adding nickel, cobalt, chromium, manganese, molybdenum, silicon and other elements to the alloys composition. Along with smelting processes hard metal heat treatment is very important [5]. However, structural features of steels and alloys which above all are determined by the selectivity of hardening phases and accompanied by the segregation of elements have a considerable influence on the properties of metal products.

Main part. The production of nanocrystalline materials with particle size from 1 to 10 nm and soft magnetic amorphous alloys improved special characteristics of products considerably.

The easiest method to obtain nanocrystalline powders is to evaporate material at a controlled temperature either under inert gas with subsequent condensation of steam or in the reaction zone with the formation of spherical particles, or on surfaces, where particles have faceting.

A plasma-chemical method of obtaining nanopowders is the most common chemical methods among others. The reaction is carried out in non-equilibrium conditions providing for a high rate of nucleus formation of the new phase and a low speed of their growth.

A nanoparticle deposition in colloidal solutions involves the consistent implementation of the following operations: a synthesis of starting reagents of the solution and a reaction termination at a particular point in time, after which the disperse system changes from a liquid colloidal state into a dispersed solid one. The method has high selectivity and allows producing particles with a very narrow size distribution.

Quite an effective way of obtaining nanocrystalline powders is mechanical abrasion. During the process of the metal synthesis grinding and plastic deformation of substances take place, the mass transfer is accelerated, as well as mixing of mixture components is performed at an atomic level, and chemical interaction of solid reagents is activated.

The way of obtaining nanopowders by the controlled crystallization of the quenched amorphous state is structurally simple and the most common. The amorphous structure is formed during fast (speed $\geq 10^6$ deg/sec.) cooling of the smelt. A jet of liquid metal with a certain speed is sent to the surface of a fast-rotating cylinder (or a disk), made of a material with high thermal conductivity (such as copper).

The structure of nanocrystalline alloy is a two phase system, in which one of the phases is nanocrystals, and the other one is a remaining amorphous matrix. Nanoalloy properties depend on composition, size and number of nanocrystals, as well as their relationship to the amorphous phase. The main advantage of such materials over crystalline ones is a favorable combination of low coercive force H_c and high relative magnetic permeability μ_{max}. Besides such main components as iron, nickel, cobalt nanoalloys consist of amorphizing elements such as boron, silicon, phosphorus, and others, which lower amorphization temperature t_a to the balance $t_a/t_L = 0.45$, and alloying components involved in the formation of special physical properties and performance characteristics of the finished product.

It is known that equilibrium metallic liquid is an analogue to a metastable amorphous state. Results of studies conducted by different authors (see, in particular, [1-4, 6,7] prove that there are nano-sized clusters or local microassotiations in the structure of these materials, which shall have the same meaning for the theory of amorphous and liquid states as an elementary cell for the theory of crystals [8]. According to the authors [9-11] a local order means that a geometric shape of a cluster is close to the FCC- and HCP- lattices and stable in time, i.e. the time of their life significantly exceeds the period of atom oscillations.

Molten metals and alloys are condensed substance, special state of which is caused by a lack of clear cluster boundaries, distortion of translational symmetry and a dynamically excited state of atoms located in an outer nuclear layer of a cluster, energetically strong interaction of atoms within the cluster, which helps to increase the electron density at the Fermi level [1].

The proportion of foreign atoms in the cluster, evaluated using experimental data, is about 15%. The role of such atoms for the formation of the properties of materials becomes even more significant, given that they have a different electronic configuration due to the change of valence and a decrease of electron density. This leads to weakening of the interatomic interaction with other atoms of the cluster and a significant increase in an amplitude of oscillations.

Another important aspect of the metal smelts structure is their non-equilibrium and heterogeneity depending not only on the composition, but also on the alloyage conditions of components. The non-equilibrium is stipulated by an uneven distribution of atoms among clusters, and the heterogeneity is stipulated by an uneven distribution of clusters differing in structure within the nanovolume of the matter.

The speed at which the equilibrium is reached is significantly less than the alteration intensity of the external environmental changes in multi-component metallic smelts. This results in the prolonged retention of the non-equilibrium states. Therefore, liquid smelts of the identical composition, in the same conditions, can be distinguished by short-range order parameters, the nature of inter-particle interactions and the way of elements distribution . This influences on the course of such processes as diffusion, crystallization and then phase transformations and distribution of atoms in cast and deformed metal [1]. Numerous experimental studies revealed the following pattern: the higher the degree of the smelt equilibrium is before solidification, the better the quality parameters of the hard metal are [1-4].

Under production conditions the equilibrium state of the smelt or a state related to it is achieved most effectively in the course of temperature-time treatment of liquid metal during smelting. Modes of such technology are developed on the basis of data on structure-sensitive properties of smelts and solid metal quality indicators [1, 3-4].

Let us consider the influence of the smelting technology with temperature-time processing of the smelt (TTP) using some parameters of the structure and properties of the alloy $Fe_{82}B_{12}Si_4C_2$ in liquid and amorphous states. In Figure 1 you can see temperature dependences of the surface tension (σ), electric resistivity (ρ), magnetic susceptibility (χ) and kinematic viscosity (ν), which show that when heated to a certain critical temperature (t_c) polytherms of heating and cooling do not match. Hysteresis represents significant structural changes taking place in the smelt at $t_{heat} \geq t_c$.

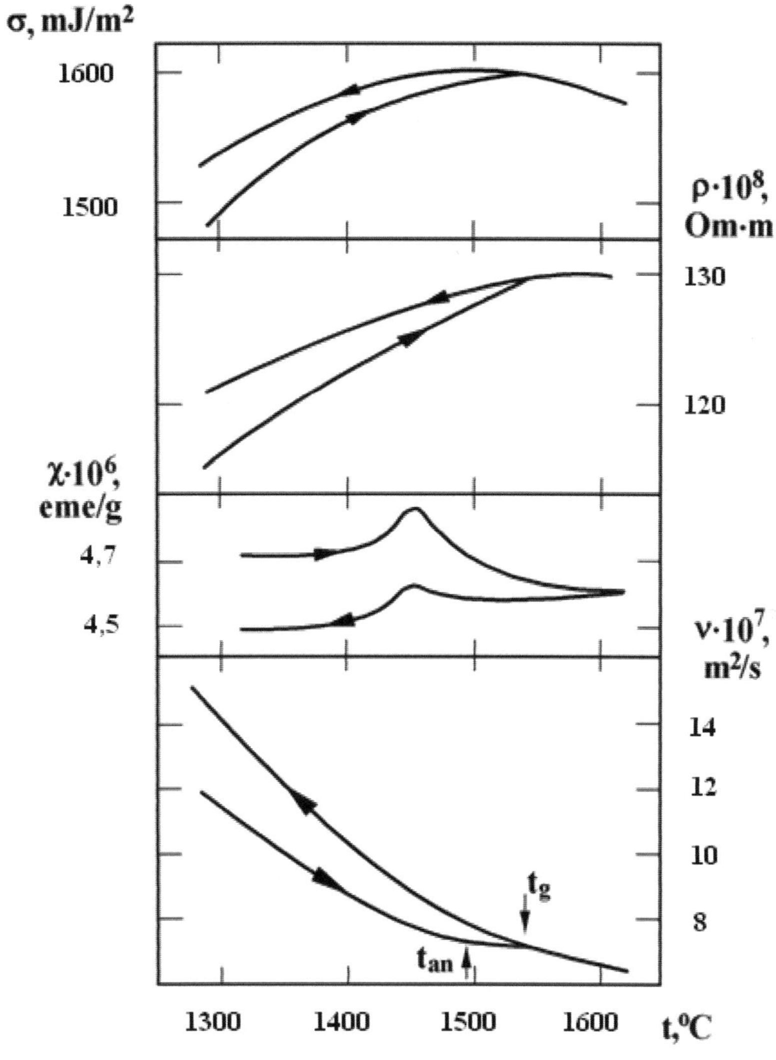

Figure1. Temperature dependences of the surface tension (σ), electric resistivity (ρ), magnetic susceptibility (χ) and kinematic viscosity (ν) of liquid $Fe_{82}B_{12}Si_4C_2$ alloy during heating and cooling

The most informative parameter for iron-based alloys is kinematic viscosity. At the temperature around 1500 °C the exponential dependence of ν(t) degenerates into a linear one, and in the interval t_{an} - t_g the thermal coefficient dν/dt is virtually zero, and for some alloys becomes positive.

The structural state of multi-component alloys is quite difficult to achieve because of the strong chemical interaction of heterogeneous atoms. The studied composition is defined by the formation of borides, silicides and carbides of different types, each of which potentially has the ability to form clusters. Therefore, in multicomponent fluids at the same time, there may be different types of clusters based on individual components or their chemical compounds.

Clusters are not conventional thermodynamic density fluctuations and are relatively stable formations with the structure close to FCC- and HCP- lattices whose lifetime τ_{cl}= (10-11)•10^{-9}s significantly exceeds the period of thermal oscillations of atoms. The local short-range order in the clusters is determined by nature and characteristics of interatomic interactions, as well as technological prehistory.

Unlike ordinary liquids, a stricter order covering several coordination spheres is naturally expected in systems with stronger interaction. In the cluster of this size the potential energy as a function of atomic positions, has a few lows. Thus allows to define this cluster in terms of a set of local states [12].

Such definition of local states is proposed in works [13,14], in which fluid is treated as a system consisting of local bulks, each of which can be in one of several configuration states. The potential polymorphism within this theory of liquids is considered in the work [15].

The smelt is approximated to the equilibrium state by a consistent positional disorder of clusters in the process of heating with the temperature over t_{an} (Figure 1). This approximation ends near the t_{cl}, i.e. the disordering temperature of the most thermally stable type of clusters and the equilibrium distribution of the elements within submicroscopic bulks of smelt [1].

In terms of the equilibrium distribution of atoms of different elements the proportion of newly formed clusters is increasing and they have energetically an advantage because of their smaller size and large crystallographic affinity between each other, which facilitates the emergence of centers of crystallization.

Cluster size reduction is accompanied by rise of intercluster volume or a disordered zone reaching to 15-45% of material volume. Compared with the clusters structural elements of the disordered zone consisting of several atoms have lower energy of the interparticle interaction, a medium-sized interatomic distance and less electronic density, contributing to alloys amorphization.

This features of the structure of liquid metals are due to changes of a structural factor and a radial distribution function of atoms at certain temperatures while preserving their general view that undoubtedly affects physical and chemical property values and their temperature dependences [1-4, 11,16].

In the process of frictional-resisted flotation the motion pulse is transferred from one atomic layer to another, without upsetting soundness and inter-particle bonds. Activation energy of the frictional-resisted flotation is the amount of energy of interatomic interaction, i.e. kinetic energy in motion.

The structure of metallic liquids is defined by two components, but they are not colloidal or emulsion systems and therefore clusters and the disordered area are involved in the frictional-resisted flotation. Strengthening of interatomic interaction in clusters increases the number of atoms in the frictional-resisted flotation and leads to input energy dissipation. In this regard, the dependence of ν (t) when cooling is characterized by elevated values of kinematic viscosity (Figure 1).

Surface tension (σ) of the alloy $Fe_{82}B_{12}Si_4C_2$ is determined by the concentration of surface active elements in the outer nuclear layer. The formation of the structure of the smelt in equilibrium distribution of atoms predetermines a more active participation of such elements in the process of cluster formation. The increase in the number of clusters leads to the decrease in the number of active elements in the surface layer, and as a result, the cooling polytherm is higher than the curve σ (t) during heating Figure 1), i.e. the surface tension increases.

The temperature-time treatment of smelt increases (Figure 1) specific electrical resistivity (ρ) and reduces magnetic susceptibility (χ). This is associated with increasing of interatomic bonds in clusters and the reduction of the number of charge carriers; with the reduction of cluster sizes and the increase in their number, i.e. the increase in scattering of conduction electrons; the increase in the density of electrons in clusters and change of the valence of atoms in the outer layer of clusters and having high amplitude of thermal vibrations due to the weak interparticle interaction with other atoms in the cluster.

Thus, the temperature-time smelt treatment contributes to the formation of the equilibrium and microhomogeneous structure of the smelt. At this the liquid metal is more capable of supercooling and flows laminarly and dense when spinning. The lack of turbulence and splashing enables high-quality appearance of an amorphous film.

With ultra-fast hardening ($V \geq 10^6$ deg/sec) the main structural parameters of the initial smelt are maintained in supercooled liquids and in the amorphous film.

Compared to the serial production experienced technology enables to obtain a thicker amorphous ribbon where the outer surface is 3-5 times less (Figure 2) under the same spinning conditions. This positively affects not only the appearance of the material, but also improves its quality indicators. Such ribbons are characterized by a homogeneous and preferred structure in a cross section, and have more ductility in bending tests.

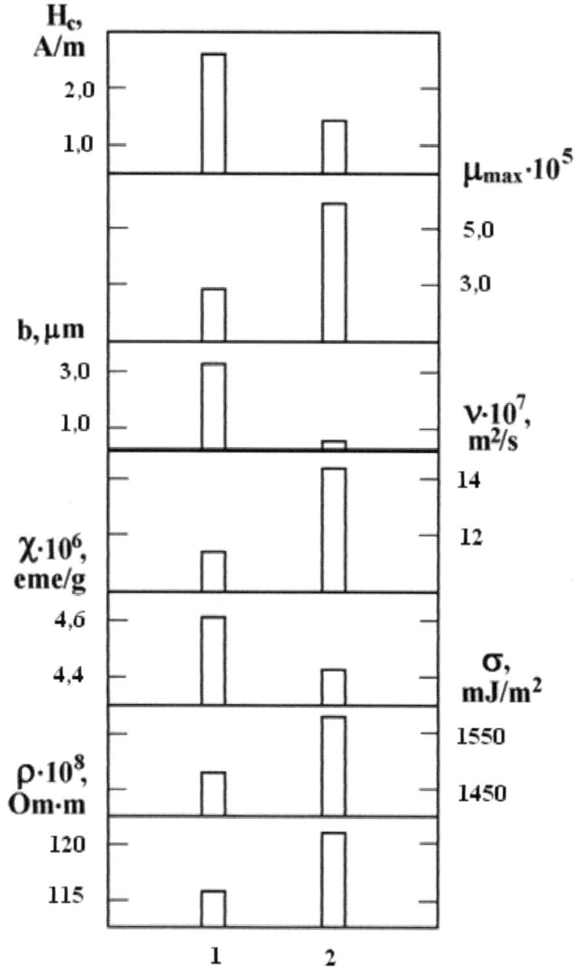

Figure 2. Effect of the smelting technology on the properties of $Fe_{82}B_{12}Si_4C_2$ alloy in the liquid (1300 °C) and amorphous states
1.- serial technology;
2.-pilot technology.

The last circumstance has important practical significance, as the amorphous ribbon is a starting material for the manufacture of various products. The workability of the ribbon increases with its plasticity, dissipation of deformation stresses in the material is improved, and high magnetic properties are maintained. [17].

The amorphous ribbons are used to obtain nano-and microcrystalline materials by annealing. To study the kinetics of the crystallization process a method of a differential thermal analysis (DTA) was used. In serial production and testing technology the crystallization process of the amorphous ribbons is characterized by a peak of heat deposition and is conducted in two stages (Figure 3).

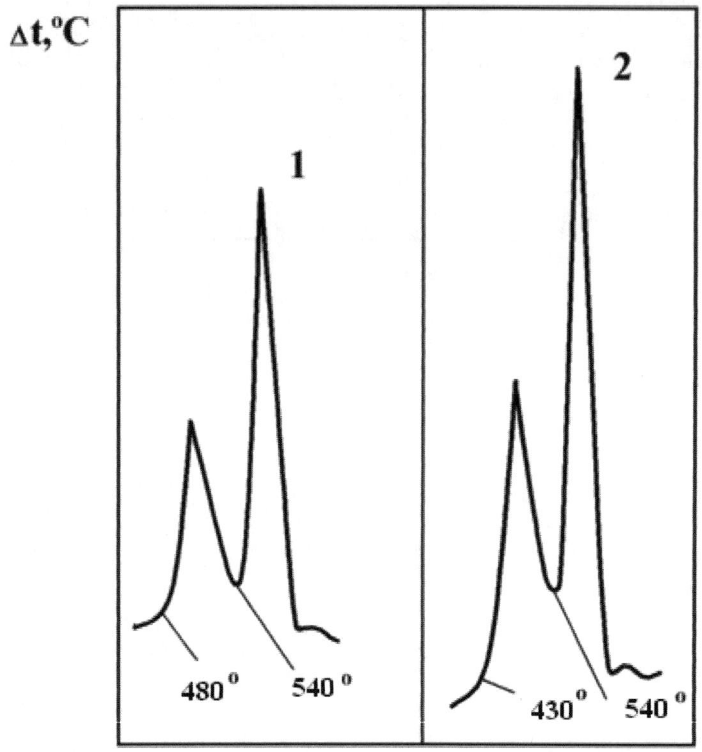

Figure 3. The results of differential thermal analysis of amorphous ribbon samples of $Fe_{82}B_{12}Si_4C_2$ alloy which have differences in the technology of smelting

1. - serial technology
2. - pilot technology

The temperature of the sample in the testing technology at which the first peak on the curve in the DTA appears is 50° higher than the temperature of the sample in a serial production, i.e. an amorphous ribbon structure which is obtained from the equilibrium smelt, has greater thermal resistance.

The comparative analysis of heat deposition peaks during crystallization process proves that the rate of amorphous state of ribbons is more stable when produced using TTP of the smelt. Such ribbons remain amorphous, with a thickness of 40 μm, and production-line ribbons when reaching a thickness of 30 μm and above become nano-and microcristalline, which negatively affects their plastic properties.

The data presented in Figure 2 shows that there is correlation dependence between structurally sensitive properties of the alloy $Fe_{82}B_{12}Si_4C_2$ in liquid state (ν, ρ, σ and χ) and quality characteristics of amorphous metal.

IOP Conf. Series: Materials Science and Engineering **133** (2016) 012001 doi:10.1088/1757-899X/133/1/012001

Summary Thus, the completed study shows that science-based technology of multi-component alloy smelting makes it possible to prepare equilibrium smelt, the structure of which has a significant effect on the properties of the amorphous ribbon before spinning and kinetics of its crystallization.
Acknowledgments. This work was supported by the Scientific researches of higher education institutions within the State task of the Russian Federation No. 2014/236.

References
[1] Baum B A and Khasin G A *et.al* 1984 *Molten steel* (Moscow: Metallurgy) p 206
[2] Elanskii G N 2006 *Structure and properties of metallic melts* (Moscow: MGHMI) p 228
[3] Tsepelev V S, Konashkov V V *et.al* 2008 *Properties of molten metals* (Yekaterinburg: USTU-UPI) p 357
[4] Tyagunov G V and Baryshev E E *et.al* 2014 *Liquid metal. Powders* (Yekaterinburg: UMC-UPI) p 192
[5] Goldshtein M I and Grachev S V *et.al* 1999 *Special steels* (Moscow:MIS&S) p 408
[6] Graebner J E and Goldling B 1986 *Phys. Rev.* B. **34** 5788
[7] Duval E A and Boukenter B 1986 *Phys. Rev.* Lett. **56** 2052
[8] Mashkovskii V K and Novikov V N *et.al* 1993 *Physics-Uspekhi* **5** 119
[9] Mitus A C Smolej F, Hahn H., Patashinski A Z 1996 *Phys.* A: *Stat. Mech. and its Appl.* **232** 662
[10] Patashinski A Z, Mitus A C, . Ratner M A 1997 *Physics Reports* **288** 409
[11] Zamyatin V M, Baum B A, Shmakova K Yu 2010 *Melts* **5** 19
[12] Son L D, Rusakov G M, Patashinski A Z ', Ratner M A 1998 *Phys.* A*: Stat. Mech. and its Appl.* **248** 305
[13] Lad'yanov V I, Vasin M G, Logunov S V, Bovin V P 2000 *Phys. Rev.*B. *Condens. Matt. and Mater. Phys.* **62** 12107
[14] Vasin M G, Lad'yanov V I 2013 *Phys. Rev.* E **68** 512021
[15] Vatolin N A and Pastuhov E A 1980 *Diffraction studies of high-temperature melts* (Moscow: Science) p 188
[16] Dubinin N E, Filippov V V, Malkhanova O G, Vatolin N A 2009 *Centr. Europ. J. of Phys.* **7** 584
[17] Gudoshnikov S A, Venediktov S N, Grebenshchikov Yu B 2012 *Measurement Techniques* **55** 329

IOP Conf. Series: Materials Science and Engineering **133** (2016) 012002 doi:10.1088/1757-899X/133/1/012002

Tribological properties of ternary nanolayers, obtained from simple/compound materials

V Jinga[1*], D Cristea[1], C Samoilă[1], D Ursuțiu[2], A O Mateescu[3], G Mateescu[3], D Munteanu[1]

[1] Department of Material Science, Transilvania University of Brașov, 1 Colina Universității street, Brașov, Romania
[2] Department of Electronics and Computers, Transilvania University of Brașov, 1 Politehnicii street, Brașov, Romania
[3] Horia Hulubei National Institute of Physics and Nuclear Engineering, Reactorului street, Bucharest, Romania

E-mail: jingavlad@yahoo.com

Abstract. Numerous recently investigations are oriented towards the development of new classes of thin films, having dry-lubrication properties. These efforts were determined by the enormous energy losses generated by friction, and due to technical complications determined by the systems used for classic lubrication. This paper presents our results concerning a new class of nanomaterials, with ternary composition deposited from simple/compound materials (Ti/Ti_xN_y, $TiB_2/Ti_xBi_yN_z$, $WC/W_xC_yN_z$). The films were deposited by magnetron sputtering, with varying sputtering parameters (sputtering power, reactive gas) on stainless steel substrates - ultrasonically and glow discharge cleaned before the deposition process. The influence of the deposition parameters on the mechanical and wear properties was assessed by nanoindentation, scratch resistance (to quantify the adhesion of the films to the steel substrate) and by pin-on-disk wear tests. The general conclusion was that the sample deposited at 550^0 C, with N_2 as reactive gas and 0.5 kV for substrate polarization, has the best mechanical characteristics (hardness and elastic modulus) and lubricant properties (represented by $\mu_{average}$), when compared to the remaining samples.

1. Introduction

In recent times friction increased its influence and importance from both technical and economical point of view. Friction generates energy losses causing energy waste, increased costs for production and has a significantly negative environmental impact.

According to the studies done in the industrialized countries, friction is the main cause for energy losses summing a total more than 5% of the country`s Gross Domestic Product [1] – a considerable value that should be minimized as much as possible.

Friction must be avoided if possible, or at least reduced. Generally, if the friction is reduced due to external lubrication, one has to consider the negative environmental impact caused by the production and waste disposal of these lubricants. A new approach is taken into consideration for this matter: the development of coatings with dry lubrication properties, meaning a reduced friction coefficient, and consequently, better wear performance.

* 11 Cibinului street, Brașov – Romania, e-mail: jingavlad@yahoo.com

This subject was already of interest for other researchers in material science. Hall – Petch and Frank – Read theories regarding surface`s tribological behaviour are slightly less significant if the thickness is lower than the critical value. When layer`s thickness decreases [2] additional layer`s coherence can be totally absorbed by the layers. If the thickness increases, the energy of surfaces being in contact is generating dislocations that don`t fit. The border separating these two trends is called critical thickness [3]. Misra, Hirth and Hoogland [4] explained the increase effect of multilayers systems: for layers having thickness larger than 100 nm, variation of increasing tensions is defined by Hall – Petch relation, meaning that tension increasing in the deposited layer is proportionally reversed with the layer`s thickness square root, $\sigma = 1/\sqrt{h}$ where σ is the yield stress in the film and h the layer thickness. The decreasing of the layer thickness less than tens of nanometers was noticed that Hall – Petch relation break down and the equation is transformed in logarithm dependence $\sigma = \ln(h/b_B)/h$ where b_B is the Burgers vector of metal law dislocation line energy.

Tambe and Busham [5] established that there is an essential connection between material`s Young modulus, friction coefficient and adherence. The rule is that a lower Young modulus implies a higher friction coefficient and a higher Young modulus results in a lower friction coefficient.

Bowden and Tabor’s model [6] from microtribology considers generation of pinning surfaces due to cold welding junctions appearance, but at nanometric scale it is stated that stick-slip is a result of overcoming energy barriers necessary for surpassing leaps over atom`s concentrations on probe`s surface.

The authors will thoroughly describe the manner in which depositions were made, the tests that were done (not only scratch tests, but also friction coefficient determinations, elastic modulus and hardness testing) and will comment on obtained results from the tests, concluding with influence of deposition parameters regarding the coating`s properties.

These contributions describe a new approach in designing tribological coatings based on mono-layered structures made of $Ti/TiB_2/WC$ in presence of N_2 as a reactive gas or not. Their properties and perspectives for application are taken into account based on author`s contributions in several patents [7].

2. Experimental procedures

2.1. Materials and process parameters

Stainless steel plates were used as substrates, and before each deposition were ultrasonic and glow discharge cleaned. During the deposition process, substrates were in a rotation movement (20 rpm), to increase the coating homogeneity. The substrates were heated at 300°C or 550°C and the bias voltage was set at 0.5 kV during the reactive deposition processes.

For the film deposition was used a multifunctional vacuum thin film deposition system. The chamber is equipped with three DC sputtering magnetrons (guns, all with maximum power of 600 W). Ar was used as a bombardment gas and N_2 as a reactive gas. The distance between the substrate holder and the sputtering target was set at approximately 12 cm.

The working process parameters for the DC magnetron sputtering deposition method used to obtain the samples described in this paper are presented in Table 1. One can notice from Table 1 that all the films were deposited by simultaneously sputtering with three targets, resulting in a quasi-composite film, each component having its distinctive role: Ti for having great adherence to stainless steel and an increased hardness, TiB_2 and WC for their hardness and for dry-lubrication.

2.2. Characterization methods

The mechanical and tribological properties of the coated samples described above were investigated by the following methods: Scratch Test Methods, Hardness Test Method and Pin-on-disk Tribometer.

Mechanical and tribological parameters: analyses were done using CSM Table Top Platform which includes the standard Micro/Nanoindentation head (NHT2) and the standard Micro scratch tester head (MST) into a small and simple-to-use instrument.

Table 1. Working process parameters for deposition of samples no. 1, 2, 3 and 4.

Sample No.	Materials	Working gases		Power injected in the gun's plasma [%P_{max}]	Deposition time [min]	Working pressure [mbar]	Temperature [^0C]	Bias Voltage [kV]
	Deposited layers	Ar [sccm]	N_2 [sccm]					
1	1x(Ti +TiB_2+WC+N_2)	100	40	9 for Ti 50 for TiB_2 22 for WC	13	2.7×10^{-3}	300	0.5
2	1x(Ti +TiB_2+WC)	100	0	5 - 3 for Ti 30 for TiB_2 25 for WC	10	2.9×10^{-3}	300	-
3	1x(Ti +TiB_2+WC+N_2)	150	40	15 for Ti 30 for TiB_2 30 for WC	30	3.5×10^{-3}	550	0.5
4	1x(Ti +TiB_2+WC)	150	0	15 for Ti 30 for TiB_2 30 for WC	30	3.4×10^{-3}	550	-

Scratching method: the micro scratch tester head (MST) has a Rockwell type indenter. The Micro Scratch Tester (0 – 30 N) is suited for analyzing thin hard or soft coatings. It contains a full software package for data acquisition and analyses of mechanical properties. The Nanoindentation Tester (0 – 500 mN) provides low loads with depth measurement in the nanometric scale. The system can be used to characterize organic, inorganic, hard and soft materials.

Tests were done with the Rockwell indenter (100 Cr6 material) having a 100 μm range. The linear scratch was progressive, having the begin load 0.03 N and the end load 10 N for all samples, the loading rate constant at 5 N/min for samples no. 2 to 4 and 8 N/min for sample no. 1. AESensitivity was 9 and the scratch length was 5 mm for samples no. 1 and 2, 2mm for sample no. 3 and 4 mm for sample no. 4. Speed varied from 4 mm/min for sample no. 1 to 2.5 mm/min for sample no. 2, 2mm/min for sample no. 3, sample no. 4 having a speed of 4 mm/min.

For samples no. 2 to 4, three scratch tests were done using the same settings for each measurement, sample no. 1 having only two scratch tests performed on. After carrying out the tests, critical loads could be established: Lc_1 – load observed for first crack occurrence, Lc_2 – load necessary for first delamination (first section of film removed from the substrate), Lc_3 – load necessary for removal of more than 50% of film from the scratch track.

Nanoidentation technique: a diamond tip –NHT2 – the standard Nanoindentation head, mounted on a stainless steel cantilever is used in order to obtain good statistics. Nanoindentation measurements were performed with a diamond tip Berkovich indenter, penetration depth being constant (no greater than 20% of coating thickness in order to minimize the effect of the substrate on the results). Thickness measurements were done using Rutherford Backscattering Spectrometry method and it was established that the film thickness is ~220nm.

Several measurements were performed on each sample, to increase the statistical relevance of the results, with the following protocol: loading/unloading speeds of 200 nm/min with a pause of 10 s between loading and unloading stages for samples no. 1 and 2 and loading/unloading speeds of 100 nm/min with a pause of 2 s between loading and unloading stages for samples no. 3 and 4.

Tribometer method: a sphere, a pin or a flat section is loaded onto the test sample with a precisely known force. The probe is mounted on a stiff lever, designed as a frictionless force transducer. The friction coefficient is determined during the test by measuring deflection of the elastic arm.

On each sample tests were performed using 6 mm 100 Cr 6 steel balls as counterparts, in rotation mode, normal load being 1 N. Atmospheric conditions were constant during all tests, 24°C temperature, and 32% humidity. Variable parameters were the Radius [mm], the Linear Speed [cm/s] and the Stop Condition [m] meaning the length of the wear track. No lubricating solutions were used.

3. Results and discussions

3.1. Adherence: scratch tests

For sample no. 1, Lc_1 values varied from 0.458 N to 0.5 N, mean value being calculated at 0.479 N ± 4.5% measurement error. Lc_2 was determined at 1.198 N and Lc_3 at 2.906 N.

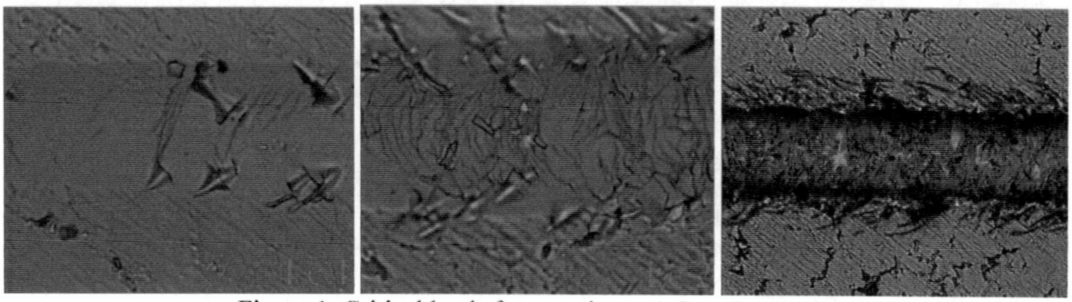

Figure 1. Critical loads for sample no. 1 (Lc_1, Lc_2, Lc_3).

For sample no. 2, Lc_1 values varied between 0.266 N and 0.404 N, having a mean value of 0.3186 ± 27%, Lc_2 values varied from 0.474 N to 0.892 N, mean value being 0.715 ± 30% determined measurement error and Lc_3 had a value of 0.933 N.

For sample no. 3, Lc_1 had a variation from 0.44 N up to 0.74 N, mean value being calculated at 0.616 ± 29%. Lc_2 varied between 1.69 N and 2.46, with a mean value of 1.986 N ± 24%. Lc3 varied from 8.02 N up to 8.29 N, mean value being calculated at 8.163 N ± 2%.

For sample no. 4, Lc_1 values varied from 0.26 N to 0.35 N, mean value being calculated at 0.313 N ± 17% measurement error. Lc_2 variation started at 0.73 N up to 0.89 N, mean value being 0.826 ± 12% and Lc_3 from 5.39 N up to 6.45 N, with a mean value of 5.973 N ± 10% error.

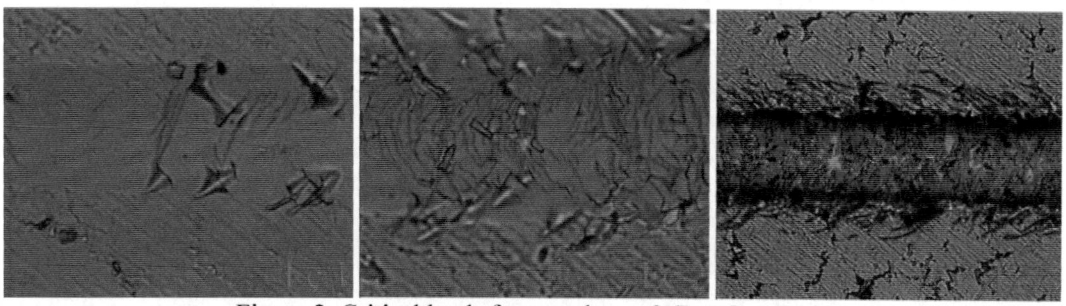

Figure 2. Critical loads for sample no. 2 (Lc_1, Lc_2, Lc_3).

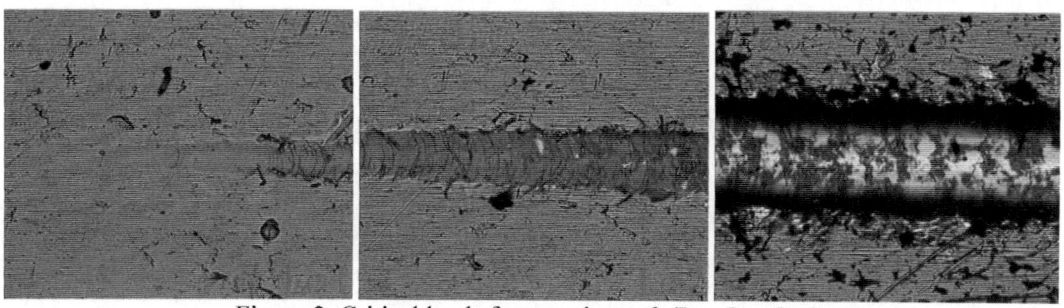

Figure 3. Critical loads for sample no. 3 (Lc_1, Lc_2, Lc_3).

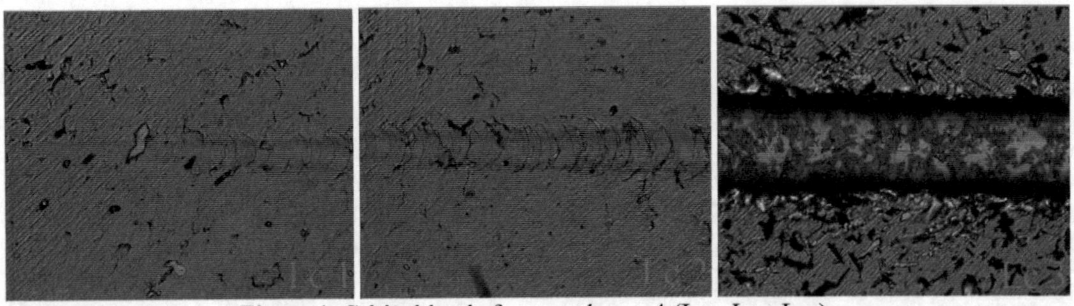

Figure 4. Critical loads for sample no. 4 (Lc_1, Lc_2, Lc_3).

3.2. Hardness indentation and elastic modulus

Five indentations were performed on sample no. 1. Hardness values varied from 1703.2 MPa to 2274.1 MPa, mean value being 1938.44 MPa ± 17% error. Elastic modulus had values starting at 163.47 GPa up to 212.67 GPa, its mean value being calculated at 185.692 GPa ± 14.5%.

For sample no. 2, seven indentations were done. Hardness varied from 1842 MPa to 2525.1 MPa, having a mean value of 2196.329 MPa ± 16%. Values for elastic modulus started at 115.55 GPa and ended at 200.2 GPa, mean value being 167.0471 GPa ± 30% error.

Seven indentations were performed for sample no. 3 also. Hardness variation was between 12095 MPa and 16841 MPa, mean value being 14627.14 MPa ± 17%. Elastic modulus values varied from 184.14 GPa up to 270.36 GPa, mean value being calculated at 217.7143 GPa ± 25% error.

For sample no. 4, seven indentations were performed. Hardness started to vary from 8308 MPa up to 10328 MPa, having a mean value of 8996.8 MPa ± 8%. Values for elastic modulus started at 128.46 GPa and ended at 187.93 GPa, mean value being 148.3 GPa ± 14% error.

3.3. Friction coefficient: wear tests

On sample no. 1 three wear tests were performed. μ_{min} coefficient had values between 0.07 and 0.245, it`s mean value being 0.1746 ± 60% error. $\mu_{max}$ coefficient varied between 0.787 and 0.829, having a mean value of 0.815 ± 3.5%. $\mu_{average}$ coefficient started at 0.526 up to 0.555, it`s mean value being calculated at 0.54 ± 3%.

Three wear tests were done for sample no. 2 also. μ_{min} coefficient values varied from 0.206 to 0.248, it`s mean value being 0.2213 ± 12%. $\mu_{max}$ coefficient varied between 0.682 and 0.752, mean value being 0.712 ± 10% error. $\mu_{average}$ coefficient had values starting at 0.397 up to 0.474, it`s mean value being calculated at 0.445 ± 10%.

On sample no. 3 was performed only one wear test. μ_{min} coefficient had a value of 0.262, μ_{max} coefficient`s value was 0.399 and 0.317 for $\mu_{average}$ coefficient.

Four wear tests were done for sample no. 4. μ_{min} coefficient values varied between 0.209 and 0.231, it`s mean value being calculated at 0.219 ± 5%. $\mu_{max}$ coefficient varied from 0.789 to 0.867, mean value being 0.825 ± 5% error. $\mu_{average}$ coefficient had values starting at 0.572 up to 0.675, it`s mean value being calculated at 0.622 ± 8.5%.

4. Comparative discussions

Analyzing the critical loads Lc1, Lc2 and Lc3 for all sample (figure 5) it can be stated that the best adhesion (that means the highest value for Lc1/ Lc2/ Lc3) was obtained for sample no. 3 (obtained at high temperature and by using of N_2 as reactive gas and bias voltage at 0.5 kV) and the lowest adhesion was obtained for sample no. 2 (achieved at low temperature and without bias voltage and reactive gas).

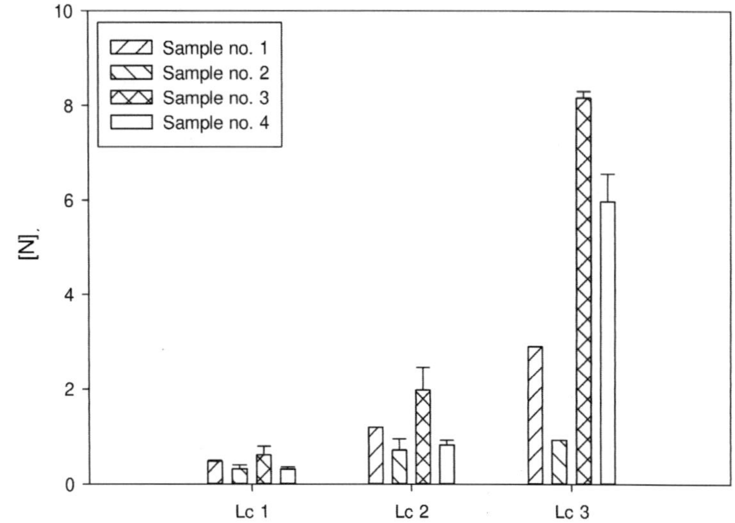

Figure 5. Scratch test`s comparative results.

The highest value of the hardness was obtained for sample no. 3 (obtained at high temperature and by using of N_2 as reactive gas and bias voltage at 0.5 kV) and the worst hardness value was obtained for sample no. 2 (achieved at low temperature and without bias voltage and reactive gas).

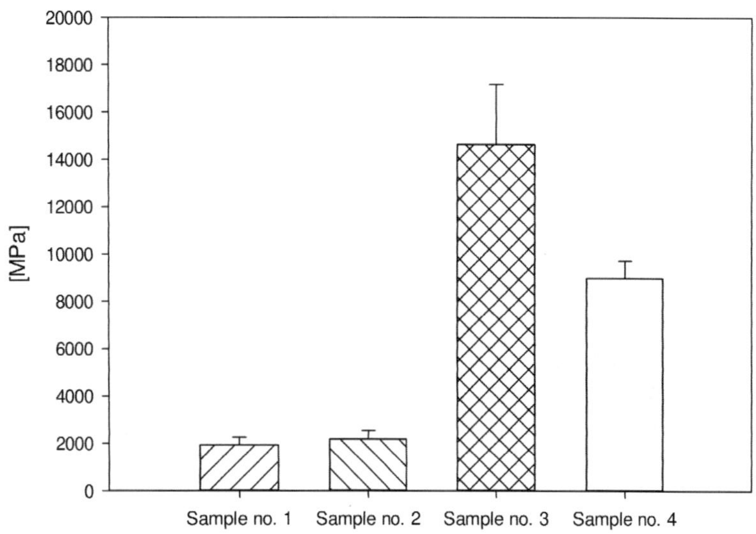

Figure 6. Hardness test`s comparative results.

Also, the highest value of the elastic modulus was obtained for the same sample no. 3 (obtained at high temperature and by using of N_2 as reactive gas and bias voltage at 0.5 kV) and the worst elastic modulus value was obtained for sample no. 2 (achieved at low temperature and without bias voltage and reactive gas) and for sample no. 4 (obtained without bias voltage for substrate and reactive gas).

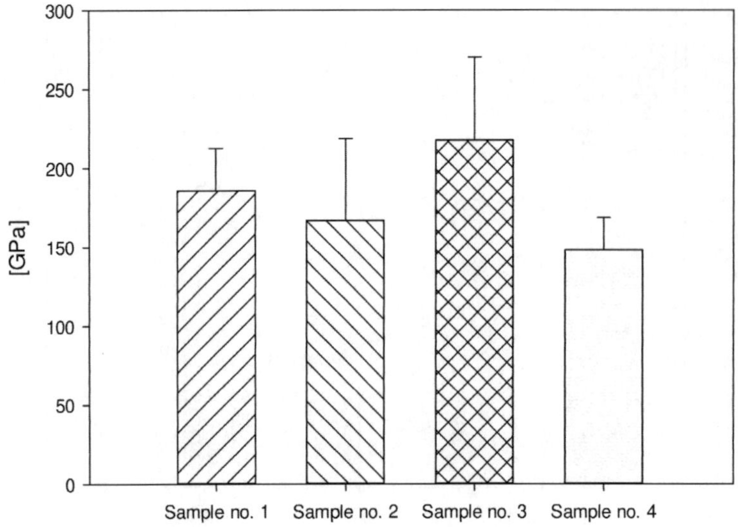

Figure 7. Elastic modulus comparative results.

All the samples have average friction coefficients with almost the same values (close to 0.4). The lowest friction coefficient ($\mu_{average}$) was obtained for sample no. 3 that was achieved with N_2 as reactive gas and 0.5 kV as bias voltage.

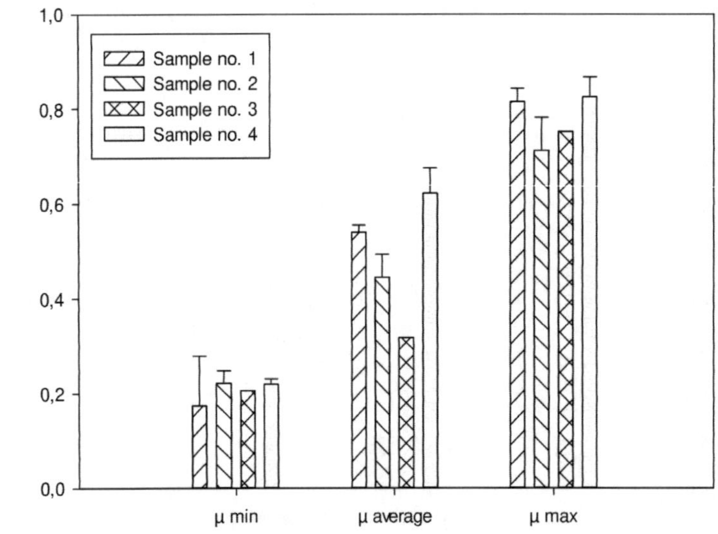

Figure 8. Friction coefficients comparative results.

5. Conclusions

Sample no. 3, deposited at 550 ^{0}C, with N_2 as reactive gas and 0.5 kV for substrate polarization, has the best mechanical characteristics (hardness and elastic modulus) and also lubricant properties (represented by $\mu_{average}$).

In theory using N_2 as a reactive gas during the deposition process would mean that N_2 gas atoms deposit on the surface and create hard particles that don`t help in achieving dry lubrication. But if the used gas concentration is not so large, the N_2 deposited particles can create on the sample a wear resistant layer without degrading the dry lubrication properties of the surface (just like in case of sample no. 3), resulting as a plus for the deposited sample (the film is not so easily destroyed in case of wear tests).

Hardness and elastic modulus values are varying quite much for each sample from one test to another, meaning that the used technique for deposition is not perfect and should be improved for the deposited thin films to be more homogeneous along all the sample`s surface. For this reason, the deposition time should be considerably increased, having in mind though not to increase a lot the thickness of the thin films.

By increasing the deposition time and the thickness of the deposited layer, the film should become also more wear resistant (gaining cumulative properties for industrial usage), so this hypothesis should be put to test as soon as possible for finding the best recipe.

Acknowledgements

This work was supported by a grant of the Romanian National Authority for Scientific Research, Program for research - Space Technology and Advanced Research - STAR, project number 68/2013.

We hereby acknowledge the structural founds project PRO-DD (POS-CCE, O.2.2.1., ID 123, SMIS 2637, ctr. No 11/2009) for providing the infrastructure used in this work and the project.

References

[1] B. Bhushan, 1999 Principles and applications of tribology John Wiley & Sons, New York.
[2] O. Marti, H. U. Krotil, 2001 Dynamic Friction Measurement Fundamentals of Trybology and Bridging the Gap Between the Macro and Micro/Nanoscales-Kluiwer Academic Publisher, Dordrecht, Netherlands.
[3] B. Bhushan, B. K. Gupta, 2004 *Springer Handbook of Nano Tribology*, New York.
[4] I. Iliuc, 1980 *Elsevier*, Amsterdam, 64.
[5] N. S. Tambe, B. Bhushan, 2005 *Appl.Phys.Lett*, 86 061906, 1..3.
[6] F. R. Bowden, D. Tabor, 1986 *The friction and Lubrication of solids Clarendon Press-Oxford*, vol I, 1.
[7] A. O. Mateescu, G. Mateescu, 2014 Coating multilayer material with tribological properties and procedure for its achievement Patent No. 128094.

IOP Conf. Series: Materials Science and Engineering **133** (2016) 012003 doi:10.1088/1757-899X/133/1/012003

A study on the soil characteristic and properties of riverbank soil samples from Sungai Perak, Kota Lama Kiri, Kuala Kangsar, Malaysia

[1]M N Adlan, [1,2]Mohd Remy Rozainy M A Z*, [1]M F Ghazali, [1]M R Selamat and [1]S Z Othman

[1]School of Civil Engineering, Universiti Sains Malaysia
14300 Nibong Tebal,Seberang Perai Selatan,Penang,Malaysia
[2]Center of Excellence Geopolymer and Green Technology (CEGeoTech), School of Materials Engineering, University Malaysia Perlis (UniMAP), P.O. Box 77, Kangar, Perlis, MALAYSIA

*ceremy@usm.my

Abstract. Riverbank filtration (RBF) technology is new in Malaysia and only a few efforts have been made to understand the RBF mechanisms and processes. Soil characteristics and properties play important roles in determining the suitability of the site for the RBF application. A research has been carried out in Kota Lama Kiri, Kuala Kangsar, Perak, Malaysia to identify the characteristics of the riverbank soil for different layers of the pumping well (PW) and three adjacent monitoring wells namely MW2, MW3, and MW5. Particle size distributions and hydraulic conductivities of the soils were obtained from sieve analyses and constant head permeability tests. The subsurface soils of the study site consisted of medium sand, fine sand, coarse sand and gravel but them medium sand was highest in percentage over the other types of soil. The aquifer extended down to 8 m. The highest hydraulic conductivity value for the PW was 0.91 cm/s and obtained for sample taken from 6 m deep. The highest hydraulic conductivity value for the monitoring wells was 5.03 cm/s and obtained for sample taken from 2.20 to 3.20 m of MW5. The overall well production capacity determined from the pumping test was 112.10 m^3/hr.

1. Introduction

RBF is a natural first-step process of water treatment. The technology is commonly employed in alluvial aquifers with complex hydrologic system that exhibits both physical and geochemical heterogeneity. Alluvial aquifers that are hydraulically connected to surface water bodies have been used for water production throughout the world. Sites have been selected because of the significant amount of groundwater that can be extracted at shallow depths [1]. River water flowing through the riverbed on its way to the underlying aquifer and subsequently to the production wells can undergo RBF. Riverbed and aquifer material act as filter media to the contaminants [2]. During the process, various contaminants such as trace organic pollutants, bacteria, viruses, and inorganic compounds can be removed [3].

The design of a RBF system usually requires detailed hydrogeological site investigation and knowledge about the hydrological characteristics of the catchment. The hydrological characteristics such as hydraulic conductivity, transmissivity, storage coefficient, and dispersivity affecting groundwater flow and contaminant transport can be characterized by various laboratory and field tests such as column tracer tests, grain size analyses, pumping tests, slug tests, and field tracer tests [4]. Alluvial aquifers near a river are typically multilayered due to the deposition processes resulting in vertical heterogeneities in hydraulic properties [5,6,7,8]. A suitable site for a RBF would involve sand and gravel aquifers with hydraulic conductivity, $k_f > 10^{-4}$ m/s, a minimal thickness of 5 m and a good hydraulic connection to the adjacent surface water [9]. Studies on the characteristics of a feasible bank filtration site based on examples from Europe and North America would mention the relevance aquifer thickness, hydraulic conductivity, flow velocity gradient, river discharge, and clogging of the riverbed [10,11,12]. The thicknesses of some existing RBF sites in Germany range from 3-5 m at Böckingen, by the Necker River to 40–55 m at Torgau-Ost, by the Elbe River, with hydraulic conductivity ranging between 1×10^{-2} and 7.5×10^{-5} m/s [10].

The RBF technology is very new in Malaysia. The characteristics of the alluvial soils need to be identified in determining the suitability of a site. The aim of this study was to describe the characteristics of the alluvial soils of the selected site in Perak and evaluate the suitability of the site for RBF application. This paper focuses on soil classification, particle size distribution (PSD), and hydraulic conductivity values.

2. Materials and methods

2.1 Study area

The study area is located in Kg Kota Lama Kiri, Kuala Kangsar, Perak. Kuala Kangsar site was selected for RBF application after considering the soil profile near the Sungai Perak. Figure1 shows the location of the study site at Kota Lama Kiri, Kuala Kangsar, Perak, Malaysia. As shown in Figure 1, the distance from the MW1 to the river was 10 meter. The distance between MW1 and MW2 was 8 meter, where the distances were applied similar to all of the monitoring wells, except for MW5 and MW6. The distance of MW5 from the PW was 25 meter, and 50 meter for the MW6, respectively. This paper will focus on the investigation of the soil samples from pumping well (PW) and the adjacent monitoring wells including MW2, MW3 and MW5.

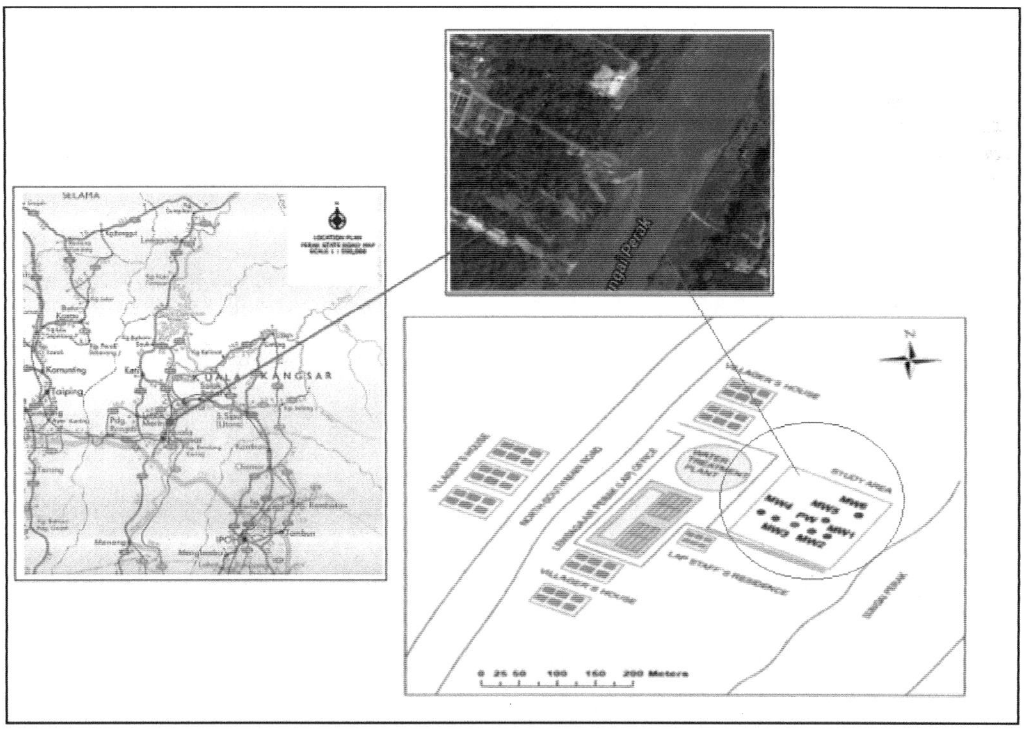

Figure 1. Location of the study site with boreholes

2.1. *Soil sampling and soil investigation*

Soil samples were collected during the construction and development of the PW and MW's in Kuala Kangsar, Perak. The coordinate of the PW is 04°48'08.5''N and 100°57'06.9''E. The samples were collected for each meter depth as the drill penetrates deep into the ground. The soils sample in each layer were collected and labeled. The maximum depth was 8.20 meters. Figure 2 shows the borelog and schematic well design of PW. The soil samples were transported to the Geotechnical Engineering Laboratory in School of Civil Engineering, Universiti Sains Malaysia (USM) for the soil characteristics investigation.

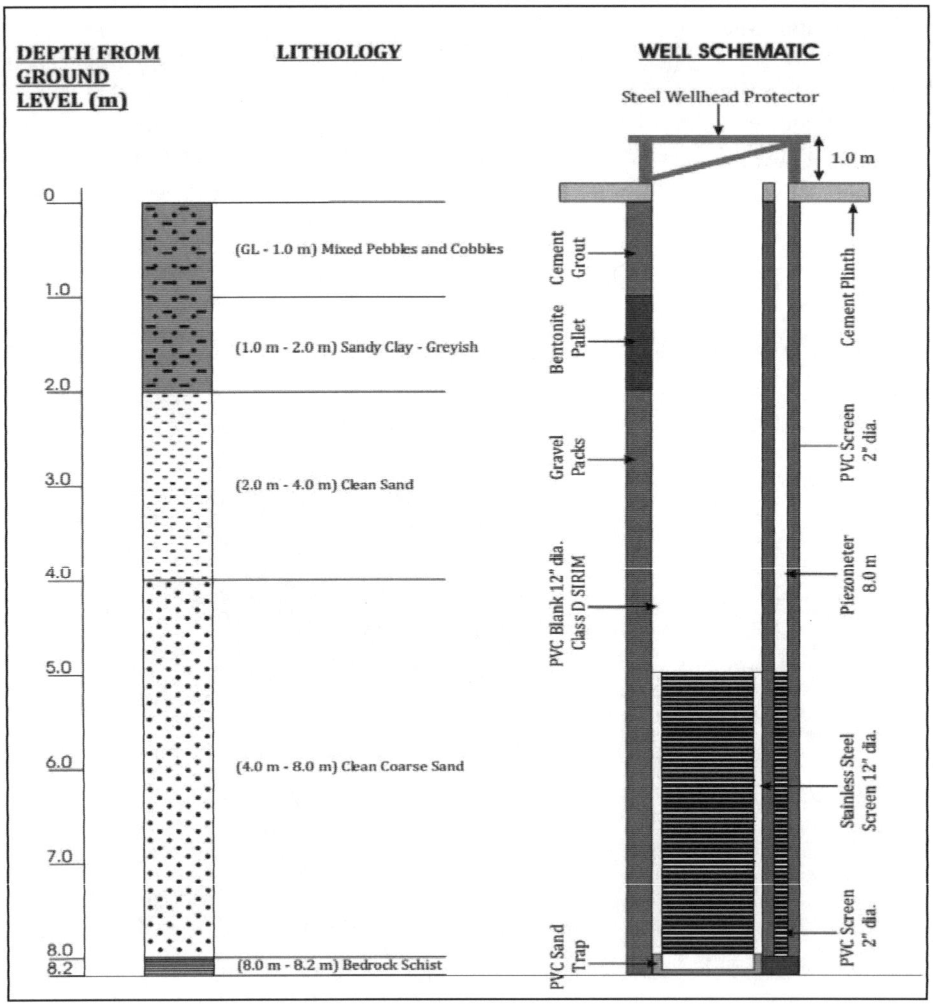

Figure 2. Borelog and schematic well design of PW

2.3. *Sieve analyses*

The sieve analysis was performed with reference to British Standard 1377: Part 2:1990. Sieve analyses were conducted by using a mesh wire woven sieve size of 14.00 mm at the top and

0.063 mm at the bottom. Soil samples were oven dried at 105 °C for 24 hours. 100g of soil samples were washed through the sieve size 0.063 mm. The soil sample must be washed carefully to avoid losing the soil material. Washed sample will be used for the mechanical sieving and the washed samples were oven dried at 105 °C for 24 hours. The dry samples were weighed again and were put on the top sieve having the largest opening which is 14.00 mm. The mechanical shaker was set to operate for 10 to 15 minutes. The soils retained on top of each sieve were weighed using the balance readable to 0.01g.

2.4. Permeability (Constant Head Test)

The constant head test was done in accordance with British Standard 1377: Part 1-9:1990. Permeability of a soil is a measure of its capacity to allow the flow of the water through the pore spaces between solid particles. The degree of permeability is determined by applying a hydraulic pressure gradient in a sample of saturated soil and measuring the flow rate of the water. The ancillary apparatus were prepared in the Geotechnical Laboratory. The internal diameter the of the permeameter cell were measured and recorded. The length and the weight of the specimen were also measured. The permeameter were placed at the test equipment and the water valves were opened. The water was allowed to flow for several minutes to let all of air out from the permeameter. A measuring cylinder was placed under the outlet of permeameter and the timer was started simultaneously. The levels of the water in the manometer tubes which is H_1 and H_2 were recorded. The quantity of water collected in the cylinder was measured. The temperature of the water was recorded.

2.5. Pumping test

The pumping test was carried using a DT 95-10 Dynatech Submersible Pump. A Submersible Pump was installed using a crane and 125 mm diameter GI riser pipe was connected to the submersible pump. A 125 mm diameter gate valve was connected to the riser pipe to regulate the flow rate. The riser pipe was then directed to a 90 V Notch tank to measure the flow rate of pumping. After the installation of the pump and all other necessary setup was completed, a calibration test was carried out for 2 hours to determine the capacity of the pumping wells, and also to determine the pumping rates for the Step Drawdown Test. The pumping test program consists of step drawdown and a 72 hours constant discharge test and recovery test.

3. Results and discussions

3.1. Sieve analyses for soil samples

Grain size analyses were carried out in the Geotechnical Engineering Laboratory, School of Civil Engineering, Universiti Sains Malaysia (USM). From particle size distribution (PSD), the types and class of the soil can be determined. Gravel, coarse sand, medium sand, fine sand, silt and clay can be found within soil sample. D_{10} represents the 10% of the particles are finer and 90% of the particles are coarser than that particular size of D_{10}. D_{30} means 30% of the particles are finer and 70% of the particles are coarser than that particular size of D_{30}. The uniformity coefficient, termed as C_u, is the ratio of D_{60} to D_{10}. C_c is another measurement of gradation, which is coefficient of gradation. Table 1 shows the classification of the soil samples within PW, MW2, MW3 and MW5 at different depth. Table 2 shows the result of sieve analyses for PW, MW2, MW3 and MW5.

Table 1 and table 2 show the results of sieve analyses for PW, MW2, MW3 and MW5 for different depths. Table 1 indicates the proportions of clay, silt, fine sand, medium sand, coarse sand, and gravel for the PW as 0 %, 0 %, 16.77 to 72.70 %, 45.12 to 77.90%, 1.18 to 18.98%, and 0.06 to 10.20 %, respectively, thus the medium sand has the most percentage. Soil samples of MW2, MW3, and MW5 indicate presence of medium sand ranging from 30.16 to 59.35 %, 0.70 to 54.51 % and 0.32 to 81.34 %, respectively. Table 2 indicates that the soils from PW had D_{10} ranging from 0.15 to 0.38 mm, D_{30} ranging from 0.40 to 0.75 mm, and D_{60} ranging from 0.60 to 1.50 mm. For MW2, MW3, and MW5, the $D_{10,}$ $D_{30,}$ D_{60} had ranged from 0.30 to 4.00 mm, 0.45 to 11.00 mm, and 0.60 to 15.00 mm, respectively. The coefficient of uniformity, C_u, for PW, MW2, MW3 and MW5 ranged from 2.0 to 14.0 while the coefficient of gradation, C_c, ranged from 0.05 to 2.75. The grain size distributions of soils from PW, MW2, MW3, and MW5 are given in Figures 3, 4, 5, and 6.

The grain size distributions for PW indicate that down to a depth of 8 m, the soils have a high percentage of medium sand, fine sand, coarse sand and gravel; and more gravel beyond the 8 m depth, thus making it suitable for the purpose of a pumping well. Nevertheless, the main aquifer was considered to reside above the 8 m mark due to the general presence of bedrocks at 8 m. The soils of MW2, MW3 and MW5 were predominantly medium and fine sands from surface down to 4.8 m. There was no silt or clay found in the soil samples of all boreholes.

Table 1. Soil Classification for PW, MW2, MW3 and MW5

		Soil Types					
Boreholes	Depth (m)	Gravel (%)	Coarse Sand (%)	Medium Sand (%)	Fine Sand (%)	Silt (%)	Clay (%)
PW	0.0-1.0	6.81	18.98	45.12	21.85	0	0
	1.0-2.0	10.20	16.60	51.27	21.32	0	0
	2.0-3.0	2.14	14.23	66.42	16.77	0	0
	3.0-4.0	2.42	9.40	55.70	31.45	0	0
	3.0-5.0	0.63	12.03	66.20	20.58	0	0
	4.0-5.0	0.06	1.18	64.82	33.16	0	0
	5.0-8.0	0.09	3.22	77.90	18.22	0	0
	6.0	7.07	13.75	66.13	72.70	0	0
	7.0-8.0	0.92	4.09	63.73	30.59	0	0
MW2	0.0-1.0	6.56	16.46	53.63	22.03	0	0
	1.0-3.0	4.72	41.59	47.56	6.10	0	0
	3.0-4.0	6.10	33.34	59.35	1.19	0	0
	4.0-4.5	22.01	47.49	30.16	0.29	0	0
MW3	0.0-1.0	34.11	30.76	11.19	2.07	0	0
	1.0-2 .0	1.85	30.78	54.51	12.75	0	0
	2.0-3.0	12.96	15.46	36.55	3.03	0	0
	3.0-4.8	51.27	10.03	7.86	0.11	0	0
	>4.8	32.94	6.23	0.70	0.30	0	0
MW 5	0.0-1.0	15.59	14.63	51.36	9.04	0	0
	1.0-2.2	40.89	16.46	38.27	4.31	0	0
	2.2-3.2	0.12	37.97	55.18	6.64	0	0
	3.2-4.0	0.061	10.76	81.34	7.73	0	0
	>4.0	52.18	4.36	0.32	0.04	0	0

Table 2. Sieve analyses results for PW, MW2, MW3 and MW5

Boreholes	Depth (m)	D_{10} (mm)	D_{30} (mm)	D_{60} (mm)	C_u	C_c	Types of samples
PW	0.0-1.0	0.15	0.50	1.50	10	1.11	Well -graded
	1.0-2.0	0.25	0.50	1.10	4.4	0.91	Poorly/uniformly graded
	2.0-3.0	0.35	0.55	1.10	3.14	0.79	Poorly/uniformly graded
	3.0-4.0	0.30	0.40	0.80	2.67	0.06	Poorly/uniformly graded
	3.0-5.0	0.25	0.40	0.60	2.40	0.07	Poorly/uniformly graded
	4.0-5.0	0.38	0.70	1.20	3.16	0.15	Poorly/uniformly graded
	5.0-8.0	0.30	0.55	0.85	2.83	1.19	Poorly/uniformly graded
	6.0	0.37	0.75	1.30	3.52	0.16	Poorly/uniformly graded
	7.0-8.0	0.20	0.40	0.75	3.75	0.14	Poorly/uniformly graded
	8.0-8.5	0.60	0.80	1.20	2.00	0.89	Poorly/uniformly graded
	13.0	0.69	1.0	2.00	2.90	0.35	Poorly/uniformly graded
MW2	0.0-1.0	0.30	0.50	1.50	5.00	0.05	Poorly/uniformly graded
	1.0-3.0	0.56	1.40	2.30	4.10	0.48	Poorly/uniformly graded
	3.0-4.0	0.55	1.40	2.00	3.63	0.54	Poorly/uniformly graded
	4.4-4.5	1.20	2.00	3.40	2.83	1.41	Poorly/uniformly graded
MW3	0.0-1.0	0.55	2.80	6.00	10.91	0.72	Poorly/uniformly graded
	1.0-2.0	0.45	0.80	1.80	4.00	0.16	Poorly/uniformly graded
	2.0-3.0	0.68	1.20	10.00	14.00	0.10	Poorly/uniformly graded
	3.0-4.8	0.50	0.55	0.60	1.20	0.25	Poorly/uniformly graded
	>4.8	4.00	11.00	15.00	3.75	2.75	Poorly/uniformly graded
MW 5	0.0-1.0	0.41	1.20	2.00	4.88	0.30	Poorly/uniformly graded
	1.0-2.2	0.48	0.80	5.10	10.63	0.06	Poorly/uniformly graded
	2.2-3.2	0.60	1.40	2.00	3.33	0.59	Poorly/uniformly graded
	3.2-4.0	0.41	0.45	1.10	2.69	0.08	Poorly/uniformly graded
	>4.0	0.00	0.00	8.50	-	-	Poorly/uniformly graded

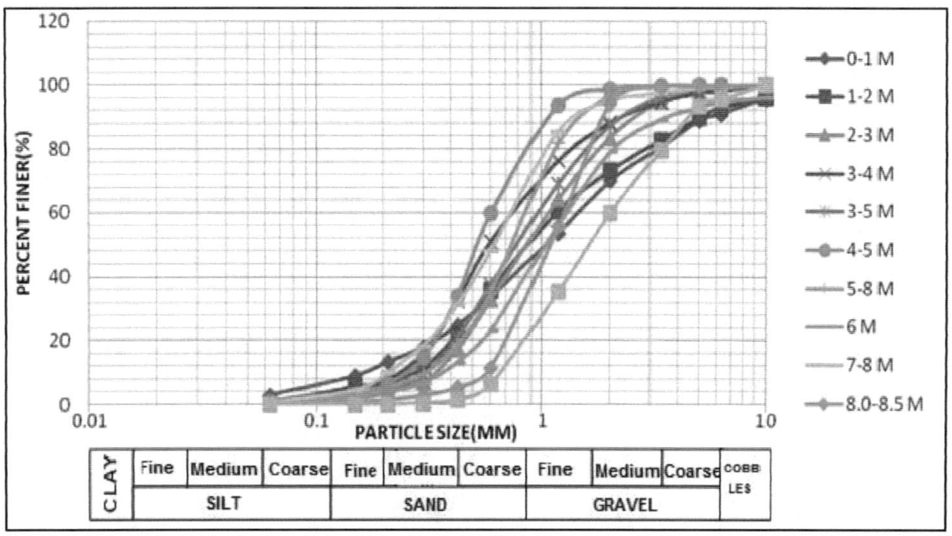

Figure 3. Graph of grain size analysis for soil samples at PW borehole

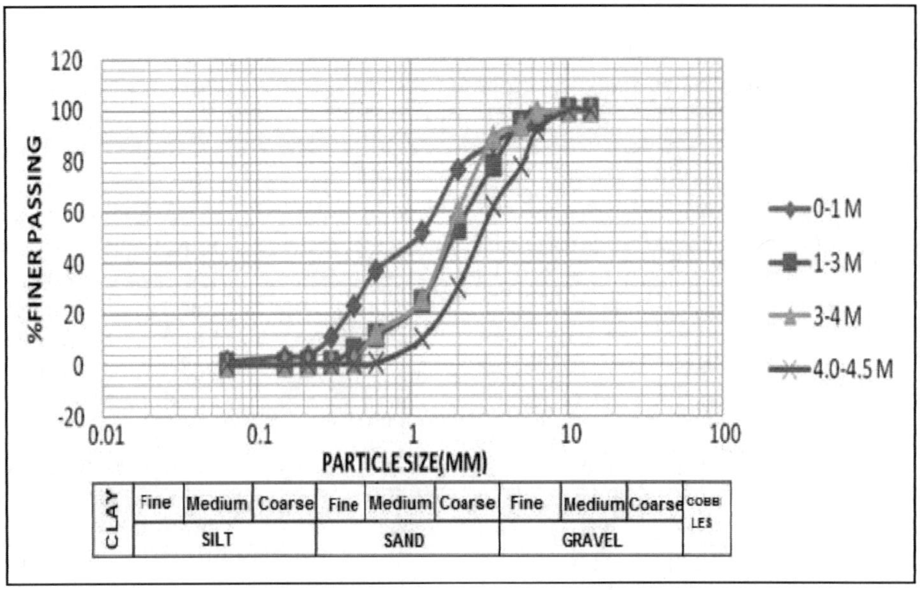

Figure 4. Graph of grain size analysis for soil samples at MW2 borehole

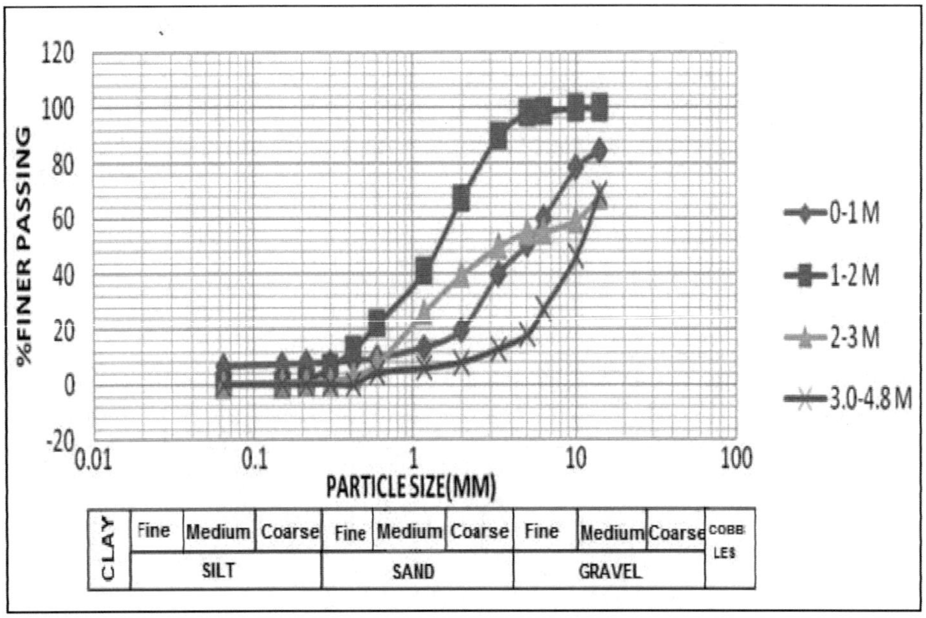

Figure 5. Graph of grain size analysis for soil samples at MW3 borehole

IOP Conf. Series: Materials Science and Engineering **133** (2016) 012003 doi:10.1088/1757-899X/133/1/012003

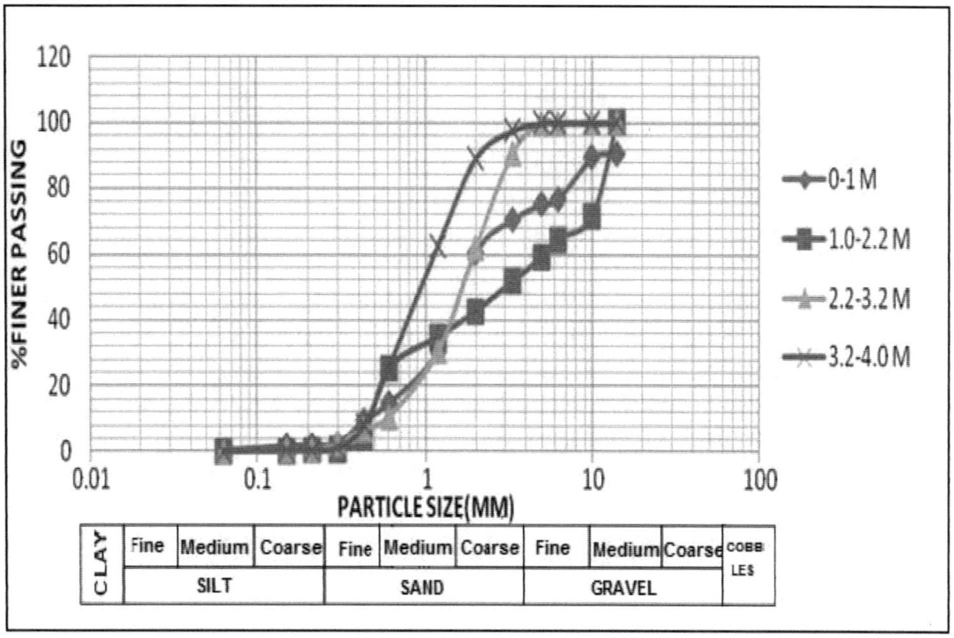

Figure 6. Graph of grain size analysis for soil samples at MW5 borehole

3.2. Permeability test for soil samples

Constant head permeability test were conducted for all the soil samples within PW and MW`s in the Geotechnical Engineering Laboratory, School of Civil Engineering, Universiti Sains Malaysia (USM). Permeability or hydraulic conductivity, k was a measurement of flow within a soil sample. Table 3 shows the results of permeability of the soil samples.

Table 3. Permeability test results for soil samples within PW, MW2, MW3 and MW5

Boreholes	Depth (m)	Hydraulic conductivity, k (cm/s)
PW	3.0-4.0	0.15
	3.0-5.0	0.15
	5.0-8.0	0.26
	6.0	0.91
	7.0-8.0	0.10
MW2	1.0-3.0	1.90
	3.0-4.0	0.90
	4.0-4.5	4.70
MW3	0-1.0	0.53
	2.0-3.0	0.47
	3.0-4.8	1.47
MW5	1.0-2.0	0.38
	1.0-2.2	0.43
	2.2-3.2	5.03
	3.2-4.0	1.15

The table indicates that the hydraulic conductivity, k, of soil samples from the PW to range between 0.10 and 0.91 cm/s. The values of k for MW2, MW3 and MW5 ranged between 0.38 and 5.03 cm/s or between 3.8×10^{-3} and 5.03×10^{-2} m/s. The literature indicates that an alluvial aquifer near a river such as the Rhine River often have a hydraulic conductivity of 10^{-3} to 10^{-2} m/s [13,14]. In previous RBF application, the hydraulic conductivity, k, of Illinois River, USA had ranged between 2×10^{-3} and 3×10^{-3} m/s. The Ohio River in Louisville, Kentucky, USA had a k value of 6×10^{-4} m/s, while the value for Great Miami River in Cincinnati, USA ranged between 8.8×10^{-4} and 1.5×10^{-3}m/s [15].

These result shows that hydraulic conductivity at field site and k obtained from constant head permeability test were quite similar with the natural materials from Sungai Perak, Kota Lama Kiri, Kuala Kangsar, Perak. Based on the screen depth between 5.00 to 8.00 metre witihin PW, soil samples hydraulic conductivity of 0.26 cm/s, 0.91 cm/s, 0.10 cm/s for depth 5.00 to 6.00 metre, 6.00 to 7.00 metre, 7.00 to 8.00 metre, respectively, and the result of pumping test 112.10 m^3/h, it can be inferred that the average value of hydraulic conductivity of 0.42 cm/s, has the possibility to yield the abstraction of 112.10 m^3/h.

3.3. Pumping test
Pumping test was conducted from 27 September 2013 to 30 September 2013 for 72 hours. Constant discharge rate for PW was 112.10 m^3/h. The recovery percentage for PW was 84.67% for duration time of 180 minutes. The static water level within PW before conducting the pumping test was 2.60 metre. The measuring drawdown during pumping test was 0.79 metre.

4. Conclusion
A typical aquifer of a riverbank alluvial deposit considered for RBF application was investigated for the characteristics of its soil samples. The grain size analysis of PW, MW2, MW3 and MW5 showed that soils consisted of medium sand, fine sand, coarse sand and gravel with the medium sand having the highest percentage. The percentages of medium sands at various depths of PW, MW2, MW3, and MW5 were 45.12 to 77.90, 30.16 to 59.35, 0.70 to 54.51, and 0.32 to 81.34 respectively. The aquifer extended down to 8 meter deep. The maximum hydraulic conductivity for the PW was 0.91 cm/s at 6 m deep. The highest permeability, k was 5.03 cm/s, which is for the depth of 2.20 to 3.20 m of MW5. The constant discharge pumping rate of 112.10 m^3/h verified the suitability of the site for RBF application.

References
[1] Hiscock KM and Grischek T 2002 Attenuation of groundwater pollution by bank filtration *Journal of Hydrology* **266**:139–144
[2] Samuel M and Jonathan L 2010 Using temperature modeling to investigate the temporalvariability of riverbed hydraulic conductivity during storm events *J Hydrol* **388**: 321-334
[3] Ray C, Melin G, Linsky RB 2002a Riverbank filtration improving source-water quality Kluwer Academic Publishers, Fountain Valley, CA, USA; 364
[4] Vandenbohede A, Lebbe L 2003 Combined interpretation of pumping and tracer tests: theoretical considerations and illustration with a field test. *J Hydrol* **277**:134–149
[5] Kim J-W, Choi H, Lee J-Y 2005 Characterization of hydrogeologic properties for a multi-layered alluvial aquifer using hydraulic and tracer test and electrical resistivity

IOP Conf. Series: Materials Science and Engineering **133** (2016) 012003 doi:10.1088/1757-899X/133/1/012003

survey *Environ Geol* **48**:991-1001

[6] Pickens JF and Grisak GE 1981Scale dependent dispersion in a stratified granular aquifer *Water Resource* **17**:1191–1211

[7] Mas-Pla J, Yeh T-CJ, McCarthy JF, Williams TM 1992 A forced gradient tracer experiment in a coastal sandy aquifer, Georgetown site, South Carolina *Ground Water* **30**:958–964

[8] Thorbjarnarson KW, Mackay DM 1997 A field test of tracer transport and organic contaminant elution in a stratified aquifer at the Rocky Mountain Arsenal (Denver, Colorado, USA) *J Contam Hydrol* **24**:287–312

[9] TZW 2006a Exportorientierte F&E auf demGebiet der Wasserver-und–entsorgung, Teil II: Trinkwasser, Leitfaden, Publisher: DVGW Technologie ZentrumWasser (TZW), Karlsruhe, Germany; 249 p

[10] Grischek T, Schoenheinz D, Ray C 2002 Sitting and design issues for river bank filtration schemes. In: *Riverbank filtration: improving source water quality.* Ray C *et al* (eds) Kluwer, Dordrecht, The Netherlands; 291–302

[11] Ray C, Grischek T, Schubert J, Wang JZ and Speth, TF 2002 A perspective of riverbank filtration *Journal of American Water Works Association* **94**:149–160

[12] Schubert J 2002 Hydraulic aspect of river bankfiltration-field studies *J Hydrol* **266**:145–161

[13] Schubert J 2006 Experience with riverbed clogging along the Rhine River In: *Bank filtration Hydrology – Impacts on system capacity and water quality* Hubbs, S.A. (eds.) NATO Science Series IV Earth and Environmental Sciences 60:221-242

[14] Price ML, Flugum J, Jeane P, Tribbet-Peelen L 1999 Sonoma County finds groundwater under direct influence of surface water depends on river conditions. In: *Proceedings Int. Riverbank Filtration Conf.*, National Water Research Institute, Fountain Valley, CA, USA; 25–27

[15] Dash RR, Mehrotra I, Kumar P, Grischek T 2008 Lake bank filtration at Nainital, India: water quality evaluation *Hydrogeol J*, **16**:1089–1099. DOI: 10.1007/s10040-008-0295-0

Acknowledgments

The authors would like to acknowledge the Ministry of Education Malaysia for providing financial support through LRGS Grant on Water Security entitled Protection of Drinking Water: Source Abstraction and Treatment (203/PKT/6726001).

IOP Conf. Series: Materials Science and Engineering **133** (2016) 012004 doi:10.1088/1757-899X/133/1/012004

Interrelationship of Kaolin, Alkaline Liquid Ratio and Strength of Kaolin Geopolymer

Shamala Ramasamy[1,a], Kamarudin Hussin[1,2,b], Mohd Mustafa Al Bakri Abdullah[1,2,c], Che Mohd Ruzaidi Ghazali[1,2,d], Mohammed Binhussain[3,e] Andrei Victor Sandu[4, f]

[1] Center of Excellence Geopolymer and Green Technology, School of Materials Engineering, Universiti Malaysia Perlis (UniMAP), 01007, P.O Box 77, D/A Pejabat Pos Besar, Kangar, Perlis, Malaysia.
[2] Faculty of Engineering Technology, Universiti Malaysia Perlis (UniMAP), 01007, P.O Box 77, D/A Pejabat Pos Besar, Kangar, Perlis, Malaysia.
[3] King Abdul Aziz City Science & Technology (KACST), P.O. Box 6086, Riyadh 11442, Kingdom of Saudi Arabia.
[4]Faculty of Materials Science and Engineering, Gheorge Asachi Technical University of Iasi, Blvd. D. Mangeron 71, Iasi 700050, Romania

E-mail: [a]shamala_ramasamy@yahoo.com,[b]vc@unimap.edu.my, [c]mustafa_albakri@unimap.edu.my,[d]ruzaidi@unimap.edu.my, [e]bnhusain@kacst.edu.sa, [f]sav@tuiasi.ro

Abstract. Geopolymer is an incredible alternative green cementitious material which has ceramic-like properties, but does not require calcining that leads to reduction in processing energy usage. The purpose of this research is to study the correlation between kaolin: liquid ratio with the performance of kaolin geopolymer. Kaolin, a prominent raw geopolymer material was used to prepare enhanced geopolymer paste by mixing with alkaline activator solution. Interrelationship of kaolin to alkaline liquid ratio with hardness and flexural strength was the focus of this work. Therefore kaolin geopolymer paste with varying solid to liquid ratio ranging from 0.7 to 1.1 was prepared. Geopolymer paste was coated on low grade wood substrate prior to Vickers hardness and flexural strength. X-ray diffraction was conducted on geopolymer paste itself after 7 days to analyze the change in phase identification at early age. Kaolin geopolymer coating on wood with solid/liquid(S/L) ratio of 0.7 shows the most promising hardness and flexural strength of 15.3 Hv and 94.73MPa. X-ray diffraction test showed high existence of kaolinite on higher S/L ratio where as sodalite was observed in S/L ratio of 0.7. Microstructural studies also compliments our finding which further proves the positive dependency of S/L ratio and kaolin geopolymer strength.

1. Introduction

The Geopolymer was first introduced by Joseph Davidovits in 1978 were it was rarely known to many [1-4]. However ever since various research development took place, geopolymers is now prevalent materials that is being rapidly developed world wide. This is mainly because of its environmental friendly nature, specific performances of high temperature resistance, high compressive strength, low density and low permeability [5]. Geopolymer source materials are materials rich in aluminosilicate such as kaolin, fly ash, metakaolin,and slag [6-9]. Geopolymers are class of largely X-ray amorphous aluminosilicate materials that is produced when solid source material powders are mixed with concentrated alkali metal silicate or hydroxide solution [10]. Geopolymers are inorganic binder that exhibits promising mechanical properties. Few works has attempted to use many materials that are rich in silica and alumina to synthesize geopolymers and succeeded [11-15]. Usage of kaolinite clays and sintered kaolin has also gained interest of multiple researchers worldwide [5, 8, 16-18]. Generally Kaolin consists of alumina octahedral sheets and silica tetrahedral sheets stacked alternately and has the theoretical formula $(OH)_8Si_4A1_40_{10}$ and the theoretical composition 46.54% SiO_2, 39.5% Al_20_3, 13.96% H_20[15].

Wood regardless of hard wood or low grade (LG) wood consist of mainly lignin, cellulose and hemicelluloses [19]. The structural and non-structural grade woods contains different grain orientation and texture, therefore their strength will also differ [20-23]. Apart from that, processing and multiple types of inherent defects also contributes to the strength of wood products. Low grade wood are natural polymeric material with lower mechanical properties [24]. LG woods are also cost competitive, architecturally attractive, natural resource and a recyclable material [25]. LW wood was used in this work to analyse if there are any improvements in terms of its mechanical properties upon being coated with kaolin geopolymer paste.

Kaolin has been used for various application but never as a coating geopolymer paste. To best of our knowledge, apart from our research group, CEGeoGtech, there are no active research on kaolin based geopolymer coating. This is an area of study that are still not studied in detailed. Focus of this work is to find kaolin based geopolymer coating paste with optimum kaolin to alkaline activator liquid ratio that exhibits enhanced mechanical properties. Thus, for mechanical testing, kaolin geopolymer paste was coated on low grade wood substrate to observe the change in properties.

2. Materials and Methods

Kaolin, raw material for this work was obtained from Associated Kaolin Industries Sdn. Bhd, Malaysia. Kaolin was the sieved using desired mesh size and stored in an air tight container at our laboratory. Sodium chloride flakes were purchased from Formosa Plastic Corporation, Taiwan. Technical grade liquid sodium silicate was supplied by South Pacific Chemicals Industries Sdn. Bhd. (SPCI) Malaysia. Sodium chloride flakes were then used to prepare stock solutions of NaOH solution with 8 M.

Liquid Na_2SiO_3 and NaOH solution are needed for preparation step. The alkaline activator solution was prepared by adding NaOH solution into liquid Na_2SiO_3 and continuously stirring until a clear solution was obtained. However the alkaline activator mixture must be prepared and kept aside minimum 24 hours prior to use to achieve equilibration and a homogenous solution. The ratios of Na_2SiO_3/NaOH used in this work is 0.45 which is the optimum range based on our findings and previous literature [26].

Sieved kaolin was mixed with alkaline activator solution at desired various solid/liquid ratio ranging from 0.7 to 1.10. The mixture was then stirred via mechanical stirrer manufactured by TPG Motors & Drives Corporation type ML712-4 at a speed of 20 rpm for 5 minutes until homogenous slurry was obtained. .

Upon mixing, kaolin based geopolymer paste is obtained. For the Vickers hardness and flexural strength test, geopolymer paste was brushed evenly onto low grade wood subsrates with a stabilo painting brush. After the first coating layer, the wood substrates were allowed to dry for 60 seconds. Next, subsequent layers were brushed onto the low grade wood until desired thickness is obtained. For the X-ray diffraction test, kaolin geopolymer paste was analysed.

After coating process, the coated LG wood and geopolymer paste were placed into hot air Table Top Oven by KCG for drying and curing process. The curing time was maintained at 72 hours. Upon curing in oven, samples were removed, allowed to cool to room temperature and sealed in thin plastic bags to prevent dust and impurity in surrounding prior to testing.

2.1. Vickers Hardness

Vicker hardness test is performed to obtain the resistance of sample towards plastic deformation. Kaolin geopolymer coated LG wood samples were cured at 60°C for 3 days for complete geopolymerisation to occur.This test will measure the geopolymer coating resistance to a localised load that leads to plastic deformation. Vickers hardness test was conducted under the test procedure of ASTM E-384 which specifies a light load range with a diamond indenter that causes the indentation to be measured in hardness value. Benchmarking previous studies and ASTM standard, for kaolin geopolymer coating, load of 1.0 kg (9.807 N) was used to provide a reasonable indent for geopolymer coating hardness determination.Vickers hardness number, H_v is a number representing hardness value related to the applied force (P) and the area of an indent is given by a square-based pyramidal diamond indenter having included face angle of 136° as shows in Figure 1. Vickers hardness value, Hv can also be calculated in MPA by the equation 1 shown as per below. For this test, five samples were tested and average value was obtained.

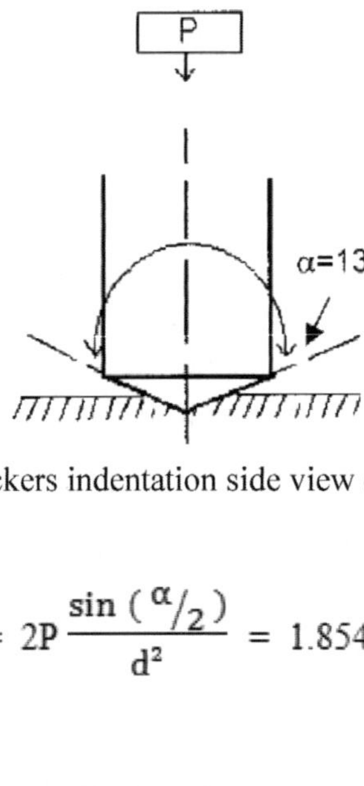

Figure 1: Vickers indentation side view of diamond indenter.

$$H_v = \frac{P}{As} = 2P\frac{\sin(\alpha/2)}{d^2} = 1.8544\frac{P}{d^2} \qquad \text{(Eq 1)}$$

Where by, P= load (N),

As= Surface area of indentation (mm2),

d= mean diagonal of identation(mm)

α = face angle of indenter (136°)

2.2 *Flexural Strength*

Flexural strength test is performed to analyse the behavioural change of material subjected to a simple beam loading Kaolin geopolymer coated low grade wood was prepared as per the standard testing size of 150mm*15mm*3mm rectangular shape. The flexural test of these coated samples were conducted

IOP Conf. Series: Materials Science and Engineering **133** (2016) 012004 doi:10.1088/1757-899X/133/1/012004

under the ASTM D709 using Automatic Max (Instron, 5569 USA).This test also determines the properties of control and reinforced materials , which allows the comparison of control and coated low grade substrates.

2.3 *X-Ray Diffraction (XRD) Test*

X-Ray Diffraction (XRD) measurements were performed for phase analysis of the starting materials and to investigate the degree of crystallinity of the resulting geopolymer coat. The sampling for this analysis was made by grinding the sample with pestle and mortar to obtain very fine powder before the sample spread on the surface of the sample holder using pieces of glass to minimize the effect of preferred orientation on the XRD pattern. Random powder samples of kaolin were prepared by lightly pressing powder samples into aluminium holders. XRD powder diffractograms of geopolymer specimens are collected on a XRD-6000 Shimadzu X-Ray Diffractometer with CuKα radiation generated at 30 mA and 40 kV. Step scans are performed from 10 - 70° 2θ at 0.02° 2θ steps, integrated at a rate of 1.2 second per step.

3. Results and discussions

Vickers hardness is conducted to measure the resistance to permanent deformation or damage layer of kaolin geopolymer coating on wood substrate. The applied load used for geopolymer is in range of 0.1 – 1.5 kg appropriate to the hardness of cementitious materials in geopolymer coating [27]. Benchmarking previous studies, for kaolin geopolymer coating, load of 1.0 kg (9.807 N) was used to provide a reasonable indent for geopolymer coating hardness determination. Table 1 shares the hardness values of various kaolin to alkaline liquid ratio of kaolin geopolymer coating on LG wood substrates. As per shown in Table 1 and Figure 2, hardness value of kaolin geopolymer coated wood substrate with S/L of 0.7 has the highest value. This is due to its good workability that enhances the dissolution of aluminosilicates in kaolin. As the S/L ratio is increased, the hardness value drops gradually. When geopolymer paste with S/L ratio below 0.6 was prepared, the geopolymer paste failed at curing stage as sever crack were observed. Thus, it can be concluded that kaolin geopolymer with S/L ratio of 0.7 is the most suitable ratio in terms of the hardness value.

Table 1. Parameters of kaolin geopolymer coating at different solid/liquid ratio tested for Vickers hardness

Sample	Solid/Liquid	$Na_2SiO_3/NaOH$	Vickers Hardness (Hv)	Hv (MPa)	Hv (GPa)
	Control	-	4.5	44.13	0.04
Kaolin Geopolymer coated Wood	0.7	0.45	15.3	150.05	0.15
	0.8	0.45	13.6	133.38	0.13
	0.9	0.45	10.3	101.02	0.10
	1.0	0.45	8.8	86.31	0.09
	1.1	0.45	6.3	61.79	0.06

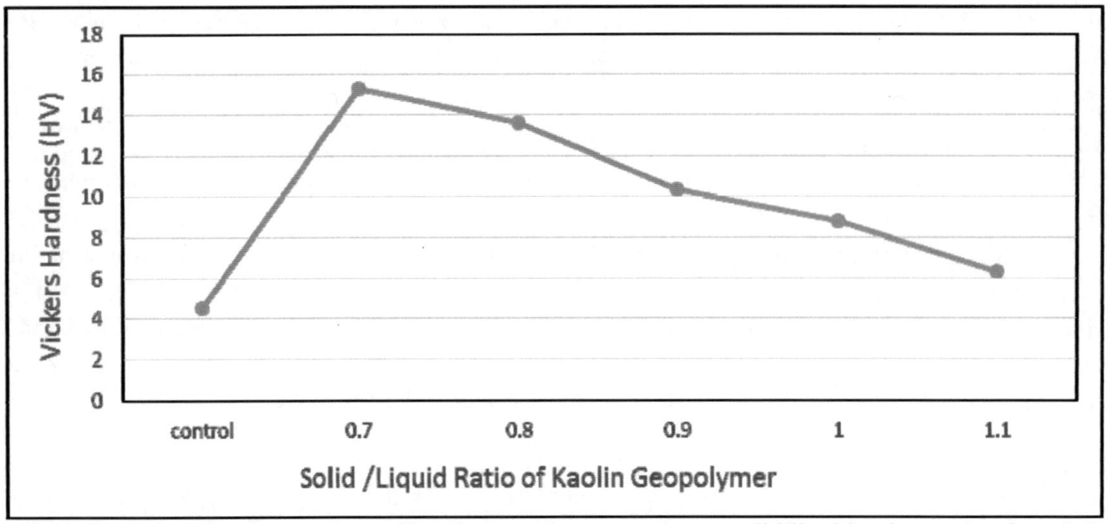

Figure 2: Vickers hardness values of kaolin geopolymer coating at solid/liquid ratio ranging from 0.7 to 1.1.

The relationship between hardness results with flexural strength is displayed in Figure 3. The line graph exhibit similar features which has enhanced result at S/L ratio of 0.7 for kaolin geopolymer coated wood substrates. These results may lead to the successful optimization of the mix design of kaolin geopolymer coating thus obtaining good performance in term of mechanical strength.

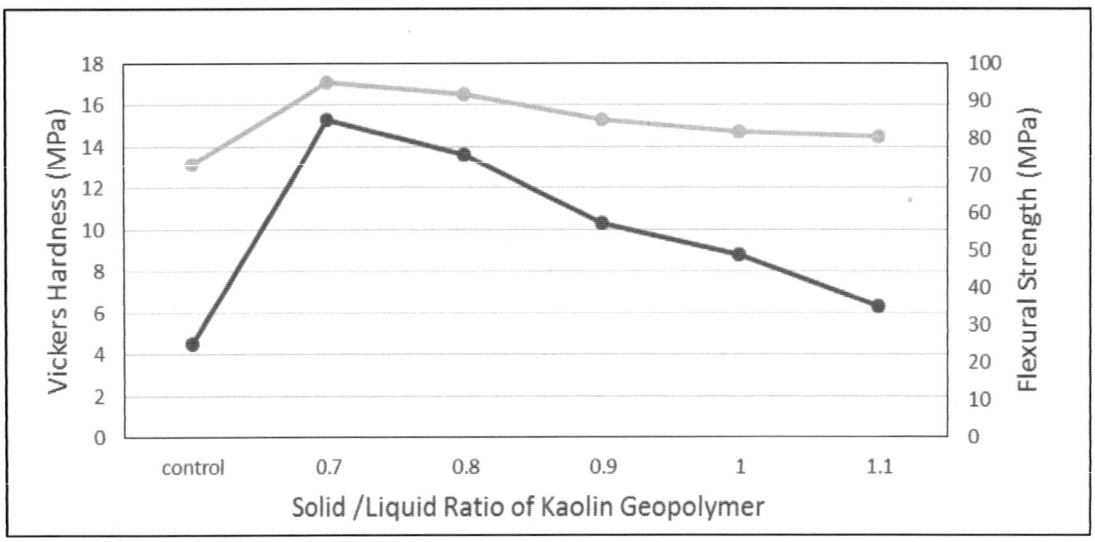

Figure 3: Vickers hardness strength versus flexural strength of kaolin geopolymer coating at solid/liquid ratio ranging from 0.7 to 1.1.

X-Ray diffraction (XRD) analysis were performed to classify phase indentification, degree of crystallinity and monitor the transformation in amorphous peak from raw kaolin materials into geopolymer paste. Selected XRD patterns of kaolin geopolymer with S/L of 0.7, 0.9 and 1.1(Figure 4) was compared and analysed. As per illustrated by Figure 4, major mineral contribution in geopolymer paste is kaolinite peaks (K) at 2θ values of 12.5°, 20.0°, 24.51°,32.1° and 35.03° in XRD patterns of kaolin geopolymers tested after 7 days. Lower intensity of kaolinite in XRD patern of kaolin geopolymer with S/L of 0.7 shows the higher dissolution of kaolin as compared to the rest. Kaolinite peaks shows the unreacted kaolin that remained in geopolymer paste [30-31]. Another major mineral quartz (Q) at 2θ values of 26.76°, 43.4 °, 46.1°, 53.09°, 55.28°, 59.23°, 62.7° and 79.97°, is the mineral compound that is not involved in the geopolymerisation process[32]. New peak of sodalite that was narrow at 2θ value of 17.70°, 21.59°, 28.1° and 33.39°. The patterns show the existence of sodalite peak indicates the geopolymerization reaction between Na, SiO_2 and Al_2O_3. The low counts in XRD pattern of geopolymer products between 25° and 35° (2θ) attributed the geopolymer products have amorphous structure [33-34].

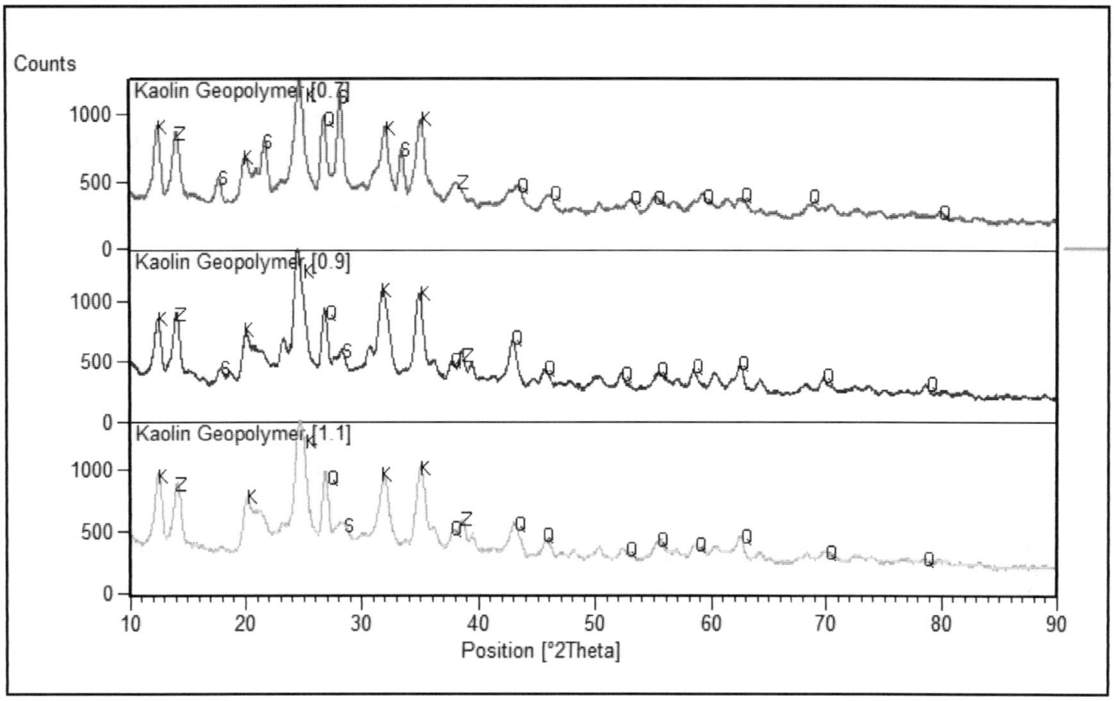

Figure 4: X-Ray Diffraction pattern of solid/liquid ratio of 0.7, 0.9 and 1.1.

4. Conclusions

Kaolin geopolymer paste showed promising enhancement which makes it a very reliable repair or coating material. Mechanical properties such as Hardness and flexural strength of these kaolin geopolymer that was coated on LG wood substrates gave the best results at kaolin/ alkaline activator liquid ratio of 0.7. This finding was further supported by the X-Ray diffraction results also gave positive reults for the solid/liquid ratio of 0.7 Thus, our findings can be summarize that with the optimum kaolin to alkaline liquid ratio, strength of kaolin geopolymer can be further enhanced.

Acknowledgments

References

[1] A.M. Mustafa Al Bakri, H. Kamarudin, M. Bnhussain, I. Khairul Nizar, Y.Zarina and A.R. Rafiza.(2012). Fly Ash Based Geopolymer Lightweight Concrete Using Foaming Agent. International Journal of Molecular Science 13, p.7186-7196.

[2] Zhang Yunsheng, Sun Wei, and Li Zongjin. (2010). Composition design and microstructural characterization of calcined kaolin-based geopolymer cement, Applied Clay Science ,47, p.271–275.

[3] Zhang, Wei and Li (2009).Preparation and microstructure of K-PSDS geopolymeric binder. Colloids and Surfaces A: Physicochemical and Engineering Aspects, 302, 473-482.

[4] Heller-Kallai L and Lapides I.(2007). Reactions of kaolinites and metakaolinites with NaOH comparison of different samples (Part 1), Appl Clay Sci 35, p.99–107

[5] Alonso and Palomo (2001). Alkaline activation of metakaolin and calcium hydroxide mixtures: Influence of temperature, activator concerntration and solids ratio, Materials Letters, 47(1-2), 55-62.

[6] Fernandez-Jimenez and Palomo (2007). Factor affecting early compressive strength of alkali activated fly ash (OPC-free) concrete. Materials de Construction, 57, 7-22.

[7] Chindaprasirt, Chareerat and Sirivivatnanon (2007). Workability and strength of coarse high calicium fly ash geopolymer. Cement and concrete composites, 29(3), 224-229.

[8] Wang, Li and Yan (2005).Synthesis and mechanical properties of metakaolinite-based geopolymer.Colloids and Surfaces A: Physicochemical and Engineering Aspects, 268(1-3),1-6

[9] Y.M. Liew, H. Kamarudin, A.M. Mustafa Al Bakri, M. Binhussain, M. Luqman, I. Khairul Nizar, C.M. Ruzaidi, and C.Y. Heah.(2011). Influence of Solids-to-liquid and Activator Ratios on Calcined Kaolin Cement Powder, Physics Procedia ,22, p.312 – 317.

[10] M.M.A. Abdullah, H. Kamarudin, H.Mohammed, I. Khairul Nizar, A.R. Rafiza, and Y.Zarina.(2011). The relationship of NaOH Molarity, Na2SiO3/NaOH Ratio, Fly Ash/Alkaline Activator, and Curing Temperature to the strength of Fly Ash –Based Geopolymer, Advanced Materials Research 328-330, p 1475-1482.

[11] A.M. Mustafa Al Bakri, H. Kamarudin, M. Binhussain, I. Khairul Nizar, A.R. Rafiza, and Y.Zarina.(2012).The Processing, Characterization, and Properties of Fly Ash Based Geopolymer Concrete, Rev.Adv.Mater. Sci 30, p90-97

[12] Duxson, Lukey and Van Deventer (2007).Physical evolution of Na-geopolymer derived from metakaolin up to 1000°C.Journal of Materials Science, 42,3044-3054.

[13] Tempest, Sanusi, Gergely, Ogunro and Weggel (2009). Compressive strength and embodied energy optimization of fly ash based geopolymer concrete. Paper presented at the 2009 World of Coal Ash (WOCA).

[14] Hardjito and Rangan (2005). Development and properties of low-calcium fly ash based geopolymer concrete.Research report GCI. Faculty of Engineering Curtin University of Technology, Perth, Australia.

[15] Y.Zarina, A.M. Mustafa Al Bakri, H. Kamarudin, Khairul Nizar, and A.R. Rafiza, (2013).Review on The Various Ash From Palm Oil Waste As Geopolymer Material, Rev.Adv.Mater. Sci 34, p37-43.

[16] Yip, Lukey and Van Deventer (2005).The coexistence of geopolymeric gel and calcium silicate hydrate at the early stage of alkaline activation. Cement and concrete research, 35(9), 1688-1697.

[17] C.Y. Heah, H. Kamarudin, A.M. Mustafa Al Bakri, M. Bnhussain, M. Luqman, I. Khairul Nizar, C.M. Ruzaidi, Y.M. Liew. (2012). Study on solids-to-liquid and alkaline activator ratios on kaolin-based geopolymers, Construction and Building Materials 35, p.912–922.

[18] A.M. Mustafa Al Bakri, J.Liyana, H. Kamarudin, M. Bnhussain, C.M. Ruzaidi,and M.A. Izzat.(2012). Fly Ash Porous Material using Geopolymerization Process for High Temperature Exposure. International Journal of Molecular Science 13, p.4388-4395.

[19] Łukasz Czajkowski, Wiesław Olek, Jerzy Weres, and Ryszard Guzenda,(2016) Thermal properties of wood-based panels: specific heat determination, Wood Science and Technology , 21,1-9.

[20] Ataollah Haddadi, Brigitte Leblon, Zarin Pirouz, Joseph Nader, and Kevin Groves (2016)

Prediction of wood properties for thawed and frozen logs of quaking aspen, balsam poplar, and black spruce from near-infrared hyperspectral images, Wood Science and Technology, 50(2), p 221-243.

[21] J. Fernández-Moya, A. Alvarado, W. Forsythe, L. Ramírez , N. Algeet-Abarquero,and M. Marchamalo-Sacristán.(2014).Soil erosion under teak (Tectona grandis L.f.) plantations: General patterns, assumptions and controversies, Catena 123, p. 236–242.

[22] Stephen Adu-Bredu, Alphonse Foua Tape Bi, Jean-Pierre Bouillet, Martial Kouame´ Me´, Samuel Yamoah Kyei, and Laurent Saint-Andre. (2008). An explicit stem profile model for forked and un-forked teak (Tectona grandis) trees in West Africa, Forest Ecology and Management 255, p.2189–2203.

[23] Narong Koonkhunthod, Katsutoshi Sakurai, and Sota Tanaka. (2007) Composition and diversity of woody regeneration in a 37-year-old teak (Tectona grandis L.) plantation in Northern Thailand, Forest Ecology and Management 247, p. 246–254.

[24] William D. Greason, (2013). Triboelectrification of wood with PTFE, Journal of Electrostatics, 71,140–144.

[25] Magnus Wiman, Benny Palmqvist, Eva Tornberg, and Gunnar Lidén (2011). Rheological characterization of dilute acid pretreated softwood, Biotechnology and Bioengineering, 108(5),1031-1041.

[26] Rovnanik (2010).Effect of curing temperature on the development of hard structure of metakaolin-based geopolymer. Construction and Building Materials, 24, 1176-1183.

[27] Rattanasak and Chindaprasirt (2009). Influence of NaOH solution on the synthesis of fly ash geopolymer.Minerals Engineering, 22(12), 1073-1078.

[28] Tempest, Sanusi, Gergely, Ogunro and Weggel (2009). Compressive strength and embodied energy optimization of fly ash based geopolymer concrete. Paper presented at the 2009 World of Coal Ash (WOCA).

[29] Sami Ullah, Faiz Ahmad, A.M Shariff and M.A. Bustam (2014).Synergistic effects of kaolin clay on intumescent fire retardant coating composition for fire protection of structural steel substrate.

[30] Y.M. Liew, H. Kamarudin, A.M. Mustafa Al Bakri, M. Bnhussain, M. Luqman, I. Khairul Nizar, C.M. Ruzaidi, and C.Y. Heah.(2012).Optimization of solids-to-liquid and alkali activator ratios of calcined kaolin geopolymeric powder, Construction and Building Materials 37, p. 440-451.

[31] Prasad, M.S., Reid, K.J., and Murray, H.H. (1991).Kaolin: processing, properties and applications, Applied Clay Science, 6, p 87-119.

[32] Provis, Yong, Duxson and Van Deventer (2009). Correlating mechanical and thermal properties of sodium silicate-fly ash geopolymers. Colloids and surfaces A: Physicochemical and Engineering Aspects, 336(1-3), 57-63.

[33] Zuhua, Xiao, Huajun and Yue (2009).Role of water in the synthesis of calcined kaolin based geopolymer. Applied Clay Science, 43,218-223.

IOP Conf. Series: Materials Science and Engineering **133** (2016) 012005 doi:10.1088/1757-899X/133/1/012005

Modeling and simulation of temperature effect in polycrystalline silicon PV cells

M Marcu, T Niculescu, R I Slusariuc*, F G Popescu

Department of Control Engineering, Computers, Electrical Engineering and Power Engineering, University of Petrosani, 20 Univerisitatii str., Petrosani, Romania

E-mail: razvansir@yahoo.com

Abstract. Due to the human needs of energy, there is a need to apply new technologies in energy conversion to supply the demand of clean and cheap energy in the context of environmental issues. Renewable energy sources like solar energy has one of the highest potentials. In this paper, solar panel is the key part of a photovoltaic system which converts solar energy to electrical energy. The purpose of this paper is to give a MATLAB/ Simulink simulation for photovoltaic module based on the one-diode model of a photovoltaic cell made of polycrystalline silicon. This model reveals the effect of the ambient temperature and the heating of the panel due to the solar infrared radiation. Also the measurements on the solar cell exposed to solar radiation can confirm the simulation.

1. Introduction

The most important device which converts solar radiation to electric energy is photovoltaic cell.

The International Energy Agency said that "the development of affordable, inexhaustible and clean solar energy technologies will have huge longer-term benefits. It will increase countries energy security through reliance on an indigenous, inexhaustible, enhance sustainability, reduce pollution, lower the costs of mitigating climate change, and keep fossil fuel prices lower than otherwise.

Solar cells are made of semiconductors, and every type of semiconductor has a property called a band gap that is different from that of other semiconductors. The band gap defines the longest wavelength of light a semiconductor can absorb (it is transparent to longer wavelengths). It also fixes the maximum amount of energy that can be captured from photons of shorter wavelength. The result is that long-wavelength photons are lost and short-wave ones incompletely utilized.

Light shining on the solar cell produces both a current and a voltage to generate electric power. This process requires firstly, a material in which the absorption of light raises an electron to a higher energy state, and secondly, the movement of this higher energy electron from the solar cell into an external circuit. The electron then dissipates its energy in the external circuit and returns to the solar cell.

A variety of materials and processes can potentially satisfy the requirements for photovoltaic energy conversion, but in practice nearly all photovoltaic energy conversion uses semiconductor materials in the form of a *p-n* junction.

2. Analysis of circuit

The equivalent circuit of a solar cell consists of a diode and a current source connected in parallel. The current source produces the photocurrent I_{ph}, which is directly proportional to solar irradiance G.

The main parameters used to characterize a PV (photovoltaic) cell are its short-circuit current and its open-circuit voltage [1],[2].

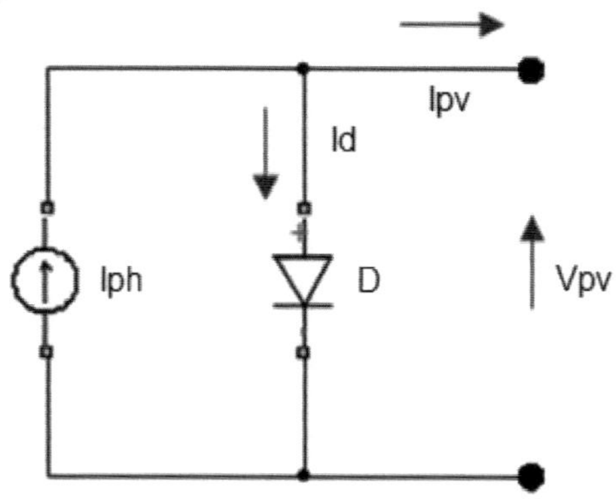

Figure 1. Simplified equivalent circuit of solar cell.

The equation of the current voltage I_{pv}–V_{pv} simplified equivalent circuit is derived from Kirchhoff's law:

$$I_{pv} = I_{ph} - I_d \tag{1}$$

Where

$$I_d = I_0 \left[e^{q\frac{(V_{pv})}{NKT_j}} - 1 \right] \tag{2}$$

So it becomes:

$$I_{pv} = I_{ph} - I_0 \left[e^{q\frac{(V_{pv})}{NKT_j}} - 1 \right] \tag{3}$$

Where:

I_{ph} - is the photocurrent that is equal to short-circuit current, I_0 - is the reverse saturation current of the diode, q - is the electron charge, k - Boltzman constant, N - is diode ideality factor, T_j - is junction temperature of the panels, I_d - is the current shunted through the intrinsic diode, V_{pv} - is the voltage across the photovoltaic cell.

$$I_{pv} = I_{sc} - I_0 \left[e^{q\frac{(V_{pv})}{NKT_j}} - 1 \right] \tag{4}$$

We can determine the reverse saturation current I_0 by setting I_{pv} =0. [2]

$$I_{pv} = 0 \tag{5}$$

$$V_{pv} = V_{oc} \tag{6}$$

$$0 = I_{ph} - I_0\left[e^{q\frac{(V_{pv})}{NKT_J}} - 1\right] \tag{7}$$

Thus we obtain, taking into account the fact that, with this model, the photocurrent is equal to the short-circuit current: [2]

$$I_0 = \frac{I_{sc}}{\left[e^{q\frac{(V_{0c})}{NKT_J}} - 1\right]} \tag{8}$$

2.1. Ohmic losses model

To obtain a better representation of the electrical behaviour of the cell of the ideal model, the second model takes account of material resistivity and the ohmic losses due to levels of contact. These losses are represented by a series resistance R_s in the equivalent circuit.[3],[4],[5]

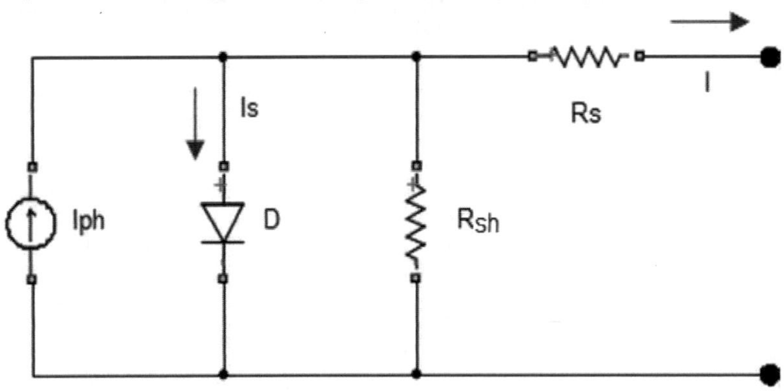

Figure 2. Simplified equivalent circuit of solar cell with R_s..

The coresponding equations are:

$$I = I_{ph} - I_S\left(\exp\frac{q(V + R_s I)}{N_1 KT} - 1\right) - \frac{(V + R_s I)}{R_{sh}} \tag{9}$$

Where:

I_{ph} - is the photocurrent that is equal to short-circuit current, I_s - is the reverse saturation current of the diode, q - is the electron charge, k - Boltzman constant , N - is diode ideality factor, T - is junction temperature of the panels, V - is the voltage across the photovoltaic cell, R_s and R_{sh} are the series resistor respectively shunt resistor [6].

2.2. Effects of Solar Radiation Variation

IOP Conf. Series: Materials Science and Engineering **133** (2016) 012005 doi:10.1088/1757-899X/133/1/012005

The above model includes two subsystems: one that calculates the PV cell photocurrent which depends on the radiation and the temperature according to next equation: [3],[7],[8]

$$I_{ph} = [I_{sc} + K_i(T-298)]\frac{\beta}{1000} \tag{10}$$

2.3. Effect of Varying Cell Temperature

The diode reverse saturation current varies as a cubic function of the temperature and it can be expressed as: [54]

$$I_s(T) = I_s\left(\frac{T}{T_{nom}}\right)^3 \cdot \exp\left[\left(\frac{T}{T_{nom}}-1\right)\frac{E_q}{NV_t}\right] \tag{11}$$

Where:

I_s is the diode reverse saturation current, T_{nom} is the nominal temperature, E_g is the band gap energy of the semiconductor and V_t is the thermal voltage. [9],[10],[5]

As a result, the complete physical behaviour of the polycrystalline photovoltaic cell is dependent on I_{ph}, I_s, R_s and R_{sh} on one hand, and on the other hand dependent on 2 external parameters: temperature and solar radiation. .[11],[12]

Based on the following equation the SIMULINK model was developed in the figure 3.

$$I = I_{ph} - I_o\left(\exp\frac{q(V+R_sI)}{N_1KT}-1\right) - \frac{(V+R_sI)}{R_{sh}} \tag{12}$$

Where:

Table 1. The values used in the block diagram of the solar cell.

Parameter	Value
I_s	0.1nA
q	1.602910^{-19} C
K	1.381910^{-23} J/K
N	1.5
T_{nom}	25^0C
I_{sc}	5.1A
K_i	0.0017 A/^{0}C
V_T	2.57 mV
E_g	1.12

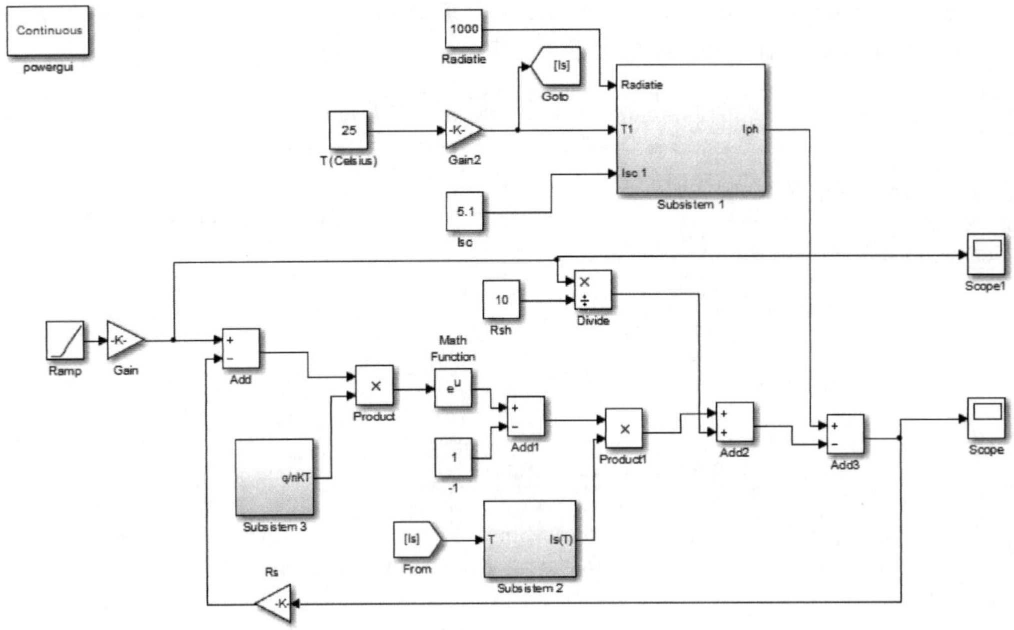

Figure 3. Block diagram of the solar cell model used in Matlab/Simulink software.

For a given value of the solar radiation, of temperature, R_s and R_{sh}, the following characteristics were generated based on the model.

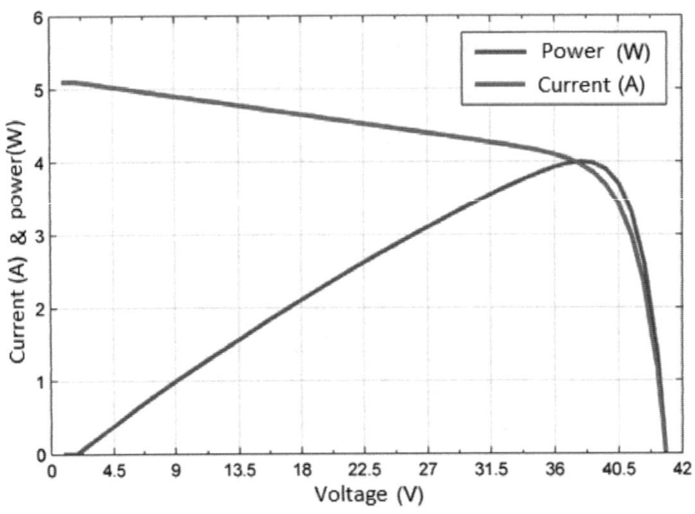

Figure 4. P-V and I-V characteristics of the solar cell.

In the figure 4 we can observe the dependency of the generated current respectively the power depending on the voltage of the polycrystalline solar cell.

In the following figure the curves of the current depending on temperature values of 25 ℃, 50 ℃, 75℃ si 100℃

For the figure 5 the value of the solar radiation was maintained constant at 1000W/m^2 and the temperature of the polycrystalline cell was varied by 25 degrees Celsius. Therefore the curves of the

IOP Conf. Series: Materials Science and Engineering **133** (2016) 012005 doi:10.1088/1757-899X/133/1/012005

current depending on voltage and the power depending on voltage were generated by the Simulink software.

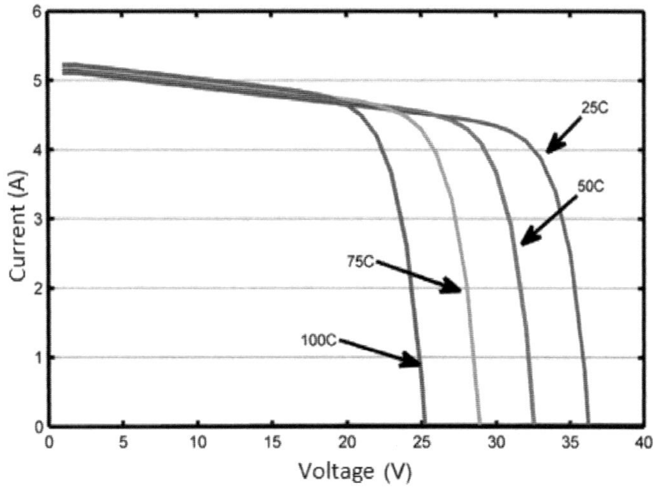

Figure 5. I-V characteristics for the temperature variation between 25 and 100 °C.

It is obvious that the generated voltage by constant solar radiation depends inversely proportional with the temperature growth, thus effecting the generated power of photovoltaic cell. To facilitate the observation of the descending current I introduced an exponential regresion trendline.

The decrease of the generated current is because of the solar movement on the sky and implicitly the changing angle of the Sun to the normal line of the solar pannel.

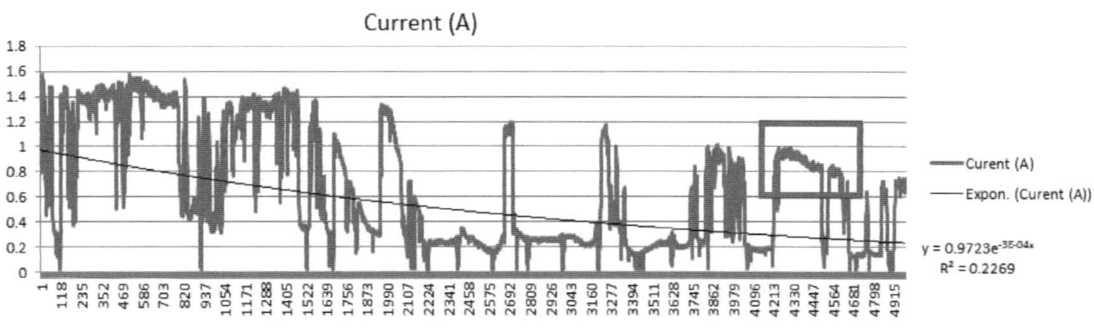

Figure 6. A selection of the measured current.

To analyse the effect of the temperature upon the generated current we will zoom in on a representative section of the measurements made in partly cloudy weather.

On the selected portion a linear regresion curve was added, from which we can see that there is an decrease in the generated current of the solar cell. The decrease is about 150mA, and is according to the simulation. If the temperature increases by roughly 10 degrees celsius the current is inversly proportional and decreases by 150mA. As the temperature increases the current drops.

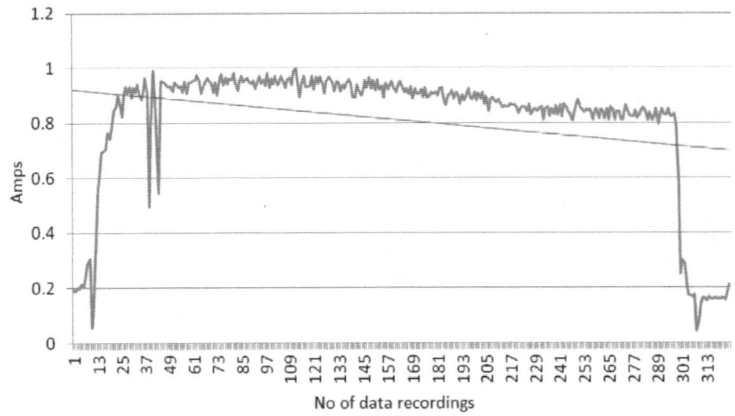

Figure 7. A zoom-in on area of the measurements along with a trendline.

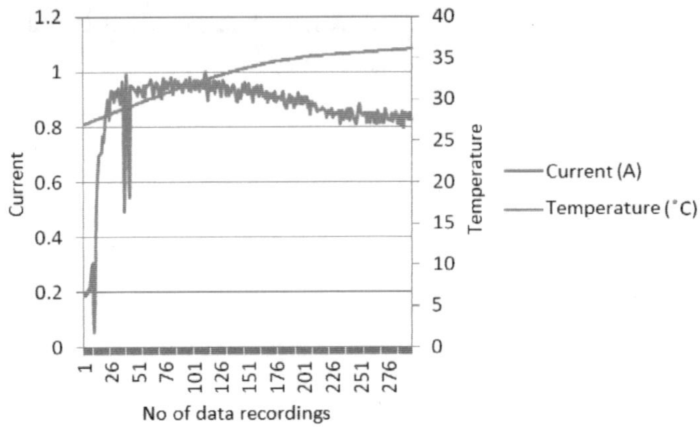

Figure 8. Current dependency on the temperature.

This decrease of the generated current is due to the exposure of the polycrystalline silicon to the mainly infrared radiation in the solar spectrum.

3. Conclusions

Using the Matlab/Simulink software it has been demonstrated the influence of temperature cell's junction and solar radiation upon the performances and therefore the efficiency of solar cell.

The studied effects are: temperature dependence, solar radiation change and series resistance influence. The solar cell with series resistance model offers a more realistic behavior for the photovoltaic systems, and the simulation give us a good perception of the influence of external factors like solar radiation and temperature.

The measurements we made in partly cloudy weather indicates that after a period of time while the Sun is covered by clouds the silicon cell starts to cool down so the effect of temperature upon the polycrystalline cell is easily visible when the Sun begins to shine.

References

[1] T. Markvart, L. Castaner, 2003 Practical Handbook of Photovoltaics: Fundamentals and Applications, *Elsevier*.
[2] D. L. Evans, 1981 Simplified method for predicting photovoltaic array output, *Sol Energy*.
[3] M. Marcu, F. Popescu, I. R. Slusariuc, 2013 Simulation of a solar cell with simple diode model, *Analele Universității din Petroșani*.

[4] T. T. Chow, 2003 Performance analysis of photovoltaic-thermal collector by explicit dynamic model, *Solar Energy*.

[5] E. E. Van Dyk et al., 2000 Temperature-dependent of performance of crystalline silicon photovoltaic modules, *S Afr J Sci*.

[6] M. A. Hamdy, R. L. Cell, 1987 The effect of the diode-ideality factor on the experimental determination of series resistance of solar cells *Sol cells*.

[7] M. A. Green, 2002 Third generation photovoltaics: solar cells for 2020 and beyond 65-70.

[8] T. A. Hove, 2000 Method for predicting long-term average performance of photovoltaic systems, *Renew Energ*.

[9] M. J. Kerr, A. Cuevas, 2003 Generalized analysis of the illumination intensity vs. open-circuit voltage of PV modules, *Sol Energy*.

[10] E. Radziemska, E. Klugmann, 2002 Thermally-affected parameters of the current–voltage characteristics of silicon photocell, *Energ Convers Manage*.

[11] M. Marcu, F. Popescu, L. Pană, I. R. Slusariuc, 2014 Modeling and simulation of solar radiation, *3rd International Conference on Nanotechnology, Technology and Advanced Materials ICNTAM*.

[12] M. Marcu, F. Popescu, L. Pană, I. R. Slusariuc, 2014 Modeling and simulation of a photovoltaic module using the two diode model, *SGEM International Conference*.

[13] M. D. Marcu, F. Popescu, L. B. Samoila, 2011 Modeling and Simuling Power Active Filter Using Method of Generalized Reactive Power Theory, *Proceedings of IEEE International Conference on Computer Science and Automation Engineering*, pp. 213-218.

[14] B. L. Samoila, M. D. Marcu, F. G. Popescu, 2013 Equipment Designed to Control a Heating Hybrid System with Solid Fuel Boiler and Solar Panels, *Proceedings of the 7th International Conference on Renewable Energy Sources (RES '13)*, pp. 106-111.

[15] M. D. Marcu, F. G. Popescu, T. Niculescu, L. Pana, A. D. Handra, 2014 Simulation of power active filter using instantaneous reactive power theory. Harmonics and Quality of Power *IEEE 16th International Conference*, pp. 581 – 585.

[16] P. Vizureanu, C. Samoilă, D.Cotfas, 2009 *Environmental Engineering and Management Journal*, Vol.8, No.2, pp. 301-306.

[17] M. D. Marcu, F. G. Popescu, L. Pana, 2014 Modeling and simulation of power active filter for reducing harmonic pollution using the instantaneous reactive power theory *Environmental Engineering and Management Journal*, pp. 1377-1382.

[18] C. Barbu, 2011 *Revista Minelor*, 2-5.

[19] I. C. Barbu, E. Pop, M. Leba, 2009 Microsisteme energetice durabile regenerative, *Editura Didactică și Pedagogică.*

IOP Conf. Series: Materials Science and Engineering **133** (2016) 012006 doi:10.1088/1757-899X/133/1/012006

Investigation of the mechanical properties of FeNiCrMnSi high entropy alloy wear resistant

G Buluc[1*], I Florea[1], R Chelariu[1], G Popescu[2], I Carcea[1]

[1]"Gheorghe Asachi" Technical University of Iasi, Romania, Department of Materials Science and Engineering, Blvd. Mangeron, No. 59A, 700050, Iasi, Romania
[2]Department of Materials Science and Engineering, Polytechnic University of Bucharest, Romania

E-mail: gheo_u@yahoo.com

Abstract. In this paper we investigated microstructure, hardness and wear resistance for FeNiCrMnAl, high entropy alloy. The FeNiCrMnSi, high entropy alloy was elaborated in a medium induction furnace, by choosing the silicon, as an alliance element within the equi-atomic high entropy alloy, we managed to obtain a dendritic structure, the formation of intermetallic compounds or separated silicon. The medium hardness value of the investigated alloy was 948.33 HV and the medium value of the friction coefficient was 0.6655 in the first 20 seconds and 0.5425 for 1667 seconds. The volume loss of the high entropy alloy FeNiCrMnSi was 0.0557 mm^3.

1. Introduction

High entropy alloys were discovered in 1995 by the researcher Yeh [1]. These sorts of alloys have appeared as a new type of material that has captivated almost instantly material science researchers. Unlike the traditional alloys, the high entropy alloys show a series of excellent physical and mechanical proprieties (elevated hardness, resistance to usage, resistance to corrosion) that recommends them for a wide range of functions [1-3]. The main principle that basses the existence of high entropy alloys (HEA) is that by increasing the number of principal elements of alliance, solid solutions are formed in a preferential way. This principle is derived from Boltzmann's hypothesis on the connection between the entropy and the complexity of a system. We can deduct from here that for an equi-molar alloy that contains at least 5 main alliance elements, the configurational entropy change during the forming of the solid solution is bigger than the entropy change necessary for the melting of the majority of metals [4]. Starting from this principle, in the last two decades, the researches for the evaluating method of the forming capacity of the HEA, for varied alloy systems, has intensified itself and also the investigation interest of a number of over 100 HEA compositions has grown. These alloys made themselves conspicuous by special functional proprieties, such as: elevated hardness, resistance to usage, resistance to high temperature, resistance to corrosion [5]. These properties are considered to be the result of 4 effects: a) the high entropy effect, b) the slow diffusion effect, c) the deformation of the crystalline network effect, d) the "cocktail" effect, which indicates the procurement of special properties by mixing a sufficiently big number of alliance elements [6].

Because of the great number of possible metal combinations that can be used to synthesize the HEA, a series of key selection criteria for the metals has been defined, they can be used for a HEA with certain properies [7-9]:

- The configurational entropy (Δ Smix), that has to be higher than 1.61R
- The difference between the atomic radius, that has to be equal or less than 6.6% in order to form solid solutions
- The concentration of the valence electrons (VEC-Valence Electron Concentration), indicates the type of solid solution that can be formed: for VEC<6.87 cubic solid solution with a centered volume, for 6.87<VEC<8 mixed structure CVC and CFC and for VEC>8 cubic solid solution with centered features
- The electronegativity difference, X, that has to be between 3 and 6% in order to form solid solutions

2. Materials and methods

To obtain FeNiCrMnSi, high entropy alloy, we used a medium induction furnace 8000 Hz. We used metallic elements like: Fe, Ni, Cr, Mn and Si. The alloy obtained was poured into molds in the form of bars with a diameter of 12 mm and length of 100 mm. Grinding abrasives samples was executed with finesse abrasive grains increasingly smaller. Polishing was done by coating the disc with a cloth that has been moistened with continuous drip solution it Al2O3. To be analyzed sample was chemical attack with Nital. To investigate mechanical properties and wear resistance of FeNiCrMnAl, high entropy alloy, we used different methods like: SEM (scanning electron microscope SEM Vega Tescan LMH II, structural analysis 2D and 3D), optical microscope (Optical Microscope Microphot FXA with a camera Hitachi HV-C201A 3CCD), chemical analysis EDAX (Equipment Bruker), hardness (hardness tester Vickers Tukon 2100 B, hardness tester Rockwell (Roc 4JR, Instron B2000), and reciprocating sliding test machine Pin-on-disc contact. Friction coefficient was monitored continuously during testing and wear of the disk measured after the disk with 3D topography analysis (confocal microscope ALICONA FOCUS G4). Parameters for wear resistance have been: F_N (normal load) =30 N, p_H (contact pressure)=1 GPa, a (amplitude)=4 mm, f(frequency)=15 Hz, v_s (average sliding speed)= 0.12 m/s, t(time)=1667 s.

3. Results and discussions

Table 1. Characteristics of metal components of the FeNiCrMnAl high entropy alloy.

Metal	Aromic radius, [pm]	Electronegativity		Valence Electron Concentration
		Pauling	Allen	
Fe	127	1,83	1,800	8
Ni	125	1,91	1,880	10
Cr	128	1,66	1,650	6
Mn	126	1,55	1,750	7
Si	132	1,90	1,916	3

3.1. Chemical composition

The chemical composition of the FeNiCrMnSi high entropy alloy is shown in table 1.

Table 2. Chemical composition for FeNiCrMnSi high entropy alloy.

Element	(at %)	(m %)
Iron	20	22.4
Nickel	20	23.5
Chromium	20	20.8
Manganese	20	22.0
Silicon	20	11.3
Total	100%	100%

The chemical composition is determined using SEM-EDAX.

3.2. Optical microscopy for FeNiCrMnAl high entropy alloy

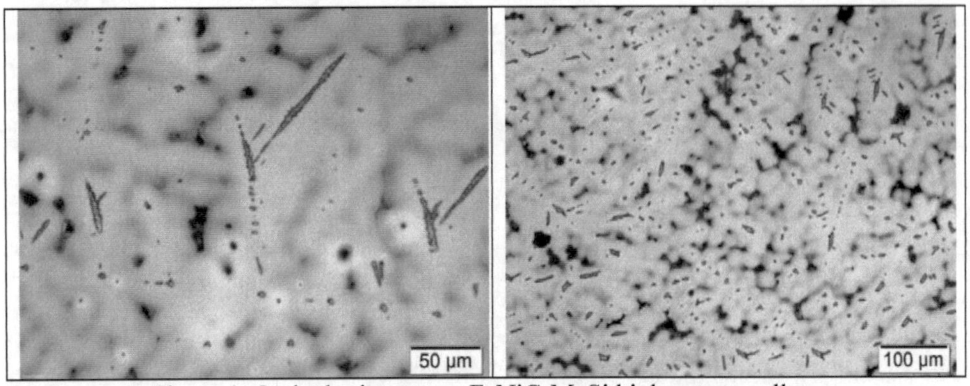

Figure 1. Optical microscopy FeNiCrMnSi high entropy alloy.

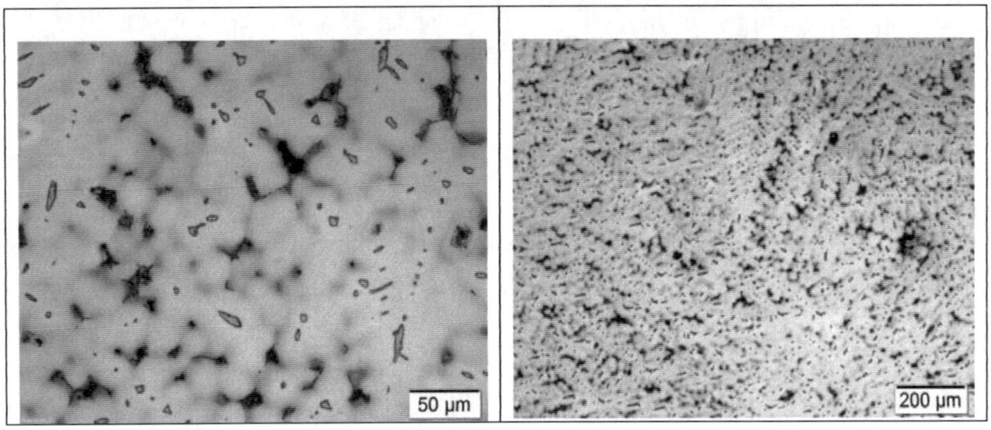

Figure 2. Optical microscopy FeNiCrMnSi high entropy alloy.

3.3. SEM analyse for FeNiCrMnSi high entropy alloy

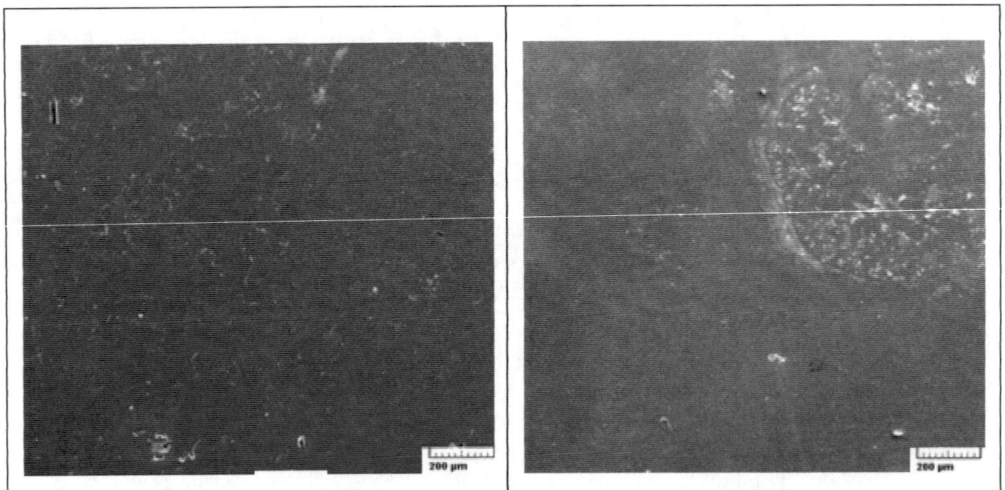

Figure 3. SEM microscopy for FeNiCrMnSi high entropy alloy.

Figure 4. SEM microscopy FeNiCrMnSi high entropy alloy.

Figure 1 and figure 2 show the microstructural analysis which renders the fact that high entropy alloy FeNiCrMnSi has a dendritic structure. Inside the dendritic branches you can notice breaks on the secondary phase, possible silicon, due to the acicular shape or an intermetallic compound with silicon. The high entropy alloy, FeNiCrMnSi, gravitates towards a structure made out of a solid solution.

In figure 3 and figure 4, can see also notice the distribution of elements in the high entropy alloy FeNiCrMnSi that shows us that the investigated alloy has a homogeneous structure.

3.4 Hardness tests

Hardness is the ability an object to resist the tendency to destroy the superficial layers by another object, the pressure acting on its location on very small areas and do not get permanent deformations.

Table 3. Hardness value for high entropy FeNiCrMnSi.

High entropy alloy	Parameters		Vickers Hardness HV	Rockwell Hardness HRC
	F, [kgf]	t, [s]		
FeNiCrMnSi	1	15	948,33	68

The silicon influence as a principal alliance element in the high entropy alloy shows that it can elevate the hardness value to significant one.

The structure of the investigated alloy has led to a growth in hardness but to fragility in the alloy due to a pronounced deterioration of the mechanical proprieties.

3.5 Wear resistance for FeNiCrMnSi high entropy alloy

Wear resistance tests have been made with reciprocating sliding test machine Pin-on-disc contact, in mechanical testing laboratory Institute of Metals and Technology from Ljubljana, Slovenia.

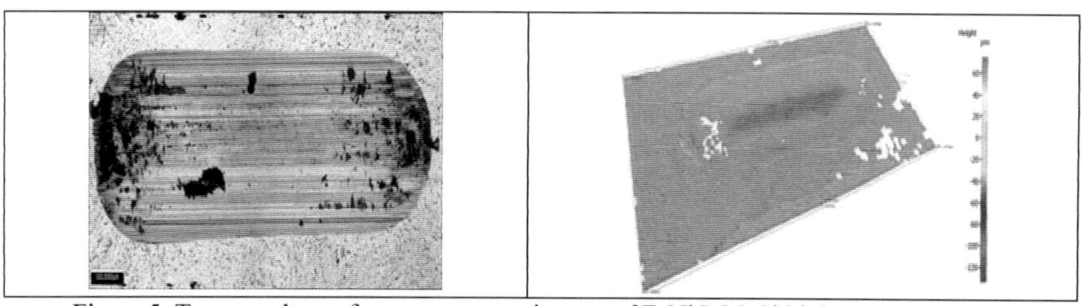
Figure 5. Trace on the surface test wear resistance of FeNiCrMnSi high entropy alloy.

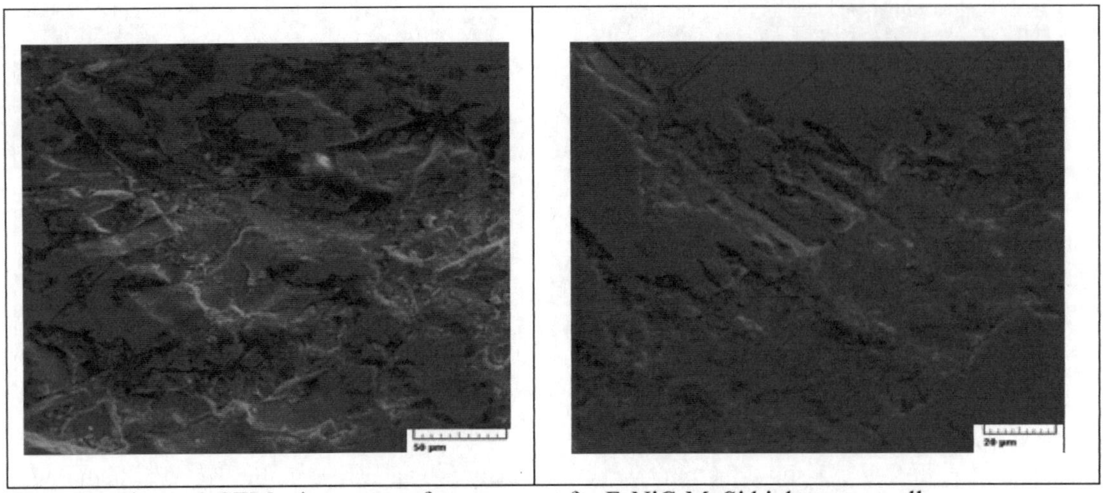

Figure 6. SEM microscopy after wear test for FeNiCrMnSi high entropy alloy.

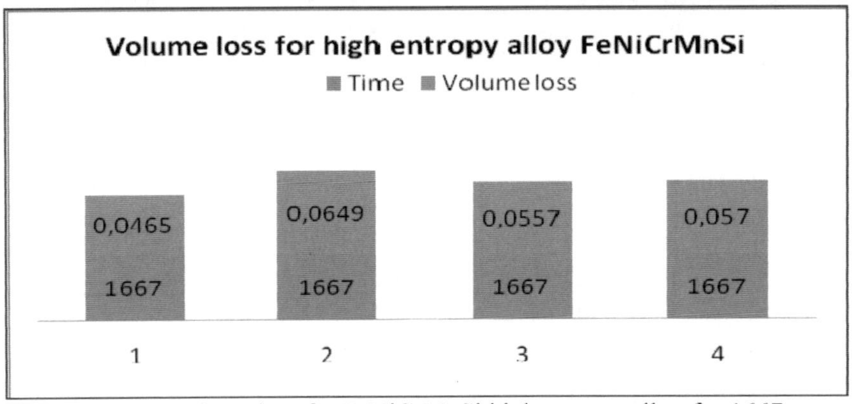

Figure 7. Volume loss for FeNiCrMnSi high entropy alloy for 1667 s.

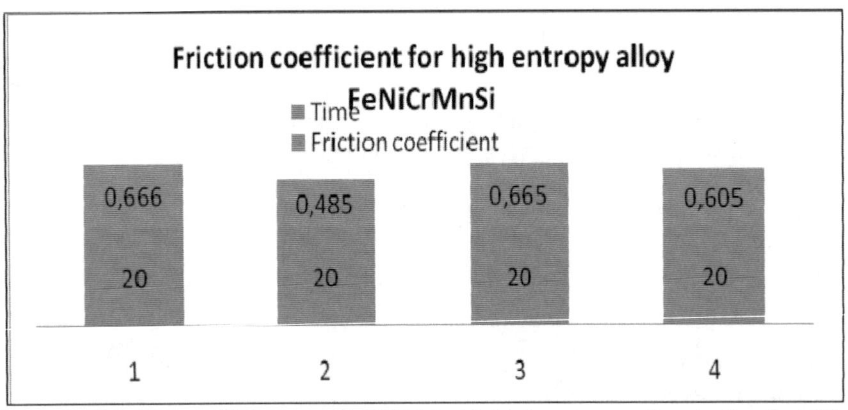

Figure 8. Friction coefficient for FeNiCrMnAl high entropy alloy in the first 20 s.

In figure 6 we can see that after a cycle of 1667 seconds we have a volume loss of 0.057 mm^3 shown in figure 4. You can notice a volume loss of 0.0465 mm^3 for the first time, represented in figure 1 and a 0.0649 loss for the second time, represented in figure 2 and a 0.0557 mm^3 loss for the third time.

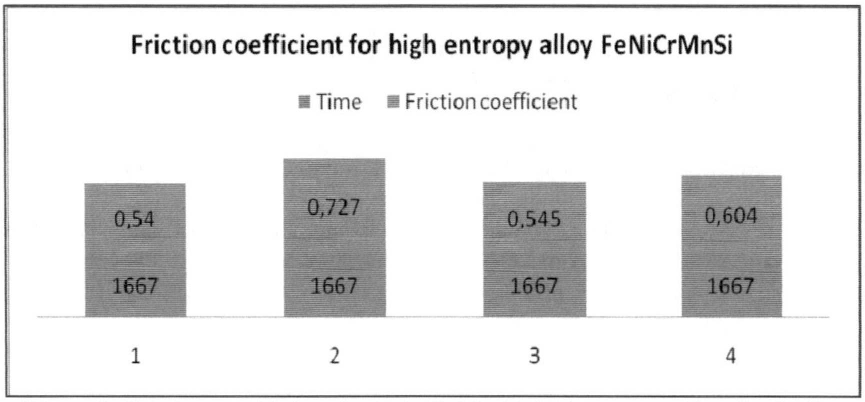

Figure 9. Friction coefficient for FeNiCrMnAl high entropy alloy for 1667 s.

Figure 7 and figure 8 represent friction coefficient for the high entropy alloy FeNiCrMnAl. The sample was performed by 3 attempts, represented in diagrams with 1, 2 and 3. No.4 represents the average of the 3 attempts. It has been observed that in the first 20 seconds friction coefficient was very high then up to the end of 1667 seconds was observed constant movement.

4. Conclusions

Microstructural analysis shows that the FeNiCrMnSi high entropy alloy has a dendritic structure closer to solid solution structure.

Hardness value for FeNiCrMnSi high entropy alloy was 948 HV.

Medium value for friction coefficient of FeNiCrMnSi high entropy alloy was 0,604 for 1667 seconds and 0,605 in the first 20 seconds.

After a cycle of 1667 seconds there is a volume loss of 0.0557 in the FeNiCrMnSi high entropy alloy.

Acknowledgments

This work was supported by a grant of the Romanian National Authority for Scientific Research, CNCS-UEFISCDI project number PN-II-PT-PCCA-2013-4-1048.

References

[1] B.S Murty, J.W. Yeh, S. Ranganathan, 2014, *Elsevier.*
[2] S. Dávid Molnár, 2015, *ELTE.*
[3] B. Cantor, 2007, *Ann. Chim. Sci.Mat.*, 32, pp. 245-256.
[4] J.W. Yeh, S. K.Chen, J. Y. Gan, S. J. Lin, T. S. Chin, T. T. Shun, et al., 2004 a. *Metall. Mater. Trans.* A, 35.
[5] J. W. Yeh, 2006 *Chim. Sci. Mat.*, 31, pp. 633-648.
[6] Y. Lu, et al., 2014 *Sci. Rep.*, 4, pp. 6200.
[7] T. Ming-Hung and Y. Jien-Wei, 2014 *Materials Research Letters*, 2:3, pp. 107-123.
[8] J.W. Yeh, 2006, *Ann. Chim. Sci. Mat.*, 31, pp. 633-648.

IOP Conf. Series: Materials Science and Engineering **133** (2016) 012007 doi:10.1088/1757-899X/133/1/012007

Effect of Montmorillonite Modification on Ultra Violet Radiation Cured Nanocomposite Filled with Glycidyl Methacrylate Modified Kenaf

A R Rozyanty[1]*, H D Rozman[2], S F Zhafer[1], L Musa[1], A Zuliahani[3]

[1]Centre of Excellent Geopolymer & Green Technology (CEGeoGTech), Universiti Malaysia Perlis (UniMap), 02600 Arau, Perlis, Malaysia.
[2]School of Industrial Technology, Universiti Sains Malaysia (USM), 11800, Penang Malaysia.
[3]Faculty of Applied Sciences, Universiti Teknologi MARA (UiTM) Perlis, 02600 Arau, Perlis, Malaysia.

Email: rozyanty@unimap.edu.my

Abstract. In this study nanocomposite cured by ultra violet radiation, were produced using modified montmorillonite (MMT) as reinforcing agent, chemically modified kenaf bast fiber as filler and unsaturated polyester as the matrix. Kenaf bast fiber was chemically modified with glycidyl methacrylate (GMA) whilst MMT were modified with cetyl trimethyl ammonium bromide (CTAB) and glycidyl methacrylate (GMA). Fixed 12 percent of GMA modified kenaf bast fiber with different percentage (i.e., 1, 3 and 5) of unmodified and modified MMT loading was used to produce the composite. The performed of GMA reaction with hydroxyl group of cellulose in kenaf bast fiber was evaluated using Fourier Transform infrared (FTIR) spectroscopy. GMA-MMT filled composite showed higher mechanical properties than MMT and CTAB-MMT filled composite. However, the increase of MMT, CTAB-MMT and GMA-MMT loading resulted in the reduction of mechanical properties. Scanning electron microscopy (SEM) analysis showed the evidence of compatibility enhancement between MMT and kenaf bast fiber with unsaturated polyester matrix.

1. Introduction

High demand and interest of lignocellulosic reinforced polymer composites is due to the cost rising, inadequate and non-renewability of synthetic materials. The use of lignocellulosic fibers as reinforcement material in polymer composites is based on several advantages such as renewable source, environmentally friendly, economical and mechanical performance that are comparable to those of synthetic fibers. In addition, using lignocellulosic fibers in polymer composites is known to be an alternative to solve the problem of decreasing petroleum resources [1].

Natural fiber particularly bast derived from flax, hemp, jute and kenaf have been applied by many researchers as reinforcement in polymer composite [2-4]. Besides providing good mechanical properties of the produced composite, the composition of the polymer matrix can be reduced through the inserting of natural fiber thus reducing the manufacturing costs and the use of non-renewable resources materials. However, high content of polysaccharides (cellulose and hemicellulose) and lignin in natural fiber has restricted its use in the production of composite [5]. Hydrophilic properties exhibited by natural fiber cause incompatibility with the hydrophobic polymer matrix which in turn weaken the mechanical properties of the composite. Therefore, natural fiber treatment or modification has been studied by many parties and it was proven to improve the compatibility between the polymer matrix and natural fibers [6, 7]. Based on a report, maleic anhydride treated kenaf dust composites have shown better tensile properties in comparison to those untreated one as a result of esterification where the hydroxyl group of kenaf reacted with the anhydride group of maleic anhydride to form an ester linkage [8, 9].

Based on a number of previous studies, the utilization of nano filler especially montmorillonite (MMT) in polymer matrix has managed to increase the strength and modulus of the composite [10,

11]. Small sized MMT provide high surface area and aspect ratio that mend the adversarial effect of irregularities of natural fiber in polymer composite. However, the performance of the nano filler filled polymer composite is depends on the interfacial interaction between the polymer matrix and filler because it determine the stress transfer efficiency of the components. Moreover, nano filler dispersion also played an important role in ensuring a uniform distribution of stress during load imposed. Appropriate modification of MMT is expected to increase the nano filler interaction with the polymer matrix and also to improve the dispersion of nano filler in the polymer matrix. Thus, this study focuses on the effect of MMT modification on mechanical and physical performance of the ultra violet (UV) radiation cured unsaturated polyester nanocomposite reinforced with glycidyl methacrylate (GMA) modified kenaf bast fiber.

2. Materials and Methods

Kenaf bast fibers were supplied by Forest Research Institute Malaysia (FRIM). Unsaturated polyester (UPE) resin (Reversol P-9728 P) was obtained from Syntomer, plc. Photoinitiator IRGACURE 1800 was purchased from Ciba Specialty (Singapore) Pte. Ltd. Montmorillonite (MMT) was obtained from Nanocor, USA. Hexadecyl trimethyl ammonium bromide also known as cetyl trimethyl ammonium bromide (CTAB) was supplied by Sigma-Aldrich. Glycidyl methacrylate (GMA) was acquired from Fluka Chemika.

2.1 Chemical modification of kenaf bast fibers

GMA was diluted in *N,N*-dimethylformamide (DMF) at a weight ratio of 3:7. To prevent the occurrence of copolymerization of GMA during reflux, 5 wt% (based on GMA) of hydroquinone was added as an inhibitor. Oven-dried kenaf bast fibers were fed into the reaction flask containing the GMA solution. The reaction was carried out for 2 hours at a temperature of 90°C. The modified kenaf bast fibers were then filtered and rinsed with acetone. The fibers were further washed by refluxing with excess fresh acetone for 3 hours to ensure the solvent, unreacted reagent, and any non-bonded polymer formed during the reaction were eliminated.

2.2 Ion substitution of MMT

CTAB solution of 0.225 mol based on the MMT cation-exchange capacity (CEC) (145 meq/100g) was prepared by dissolving CTAB into a mixture of ethanol and distilled water. Ethanol to distilled water ratio was 1:1. MMT was incorporated into distilled water and stirred using a high speed motor stirrer. CTAB solution was then added to the MMT and stirred at the speed of 500 rpm in a water bath at 80°C for 12 hours. Modified MMT was filtered and washed repeatedly with ethanol and distilled water to remove excessive CTAB. It was then oven dried for 48 hours.

2.3 Chemical modification of MMT

GMA was diluted in DMF at a volume ratio of 5:3. Hydroquinone of 5wt% based on GMA was added as an inhibitor. Oven dried MMT was then added into the GMA solution and the reaction was carried out for 3 hours at a temperature of 90°C. Reacted MMT was then filtered and refluxed with acetone for 2 hours before being dried in an oven for 24 hours.

2.4 Composite preparation

Kenaf bast fibers were cut into 5 mm long segment. The fiber mat was prepared by soaking the weighted fibers in distilled water for 10 minutes and arranged randomly in a deckle box with nylon mesh covered at the bottom to drain out water. The mat was oven dried at 105°C for 24 hours. The fiber mat was then hot-pressed for 1 hour at 100°C to give the thickness approximately of 1 mm. MMT, CTAB-MMT and GMA-MMT nano filler was poured into the UPE at 1, 3 and 5% separately. The mixtures were blended using a high-powered mixer at 600rpm and further mixed using three-roll mill for 1 hour. Photoinitiator IRGACURE 1800 of 3 wt% was then added into the mixture and stirred. Mix UPE/nano filler/IRGACURE 1800 was applied to the fiber mat using hand lay-up process in a steel mold. Curing process took place in a UV 1st ultraviolet machine at a conveyor speed of 10 m/min for 20 cycles for both sides of the composite.

2.5 Testing and characterization

2.5.1 Fourier transforms infrared (FTIR) analysis. Infrared spectra were obtained using Nicolet Avatar 360. The purpose of using FTIR was to determine the functional groups formed from the modification of kenaf bast fiber with GMA.

2.5.2 Tensile test. A number of specimens were cut out from the composite sheets. A tensile test was performed on an Instron Merlin Series in accordance to ASTM D 638.

2.5.3 Scanning electron microscope (SEM) analysis. Surface morphologies of composite fractured tensile specimens were examined using Jeol JSM-7000F scanning electron microscope (SEM).

2.5.4 Water absorption and thickness swelling. The test was conducted in accordance to ASTM D 579. Rectangular specimens were cut out and oven dried at 80°C for 24 hours before it were cooled at room temperature in a desiccator and weighted. The specimens were immersed in distilled water at room temperature for 24 days. Weight and thickness were measured every 24 hours.

3. Results and Discussions

3.1 FTIR Analysis

Figure 1 shows the FTIR spectra of untreated and GMA treated kenaf bast fiber. It can be seen that peaks at 3430 cm^{-1} and 3424 cm^{-1} had appeared for unmodified and modified kenaf bast fiber, respectively. It indicates the presence of hydroxyl (OH) group in the fiber. However, the decreased intensity of OH peak of GMA modified kenaf bast fiber compared to the unmodified kenaf bast fiber is clearly be seen, which designated to the reduction of OH group in the fiber [12].

The peaks located at 1736 cm^{-1} for unmodified kenaf bast fiber and 1718 cm^{-1} for modified kenaf bast fiber were attributed to the carbonyl (C=O) group. Modified kenaf bast fiber shows higher absorption as compared to unmodified kenaf bast fiber which indicates the merged between the fiber and GMA had occurred. High frequency of carbonyl group at 1700 cm^{-1} region for unmodified kenaf bast fiber is attributed to the presence of ester wax [13].

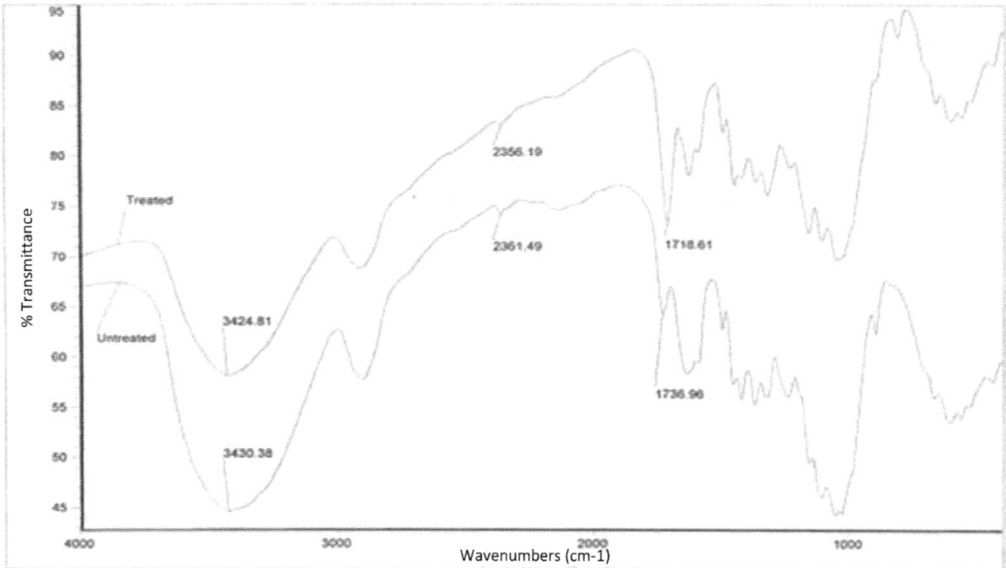

Figure 1: FTIR spectra of untreated and GMA treated kenaf bast fiber.

3.2 Tensile strength and modulus

Figure 2 shows the effect of MMT modification and loading on tensile strength of 12% of GMA modified kenaf bast fiber reinforced composite. It can be seen that composite with 1% loading of GMA-MMT shows the highest tensile strength. However, at 5% loading of GMA-MMT, the composite shows the lowest tensile strength value. CTAB-MMT shows lower tensile strength compared to that MMT and GMA-MMT at 1% loading. Figure 3 shows the results of the tensile modulus of the same samples. The similar pattern of outcomes as tensile strength was observed.

The existence of reactive sites resulted from chemical modification on the surface of MMT with GMA allows the occurrence of crosslinking between the matrix and MMT. It has improved the interfacial adhesion thus enhance the compatibility between the polyester matrix and the MMT. Besides, micron sized GMA-MMT yielded large surface area that subsequently assists good interaction with polymer matrix. The strong interaction has enabled the modified filler to be a reinforcing agent in the composite that can resist well when tensile force is applied [14]. Moreover, the presence of modified kenaf bast fiber that has good interfacial interaction with polymer matrix has augmented the tensile strength and modulus of the composite.

The decline in tensile strength and modulus with the addition of GMA-MMT at 3 and 5% is caused by the poor penetration of UV radiation in the composite. This is due to the dark colour given by MMT and kenaf bast fiber after being modified with GMA. With the increasing of GMA-MMT percentage, the composite dark colour is raised and limits the penetration of UV radiation only to the surface of the composite. Furthermore, the decrease of tensile strength and modulus also contributed to the dispersion of the filler in the polymer matrix. The separation of platelets is important in providing good dispersion and to avoid agglomeration of MMT in the polymer matrix [15]. However, at higher filler loading, shear forces exerted by high power mixer and two-roll mill was not able to well separate the MMT platelets. The same situation applies to the MMT and CTAB-MMT which reflected by the decreased of tensile strength and modulus with the increasing of filler loading.

Lower tensile strength and modulus of CTAB-MMT compared to MMT and GMA-MMT at 1% loading was due to the nano size of the filler. Although CTAB-MMT has good interaction with the polymer matrix, the addition of this nano filler in the composite is less effective because of their smaller size [16]. This causes the tensile strength and modulus of the composite is much depends on the reinforcing effect given by the GMA modified kenaf bast fiber.

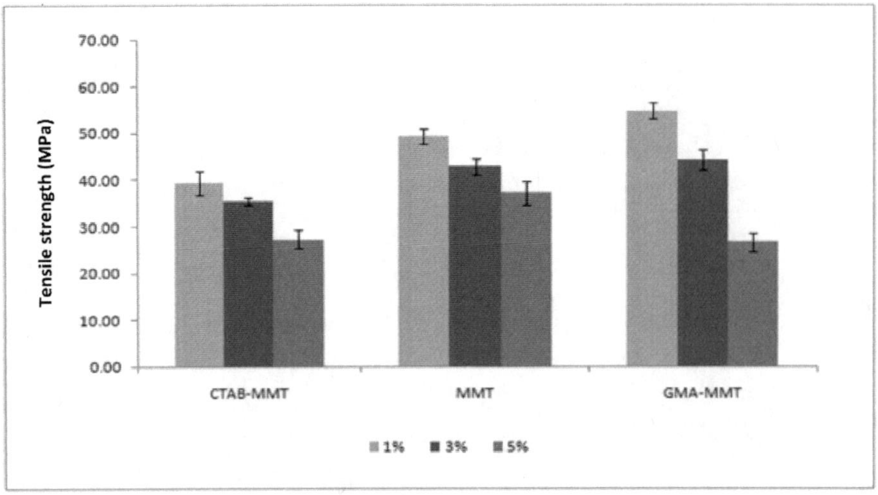

Figure 2: Tensile strength of 12% of GMA modified kenaf bast fiber reinforced composite with different loading of MMT, CTAB-MMT and GMA-MMT

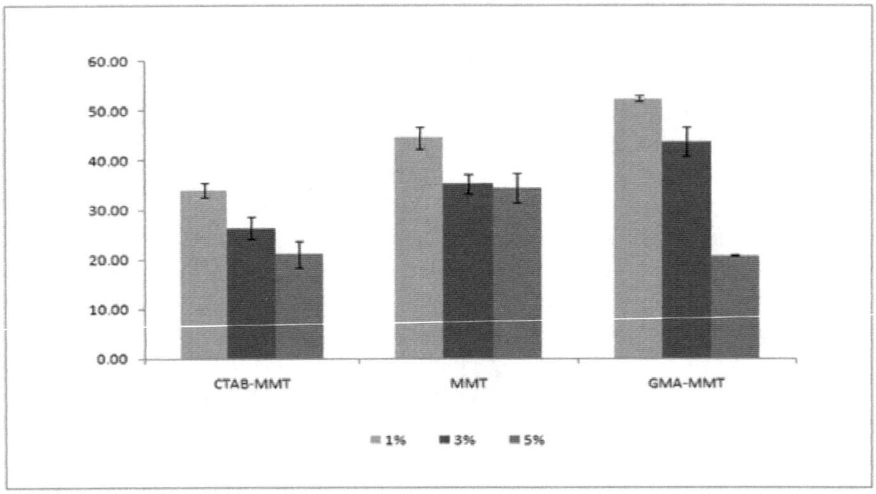

Figure 3: Tensile modulus of 12% of GMA modified kenaf bast fiber reinforced composite with different loading of MMT, CTAB-MMT and GMA-MMT

3.3 SEM micrographs

Figures 4(a), (b) and (c) show the SEM micrograph of unmodified MMT, CTAB-MMT and GMA-MMT contained in 12% of GMA modified kenaf bast fiber reinforced composite, respectively. It can be seen that the composite contain unmodified MMT has smooth surface of polymer matrix with non-agglomeration observed which implies that MMT is well dispersed. The figure also shows the occurrence of good interfacial bonding between GMA modified kenaf bast fiber and polymer matrix. In contrast, CTAB-MMT and GMA-MMT contained composite shows rough surface of the polymer matrix. Agglomeration can be observed on CTAB-MMT composite indicates that the poor dispersion of nano filler in polymer matrix. Fiber breakage spotted on GMA-MMT composite signified good matrix-fiber adhesion strength.

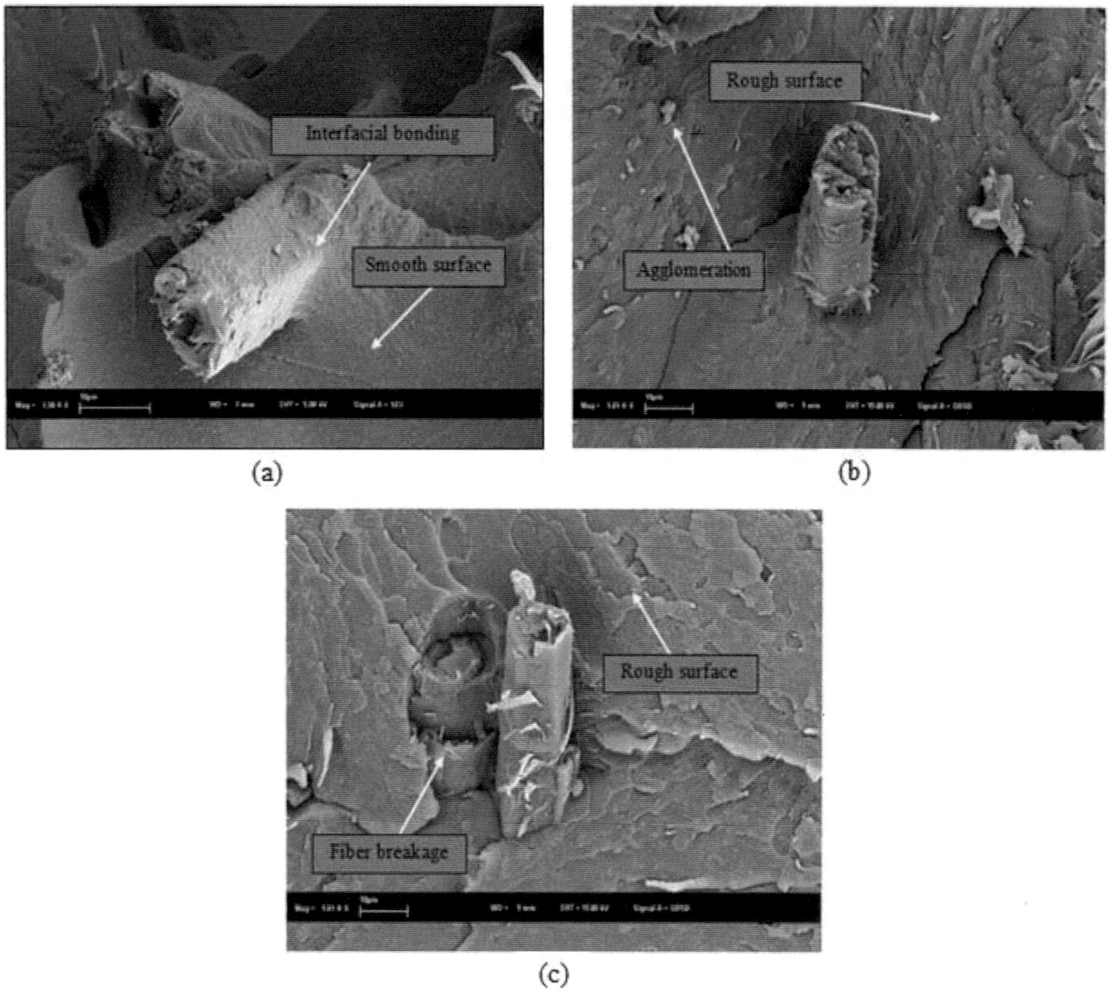

Figure 4: SEM micrographs of (a) unmodified MMT (b) CTAB-MMT and (d) GMA-MMT filled 12% of GMA modified kenaf bast fiber reinforced composite

3.4 Water absorption and thickness swelling properties

Figures 5 and 6 show the effect of MMT modification and loading on the percentage of water absorption and thickness swelling of 12% of GMA modified kenaf bast fiber reinforced composite, respectively. Both figures show a similar pattern where composite contain 5% CTAB-MMT exhibit the highest percentage of water absorption and thickness swelling. It was then followed by composite contain 5% MMT, 5% GMA-MMT, 1% CTAB-MMT and 1% MMT. The lowest percentage of water absorption and thickness swelling is indicated by 1% GMA-MMT.

Naturally, MMT contains hydroxyl group (OH) on the surface and in the gallery that attract to moisture. Modification of MMT with CTAB expanded the gallery space thus revealed more OH group to allow the interaction with water molecules. This phenomenon occurs when the layer of MMT silicate intercalated or exfoliated. With higher loading (5%) of CTAB-MMT, more hydrophilic material contained in the composite hence increases the tendency of the composite to absorb water. This directly destabilized the dimensional of the composite by increasing the thickness of the composite.

In contrast, modification of MMT with GMA has managed to make the MMT particles more hydrophobic. This is the result of OH group on MMT particle has been replaced with methacrylate groups. This causes the percentage of water absorption demonstrated by GMA-MMT contained composite is the lowest, consequently lowering the percentage of thickness swelling of the composite. However, the increase of water absorption and thickness swelling percentage at high unmodified and modified MMT loading might be caused by the agglomeration of MMT on the surface of the composite which cause concentration of water uptake at certain points along the composite [17].

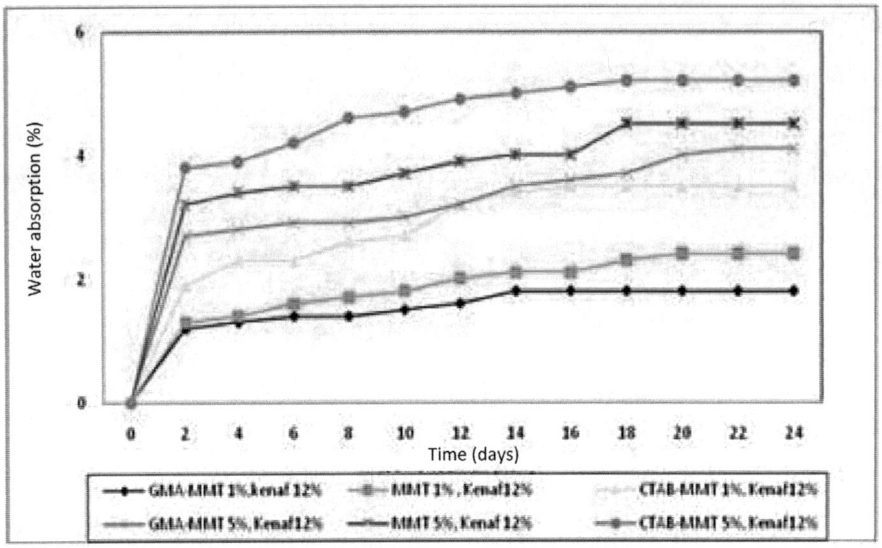

Figure 5: Water absorption of 12% of GMA modified kenaf bast fiber reinforced composite with different loading of MMT, CTAB-MMT and GMA-MMT

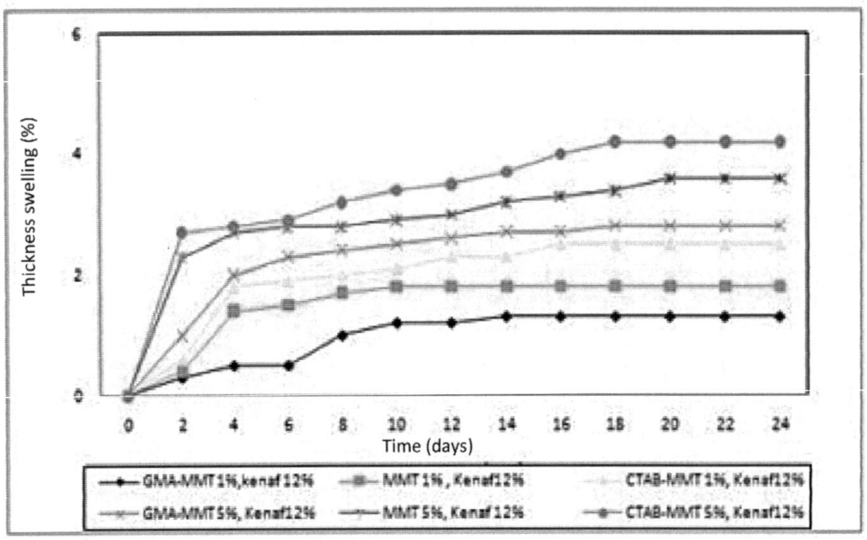

Figure 6: Thickness swelling of 12% of GMA of kenaf bast fiber reinforced composite with different loading of MMT, CTAB-MMT and GMA-MMT

4. Conclusions

From this study, it can be concluded that the modification of kenaf bast fiber with GMA has reduced the OH content which promotes good fiber-matrix interfacial bonding. Modification of MMT with GMA has improved the tensile properties of the composite at 1% loading. An observation attributed to the improvement of interfacial adhesion between the polyester matrix and GMA-MMT. CTAB modified MMT demonstrate the lowest tensile properties at 1% loading due to the small size of the filler that less effective to the composite. Further loading of nano filler decreases the tensile properties of the composite cause by the poor dispersion of the filler in the polymer matrix. SEM micrographs show the interaction of the fiber to matrix and the filler to matrix. GMA-MMT at 1% loading composite shows the lowest percentage of water absorption and thickness swelling due to the OH group on MMT has been replaced with methacrylate groups.

Acknowledgement

The authors would like to extend the appreciation to the Center of Excellence Geopolymer & Green Technology (CEGeoGTech), Universiti Malaysia Perlis (Unimap) and Universiti Sains Malaysia (USM) for the facilities and support.

References

1. Dhal J P and Mishra S C 2012 *Processing and Properties of Natural Fiber-Reinforced Polymer Composite.* J. Mater. 2013.
2. Kikuchi T, Tani Y, Takai Y, Goto A and Hamada H 2014 *Mechanical Properties of Jute Composite by Spray up Fabrication Method.* Energy Procedia. **56** p 289-297.
3. Saba N, Paridah M, and Jawaid M 2015 *Mechanical properties of kenaf fibre reinforced polymer composite: a review.* Constr. Build. Mater. **76** p 87-96.
4. Yan L, Chouw N, and Jayaraman K 2014 *Flax fibre and its composites–A review.* Compos. Part B-Eng. **56** p 296-317.
5. Koronis G, Silva A, and Fontul M 2013 *Green composites: a review of adequate materials for automotive applications.* Compos. Part B-Eng. **44**(1) p 120-127.
6. Mahjoub R, Yatim J M, Sam A R M and Hashemi S H 2014 *Tensile properties of kenaf fiber due to various conditions of chemical fiber surface modifications.* Constr. Build. Mater. **55** p 103-113.
7. Shukor F, Hassan A, Islam M S, Mokhtar M and Hasan M 2014 *Effect of ammonium polyphosphate on flame retardancy, thermal stability and mechanical properties of alkali treated kenaf fiber filled PLA biocomposites.* Mater. Des. **54** p 425-429.
8. Chang S Y, Ismail H, and Ahsan Q 2012 *Effect of maleic anhydride on kenaf dust filled polycaprolactone/thermoplastic sago starch composites.* Bioresources. **7**(2): p. 1594-1616.
9. Yang X, Huang L, Cheng L and Yu J 2012 *Studies of Moisture Absorption and Release Behaviour of Akund Fiber.* Adv. Mech. Eng. **4** p 356548.
10. Tay G S, Azniwati A A, Azizah A B, Rozman H D and Musa L 2012 *The flexural and impact properties of kenaf-polypropylene composites filled with montmorillonite filler.* Polym.-Plast. Technol. Eng. **51**(2) p 208-213.
11. Salehiyan R, Yussuf A A, Hanani N F, Hassan A and Akbari A 2015 *Polylactic acid/polycaprolactone nanocomposite Influence of montmorillonite and impact modifier on mechanical, thermal, and morphological properties.* J. Elastomers Plast. **47**(1) p 69-87.
12. Khalil H A, Suraya N L, Atiqah N, Jawaid M and Hassan A 2012 *Mechanical and thermal properties of chemical treated kenaf fibres reinforced polyester composites.* J. Compos. Mater. p 0021998312465026.
13. Fiore V, Di Bella G and Valenza A 2015 *The effect of alkaline treatment on mechanical properties of kenaf fibers and their epoxy composites.* Compos. Part B-Eng. **68** p 14-21.

14. Lu H, Lei M, Yao Y, Yu K and Fu Y Q 2014 *Shape memory polymer nanocomposites: nano-reinforcement and multifunctionalization.* Nanosci. Nanotechnol. Lett. **6**(9) p 772-786.
15. Mirjalili F, Chuah L and Salahi E 2014 *Mechanical and Morphological Properties of Polypropylene/Nano α-Al 2 O 3 Composites.* Scientific World J. **2014**.
16. Fu S Y, Feng X Q, Lauke B and Mai Y W 2008 *Effects of particle size, particle/matrix interface adhesion and particle loading on mechanical properties of particulate–polymer composites.* Compos. Part B-Eng. **39**(6) p 933-961.
17. Liu R, Peng Y, Cao J and Luo S 2013 *Water absorption, dimensional stability, and mold susceptibility of organically-modified-montmorillonite modified wood flour/polypropylene composites.* BioResources. **9**(1) p 54-65.

Synthesis and characterization of a new high entropy composite matrix

G Popescu[1], M A Matara, I Csaki[1]*, C A Popescu[1], R Trușcă[2]

[1]University Politehnica of Bucharest, 313 Splaiul Independenței, 6 District, 060042 Code, Bucharest, Romania
[2]METAV CERCETARE-DEZVOLTARE, 31 C.A.Rosseti Street, 2 district, Buchrest, Romania

E-mail: * ioana.apostolescu@upb.ro

Abstract. Even if high entropy alloys were not reported in a scientific journal till 2003, these new alloys have been investigated since 1995 due to their high temperature properties. In the last years the synthesis of these alloys has been widely investigated. Thus, the present work has been carried out to produce a high entropy composite using an equiatomic AlCrFeMnNi high entropy alloy (HEA) matrix and graphite particles (Gr) as reinforcing material. The high entropy composite was obtained by powder metallurgy route using a planetary ball mill. The mechanically alloyed mixture was investigated by scanning electron microscopy (SEM). Microstructural investigation realized by SEM revealed the homogenous structure of the composite, with multiple phases and decreasing particles size, mostly reaching nanometric scale.

1. Introduction

Conventional alloys are based on one principal element with different elements added to improve their properties [1]. Based on the existing research and literature on classical alloys, Jien-Wei Yeh explored the multicomponent world and after years of research he developed the HEA concept providing experimental results and related theory. He defined HEAs as those alloys containing at least 5 principal elements with the atomic percentage of between 5% and 35%, with the atomic percentage of each minor element, if any, less than 5% [2].

Due to the combinations of composition and process for producing HEAs and each HEA having a particular microstructure and properties to be identified and understood, the research work is almost limitless [3, 4]. HEAs have shown interesting properties such as high hardness and high strength, good thermal stability wear and oxidation resistance, properties which provide great potential for engineering applications [4 – 9]. The high entropy alloys have been produced by different routes, the usual route being liquid processing.

During the last years, mechanical alloying technique has been more often approached due to the good homogenous chemical distribution and near net shape technology provided by the respectively process [9, 10, 11]. Chao Wang, Wei Ji and Zhengyi Fu obtained simple structured solid solution after 30 h of milling, with a grain size of less than 20 nm while B.S.Murty et al. succeeded to obtain a HEA with BCC structure and grain size of approximately 10 nm [11,12,13].

This report aims to reveal the synthesis method for a composite material using a HEA matrix and Gr particles as reinforcing material. Based on this objective, an equiatomic AlCrFeMnNi HEA matrix was fabricated by MA process and Gr was added after the HEA was obtained. The alloy microstructure was investigated and the XDR analyses were performed to see the moment when the mechanical alloying was produced.

2. Materials and Methods

Al, Cr, Fe, Mn, Ni elemental powders of >99% purity and ≤45 mμ, used as matrix, were mechanically alloyed in a high energy planetary ball mill (Retsch 400 PM) for 40 h in Ar atmosphere at 300 rpm with a ball to powder ratio (BPR) of 10:1. Graphite was added to obtain a new HEA composite. The graphite particles were added 3% (volumetric) to improve the powder flow and weldability. Stainless steel vials and balls have been used. Heptane as process control agent (PCA) was used to prevent powder deposition on the vials walls and milling balls. Samples were extracted at intervals of 20, 25, and 40 hours.

The mechanically alloyed powder was microstructural characterized. QUANTA INSPECT F scanning electron microscope (SEM) foreseen with a field emission electron gun, and equipped with analysis dispersive system in X-radiation energy was used to observe the morphology and particle reduction. The samples were milled at different times in order to assess the best milling time for a planetary ball mill, with 300 rpm, 10:1BPR to process a high entropy composite material. Samples taken at different intervals revealed that after 20 hours the alloying process began.

3. Results and discussions

Figure 1 reveals the AlCrFeMnNIGr HEA composite microstructure under different milling times.

The particles in the primitive powder are roughly between 5 and 45 μm and of irregular shape, with sharp edges, as seen In Figure1 (a). The ball to powder ratio chosen was 10:1 and the ball mill speed was 300rpm. The idea of the experiment was to assess the milling time necessary for the alloying process to be produced. The authors tried to decrease as much as possible the milling time using a planetary ball mill with an optimum speed of 300rpm. Previously the milling time reported was of 60 hours [9] but with a BPR of 7:1. With the increase of the milling time the particles have reduced their size exponentially with time and cold weld together to form agglomerations of elliptical shape after 20 and 25 h as seen in figure 1 (b) and (c). Figure 1 (d) reveals that at 40 h milling time further decrease in particles size with grain size reaching nanometric scale as it will be further described by the diffraction pattern. The microstructure of the 20 h milled sample shows a good homogeneity and the welding process of the particles began.

The powder further processed present agglomeration for the particles of different powder. The alloyed particles have a light grey colour, uniform and the particles are so small that the agglomeration could not be avoided. The percentage of graphite particles added improved the milling process and the powder was alloyed after 40 hours milling. In further papers the properties of the new composite material obtained will be investigated. The graphite particles were added from the beginning to obtain a homogeneous structure.

The microstructure study is important for revealing the sample homogeneity. To establish if the component of the powder were alloyed after the process an EDS analysis was performed.

In figure 2 EDS analyses results show the homogeneity of the mentioned HEA composite during the mechanically alloying process. In figure 2a the distribution of the particles reveals all elemental powder used to create the high entropy composite. Figure 2b shows a mixture of all these elements showing that the alloying process began.

IOP Conf. Series: Materials Science and Engineering **133** (2016) 012008 doi:10.1088/1757-899X/133/1/012008

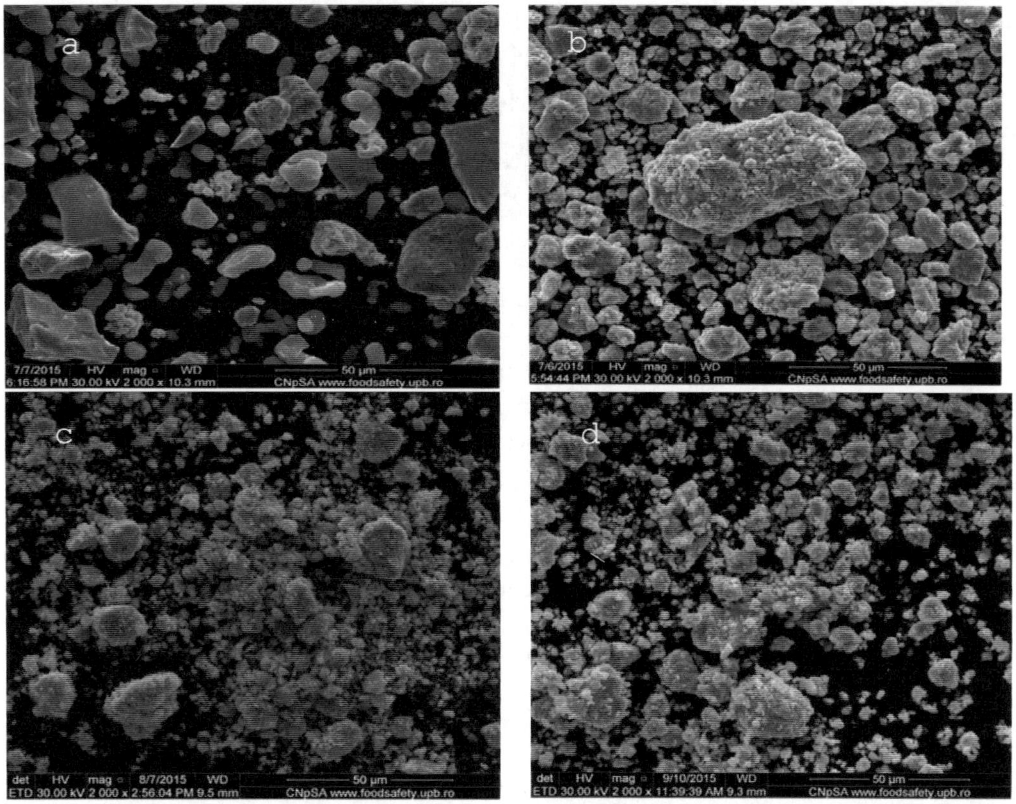

Figure 1. SEM image of AlCrFeMnNiGr HEA composite at (a) 0 h; (b) 20 h; (c) 25 h; (d) 40 h.

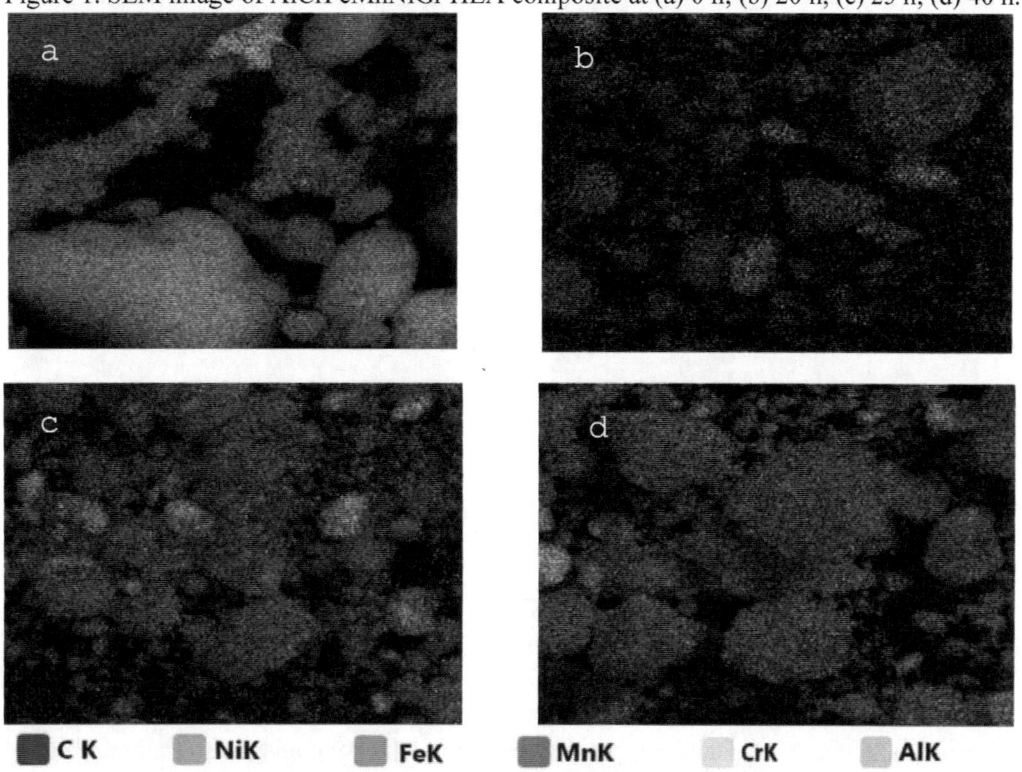

Figure 2. EDS of AlCrFeMnNiGr HEA composite: a) 0 h, b) 20 h, c) 25 h, d) 40 h.

The authors assessed the microstructure of the samples after 25 hours (figure 2c). The EDS analyses revealed the alloying process in a better phase than after 20 hours but still Mn particles could be observed unalloyed. After 40 h milling time a homogenous structure can be observed with the elements evenly distributed.

Figure 3 shows the XRD pattern of AlCrFeMnNiGr HEA composite before and after milling time. The peaks of all the pure elements in the powder before alloying could be observed in figure 3, in the initial mixture. After 40 h of milling time most of the peaks disappear revealing the solid solution formation. The transformation of the phases from FCC to BCC can be observed in the diffraction pattern. The blue line indicates also the absence of the other elemental peaks proving the fact that the alloying process was realized.

Table 1 exhibits crystallite size and lattice strain for 40 h milling time, calculated with Scherrer's formula

$$D = (K\lambda)/(\beta\cos\theta) \quad (1)$$

Where λ is the X-ray wavelength in nanometres (nm), β is diffraction broadening and K is a constant. The θ can be in degrees or radians [14]. The Scherrer formula was used to determine the crystallite size and lattice strain which is one of the four core effects of HEAs. Because of the atomic size difference lattice strain occurs increasing hardness and strength but also giving a higher resistance of the material properties at high temperatures [2].

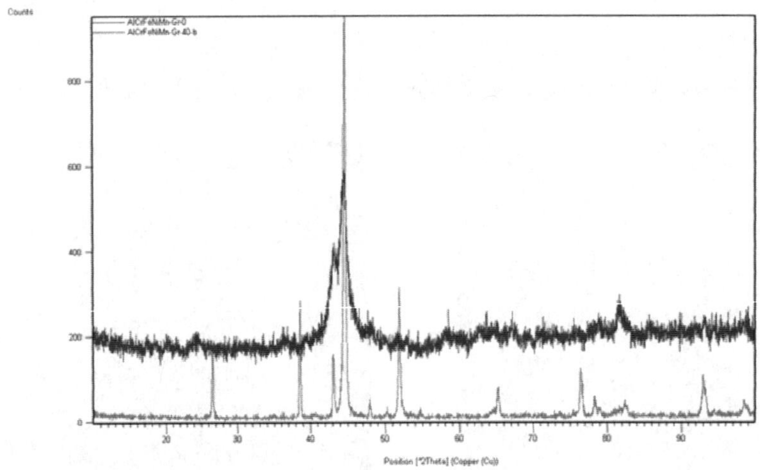

Figure 3. XRD patterns at 0 and 40 h milling time of AlCrFeMnNiGr HEA composite.

Table 1. The crystallite size and lattice strain of AlCrFeMnNiGr HEA at 0 and 40 h milling time

Milling time/h	Crystallite size/nm	Lattice strain/%
40	39.28	2.59

4. Conclusions

A high entropy alloy was syntheses starting from the pure elemental powders from Al, Cr, Fe, Ni, Mn. To enhance this high entropy alloy properties graphite particles were added as reinforcing elements. The graphite particles were added from the beginning to improve the powders properties.

The high entropy composite was alloyed for 20, 25 and 40 hours. The microstructure studies revealed that the composite obtained was homogeneous and EDS analyses results showed that the elements are uniformly distributed.

The XRD analysis revealed the solid solution formation and the nanometric size of the particles.

The high entropy composite was successfully synthetized after 40 hours of milling time.

The lattice strain calculate with the aid of Scherrer formula shows a value characteristic for high entropy alloys, providing a better behaviour for these type of alloys at elevated temperatures.

Acknowledgments
This research was supported by a grant of the Romanian National Authority for Scientific Research, CNDI – UEFISCDI, project number PCCA 270/2014, HEADURCOR
The work has been funded by the Sectoral Operational Programme Human Resources Development 2007-2013 of the Ministry of European Funds through the Financial Agreement POSDRU/159/1.5/S/132395.

References

[1] K. H. Huang, J. W. Yeh, 1996 *A study on multicomponent alloy systems containing equal-mole elements*, M.S. thesis, Hsinchu: National Tsing Hua University.
[2] B.S. Murty, J.W. Yeh, S. Ranganathan, 2014 *High Entropy Alloys. Elsevier.*
[3] Y. Jien-Wei, 2013 *JOM,* Vol. 65, No. 12.
[4] R. Ştefănoiu, V. Geantă, I. Voiculescu, I. Csaki, N. Ghiban, 2014 *Revista de Chimie*, Vol. 65, No. 7.
[5] P. K. Huang, J. W. Yeh, T. T. Shun, S K. Chen, 2004 *Adv Eng Mater*, 6(1–2), pp.74–78.
[6] S. Varalakshmi, G. A. Rao, M. Kamaraj, B. S. Murty, 2010 *J Mater Sci*; 45(19), pp. 5158–5163.
[7] S. Varalakshmi, M. Kamaraj, B. S. Murty, 2010 *Mater Sci Eng A;* 527(4–5), 1027–1030.
[8] C. J. Tong, M. R. Chen, S. K. Chen, J. W. Yeh, T. T. Shun, S. J. Lin, et al., 2005 *Metall Mater Trans A;* 36(5), pp. 1263–1271.
[9] J. M. Wu, S. J. Lin, J. W. Yeh, S. K. Chen, Y. S. Huang, 2006 *Wear*, 261(5–6), pp. 513–519.
[10] F.J. Baldenebro-Lopez, J.M. Herrera-Ramírez, S.P. Arredondo-Rea, C.D. Gómez-Esparza, R. Martínez-Sánchez, 2015 *Journal of Alloys and Compounds 643*, pp. S250–S255.
[11] Y. Shaofeng, Z. Yan, C. Jialin, Z. Chen, C. Weiping, 2014 *Rare Metal Materials and Engineering*, Vol. 43, Issue 12.
[12] C. Wang, W. Ji, Z. Fu, 2014 *Advanced Powder Technology*, 25, pp. 1334–1338.
[13] S. Varalakshmi, M. Kamaraj, B.S. Murty, 2008 *Journal of Alloys and Compounds*, 460, pp. 253–257.
[14] L. Alexander and H. P. Klug, 1949 *Determination of Crystallite Size with the X-Ray Spectrometer*. Department of Research in Chemical Physics, Mellon Institute, Pittsburgh, Pennsylvania.

Friction Behaviour of Polymeric Composite Materials Mixed with Carbon Fibers Having Different Orientations Layout

R Caliman

Calea Marasesti 157, "Vasile Alecsandri" University of Bacau, Romania

E-mail: rcaliman@ub.ro

Abstract. This paper presents a study of the friction properties of polymeric composite materials reinforced with unidirectional carbon fibers having different stratified structure. So, the composites are complex and versatile materials but their behaviour in practice is not fully studied. For instance, these polymeric composite materials mixed with carbon fibers after being investigated in terms of wear, did not elucidate the effect of fiber orientation on wear properties. Is therefore necessary to investigate the effect of carbon fibers orientation on the friction-wear properties of the reinforced composite materials tested to abrasive and adhesive friction. Research work has been done with unidirectional composite materials having overlap 18 successive layers made from a polymeric resine and 60% of carbon fibers. The stratified structure was obtained by compressing multiple pre-impregnated strips, positioned manually. During this experimental work, three types of test samples were investigated: parallel, normal and anti-parallel, taking in consideration the carbon fibre orientation with respect to the sliding direction. The friction coefficient is computed function to the friction load and loading value. Also, the specific wear rate was calculated according to: the mass loss, density, the normal contact surface, the sliding distance and load rating.

1. Introduction

The importance of composite materials plays an increasing role in everyday life. Their superior combination of stiffness, strength and low density has given composite materials a competitive advantage in aerospace, automotive, civil infrastructure and wind energy industries [1-3].

Composite materials are made of a combination of two or more constituents to make one heterogeneous material. The advantage of the resultant material is that it is designed to take advantage of the beneficial properties of each constituent without being limited as much by the detrimental properties of the original constituents. Composite materials consist of a matrix and filler. The matrix usually consists of an epoxy type polymer and the filler is usually particles or fibers. Thin layers can then be stacked into a resultant called a laminate [4,5].

Although composite laminates have obvious advantages to traditional metals, the composites are much more susceptible to damage induced by impact from foreign objects [6].

The use of composite materials in aerospace, electronics and wind industries has become increasingly common and the composite components are required to carry mechanical, electrical, and thermal loads simultaneously. For instance, carbon fiber epoxy matrix composites are expected to disperse high electric currents during lightning strike events, which has sparked research interest into the electrical properties of composites [7-10].

At present, because of easy adaptability to conventional manufacturing techniques and low cost of fabrication, short fiber reinforced composites will increasingly be used in a wide range of application. So it is more important to understand short fiber reinforced composites mechanical properties, which depend strongly on the interfacial bonding between fiber and matrix. Compare with continuous fiber reinforcement, the composites with short carbon fibers can be prepared by conventional methods and lead to the homogenous properties because of the uniform distribution of short carbon fiber in the matrix [10-13].

In this paper, the effect of carbon fiber orientation function to the sliding direction on the tribological behaviour of the resulted composites, especially the friction and wear behaviours of the composites sliding under different sliding speed and constant applied load and time, were carefully investigated in order to provide some practical guidance for the use of polymer-based composites.

The objective of this research work it was to investigate a polymeric composite material reinforced with unidirectional carbon fibers having stratified structure.

2. Experimental procedure

Investigations were carried out on unidirectional composite materials having stratified structure with 18 successive layers made from a polymeric resin and 60% of carbon fibers. The experiments have involved the realization of the composite materials reinforced with short carbon fibers. Technological flow related to preliminary experiments is shown in figure 1.

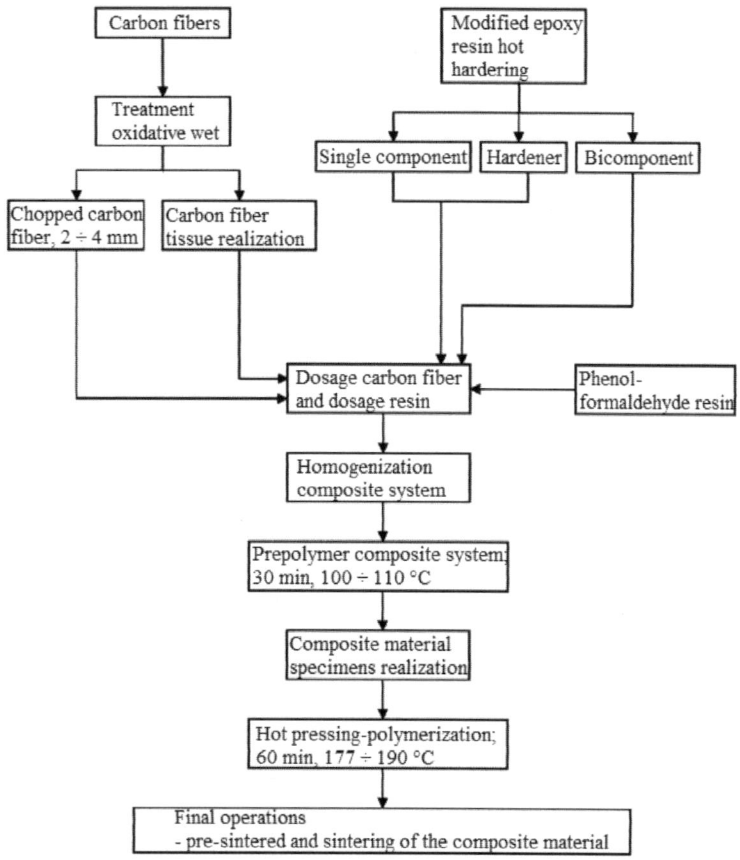

Figure 1. Technological flow for obtaining the polymeric composite materials sintered with the carbon fiber.

Depending on the type of resin used, the temperature changes of the pre-polymerization and the polymerization for the composite systems [14].

In these experiments were carried out the tests by realization of the polymer composite material reinforced with short carbon fibers, using the polymers in the solid state. The formulations used in these tests are shown in table 1.

Table 1. The formulations used in the production of the composite materials with short carbon fibers and the polymer matrix.

Tested type	Composition, %		
	Resin	Fiber	Filling material (graphite)
T_1(P)	22.62	61.42	6.56
T_2(AP)	19.25	58.15	13.34
T_3(N)	36.40	62.48	-

The physic-mechanical characteristics determined for the composite materials tested are shown in table 2.

Table 2. Characteristics of polymer composite materials made with short carbon fiber polymer in the solid state.

Tested type	Characteristics			
	Mechanical resistance (daN/mm^2)	Density (g/cm^3)	Porosity (%)	Expansion coefficient (1/°C)
T_1(P)	0.71	1.827	14.95	4.21
T_2(AP)	0.53	1.757	10.97	9.68
T_3(N)	4.52	1.792	14.88	10.38

There have been made investigations by scanning electron microscopy on the samples subject to wear, in order to elucidate the wear process.

In principle, a wear process is a phenomenon of the cracks propagation which lead to the formation of the very fine particles. Regarding to the carbon-fiber-reinforced polymers, can be detected some differences in the wear mechanism, related to the fibers behavior in the matrix.

Thus, the wear phenomenon in this case is a sum of [15]:

- matrix wear;
- sliding wear of the fibers;
- cracking wear of the fibers;
- fiber separation wear from the matrix at the interface.

In case of polymer composites reinforced with short fiber, characteristic are the traces of the wear from the matrix material and spraying of the fibers. The presence of fibers tends to inhibit the formation of the large chips from the matrix material, by accumulating layers of the sheared polymer material. These mechanisms are more probable for the oriented fibers.

Characteristic values determined for the polymer composite are shown in table 3.

Table 3. Tribological characteristics for polymeric composite materials reinforced with short carbon fibers.

Tested type	Parameters			
	N = 0.25 daN		N = 0.85 daN	
	v = 6.5 mm/s	v = 41.5 mm/s	v = 6.5 mm/s	v = 41.5 mm/s
	μ	μ	μ	μ
T_1(P)	0.65	0.53	0.66	0.55
T_2(AP)	0.62	0.78	0.68	0.61
T_3(N)	0.66	0.79	0.78	0.77

Investigations by electron microscopy have revealed, in this case, a particular behavior in the friction tests, generated by the presence of the short fibers, and the areas of the resin epoxy polymerized.

3. Results and discussion

The influence of the carbon fiber orientation on the friction-wear properties in the case of fiber polymer composites reinforced can be studied by using of the mechanical wear tests and the microscopy, and by using of the phenomenological models.

For the polymer composites reinforced with fiber investigated in this study from the point of view of wear, it was able to elucidate the effect of the fiber orientation on the properties of wear. Therefore, it is necessary to investigate the effect that it has the fiber configuration on the friction-wear properties presented by polymeric reinforced carbon fiber, in case of adhesive and abrasive wear [16,17].

There have been made some investigations on the uniaxial multi-layered composite with 18 layers and a percentage of 60% carbon fiber, impregnated with polymeric resin. The multi-layered was made by pressing of the sheets pre-impregnated type filed manually. There are analyzed three types of samples: normal, parallel and antiparallel (figure 2).

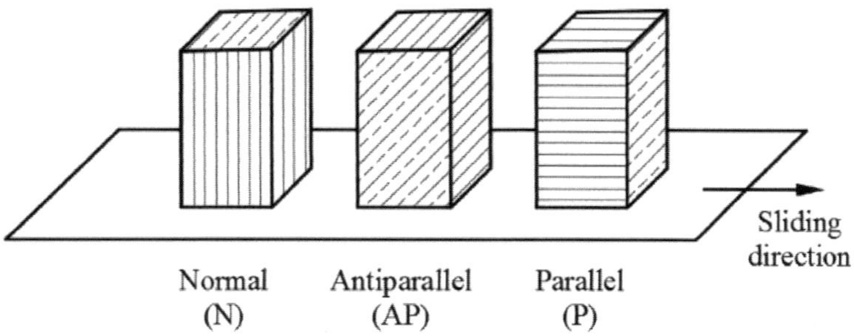

Figure 2. Types of samples used in friction tests.

Specific wear rate, W_S, is calculated by the following equation [18]:

$$W_{S.} = \frac{\Delta M \cdot P}{\rho AS} = \frac{\Delta M}{\rho SL} \quad (1)$$

where M - mass loss; ρ - density; A - nominal area of contact; S - sliding distance; P = L/A; N - normal load.

The coefficient of friction is calculated by the equation:

$$\mu = F / N \quad (2)$$

where F is the friction force.

The dates corresponding to the friction coefficient are given in figures 3 and 4.

It is observed that:

- the wear rate increases in the order N → AP → P, indifferent of the sliding speed or variation of the normal load;
- the wear rate is almost insensitive to the variation of the sliding speed, while the wear rate tends to decrease with the increase of the normal load;
- coefficient of friction under high load increases in the order P → AP → N with increasing of the sliding speed;
- coefficient of friction at the high speed decreases with increasing of the normal load;
- generally, the friction coefficient is insensitive to sliding speed variation under the heavy load;
- correlation between the wear rate and the friction coefficient is reduced.

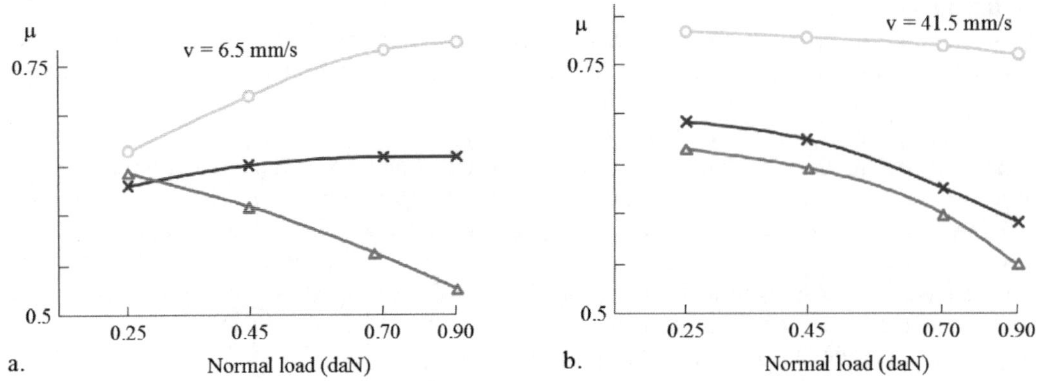

Figure 3. a) The coefficient of friction according to the normal load, to the abrasive wear, for the low sliding speed (6.5 mm/s); b) coefficient of friction according to the normal load for the abrasive wear at the high sliding speed (41.5 mm/s); o-N type; Δ-P type; x-AP type.

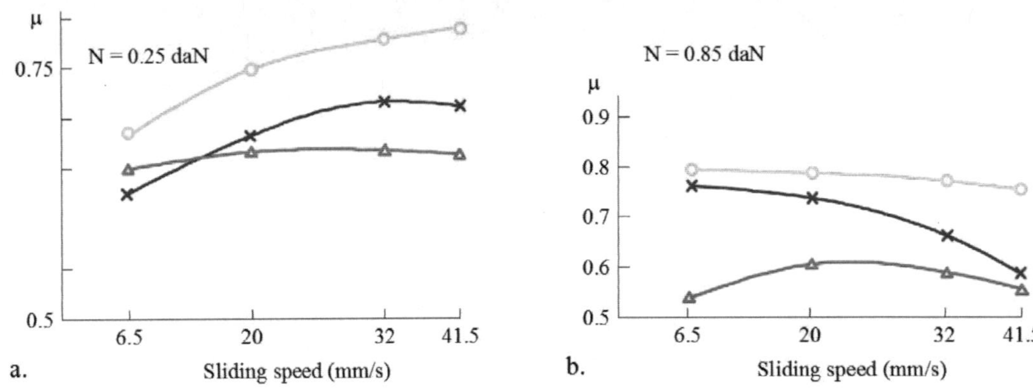

Figure 4. a) Abrasive friction coefficient, according to the sliding speed under to the small normal load (0.25 daN); b) abrasive friction coefficient, depending on the sliding speed, under to the high normal load (daN 0.85); o-N type; Δ-P type; x-AP type.

A similar type of behavior is observed on the curve of the friction coefficient as a function of the sliding speed (figure 5). The effect of the fiber orientation is unclear both case of the friction and the wear.

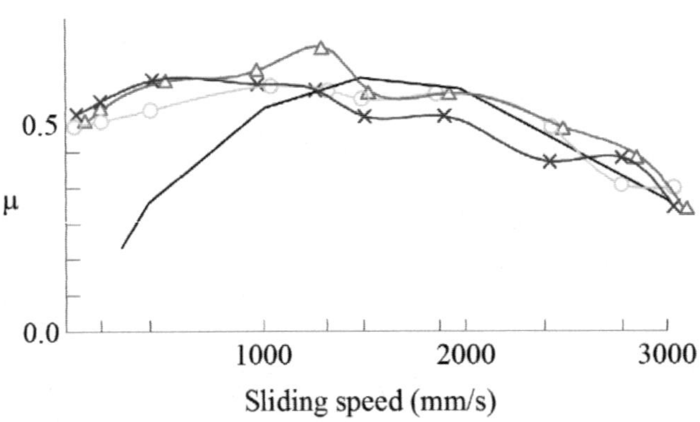

Figure 5. The coefficient of friction on the sliding speed for abrasive wear of the polymer laminates reinforced with the carbon fiber; o-N type; Δ-P type; x-AP type.

As the normal load increases, the friction coefficient tends to decrease as is shown in figure 6 a, indicate an increase in the coefficient of friction, in order AP → P → N, under a constant sliding speed. This behavior has been observed with the abrasive wear. Under a very low sliding speeds, the wear rate appears to be insensitive to the normal load variation, figure 6 b.

The resulting material by the wear is formed by fracturing and deformation of the matrix. The wear particles, powders, were easily observed at very low sliding speeds, appearing to involve the abrasive wear, prevalent when the thermal effect is negligibly for the low sliding speeds.

The values of the friction coefficient can be seen in figures 3 and 4, differ by orientations N and P about 10% (μ_N, μ_P), and therefore the friction coefficient cannot explain the wear rate. Therefore, the specific wear rate seems to be under control by three factors: E, E_W and I_S.

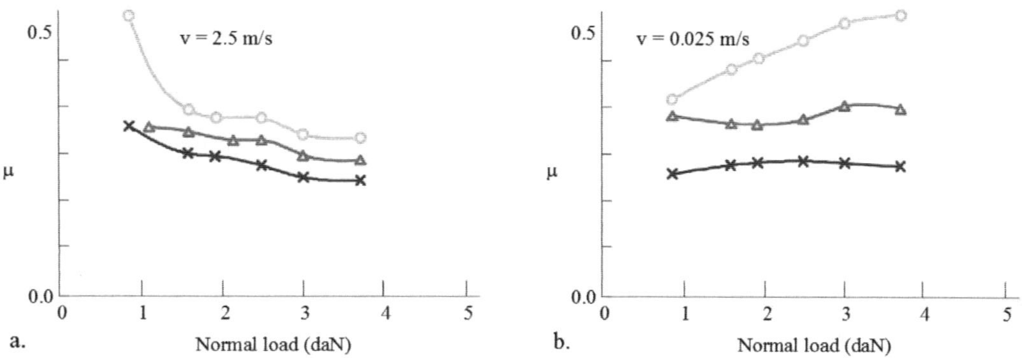

Figure 6. a) The coefficient of the abrasive friction according to the normal load, for the polymeric laminates with the carbon fiber at a high sliding speed (2.5 m/s); b) the coefficient of the abrasive friction according to the normal load, for the polymeric laminates with the carbon fiber at a low sliding speed (0.025 m/s); o-N type; Δ-P type; x-AP type.

The previously made calculations for the composites with short fiber indicated that the E_N is greater than E_P by about 200% and that ε_N is greater than ε_P about 20%. It should be noted that in the case of a short fiber composite, the percentage of the existing fiber is 45% (weight percent), without an orientation of the fibers.

In case of the layered polymer composites with a fiber content of 60%, E and ε values rise. Also, $(I_S)_N$ is greater than $(I_S)_P$, because the carbon fiber obtained from PAN precursor has a structure with layers of graphite circumferentially oriented. Due to the fact that the connection between the two layers is poor, the inter-laminar shear strength tends to be small in the direction of the fiber axis, while the shear strength normal to the axis of the fiber tends to be greater, due to a stronger strength of the covalent bond of the carbon atoms are along the carbon chain, i.e. $(I_S)_N > (I_S)_P$.

The net effect of these factors (E, μ and I_S) makes $(W_S)_N$ to be less than $(W_S)_P$. The wear rate difference between the samples P and AP types comes from the difference between the I_S factors, given that there is no difference for E and μ between P and AP types. Such a difference in I_S seems to be generated by the fiber morphology specific for the PAN layer and the morphology of the cylindrical envelope of the lamellar plates by semi-crystalline polymer matrix. The orientation of the plates is parallel to the sliding direction across the sample P, while the lamellar morphology of the AP specimen can be parallel or perpendicular to the sliding direction, as is shown in figure 7.

Briefly, (I_S) AP> (I_S) OP due to the morphology of the lamellar layer of the graphite sheet and, in addition, the crystalline lamellar plates which surrounds, like a cylinder, a fiber from to a polymer matrix. In addition, if the case of N-type specimens (I_S) AP < (I_S) N (W_S) N < (W_S) AP < (W_S) P. Another possible source of low wear rate of the N type is the ease with which can be breaked and removed from the wear surface the fibers from the higher parts.

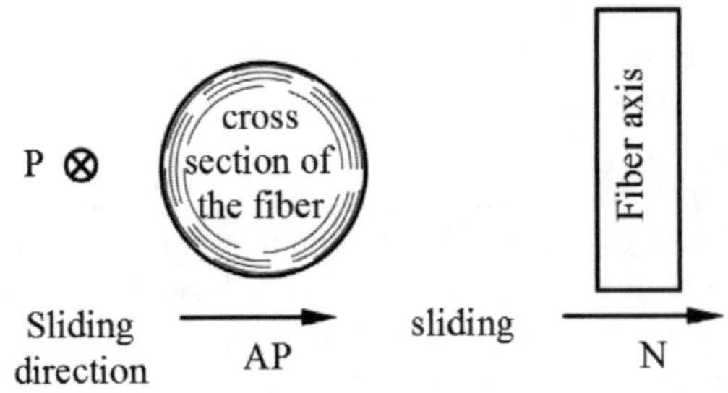

Figure 7. Schematic presentation of the relationship between the sliding direction and the lamellar graphite plate a PAN carbon fiber.

The difference between the friction coefficients in N and AP direction decreases with the load, is substantially constant, as seen in Figures 3 and 4. Mechanical component of the friction force may explain the large difference between the coefficients of the abrasive wear or the adhesive wear, ie: $\mu_{ab} > \mu_{ad}$. In case of the adhesive wear rate, the temperature rise in the contact area and is forming a third layer between the two components of friction tend to reduce the gap between the different fiber orientations. All these factors seem to contribute to the impossibility to establish a clear relationship between the tribological parameters in case of polymer composites reinforced with the carbon fiber.

For the friction coefficient of the adhesive wear may involve a similar reasoning to that in case of the abrasive wear, to conclude that $\mu_N > \mu_P > \mu_{AP}$.

The specific PAN carbon fiber morphology ("leaf onion" type) was correlated with the friction wear behavior to explain the difference between the specimens' P and AP type.

The wear adhesive rates for different orientations of the fibers are still very difficult to correlate with the properties of the mechanical strength or with the microstructure, only that the effect of fiber orientation is relatively weak and almost nonexistent, which is probably due to a third layer newly formed as a lubricant between the two surfaces which is sliding one another (sample plate of stainless steel).

5. Conclusions

Unidirectional reinforces carbon fiber materials are more effective if refer to specific properties per unit volume compared to conventional isotropic materials. Some benefits of carbon fibers composite materials are: low density and high resistance to wear; low or high friction coefficient; resistance to high temperatures; high resistance to breakage and high value ratios strength/density.

As can be seen from the analysis conducted in this paper, due to morphology of the laminated plates of the graphite plates layer and the crystalline lamellar which surrounding a fiber from a polymer matrix and through a low wear rate N type is eased the way which can be broken and removed from the wear surface, fibers from the larger pieces.

In this paper was analyzed the formulations used in the production of the composite materials with short carbon fibers and the polymer matrix, the characteristics of polymer composite materials made with short carbon fiber polymer in the solid state and the tribological characteristics for polymeric composite materials reinforced with short carbon fibers for a better understanding of the phenomena.

The experiments were conducted on the composite material reinforced with short carbon fibers, with polymers in the solid state, and by means of microscopic investigation on the samples subject to wear can be detected some differences in the wear mechanism, related to the behavior of matrix fibers.

The presence of fibers tends to inhibit the formation of the large chips of the matrix material, by accumulating of the polymer layers from sheared material generally being present in these fibers oriented. Electron microscopy investigations have revealed, in this case, a particular behavior of the friction tests, generated by the presence of short fibers, but also because the areas with polymerized epoxy resin.

References

[1] P. C. Verma and others, 2016 *Wear*, Volumes 346–347, pp. 56-65.
[2] R. M. German, 2004 *Powder Metallurgy*, no 8.
[3] J. P. Davim and R. Cardoso, 2009 *Wear*, Vol. 266, Issues 7–8 pp. 795-799.
[4] K.K. Chawla, 2005 *Journal of Minerals, Metals & Materials*, pp. 47.
[5] A. Molazemhosseini, H. Tourani, A. Khavandi, and Y. B. Eftekhari, 2013 *Wear*, Vol. 303, Issues 1–2, pp, 397-404.
[6] J. P. Davim, N. Marques and A. M. Baptista, 2001 *Wear*, Elsevier Sc., 251, pp. 1100-1104.
[7] Z-Z. Zhang, H-J. Song, X-H. Men and Z-Z. Luo, 2008 *Wear*, Vol. 264, Issues 7–8, pp. 599-605.
[8] M. Minus and S. Kumar, 2005 *Journal of Minerals, Metals & Materials*, pp. 52-58.
[9] N. Chand and M. K. Sharma, 2008 *Wear*, Vol. 264, Issues 1–2, pp. 69-74.
[10] K-J. Lee, H-Z. Cheng and J-S. Chen, 2006 *Wear*, Vol. 260, Issues 1–2, pp. 99-108.
[11] S. Rawal, 2001 *Journal of Minerals, Metals & Materials*, pp. 14-17.
[12] J. M. Kunze and C. C. Bampton 2001 *Journal of Minerals, Metals & Materials*, pp. 22-25.
[13] J. P. Davim 2000 *Comparative study of metal matrix composites machinability with diamond cutting tools*, Diamond 2000-11th Europeen Conference, Elsevier Sc, Porto, Portugal 15.7.19.
[14] T. Policandriotes and P. Filip, 2011 *Wear*, Vol. 271, Issues 9–10, pp. 2280-2289.
[15] B. S. S. Kishore, S. Seetharamu and P. S. Kumaran, 2009 *Wear*, Vol. 267, Issues 9–10, pp. 1405-1414.
[16] I. Dutta, K. A. Peterson and C. Park, 2003 , *Journal of Minerals, Metals & Materials*, pp. 38-43.
[17] J. Grácio, J. P. Davim, F. Q. Hua and N. Ali, 2002 *New Developments on Tribology: Theoretical Analysis and Application to Industrial Process*, Proceedings 8th Portuguese Conference of Tribology, University of Aveiro.
[18] S. S. Kim, M. W. Shin and H. Jang, 2012 *Wear*, Vol. 274–275, pp. 34-42.

IOP Conf. Series: Materials Science and Engineering **133** (2016) 012010 doi:10.1088/1757-899X/133/1/012010

Influence of ZrO_2-Y_2O_3 and ZrO_2-CaO coatings on microstructural and mechanical properties on Mg-1,3Ca-5,5Zr biodegradable alloy

B Istrate[1], C Munteanu[1], MN Matei[2], B Oprisan[3], D Chicet[4] and K Earar[2]

[1] "Gheorghe Asachi" Technical University of Iasi, Faculty of Mechanical Engineering, 43 "D. Mangeron" Street, 700050, Iasi, Romania
[2] "Dunarea de Jos" University of Galati, Faculty of Medicine, Dental Medicine, Al. I. Cuza 35 Street, Galati, Romania
[3] "Gr.T. Popa" Medicine University of Iasi, Faculty of General Medicine, University street nr. 16, Iasi, Romania
[4] "Gheorghe Asachi" Technical University of Iasi, Faculty of Material Science and Engineering, 61-63 "D. Mangeron" Street, 700050, Iasi, Romania

E-mail: cornelmun@gmail.com

Abstract. Zirconia (ZrO_2) as a ceramic biomaterial facilitates the osteoconductivity in new bone formation around implant. In order to improve the degradation and the surface properties, it is necessary to apply a surface film to satisfy multiple clinical requirements such as mechanical strength, biocompatibility, and degradation rate. Therefore, surface changing to form a tenacious, biocompatible and corrosion resistant modified layer has become a necessary study in biodegradable materials. The aim of the study is to observe the morphology, structural and scratch analysis for some coatings of ZrO2-CaO and ZrO2-Y2O3 having similar thickness deposited with an atmospheric plasma spraying facility, Sulzer Metco 9MCE, using scanning electron microscopy and X-Ray diffraction. Some mechanical aspects were highlighted during the scratch test. Comparative scratch tests were carried out to study the bonding properties between the coatings and the substrates.

1. Introduction

Mg and most of its alloy are considered to be relatively safe materials for human body. Pure Mg and Mg-X alloys (X:Ca, Si, Zn, Zr) seems attractive for medical use. Some impurities like: Fe, Ni, Cu in magnesium give low corrosion resistance. Qiao et al. [1] investigated that magnesium alloys with Fe concentration of 35 ppm accelerates the corrosion up to 60 times.

Ca belongs to second group of the periodic table. Calcium's solubility in magnesium is relatively low, about 1.34 wt.% under equilibrium conditions. Supersaturated Ca, 1.34 wt.% conducts to formation of Mg2Ca compound along the grain boundaries. Alloying calcium in magnesium alloys helps refining the microstructure of Mg and gives mechanical strength and creep properties.Wan et al. [2] investigated that introducing 0.6 wt.% of Ca addition improves compressive strength of pure Mg.

Reviewing the corrosion behavior of the mageniusm alloys, the most suitable calcium content should be between of 0.6–1.2 wt. % [3-5].Gu et al. [5] estimated that 1-1.5 wt.% of Zr addition in magnesium alloys improves the tensile strength and materials ductility, having values of ultimate tensile strength of 171.87 MPa, elongation of 27.2%, and the corrosion rate is decreased by 40-50%. Zhou et al. [6] obtained the Mg–Ca–Zr alloys which higher values for tensile strength and elongation when zirconium concentration is between 0.5 wt. % to 1.2 wt. %. The as-cast Mg–1Ca–1Zr alloy reveals a value of 125 MPa for ultimate tensile strength and 8% in elongation to failure. [7].A way to improve corrosion resistance on magnesium alloys is surface coatings. Piconi et. al. [8] reveals that cytotoxicity of ZrO_2 is lower than TiO_2. Bozzini et al. [9] demonstrated that ZrO_2 has a satisfactory biocompatibility and could be used as coating for biodegradable materials. Plasma spraying ZrO_2 layer is a ceramic coating used for corrosion resistance of metallic components [10, 11]. Li et al. [12] obtained a polylactic-co-glycolic acid (PLGA) coating on Mg6Zn alloy. He realized two coatings of 33 μm thickness, respectively 72 μm thicknesses. The 72 μm coating exhibits a lower corrosion resistance than the thiner coating. In vitro analysis with cell culture tests (MC3T3-E1) cells on coatings resulted good adhesion of cells on the PLGA coating. The purpose of this work was to obtain two ceramic coatings (ZrO_2-Y_2O_3 / ZrO_2-CaO) using atmospheric plasma spraying and to investigate the morphology, phase analysis and some mechanical characteristics.

2. Materials and methods

Mg-1.3Ca-5.5Zr alloy was obtained by melting in a induction current with controlled atmosphere facility, CTC50K15 type, using zirconia crucibles at Faculty of Materials Science and Engineering, University "Politehnica" of Bucharest, Laboratory of casting and refining metal alloys [13]. For casting it was used as raw materials pure magnesium (99.7%), pure calcium particles and a master Mg-25Zr alloy. Chemical composition correction was made by repeated casting until final composition of 1.3% Ca(wt.%) and 5.5% Zr(wt.%).

The chemical compositions has been determined by EDX analysis using a scanning electron microscope SEM Quanta 200 3D DUAL BEAM equipped with EDS-EDAX detector. The above mentioned percentages of alloying elements have been calculated as average of theree values. Ceramic coatings of ZrO2-Y2O3 / ZrO2-CaO on Mg-Ca-Zr alloy were obtained using athmospheric plasma spray facility (SPRAYWIZARD-9MCE, Sultzer-Metco, USA). For phase analysis, XRD analysis was carried out using a X'Pert Pro MPD diffractometer. Aparrent coefficient of friction, hardness and elastic modulus were measured using CETR UMT-2 Tribometer. For micro-scratch analysis it was used a constant load method with a load of 5N on a distance of 4 mm, for a single determination. In figure 1 is shown the EDS analysis and the chemical composition of the ceramic coatings on Mg-1,3Ca-5,5Zr alloy after deposition. In table 1 are highlighted the technological parameters used for the coating process.

Table 1. Technological parameters used for coating process.

Powder	Gun	N_2		H_2		Electric		Powder feeder 9MP			Spray distance (mm)
		Pressure (bar)	Gas flow (NLPM)	Pressure (bar)	Gas flow (NLPM)	DC (A)	DC (V)	Carrier gas flow (NLPM)	Air Preasure (bar)	Quantity (g/min)	
ZrO2 – CaO	9MB	3,4	44	3,4	6,6	400	70-80	5.3	1.4	144	127
ZrO2 – Y2O3		3,6	39	3,4	6,6	400	70-80	5.1	1.4	126	127

IOP Conf. Series: Materials Science and Engineering **133** (2016) 012010 doi:10.1088/1757-899X/133/1/012010

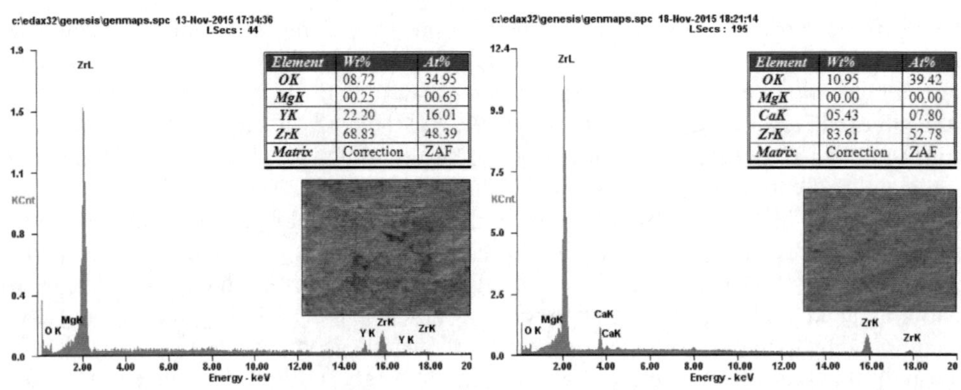

a) b)

Figure 1. EDS spectrum and chemical composition of : a) ZrO_2-Y_2O_3; b) ZrO_2-CaO.

3. Results and discussions

3.1. Structural analysis

3.1. SEM microstructure

Morphology of the ceramic coatings is presented using SEM Quanta 200 3D images. In figure 2 and 3 are shown cross-section and surface images (a-d) for ZrO_2-Y_2O_3 / ZrO_2-CaO layers. Ceramic coatings have the appearance of columnar grains called "splats" having an average roughness with the presence of micropores and some unmelted particles from the thermal deposition process. ZrO_2-Y_2O_3 layer presented some cracks due the recovery process. Average thickness of coatings are presented in figure 2(a) and 3(a), 124.72 μm for ZrO_2-Y_2O_3 coating, respectively 100.05 μm for ZrO_2-CaO.

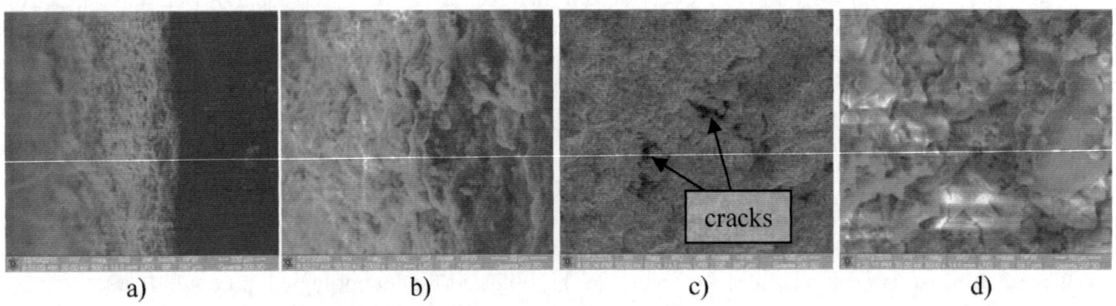

a) b) c) d)

Figure 2. SEM images of ZrO_2-Y_2O_3 coating morphology: a) cross-section - 500X; b) cross-section - 2000X; c) surface - 500X; d) surface - 5000X.

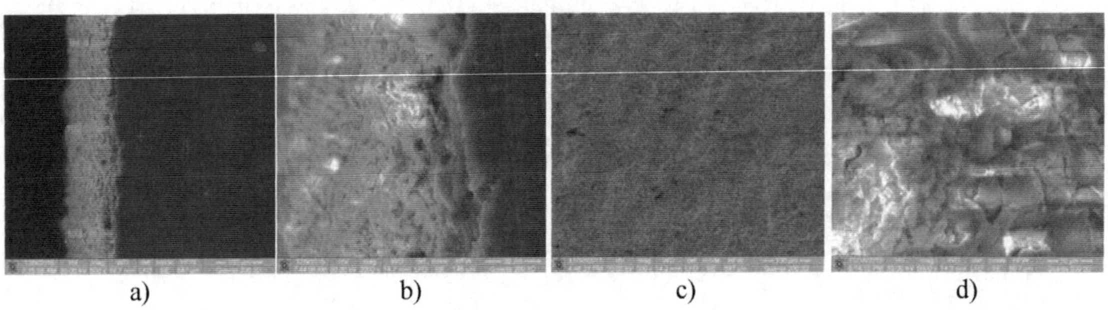

a) b) c) d)

Figure 3. SEM images of ZrO_2-CaO coating morphology: a) cross-section - 500X; b) cross-section - 2000X; c) surface - 500X; d) surface - 5000X.

Studying the morphologies of the ceramic layers it can be noticed that ZrO_2-CaO seems to be a more compact layer than ZrO_2-Y_2O_3 due the lack of micropores presented in the microstructure.

3.2. XRD analysis

XRD diffraction analysis are shown in figure 4. The diffraction analysis highlighted an α-ZrO_2 predominant phase around 2θ angle: 30° and 50° having a tetragonal crystal system. Y_2O_3 compound is found most at 2θ angles: 73,47°, 94,27° and 117,95°with a crystal structure of cubic type. Calcium phase was identified at 2θ = 73,92° and 102,69° with a cubic structure type. Lattice parameters of all identified compounds are shown in table 2.

Tabel 2. Lattice parameters of ZrO_2-Y_2O_3 / ZrO_2-CaO coatings.

	Compound	**Space Group**	**Crystal system**	**a (Å)**	**b (Å)**	**c (Å)**	**α (°)**	**β (°)**	**γ (°)**	**Cell volume (10^6 pm^3)**
ZrO_2-Y_2O_3 coating	Mg	P63/mmc	Hexagonal	3,2089	3,2089	5,2101	90	90	120	46,46
	ZrO_2	P42/nmc	Tetragonal	3,6292	3,6292	5,1973	90	90	90	68,45
	ZrO_2	P42/nmc	Tetragonal	3,6055	3,6055	5,1797	90	90	90	67,33
	Y_2O_3	Ia-3	Cubic	10,6080	10,6080	10,6080	90	90	90	1193,70
	$YO_{1.335}$	P-3m1E	Hexagonal	3,8750	3,8750	6,0340	90	90	120	78,47
ZrO_2-CaO coating	ZrO_2	Fm-3m	Cubic	5,1280	5,1280	5,1280	90	90	90	134,85
	ZrO_2	P42/nmc	Tetragonal	3,5960	3,5960	5,1770	90	90	90	66,94
	Ca	Fm-3m	Cubic	5,5884	5,5884	5,5884	90	90	90	174,53
	Ca	Fm-3m	Cubic	5,5886	5,5886	5,5886	90	90	90	174,55

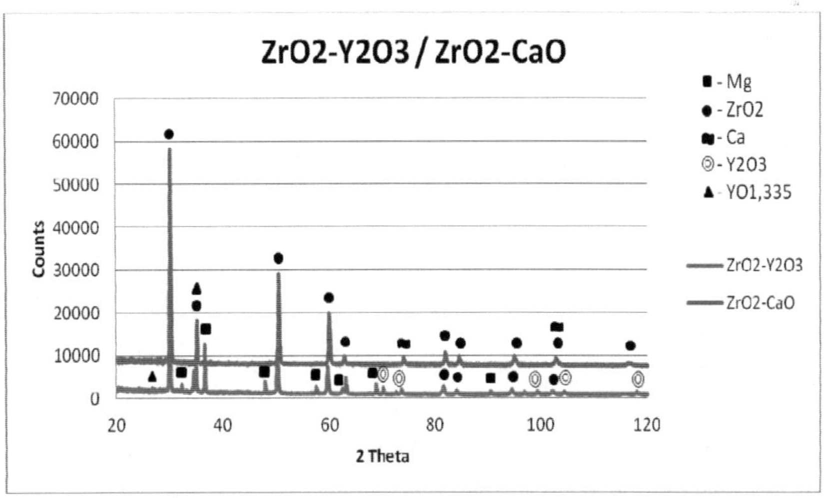

Figure 4. XRD patterns of ZrO_2-Y_2O_3 / ZrO_2-CaO coatings.

Using the the Scherrer equation [14], it was possible to calculate the crystallite size for each phase determinated and show the results in table 3.

Table 3. Crystalite size of compounds (nm).

	ZrO_2	Y_2O_3	$YO_{1,335}$	ZrO_2	Ca
$ZrO_2 - Y_2O_3$	23,94738	30,09688	24,79977	-	-
$ZrO_2 - CaO$	-	-	-	22,4809	15,3444

3.3. Scratch and micro-indentation analysis

Figures 5 and 6 presents SEM images (a-c), 2-D (e) and 3-D (d) profile surface of the "scratch" analysis. Figure 5 a), b) is presented the head-ends of scratch marks, then a SEM image at 500X magnification, highlighting the fracturing of the ZrO_2-Y_2O_3 coating. The apparent coefficient of friction for the ZrO_2-Y_2O_3 layer is 0.868. Same analyses were carried out for the second layer, respectively ZrO_2-CaO, which presented an apparent coefficient of friction of 1.042.

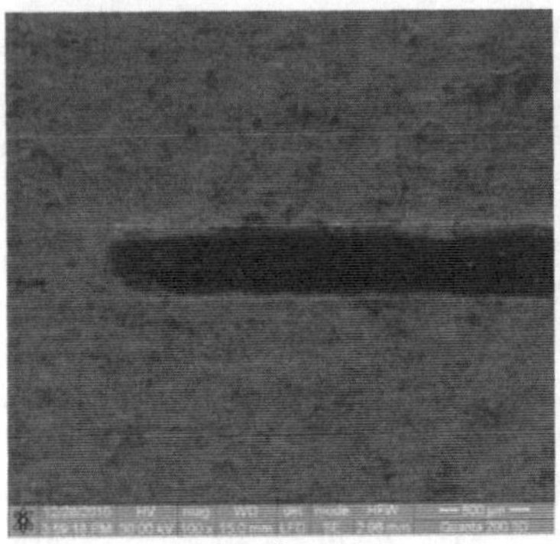

a)

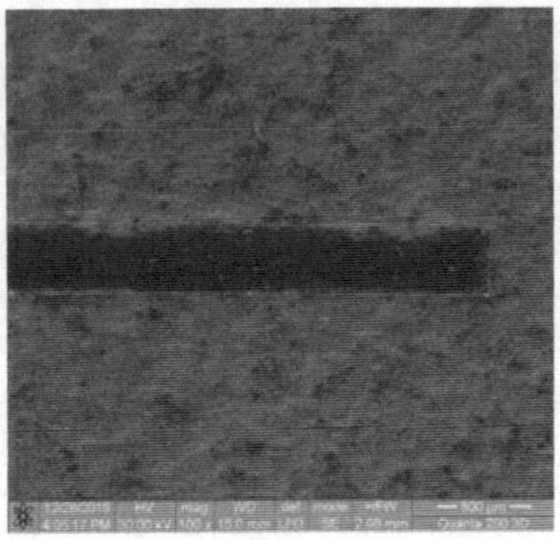

b)

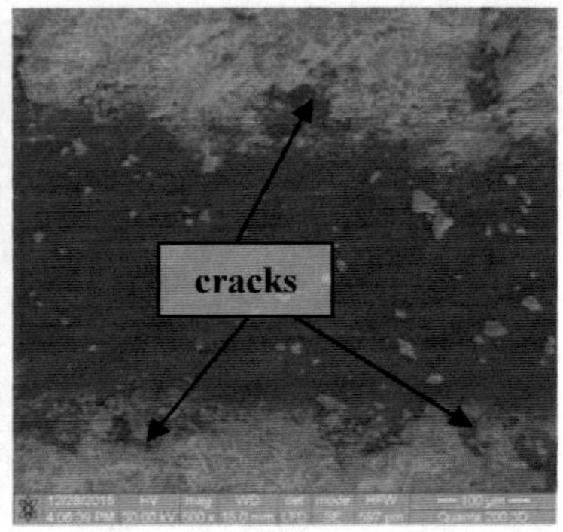

c)

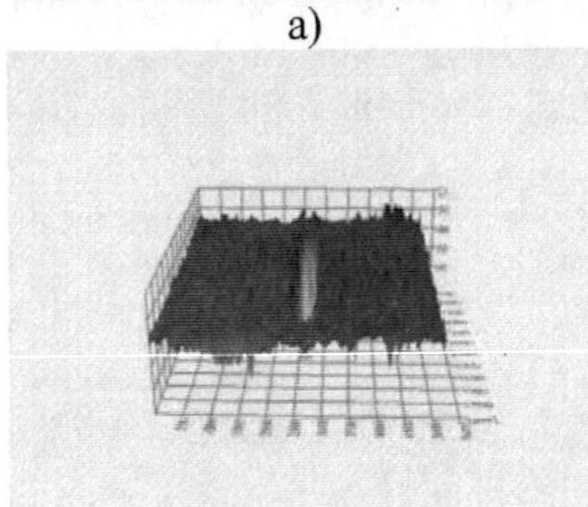

d)

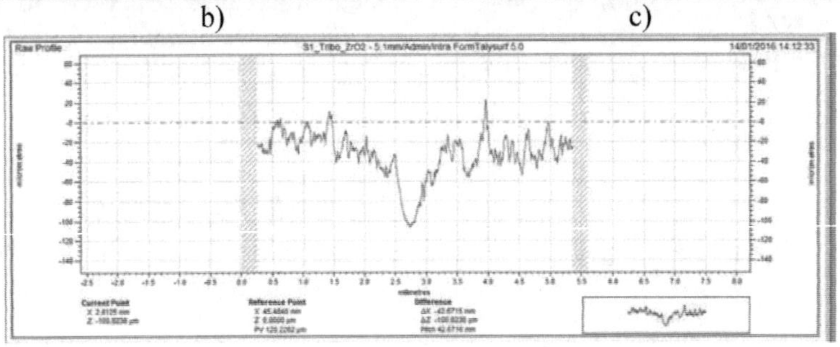

e)

Figure 5. SEM images of scratch test (a-c) and 2-D/3-D profile (d-e) on ZrO_2-Y_2O_3.

a)

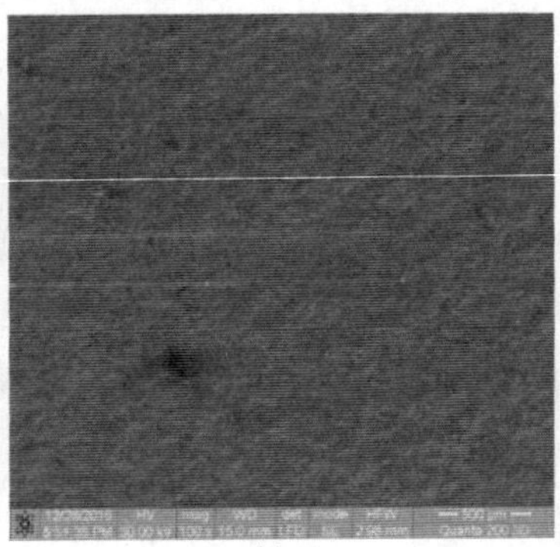

b)

c)

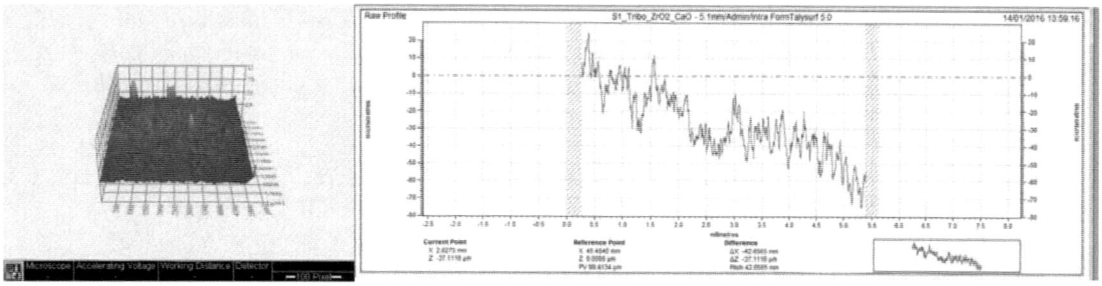

d) e)

Figure 6. SEM images of scratch test (a-c) and 2-D/3-D profile (d-e) on ZrO_2-CaO.

SEM images showed an adherent layer (ZrO_2CaO) to the base material, a much harder coating (table 4) than ZrO_2-Y_2O_3 layer. Also it was noticed a lack of cracks in the layer of ZrO_2CaO, resulting a less fragile layer. The results have shown a depth of 65 microns for scratch of ZrO_2-Y_2O_3 layer with layer coated fracturing and 25 microns depth for scratch of ZrO2-CaO layer, without fracturing. it can be concluded that the layer of ZrO2-CaO has a high resistance to fracturing, a fact confirmed by hardness values and coefficient of friction apparently values presented in table 4.

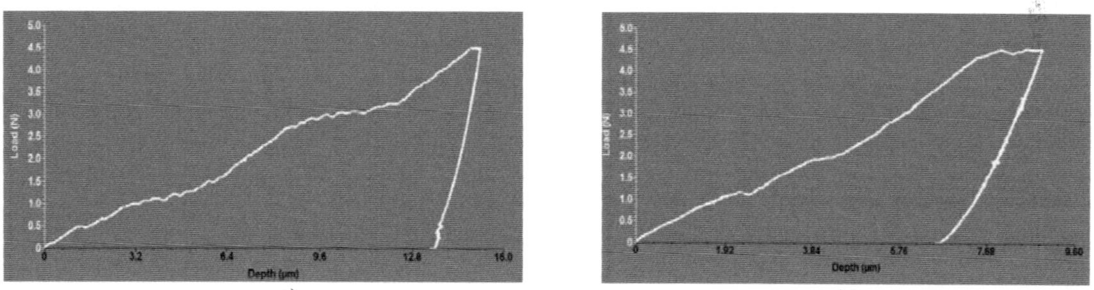

a) b)

Figure 6. The variation curve of the micro-indentation test of ZrO_2-Y_2O_3 (a) / ZrO_2-CaO (b) coatings.

ZrO_2-Y_2O_3 layer presented a strain load of approximately 14,6 μm and a strain release of approximately 13,2 μm resulting an elastic deformation of 1,4 μm. ZrO2-CaO layer showed a strain load of approximately 11,5 μm and a strain release of approximately 8,2 μm resulting an elastic deformation of 3,3 μm. These values indicate that ZrO2-CaO layer has an elastic behaviour towards ZrO2-Y2O3 coating that is more fragile. Micro-indentation tests results are presented in table 4, showing the elastic modulus, hardness and stiffness for the ceramic layers.

Table 4. Some mechanical properties of ZrO_2-Y_2O_3 / ZrO_2-CaO coatings.

	COF	Hardness (HV)	Young Modulus (GPa)	Stifness(N/μm)
ZrO_2-Y_2O_3	0.868	20.053	20.541	3.733
ZrO_2-CaO	1.042	41.229	17.903	2.377

4. Conclusions

In this article was studied the morphological appearance, X-ray diffraction, micro-scratch and micro-indentation aspects on ceramic coatings ZrO2-Y2O3, respectively ZrO2-CaO. These ceramic coatings have the appearance of columnar grains called "splats" having an average roughness with the presence of micropores and some unmelted particles from the thermal deposition process. X-Ray analysis showed the presence of α-ZrO_2 as predominant phase having a tetragonal crystal system. Also was identified as secondary phase Y_2O_3 and calcium with a crystal structure of cubic type. ZrO2-CaO presented a higher fracturing resistance then ZrO2-Y2O3, having a lower elastic modulus, with a higher hardness value. In conclusion ZrO2-CaO coating behaved superior to mechanical tests than ZrO2-Y2O3 coating which makes it suitable for further research in order to use in the medical field.

References

[1] Qiao Z, Shi Z, Hort N, Zainal N, Abidin I and Atrens A 2012 Corrosion behavior of a nominally high purity Mg ingot produced by permanent mould direct chill casting *Corrosion Science* **61** pp 185–207

[2] Wan Y, Xiong G, Luo H, He F, Huang Y and Zhou X 2008 Preparation and characterization of a new biomedical magnesium–calcium alloy *Materials & Design* **29** pp 2034–2037

[3] Kim W C Kim J G Lee J Y and Seok H K 2008 Influence of Ca on the corrosion properties of magnesium for biomaterials *Materials Letters* **62** pp 4146–4148

[4] Kirkland N T, Birbilis N, Walker J, Woodfield T, Dias G J and Staiger M P 2010 *Journal of Biomedical Materials Research Part B: Applied Biomaterials* **95B** pp 91–100

[5] Gu X N, Zheng Y F, Cheng Y, Zhong S P and Xi T F 2009 In vitro corrosion and biocompatibility of binary magnesium alloys *Biomaterials* **30** pp 484–498

[6] Zhou Y L, An J, Luo DM, Hu W Y, Li Y C, Hodgson P and Wen P C 2012 Microstructures and mechanical properties of as cast Mg–Zr–Ca alloys for biomedical applications *Materials Technology: Advanced Performance Materials* **27** pp 52–54

[7] Zhang W, Li M, Chen Q, Hu W, Zhang W and Xin W 2012 Effects of Sr and Sn on microstructure and corrosion resistance of Mg–Zr–Ca magnesium alloy for biomedical applications *Materials & Design* **39** pp 379–383

[8] Piconi C and Maccauro G 1999 Zirconia as a ceramic biomaterial-rev *Biomaterials* **20** 1-25

[9] Bozzini B, Carlino P and Mele C 2011 Electrochemical behaviour and surface characterisation of Zr exposed to an SBF solution containing glycine, in view of dental implant applications, *Journal of Materials Science: Materials in Medicine* **22** pp 193-200

[10] Morks M F, Gao Y, Fahim N F and Yingqing F U 2006 Microstructure and hardness properties of cermet coating sprayed by low power plasma *Materials Letters* **60** pp 1049-1053

[11] Morks M F, Gao Y, Fahim N F, Yingqing F U and Shoeib M A 2005 Influence of binder materials on the properties of low power plasma sprayed cermet coatings *Surface & Coating Technology* **199** pp 66-71

[12] Li J N, Cao P, Zhang X N, Zhang S X and He YH 2010 In vitro degradation and cell attachment of a PLGA coated biodegradable Mg–6Zn based alloy *Journal of Materials Science* **45(22)** pp 6038-6045

[13] *** www.eramet.ro

[14] Patterson A 1939 The Scherrer formula for X-ray particle size determination *Physical Review* **56** pp 978-982

Sodium silicate as source of silica for synthesis of mesoporous SBA-15

Norhasyimi Rahmat, Fazlena Hamzah, Norsuraya Sahiron, Marissa Mazlan and Muhammad Mukmin Zahari
Faculty of Chemical Engineering, Universiti Teknologi MARA, 40450 Shah Alam

E-mail: norhasyimi@salam.uitm.edu.my

Abstract. Ordered mesoporous silica SBA-15 was prepared using hydrothermal synthesis using sodium silicate (Na_2SiO_3) as the silica source and amphiphilic block copolymer Pluronic P123 as the structure directing agent. The influence of the mass Na_2SiO_3, ripening duration, aging time and calcination temperature on the structural and mesoporous properties of silica was studied. X-ray diffraction (XRD), Fourier transform infrared (FTIR), Scanning electron microscopy (SEM) and the nitrogen adsorption desorption using Brunauer Emmett Teller (BET) are some instruments used to characterize the results of investigation. From XRD analysis, SBA-15 synthesized from sodium silicate yield 2D-hexagonal symmetry (p6mm). From FTIR analysis, functional group Si-O-Si symmetric stretching modes and asymmetric Si-O-Si stretching modes were present. The sample with the highest mass of Na_2SiO_3 and the shortest aging time exhibited the largest surface area and large pore size. The results also showed the morphological structure could be tuned during ripening stage.

Keywords: Mesoporous, SBA-15, sodium silicate, characterization

1. Introduction

Mesoporous materials are made up of inorganic metal oxides such as silica or alumina. Silica mesoporous materials are synthesized by applying the use of surfactant to form regularly aligned assemblies that act as a template for the metal oxide followed by the removal of the template. The surfactant being used is the amphiphilic block copolymers, poly (ethylene oxide) – poly (propylene oxide) – poly (ethylene oxide). Its properties such as its amphiphilic character, biodegradability, low-cost commercial availability and mesostructural ordering action make it desirable to be used for SBA-15 synthesis [1]. It can be made up of pores ranging from 2 to 50 nm in width [2] and disordered micropores that allows interconnection between mesopores. SBA-15 products were found to be more hydrothermally stable as it has thicker walls. Its high internal surface area typically in the range between 400 and 900 m^2g^{-1} makes SBA-15 very versatile. The physico-chemical properties of SBA-15 materials can be possible tuned by varying the synthesis parameters which includes the pH solutions, reactant, additives and solvents, aging temperature, stirring and aging durations and temperature.

SBA-15 has been used as support in catalytic reactions where the high surface area enables better dispersion of particles [3]. It has also been known that SBA-15 is used in environmental analytics for separation and adsorption, advances optics and many more uses [4]. Mesoporous materials are usually synthesized by using tetraethylorthosilicate (TEOS) as the silica source. However, high cost of complex TEOS processing motivates the development of alternative silica source [5]. In this study, sodium silicate (Na_2SiO_3) was investigated to synthesize mesoporous SBA-15. It was also conducted to determine the effects of different mass of Na_2SiO_3, ripening duration, aging time and calcination temperature.

2. Methodology

2.1 Synthesis of SBA-15 samples

SBA-15 samples were synthesized using Pluronic P123 (non-ionic triblock copolymer, $EO_{20}PO_{70}EO_{20}$) as a surfactant and Na_2SiO_3 as the silica source. 4.0 gram of Pluronic P123 was dissolved in 80 gram of 2 M hydrochloric acid and 20 gram of distilled water with constant stirring for 1 hour at 40°C until homogeneous. Then 8.0 gram of Na_2O_3Si is added to the solution drop wise with vigorous stirring at the same time and temperature.

2.2 The effect of amount of Na_2SiO_3 and aging time experimental procedure

The samples are aged at 40°C for 24 hours. The aging time is varied at 3, 6 and 16 hours. The solution is then treated hydrothermally at 100°C for 4 hours. The final product derived was filtered, washed and dried at 40°C. The samples were then calcined in a furnace at 550°C for 5 hours to decompose triblock copolymer [6].

2.3 The effect of ripening duration and calcination temperature experimental procedure

4 gram of sodium silicate was slowly added into Pluronic123 – HCl solution under stirring condition and continuously stirred at 40°C for different duration for ripening process. After different hours of ripening step is completed, the solution was then transferred to the oven for hydrothermal aging process at 100°C for 4 hours. After 4 hours, the precipitate formed was filtered and washed with distilled water. Filtered precipitates are then dried in oven at 100°C for 4 hours. Dried sample are then divided into two and calcined at 550°C and 750°C respectively in a furnace.

2.4 Characterization of SBA-15 samples

The SBA-15 samples crystalline structure is characterized using X-ray Diffraction (XRD) patterns using the Rigaku Diffractometer. The 2θ angular regions between 0 and 5° were done at a scan rate of 0.004 °/min (40 kV, 40 mA).

Adsorption and desorption isotherms of N_2 was measured of the SBA-15 samples at -196°C using Micromeritics ASAP 2020. From the isotherms the BET surface area, A_{BET} pore size, and micropore area, A_{mic}. The mesoporous area, A_{meso} was calculated by subtracting the A_{mic} from A_{BET}.

Fourier Transform Infrared (FTIR) analysis of the SBA-15 samples was carried out using Perkin Elmer Spectrum One. The Field Emission Scanning Electron Microscope (FESEM) was used to capture the images of SBA-15 sample. The electron microscopy images of SBA-15 were taken on Carl Zeiss Supra 40 VP at acceleration voltage of 5 kV and magnification of x5000 and x20000.

3. Results and discussion

3.1 Synthesis of SBA-15 from sodium silicate as silica source

Figure 1 shows the XRD analysis for SBA-15 synthesized from sodium silicate. For different amount Na_2SiO_3 and aging time imposed on the synthesis process, it can be seen diffraction peaks at almost the same intensities were observed. The peaks range from 0.32-0.38°, 0.70-0.72° and 1.04-1.06°. All samples exhibit very intense diffraction peaks at 2θ=1.048-1.06° which are characteristic of ordered SBA mesoporous materials with a 2-D hexagonal symmetry (p6mm). The

XRD patterns show that the samples synthesized have pattern characteristics of hexagonal mesostructures.

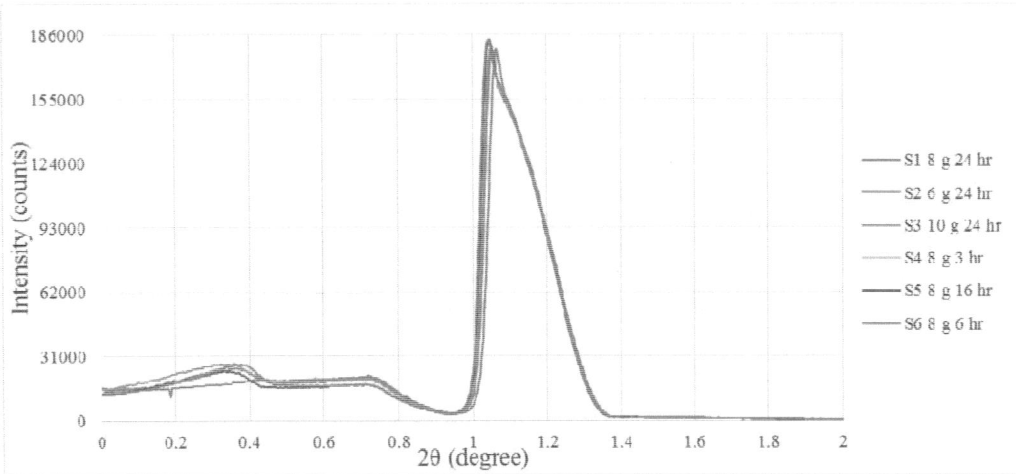

Figure 1. XRD patterns of SBA-15 sample at different sodium silicate mass (S1) 8g, (S2) 6g, (S3) 10 g and at different aging time (S1) 24 hours, (S4) 3 hours, (S5) 16 hours and (S6) 6 hours.

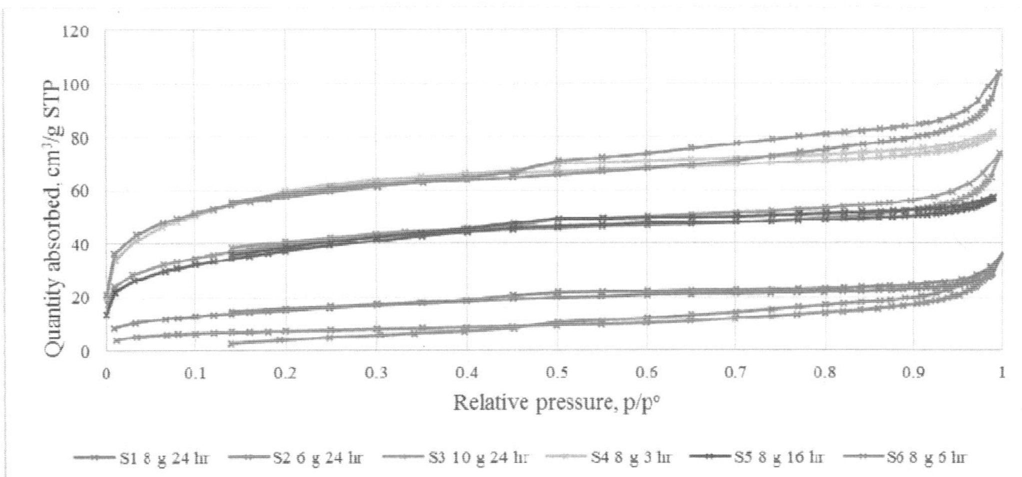

Figure 2. N_2 adsorption-desorption isotherms of SBA-15 sample at different sodium silicate mass (S1) 8g, (S2) 6g, (S3) 10 g and at different aging time (S1) 24 hours, (S4) 3 hours, (S5) 16 hours and (S6) 6 hours

Nitrogen adsorption desorption of the samples using the Micromeritics ASAP 2020 enables the determination of the surface area of the samples and the pore distribution whether it is microporous, mesoporous or macroporous. Figure 2 is the N_2 adsorption desorption isotherm of the SBA-15 samples. All samples exhibit the same adsorption isotherm trend that is the Type IV isotherm. Type IV isotherms indicates that the samples are mesoporous. Mass of Na_2O_3Si in the synthesis of the SBA-15 samples S1, S2 and S3 yields 54, 26 and 140 m^2/g surface area. It can be inferred that the greater the mass of the sodium silicate used, the greater the surface area developed from the synthesis. Even though the surface area varies with each sample with different sodium silicate mass, all three isotherms correspond to Type IV mesoporous materials.

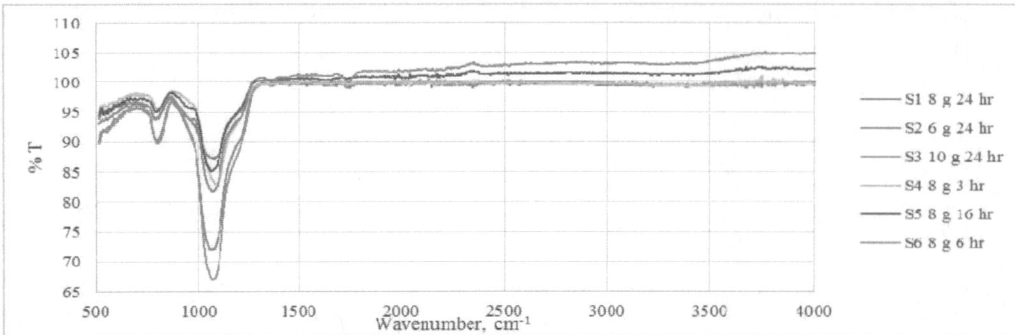

Figure 3. Fourier transform infrared (FTIR) peaks for samples at different amount of Na_2O_3Si and aging time.

The FTIR adsorption data was collected using the Perkin Elmer Spectrum One FTIR spectrometer. Figure 3 displays the frequency band of samples S1 to S6 of SBA-15 synthesized by sodium silicate characterized using FTIR to identify the functional groups present in those samples. It can be observed that all the samples exhibit peaks of almost the same frequency. Each peak is characteristic of a specific functional group. The peaks ranging between 785 to 801 cm^{-1} are due to the Si-O-Si symmetric stretching modes. The peaks located between 1063.05 to 1077.49 cm^{-1} are apparent due to the asymmetric Si-O-Si stretching modes. The silica from the sodium silicate on the surface of the sample is evident from the frequency peaks detected.

3.2 *The effect of amount Na_2O_3Si and aging time*

For the variation of aging time, the mass of the sodium silicate used was the same for the samples, which are 8g. In the terms of the aging time as the synthesis parameters, the shorter the aging time the greater the BET surface area of the samples. The samples S4, S6, S5 and S1 arranged according to ascending aging time which is 3, 6, 16 and 24 hours respectively The surface areas of the samples S4, S6, S5 and S1 are approximately 208, 204, 132 and 53 m^2/g respectively. It can be inferred that the longer the aging time, the surface area reduces. This may happen due to the degradation of pores and breakage of fiber segments formed due to prolonged aging time. The samples however still correspond to Type IV isotherms classifying it as mesoporous materials.

All samples exhibit small micropore area. The micropore area makes up approximately 9-35% of the total surface area. The samples are made up mostly of mesopores. The SBA-15 samples prepared have low microporosity have higher thermal stability (Barrabino, 2011). The pore width of all samples is also characteristic of mesopores that is in the range of 2-50 nm.

Table 1. Porous properties of SBA-15 samples with varying sodium silicate mass and aging time

Parameter			a	b	c	Pore size, nm				Average pore width, nm	
						d	e	f	g	h	i
^{j}(g)	S1	8	53.995	5.276	0.043	3.150	3.2338	4.2402	4.4338	3.1507	3.234
	S2	6	25.961	6.212	0.042	6.619	6.619	11.042	7.196	6.619	6.619
	S3	10	140.42	36.062	0.100	2.794	2.933	5.449	5.226	2.9944	2.9330
^{k}(hr)	S1	24	53.995	5.2760	0.043	3.151	3.234	4.2402	4.4338	3.1507	3.2338
	S4	3	208.23	36.616	0.124	2.376	2.3801	3.3615	3.2504	2.3756	2.3801
	S5	16	132.38	18.756	0.087	2.618	2.6384	3.4435	3.2855	2.6177	2.6385
	S6	6	204.58	71.514	0.144	2.791	2.8315	5.6145	5.1614	2.7908	2.8315

a- BET surface area, m^2/g
b- Micropore area, m^2/g

c- Pore volume, cm^2/g
d- Adsorption average pore width
e- Desorption average pore width
f- BJH adsorption average pore width
g- BJH desorption average pore width
h- Adsorption
i- Desorption
j- Mass of sodium silicate , Na_2O_3Si
k- Aging time

3.3 The effect of ripening duration and calcination temperature

The ripening duration were varied at 2h, 6h 16h and 24h and the effect on sample characteristics were studied. All samples exhibited similar nitrogen adsorption/desorption isotherm pattern but different isotherm values. The N_2 adsorption/desorption isotherms of all samples indicate different duration of ripening could change the mesostructural characteristic of SBA-15. Figure 4 shows the N_2 adsorption /desorption isotherm of a sample synthesize at 2h, 6h, 16h, and 24h ripening duration and calcined at 550°C.

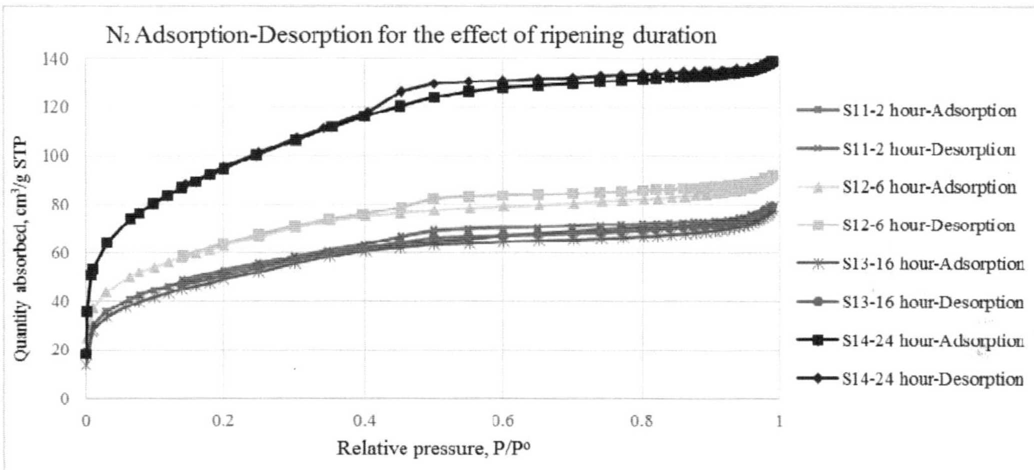

Figure 4. Nitrogen adsorption/desorption isotherm of sample ripened for (a) 2h (b) 6h (c) 16h and (d) 24h, which calcined at 550 degree Celcius.

The SEM image of the samples in Figure 5 shows that spherical like aggregates was formed. As the ripening period increased, the sizes of the aggregates also increased. The aggregates alignment was found to be better aligned as the period of ripening increased. The large aggregates was formed in favor of the ripening conditions was also reported by Benamor et al. The trend of structural and textural properties of SBA-15 synthesized is in agreement with investigation by Chareonpanich et al.

(a)

(b)

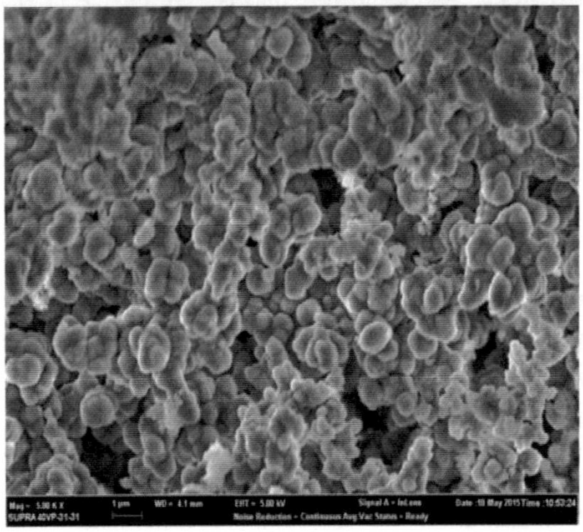

(c)

Figure 5. SEM images of SBA-15 ripened for (a) 2h (b) 16h (c) 24h and calcined at 550°C

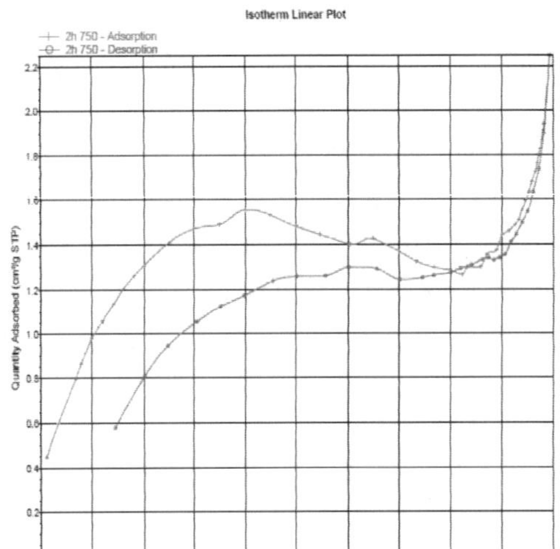

Figure 6: Nitrogen adsorption/desorption isotherm of sample ripened for 2h calcined at 750°C.

Figure 6 shows the SBA-15 sample ripened for 2h and calcined at 750°C. The Nitrogen adsorption/desorption pattern of samples calcined at 750°C showed a much smaller value compared to samples calcined at 550°C. All the samples calcined for both temperature 550°C and 750°C exhibits the type H1 hysteresis loop.

The SEM images SBA-15 samples calcined at 750°C in Figure 7 shows a significant difference compared to SBA-15 samples calcined 550°C in Figure 5. The particles has smaller surface area and pore volume at 750°C than that of 550°C. However, the pore diameter of mesoporous structure of SBA-15 have higher value compare to pore diameter calcined at 550°C. Table 2 displays the value of surface area and total pore volume is higher for samples calcined at 550°C but the higher pore diameter value for samples calcined at 750° As reported by Sahu et al. the surface area and total pore volume of SBA-15 samples would reduce as the calcined temperature increased.

(a)

(b)

(c)

Figure 7: SEM images of SBA-15 samples ripened for (a) 2h (b) 16h (c) 24h and calcined at 750°C

Table 2: Structural and textural properties of ripened at 40°C for various ripening period and calcined at 550°C and 750°C.

Sample	S_{BET} [a] (m^2/g)	V_P [b] (cm^3/g)	D_{BJH} [c] (Å)
2h 550	183.2982	0.080441	31.286
6h 550	222.561	0.091274	30.235
16h 550	172.7141	0.087265	32.86
24h 550	332.8657	0.147502	28.922
2h 750	5.1029	0.00267	36.481
6h 750	1.3588	0.002104	127.5
16h 750	1.2392	0.002221	370.504
24h 750	3.0783	0.004184	84.147

[a] BET surface area

[b] Total pore volume deduced from nitrogen adsorbed

[c] Pore diameter determined by the BJH method from the nitrogen desorption branch

4. Conclusion

Mesoporous silica SBA-15 was successfully synthesized under acidic conditions using sodium silicate as the silica source. The characterization showed that the materials have large specific surface area and developed mesoporosity. The surface area however is smaller as compared to the standard characteristics of SBA-15, which is around 500-1000 m^2/g. The synthesis parameters of concern were the mass of the sodium silicate and aging time. From the result, the conditions for the synthesis of SBA-15 using sodium silicate with the highest mass of sodium silicate and the shortest aging time which is 10g and 3 hours respectively. Modifications with the mass and time can result in greater surface area of the SBA-15 product. Ripening duration due to different period of stirring and different calcination temperature also affect the structural properties of SBA-15 synthesized from sodium silicate.

ACKNOWLEDGMENT

Appreciation goes to Malaysia of Higher Education for funding this study through Fundamental Research Grant Scheme (FRGS/2/2013/TK05/UITM/03/2). Special thanks to Research Management Institute of Universiti Teknologi MARA for assisting the process of this investigation.

References

[1] Zhao, D., Feng, J., Huo, Q., Melosh, N., Fredrickson, G.H., Chmelka, B.F. and Stucky, G.D. (1998). Triblock copolymer syntheses of mesoporous silica with periodic 50 to 300 angstrom pores. *Science Vol 279.*

[2] Barrabino, A. (2011) Synthesis of mesoporous silica particles with control of both pore diameter and particle size. *Chalmers University of Technology.*

[3] Zhu, J., Wang, T., Xu, X., Xioa, P., and Li, J. (2013). Pt nanoparticles supported on SBA-15: synthesis, characterization and applications in heterogeneous catalysis. *Applied Catalysis B: Environmental 130-131 (2013) 197-217.*

[4] Thielemann, J.P., Girgsdies, F., Schlögl, R., and Hess, C. (2011). Pore structure and surface area of silica SBA-15 : influence of washing and scale-up. *Beilstein Journal of Nanotechnology.*

[5] Meléndez-Ortiz, H.I., Mercado-Silva, A., García-Cerda, L.A., Castruita, G. and Perera-Mercado, Y.A. (2013). Hydrothermal synthesis of mesoporous silica MCM-41 using commercial sodium silicate.

[6] Björk. E.M. (2013). Mesoporous building blocks – synthesis and characterization of mesoporous silica particles and films. *Linköping Studies in Science and Technology Dissertation No.1542. Linköping University.*

[7] Erdem, S., Erdem, B., Öksüzoğlu, R.M., and Çitak, A. (2013). Bifunctional Fe-SBA-15-SO_3H mesoporous catalysts with different Si/Fe molar ratio: synthesis, characterization and catalytic activity. *Bull. Korean Chem. Soc (2013) Vol.34, No.5 (1481).*

IOP Conf. Series: Materials Science and Engineering **133** (2016) 012012 doi:10.1088/1757-899X/133/1/012012

Metal Chloride Induced Formation of Porous Polyhydroxybutyrate (PHB) Films: Morphology, Thermal Properties and Crystallinity

W.L.Tan*, N.N. Yaakob, A. Zainal Abidin, M. Abu Bakar and N.H.H. Abu Bakar

Nanoscience Research Laboratory, School of Chemical Sciences, Universiti Sains Malaysia, 11800 Penang, Malaysia

E-mail: weileng_tan@usm.my; weileng728@gmail.com

Abstract. Polyhydroxybutyrate (PHB) films with highly porous structures were synthesized using a one phase system comprising of metal chloride/methanol/PHB/chloroform (MCl_2/CH_3OH/PHB/$CHCl_3$). SEM analyses confirmed that the MCl_2 (where M = Cu^{2+} or Ni^{2+}) induced porous structures with pore sizes ranging from 0.3 – 2.0 µm. The average pore size increased with the increasing MCl_2 content. There existed weak physical interactions between the PHB chains and MCl_2 as revealed by FTIR and NMR spectroscopies. The residue of MCl_2 in the porous PHB film does not exert significant influence on the thermal stability of PHB. Nevertheless, the crystallinity of the prepared film is enhanced, as MCl_2 acts as the nucleation sites to promote the growth of spherullites.

1. Introduction

Nowadays, the ever increasing environmental issues and the increasing cost and limitations of petroleum resources have driven researches to focus on biopolymers. Polyhydroxybutyrate (PHB) a biopolymer belonging to the family of polyesters, polyhydroxyalkanoate (PHA). It is derived from renewable resources and shows excellent biocompatibility and biodegradability. It is also non-toxic and harmless to the environment.

The fabrication of porous PHB is of great interest among scientists as this material has shown a wide variety of applications. One of the most common applications of porous PHB is in the biomedical area. The porous structure of PHB not only acts as a support to adhere the cells, but also enables the transport of nutrient, drugs or bioactive molecules to the tissues or cells [1]. As such, it has been applied as nerve guides [2], tissue scaffolds [3] and in drug delivery [4]. Another prominent field of porous PHB is water treatment. Zhang and co-workers [5, 6] has demonstrated the removal of chlorobenzene and o-nitrochlorobenzene using porous PHB as a biomimetic adsorbent. The maximum adsorption capacity for chlorobenzene and o-nitrochlorobenzene is 125.99 and 39.56 mg g^{-1} respectively.

Solvent casting-particulate leaching method, a conventional technique to produce scaffold/highly porous biopolymer was first developed by Mikos et al. [7]. This technique was adopted by Zubairi [8] and Misra et al. [9] to prepare PHB scaffolds using sodium salt crystals or sugar cubes as porogens. The porous structure was formed upon leaching of the porogen using deionized water. Porous PHB can also be prepared via enzymatic hydrolysis technique where PHA depolymerase enzyme from Acidovorax sp. is used to degrade the PHB film [10]. Emulsion is another method to create porous PHB [5, 6, 11]. In this method, water droplets template the porous structure of PHB during evaporation. Other techniques including freeze-drying [3], laser sintering [12] as well as electrospinning [13], have also been reported.

Most of the techniques that were used to create porous PHB are either time consuming, involves complicated procedures or needs specific instruments. Furthermore, the homogeneity as well as the inter-connectivity of the pores are hardly achieved. In this study, we report a simple and low cost method that is capable to produce porous PHB film with homogenous inter-connected pores. Unlike the emulsion and particulate-leaching techniques that employs a bi-phasic system, the method described here involves only one phase. The pore textures, thermal properties and crystallinity of the PHB film prepared in a singular phase (MCl_2/methanol/PHB/chloroform, where M = Ni^{2+} or Cu^{2+}) is thus investigated. The role of metal chloride and methanol on the formation of the porous structure is also discussed. These porous PHB films may have potential applications in the field of waste water treatment as an adsorbent or separation membrane [5,6].

2. Materials and Methods

2.1 Materials

Poly(3-hydroxybutyrate) (PHB) was supplied by BIOCYCLE, Brazil and purified before used [14]. Other chemicals were commercially obtained and used without further purification: 98% nickel (II) chloride hexahydrate ($NiCl_2.6H_2O$) from UNIVAR, Ajax Chemicals, Australia, 99.99% copper (II) chloride dehydrate ($CuCl_2.2H_2O$) from Aldrich, US, chloroform from Systerm, Malaysia and methanol from QRec, Malaysia.

2.2 Preparation of porous PHB films

Three different moles of MCl_2 (where M = Cu^{2+} or Ni^{2+}) which are 5.0 x 10^{-6}, 2.5 x 10^{-5} and 5.0 x 10^{-5} mol were used against a constant mole of PHB (5 x 10^{-4} mol). In a typical preparation, 10 mL of 0.05 M PHB in chloroform was added to a stipulated amount of 0.05 M MCl_2 in methanol and topped up with methanol where necessary (Table 1). The mixture was stirred for 10 min. Then, the solvent was evaporated via rotary evaporator at room temperature until a viscous solution was obtained (~ 10 minutes). A porous PHB film was obtained upon further air drying the viscous solution at ambient condition. For comparison purposes, a sample with solely PHB mixed with equivalent volume of methanol was also prepared.

Table 1. Preparation of PHB film with various amount of MCl_2 salt

MCl_2 (mol)	0.05 M MCl_2/CH_3OH (mL)	Methanol (mL)
5.0 x 10^{-6}	0.1	0.9
2.5 x 10^{-5}	0.5	0.5
5.0 x 10^{-5}	1.0	-
NIL	-	1

2.3 Characterizations

The possible structural changes of the samples were determined by Fourier Transform Infrared (FTIR) spectroscopy and nuclear magnetic resonance (NMR) spectroscopy. The samples were scanned from 400 – 4000 cm-1 using a Frontier Universal ATR Perkin Elmer FT – NIR Spectrometer. The 1H NMR and 13C NMR spectra were analyzed using a BRUKER AscendTM 500 NMR spectroscopy. Samples were prepared by dissolving ~ 10 mg of the samples in deuterated chloroform. The percentage of metal content in the samples (expressed as wt % of M) was analyzed using a Perkin Elmer AAnalyst 400 atomic absorption spectrometer. The samples were digested using aqua regia solution and diluted to 50 mL with distilled water. The surface morphology of the samples were characterized by a LEO SUPRA 50 VP scanning electron microscope (SEM) operating at 15 kV. The samples were coated with a layer of gold prior to analysis. The thermal degradation of the samples were analyzed on ~7 mg sample using a Perkin Elmer STA 6000 simultaneous thermal analyzer at a heating rate of 20 °C min-1. The samples were heated from 30 °C to 920 °C under nitrogen atmosphere. The thermal behaviour of the samples were investigated using a Perkin Elmer DSC 800. Approximately 7 mg of sample were first heated from 30 °C to 190 °C at a heating rate of 20 °C min^{-1} under nitrogen atmosphere and held for 1 min. The samples were then quenched to -40 °C at a rate of 100 °C min^{-1} and held for another 3 min. Finally, the samples were subjected to the second heating scan where they were again heated from -40 °C to 190 °C at a rate of 20 °C min^{-1}. The growth of PHB spherulites was observed under a polarized optical microscope, Olympus EX53 equipped with Linkam THM5600 hot stage and thermal programmer. A thin layer specimen was heated from room temperature to 190 °C in the hot stage at a rate of 20 °C min-1. It was annealed for 1 min before cooling to room temperature. PHB spherulites were allowed to grow during the cooling stage and the images of the growing spherulites were acquired using an Infinity Analyze software.

3. Results and Discussion

3.1 Surface Morphology and Porous Texture

The surface morphology and textural properties of the samples are studied by SEM and BET analyses. Typical SEM images of pristine PHB film, PHB film prepared in methanol and films prepared from the various concentration methanol solution of MCl_2 are shown in Figure 1. The surface of pristine PHB (Figure 1(a)) is almost smooth but rougher when methanol is employed during the film formation (Figure 1(b)). There is also no apparent porous texture observed in both samples. However, the interconnected porous structure can be observed when MCl_2/methanol was added during the preparation of PHB film (Fig. 1(c)-(h)). The pores were homogenously distributed throughout the film with size ranging from 0.3 – 2.0 μm. It is also observed that the porosity and pore size of the film increases with the increased in MCl_2 concentration. For instance, the average pore size for PHB film prepared with 5.0 x 10^{-6} mol $CuCl_2$ in methanol was 0.80 ± 0.40 μm. This was increased to 1.29 ± 0.43 μm for 2.5 x 10^{-5} mol $CuCl_2$ and continued to increase to 1.46 ± 0.65 μm for 5.0 x 10^{-5} mol $CuCl_2$. Furthermore, the pore size of the PHB prepared with $NiCl_2$ was generally smaller as compared to $CuCl_2$ analogues. This may be due to the larger size of $CuCl_2$ as compared to $NiCl_2$. It is also worth to point out that no serious aggregation of MCl_2 salt particles occurred in these samples.

Table 2. Preparation of PHB film with various amount of MCl_2 salt.

Additive	MCl_2 (mol)	#Pore size (μm)	*MCl_2 content (% wt)	
			Before washing	After washing
CH_3OH	-	N/A	-	-
$NiCl_2/CH_3OH$	5.0 x 10^{-6}	0.40 ± 0.14	4.86	2.47
	2.5 x 10^{-5}	0.62 ± 0.24	12.31	0.51
	5.0 x 10^{-5}	0.73 ± 0.29	20.02	0.44
$CuCl_2/CH_3OH$	5.0 x 10^{-6}	0.80 ± 0.40	2.29	1.40
	2.5 x 10^{-5}	1.29 ± 0.43	6.99	2.43
	5.0 x 10^{-5}	1.46 ± 0.65	11.96	1.82

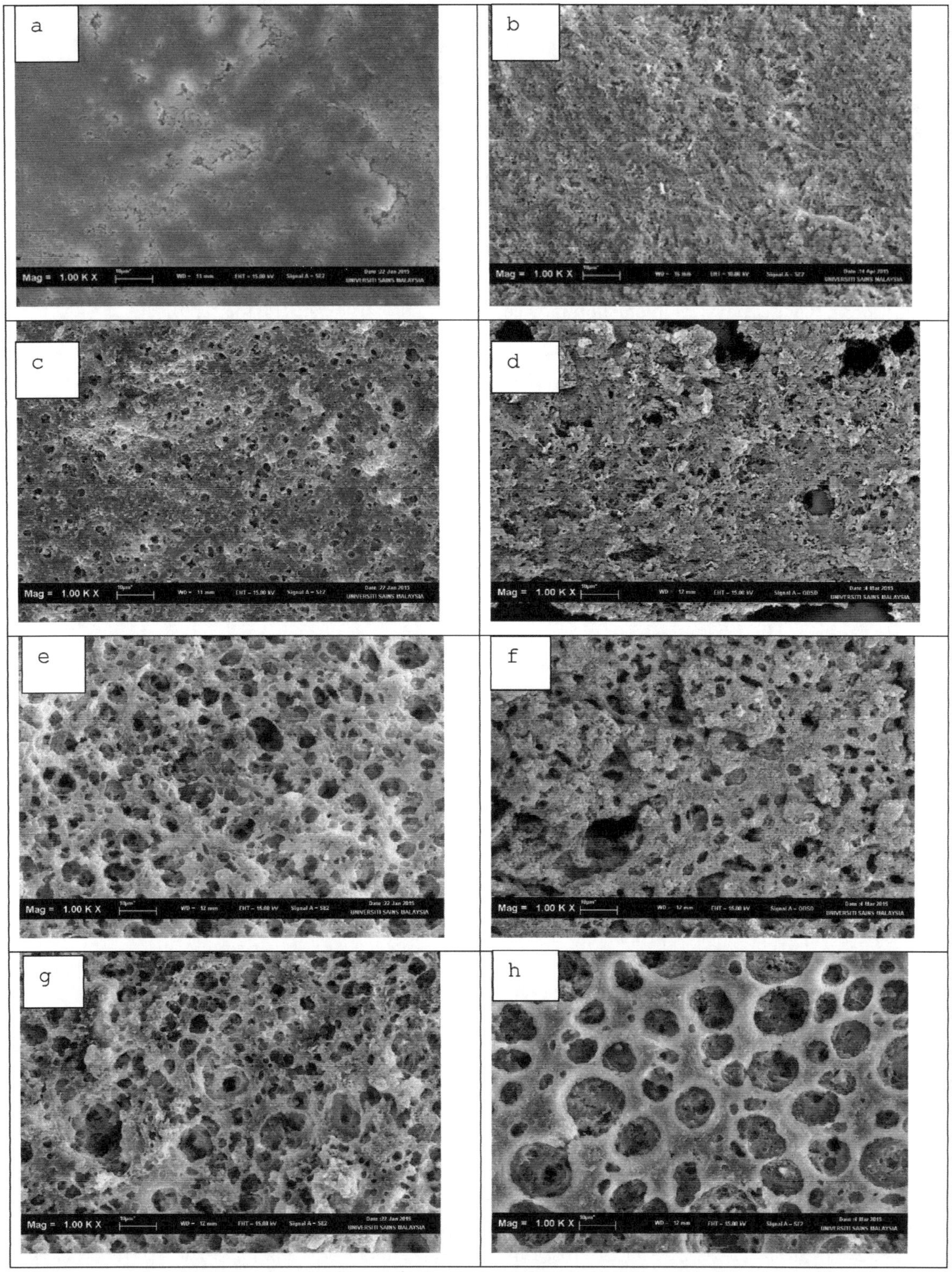

Figure 1: Typical SEM images for films of (a) pristine PHB and PHB prepared in (b) methanol, (c) 5.0 x 10^{-6} mol $NiCl_2/CH_3OH$ and (d) $CuCl_2/CH_3OH$, (e) 2.5 x 10^{-5} mol $NiCl_2/CH_3OH$ and (f) $CuCl_2/CH_3OH$, (g) 5.0 x 10^{-5} mol $NiCl_2/CH_3OH$ and (h) $CuCl_2/CH_3OH$ solutions

3.2 Surface interactions

NMR analyses was conducted to study the structural changes in PHB if any. Figures 2 and 3 show respectively the typical ^{1}H and ^{13}C NMR spectra of film of pristine PHB as well as porous PHB prepared in $CuCl_2/CH_3OH$. The ^{1}H spectrum of pristine PHB shows the doublet signals at 1.25 ppm that was assigned to $-CH_3$. Mirror quartet signals appeared at 2.43 and 2.61 ppm indicate the CH_2 (bonded to C=O). Multiplet signals of CH (bonded to C-O) are at 5.20 – 5.27 ppm, whereas singlet signal of H_2O was at 1.49 – 1.56 ppm. Similar results was reported in the literature [15]. The ^{13}C NMR spectrum of pristine PHB displayed four signals for four types of carbon. These are situated at 19.77, 40.80, 67.62 and 169.14 ppm which represents the CH_3, CH_2, C-O and C=O (due to ester group) respectively. The NMR spectrum for porous PHB prepared in MCl_2/CH_3OH solution are similar to pristine PHB. Nevertheless, the signals are broader and slightly shifted as compared to pristine PHB indicating only weak physical interaction between MCl_2 and PHB existed in the samples.

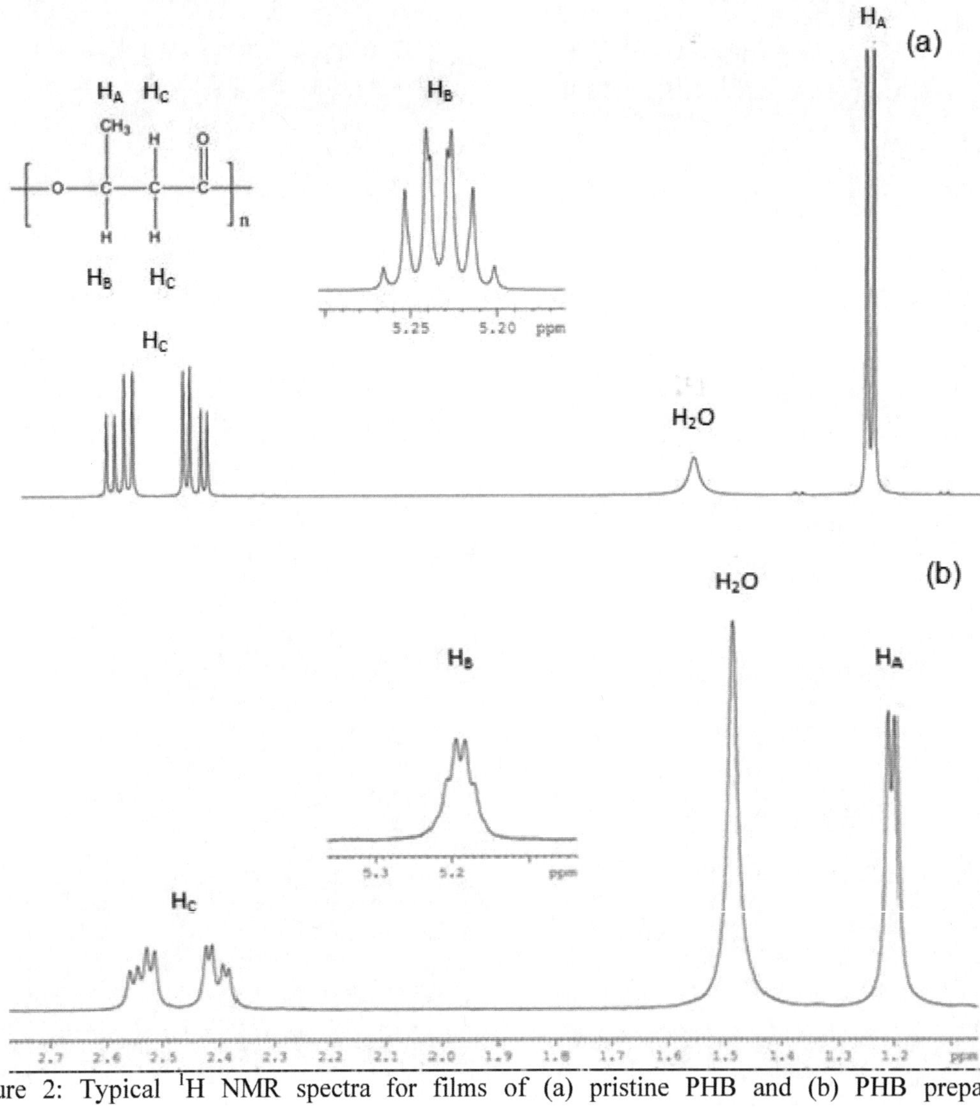

Figure 2: Typical ^{1}H NMR spectra for films of (a) pristine PHB and (b) PHB prepared in $CuCl_2/CH_3OH$

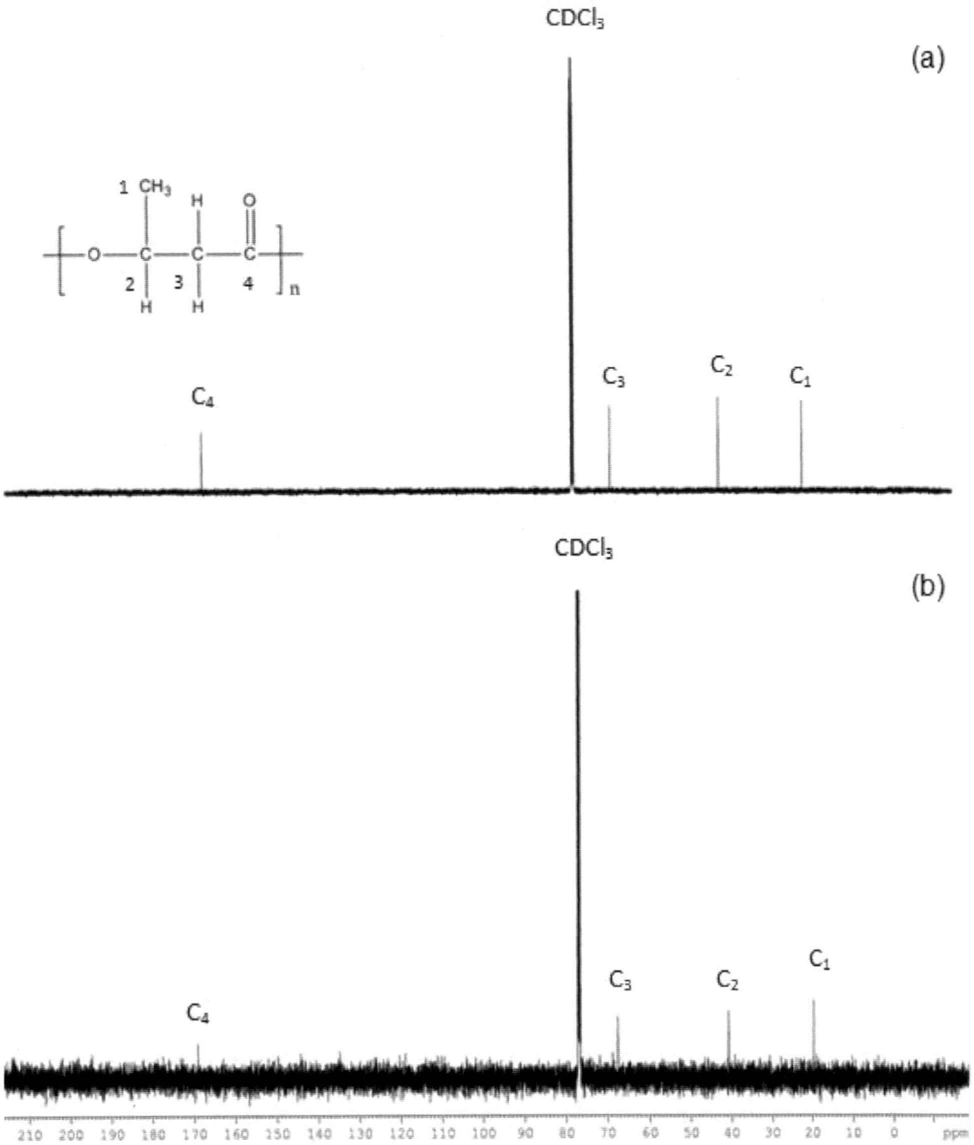

Figure 3: Typical ^{13}C NMR spectra for films of (a) pristine PHB and (b) PHB prepared in $CuCl_2/CH_3OH$ solution.

Figure 4 shows the typical FTIR spectra for films of pristine PHB and PHB prepared in $NiCl_2/CH_3OH$ and $CuCl_2/CH_3OH$ solutions. Pristine PHB exhibits characteristic peaks similar to that reported in the literature [16]. The –CH stretching are situated at 2976 and 2933 cm^{-1} whereas the –CH_2 and –CH_3 bending are located at 1452 and 1379 cm^{-1} respectively. The intense peak at ~1720 cm^{-1} indicates the C=O stretching. The stretching and wagging of the C-O due to the ester group of PHB chain are at ~1276, ~1228 and 1181 cm^{-1}. The MCl_2 in PHB films showed no significant difference (i.e. the position of peaks) in the FTIR spectrum as compared to pristine PHB. Hence, it is envisaged that no chemical interactions occurred between the MCl_2 and PHB. The MCl_2 is only physically interacted with PHB. The FTIR results are in accordance to the NMR analyses as discussed above.

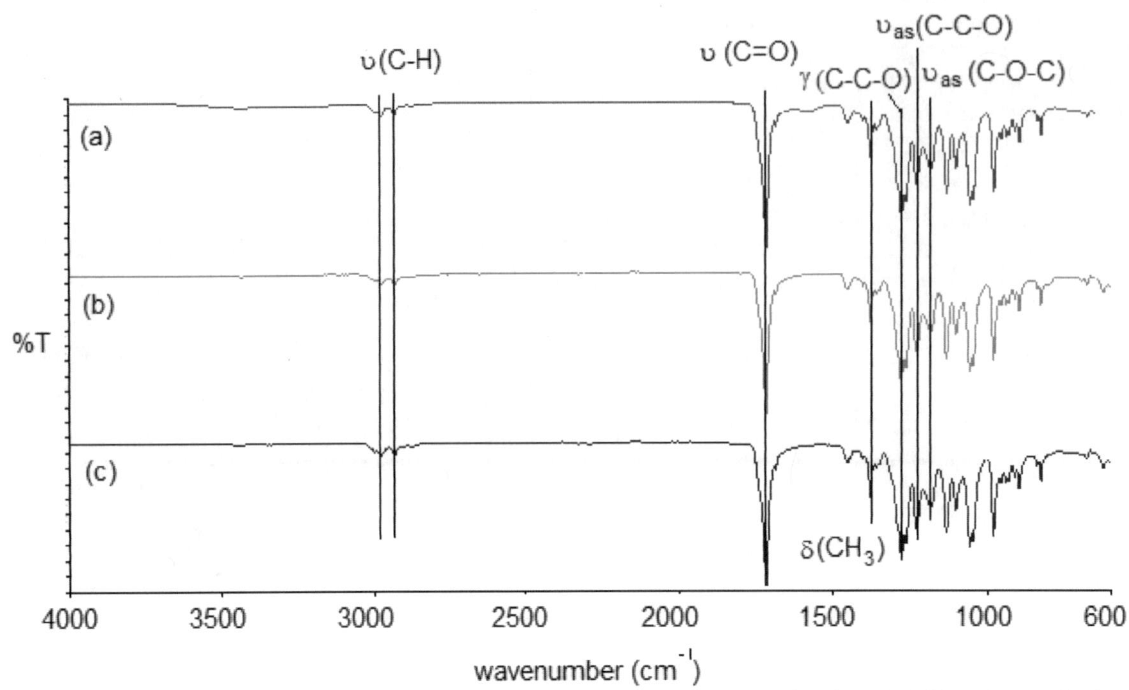

Figure 4: Typical FTIR spectra of (a) pristine PHB and PHB prepared in (b) $NiCl_2/CH_3OH$ and (c) $CuCl_2/CH_3OH$ solutions.

Some FTIR peaks in the spectrum are sensitive to the changes in the crystallinity of PHB. For instance, the peaks at 1228, 1278 as well as 1724 cm^{-1} are indicative for the crystalline phase in PHB [17]. On the other hand, the peak at ~1181 cm^{-1} is assignable to the asymmetric stretching of C-O-C backbone in the amorphous phase of PHB [17]. By monitoring the relative peak height of these peaks to a reference peak, the degree of crystallinity/amorphocity of PHB in a sample can be obtained. According to Porter and Yu [17], the C-O-C backbone stretching (*viz.* at 1181 cm^{-1}) is the peak of interest while $-CH_3$ bending (*viz.* at 1379 cm^{-1}) was selected as the reference peak. In this study, the ratio of $[C\text{-}O\text{-}C_{amorphous}/\text{-}CH_3]$ for pristine PHB film is 1.20. The porous PHB film prepared in 5 x 10^{-6} mol MCl_2/CH_3OH solution showed similar values (i.e. 1.20-1.21) to PHB. However, these are respectively 1.10 and 1.14 for the porous film samples prepared in 5 x 10^{-5} mol $NiCl_2$ and $CuCl_2/CH_3OH$ solutions. Obviously, the amorphocity of PHB decreases swith the increase of MCl_2 in the samples. In other words, the crystalline phase increases as more MCl_2 is present during the preparation of PHB porous films.

3.3 Thermal Properties

Figure 5 shows the typical TG and DTG curves of pristine PHB and porous PHB films prepared in MCl_2/CH_3OH. The onset of decomposition temperature (T_{onset}) and temperature at maximum decomposition rate (T_{max}) of the samples are presented in Table 2. Pristine PHB was stable up to ~ 280 °C, beyond which, it starts to decompose [18]. The maximum mass loss of purified PHB occurred at 292 °C. The porous PHB films prepared in MCl_2/CH_3OH exhibit a low mass loss of < 5% of the total mass loss at the initial stage (30-150 °C). This relates to the evaporation of adsorbed moisture or trapped solvent in the samples. The major mass loss that attributed to the decomposition of PHB chains is observed at ~280-300 °C. Also, the higher residue as compared to pristine PHB observed (~10 %) is due to the presence of MCl_2 in the samples. As tabulated in Table 3, the values of T_{onset}, T_{50} and T_{max} for porous PHB films prepared in MCl_2/CH_3OH are comparable to that of pristine PHB. This depicts that the presence of MCl_2 in PHB does not exert significant influence on the thermal stability of PHB.

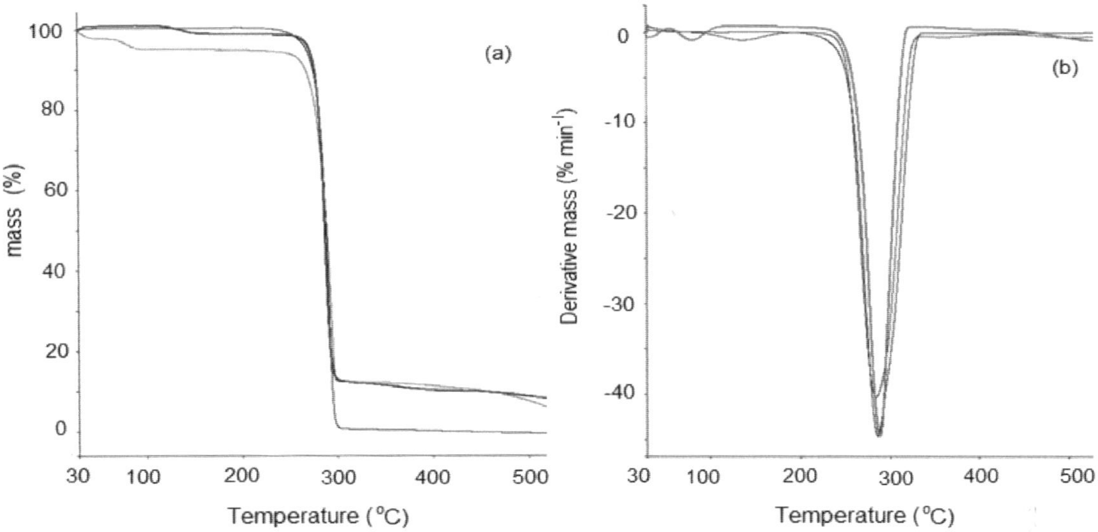

Figure 5: Typical (a) TG and (b) DTG curves for pristine PHB (black) and porous PHB films prepared with $NiCl_2/CH_3OH$ (blue) or $CuCl_2/CH_3OH$ (red)

Table 3: Thermal data of pristine PHB film and PHB films incorporated with $NiCl_2$ or $CuCl_2$.

Sample	MCl_2 (mol)	T_{onset} (°C)	T_{50} (°C)	T_{max} (°C)	T_c (°C)	T_m (°C)	χ_c (%)
PHB	-	281	288	292	53	168	41.02
PHB-$NiCl_2$	5.0 x 10^{-6}	271	287	290	N/D	168	39.12
	2.5 x 10^{-5}	285	289	291	N/D	169	38.85
	5.0 x 10^{-5}	278	287	288	N/D	169	51.12
PHB-$CuCl_2$	5.0 x 10^{-6}	277	290	294	54	167	31.63
	2.5 x 10^{-5}	281	293	289	N/D	169	37.36
	5.0 x 10^{-5}	281	287	287	N/D	169	43.42

Typical DSC thermograms for pristine PHB and porous PHB film prepared in $CuCl_2/CH_3OH$ are depicted in Figure 2. The thermal data including the cold crystallization temperature (T_c), melting temperature (T_m) as well as the calculated polymer's crystallinity (χ_c) for all the prepared samples are tabulated in Table 3. As can be seen in Figure 2(a), pristine PHB clearly showed three stages of thermal transitions. The glass transition of pristine PHB is found at 6 °C, whereas the cold crystallization and melting temperature are respectively located at ~53 °C and ~168 °C [18]. The appearance of cold crystallization peak in the second heating scan of pristine PHB indicates that the crystallization process was incomplete during the cooling step. Some crystallizable PHB segments are thus reorganized to form folded lamellar crystals during the second heating scan [19,20].

It is interesting to note that the crystallization peak are undetectable in all the samples prepared in MCl_2/CH_3OH. Only an endothermic peak that relates to the melting of PHB is observed in their thermograms. As much as 2-20 wt% of metal salt are present in the porous PHB samples as determined by AAS analyses (see Table 2). It is envisaged that the presence of MCl_2 interrupts the molecular attraction between PHB chains. Hence inhibits cold crystallization to occur. Similar phenomenon is reported elsewhere [19]. There are small differences in T_m value ($\Delta = \pm$ 1 °C) between the pristine PHB and porous PHB films prepared in MCl_2/CH_3OH. Nevertheless, the pristine PHB exerts a broader range of melting, with a shoulder at ~171 °C as compared to porous PHB film. The peak is attributed to the melting of crystals formed during the cooling step. However, the shoulder observed in pristine PHB is associated to the melting of crystals formed during the second heating scan.

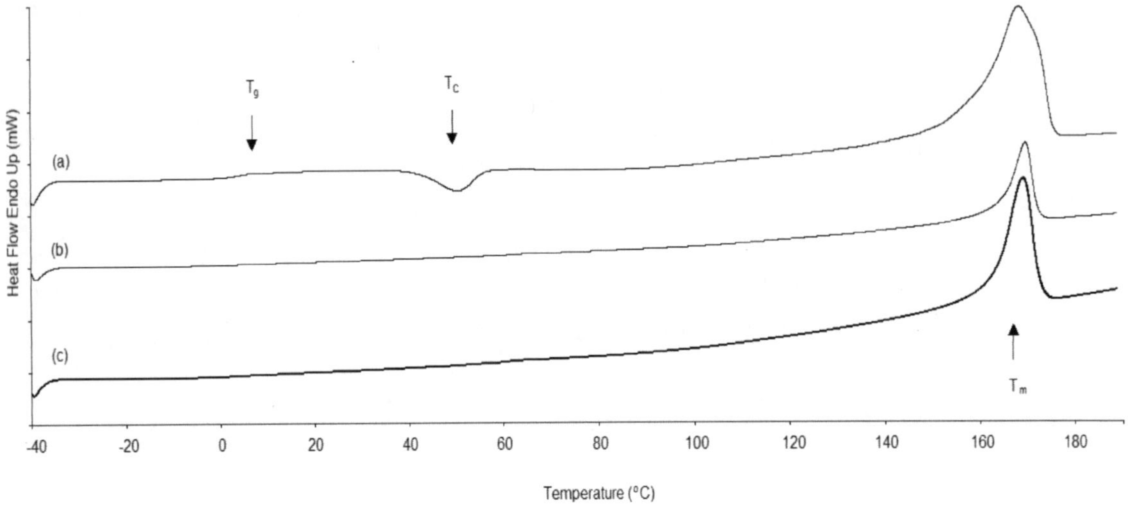

Figure 6: Typical DSC thermograms (second heating scan) of (a) pristine PHB and porous PHB film prepared in (b) $NiCl_2$ or (c) $CuCl_2/CH_3OH$ solutions.

Degree of crystallinity (χ_c) of the pristine PHB and porous PHB containing MCl_2 is calculated according to equation (1) [21]:

$$\chi_c = \Delta H_m/\Delta H_{100} \cdot W_{PHB} \text{ x } 100\% \qquad (1)$$

where ΔH_m and W_{PHB} respectively represents the melting enthalpy and weight fraction of PHB in the sample. The ΔH_{100} is the melting enthalpy of a 100 % crystalline PHB with the value of 146 Jg^{-1} [18]. Pristine PHB exerts 41 % crystallinity. On the other hand, χ_c for porous PHB films falls into the range of 32-51 %. Similarly to the result obtained by FTIR, the degree of crystallinity increased with an increase in MCl_2 content in the samples regardless of the type of metal. For instance, porous PHB films with the highest salt loading (20.02 wt% Ni) displayed the highest χ_c of 51.12 %. This was decreased to 39.12 % when the sample contains only 4.86 wt% of Ni salt. It is also interesting to note that the χ_c value for sample with lower MCl_2 content (i.e. prepared with $\leq 2.5 \times 10^{-5}$ mol salt) is lower than pristine PHB. This may due to the adverse effects that is discussed in the following section.

The growth of PHB spherullite is further investigated using a polarized optical microscope (POM). As shown in Figure 7(a), the POM images of purified PHB features large spherulites with closely ring bands. In the presence of MCl_2, the size of spherulites are much smaller and imperfect, especially at high MCl_2 content (Figure 7(c-d)). This indicates that MCl_2 can act as the heterogeneous nucleation sites to increase the growth rate of spherulites. Similar observation is reported elsewhere [22].

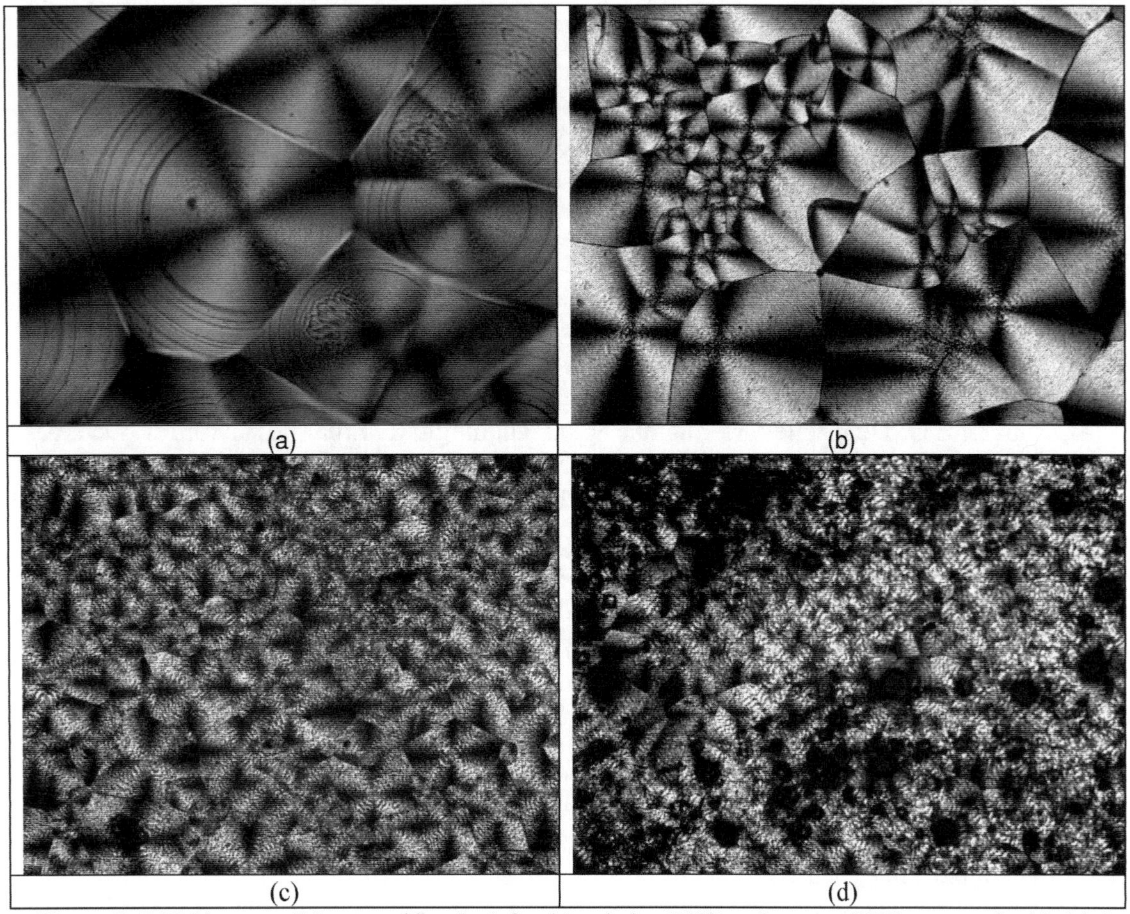

Figure 7: POM images (10x magnification) for (a) pristine PHB and porous PHB prepared with (a) 5 x 10^{-6} mol, (b) 2.5 x 10^{-5} mol and (c) 5.0 x 10^{-5} mol $CuCl_2/CH_3OH$

The presence of salt as an additive in PHB sample may influence the crystallinity of PHB in two adverse ways – (1) as the nucleation agent to promote the crystal growth or (2) decrease the mobility of PHB segments and retard the growth of crystal [23]. Generally, the additive with good dispersion provides heterogeneous nucleation sites to initiate the formation of polymer crystallite. In contrast, highly aggregated additive or additive that interacted strongly with polymer reduces the growth of polymer crystallite. In lower MCl_2 content, the contribution of (2) maybe higher than (1), hence the χ_c is slightly lower than pristine PHB. Nevertheless, at high MCl_2 content (*viz.* 5.0 x 10^{-5} mol MCl_2), the presence of MCl_2 salt encourages more reorganization of PHB chains. This can be clearly seen from POM results. Hence in this study, (1) is the dominant attributes and the crystallinity of the PHB samples increase with the increased in MCl_2 content in the samples.

4. Conclusion

PHB films with porous structure have been successfully formed via simple MCl_2 induced solvent casting method. The pore size of the films are fall into the range of 0.3-2.0 μm and it increases with the increased in MCl_2 concentration. Generally, PHB prepared with $NiCl_2$ gave smaller pore size as compared to $CuCl_2$ analogues. The presence of MCl_2 exerts no significant influence on the thermal stability of PHB but greatly enhance the crystallinity of PHB. FTIR and NMR analyses shown that there is only weak physical interaction occurred between the PHB chains and MCl_2 salt. Thus most of the MCl_2 can be easily removed by simple washing. AAS results revealed that there exist < 2.5 wt% of MCl_2 in the PHB after washing step.

Acknowledgement
The authors wish to thank the Universiti Sains Malaysia for financial support in the form of RUI grant: 1001/PKIMIA/811311

References

1. Bergstrand A, Uppström S and Larsso A 2014 *Int. J. Polym. Sci.* 958975
2. Sangsanoh P, Waleetorncheepsawat S, Suwantong O, Wutticharoenmongkol P, Weeranantanapan O, Chuenjitbuntaworn B, Cheepsunthorn P, Pavasant P and Supaphol P 2007 *Biomacromolecules* **8** 1587
3. Guven E O, Demirbilek M, Saglam N, Karahaliloglu Z, Erdal E, Bayram C, Denkbaş E B 2008 *Hacettepe J. Biol. Chem.* **36** 305
4. Artsis M I, Bonartsev A P, Iordanskii A L, Bonartseva G A and Zaikov G E 2010 *Mol. Cryst. Liq. Cryst.* **523** 593
5. Zhang X, Wei C, He Q and Ren Y 2010 *J. Env. Sci.* **22** 1267
6. Zhang X, Wei C, He Q and Ren Y 2010 *J. Hazardous Mater.* **177** 508
7. Mikos A G, Thorsen A J, Czerwonka L A, Bao Y, Langer R, Winslow D N and Vacanti J P 1994 *Polymer* **35** 1068
8. Zubairi S I 2014 *Adv. Env. Bio.* **8** 925
9. Misra S K, Ansari T I, Valappil S P, Mohn D, Phillip S E, Stark W J, Roy I, Knowles J C, Salih V and Boccaccini A R 2010 *Biomater.* **31** 2806
10. Ansari N F and Amirul A A 2013 *Appl. Biochem. Biotechnol.* **170** 690
11. Bergstrand A, Andersson H, Cramby J, Sott K and Larsson A 2012 *J. Biomater. Nanobiotechnol.* **3** 431
12. Pereira T F, Oliveira M F, Maia I A, Silva J V L, Costa M F and Thire R M S M 2012 *Macromol. Symp.* **319** 64
13. Ying T H, Ishii D, Mahara A, Murakami S, Yamaoka T, Sudesh K, Samian R, Fujita M, Maeda M and Iwata T 2008 *Biomater.* **29** 1307
14. Lee H K, Ismail J, Kammer H W and Abu Bakar M 2005 *J. Appl. Polym. Sci.* **95** 113
15. Zhu C, Noruma C T, Perrotta J A, Stipanovic A J and Nakas J P 2010 *Biotechnol. Prog.* **26** 424
16. Diez-Pascual A M and Diez-Vicente A L 2014 *Int. J. Mol. Sci.* **15** 10950
17. Porter M and Yu J 2011 *J. Microbio Methods* **87** 49
18. Yeo S Y, Tan W L, Abu Bakar M and Ismail J 2010 *Polym. Degrad. Stab.* **95** 1299
19. Pachekoski W M, Agnelli J A M, Belem L P 2009 *Mater. Res.* **12** 159
20. Tri P N, Domenek S, Guinault A, Sollogoub C 2013 *J. Appl. Polym. Sci.* **129** 3355
21. Gunaratne L M W K, Shanks R A and Amarasinghe G 2004 *Thermochim. Acta* **423** 127
22. Yu H-Y, Qin Z-Y and Zou Z 2011 *Prog. Nat. Sci.* **21** 478
23. Im S and Lee W 2010 *Chapter 9: Polyester nanocomposites* (Recent advances in polymer nanocomposites: Synthesis and characterization) Ed Thomas S *et al.* (CRC Press, US)

IOP Conf. Series: Materials Science and Engineering **133** (2016) 012013 doi:10.1088/1757-899X/133/1/012013

Damage detection of carbon reinforced composites using nondestructive evaluation with ultrasound and electromagnetic methods

A Savin [1*], P D Barsanescu[2], P Vizureanu[3], M D Stanciu[4], I Curtu[4], N Iftimie[1], R Steigmann [1]

[1] National Institute of Research and Development for Technical Physics, Nondestructive Testing Department, Iasi, Romania
[2]Gheorghe Asachi Technical University, Department of Mechanical Engineering, Mechatronics and Robotics, Iasi, Romania
[3] Gheorghe Asachi Technical University, Department of Technologies and Equipment for Material Processing, Iasi, Romania
[4]Transilvania University, Department of Mechanical Engineering, Brasov, Romania

*E-mail: asavin@phys-iasi.ro

Abstract. CFRP have applications among most different domains due their low density, high elastic modulus and high ultimate strength along the carbon fibers direction, no fatigue and the expansion coefficient is small. This paper presents the behavior of carbon fiber woven-PPS composites at low velocity impacts. The transversal electrical conductivity is modified due to the plastic deformation following the impacts, and thus electromagnetic procedures can be used for assessment of CFRP using a high resolution sensor with metamaterials lens and comparing the results with those obtained from ultrasound testing with phased array sensor. The area of the delamination is overestimated when the method of phased array ultrasound is used and substantially underestimated by the electromagnetic testing. There were a good agreement between the simulations with finite element method and experimental measurements.

1. Introduction

Carbon fiber reinforced plastics composites are lightweight, strong materials used in the manufacturing of numerous products, from the sports goods to aerospace industry components. CFRP allow the production of goods without scrap parts, becoming a cost effective technology [1]. In general, CFRP composites use thermosetting resins such as epoxy, polyester, or vinyl ester. High performance thermoplastic resins and more particularly Polyphenylene sulfide (PPS) [2] represent a promising alternative to thermosetting resins problems because it offers a number of advantages over epoxy resins (chemical and impact resistance over a wide range of temperatures, high hardness and stiffness and superb long-term creep under load properties, extremely low moisture absorption).

For 2020, it is estimated that the carbon fiber demand will outstrip supply, doubling the production of 2015 (tripling the production of 2010). In order to avoid the incurring of all costs within selection between good and bad parts, NDE is required from the manufacturing stage of composites, from base fiber and production of matrix. Nondestructive evaluation is no longer about "pass/fail" [3]. NDE should provide validation of ply sequences and fiber/woven orientation, the state of the matrix, parameters of the final materials. Low velocity impacts (LVI), that can be defined by quasi-static

IOP Conf. Series: Materials Science and Engineering **133** (2016) 012013 doi:10.1088/1757-899X/133/1/012013

events of which superior limits can be in the range of tens of m/s [4], depending on material properties as stiffness, can produce delaminations, carbon fiber breaks, cracking of matrix [5] etc.

Usually, CFRP are tested through C-scan or top view ultrasound [6], Lamb waves using noncontact transducers [7], with Hertzian contact [8] or compression waves generated by normal transducers. Also, microwave [9], electrical resistivity measurement [10], shearography [11], acoustic emission [12] can be used in testing CFRP. Carbon fiber have a low electrical conductivity [13], allowing CFRP to be tested using electromagnetic methods. The electromagnetic investigation of CFRP can be done using sensors with orthogonal coils [14], eddy current sensor with 16 individual sensor pairs [15], sensors with metamaterials lens [16], etc.

This paper present the possibility to use a high resolution sensor based on metamaterials lens to assess the delaminations in CFRP with Polyphenylene sulphide matrix reinforced with carbon fibers woven fabric, created by LVI and comparing the results with those obtained from top scan ultrasound using phased array method.

2. Studied samples

The plates taken into study have the dimensions of 150mm×100mm×4.2mm [17]. The composite, named CETEX, is made from PPS matrix reinforced with 12 layers of carbon fibers 5Harness satin fabric type. The carbon fibers are T300JB with 1.75 g/cm^3 density, fabric surface density is 285g/cm^2 and overall fiber volume fraction 50 ± 3% [18]. In figure 1 are presented the studied samples, made by Tencate, The Netherlands and the layout of 5 Harness satin fabric.

a. b.

Figure 1. Studied samples: a) composite plates; b) 5 Harness satin fabric layout.

The samples were impacted with energies of 2 J, 4 J, 6 J, 8 J, 10 J, and 12 J at room temperature using equipment FRACTOVIS PLUS 9350-CEAST-Instron USA with a hemispherical bumper head having 20 mm diameter and 2.045 kg weight, according to ASTM [17]. The intermediate energy of 10J has been taken into consideration before the failure of the fibers that have taken place at 12J.

3. Experimental setup

3.1. Mechanical tests

The conditions of impacts were established by simulating the Drop Weight impact method [19] using Finite Element Method (FEM). Drop weight testing is the preferred method when using low velocity impact testing and the results may also be enhanced with post impact testing. The energy of impact can be calculated knowing the mass of the impactor m and the height of dropping h as $E = m \cdot g \cdot h$, where g is the gravitational acceleration. The geometry of the impact is presented in figure 2, where the displacement of the impactor δ is the sum of the displacement of the impacted plate median plane ω and the plastic deformation of the plate after impact α, as $\delta = \omega + \alpha$. The displacement of the composite plate during the impact is

$$\delta = \iint_i \frac{F(t) - gM_{total}}{M_{total}} dtdt \qquad (1)$$

with δ the displacement of the plate in point i; $F(t)$ is recorded by the data acquisition system, M_{total} is total mass of the impactor. The absorbed energy until the i point is calculated as the area described by the curve in the Force-Displacement diagram

$$E_i = \int_i F(\delta) d\delta \tag{2}$$

The speed till point *i* is calculated by simple integration of Force-Time diagram

$$v_i = \int_i \frac{F(t) - gM_{total}}{M_{total}} dt \tag{3}$$

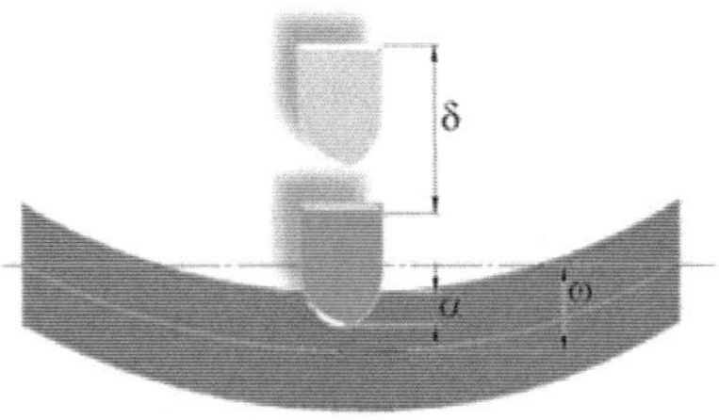

Figure 2. Geometry of the impact of plate.

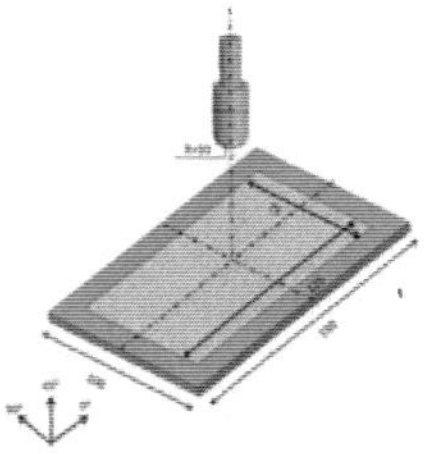

Figure 3. Dimensions of sample for the impact tests [17].

3.2. Electromagnetic testing

The plates were examined using the sensor with metamaterial lens for detection. The sensor has the emission part made from a 20mm diameter coil, 2 turns of Cu wire (1.2mm diameter). The reception coil has a 1mm diameter, 2 turns of Cu wire (0.1mm diameter). In front of the reception coil is placed a conductive screen made from a flexible foil LONGLITE^{TM}200 produced by ROGERS CORPORATION USA, having a circular aperture, with 0.1mm diameter, coaxially placed at 0.05mm distance from the reception coil. The screen was grounded. The circular aperture serves for the diffraction of the evanescent waves, appeared in the slits. A conical Swiss roll [20] is placed above the reception coil and acts as metamaterials allowing the transmission of the evanescent waves [20] from the region between carbon fibers towards the reception coil (Figure 4). The transducer was fixed, being connected to the Network/Spectrum/Impedance Analyzer 4395A Agilent USA. The plate is displaced by a motorized XY stage Newmark USA, commanded by PC through RS232 interface. The command and the data acquisition were made through a GPIB IEEE 488 interface, using programs elaborated in Matlab 2014a. The surface was raster scanned using 1mm steps in both directions.

3.3. Ultrasound testing

The testing using ultrasound phased array sensors has been effectuated using Phasor XS - GE USA. The sensor contains a linear array of 32 sensors having the pitch of 0.5mm and the central frequency 5MHz. The sensor is placed on an edge of 36° and the displacement has been assured by an axial scanned ENCSTD with linear encoder (Figure 5). The coupling is a gel with average viscosity.

Figure 4. Electromagnetic testing.

Figure 5. Ultrasound testing.

4. Experimental results and discussions

The composite specimen prepared for mechanical tests, is cropped from the central region of a sample impacted with 10J. This impact only produce a dent. The tensile tests were done using Instron E8801 with a displacement speed of 0.2mm/min, using the pressure of 70barr [21]. In figure 6 is presented the residual stress at tensile test after impacts. Approximating the plot as a parabola, it can be shown that the residual stress of the sample impacted with 12 J energy is 50% from the non-impacted samples.

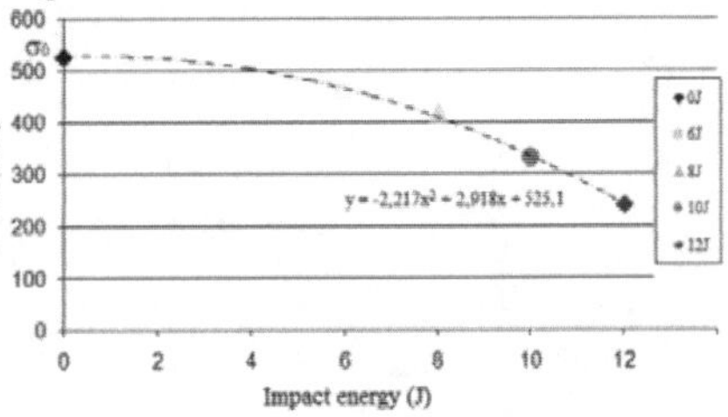

Figure 6. Residual stress vs impact energy.

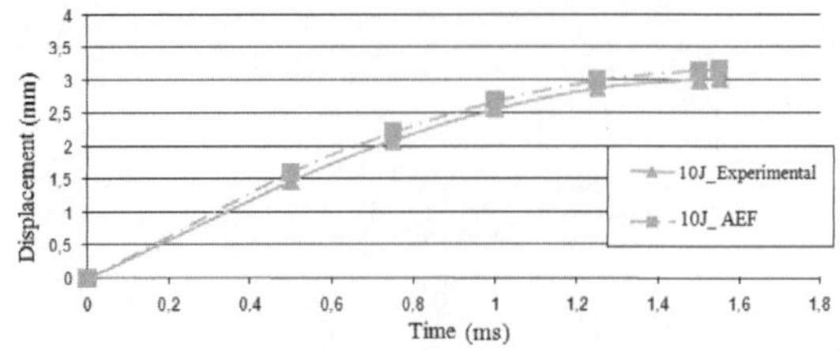

Figure 7. Displacement of the center of the plate during the impact vs time at 10J energy impact.

Taking into account the values of displacement at impact simulated by FEM in the conditions specified above, there were found a good concordance with the values experimentally determined during the impact tests (Figure 7). The tests were determined on the sample impacted with 10J, energy below the energy where breaking of the fiber appears [22].

Microscopic analysis using Olympus Microscope BX51 of the composite specimens impacted with 10J has been effectuated for evaluation of the the damage initiation and propagation of delamination. The specimen has been cut on the middle of the impacted zone and prepared for optical inspection. The damage developed in the 5-HS fabric under impact takes the form of weft yarn cracking, which is perpendicular to the loading direction. The damage is developed at the yarn crossover positions, and then propagates into the matrix (Figure 8).

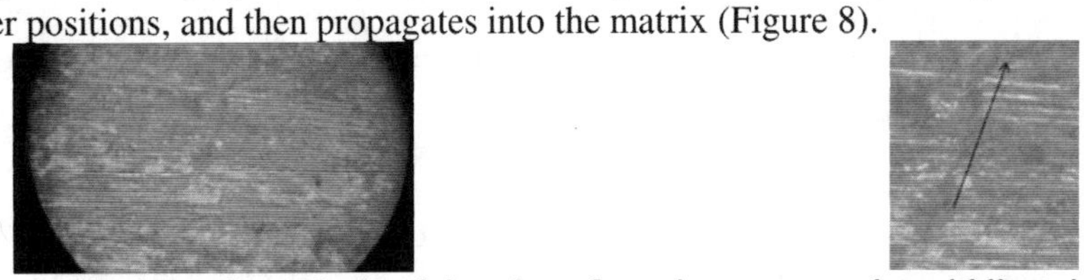

Figure 8. Microscopic analysis of the edge of specimen cut on the middle axis of the impacted zone.

From the point of view of nondestructive evaluation, the two complementary methods presented above were involved in determination of the area of delamination due to impact with energy of 10J, the energy below the value where the breaking of fibers occurs. Both types of measurements were effectuated on the impacted face of the plate. In figure 9a is presented the ultrasound image of top scan obtained at the scanning of the plate impacted with 10 J energy meanwhile in figure 9 b is obtained the electromagnetic image.

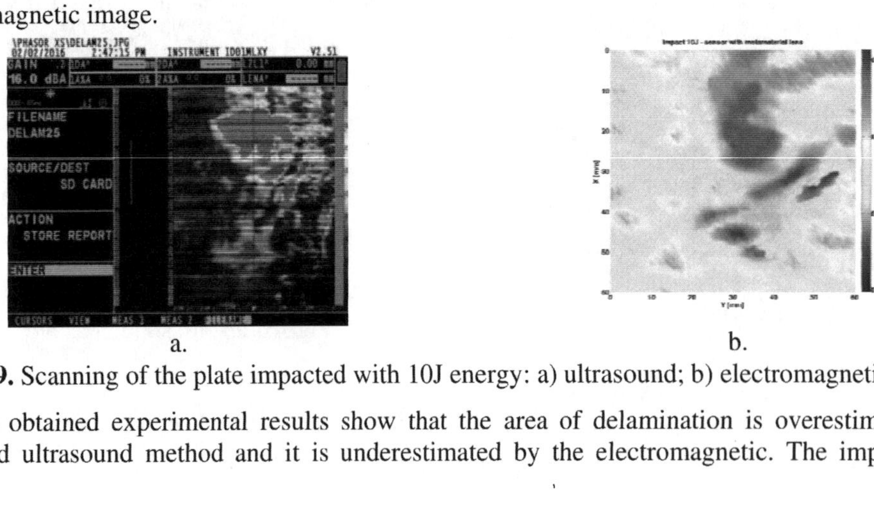

a. b.

Figure 9. Scanning of the plate impacted with 10J energy: a) ultrasound; b) electromagnetic image.

The obtained experimental results show that the area of delamination is overestimated by the proposed ultrasound method and it is underestimated by the electromagnetic. The impact damage exists at different levels through thickness. The delamination near the back wall is larger than the ones near the front wall (where the impact took place), there is rapid local bending at the impact point and at the face of the delamination, almost all the energy will be reflected back to the sensor as an echo. It is not possible to penetrate past the first incident delamination to those in its shadow; all the energy is then confined between the first incident delamination and the front wall to create repeat echoes. Multiple level delaminations are detected by virtue of the conical damage feature (those near the back wall being larger in area).The underestimation using electromagnetic method is given by modification of transversal electrical conductivity of the composite material due to the impact, that took place in the central region of the impact, where the material suffers a plastic deformation. The electromagnetic sensor "senses" only the subsurface delamination.

5. Conclusions

CFRP plates made from 12 layers of carbon fibers 5Harness satin fabric embedded into PPS matrix have been impacted with different energies in order to study their damage using nondestructive evaluation methods by means of ultrasound phased array and electromagnetic sensors. The residual stress of the sample impacted with 12 J energy is 50% from the non-impacted samples. The area of delamination is overestimated by phased array ultrasound method because the delamination on the back face is larger than the ones on the impacted face and due to repeated back end echoes and respectively is underestimated by the electromagnetic method due to plastic deformation that modify local conductivity. In order to obtain more eloquent results, a fusion of data obtained by both nondestructive evaluation methods using Dempster Shafer theory of evidence is required and will be further studied.

Acknowledgments

This work was supported by the Ministry of National Education, CNCS–UEFISCDI, projects PN-II-ID-PCE-2012-4-0437, PN-II-PCCA-2013-4-0656 and Nucleus PN 16 37 01 01.

References

[1] Morgan P 2005 Carbon fibers and their composites *CRC press*

[2] Vielle B and Taleb L 2011 About the influence of temperature and matrix ductility on the behavior of carbon woven-ply PPS or epoxy laminates: Notched and unnotched laminates *Composite Science Technology* **71** pp 998-1007

[3] Brosius D 2016 Advanced composites – An idealized future *Composites World*

[4] Richardson M O W and Wisheart M J 1996 Review of low-velocity impact properties of composite materials *Composites Part A: Applied Science and Manufacturing* **27(12)** pp 1123-1131

[5] Abrate S 1991 Impact on Laminated Composite Materials *Appl Mech Rev* **44(4)** pp 155-190

[6] Hosten B and Castaings M 2008 Comments on the ultrasonic estimation of the viscoelastic properties of anisotropic materials *Composites Part A: Applied Science and Manufacturing.* **39(6)** pp1054-1058

[7] Su Z, Ye L and Lu Y 2006 Guided Lamb waves for identification of damage in composite structures: A review *Journal of sound and vibration* **295(3)** pp 753-780

[8] Grimberg R, Savin A, Steigmann R, Bruma A and Barsanescu P 2009 Ultrasond and eddy current data fusion for evalution of carbon-epoxy composite delaminations *Insight-Non-Destructive Testing and Condition Monitoring* **51(1)** pp 25-31

[9] Sediq A S B and Qaddoumi N 2005 Near-field microwave image formation of defective composites utilizing open-ended waveguides with arbitrary cross sections *Composite Structures* **71(3)** pp 343-348

[10] Park J M, Lee S I and DeVries K L, 2006 *Composites Part B: Engineering* **37(7)** pp 612-626

[11] Hung Y Y 1999 Applications of digital shearography for testing of composite structures *Composites Part B: Engineering* **30(7)** pp 765-773

[12] Park J M, Kim P G, Jang J H, Wang Z, Kim J W, Lee W I, Park J G and DeVries K L 2008 Self-sensing and dispersive evaluation of single carbon fiber/carbon nanotube (CNT)-epoxy composites using electro-micromechanical technique and nondestructive acoustic emission *Composites Part B: Engineering* **39(7)** pp 1170-1182

[13] Mook G, Lange R and Koeser O 2001 Non-destructive characterisation of carbon-fibre-reinforced plastics by means of eddy-currents *Composites science and technology* **61(6)** pp 865-873

[14] Grimberg R, Premel D, Savin A, Le Bihan Y and Placko D 2001 Eddy Current Holography Evaluation of Delamination in Carbon – Epoxy Composites *Insight* **43(4)** pp 260-264

[15] Schulze M H, Heuer H, Küttner M and Meyendorf N 2010 High-resolution eddy current sensor system for quality assessment of carbon fiber materials *Microsystem technologies* **16(5)** pp 791-797

[16] Grimberg R and Savin A 2011 Electromagnetic transducer for evaluation of the structure and integrity of composite materials with polymer matrix reinforced with carbon fibers *(Patent no. RO126245-A0)*

[17] ASTM D7136 / D7136M - 15 Standard Test Method for Measuring the Damage Resistance of a Fiber-Reinforced Polymer Matrix Composite to a Drop-Weight Impact Event

[18] ***www.tencate.com

[19] ASTM D 3763, High Speed Puncture Properties of Plastics Using Load and Displacement Sensors 1997

[20] Grimberg R, Savin A and Steigmann R 2012 Electromagnetic imaging using evanescent waves *NDT & E International* **46** pp 70-76

[21] ***www.vishaypg.com/docs/11331/125tm.pdf

[22] Savin A, Steigmann R, Bruma A and Šturm R 2015 An Electromagnetic Sensor with a Metamaterial Lens for Nondestructive Evaluation of Composite Materials *Sensors* **15(7)** pp15903-15920

IOP Conf. Series: Materials Science and Engineering **133** (2016) 012014 doi:10.1088/1757-899X/133/1/012014

Chemical elements diffusion in the stainless steel components brazed with Cu-Ag alloy

I Voiculescu[1], V Geanta[1], I M Vasile[1], E F Binchiciu[2] and R Winestoock[3]

[1]University Politehnica of Bucharest, Romania 313, Splaiul Independentei, sector 6, 060042, Bucharest, Romania
[2]S.C. Sudotim As SRL, 30, Bv. Mihai Viteazu, Timisoara, Romania
[3]Ben Gurion University of the Negev, Beer-Sheva, Israel

E-mail: ioneliav@yahoo.co.uk

Abstract. The paper presents the study of diffusion of chemical elements through a brazing joint, between two thin components (0.5mm) made of stainless steel 304. An experimental brazing filler material has been used for brazing stainless steel component and then the diffusion phenomenon has been studied, in terms of chemical element displacement from the brazed separation interface. The filler material is in the form of a metal rod coated with ceramic slurry mixture of minerals, containing precursors and metallic powders, which can contribute to the formation of deposit brazed. In determining the distance of diffusion of chemical elements, on both sides of the fusion line, were performed measurements of the chemical composition using electron microscopy SEM and EDX spectrometry. Metallographic analysis of cross sections was performed with the aim of highlight the microstructural characteristics of brazed joints, for estimate the wetting capacity, adherence of filler metal and highlight any imperfections. Analyzes performed showed the penetration of alloying elements from the solder (Ag, Cu, Zn and Sn) towards the base material (stainless steel), over distances up to 60 microns.

1. Introduction

Nowadays, stainless steels like 304 and 316 are used extensively for steam pipes and exhaust systems, due their resistance to elevated temperature, oxidation, and corrosion, being the choice for food preparation equipment, including steam-heated boilers and storage tanks. The addition of chromium and nickel to the iron matrix creates a significant percentage of chromium and nickel atoms at the surface [1].

The corrosion resistance of stainless steel depends on the chromium content. Long maintenance at higher temperatures, in the range of 425-870 °C, causes the atoms diffusion and changes the metal's properties. Such high temperatures allow the chromium to diffuse away from the grain boundaries to form chromium carbides, its preferred crystalline structure at this temperature range [2]. During brazing, the assembly is heated from room temperature to the filler material melting temperature, and significant diffusion phenomena of chemical elements occur between the parent material and the filler metal. This process is particularly important when the filler metal layer is very thin [3].

[1] To whom any correspondence should be addressed.

By dissolution of the base material into the molten filler, occurs the increase of width of the liquid zone. At the same time, the temperature increases from the melting point to the joining temperature. By lowering the temperature, isothermal solidification occurs, as a result of diffusion of one or more constituents of the filler into the base metal (and vice versa).

The minimum suitable temperature for a diffusion brazing of the stainless steels is not simply determined by the melting point of the filler material or that of the alloys generated by reaction. If it is desired that the end product of reaction is a solid solution across the joint and that there are no residual interfacial inter-metallic phases present, then the solubility limit of the minor constituents in the primary metal phase is also crucially important [4].

The brazed joints made in stainless steel can be frequently exposed to water or vapour during service. This specific medium can promote interfacial corrosion of failure of the joints. The special silver-copper brazing filler metal allows to avoiding the problems of interfacial corrosion and improves the strength of the joint. Most stainless steel types, with the exception of titanium or niobium stabilized grades, can be brazed. For brazing stainless steels in air, low-temperature silver brazing alloys are generally used [5].

One of the basic conditions to achieve a quality joint is to ensure proper wetting of the brazed surfaces. The stainless steel surface is difficult wetted by brazing alloys, due to the surface oxide that protect it from corrosion. The chemical composition of flows used for brazing must ensure rapid dissolution of oxide surfaces and to ease wetting by molten filler material [6]. Brazing alloy must be sufficiently fluid, in order to penetrate into gaps between components on large distances to ensure a good resistance of the joint. Melting temperature of brazing alloy that contains silver and copper must be well correlated with the type of base material. In the case of austenitic stainless steels, the effective maintenance in the 600- 800°C temperature range is critical to avoid the inter-crystalline corrosion. In the case of very thin components, the effects of high temperatures are more critical [6].

The aim of the research paper was to establish the brazing behaviour of an experimental brazing filler material, named VIAg25SnSiPR. The filler materials used in the experimental study were produced through environmentally friendly technologies, without the use of highly toxic materials (cadmium free) and based on an optimized technology, with differences between the melting temperature up to 50°C, between metal rod and coating layer [7, 8, 9]. The increasing of tin content can improve the wettability and mechanical properties of the joints. However, when the tin content exceeds 5%, the shear strength of the joints decreases [5]. Following this observation, in the experimental filler material the tin content was kept lower than 2%Sn [10].

2. Experimental details

To be able to enter in the gap of the brazed joint, the brazing filler material contains a special covering that allows obtaining a difference between the melting temperature of coating and of the metallic rod. The filler material is in the form of a metal rod coated with ceramic slurry mixture of minerals, containing precursors and metallic powders, which can contribute to the formation of deposit brazed (Figure 1).

Figure 1. The production line of coated rods and the filler materials used in the experimental program.

At the low temperature (650-700°C) first ceramic mixture melting occurs, for the base material surface deoxidation. At higher temperatures, between 700-800°C, the second ceramic mixture and silver particles melting occur, completing the final chemical composition of filler material. In this way, the filler metal can spread rapidly and fill completely the gap. This alternance in melting that appears at filler material coating was evidenced by differential thermal analysis performed with the device Netzsch STA 449 F3, in Ben-Gurion University of Negev, Israel. Composite coatings contain 60 to 65% deoxidation flows, in accordance with EN 14045: 1999, 15 to 20% plasticizers and 10 to15% binder. They also contain nano-powders with suitable properties, such as wetting and deoxidation agents, to improve the diffusion and to increase the strength tear.

In order to study the distance of filler material penetration, two type of samples were joined (figure 2), using flame brazing process. The average value of the gap between the 2 components was established in the range of 0.1 - 0.4 mm. Before brazing, the edges of the samples were mechanical cutted, and the surfaces was cleaned with propane, for eliminate all the contaminants (scale, oil/grease) and then were dried with hot air, for eliminate moisture.

An oxidizing flame type for brazing was chosen, for avoiding the evaporation effect of the chemical elements of the filler material. The total duration of the process was of 30 seconds, including heating the components and brazing period. This short heating period ensures a low deformation effects, considering the low thickness of the samples (0.5mm).

a

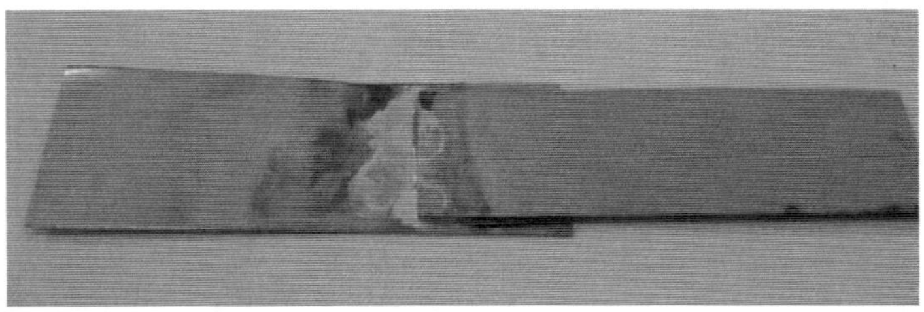
b

Figure 2. Samples brazed with experimental filler material VIAg25SnSiPR:
a) Corner joint; b) Overlap joint.

After the cooling period, the brazing zones were washed in hot water and were dried with hot air. The representative area from each brazed sample was removing using precision cutting machine with cooling agent, to avoid the alteration of the microstructure. To ensure the correct edge of samples during the polishing, those were embedded by compression in molten phenolic resin, at 175 °C, then it was applied the metallographic rough polishing procedure [1] using abrasive grit paper (400, 600, 800, 1000, 1500 grit) followed by final polishing using alumina alpha powder (Topol1, Topol, Topol 2 and 3, the grain size of 3 to 0.1 μm).

Metallographic analysis was performed with the aim of highlight the microstructural characteristics of brazed joints, for measuring the geometrical characteristics and to highlight any imperfections. The analysis was conducted using scanning electron microscope FEI QUANTA INSPECT F provided with the electron gun with field emission - EGF with a resolution of 1.2 nm and X-ray spectrometer energy dispersive (EDS) with resolution of 133 eV at MnK. Measuring conditions was: temperature + 24°C (reference temperature + 23 ± 5 °C); Humidity 50%.

3. Microstructural analysis

Metallographic analysis of brazed samples has targeted viewing of the distribution of the molten alloy in the joint gap, the correct sliding of the molten filler material and the absence of imperfections. For this purpose, some geometric measurements of joining areas have been done (joint width and stretching distance of molten metal). For "L" joint (Figure 3, 4 and 5) geometrical dimensions were:

penetration aperture of 42μm in the root and maximum width of 3.14mm, without misalignment of components, corresponding to the quality conditions imposed on this type of joint.

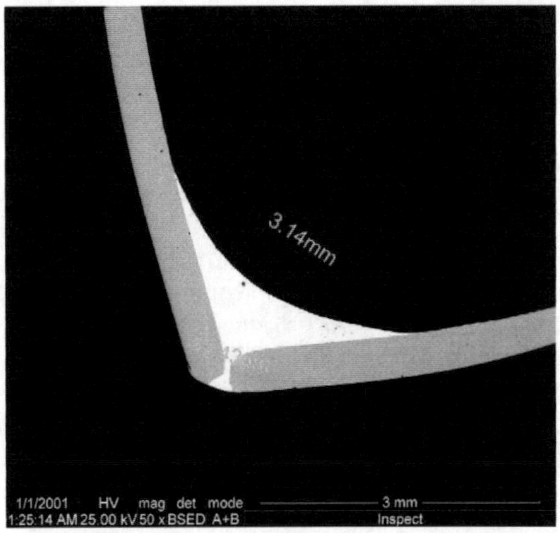

Figure 3. Macrographic aspect in cross-section of corner joint (sample "L") (50x).

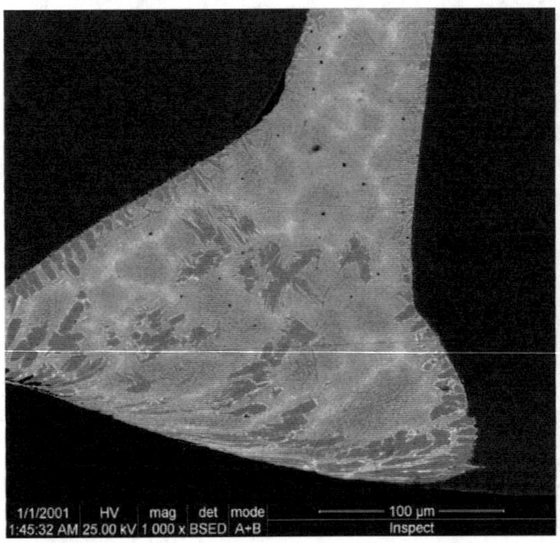

Figure 4. Brazed seam in root zone. Filler metal has an acicular microstructure (1000x).

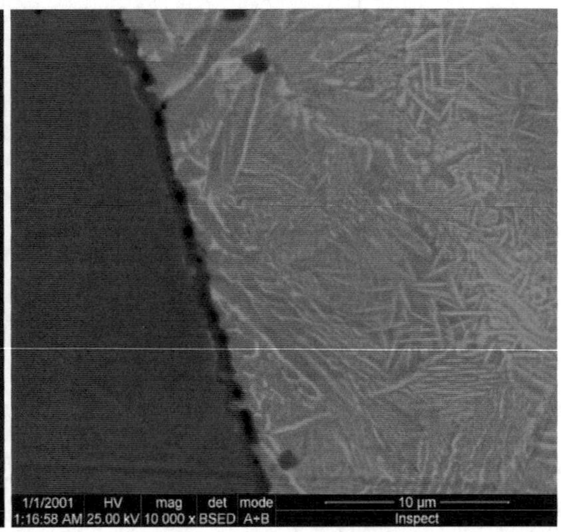

Figure 5. Detail on the interface between the base material and deposed material. Good adhesion and small pores(10000x).

At higher magnifications (1000 x, Figure 4) it can see the correct geometry of the root and the absence of secondary fragile phases. Some pores are small (diameter of 1-2μm) with random spreading that do not endanger the bond strength (0.5mm diameter are admitted for imperfections according to EN 12799-2002). Also it can observe a good adhesion of molten material from the surface of the stainless steel component (Figure 5).

For the "O" joint (Figure 6) the average size of the gap was 412μm (Figure 6), with a continuous spreading distance of 4mm (Figure 7), without excess of molten material and correct shape of concavity , what meet the conditions imposed for this type of joint. The quality condition for required

IOP Conf. Series: Materials Science and Engineering **133** (2016) 012014 doi:10.1088/1757-899X/133/1/012014

maximum length of penetration is 5s, where s is the thickness of components. In this case, the effective length of penetration was 4mm, which is higher than the required minimum value of 2.5mm.

On examination with higher magnifications it can observe a good adhesion of brazing filler metal to components of stainless steel. The constituents that are present in molten metal have fine shape needle-like appearance of (Figure 8). Expanding heat affected zone (HAZ) is quite restricted (270-315μm) given the very small component thickness (0.5mm) (Figure 9).

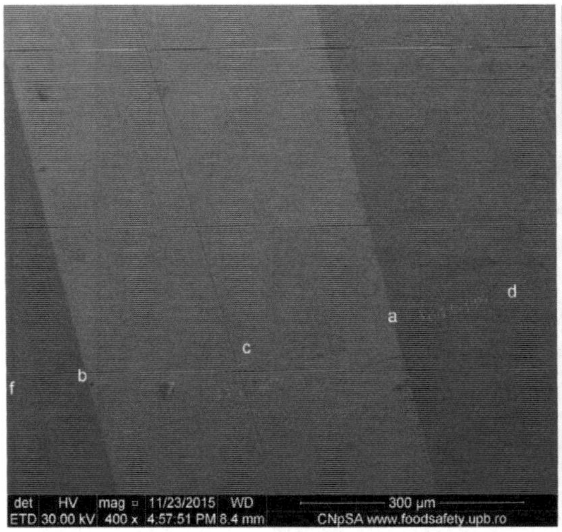

Figure 6. Cross section from "O" shape lap joint (400x).

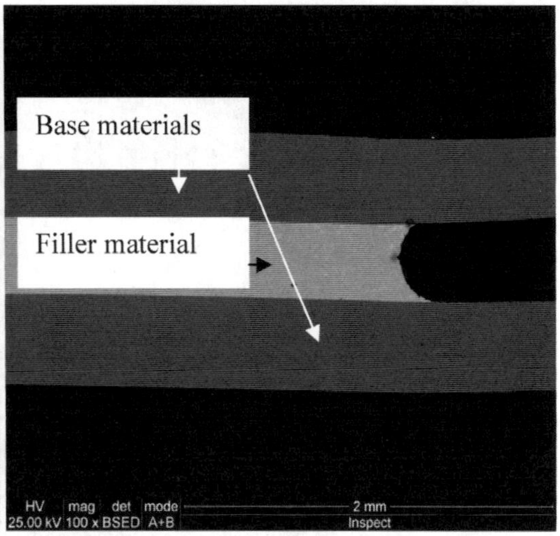

Figure 7. Aspect of lap brazed joint with very good penetration and adherence (100x).

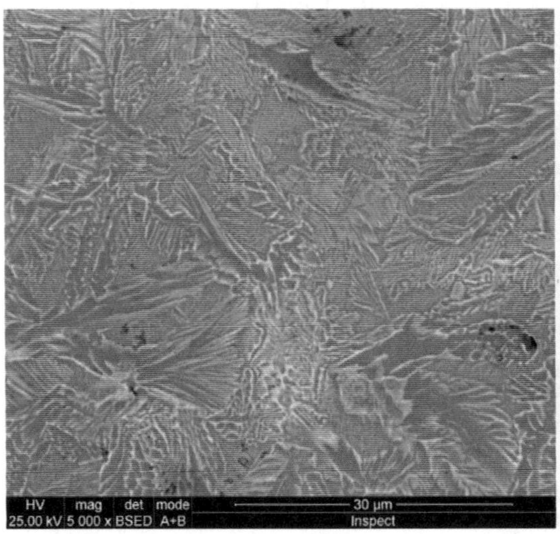

Figure 8. Details of the molten brazing alloy (5000x).

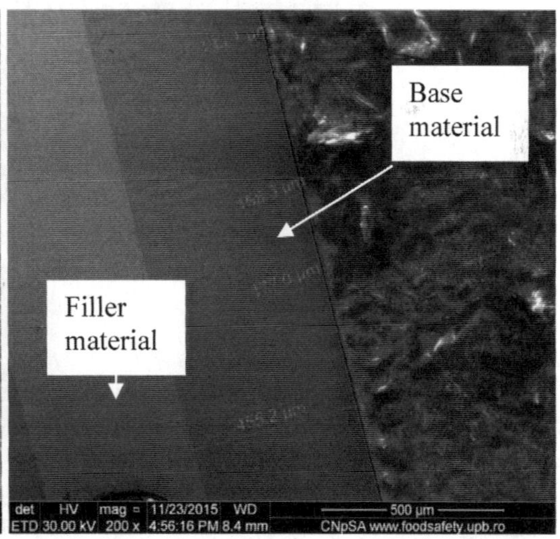

Figure 9. Heat affected zone with expansion in the range of 270 - 315μm (200X).

3.1. EDAX analysis on the cross section of "L" joint

To evaluate the effects of diffusion, some analysis of chemical composition was made, performed using scanning electron microscope. The analyze was carried out in 105 different points, located the area of the root and the seam, following a line that crossing successively the base material, the interface and filler material, then again in the base material (Figure 10).

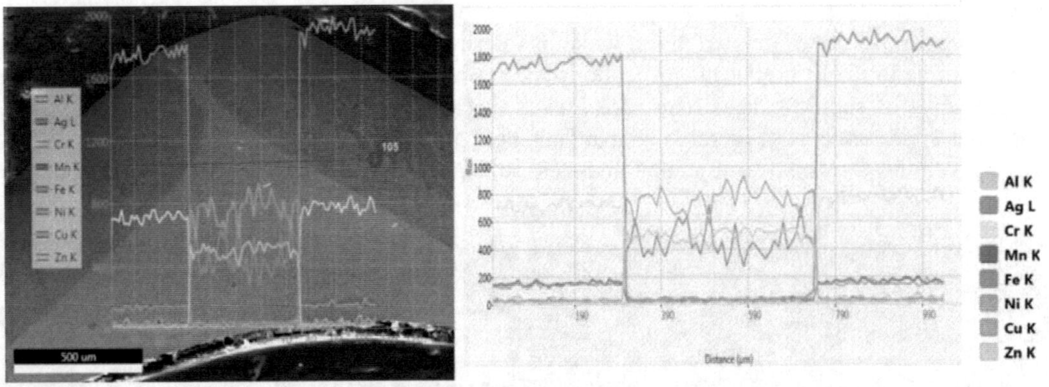

Figure 10. Evolution of the chemical composition in 105 different points, on the line that cross by the joint between stainless steel components and filler material.

The measurements were carried out on the straight line, from left to right, between points 1 and 105, over a distance of 10 microns. The extention in the base material was of maximum 240 microns on both sides of the line of fusion (Figure 11).

Figure 11. Overall distribution of elements in the root zone of brazing joint, evidenced by EDAX analysis.

3.2. EDAX analysis on the cross section of "O" joint

Measurements of the chemical composition have been done in order to determine the distance of diffusion of chemical elements in "O" joint. The chemical composition values for each point are presented in table 1.

The measurements were performed in the base material and HAZ, in 5 points located at different distances from the brazing surface (Figure 12a). As results from the chemical composition values in table 1, diffusion effects are detected only at very small distances from the molten metal surface, at approximately 20μm length. It can be observed the diffusion of chromium, iron and nickel from the base metal towards the molten material, and diffusion of silver, zinc and copper from the filler metal towards the base material.

To estimate the effects of diffusion in the bulk of molten filler material the chemical composition was analyzed in other 5 points of the cross section of joining "O" (Figure 12b). Points "f" and "d" are placed outside of the diffusion zones, while "a" and "b" points are located in the transition zone, between base materials (overheated) and melted deposit (filler material).

Only at very small distances from the molten metal surface (points a and b), at approximately 20μm length, some diffusion effects of elements like chromium, iron, manganese and nickel can be observed, in direction towards filler material. In opposite direction, elements like silver, copper, zinc and tin diffuses from filler metal towards base material, the biggest diffusion effect being for copper (25%).

a) base material b) filler material

Figure 12. The points location for determination of the chemical composition in "O" joint (400x).

Table 1. Chemical composition measured in 5 points in cross section of base material.

Measurement zone	Chemical composition, %wt									
	Si	Cr	Mn	Fe	Ni	Al	Ag	Zn	Sn	Cu
1- base metal	0.23	18.82	2.5	70.94	7.5	-	-	-	-	-
2 - HAZ	0.2	18.08	2.27	69.34	6.92	3.18	-	-	-	-
3 - HAZ	0.19	18.16	2.46	69.53	7.59	2.06	-	-	-	-
4 - HAZ	0.23	18.48	2.49	69.83	7.14	1.83	-	-	-	-
5 - near fusion line (20 μm)	0.3	13.45	1.56	49.43	4.63	-	3.95	10.97	0.43	15.28

The chemical composition values for each point located in filler material zone are presented in table 2. As results from table 2 values, diffusion effects are not detected in the center of filler material.

Table 2. Chemical composition measured in 5 points through filler material and HAZ.

Measurement zone	Chemical composition, %wt								
	Si	Cr	Mn	Fe	Ni	Ag	Zn	Sn	Cu
a- base metal	0.28	15.12	1.58	56.99	4.92	3.19	6.99	0.21	10.7
b- base metal	0.09	14.11	1.54	56.75	5.71	1.96	7.57	0.13	12.15
c – center zone of filler metal	-	-	-	-	-	21.36	34.5	1.16	42.98
d- HAZ (164μm)	0.23	18.43	2.74	70.33	8.1	-	-	-	-
f - HAZ (60 μm)	0.22	18.45	2.33	69.24	7.28	-	-	-	-

4. Conclusion

The experimental alloy used in the study has a good spreading behaviour and do not produce any cracking tendency. The only types of imperfections observed in the brazed area were: un-melted flux inclusions and pores (gas) resulted from the coating of the filler material.

In the brazed joints made by stainless steel components, at the interface between the base material and melted filler material, diffusion phenomenon can occurs, by migration of some chemical elements through the interface, even the time of maintain at high temperatures is short (30 minutes).

At approximately 20μm length from melting line, some diffusion effects of elements like chromium, iron, manganese and nickel can be observed, in direction towards filler material.

The elements with greater participation in the filler material (copper, zinc and silver) migrate by diffusion towards the base material, the biggest rate of diffusion being for copper (25%). These exchanges are extending only in a small area, of 20 to 60 microns form de interface, and do not affect the mechanical resistance or corrosion resistance of the base material.

The analyzes performed to determine the diffusion effects in the case of two types of brazed joints showed the penetration of alloying elements from the solder (Ag, Cu, Zn and Sn) towards the base material (stainless steel), over distances up to 20 microns. Concentration of alloying elements in base material was 3% Ag, 10% Cu and 7% Zn at 10 microns distance from the interface, and 2% Ag, 12% Cu and 7.5% Zn at 20 microns from the interface.

Acknowledgements

The research work was financially supported by the Romanian National Program for Research in the framework of the Project **– PCCA 188/2012** -„Materials and performant technologies designed to realize knives for milling the asphalt – MATFREZ.

References

[1] Millspaw M 1994 The Care and Feeding of Stainless Steel *Brewing Techniques* **2(4)** pp 44-47
[2] Diffusion Brazing, Principle of Brazing *ASM International* 2005 pp 207-216
[3] ***An Introduction to Brazing 2014 *Oerlikon Metco* **4** pp 6 – 10
[4] Jhonson M Metal joining, Silver Brazing Alloys and Fluxes, pp 4 - 39 http://www.jm-metaljoining.com
[5] Xing F, Qiu X M, Li Y D 2015 Effects of Sn element on microstructure and properties of Zn–Cu–Bi–Sn high-temperature solder *Trans. Nonferrous Met. Soc. China* **25** pp 879 – 884
[6] Stainless Steel Cladding and Weld Overlays, ASM Speciality Handbook: Stainless steels, *ASM International* 1994 pp 109-110
[7] Voiculescu I, Geantă V, Binchiciu H, Binchiciu E et. all, Research Project – PCCA 188/2012 - Materials and performant technologies designed to realize knives for milling the asphalt – MATFREZ
[8] Binchiciu A, Voiculescu I, Geantă V, Binchiciu H, Ştefănoiu R and Binchiciu E 2010 Coated silver alloy rods of reduced hygroscopicity for brazing, comprises silver and tin deposited by extrusion deoxidizing flux made of synthetic vitreous precursors and metal powder additions (Patent Number: RO125836)
[9] Binchiciu A, Voiculescu I, Geantă V, Binchiciu H, Ştefănoiu R, Negriu R M and Binchiciu E 2012 Ecological coated core wires of silver alloys for brazing, made by extrusion coating with deoxidizing flux with metal powder addition (Patent Number RO125835)
[10] Voiculescu I, Geanta V, Binchiciu E, Moisa S, Vasile I M and Trusca R 2016 Study of Dilution Effects in the Joint Brazed with Cu-Ag Alloy for Stainless Steel Components, *IMEC 17 The 17th Israel Materials Engineering Conference*

IOP Conf. Series: Materials Science and Engineering **133** (2016) 012015 doi:10.1088/1757-899X/133/1/012015

Study on structure and properties of CuZn40Pb alloy

D C Achiței[1,3*], M G Minciună[1,3], P Vizureanu[1,3], A V Sandu[1,3,5], R Cimpoeșu[2], B Istrate[4]

[1]Department Technologies and Equipments for Materials Processing, Gheorghe Asachi Technical University of Iasi, D Mangeron Street, no 41A, Iasi
[2]Department Materials Science, Gheorghe Asachi Technical University of Iași, D Mangeron Street, no 41A, Iasi
[3]Center of Excellence Geopolymer & Green Technology (CEGeoGTech), School of Materials Engineering, Universiti Malaysia Perlis, 01000 Kangar, Perlis, Malaysia
[4]Department of Mechanical Engineering, Mechatronics and Robotics, Gheorghe Asachi Technical University of Iași, D Mangeron Street, no 41A, Iași
[5] Faculty of Technology, Universitas Ubudiyah Indonesia (UUI), JlnAlue Naga DesaTibang, Banda Aceh, Indonesia

E-mail: dragos_adc@tuiasi.ro, mirabela.minciuna@yahoo.ro, peviz2002@yahoo.com, euroinvent@yahoo.com, ramonacimpoesu@gmail.com, bogdan_istrate1@yahoo.com

Abstract. The paper shows aspects about the behavior of Cu-Zn-Pb alloys a subjected to the temperatures variation and corrosion resistance in saline medium (sea water). The chemical composition was determined by spectral analysis on optical spectrometer, type Foundry Masters. The experiments are completed by a microstructure analysis made on scanning electronic microscope.

1. Introduction

The nonferrous alloys are used in the domain of manufacturing of cast parts. At the elaboration of nonferrous alloys can be use blocks of cast alloys, pure metals, pre-alloys, metallic waste, recycling materials, fondants and modifying admixtures. The calculus of load is made in function of the available materials, the burnt during elaboration and the final alloys which are achieved.

To obtain the superior characteristics of cast parts from nonferrous alloys, the limits of impurities admitted are provided in standards, being necessary their diminution. Due to his properties, the copper have a large utilization in industry: 50 % in electrotechnique parts and 30 ... 40 % for elaboration of alloys.

The brasses are copper alloys with zinc and contain at least 55 % copper. The alloyed brasses can contain and additions by aluminum, tin, manganese, iron, nickel, lead. The casting temperature for binary brasses is 1100 ... 1200 °C, in function of cast part and shape. If the load is composed by unpurified waste or metals, it is necessary to refining of metallic bath after the complete melting under a flux layer. The brasses with a high content of lead, can replace the bronzes with tin. At their elaboration is recommended initial the melting of copper, then simultaneous introduction of zinc and

* dragos_adc@tuiasi.ro

lead. Before casting is added 0.1 ... 0.2 % aluminum for fluidity and dezoxidation, and the casting temperature is between 1000 ... 1080°C.

2. Results and discussions

2.1. Determination of chemical composition

The spectral analysis was made on Foundry Masters spectrometer, type 01J0013, using copper base for analysis. The determinations were made on samples, obtained by cutting at optimal dimensions, using METACUT M250 machine, with abrasive disk and a cooling system. After 3 determinations, made on samples surface, it was obtained the average value for the alloying elements (table 1). The alloy is part of copper-tin alloys (brasses), cast in blocks, type CuZn40Pb.

Table 1. Chemical composition for the studied alloy

Cu	**Zn**	**Pb**	**Sn**	**P**	**Mn**	**Fe**	**Ni**	**Si**
58.5	38.8	1.88	0.186	0.003	0.0091	0.282	0.146	0.0063
Mg	**Cr**	**Al**	**S**	**As**	**Be**	**Ag**	**Co**	
0.0014	0.0053	0.0147	0.002	0.0067	0.005	0.0207	0.015	

Table 2. Mechanical characteristics and utilization domains for the CuZn40Pb alloy [1]

Type	**Cast mode**	**R_m [N/mm^2]**	**A_5 [%]**	**HB [daN/mm^2]**	**Utilization domain**
CuZn40Pb	Cast parts under pressure	280	7	80	fittings, elements subjected to usage, elements with good electrical and thermal conductivity
	Cast parts in metallic matrix	280	7	80	

Table 3. Physical properties[6]

Properties	**Units**	**Value**
Density at 20°C	[g/cm^3]	8.4
Specific heat	[cal/g·°C]	0.09
Modulus of elasticity	[kg/mm^2]	10000
Modulus of rigidity	[kg/mm^2]	3700

Table 4. Manufacturing properties[6]

Properties	**Units**	**Value**
Melting interval	[°C]	895 – 900
Melting temperature	[°C]	1000 – 1050
Annealing temperature	[°C]	450 – 650
Hot processing interval	[°C]	650 – 750
Hot plasticity	-	Good
Cold plasticity	-	Limited

2.2. Dilatometer analysis

The dilatometer analysis was made on Linseis L75H/1400 and is used samples with small dimensions. By dilatometer analysis is highlighted the structural transformations triggered at heating in linear regime, for temperatures domain between ambient up to 200°C.

Table 5. Measured values for elongation by expansion at heating

Heating speed 2.5 [°C/min]	**T [°C]**	17.7	36.2	55.8	71.3	87.8	104.3	119.3	137.6	156.5	172.4	186.8	200.0
	Δl [mm]	0	2.7	7.0	11.2	16.0	20.7	24.7	29.5	34.3	38.5	42.2	45.6
Heating speed 5 [°C/min]	**T [°C]**	28.1	51.9	65.6	89.2	108.9	126.7	144.4	160.9	171.2	177.0	191.6	200.0
	Δl [mm]	0	0.9	1.7	4.5	8.9	13.6	18.7	23.6	25.8	28.2	32.1	34.3
Heating speed 7.5 [°C/min]	**T [°C]**	28.2	53.3	74.7	96.5	115.3	131.7	148.1	164.4	178.9	183.4	190.9	200.0
	Δl [mm]	0	0.10	1.95	2.92	6.35	9.75	13.75	18.20	22.44	24.85	26.15	29.0
Heating speed 10 [°C/min]	**T [°C]**	32.9	59.9	81.1	99.5	116.2	129.9	146.2	159.2	169.3	182.3	193.8	200.0
	Δl [mm]	0	2.19	1.42	2.78	3.80	6.09	9.24	12.04	14.41	17.61	20.63	22.36

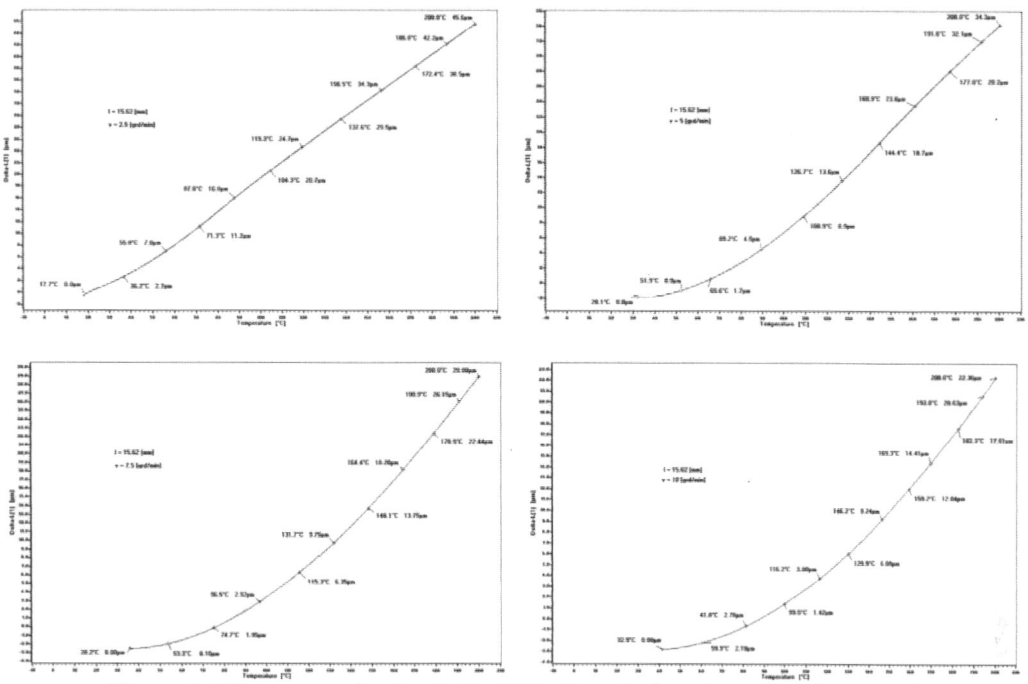

Figure 1. Dilatograms for the CuZn40Pb alloy, in function of heating speed.

2.3. Study on corrosion resistance

The electrochemical corrosion represent a ensemble of physical – chemical processes after which the alloys pass to metallic compounds form (oxides, hydroxides, salts), as a result of the formation of micropores in which the metal is subjected to anodic oxidation. In this time a compounds of the solution, named depolarizing, suffer a complementary reaction by reduction. Between the principal causes which determine the appearance of locally elements can be mentioned:

- Inpurifying with non-noble metals, oxides of metals;
- Chemical heterogeneity;
- Physical heterogeneity, due to mechanical treatment of irregular thermal treatment.

To appearance of this corrosion type it is necessary to exist an anode, a cathode, an electrolyte and a conductor. If one of these conditions not exists, the electrochemical corrosion is not produced,

To study the processes of electrochemical corrosion it is used the potentio-dynamic method (linear and cyclic) [12].

2.3.1. Characterization by linear polarization studies. Electrochemical corrosion tests were conducted using a three electrode electrochemical cell connected to a Volta Lab 21 (Radiometer, Copenhagen) potentiostat.

A saturated calomel electrode (SCE) was used as the reference electrode. The auxiliary electrode was a platinum wire [10,11]. The working electrode was realized in square shape and was assembled in teflon. Exposed area of the sample to the electrolyte was 1.0 cm^2.

The tests were conducted at room temperature (20°C). The solution used for the tests was orange juice,

The curves of linear polarization were recorded at a scanning speed of electrode potential by 3 mV/s, in a potential domain ± 545 mV, in a open potential circuit. The curves of cyclic polarization were recorded with 10 mV/s speed, in the potential domain between -1000...+2000 mV.

The corrosion potential at corrosion current $E_0 \equiv E$ (I = 0), the Tafel branches (b_a şi b_c), the polarization resistance R_p, density of corrosion current J_{cor} and corrosion speed V_{cor}, were evaluated with Volta Master 4 software.

The measurements were made at potential, in open circuit, using sea water. The sea water is considered the most aggressive natural agent of corrosion [9]. By his content in salt, gases and the fact it is in continuous moving, the sea water corroded all the metallic construction which are immersed or which are at surface. The principal form of corrosion is locally, type pitting.

The principal parameters of corrosion process (E_0 and J_{cor}) obtained by processing the curves of linear polarization for Cu-Zn alloy, in cast state are centralized in table 2.

Table 6. The parameters of electrochemical process

Sample	Corrosive medium	E_0	b_a	b_c	R_p	J_{cor}	V_{cor}
		mV	mV	mV	kΩ/cm²	µA/cm²	mm/Y
Cu-Zn	Sea water	- 961.1	528.1	- 306.1	1.77	0.0272	317.5

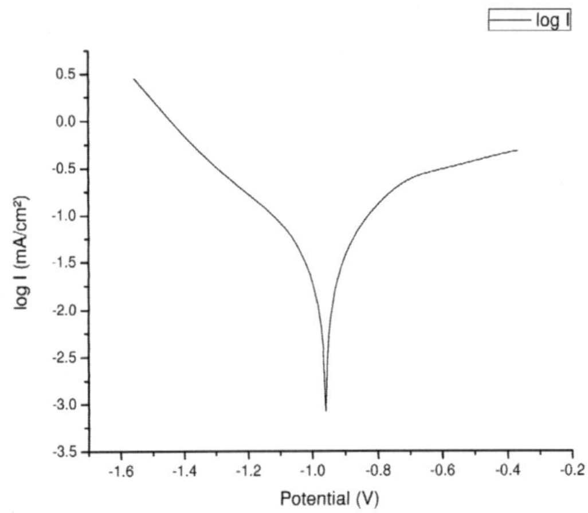

Figure 2. The curve of linear polarization for Cu-Zn alloy, immersed in sea water, obtained by sweep with 3 mV/s speed.

In order to check the passivation effect of the sea water to Cu-Zn alloy, is drawing the polarization curves for samples cu a polished surface [7, 8].

The copper forming compounds in which can be monovalent (Cu^+), divalent (Cu^{+2}) or rarely trivalent (Cu^{+3}). The divalent copper combination is colored, and for our case present an greenish color. The corrosion potential present negative value, relatively high, this fact highlighted the natural tendency of this alloy to spontaneous corrosion.

The corrosion current (J_{cor}) present a positive value by 25 µA/cm² and representative for the degree of material deterioration.

2.3.2. Characterization by cyclic polarization studies. The analysis of cyclic diagram presents the pitting character of corrosion, being by hysteresis type with return branch on external. The surface of hysteresis branch is a value of intensity of corrosion process with pitting character.

The pitting corrosion represents a dangerous behavior of materials at the interaction with external medium, due to the material quantity which pass in the electrolyte medium.

The curve of cyclic polarization obtained Cu-Zn alloy immersed in sea water present the form of pitting corrosion, process which is possible at a small potential, but the corrosion intensity is high.

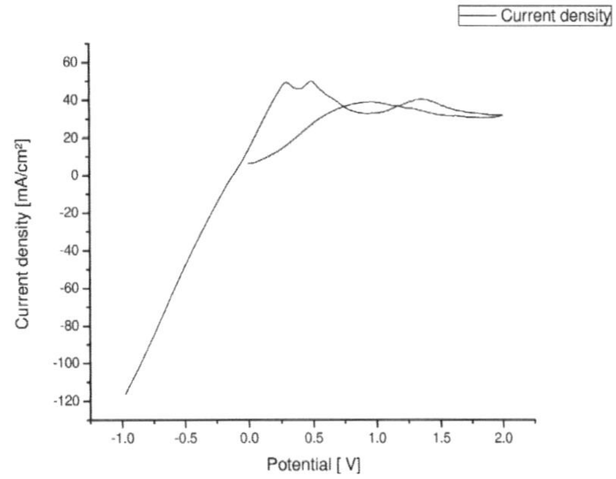

Figure 3. The curve of cyclic polarization for Cu-Zn alloy immersed in sea water, obtain by sweep with 10 mV/s speed.

The obtained layer is not adherent because can be removed by simply deleting with filter paper.

2.4. Structural analysis by electronic microscopy
It was made on a scanning electronic microscope, type VEGA II LSH, manufactured by TESCAN Cehia, coupled with a EDX detector, type QUANTAX QX2, manufactured by ROENTEC Germany.

a.

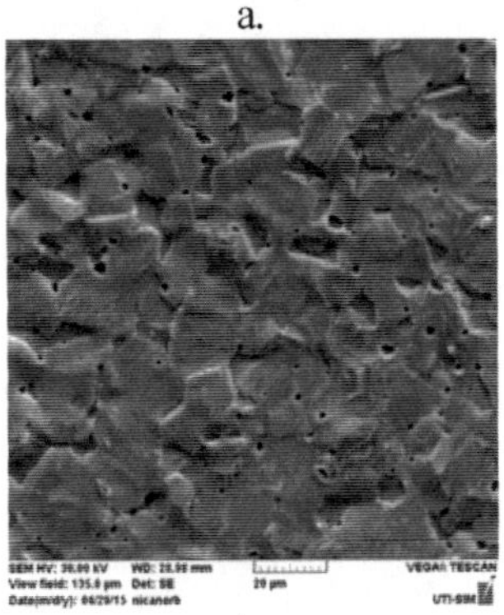

b.

Figure 5. SEM microstructures in cast state: a) 500x; b) 2000x.

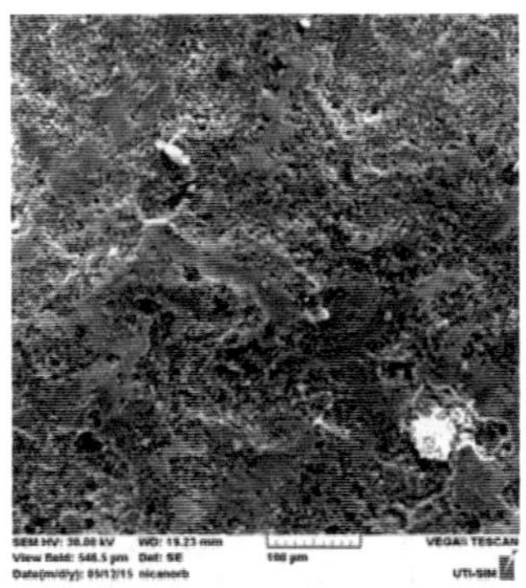

a.

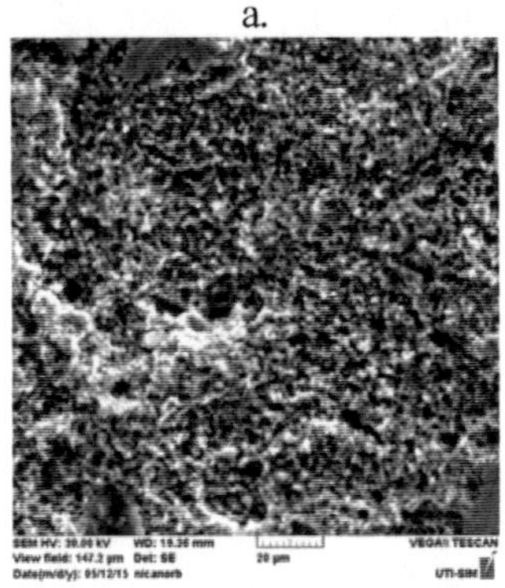

b.

Figure 6. SEM microstructures after corrosion in water: a) 500x; b) 2000x.

a.

b.

Figure 7. SEM microstructures after corrosion in sea water: a) 500x; b) 2000x.

3. Conclusions

The nonferrous alloys can be used in large domain of application for manufacturing the cast parts.

The elaboration of nonferrous alloy can be used cast alloys in blocks, pure metals, pre-alloys, metallic waste, recycle material, fondants and modifying additions.

Brasses are copper zinc alloys. These alloys have good strength and corrosion resistance, although their structure and properties are a function of zinc content. Alloys containing up to approximately 35% zinc are single phase alloys, consisting of a solid solution of zinc and alpha copper.

Brasses are used in applications such as blanking, coining, drawing, piercing, springs, fire extinguishers, radiator cores, lamp fixtures, ammunition, flexible hose, and the base for gold plate.

Brasses have excellent castability, and a good combination of strength and corrosion resistance.

The cast brasses are used in applications such as plumbing fixtures, fittings and low pressure valves, gears, bearings, decorative hardware and architectural trim.

We investigated the effects of natural seawater on the corrosion of a Cu-Zn alloy. In this study, the corrosion behavior of the alloy has been investigated using the open circuit potential measurements,

Tafel curves and anodic potentiodynamic curves.

By comparing different samples of marine water indicates that salinity is the primary influence on corrosion behavior near the open circuit potential. At higher anodic potentials, the impact of mineral composition becomes significant.

References

[1] G. Amza, et.al., 2002 *Tratat de tehnologia materialelor*, Editura Academiei Romane, București.

[2] N. Geru, et.al., 1985 *Materiale metalice – structură, proprietăți, utilizări*, Editura Tehnică, București.

[3] C. Ștefănescu, et.al., 1985 *Îndrumatorul proiectantului de tehnologii in turnatorii*, Editura Tehnică, București.

[4] I. Rusu, 2011 *Tehnici avansate in ingineria materialelor. Aplicații practice*, Editura PIM, Iasi.

[5] http://admin.copperalliance.eu

[6] L. Suarez, P. Rodriguez - Calvillo, J.M. Cabrera, 2015 *Materials Science and Engineering A - Structural Materials Properties Microstructure and Processing*, vol. 627, pp. 42-50.

[7] M. M. Nandi, R. Banerjee, 2014 *Journal of the Indian Chemical Society*, vol. 91, issue 6, pp. 977-989.

[8] S. Bruschi, A. Ghiotti, M. F. Novella, 2013 *Computational Plasticity XII: Fundamentals and Applications*, pp.332-342.

[9] K. Holler, B. Reetz, K. B. Muller, 2003 *PTS 1-5 Book Series: Materials Science Forum*, vol. 426-4, pp. 3667-3672.

[10] B. Reetz, W. Reimers, 2008 *International Journal of Materials Research,* vol. 99, issue 7, pp. 787-794.

[11] A. I. Toulfatzis, G. A. Pantazopoulos, A. S. Paipetis, 2014 *Journal of Materials Engineering and Performance*, vol. 23, issue 9, pp. 3193-3206.

[12] A. Nawaz, K. M. Deen, A. Farooq, 2015 *International Journal of Materials Research*, vol. 106, issue 1, pp. 92-96.

[13] I. Carcea, M. Gherghe, 2009 *Aliaje neferoase de turnatorie*, Editura Performantica.

IOP Conf. Series: Materials Science and Engineering **133** (2016) 012016 doi:10.1088/1757-899X/133/1/012016

Hafnium influence on the microstructure of FeCrAl alloys

V Geanta[1], I Voiculescu[1], E-M Stanciu[2]

[1]University Politehnica of Bucharest, Splaiul Independenței 313, sector 6, 060042, Bucharest, Romania
[2]S.C. Optoelectronica 2001S.R.L, str. Atomistilor, 409, 077125, Magurele, Romania

E-mail: victorgeanta@yahoo.com

Abstract. Due to their special properties at high temperatures, FeCrAl alloys micro-alloyed with Zr can be regarded as potential materials for use at nuclear power plants, generation 4R. These materials are resistant to oxidation at high temperatures, to corrosion, erosion and to the penetrating radiations in liquid metal environments. Also, these are able to form continuously, by the self-generation process of an oxide coating with high adhesive strength. The protective oxide layers must be textured and regenerable, with a good mechanical strength, so that crack and peeling can not appear. To improve the mechanical and chemical characteristics of the oxide layer, we introduced limited quantities of Zr, Ti, Y, Hf, Ce in the range of 1-3 %wt in the FeCrAl alloy. These elements, with very high affinity to the oxygen, are capable to stabilize the alumina structure and to improve the oxide adherence to the metallic substrate. FeCrAl alloys microalloyed with Hf were prepared using VAR (Vacuum Arc Remelting) unit, under high argon purity atmosphere. Three different experimental alloys have been prepared using the same metallic matrix of Fe-14Cr-5Al, by adding of 0.5%wt Hf, 1.0%wt Hf and respectively 1.5%wt Hf. The microhardness values for the experimental alloys have been in the range 154 ... 157 HV0.2. EDAX analyses have been performed to determine chemical composition on the oxide layer and in the bulk of sample and SEM analyze has been done to determine the microstructural features. The results have shown the capacity of FeCrAl alloy to form oxide layers, with different texture and rich in elements such as Al and Hf.

1. Introduction

FeCrAl alloys along with the ZrCrAl alloys class can be used for realizing the metallic structures of the reactor in nuclear power plant, generation 4R. The components that can be made from these modern alloys are: the outer jacket of the reactor, recirculation pipes for the liquid metal, pumps for distribution of the cooling medium etc. Modern type LFR reactors working in closed cycle with optional fuel flexibility and fast neutron spectrum are cooled with metallic medium like Pb or Pb-Bi alloy. This particularity ensures a much higher level of security at a lower overall volume compared to the previous generation reactors. Such materials must provide excellent mechanical and technological characteristics, among which: plasticity and ductility, mechanical workability, weldability, thermal shock resistance and mechanical strength, creep resistance at high temperature, high chemical stability in specific environmental conditions (temperatures above 400 – 800°C), corrosion and erosion resistance, high thermal conductivity to ensure optimum heat transfer.

[1] To whom any correspondence should be addressed.

Special FeCrAl alloy is similar to ferritic stainless steels, but contains a higher aluminum percent and some microalloying elements like: Hf, Ta, Ti and Y. This alloy is designed to be used at temperatures up to 1400C and exhibits outstanding electro-thermal characteristics: high resistivity, low temperature coefficient, long life at high temperatures and high resistance to oxidation [1, 2, 3, 4, 5].

Oxidation resistance is given by the presence of a special and adherent layer at the surface level of the alloy, a layer that contains complex oxides of aluminum and chromium, which prevent deep oxidation at high temperatures. The Al2O3 layer acts like a ceramic material and a barrier against the effects of corrosive or erosive environment generated by metal coolant and provide optimum protection, if it keeps a high density and if the oxide does not peel over time.

Class of alloys that possess necessary characteristics for the operation in the new 4R nuclear power plant is the alloy FeCrAl, known in the literature as the Fecralloy trade mark. It is known that the addition of chromium can increase the ductility at room temperature, but it does not have a substantial effect on the flow resistance at 600 °C for alloys with up to 25% Al. At high contents of aluminum, the ability to form at the surface level of a textured and uniform aluminum oxide is diminished. Thus, in excess of 8% Al, the aspect of the oxide is not uniform, adherent and compact, the superficial layer containing very thin alumina whiskers. The choice of a correct aluminum content of the FeCrAl alloys must be correlated with the content of chromium, according to the equation:% Al = (40 -% Cr)/6 [6,7]. At the same time, the chromium content below 12% does not offer sufficient resistance to oxidation. The addition of rare earth elements with high affinity for oxygen (yttrium, zirconium, hafnium, titanium and cerium) leads to the formation of highly stable oxides on the surface of alloys FeCrAl [8, 9].

The protective oxide layer, with self-generation effect and with high adhesion to the substrate of alloy FeCrAl, must be textured and must have a good mechanical strength at cracking process or at dissolving effects. There are various known technologies for producing oxide layers on the surface of the alloy, such as the deposition of layers by plasma spraying in a vacuum (PVD), re-melting the surface with a jet of electron irradiation, laser deposition, plasma metalizing [10, 11, 12, 13, 14]. These methods aim to eliminate the porosity of the surface, by remelting of the superficial layer of oxide and obtaining a roughness as low as possible, for avoid the mechanical anchoring effect on the metallic surface of the molten lead. For designing the correct chemical compositions of FeCrAl alloy (micro-additions of rare earth elements), the influences of all alloying elements on the properties of these materials must be taken into account.

Alumina forming alloys exhibit interesting anti-corrosion properties under high temperature oxidizing conditions since aluminum allows the formation of an α-alumina scale. Under thermal cycling conditions, the oxide scale may undergo severe spallation. The addition of reactive elements is known to improve the high temperature oxidation performance of alumina forming alloys. FeCrAl – type alloys are of interest for applications at high temperatures, also in molten lead or in a mixture of lead and bismuth. Small additions of reactive elements, like hafnium, alter the morphology of the oxides, leading to the formation of an adherent alumina scale. Utilizing the rare elements in these alloys has a beneficial effect to increase the chemical bonding between metal and oxide, and/or to the formation of pegs anchoring the oxide scale. Reactive elements like hafnium are also believed to act as vacancy sinks to suppress the formation of voids.

Hafnium is used for nuclear reactor control rods because of its ability to absorb neutrons and its good mechanical and corrosion resistance qualities. Its neutron-capture cross-section is about 600 times that of zirconium. Hafnium resists to corrosion due to the formation of an oxide film on exposed surfaces [15]. In FeCrAl alloys micro-alloyed with hafnium, a complex aluminum oxide that contains both aluminum and hafnium forms on the metallic surface. The physical properties of hafnium metal samples are markedly affected by zirconium impurities, especially the nuclear properties, as these two elements are among the most difficult to separate because of their chemical similarity [16].

Hafnium is a very reactive element and reacts with oxygen according to the formula:

$$Hf + O_2 \rightarrow HfO_2 \qquad (1)$$

Irradiation by high energy particles (electrons, ions, neutrons) may induce the redistribution of impurities and solutes near sinks, such as point defects, dislocations and boundaries. The enrichment and/or depletion of the elements at grain boundaries will influence the mechanical properties of materials. The enrichment of undersized impurity atoms, e.g. phosphorus in ferritic reactor pressure vessel steels, reduces the grain boundary cohesion, causing the materials to fail through inter-granular embrittlement. Radiation induces undersized atom (P) enrichment and oversized atom (Cr) depletion at grain boundaries in E911 steel. This phenomenon can be suppressed by the addition of an oversized atom like Hf. The addition of Hf suppresses radiation-induced undersized atom enrichment and oversized atom depletion. Hf is added to ferritic steel for induce the redistribution of impurities at grain boundaries with influence of the mechanical properties under irradiated effect [16,17].

2. Obtaining experimental FeCrAl alloys micro-alloyed with Hafnium

The research paper aims to design and manufacture FeCrAl alloys as mini-ingot, using VAR furnace (vacuum remelting arc furnace), documenting the procedure and analyzing the microstructure in order to estimate the effect of hafnium addition on the alloys' characteristics. The samples were designed and manufactured at ERAMET laboratory, at the Politehnica University of Bucharest, Material Science and Engineered Faculty (www.eramet.wix.com/eramet), using a MRF ABJ 900 furnace (55 kVA, min. 650 A @ 60% DS). After obtaining the stable vacuum of 1x10-4 mBar, the furnace chamber was filled with argon (Ar 5.3) to insure the electric arc stability. Each mini-ingot was remelted and solidified five times, in order to obtain the microstructural homogeneity. The assimilation coefficient for this particular alloy class was over 99.5%, due to the low vapor losses of the components in the electric arc remelting process. Mini ingots had quasi-constant weight (39,63g – 39,96 g) (Table 1).

Three experimental FeCrAl alloys were obtained, with different contents of hafnium: 0.5%wt Hf, 1.0%wt Hf and respectively 1.5%wt Hf added into the same metallic matrix of the experimental alloy Fe-14Cr-5Al. The base alloy was a high purity "extra soft" steel, mark MK3, having the chemical composition: C = 0.02%wt, Si = 0.04%wt, Mn = 0.21%wt, S = 0.02%wt, P = 0.015%wt, Ni = 0.2%wt, Cr = 0.15%wt, Mo = 0.07%wt, 0.14%wt Cu, Al = 0.12%wt and Fe= bal. For obtaining high alloyed material, high purity alloying elements were introduced into the base material: metallic chromium 99.5 % Cr; electrolytic aluminum 99.4% Al; hafnium 99.5 % Hf.

Table 1. Assimilation efficiency and the weights for each experimental batch and mini-ingots.

Sample code	Initial weight batch, g	Weight of mini-ingot, g	Assimilation efficiency, $(G_{ling.}/G_{ch.})$x100, %	Global efficiency, %
NUC 8 - Fe-14Cr-5Al-0.5Hf	40.0	39.96	99.90	
NUC 9 - Fe-14Cr-5Al-1.0Hf	40.0	39.91	99.77	99.58
NUC10 - Fe-14Cr-5Al-1.5Hf	40.0	39.63	99.07	

3. Results and discussions

To each sample sectioned from the mini-ingots we applied the metallographic rough polishing procedure using abrasive grit paper (400, 600, 800, 1000, 1500 grit), followed by a final polishing using alumina alpha powder (Topol1, Topol, Topol 2 and 3, the grain size of 3 to 0.1 μm). Metallographic analysis was performed with the aim of highlighting the microstructural characteristics of the experimental alloys, in order to estimate the aluminium and hafnium distribution in the superficial layer of oxide. The analysis was conducted using scanning electron microscope FEI QUANTA INSPECT F provided with the electron gun with field emission - EGF with a resolution of 1.2 nm and X-ray spectrometer energy dispersive (EDS) with resolution of 133 eV at MnK.

3.1. Chemical composition and microstructural analysis

The chemical composition of the samples was determined by EDAX analysis in the successive points, in accordance with figure 1. The values obtained for the three batches named NUC 8 - NUC 10 in the center thereof are shown in table 2. Concentration analysis of chemical elements in center of the samples (point 5) shows a good concordance between projected chemical composition and chemical composition performed for each batch separately. Although different amounts of Hf were introduced into the metallic matrix (0.5%Hf, 1.0%Hf and 1.5%Hf), this element is not detected in the center of the sample, due to its migration to other areas, like the peripheral oxide layer.

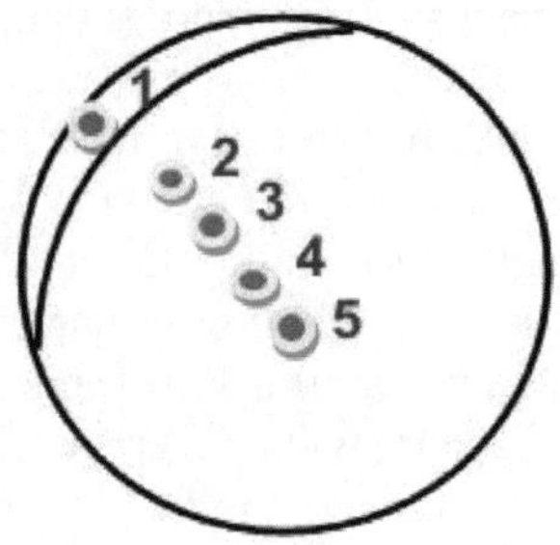

Figure 1. Radial arrangement of the points of analysis of FeCrAlHf alloy.

Table 2. Amounts of alloying elements in the center of the samples (point 5).

Sample	Chemical composition, %wt		
	Al	Cr	Fe
NUC 8	4.17	14.54	81.29
NUC 9	4.68	14.78	80.55
NUC 10	4.51	14.84	80.64

EDAX analysis of these alloys was done in several areas of the surface of the sample in radial points, as shown in figure 1. In table 4, we have shown the chemical compositions of the edge area of the samples analyzed (point 1). The chemical composition of the marginal crust of the samples micro alloyed with hafnium (point 1) indicates that the value of the Hf content is increasing with increasing amount of hafnium in the alloy. This value decreases in the direction of points 1 to 5 due to strong migration of this element to the marginal zone, where it combines with oxygen dissolved in the sample. Due to the simultaneous presence of the two elements with very high affinity for oxygen (aluminium and hafnium), the marginal crust are formed by complex oxides of Al and Hf, whose share is given by the amounts of these elements in the alloy.

Table 4. Chemical compositions on marginal crust.

Chemical composition, %wt					
NUC8 (0.5%wt Hf)		NUC9 (1%wt Hf)		NUC10 (1.5%wt Hf)	
O	22.64	O	15.51	O	13.86
Al	48.58	Al	11.09	Al	0.87
Cr	0.34	Cr	0.96	Cr	0.82
Fe	0.62	Fe	4.07	Fe	3.66
Hf	27.81	Hf	68.37	Hf	80.79

The distribution of elements in peripheral oxide layer for the alloy named NUC 8 is shown in figure 2.

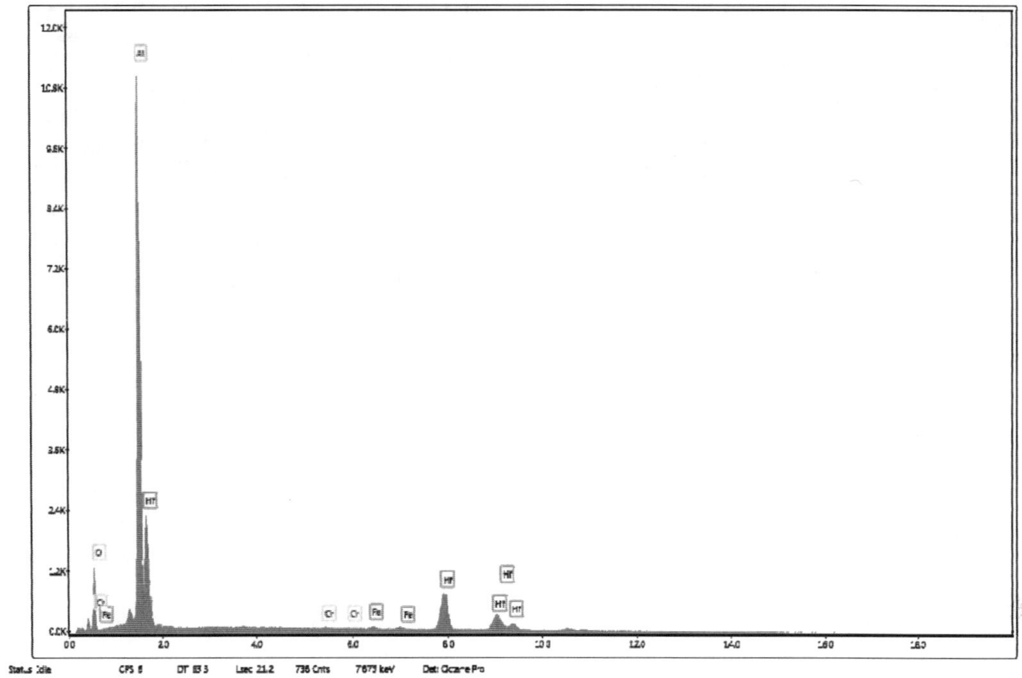

Figure 2. The distribution of elements in peripheral oxide layer for the alloy NUC 8 (0.5 %wt Hf).

The thickness of peripheral oxide, for the 3 samples analyzed, was between 9-13 µm. Their aspect is like elongated and adherent islands, rich in Al, O and hafnium, near other clean areas, as shown in figures 3, 4 and 5. Figure 5 shows the distribution of chemical elements into peripheral oxide layer for alloy NUC 8 from which it appears that, even for small amounts of hafnium in the alloy (0,5%wt Hf) the presence of this element is highlighted in this zone, along with other elements.

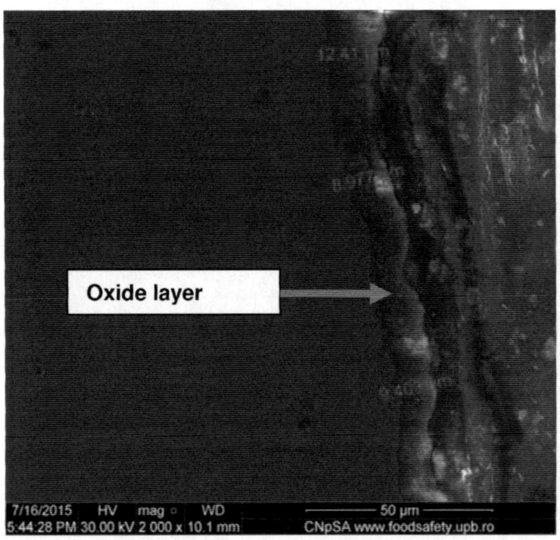

Figure 3. Oxide layers, rich in Al, O and hafnium (sample NUC 8) (2000x).

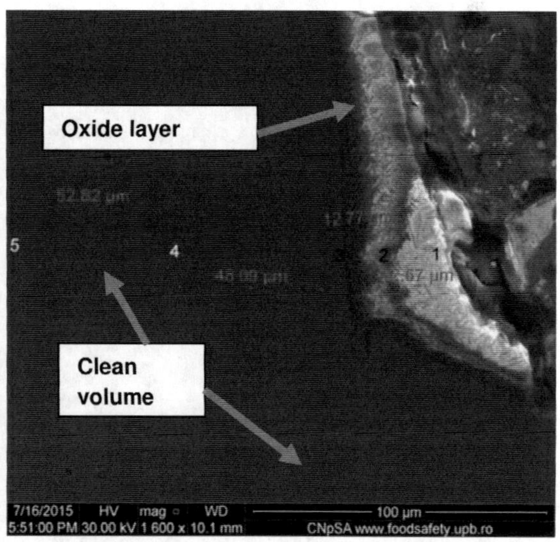

Figure 4. Peripheral oxide layer and clean volume (sample NUC 8) (1600x).

IOP Conf. Series: Materials Science and Engineering **133** (2016) 012016 doi:10.1088/1757-899X/133/1/012016

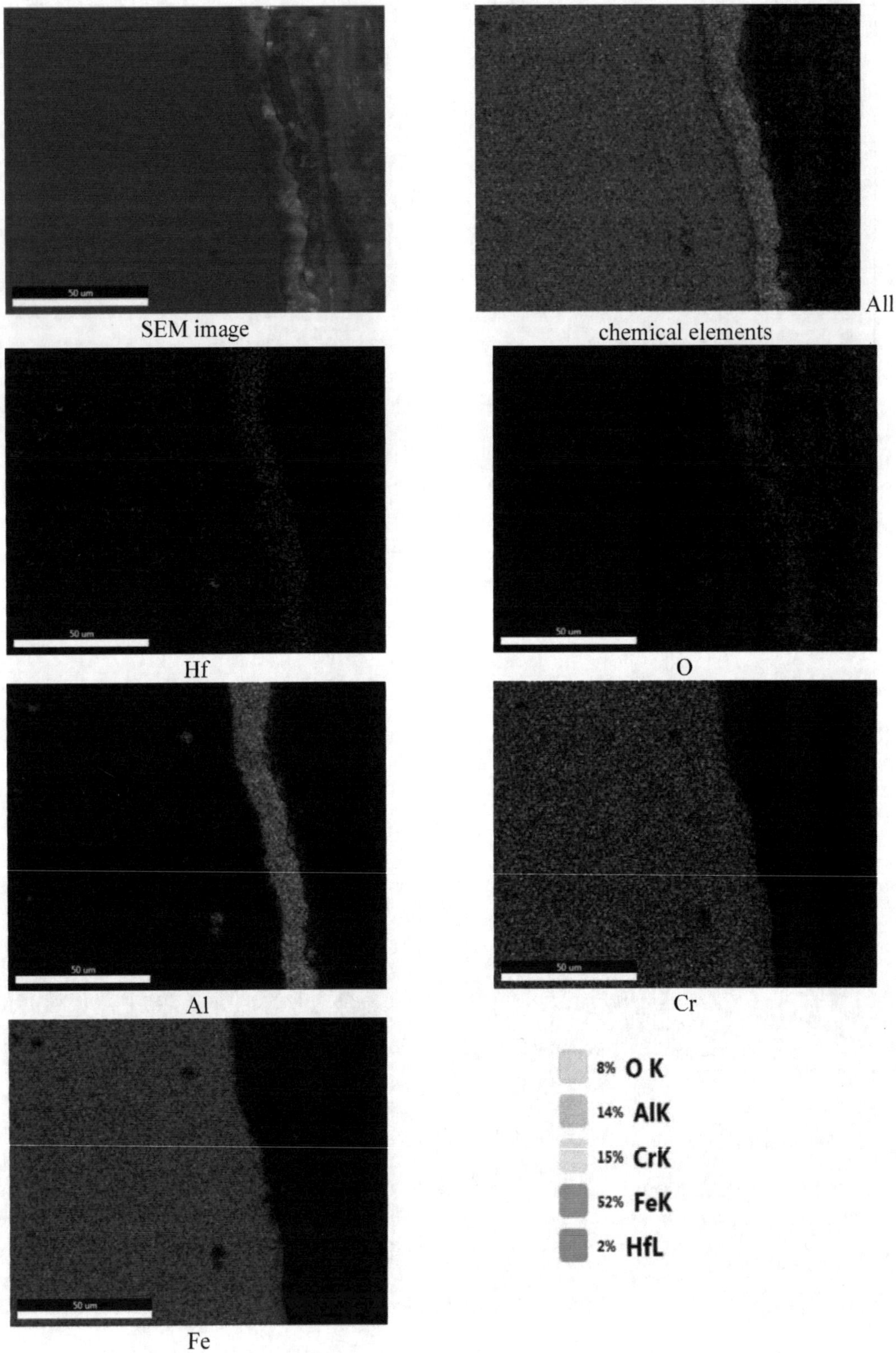

Figure 5. Distribution of chemical elements into peripheral oxide layer for alloy NUC 8.

3.2. Microhardness measurements

Microhardness measurements were made using a Shimadzu HMV 2TE testing machine, with a measuring force of 1.9614 N, a period of 10 seconds and an extended relative uncertainty of 1%. Measurement results of micro-hardness for the three samples analyzed are presented in table 3.

Table 3. Microhardness values ($HV_{0.2}$).

Sample	Individual values*					Average value, $HV_{0,2}$
NUC8	156	162	154	161	151	157
NUC9	157	158	154	148	154	154
NUC10	149	156	154	156	156	154

* The measurements were carried out on the cross sections of mini-ingots, in diagonal line through center zone, with distances between indentations of minimum 1000 microns.

Primary analysis of experimental data shows that there is a quasi-constancy of the micro-hardness values, which reflects an increased homogeneity of all samples obtained in VAR furnace. For all the samples, one can observe that there is a narrow distribution of micro-hardness values, in the range of 154-157 HV0.2, which is solely due to the influence of the alloying elements in the metal alloy composition and disposition homogeneous constituents in the metal matrix. These results are similar to other results obtained for the same class of alloys [14].

4. Conclusions

This research is aimed at obtaining and characterizing the alloy FeCrAl, microalloyed with hafnium, potentially usable in nuclear power plant (generation 4R, type LRF).

Compositional analysis performed in the central area of the samples (FeCrAl microalloyed with 0.5%wt Hf, 1%wt Hf and 1.5%wt Hf) shows a relatively similar composition comparative with technological calculations performed. However, despite the different amount of hafnium added in metallic matrix (0.5%, 1.0% and 1.5%), this element is not detected in the volume of the sample. A large amount of hafnium was detected in the peripheral layer, due to its high affinity for oxygen, by quickly migrating during the elaboration of the alloy.

The hardness values are in the range of 154-157 HV0.2, normal limits for these materials. This is exclusively due to the influence of alloying elements in the metal alloy composition and due to homogenous arrangement of the metallographic constituents in the metal matrix.

Microstructural analysis of the different areas of the samples shows that the peripheral areas contain complex oxides of elements aluminum and hafnium, whose proportion is given by the concentration of these elements in the alloy composition. Analysis of the chemical composition in the peripheral zone of the samples indicates that the content of hafnium increase with increasing this element content in the alloy. The presence of this element in the center of the sample wasn't detected, due to its migration to the peripheral zone, where the amount of oxygen dissolved in the sample is higher.

For less than 1.5%wt Hf content, this element is present only in superficial layer of oxide. Based on these observations, it can be concluded that alloys of the FeCrAl class can be used at high temperatures in an oxidizing atmosphere, due to their capacity to form a protective and adherent oxide layer, rich in chemical elements such as aluminum and hafnium.

Acknowledgments

The research work was financially supported by the Romanian National Program for Research in the framework of the Project No." PCCA 243/2014 *Advanced Metallic Materials used for the New Generation of Nuclear Power Plant* 4R – *NUCLEARMAT*.

References

[1] Allen T R and Crawford D C 2007 Lead-Cooled Fast Reactor Systems and the Fuels and Materials Challenges *Science and Technology of Nuclear Installations*, (Article ID 97486)

[2] Cedergren M and Göransson K 2004 Ferritic stainless steel for use in high temperature applications (United States Patent 6773660)

[3] Cairns R L and Benjamin J St 1976 Iron-chromium-aluminum alloys with improved high temperature properties (United States Patent 3992161)

[4] Aggen G and Borneman P R 1983 Iron-chromium-aluminum alloy and article and method therefor (United States Patent 4414023)

[5] Uehara T, Minagi Y and Inoue K 2002 Ferritic Fe-Cr-Ni-Al alloy having exellent oxidation resistance and high strength and a plate made of the alloy (United States Patent Application 20020124913)

[6] Aggen G and Borneman PR 1982 Iron-chromium-aluminum alloy and method therefor (United States Patent 4414023)

[7] Pillis M F, Correa O V, De Araújo E G and Ramanathan L V 2008 Oxidation Behavior of FeCr and FeCrY Alloys Coated with an Aluminium Based Paint *Materials Research* **11(3)** pp 251-256

[8] Allen R E 1971 Strengthening of Fe-Cr-Al-Y Oxidation Resistant Alloys *General Electric CO Cincinnati Ohio Material and Process Technology*

[9] Jianu A 2010 Corrosion barriers for in-core components of nuclear reactors *Conference "Diaspora"* Bucureşti

[11] Jianu A, Weisenburger A, Heinzel A, Fetzer R, Delgiacco M, An W, Mueller G, Voiculescu I and Geantă V 2011 Alumina scale formation on FeCrAl-alloys exposed to 400-600^{0}C in oxygen containing liquid lead (E11-P-1-12 (1961/1/1) *European Congress on Advanced Materials and Process EUROMAT 2011* France

[12] Liu C A, Humphries M J and Krutenat R C 1983 Production of FeCrAlY and CoCrAlY coatings by laser surface fusion and their oxidation behavior *Thin Solid Films* **107(23)** pp 269-275

[13] Liu C A and Humphries M J 1984 Effects of Process Parameters on Laser Surface Modification *Laser Institute of America* **38** pp 108-117

[14] Kadolkar P and Dahotre N B 2002 Variation of structure with input energy during laser surface engineering of ceramic coatings on aluminum alloys *Applied Surface Science* **199(1-4)** pp 222-233

[15] Geantă V, Voiculescu I, Ştefănoiu R and Jianu A 2011 Influence of chemical composition of FeCrAl alloys on the Microhardness *Metalurgia International* **XVI (50)** pp153-156

[16] Pilone D, Felli F and Bernabai U 2009 FeCrAl Alloys Produced by Roll Bonding and Annealing of Al (RE)-Clad Stainless Steel: How addition of Hf and Zr Affects Their Oxidation Behaviour *Trans Tech Publications* **604-605** pp 133-137

[17] Lu Z, Faulkner R G, Sakaguchi N, Kinoshita H, Takahashi H and Flewitt P E J 2006 Effect of hafnium on radiation-induced inter-granular segregation in ferritic steel *Journal of Nuclear Materials* **351** pp155–161

Hydrogenation of Liquid Styrene by Alumina Supported Nickel Catalysts: Comparison between Classical and Non-Classical Methods

Y. C. Tan, N. H. H. Abu Bakar, W. L. Tan, M. Abu Bakar
School of Chemical Sciences, Universiti Sains Malaysia, 11800 USM, Penang, Malaysia

E-mail: hana_hanif@usm.my

Abstract. Almina supported Ni catalysts (Ni/Al_2O_3) with different Ni weight percentages (wt%) were prepared via classical and non-classical methods. All samples were prepared via impregnation technique. The samples prepared via non-classical methods were reduced using KBH_4 as the reducing agent. The catalysts were tested for the hydrogenation of styrene in liquid phase. Optimum activation conditions for the hydrogenation reaction were found to be 633 K for 2 hours. Comparison of the catalytic reactivity for all catalysts at these activation conditions showed that catalysts prepared via classical methods exhibited better activity. Furthermore the 7.6wt% Ni-Al_2O_3/C showed enhanced activity when compared to the 5.9wt% and 13.8wt% Ni-Al_2O_3/C catalyst. This phenomenon is mainly attributed to the type of Ni active sites available on the catalyst. The surface properties of the catalysts investigated via H_2-temperature programmed reduction (H_2-TPR), H_2-chemisorption and H_2-temperature programmed desorption (H_2-TPD) confirm this.

1. Introduction

Olefins which are contaminants formed during the cracking process, are especially found in petrochemical and petroleum refining industries [1]. To accelerate the removal of these reactive unsaturated components in the gas stream, catalytic hydrogenation reaction is commonly employed. In laboratory research, the hydrogenation of styrene is frequently used as a catalytic test model to access the performance of hydrogenation catalysts.

Over the years, the hydrogenation of styrene over a series of catalysts consisting of metal particles such as palladium (Pd) [1, 2], nickel (Ni) [3-5], iron (Fe) [6], copper (Cu) [4], cobalt (Co) [4], ruthenium (Ru) [1], rhodium (Rh) [1] and platinum (Pt) [1] has been reported. Amongst these, Ni is the cheapest and is available in abundance. As such, studies on hydrogenation catalysts based on mono-Ni supported alumina [3], silica [4] and sepiolite [5] as well as bimetallic Ni-Pt/alumina [7], Ni-B/MWCNT [8], Ni-Cu/alumina [9] can be found in the literature. The reason behind the numerous studies available is basically because the catalytic hydrogenation of a compound can be affected by several factors such as the metal content, type of support and preparation procedures for the catalysts. As an example, different supports can exert different metal-support interactions which can influence catalytic reactivity. In addition, the method in which the catalysts are prepared can also influence the metal-support interaction differently. In general, catalysts can be synthesized through classical and non-classical methods. Nowadays, the use of non-classical methods have gained popularity as it offers a faster preparation procedure and can prevent aggregation of particles as a result of not needing high temperature pre-treatments for long durations of time. There are many types of non-classical techniques that have been proposed for the synthesis of supported metal catalysts. Among the various non-classical methods available are those employing reducing agents such as KBH_4 [10] or hydrazine [11] and the sol-gel method [12]. Although this method has its advantages, the classical methods may be a better option as it can remove poisonous elements in a catalyst via calcinations at high temperatures. Notably, many researchers are still employing this method to prepare catalysts.

Hence, in this work, a series of Al_2O_3 supported Ni catalysts have been prepared via both the classical and non-classical methods using the impregnation technique. Here, potassium borohydride, KBH_4 was used as the reducing agent for the catalysts prepared via the non-classical method. The activity of Ni-Al_2O_3 catalysts with different Ni wt% were investigated for hydrogenation of styrene in liquid phase. Catalysts were characterized using imaging and adsorption techniques to understand the phenomenon.

2. Materials And Methods

2.1 Materials

All materials were used as received. Aluminum oxide (Al_2O_3, purity 99.99%) was obtained from Aldrich Chemical Company, Inc., nickel (II)-acetate-4-hydrate, ($(CH_3COO)_2Ni{\cdot}4H_2O$, purity 99.998%) was obtained from Merck Sdn Bhd and potassium borohydride, (KBH_4, purity $\geq$ 96.5%) was acquired from Sigma Aldrich. Ethanol (AR grade, 99.7%) was purchased from Qrec (Asia) Sdn Bhd while Hydrofluoric acid (HF, 40%) was supplied by Univar. Nitric acid (AR grade, HNO_3 69% Grade) was obtained from Brightchem Sdn Bhd, 1-butanol (AR grade, $C_4H_{10}O$, purity > 99%) was acquired from Al Ilham Utama Enterprise, styrene, (C_8H_8, purity $\geq$ 99.6%) was obtained from Merck Schuchardt OHG while acetone (C_3H_6O, purity $\geq$ 99.5%) was purchased from Intrachem Sdn. Bhd.

All the gases were purchased from Malaysian Oxygen Berhad unless otherwise stated. Purified hydrogen gas (H_2, purity 99.999%), purified argon gas (Ar, purity 99.999%) and diluted hydrogen gas (H_2, 1000 ppm, purity 99.999%) from Air Liquid were used for characterization tests while nitrogen gas (N_2, purity 99.995%) was used for the preparation of catalysts *via* the non-classical method.

2.2 Methods

2.2.1. Synthesis of Ni/Al_2O_3 Catalysts via Classical Methods (Ni/Al_2O_3-C)

Typically, for the preparation of 7.6 wt% Ni-Al_2O_3/C, 5 g of Al_2O_3 was weighed and added into 17 mL of a 0.5 M $Ni(CH_3COO)_2{\cdot}4H_2O$ solution. Subsequently, 23 mL of distilled water was added. The solution was left to homogenize for 8 days. The mixture was then dried in an oven overnight at about 60°C. Similar procedures were repeated to prepare other Ni-Al_2O_3/C catalysts however the amount of the Ni salt solution and distil water was varied as in Table 1 to obtain the 13.8 wt% and 5.9 wt% Ni/Al_2O_3/C catalysts.

Upon drying, each sample was calcined. Typically, the sample was heated in air from room temperature to 373 K at the rate of 5 K min^{-1} then held for 1 hour. Subsequently, the temperature was raised to 873 K while heating at the rate of 10 K min^{-1}. Upon reaching 873 K, the temperature was held for 4 hours.

2.2.2 Synthesis of Ni/Al_2O_3 Catalysts via Non-Classical Methods (Ni/Al_2O_3-NC)

Similar procedures as the classical methods were repeated to synthesize the non-classical catalysts. However 10 mL of 2.5 M KBH_4 (Table 1) was added to the samples after the homogenizing stage. The solution turned dark immediately upon addition of KBH_4. The solution was left to stir for about 15 min before filtering. Catalysts were washed thoroughly with distilled water then dried in a vacuum oven overnight at about 348 K.

IOP Conf. Series: Materials Science and Engineering **133** (2016) 012017 doi:10.1088/1757-899X/133/1/012017

Table 1: Precursors for preparing the Ni/Al_2O_3 catalysts and the respective metal compositions.

Samples Denotation	Volume of Ni Salt Solution (0.5 M), mL	Volume of KBH_4 (2.5 M), mL	Volume of Distilled Water, mL	wt % of Ni
5.9wt% Ni-Al_2O_3/C	8.5	–	31.5	5.9
7.6wt% Ni-Al_2O_3/C	17.0	–	23.0	7.6
13.8wt% Ni-Al_2O_3/C	25.6	–	14.4	13.8
5.0wt% Ni-Al_2O_3/NC	8.5	10	31.5	5.0
8.2wt% Ni-Al_2O_3/NC	17.0	10	23.0	8.2
8.9wt% Ni-Al_2O_3/NC	25.6	10	14.4	8.9

2.3. *Characterization Technique*

The real content of the Ni element was determined by atomic absorption spectroscopy (AAS) using a PerkinElmer Analyst 400. An acid mixture (H_2O:HNO_3:HF = 5:3:8) was added to 0.05 g of a sample while heating until the catalyst was dissolved. The solution was filtered then topped up to 50 mL of distil water.

The morphology of the Ni particles in the catalysts was investigated using scanning electron microscopy (SEM). The catalysts were placed on a carbon tape. Subsequently, samples were sputtered with chromium and then subjected to an FEI Quanta 650 SEM. The samples were viewed using backscattering mode.

The H_2-TPR analysis of the catalysts was carried out using diluted hydrogen gas (a mixture of 1000 ppm H_2/Ar) as reductant with a flow rate 100 mL min^{-1}. As much as 0.2 g of the sample was measured and placed in a U-tube quartz reactor. The samples were heated at a rate of 5 K min^{-1} from room temperature to 973 K. Hydrogen consumption was measured every 2 minutes with an Agilent G3000A Micro gas chromatograph equipped with a thermal conductivity detector.

The H_2-TPD analysis was conducted on similar apparatus as the H_2-TPR analysis. As much as 0.8 g of a sample was placed in a U-tube quartz reactor. The sample was then activated in pure H_2 with a flow rate of 400 mL min^{-1} at 633 K for 2 hours. Subsequently, the samples were cooled to room temperature and switched to purified Ar at a flowrate of 100 ml min^{-1}. The sample was then heated from room temperature to 973 K at a heating rate of 5 K min^{-1}, and hydrogen desorption was measured every 2 minutes.

2.4. *Hydrogenation of Styrene*

The hydrogenation of liquid styrene was carried out to compare the activity of the Ni-Al_2O_3 catalysts prepared via classical and non-classical methods. Typically, as much as 0.2 g of Ni-Al_2O_3 sample was measured and place into a 3-necked round bottom flask. Pure H_2 was flowed into a reaction flask at a flow rate of 100mL min^{-1}. The samples were then activated at 633 K for 2 hours. Other activation conditions such as the temperature (483 to 633 K) and activation time (2 to 4 hours) were also studied to optimize activation of the catalysts.

After activation, the sample was cooled to 373 K. As much as 10 mL 1-butanol followed by 1 mL of styrene was added to the sample. The mixture was homogenized and the extent of hydrogenation was followed using a burette filled with purified hydrogen gas by water displacement. The water level in the burette was recorded every 5 min.

3. Results and Discussion

3.1. *Synthesis of Al_2O_3 Supported Ni Catalysts via Classical and Non-Classical Methods*

The Al_2O_3 supported Ni catalysts were synthesized via classical as well as chemical reduction method using KBH_4 as the reducing agent. Catalyst denotations including the metal content are presented in Table 1.

The Ni-Al_2O_3/C catalysts were green in color after drying the mixture of Ni salt with Al_2O_3. Upon calcinations in air, the samples changed to grayish in color indicating that oxidation of Ni occurred. These samples changed to black in color after activation of the catalysts in pure H_2 according to equation (1).

$$NiO + H_2 \longrightarrow Ni + H_2O \tag{1}$$

For the Ni-Al_2O_3/NC catalysts, samples were also green in color after homogenizing the Ni salt with Al_2O_3. However upon reduction with KBH_4, the samples changed to black immediately. Bubbles were produced, indicating the release of hydrogen gas according to equation (2) [13].

$$Ni^{2+} + KBH_4 + 4H_2O \longrightarrow Ni + KB(OH)_4 + 3H_2\uparrow + 2H^+ \tag{2}$$

3.2 Hydrogenation of Styrene

The optimum conditions for the hydrogenation of styrene were obtained using the 7.6wt% Ni-Al_2O_3/C catalyst. The sample was first activated to a temperature of approximately 583 K. Activation was then carried out for various durations to determine the optimum activation time. From Figure 1, it can be seen that activation in pure H_2 for 2 hours showed the best results. Generally, the H_2 uptake increased steadily for the first 100 min. After that, the uptake of H_2 slowed down reaching 60 mL after 400 min. When the sample was activated for 3 hours, the H_2 uptake raised steadily also reaching 60 mL at approximately 400 min. In contrast, when the sample was activated for 4 hours, the H_2 uptake increased for the first 80 min then slowly came to a constant at 10 ml of H_2 uptake. The profiles show that the catalyst activated for 4 hours at 583 K exhibited the lowest hydrogenation of styrene. The low activity may be attributed to the extensive heating times, which may have caused the metal particles to migrate and aggregate. This indirectly may have resulted in a decrease in the surface area of the active phase. Thus, there was low surface area available for catalytic reaction to take place.

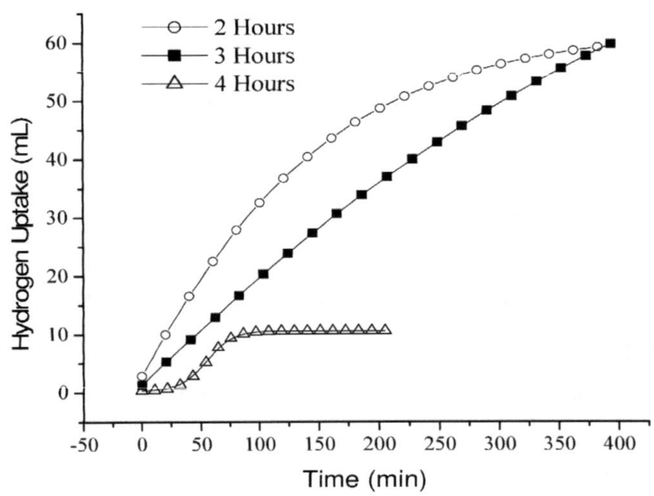

Figure 1:The effect of activation time on the hydrogenation of styrene.

Considering that the 7.6wt% Ni-Al_2O_3/C showed better catalytic activity when activated for duration of 2 hours, this activation time was employed to study the effect of activation temperature. From Figure 2, it can be seen that the catalyst activated at 663 K showed better H_2 uptake compared to catalysts activated at other temperatures. The H_2 uptake increased steadily until reaching 100 mL in approximately 150 min. After this, the uptake continues however at a much slower rate until reaching a maximum of 110 mL. The H_2 uptake for catalysts activated at 583 K and 483 K were also compared. It is seen that activation at 583 K showed a better H_2 uptake than catalysts activated at 483 K. This maybe because when the sample was heated at the temperature of 483 K, the temperature is insufficient to totally reduce the NiO particles. Here, only reduction of oxides on the surface of the Ni species occurred. When the temperature was increased to 583 K, activity was better because at higher temperature more NiO particles can be reduced. Catalysts which were activated at 663 K had the best activity because the temperature is sufficient to reduce both the surface and bulk oxides of the particles. Therefore, it has been shown that in the temperature and time range investigated, the optimum activation condition for this hydrogenation test is 583 K for 2 hours.

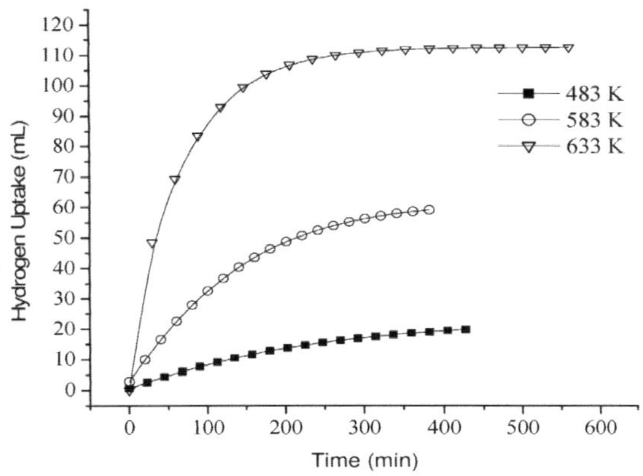

Figure 2: The effect activation temperature for the hydrogenation of styrene.

These conditions were applied to all the classical and non-classical catalysts. Results are presented in Figure 3. From Figure 3, it is obvious that the 7.6wt% Ni-Al_2O_3/C catalyst demonstrated the best catalytic activity. This is followed by the 13.8wt% and 5.9wt% Ni-Al_2O_3/C catalysts. The 13.8wt% and 5.9wt% Ni-Al_2O_3/C showed similar maximum activity whereby maximum H_2 uptake is 35 mL. However, the 5.9wt% Ni-Al_2O_3/C took a longer time to reach the maximum activity, which was at approximately 200 min while 13.8wt% reached the maximum activity at approximately 80 min. Among the samples prepared via the non-classical method, 8.2wt% Ni-Al_2O_3/NC showed the best catalytic activity. For this catalyst, H_2 uptake reached 40 mL. Other catalysts prepared via non-classical methods were almost inactive. Hence as an overall, the graph showed that the classical method had a much better catalytic activity than the non-classical method. In general, this may be due to the fact that in classical method, all samples were calcined. As a result, this removes any impurities which may poison and indirectly affect the catalysts. In contrast, catalysts prepared via non-classical methods were reduced using KBH_4. Even traces of boron left in the catalyst may have poisoned the catalysts, resulting in the extremely low activity observed.

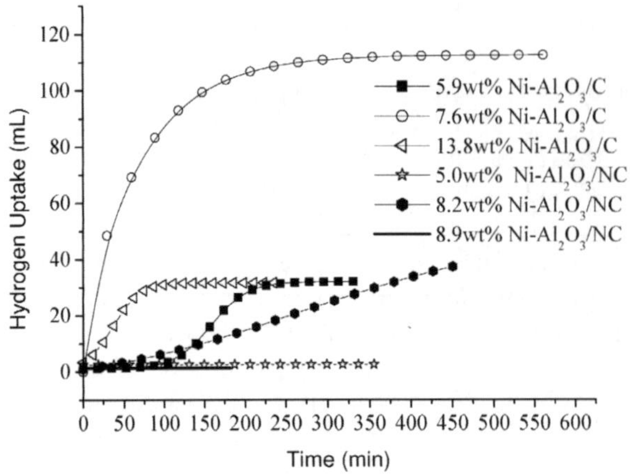

Figure 3: The effect of different wt% of Ni-Al_2O_3 catalyts prepared via classical and non-classical method for the hydrogenation of styrene.

3.3 Characterization of Ni-Al_2O_3/C Catalysts

To understand the factors which contributed to the enhanced activity demonstrated by the catalysts prepared via classical methods, this series of catalysts were characterized to gain knowledge on the morphology of the active phase and its surface properties.

The SEM analysis were conducted for the 5.9 wt%, 7.6 wt%, 13.8wt% Ni-Al_2O_3/C catalysts. The images are shown in Figure 4. The Al_2O_3 support appears as greyish particles while the Ni species are evidenced by the white particles. It can be seen that very few particles attributed to Ni species are observed in the 5.9wt% Ni-Al_2O_3/C catalyst. Furthermore the sizes of the particles seem to be very small. In contrast, larger Ni species are available in the 7.6 wt% and 13.8wt% Ni-Al_2O_3/C catalysts. In the case of the 13.8wt% Ni-Al_2O_3/C catalyst, aggregation of the Ni species was also observed. Thus this indicates that the 13.8wt% Ni-Al_2O_3/C may give rise to a lower surface area for the active phase when compared to the 5.9 wt% and 7.6wt% Ni-Al_2O_3/C catalyst.

The H_2-TPR analysis was conducted to determine the reducibility of the oxide particles and to give an idea on the size of the particles in the Ni-Al_2O_3/C catalysts. The H_2-TPR profiles are presented in Figure 5. It can be seen that the 13.8wt% Ni-Al_2O_3/C exhibits a strong peak at 400 K. According to Kim and co-workers [14], this peak can be attributed to bulk oxides which do not interact with the support. The availability of higher Ni content in this catalyst enabled the growth of the particles, which increased the size of particles [14,15]. This can result in a weaker interaction between the oxide particles with the support. Therefore, this gave rise to the sharp peak observed in the 13.8wt% Ni-Al_2O_3/C. Observation of the 5.9 wt% and 7.6wt% Ni-Al_2O_3/C catalysts on the other hand does not show a peak in this region. This may be due to the low Ni content as compared to the 13.8wt% Ni-Al_2O_3/C. Further inspection of the H_2-TPR profiles reveal that all the Ni-Al_2O_3/C catalysts showed one broad peak from 600 - 811 K and a sharp peak at around 900 K. It seems that the intensity of the broad peak is significantly lower for the 7.6wt% Ni-Al_2O_3 when compared to the 5.9 wt% and 13.8wt% Ni-Al_2O_3/C catalysts. This depicts that metal-support interaction occurs to a greater extent in the later catalysts, which reduces the capability of the Ni^{2+} ions to decrease to metallic form. In contrast, the sharp peak at higher temperatures of about 900 K can be assigned to Ni^{2+} ions in strong interaction with the alumina support.

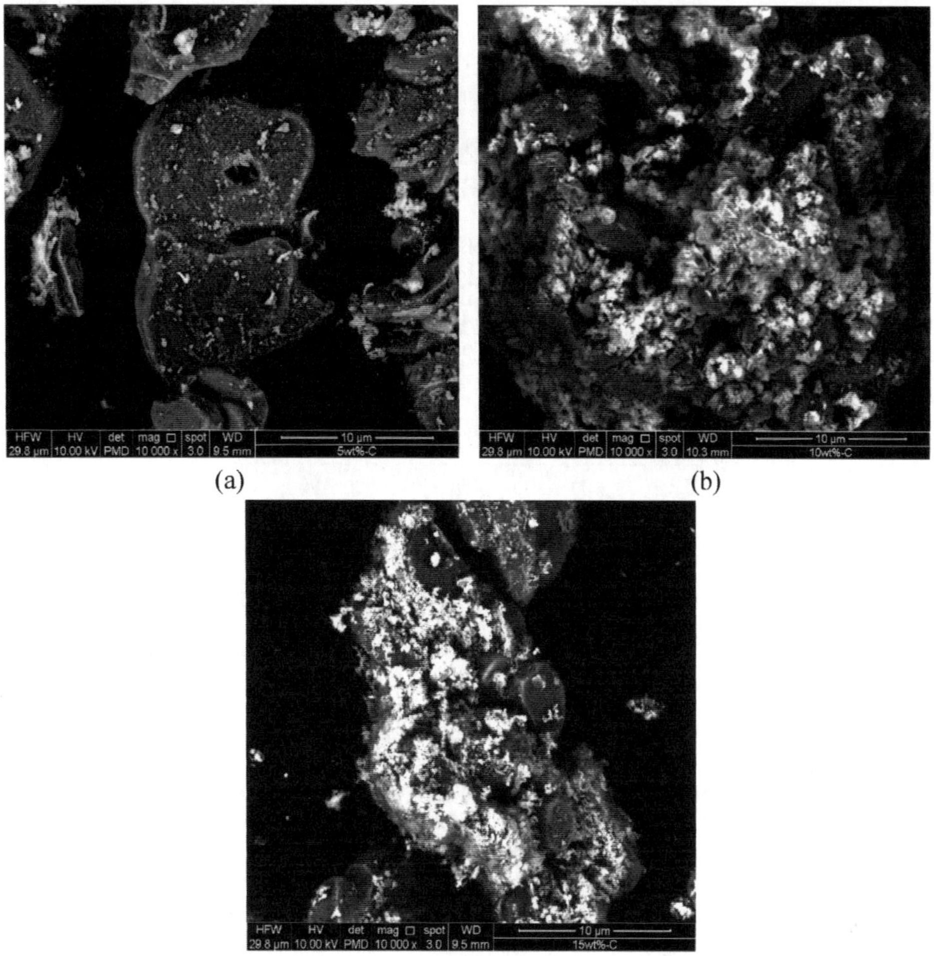

(a) (b)

(c)

Figure 4: SEM images of (a) 5.9wt% (b) 7.6wt% and (c) 13.8wt% Ni-Al_2O_3/C.

Subsequently, another interesting observation is that the 7.6wt% Ni-Al_2O_3 catalysts showed an additional prominent peak at 592 K. This peak is only observed in this catalyst and indicates some additional characteristic of the catalyst. The availability of this peak depicts that Ni species are in weaker interaction with the support or smaller NiO aggregates exists in the catalyst when compared to the 5.9 wt% and 13.8wt% Ni-Al_2O_3.

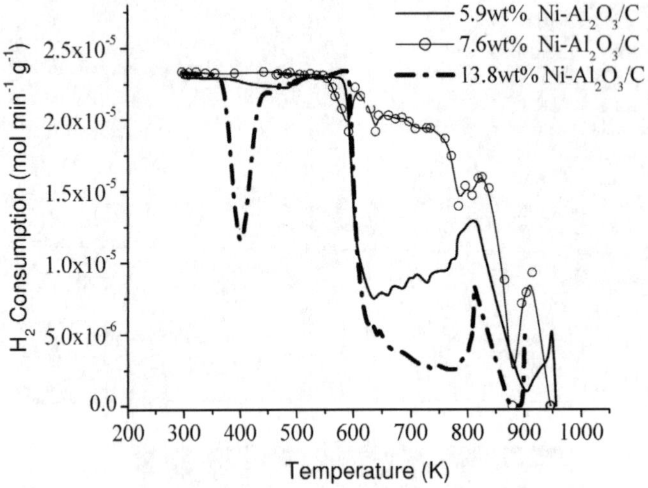

Figure 5: H_2-TPR profiles of Ni-Al_2O_3/Ccatalysts.

H_2 - chemisorption studies of the Ni-Al_2O_3/C catalysts revealed that the 5.9wt%, 7.6wt% and 13.8wt% Ni-Al_2O_3 catalysts adsorbed as much as 2.74 x 10^{-6} mol g_{act}^{-1}, 1.80 x 10^{-6} mol g_{act}^{-1} and 2.15 x 10^{-6} mol g_{act}^{-1} of H_2 correspondingly. This gives a total metallic surface area of 21.47 m^2 g_{cat}^{-1}, 14.11 m^2 g_{cat}^{-1} and 16.85 m^2 g_{cat}^{-1} for each of the catalysts respectively. The results show that the 7.6 wt% Ni-Al_2O_3/C gives rise to the lowest amount of H_2 adsorbed and lowest metallic surface area. This does not correlate with the high activity demonstrated by this catalyst and reveals that the catalytic activity is not directly influenced by the surface area of the active phase.

H_2-TPD analyses were also carried out for this series of catalysts to gain an understanding of the strength of H_2 adsorption on the metal phase. Profiles of the catalysts are presented in Figure 6. As can be seen, the 5.9wt% Ni-Al_2O_3/C gives rise to three main peaks positioned at 385 K, 442 K and 709 K respectively. Previous work has shown that the peaks at lower temperatures of 385 K and 442 K are attributed to H_2 weakly adsorbed onto Ni active sites [9]. In contrast, the peak arising at 709 K can be assigned to several factors such as H_2 strongly bonded to the Ni particles, hydrogen retained at the metal-support interphase or H_2 spillover [16]. In contrast to the 5.9wt% Ni-Al_2O_3/C, the 7.6 wt% Ni-Al_2O_3/C exhibits only two H_2 desorption peaks. The peaks have maximum temperatures at approximately 401 K and 506 K. As such, only H_2 weakly adsorbed onto the Ni active sites exists in this catalyst. Finally, the 13.6 wt% Ni-Al_2O_3/C catalyst exhibits a strong desorption peak at 414 K with a shoulder at lower temperatures of about 346 K as well as a H_2 desorption tail ranging from 435 to 780 K. Similarly to the catalysts with low Ni contents, the high desorption at low temperatures is attributed to H_2 weakly bonded to the Ni particles. Even so, the desorption tail towards higher temperatures indicates the availability of H_2 which are moderately and strongly adsorb onto the Ni active sites, hydrogen at the metal-support interphase or hydrogen spillover. Based on these results, it is obvious that the 5.9wt% and 13.6 wt% Ni-Al_2O_3/C show similar characteristics whereby H_2 desorption is seen at higher temperatures of more than 550 K. The availability of these active sites may have resulted in the lower catalytic activity observed when compared to the 7.6 wt% Ni-Al_2O_3/C where only low temperature desorption is observed. This correlates well with the H_2-TPR findings. Another interesting observation is that the 13.6 wt% Ni-Al_2O_3/C exhibits very high H_2 desorption (3.044 x 10^{-4} mol g_{cat}^{-1}) in the temperature range of 300 to 435 K. This phenomenon may have contributed to the initial rapid H_2 uptake during the hydrogenation of styrene when compared to the 5.9wt% Ni-Al_2O_3/C.

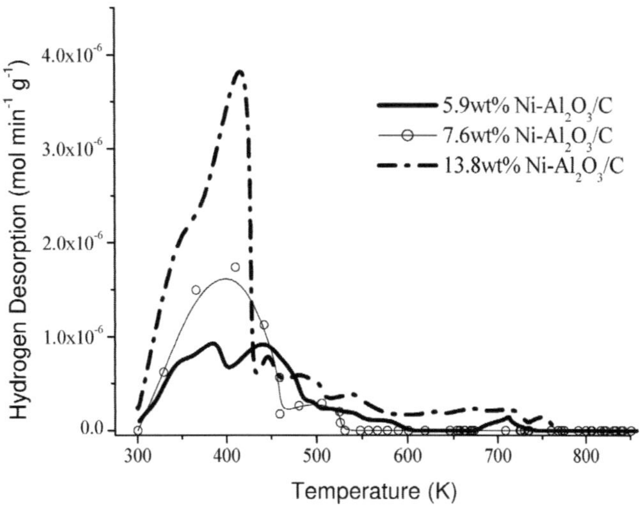

Figure 6: H_2-TPD profiles of Ni-Al_2O_3 catalysts prepared via classical methods.

4. Conclusion

A series of studies was conducted to investigate the optimum activation conditions for the hydrogenation of liquid styrene. The optimum conditions in the range of this study were determined to be 360°C for 2 hours. The effect of classical and non-classical preparation methods were studied at these conditions. The Ni-Al_2O_3/C catalysts demonstrated better catalytic activity when compared to the Ni-Al_2O_3/NC catalysts. The 7.6wt% Ni-Al_2O_3 classical catalyst gave rise to the best activity because it exhibited weaker H_2 adsorption on the Ni particles when compared to the 5.9wt% and 13.8wt% Ni-Al_2O_3/C catalysts. Furthermore, results show that the activity of this catalyst is largely dependent on the type of active site available and not the surface area of the catalyst.

Acknowledgements

The authors would like to show their gratitude to the Malaysian Ministry of Higher Education for the financial support in the form of the Fundamental Research Grant Scheme (FRGS) grant (203/PKIMIA/6711419).

References

1. Badano J M, Quiroga M, Betti C, Vera C, Canavese S and Coloma-Pascual F 2010 *Catal. Lett.* **137** 35
2. Harraz F A, El-Hout S E, Killa H M and Ibrahim IA 2012 *J. Catal.* **286** 184
3. Nijhuis T A, Dautzenberg F M, Moulijna J A 2003 *Chem. Eng. Sci.* **58** 1113
4. Corvaisier F, Schuurman Y, Fecant A, Thomazeau C, Raybaud P, Toulhoat H and Farrusseng D 2013 *J. Catal.* **307** 352
5. Anderson J A, Daza L, Damyanova S, Fierro J L G and Rodrigo MT 1994 *Appl. Catal. A* **113** 75
6. Paˆrvulescu V I, Filoti G, Paˆrvulescu V, Grecu N, Angelescu E and Nicolescu L V 1994 *J. Mol. Catal.* **89** 267

7. Betti C, Badano J, Maccarrone M J, Mazzieri V, Vera C, Quiroga M 2012 *Appl. Catal. A* **435-436** 181
8. Liu Z, Li Z, Wang F, Liu J, Ji J, Park K C and Endo M 2012 *Mater. Res. Bull.* **47** 338
9. Muhammad S, Tan W L, Abu Bakar N H H, Abu Bakar M and Bettahar M M 2016 *Reac. Kinet. Mech. Cat.* (DOI 10.1007/s11144-016-0980-2)
10. Yu W, Xu J, Ma H, Chen C, Zhao J, Miao H and Song Q 2010 *Catal. Comm.* **11** 493
11. Hussain M S and Haque K A 2010 *J. Sci. Res.* **2** 313
12. Gronchi P, Kaddouri A, Centola P and Del Rosso R 2003 *J. Sol-Gel Sci. Tech.* **26** 843
13. Büchner W, Niederprüm H, Bayer A G, Leverkusen, and W. Germany. Pure & Appl. Chem. 49 (1977) 733-743.
14. Kim P, Joo J B, Kim H, Kim W, Kim Y, Song I K and Yi J 2005 *Catal. Lett.* **104** 181
15. Wen G, Xu Y, Xu Z and Tian Z 2009 *Catal. Lett.* **129** 250
16. Chettibi S, Wojcieszak R, Boudjennad E H, Belloni J, Bettahar M M and Keghouche N 2006 *Catal. Today* **113** 157

IOP Conf. Series: Materials Science and Engineering **133** (2016) 012018 doi:10.1088/1757-899X/133/1/012018

Investigation on the mechanical and physical properties of TPU/PVC blend as potential material of bedpan

Hakim M.I[1], Shahnaz S. S. B.[1], Luqman M.[1], Anis N.I.[1], Shuhaida Y.[2], Zaharah S.A.[3], Baidi M.B.[3], Yu C.W.[3] and Kathiresan V. S.[3]

[1]School of Materials Engineering, Universiti Malaysia Perlis, Taman Muhibah, Jejawi, 02600 Arau, Perlis, Malaysia
[2]Schools of Mechatronic Engineering, Universiti Malaysia Perlis, Pauh Putra Campus, 02600 Arau, Perlis, Malaysia
[3]Faculty of Allied Health Professions, Aimst University, Jalan Bedong - Semeling 08100 Bedong, Kedah Darul Aman, Malaysia

E-mail: mohdhakim5888@ymail.com

Abstract. The purpose of this study is to determine the impact properties, chemical resistance and morphology of TPU/PVC blend in order to be used as potential material to fabricate bedpan. TPU and incorporation of PVC in TPU was done by melt-mixing method. The blends of TPU/PVC from melt mixing were then undergoing hot press compression moulding to produce sample with 3mm thickness. The blend sample was subjected to drop weight impact test, chemical resistance test and SEM test. Results from impact test showed that 30/70 TPU/PVC has recorded the maximum velocity impacted by 6 J of 1.8 m/s at 97ms while 50/50 TPU/PVC has the highest force (impact strength) of 0.95kN compared with other samples. The result also found that 50/50 TPU/PVC blend has lowest weight change, indicated that the blend has good resistance toward NaOH. HCl has a stronger effect on TPU/PVC blend compared to NaOH for all blend composition where 10/90 TPU/PVC showed highest resistance towards HCl. The micrograph shows that 50/50 TPU/PVC exhibited both smooth surface, indicating the brittleness of PVC and rough and the plasticity fractured surface, indicating the ductility of TPU features.

1. Introduction

In recent years, the utilization of polymers in production of medical product are being increases. The advantages of these materials consist of a low cost manufacturing processes which can led to many products of polymer materials in medical application [1]. Polymers also can be mixed or copolymerized with or without any chemical bonding between them that also called polymer blend. Polymer blending is done to achieve commercially variable products through either superior properties or lower cost as compared to the virgin material. Various advantages accomplished by developing polymer blending such as improvement of the properties, reduction of the material cost, improved processability of polymer and extended service temperature range [2]. One example of polymer blend is blending of thermoplastic polyurethane (TPU) and polyvinylchloride (PVC) which can be obtained by melt-blending the materials that have been used as a coated wire, cable, tubing, hoses, upholstery, apparel fabric, film, and others. Soft grades of TPU are often used as polymeric plasticizers in PVC. On the other hand Polyurethanes (PU) exhibit excellent mechanical properties and good low temperature properties. Thus, the versatility of PU chemistry allows thermoplastic polyurethanes (TPU) to be tailor made for various polymer blends [3].

Shashidhara et. al. (2010) compared the mechanical and thermal properties PVC/TPU blends with variable weight ratio prepared using single and twin screw extruder. Two grades of TPUs different in hardness (Shore A hardness 66 and 85) were used in making the blends. The tensile strength of PVC/TPU-1 and PVC/TPU-2 blends obtained from twin screw extruder is higher than that obtained from single screw extruder. The effect of addition of TPU to PVC which is already plasticized with DOP was expected to decrease the modulus of the blends [4]. Study by Hernandez et. al. (2002) showed that the PVC plasticized with common Dioctyl phthalate (DOP) shows a considerable higher elongation at break and lower Young modulus than PVC mixed with TPU. A remarkable reduction in the viscosity of the polymer blends observed by these authors indicates that these compounds worthy of attention from a technological point of view [5]. Pita et al studied the influence of processing condition on the mechanical properties of plasticized PVC and PVC/TPU blend. It is suggested that the structure of TPU/PVC mixtures could be maintained efficiently up to 50 phr TPU content. Processing of the corresponding melts is obviously very interesting and promising in the study of polymeric materials [6].

Pielichowski et. al. (2000) has examined the thermal behaviours of compatible polyvinyl chloride (PVC)/chlorinated polyurethane (CPU) blends. Decomposition route was found to be of a two-step process; the decisive degradation stage was found to be a result of parallel reactions of PVC and CPU decomposition. The enhanced thermal stability of several PVC/CPU blends can be explained in terms of specific dipole-dipole interactions whose intensity may, however, be lowered by diffusion conditions of volatile products through the polymer matrix [7].

The purpose of this research is to determine the potential of TPU/PVC blend as material for fabricating bedpan. The effect of TPU loading percentages on the impact properties, chemical resistance and morphology of TPU/PVC blend will be determined. The drop weight impact test can give the information on impact properties of the TPU/PVC blend from the graph of force against displacement, and velocity against time. Chemical resistance testing were using sodium hydroxide (NaOH) and hydrochloric acid (HCl) as chemical reagent to get the information of the effect of chemical to dimensions or weight of the TPU/PVC blend. The morphology of the TPU/PVC blend was study using Scanning electron microscopy (SEM).

2. Experimental Details

2.1. Materials

The materials used in this studied are commercial polyvinyl chloride (PVC) and thermoplastic polyurethane (TPU). PVC grade Kane vinyl PSM 31 (Degree of polymerization 1230 to 1430 and density 1.4 g/cm3) was obtained from Kaneka (Malaysia) Sdn. Bhd., and TPU grade Estane 58887 (hardness 87 Shore A and density 1.12 g/cm3) acquired from Lubrizol Advanced Material Malaysia Sdn. Bhd. company.

2.2. Sample preparation

TPU and incorporation of PVC in TPU was done by melt-mixing method using Brabender Internal Mixer Plastograph EC machine at different ratios of 10/90 %wt, 30/70 %wt and 50/50 %wt. of TPU/PVC at temperature of 180°C and speed of 60 rpm. Firstly TPU was charged to start the melt mixing. After 2 min PVC was added and mixing was continued for another 6 minutes. At the end of 8 min, the blend was taken out and the sample was obtained. The blends of TPU/PVC sample from melt mixing were then undergoing hot press compression moulding to produce sample with 3mm thickness. The hot press compression machine GT-7014-H model was used for compaction a sample for test. The temperature was set at 180°C and the pressure was set at 15 ton. The mould was preheated for 2 minutes before compression to remove air trapped in the sample. The compression process was done for 5 minutes and then, the sample was cool down for 5 minutes. The hollow rectangular mould was used in the compression moulding has the dimension of 170 mm x 135 mm with thickness 3 mm. The samples were then cut according to the America Society for Testing and Materials (ASTM) standard dimensions for mechanical and physical testing [8].

2.3. Drop weight impact test
A model IM10 drop weight impact testers manufactured by IMATEK was used to produce the instrumented impact loading. For this experiment, the impactor surface was hemispherical with a radius of 12.7 mm and the total mass including impactor and carriage was 9.6209 kg. The test was performed at room temperature with 6 J impact energy level. Five specimens for each of the blend were tested and mean values were calculated. The test specimen size is 150mm x 100mm flat rectangular with thickness of 3mm.

2.4. Chemical resistance test
The chemical resistance test was performed by following ASTM standard test method D543[9]. Two chemical reagents were used in this study; which are sodium hydroxide (NaOH) and hydrochloric acid (HCl). The chemical resistance towards NaOH and HCl was measured through swelling effect and weight loss. The specimens were immersed in the appropriate reagent for a specified time under the room temperature. For the swelling effect, changes in mass were reported for time duration of 24, 72 and 168 hours which represent short, intermediate and long period of the tests. Three sample tests were carried out for each blend sample.

2.5. Scanning electron microscopy(SEM) test
The morphology of the surface of samples was examined by scanning electron microscope (SEM) JEOL JSM-6460LA. The impact fractured sample was cut into small rectangular shape, so the sample can be mounted with the tape on the sample holder.

3. Result and discussion
3.1. Drop weight impact
The impact properties of TPU, PVC and TPU/PVC blend with a ratio of 10/90, 30/70, and 50/50 are used to compare the mechanical properties of TPU, PVC and their blends. Figure 1 shows the peak of velocity against time behaviour at impact energy of 6J. The results showed that the velocity was increased with increasing of time at the beginning of the test before significantly dropped after 10ms and increase back for all samples. TPU sharp drop was noted at 20ms while PVC was noted at 50ms. 50/50 TPU/PVC blend has exhibited intermediate drop between TPU and PVC at 25ms. The maximum velocity has occurred at 97 second where the maximum velocity was recorded by 30/70 TPU/PVC by 1.8 m/s.

Figure 2 shows the peak of force against displacement at impact energy of 6J. The force against displacement curves give information about the impact response of composite materials tested [10]. The displacement of impact test was conducted to determine the ability of the blend to withstand the forces applied vertically from a specified height in the drop load method [11]. The peak of force against displacement shows that the 50/50 TPU/PVC have the highest force (impact strength) by 0.95kN as compared to other samples. There are no perfect rebounding type curve was occurring of the diagram after reaching a maximum force for the test at impact energy of 6J. For 10/90 TPU/PVC and 30/70 TPU/PVC blend, the scattered peak was observed that possibly due to the damage of both specimens. This affects significantly showed that 10/90 TPU/PVC and 30/70 TPU/PVC blends have lower impact strength.

IOP Conf. Series: Materials Science and Engineering **133** (2016) 012018 doi:10.1088/1757-899X/133/1/012018

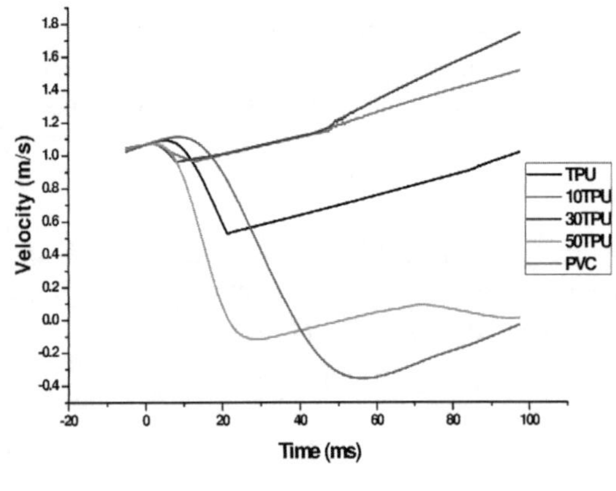

Figure 1. Velocity vs time at impact energy of 6J

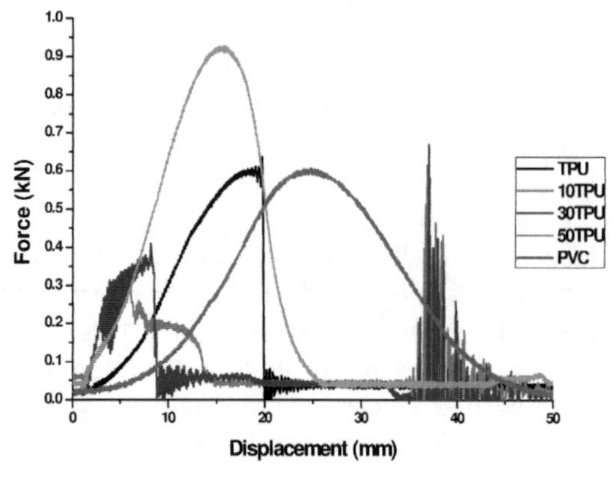

Figure 2. Force vs Displacement at impact energy of 6J

3.2. Chemical resistance

The chemical resistance of TPU/PVC blends against NaOH and HCl are shown in Figure 3 and 4. The chemical resistance effect is determined by the percentage of weight change after immersion. The chemical resistance of the TPU/PVC blends after 24 hours, 72 hours and 168 hours of immersion in NaOH as shown in Figure 3. It was found that a TPU has the highest weight gains which are 0.19% after 24 hours, 0.39% after 72 hours and 0.57% 168 hours followed by PVC for all immersion period. This implies that TPU has lower resistant ability to resist the absorption of NaOH. While, the lowest weight change is found by 50/50 TPU/PVC blend where only 0.19% weight gained over 168 hours of immersion, which indicated that 50/50 TPU/PVC has a good resistance toward NaOH.

Figure 4 representing the chemical resistance effect of the blend after immersion in HCl for 24 hours, 72 hours and 168 hours. PVC showed the highest percentage of 0.93% than other samples followed by TPU with 0.85% and 50/50 TPU/PVC blend with 0.83% of weight gained after immersion for 168 hours. Blend of 10/90 TPU/PVC has highest resistance towards HCl with only 0.69% weight increase after 168 hours. Even though the 50/50 TPU/PVC blend showed a high gain during longer immersion period (168 hours), it has displayed good properties in 24 hours immersion with the lowest value of gain with only 0.15%. The 30/70 TPU/PVC blend showed the highest weight gain recorded at 0.23% after immersing for 24 hours as compared to the other blends. The result shows that the chemical resistance of the TPU/PVC blend has lessened with increasing of TPU content after long exposure. Strong interaction between the plastic material and chemical are preferred when the chemical nature of the materials are similar. For example, if the solvent is highly polar and plastic materials also have polar groups, the plastic material may be influenced by the chemical [12]. Therefore in this case there will be a strong interaction between PVC and HCl, which are both non-polar.

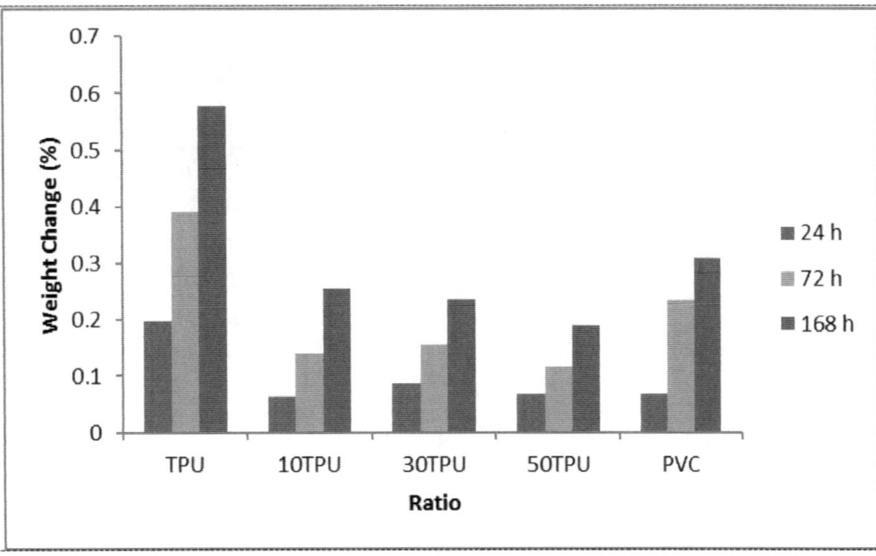

Figure 3. The swelling effect of TPU/PVC blend after immersion in NaOH

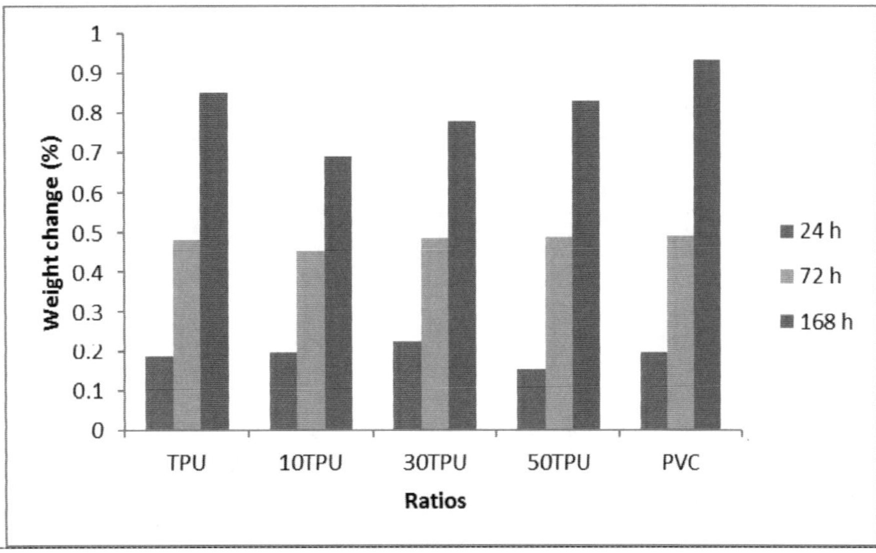

Figure 4. The swelling effect of TPU/PVC blend after immersion in HCl

3.3. Scanning electron microscopy(SEM)

Scanning electron microscope (SEM) is used to observe the fracture surface of the TPU/PVC blend sample in order to evaluate the morphology of the polymer blends. The fracture specimens were acquired from impact test. Figure 5 shows scanning electron micrographs of TPU, PVC and 50/50 TPU/PVC blends at magnification of x2000. PVC displayed smooth surface, indicating the brittle fracture has occurred as shown in Figure 5 (a). The surface of TPU was rough with small particle surface which signifies the ductile fracture while 50/50 TPU/PVC exhibited the mixture of both features with smooth cut with smaller particulate of agglomeration and pores on its surface that indicate in Figure 5 (b) and (c). From the result, it is assured certainly that by adding TPU resulted in reduction of PVC brittle fracture [13]. This suggests the good compatibility between TPU and PVC that could achieve good performance.

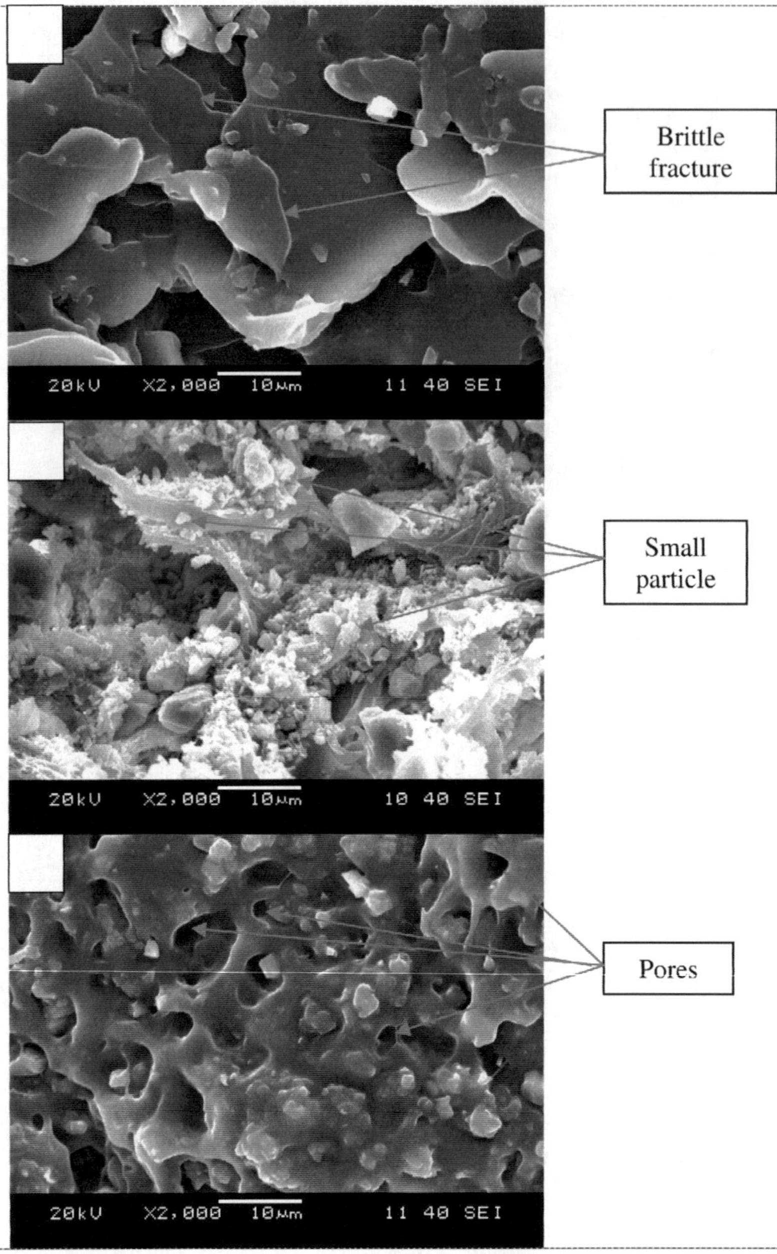

Figure 5. SEM micrograph of (a) PVC, (b) TPU and (c) 50/50 TPU/PVC

4. Conclusion

Impact test showed the maximum velocity has occurred at 97 ms by 1.8 m/s that was recorded by 30/70 TPU/PVC. The peak of force against displacement shows that the 50/50 TPU/PVC have the highest force by 0.95kN as compared to other samples. The immersion of TPU/PVC blend in NaOH found that the 50TPU has lowest weight change of only 0.07 % after being immersed for 168 hours. The highest weight change after immersion in NaOH for 72 hours of immersion was TPU by 0.39% of weight gain. The weight increased for 50/50 TPU/PVC after complete immersion in NaOH for 24 hours is 0.06 % and after immersion for 168 hours is only 0.18% as compared to other blends which indicated that 50/50 TPU/PVC blend has a good resistance toward NaOH. HCl had a stronger effect on TPU/PVC blend compared to NaOH where the weight change occurred above 0.6% for all blend composition. The SEM micrographs showed that the fracture surface of the blend of 50/50 TPU and PVC sample resulted in the adaption of both PVC brittleness and TPU ductility features. The results of this blend showed promising potential to be used to fabricate bedpan.

Acknowledgements
The author wishes to thank Universiti Malaysia Perlis (UniMAP), AIMST University and Ministry of Higher Education (MOHE) of Malaysia for supporting this work. This research was supported by MOHE Prototype Research Grant Scheme (PRGS) grants, PRGS/1/2013/SKK12/AIMST/03/1.

References

[1] Valentine Kanyanta, Alojz Ivankovic ,Neal Murphy P. M. V. e. al. 2013 *Springer-Verlag Berlin Heidelberg* Dublin, Ireland 10.1007/978-3-642-20928-4_8 p. 227-252.

[2] M. Kannan 2012 *Studied On Thermoplastic Polyurethane (Segmented PU) based Blends* Faculty of Science Doctor of Philosophy, Mahatma Gandhi University.Reference.

[3] Johanna Baena 2006 *Processing and Kinetic studies of the reactive blends of polyvinyl Chloride and thermoplastic polyurethanes* The Graduate Faculty of The University of Akron Doctor of Philosophy, University of Akron.Reference.

[4] G.M. Shashidhara, R. Goel, K.G. Pradeepa, R. Bharath, A.Dhumal Rao ,R. Abhilash 2010 Comparative study on the properties of PVC/TPU blends prepared by single screw and twin screw extruder *Material Science Research India* **7** 2 457-466.

[5] R. Hernandez, J. J. Pena, L. Irusta ,A. Santamaria 2000 The effect of a miscible and an immiscible polymeric modifier on the mechanical and rheological properties of PVC *European Polymer Journal* **36** 5 1011-1025.

[6] Victor J.R.R. Pita, E.E.M. Sampaio ,Elisabeth E.C. Monteiro 2002 Mechanical properties evaluation of PVC/plasticizers and PVC/thermoplastic polyurethane blends from extrusion processing *Polymer Testing* **21** 545-550.

[7] Krzysztof Pielichowskia ,Ian Hamerton 2000 Compatible poly(vinyl chloride)/chlorinated polyurethane blends: thermal characteristics *European Polymer Journal* **36** 171-181.

[8] Alewo Opuada Ameh, Mohammed Tijani Isai ,Ibrahim Sanusi 2005 Effect of particle size and concentration on the mechanical properties of polyester/date palm seed particulate composites *Leonardo El J Pract Technol* **26** 65-78.

[9] Astm Standard D543 2014 *Standard Practices for Evaluating the Resistance of Plastics to Chemical Reagents* ASTM International West Conshohocken, PA 10.1520/D0543-14 www.astm.org.

[10] Hawa A., M. S. Abdul Majid, M. Afendi, Haslan M., Pranesh Krishnan ,N. A. M. Amin 2015 Ageing Effect on Burst Pressure Test of Impacted Glass Fibre Reinforcement Epoxy (GRE) Pipes` *Applied Mechanics and Materials* **695** 717-720.

[11] Firuz Zainuddin, Hazizan Md Akil, Sahrim Ahmad, Rozaidi Rasid ,Syed Nuzul Fadzli Syed Adam 2014 Drop Weight Impact Analysis of Laminated Polyurethane Foam/Cloisite 30B Clay *Engineering Materials* **594-595** 1089-1093.

[12] Sani Amril Samsudin, Azman Hassan, Munirah Mokhtar ,Syed Mustafa Syed Jamaluddin 2006 Chemical Resistance Evaluation of Polystyrene/Polypropylene Blends:Effect of Blend Compositions and SEBS Content *Malaysian Polymer Journal (MPJ)* **1** 1 11-24.

[13] Yapeng Li, Xuejia Ding, Hanqing Gu ,Guoying Hu 2015 The Effect of TPU on Properties of Polyvinyl Chloride *Journal of Materials Sciences and Applications* **1** 4 131-141.

The structural characterization of some biomaterials, type AISI 310, used in medicine

M G Minciuna[1,2], P Vizureanu[1,2*], C Hanganu[3], D C Achitei[1,2], D C Popescu[4], S C Focsaneanu[1]

[1] Gheorghe Asachi Technical University, Faculty of Materials Science and Engineering, 61 D. Mangeron Blvd, 700050, Romania
[2] Center of Excellence Geopolymer & Green Technology (CEGeoGTech), School of Materials Engineering, Universiti Malaysia Perlis, 01000 Kangar, Perlis, Malaysia
[3] "Gr. T. Popa" Medicine and Pharmacy University, Department of Dental Medicine, 16 Universitatii Street 700115, România
[4] "Gr. T. Popa" Medicine and Pharmacy University, Department of Orthopedics and traumatology, 16 Universitatii Street 700115, România

Email:mirabela.minciuna@yahoo.ro,peviz2002@yahoo.com,carmenhanganu1957@yahoo.com,dragos_adc@tuiasi.ro,dragospopescu1975@yahoo.com, focsaneanu.sergiu@gmail.com

Abstract. Orthopedics biomaterials are intended for implantation in the human body and substituted or help to repair of bones, cartilage or organ transplant, and tendons. At the end of the 20th century, the availability of materials for the manufacture implants used in medicine has been the same as for other industrial applications. The most used metals for manufacturing the orthopedics implants are: stainless steels, cobalt-chrome-molybdenum alloys, titanium and his alloys. The structural researches which are made in this paper, offer a complete analysis of AISI310 stainless steels, using: optical spectrometry, X-ray diffraction and scanning electronic microscopy.

1. Introduction

The orthopedics, dentistry, plastic surgery, ophthalmology, maxillofacial surgery, cardiology, neurology and neurology, and practically all the medical specialities require the biomaterials utilisation, in different goals like: diagnosis, prevention and therapy [1-4].

Metals are one of the most used biomaterials in the case of orthopaedic implants (figure 1), and not only, being known for his high usage resistance, high ductility and hardness. Most commonly used metals, for implants manufacturing, are the stainless steels, the cobalt-chrome-molybdenum alloys, the titanium and titanium alloys [5].

First stainless steel used like material for implant manufacturing was AISI302 steel, which have a resistance bigger than vanadium and is much resistant to corrosion. The steel based on vanadium is not used in implants because, actually, his corrosion resistance is inadequate [6].

*peviz2002@yahoo.com

Further, it starts to be use the Mo 18-8 stainless steel, which contain molybdenum and present an improved corrosion resistance in salted water. This alloy is known like AISI 316 stainless steel. In 1950, the carbon content from AISI 316 steel was reduced from 0.08% to 0.03%, fact that improves the corrosion resistance in chlorides. This new alloy is named AISI 316L [7].

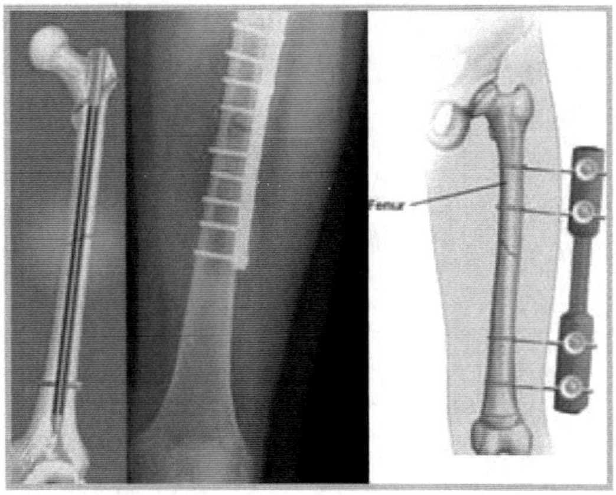

Figure 1. Orthopaedics implants [8].

The metallic biomaterials are classified, in terms of chemical composition and physical structure, in 3 principal classes [9,11]:

- Technically pure metals
- Metallic alloys
- Composites with metallic matrix (CMM)

Metals are used in medical domain, curently, like surgical implants, dentistry materials and for manufacturing of various aparatus and medical divices. Exist must 30 metallic elements used in medicine like [12-13]:

- Implants and orthopedics and dentistry prosthesses: Fe, Al, Ti, Co, Cr, Mn, Mo, Nb, Ni, Sn, Ta, V, W, Zr.
- Nonprecious dentistry alloys, aloyed with: Co, Mo,Al, B, Be, Cd, Cr, Fe, Mn, Ni, Si, Ti, V, W.
- Precious and semiprecious dentistry alloy, alloyed with: Au, Ag, Cu, Ga, In, Ir, Pd, Pt, Rh, Ru, Sn, Ti, Zn.

Alloys used like surgery implants in orthopedics and dentistry, are clasified in:

- Austenite stainless steels – represents the alloys class base on iron, with a high content by chrome and nickel;
- Cobalt base alloys, named CoCr alloys, characterized by a high chrome content (25-30%), molibdenum (5-7%) and other metals like nickel, manganese, zirconium, tin;
- Titannium alloys, with 70-90% Ti, can contain elements like: Al, V, Nb, Ta, Mn, Zr and Sn.

The stainless steel is a iron-carbon alloy, alloyed with chrome, which have a high corrosion resistance in air, in salt and acids solutions. The stainless steels proccesed by plastic deformation are more types; these are clasified in 4 classes, in term of structure (according AISI – American Institute of Steel and Iron), like is present in table 1.

Class I – martensite stainless steels, which contain chrome like principal alloying element; these can be hardened by heat treatments with rapid cooling and formation of martensite [14]. The stainless steels from this class dont have enough corrosion resistance, but presents good mechanical properties. From this reason, are used in manufacturing of surgery instruments, mostly due the hardness properties obtained by martensite quenching [15].

Class II – feritte stainless steels which have a structure formed by a Fe_{α} solid solution and dont be hardened by heat treatment [16]. The steels from this class present good properties of corrosion resistance, espacially to corrosion under load, being used in chemical and medical industry, mostly to manufacturing of recipients [17].

Table 1. Plastically deformable stainless steels and their chemical compositions [18]

Mark AISI	Chemical composition %				Mentions
	C	Cr	Ni	Other elements	
Class I – martensite steels					
410	max 0.15	11.5 – 13.5	-	-	Turbines, valvs
420	0.35 – 0.45	12 –14	-	-	Surgery instruments
431	max 0.20	15 –17	1.25 – 2.5	-	High ductility
440	0.60 – 0.70	16 –18	-	-	Extremely hard
Class II – ferrite steel unhardened					
430	max 0.12	14 –18	max 0.5	-	-
446	max 0.20	23 –27	max 0.5	max 0.25 N	Temperature resistant
Class III – Cr-Ni austenite steels					
301	max 0.15	16 –18	6 – 8	max 2% Mn	Hardned under load
304	max 0.08	18 –20	8 – 12	max 1%Si	Especial 18-8 steel
304L	max 0.03	18 – 20	8 – 12	max 1% Si	Low carbon content
310	max 0.25	24 – 26	19 – 22	max 1.5% Si	Steel 25-20 hot resistant
310X	max 0.08	24 – 26	19 – 22	max 1.5% Si	Low carbon content
314	max 0.25	23 – 26	10 – 14	1.5 – 3% Si	Steel with Si and hot resistant
316	max 0.10	1 6 – 18	10 – 14	2 – 3% Mo	Steel 18-8 with Mo used for implnts
316L	max 0.03	16 –18	11 – 14	2 – 3% Mo	Steel with low carbon content, used for surgery implants
317	max 0.08	18 – 20	11 – 14	3 – 4% Mo	Steel with a high molibdenum content
321	max 0.08	17 – 19	8 – 11	max 4% Ti	Steel stabilized with titanium
Class IV – Hardened steel by ageing					
322	max 0.07	max 17	max 7	max 0.02% Al max 0,007 % Ti	

Class III – austenite stainless steels, based on chrome-nickel, are must used in implants, especially 316 and 316L marks [19]. These steels arenot hardened by heat treatments, but can be hardned by cold plastic deformation. This steels class is nonmagnetic and have a good corroion resistance.

The austenite steels are, usually, chrome-nickel steels, which contain more then 13% chrome and more than 8% nickel. Nickel is introducing to stabilize the austenite; the addition of 2-3% molibdenum in AISI 316 and 317 steels, the corrosion resistance increase [20].

It was found that the decrease of carbon content from 0.1% to 0.03% in AISI 316 it is improved the corroion resistance in chlorine and human physiological saline. From this reasen the american standard ASTM recommend for implants to use the AISI 316L stainless steel [21].

2. Results and discussions

2.1. Spectral analysis for determination of chemical composition
From the spectral analysis effectuated on samples, using Foundry Masters Spectrometer, with iron base analysis, it is obtained the analysis bulletin present in table 1.

Table 2. Chemical composition of analyzed alloy.

Fe	C	Si	Mn	P	S	Cr	Mo
52.30	0.12	1.08	1.46	0.03	<0.01	24.10	0.12
Ni	**Al**	**Co**	**Cu**	**Nb**	**Ti**	**V**	**W**
20.10	<0.01	0.2	0.08	0.03	0.09	0.07	0.02

According the table 2, the studied alloy is part of class III, being considered an austenite steel, type AISI 310B, with iron, chrome and nickel as principal alloying elements.

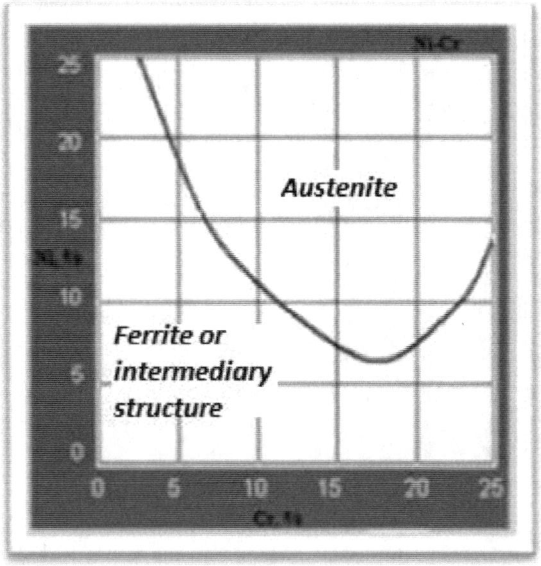

Figure 2. Austenite domain of stainless steels [10].

Chrome is the major element form AISI 310 B steel (with content by 24.10%), fact that it offers a high corrosion resistance. Nickel, in content by 20.1%, it is used in stabilizing the austenite phase at ambient temperature and, more than that, to increase the corrosion resistance. In the case of stainless steels which contain 0.12% carbon, the stability of austenite phase can be influenced by the nickel and chrome presence (figure 2).

2.2 . Structural analysis by scanning electronic microscopy
For a rigorous research in materials domain, for development and characterization of new materials or for understanding their properties it is necessary to use SEM (Scanning Electron Microscopy), in our case type Vega TESCAN LSH II

The quality of images, obtained on SEM is according by the quality of the analyzed sample and I used a detector BSE (Back Scattered Electrons).

In the images from the figure 3, obtaining by electronic microscopy, can remark details of obtained structures, for AISI 310B steel, at magnitudes between 2000- 6000 X BSE.

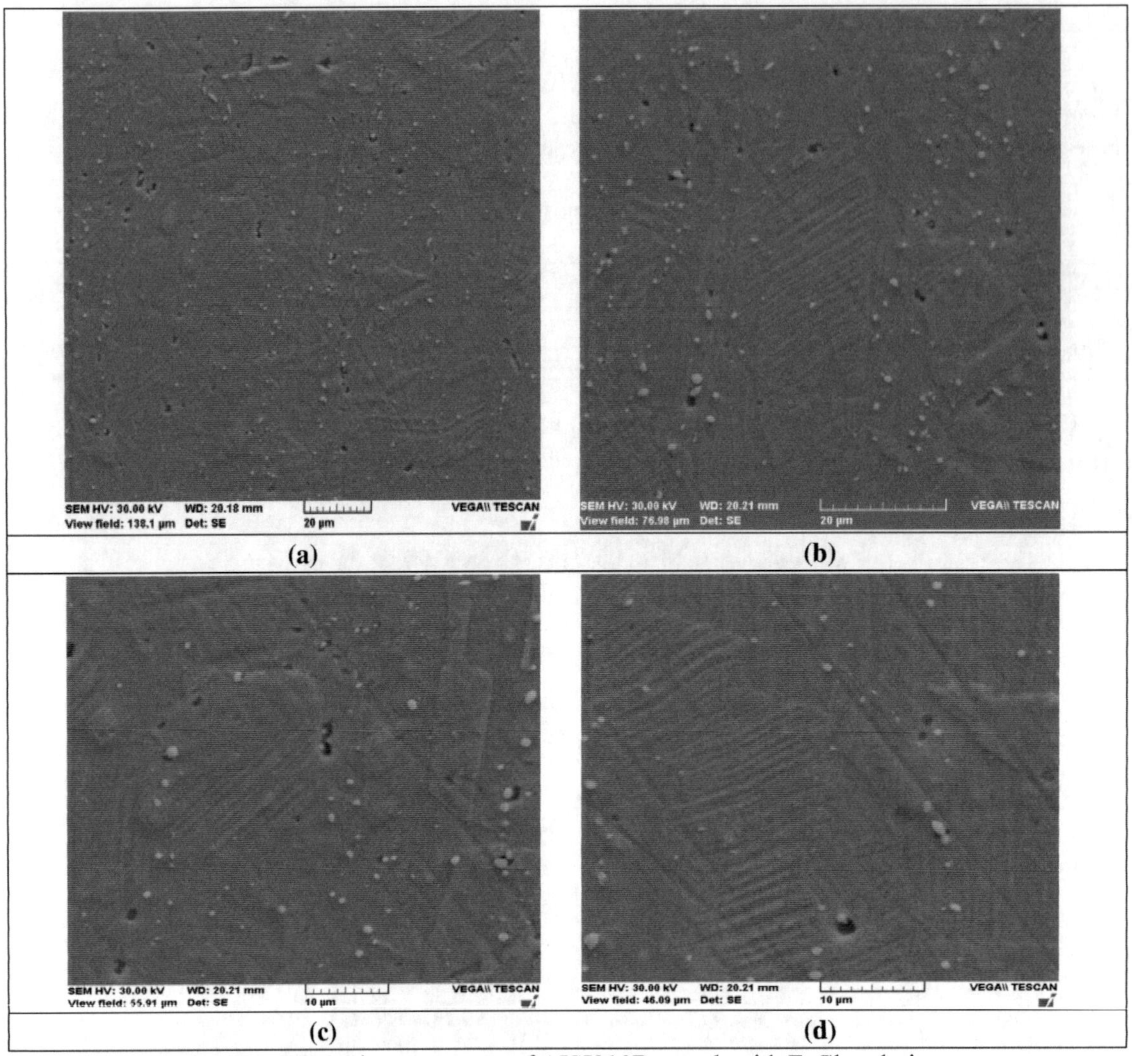

Figure 3. SEM microstructures of AISI310B, attack with $FeCl_3$ solution;
magnitude: (a) 2000X BSE; (b) 4000X BSE; (c) 5000X BSE; (d).6000X BSE.

The AISI 310B steel presents in structure an σ (FeCr) phase, which can cause cold deformations, but and also the presence of elements like: silicon, molibdenum, niobium, titanium. The σ phase causes the decrease of relative elongation, of necking and resilience, increase the harness and tensile strength. In figure 3.d, can highligth in detail, carbides (points by dark color) undisolved, in a austenite base mass and presence of ferrite crystals (white color).

The inluence of elements by AISI 310B steel composition on structure, is given by Schaeffler diagram (figure 4), drawing in coordonate: equivalent nickel – equivalent chrome, conforme relations:

$$Ni_{echiv.} = \%\,Ni + 30\%\,C + 0{,}5\%\,Mn$$

$$Cr_{echiv.} = \%\,Cr + \%\,Mo + 1{,}5\%\,Si + 0{,}5\%\,Nb$$

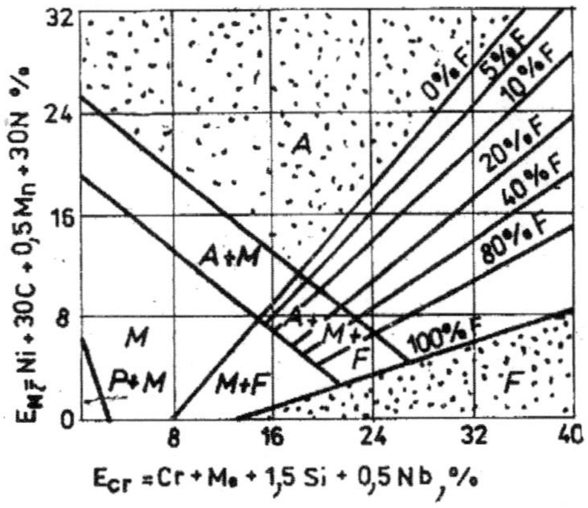

Figure 4. Stainless steels structure – Schaeffler diagram [22].

By the Schaeffler diagram, result that the structure type of AISI 310B stainless steel is in function of chemical composition and the ratio of alfagene and gamagene elements.

2.3. Qualitative phase's analysis by X-ray diffractometer investigations
Using XRD device we characterized the obtained alloys. Determination of structural constituents by diffraction is dependent to the diffracted radiation intensity and double diffraction angle.

The establish of compositional phases was made on an X-ray diffractometer, type Panalytical X'Pert PRO MPD. It is used a fascicle by X-ray – monochrome CuK_{α}, with nickel filter. The 2θ analysis domain is comprised between 20-80°, with a step value by 0.001°, and the step's time is 3 second/step. Figure 5 shows the diffraction pattern obtained from the AISI 310B steel.

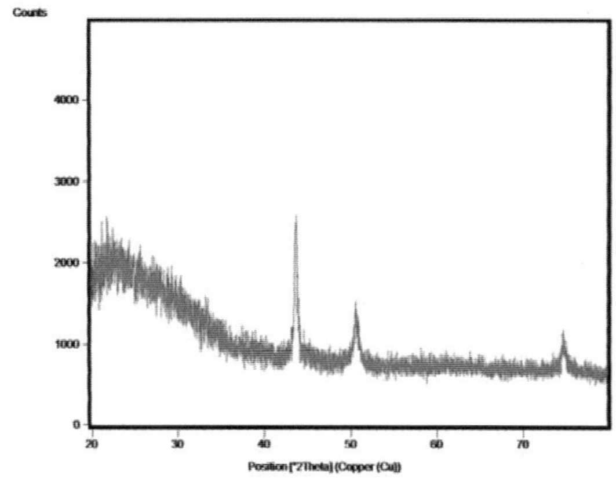

Figure 5. XRD pattern for AISI310B.

In sample are present like phases:

- Fe_3Ni_2 with cubic crystaline network, having a principal maximum at $2\theta = 43{,}76^0$ angle;
- FeCr with cubic crystaline network, having a principal maximum at $2\theta = 50{,}58^0$ angle.

Composition, the chemical state and struture of AISI 310B steel, highligthed by X-ray diffraction, certify the obtained results from spectral analysis and scanning electronic microscopy.

2.4 . Micro-hardness measurements
Micro-hardness measurements were performed on PMT3 micro-hardness device, with standard measure of hardness HV_{50}, using a force measuring 50 g and a measurement time of 15 seconds. Were

conducted a total of 3 measurements for alloy (table 3), with the following measuring conditions: temperature 28°C (reference temperature: 23 ± 5° C) and humidity of 62 %.

Table 3. Micro-hardness of the AISI310B steel

AISI310B	N [div]	P [g]	HV	The average value
1	46,0	50	472,59	438,07
2	48,5	50	425,12	
3	49,0	50	416,49	

The orthopedic implants are required to sag, torsion, and stretch and at the contact of two pieces, due to this requests, it is create tensions and in time, can appear usage. The micro-hardness, with value by 438.07 HV, obtained for AISI 310B steel, corresponding to the standard imposed by the Romanian Society of Biomaterials.

3. Conclusions

The stainless steels used in reparatory medicine cannot be eliminated, but the load to adjust these biomaterials to the integration scopes for tissues and regeneration after implant, can be a challenge for future.

The AISI 310B stainless steel is resistant to large types of corrosive agents, due to his high content in chrome (by 24.10%) and his protector oxide layer Cr_2O3, resistant to corrosion.

The structural characterization realized by X-ray diffraction certifies the obtained results by spectral analysis and scanning electronic microscopy.

References

[1]. H. Schumann, 1962 *Metalurgie Fizică*, Editura Tehnică, Bucureşti.
[2]. M. Truşculescu, A. Ieremia, 1983, *Oţeluri inoxidabile şi refractare*, Editura Facla, Timişoara.
[3]. F. Bratu, L. Benea, J. P. Celis, 2008 *Revista de chimie*, vol. 59, issue 3, pp. 346-350.
[4]. J. T. Zhao, Y. R. Zuo, W. P. Zhao, 2015 *Advanced Materials Research*, volume 1095, pp. 79-82.
[5]. L.Y. Lun You Li, H. N. Ya Nan, 2014 *Advanced Materials Research*, vol. 989-994, pp. 446-449.
[6]. S. Srikanth, P. Saravanan, P. Govindarajan, S. Sisodia, K. Ravi, 2013 *Advanced Materials Research*, vol. 794, pp. 714-730.
[7]. B. K. Yeshwanth, S. Senthil Kumar, B. Hafeez Sharief, K. Devendranath, S. Ramkumar, Narayanan, N. Arivazhagan, 2011 *Advanced Materials Research*, volumes 383-390, pp. 5863-5868.
[8]. G. Balachandran, V. Balasubramanian, 2013 *Advanced Materials Research*, vol. 794, pp. 135-158.
[9]. H. Hosseini Shah, M. Shamanian, A. Kermanpur, 2011 *Materials characterization*, vol. 62, issue 4, pp. 425-431.
[10]. S. S. M. Tavares, V. V. C. daCosta Moura, 2009 *Materials characterization*, vol. 60, issue 6, pp. 573-578.
[11]. H. K. Rafi, G. D. J. Ram, G. Phanikumar, 2010 *THERMEC2009, PTS 1-4 Book Series: Materials Science Forum*, vol. 638-642, pp. 864-869.
[12]. B. Istrate, C. I. Crimu, S. Stanciu, B. Oprişan, C. Munteanu, 2015 *Key Engineering Materials*, 660, pp. 51-56.
[13]. A. E. Kandjani, S. E. H.Amiri, M. R. Vaezi, S. K. Sadrnezhaad, 2010 *Journal Of Optoelectronics And Advanced Materials*, 12, pp. 2057.

[14]. P. Asholt, 1999 *Metal Foams and Porous Metal Structures*, ed. J. Banhart, M. F. Ashby and N. A. Fleck, Bremen, Germany, MIT-Verlag.
[15]. D. C. Achiţei, P. Vizureanu, M. G. Minciună, B. Istrate, A.V. Sandu, 2014 *AFASES*, pp. 13.
[16]. M. G. Minciună, P. Vizureanu, D. C. Achiţei, B. Ghiban, A.V. Sandu, D. Mareci, A. Bălan, 2014 *Revista de Chimie*, 65, pp. 1138.
[17]. I. Alexandru, 1997 *Alegerea şi utilizarea materialelor metalice*, Editura Didactică şi Pedagogică, Bucureşti.
[18]. A. Fraunhofer, M. M. Rogers, 2004 *Gen. Dent.*, 52, pp. 372.
[19]. http://sbecpl.com/products/stainless-steel/ss-310-310s/
[20]. http://www.acerinox.com/en/productos/tipos-de-acero/EN-1.4845---AISI-310S-00001/
[21]. http://www.outokumpu.com/SiteCollectionDocuments/Comparative-behaviour-of-specialty-austenitic-stainless-steels-in-high-temperature-environments-Acom.pdf
[22]. http://www.osti.gov/scitech/biblio/7284338.

Wetting and spontaneous infiltration: the case study of TaC/(Au, Al and Cu) compared to TiC/Cu

M Aizenshtein[1]*, N Froumin[2], O Nafman[2], N Frage[2]

[1] Materials Department, NRC-Negev, P.O. Box 9001, Beer-Sheva, 84190, Israel
[2] Department of Material Engineering, Ben-Gurion University, P.O.Box 653, Beer-Sheva 84105, Israel

E-mail: aizensht@bgu.ac.il

Abstract. Spontaneous infiltration of molten metals in to ceramic skeletons, in the course MMCs' production, is related to improved wetting of the ceramic by metals. TiC is considered a "metal-like" carbide and is supposed to be wetted well by metals through metallic bonding mechanism. Nevertheless, TiC/Cu exhibit an unusual behavior since spontaneous infiltration of molten Cu takes place, while TiC is partially wetted by Cu (θ=90°).

In this work we studied the relation between wetting and spontaneous infiltration in the TaC/Au, Al and Cu systems. TaC is also considered a "metal-like" carbide and indeed no chemical interaction was observed at the interfaces of the studied systems.

Sessile drop experiments showed almost perfect wetting in the three system but spontaneous infiltration occurred only in the first two (e.g. TaC/Au or Al). Thermodynamic calculation shows the difference between the systems which also has its' influence on the mechanical properties of the MMCs'. Further calculation clarifies the difference between TaC/Cu and TiC/Cu infiltration behavior, but is unable to explain the wetting results differences.

Correlation between wetting and spontaneous infiltration in some cases is not straight forward and more studies and calculations on the atomistic level should be done in order to clarify this matter.

1. Introduction

Spontaneous production of MMC's (Metal Matrix Composites) in a pressure less atmosphere is very attractive. Nevertheless, external pressure is usually necessary since ceramics are not wetted well by molten metals, therefore spontaneous infiltration doesn't take place [1,2]. Often improved wetting of ceramics is achieved by the addition of a reactive element which interacts with the ceramic and forms a new phase at the interface (e.g. "reactive wetting"). Unfortunately this new phase is brittle and its' width is not controlled. This might lead to production of MMC's with poor properties. It became common knowledge that successful spontaneous infiltration could occur if the contact angle (called "wetting angle"),θ, is lower than 90°. When ideal parallel pore walls of the ceramic are considered, this condition is sufficient [3]. In practice, pore walls are never parallel; they also have a certain roughness which could hinder infiltration and actually successful infiltration requires contact angles bellow 60° [3]. Trumble [4] proposed an analytical model which predicted that infiltration could take place when θ_c is lower than 50.7°. The straight forward link between contact angle and spontaneous infiltration is done since both depend on the same physical phenomenon; both wetting and infiltration are controlled by the energy gain of the system when surfaces of the ceramic and the molten metal are replace by a solid-liquid interfaces. Lack of infiltration despite improved wetting might occur in systems in which reactive wetting takes place (graphite/Si system for example [5]), in this case the growth of the new interface phase, can block pores and deny further infiltration of the molten metal. Blockage of pores was also observed in the case of CaF_2/In-Ti. Despite improved wetting (θ=30°), infiltration didn't take place probably due to the high vapor pressure of CaF_2, which practically blocked the access of the molten alloy into the open pores [6].

In previous results [7], despite the fact that the contact angle in the TiC/Cu was found to be 90° and in other reports ever higher [8,9], spontaneous infiltration took place in the frame time of second which clearly suggests that infiltration didn't proceed through a reactive wetting mechanism. Titanium carbide has a high melting point, displays extremely high hardness and shock resistance and its electrical conductivity is comparable to that of metals [10-12]. In the high temperature capillarity community it is considered a "metal-like" material and therefore could be easily wetted by liquid metals, since it is well known that metal wets other metals due to metallic bonding formed at the solid/liquid interface, without the need of chemical reaction (dissolution or new phase formation) [1,13]. The partially wetting and improved infiltration in the TiC/Cu is peculiar, therefore we decided to study the wetting and infiltration behavior in a similar system. For this purpose we chose TaC and Au, Al and Cu as infiltrating metals. Tantalum belong the V group of elements which exhibits, just like the IV group, carbides with a wide composition range. The high density of TaC (14.3 gr/cm^3) limits its use as a structural material, yet it is used in various applications due to its high melting point, chemical inertness and metallic electrical conductivity [10,14]. These properties are related to the mixed covalent-metallic bonds. Nevertheless, TaC is brittle, therefore there is an advantage in producing a MMC based on TaC and a ductile metal. Shanker et al. [15] made an attempt to develop MMCs' based on TaC for use as electrodes in plasma systems and obtained improved mechanical properties. Due to its metallic character, TaC just like TiC is an excellent candidate for production and study spontaneous infiltration processes of MMC's. The high stability of TaC suggests that little if any chemical interaction might occur at the interface, which would not compromise its' integrity. In the current study we intend to investigate the wetting properties of TaC in contact with pure metals (Al, Cu and Au) at 1423 K, followed by spontaneous fabrication of MMCs' and characterization of its' mechanical properties, in order to find more about the relations between wetting, spontaneous infiltration and properties.

2. Materials and Methods

Fully dense TaC samples were fabricated from 99% purity powder using SPS (Spark Plasma Sintering) - apparatus (type HP D5/1, FCT System. Rauenstein, Germany). The SPS consolidation was made using 20KN pressure at 2473K for 30 minutes. The final Ta:C ratio was determined by XRD analysis. The lattice parameter of the rock salt TaC after SPS process was 0.4453 nm. According to Bowman [16], the C:Ta ratio is 0.974. Sessile drop wetting experiments were performed at 1423K, under a dynamic vacuum (10^{-3} Pa). Contact angles were measured using a Nikon 990 Coolpix digital camera and derived from the magnified images using the "Image Pro 4" software. Composites based on TaC were fabricated by free infiltration of TaC preforms obtained by cold compacting under 15MPa (60% porosity) and after densification by SPS at 2073K for 30 min under 60MPa applied pressure (20% porosity). The infiltration process was carried out at 1423K, under a pressure of 10^{-3}Pa, for 30 minutes. Solidified ceramic/drops interfaces were characterized using SEM (Jeol JSM 35C) while ceramic/metal interfaces in the infiltrated composites were characterized using TEM (Jeol 2010F) analysis. Samples for TEM characterization were first sectioned and mechanically thinned to about 50 μm. Then, 3 mm in diameter samples were punched out of the thinned samples followed by dimple grinding and Ar-ion beam milling using a Gatan PIPS (Precision Ion Polishing System) apparatus. The effectiveness of the spontaneous infiltration process was determined indirectly by the bending strength of the composites. In this study we used the three point method (LLOYD Instruments LTD machine). Bars (four samples for each composite) with the length of 20 mm, 2 mm width and 1.5 mm height were tested at a strain rate of 0.2 mm/min according to ASTM 1161-94. The velocities of longitudinal and transverse acoustic waves were determined by the "Pulse-Echo" method using a 5 MHz probe and according to ASTM E494-10. The elastic modulus of the composite was calculated using the ultrasonic sound velocities and Archimedes density measurements.

3. Results and discussions

3.1. Wetting and infiltrated composites microstructure

The final contact angle was obtained during heating before reaching 1423K. In the TaC/Al system, contact angle of 10° ± 1° was measured during the 30 minutes dwell, while 10.5° ± 1° and 11° ± 1° was measured in the TaC/Cu and TaC/Au systems respectively. No new phases or signs of dissolution were found at TaC/Au, Al or Cu interfaces (figure 1). Liquid Al and Au fully infiltrate the TaC preforms (figure 2), TaC particles are homogeneously distributed in metallic matrix.

Cu didn't infiltrate in the preform and TaC/Cu MMC was not obtained. This result matches the results of Shanker et al. [8] which reported difficulties in production of TaC/Cu MMCs'. According to Shanker et al., successful infiltration was achieved over a narrow TaC range (43-50%) at temperatures higher than 1300°C in vacuum.

TEM characterization of the TaC/Al and TaC/Au MMCs' showed no interface interaction or new phases formed at the TaC/Al and TaC/Au interfaces (figure 3).

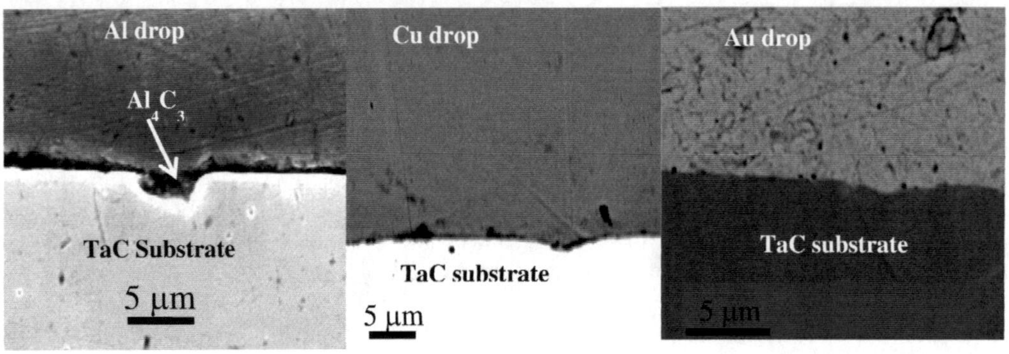

Figure 1. TaC/(Al, Cu, Au) interfaces showing no traces of interaction and dissolution and precipitation of Ta in the solidified melt. In a few discrete points Al_4C_3 particles were found. These phases are related to excess graphite remains after the SPS process.

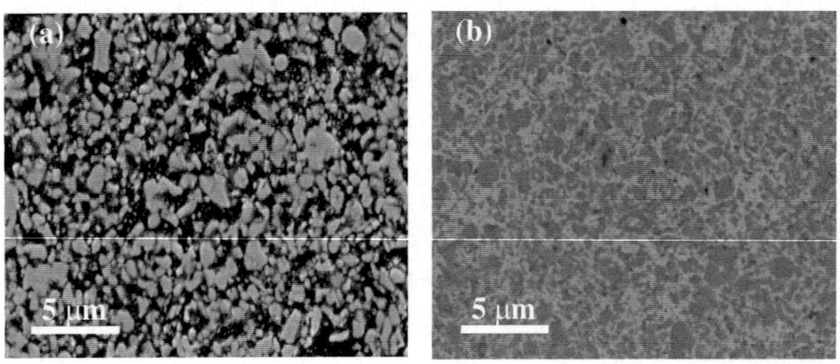

Figure 2. MMCs' bases of TaC-60% porosity, (a) infiltrated in Al (b) infiltrated in Au.

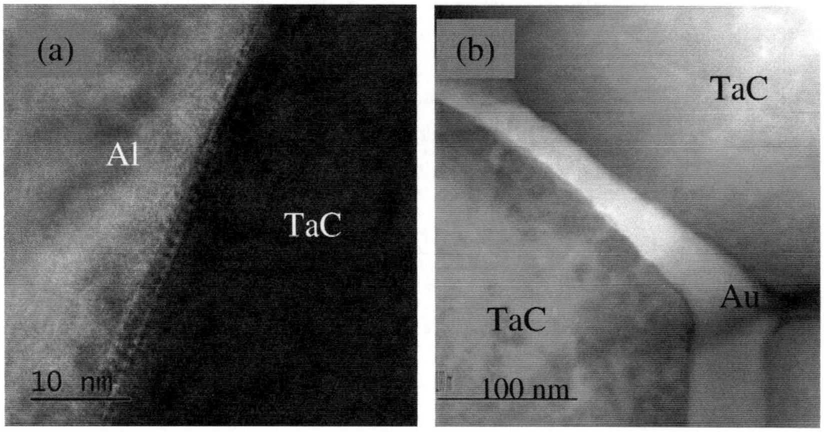

Figure 3. TEM images of representative TaC particles in contact with the infiltrating metal (a) Al (b) Au.

3.2. Mechanical properties Interpretation

The properties of the composites strongly depend on the metal to ceramic ratio. Moreover, the bending strength of the TaC-Au composites is higher than that for the TaC-Al composites (table 1). The image of the fractured surface of the TaC-60vol.%Au composite clearly demonstrates that the fracture mode is ductile and the cracks propagate rather through the ceramic phase than through the metal/ceramic interfaces (figure 4). Characteristic bending stress curves in the TaC/Al system are presented in figure 5 and support the ductile behavior of the high metal content composites. Despite the fact that no interface interaction is observed in the two studied systems, according to the results presented in table 1, TaC/Au interfaces are stronger than TaC/Al ones. This feature may be correlated with the work of adhesion, W_a, between liquid metal and solid substrate. The work of adhesion depends on contact angle θ and liquid vapor surface energy of the metal γ_{LV} (comprehensive review of the subject [17]),

$$W_a = \gamma_{LV}(1+\cos\theta) \quad (1)$$

According to Keene [18], the surface energies of liquid Au and Al at 1423K are 1.14 J/m^2 and 0.8 J/m^2, respectively and the calculated values of the work of adhesion are 2.25 J/m^2 and 1.6 J/m^2, respectively. Furthermore, thermal stresses might develop at the interface during cooling. Due to the mismatch between the thermal expansion coefficient (CTE) of the metal and TaC and the fact that the CTE of TaC is lower than the CTE of Al and Au, the ceramic at the interface is under compression stresses and the metal is under tensile stresses.

The thermal stress at the TaC/metal interface could be expressed by the following equations [19]:

$$\sigma_{TaC} = (-E'_{metal} \cdot E'_{TaC})/(E'_{metal}+E'_{TaC}) \cdot \Delta T \cdot (\alpha_{metal}-\alpha_{Tac}) = \sigma_{metal} \quad (2)$$

And $E'=E/(1-\nu)$, where E is theYoung's modulus, α = CTE and ν is the Poisson's ratio.

The data and the calculated interface stresses are summarized in table 1. The calculated values obtained show that while the compression stresses which undergoes the ceramic is beneath the yield of TaC, the tensile stresses of both metals exceeds their yield stress. Yet the stresses at the TaC/Al interface are twice higher than the ones at the TaC/Au interface, which should have a contribution to the decreasing bonding strength of the Al/TaC MMC.

Table 1. Results of three point bending strength and Young modulus of TaC/Al and TaC/Au MMCs'

Composite	Bending strength, MPa	Young modulus GPa
TaC -60vol.%Al	697 ± 50	151±10
TaC -20vol.%Al	632±45	349±14
TaC -60vol.%Au	987±48	198±12
TaC -20vol.%Au	720±35	365±12
TaC	590	537 [20]

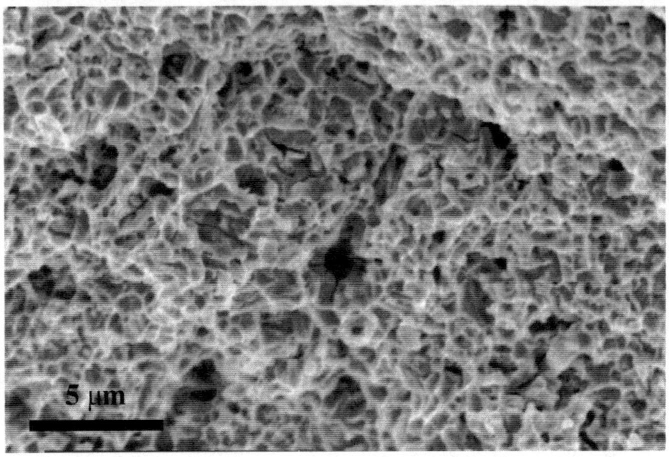

Figure 4. Fractography image of the TaC (60%por)/Au MMC after three point bending experiment showing dimples with are characteristic to ductile fracture.

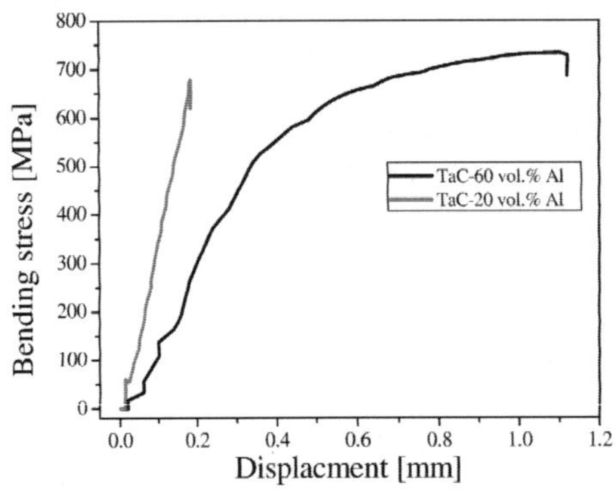

Figure 5. Characteristic three point bending stress diagram for TaC/60vol.%Al and TaC/20vol.%Al composites.

Just for comparison, the bending strength reported by Frage et al. [6] in the TiC/Cu (in the rage of 20-50% porosity) is 500 – 960 MPa.

IOP Conf. Series: Materials Science and Engineering **133** (2016) 012020 doi:10.1088/1757-899X/133/1/012020

3.3. *Thermodynamic considerations*

The difference between the three studied systems was considered in the perspective of the thermodynamic properties of the systems. The carbon solubility in the various liquid metals is negligible. Therefore the thermodynamic properties of the various metals-Ta and the composition range of TaC at 1423K were taken into account.

The equilibrium content of Ta within the liquid melt were calculated according to equations (3-5), where

$$\mu^{Ta}_{Solid} = \mu^{Ta}_{Liquid} \quad (3)$$

$$\mu_{\downarrow}0^{\uparrow}(Ta\,(S)) + RTln(a)_{\downarrow}(\llbracket Ta \rrbracket^{\uparrow}S) = \mu_{\downarrow}0^{\uparrow}(Ta\,(L)) + RTln(a)_{Ta^{L}} \quad (4)$$

$$RTln\,(a)_{\downarrow}(\llbracket Ta \rrbracket^{\uparrow}S)/(a)_{\downarrow}(\llbracket Ta \rrbracket^{\uparrow}L) = \mu_{\downarrow}0^{\uparrow}(Ta\,(L)) - \mu_{\downarrow}0^{\uparrow}(Ta\,(S)) = \llbracket \Delta G \rrbracket_{\downarrow}Fusion^{\uparrow}Ta \quad (5)$$

The activities of Ta ($\mu^{Ta}{}_{Liquid}$) in the various Al, Cu and Au melts are described by the Redlich-Kister formalism [21] using the interaction parameters, L^0, L^1 and L^2 (**Error! Reference source not found.**) and the data concerning the Ta activity within its composition range at 1423K. While the binary phase diagrams of Al-Ta and Au-Ta exhibits the existence of various stable intermetallics, Cu and Ta don't have any. This behavior implies that Ta and Cu should have a positive departure from ideality while the former systems should show a negative departure from ideality as could be seen through the interaction parameters.

Table 1. The interaction parameters of Ta-Al, Cu and Au binary liquid solutions at 1423K.

System	L^0 [J/mol]	L^1 [J/mol]	L^2 [J/mol]
Al-Ta [22]	-41521.2	-4732.84	19928.86
Cu-Ta [23]	99366.18	0	0
Au-Ta [24]	-135370	0	0

The thermodynamic properties of TaC_x were extracted from [25]. At 1423K, Ta composition in Ta_xC_{1-x} varies between $0.5<x<0.55$. The calculated content of Ta in various melts as a function of tantalum carbide stoichiometry is shown in figure 6. The calculated results express the high affinity of Au towards Ta and the repulsion between Ta and Cu. These huge differences in thermodynamic properties are not expressed in the sessile drop experiments, which mean that the spreading kinetics is faster than dissolution. Yet, the thermodynamic differences between the studied systems are qualitatively observed in the spontaneous infiltration and bending strength results. Cu was not able to infiltrate at all, at the bending strengths of TaC/Au composites were higher than those of TaC/Al.

When questioning the differences between the TaC/Cu and TiC/Cu systems, first the Ti-Cu liquid system exhibits a negative departure from ideal solution ($\ln(\gamma_0) = 1.02$ at 1423K [26]), second the formation Gibbs energy of TiC is lower than that of TaC. The thermodynamic calculation in the case of titanium carbide within its composition range is based on Frage et al. [6]. Figure 7 expresses the differences between Ta solubility in molten Cu in equilibrium with tantalum carbide compared to Ti solubility in molten Cu in equilibrium with titanium carbide at 1423K.

It could be seen that the solubility of Ti in Cu is about two orders of magnitude higher than Ta in Cu in the stochiometric state (x=0.5) and about one order of magnitude for lower content of carbon in the carbides.

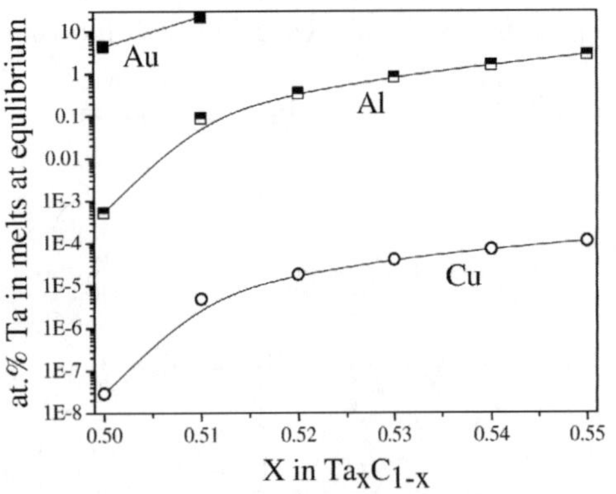

Figure 6. Calculated Ta content in various liquid metals (Au, Al, Cu) as a function of tantalum carbide stoichiometry.

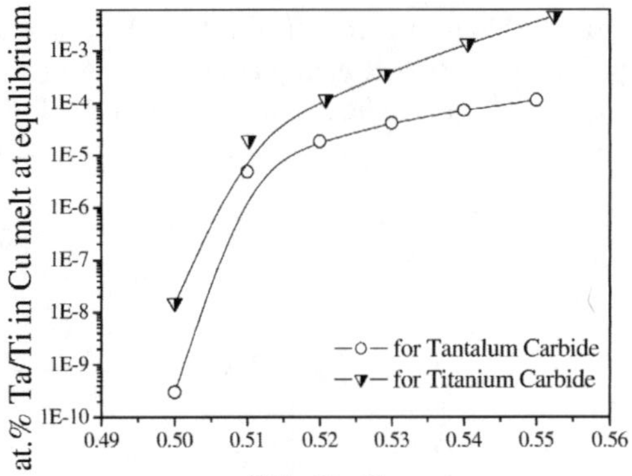

Figure 7. Calculated Ta and Ti content in liquid Cu as a function of tantalum carbide and titanium carbide stoichiometry.

4. Conclusions

The sessile drop experiments in the TaC/(Al, Cu and Au) at 1423 K systems showed almost perfect wetting and predicts that spontaneous infiltration should occur in all cases. The flat interfaces and the lack of reaction products or detected solubility of Ta, led to the conclusion that improved wetting in the TaC/(Al, Cu and Au) occurs due to the "metal-like" nature of TaC. The pressure free infiltration experiments and bending strength exhibited best results in the TaC/Au system while in the case of TaC/Cu no infiltration took place, which is in qualitative correlation with the thermodynamic properties of the studied systems. The fractography images show ductile fracture implying that the bonds at the interface are stronger than the strength of the metals. In contrast, Cu partially wets TiC ($\theta=90^{o}$) [8] but infiltrates it perfectly [7]. Although TiC is more stable thermodynamically than TaC, the interaction parameter of Cu-Ti dilute liquid phase (L^0 or $RTln\gamma^0$) at 1423K is ~ -13 kJ/mol (attraction forces) while the interaction parameter in the Cu-Ta dilute liquid phase is ~ 99 kJ/mol (repulsion forces). These differences could lead to different behavior in the spontaneous infiltration process but are unable to explain the wetting results differences. Lastly, the correlation between wetting and spontaneous infiltration in some cases as the ones showed here is not straight forward and more studies and calculations on the atomistic level should be done in order to clarify this mater.

References

[1] F. Delannay, L Froyen and A. Deruyttere, 1987 *Journal of Materials Science*, vol. 22, pp. 1-16.
[2] A. Mortensen and J. A. Cornie, 1987 *Metallurgical Transactions A*, vol. 18A, pp. 1160-1163.
[3] N. Eustathopoulos, 2015 *Metals*, vol. 5, pp. 350-370.
[4] K. P. Trumble, 1998 *Acta Materialia*, vol. 46, pp. 2363-2367.
[5] O. Dezellus, S. Jacques, F. Hodaj and N. Eustathopoulos, 2011 *Journal of Materials Science*, vol. 40, pp. 2307-2311.
[6] S. Barzilai, H. Nagar, N. Froumin and N. Frage, 2011 *Journal of Materials Science*, vol. 46, pp. 5698-5701.
[7] N. Frage, N. Froumin, L. Rubinovich and M. P. Dariel, 2001 *15th International Plansee Seminar*, vol. 1, pp. 202-216.
[8] N. Frage, N. Froumin and M. P. Dariel, 2002 *Acta Materialia*, vol. 50, pp. 237-245.
[9] P. Xiao and B. Derby, 1996 *Acta Materialia*, vol. 44, pp. 307-314.
[10] P. M. McKenna, 1936 *Industrial and Engineering Chemistry*, vol. 28, pp. 767–772.

[11] S. A. Shvab and F. F. Egorov, 1982 *Soviet Powder Metallurgy Metal Ceramic*, vol. 21, pp. 894-897.
[12] H. O. Pierson, 1996 *Handbook of Refractory Carbides and Nitrides*, Noyes Pub.
[13] N. Eustathopoulos, M. G. Nicholas and B. Devet, 1999 *Wettability at High Temperatures*, Oxford: Pergamon Materials Series.
[14] X. Zhang, G. E. Hilmas and W. G. Fahrenholtz, 2009 *Materials Science and Engineering A*, vol. 501, pp. 37-43.
[15] K. Shanker, L. T. Mavropoulos, R. A. L. Drew and P. G. Tsantrizos, 1992 *Composites*, vol. 23. pp. 47-53.
[16] A. L. Bowman, 1961 *Journal of Physical Chemistry*, vol. 65, pp. 1596–1598.
[17] W. D. Kaplan, D. Chatain, P. Wynblatt, W. C. Carter, 2013 *Journal of Materials Science*, vol. 48, pp. 5681–5717.
[18] B. J. Keene, 1993 *International Materials Reviews*, vol. 38, pp. 157-192.
[19] S. Ho, C. Hillman, F. F. Lange and Z. Suo, 1995 *Journal of the American Ceramic Society*, vol. 78, pp. 2353-2359.
[20] H. L. Brown, P. E. Armstrong and C. P. Kempter, 1966 The Journal of Chemical Physics, vol. 45, pp. 547–549.
[21] M. Hilert, 1998 Phase Equilibria, Phase Diagram and Phase Transformation—their *Thermodynamic Basis*, 1st edn. Cambridge University Press, Cambridge.
[22] V. T. Witusiewicz , A. A. Bondar , U. Hecht , J. Zollinger , V. M. Petyukh , O. S. Fomichov, V. M. Voblikov and S. Rex, 2010 *Intermetallics*, vol. 18, pp. 92–106.
[23] P. R. Subramanlan and D. E. Laughlin, 1989 *Bulletin of Alloy Phase Diagrams*, vol. 10, pp. 652-655.
[24] K. Fitzner, N. Selhaoui and O. J. Kleppa, 1992 *Metallurgical Transactions A*, vol. 23, pp. 1836-1839.
[25] E. A. Guseva, A. S. Bolgar, V. A. Gorbatyuk and V. V. Fesenko, 1974 *Refractory Carbides*, Editor: G.V. Samsonov, pp. 317-322.
[26] P. Wei and L. Jie, 1999, *Materials Engineering A*, vol. 269, pp. 104-110.

IOP Conf. Series: Materials Science and Engineering **133** (2016) 012021 doi:10.1088/1757-899X/133/1/012021

Geopolymer concrete for structural use: Recent findings and limitations

M F Nuruddin[1*], A B Malkawi[1,2], A Fauzi[1,3], B S Mohammed[1], H M Almattarneh[2]

[1] Civil and Environmental Engineering Department, Universiti Teknologi PETRONAS, 32610 Seri Iskandar, Perak, Malaysia
[2] Civil Engineering Department, Najran University, 11001 Najran, Saudi Arabia
[3] Civil Engineering Department, Politeknik Negeri Lhokseumawe, 24301 Lhokseumawe, Aceh, Indonesia

E-mail: fadhilnuruddin@petronas.com.my

Abstract. Geopolymer binders offer a possible solution for several problems that facing the current cement industry. These binders exhibit similar or better engineering properties compared to cement and can utilize several types of waste materials. This paper presents the recent research progress regarding the structural behaviour of reinforced geopolymer concrete members including beams, columns and slabs. The reported results showed that the structural behaviour of the reinforced geopolymer concrete members is similar to the known behaviour of the ordinary reinforced concrete members. In addition, the currently available standards have been conservatively used for analysis and designing of reinforced geopolymer concrete structures. On the other hand, the main hurdles facing the spread of geopolymer concrete was the absence of standards and the concerns about the long-term properties. Other issues included the safety, cost and liability.

1. Introduction

Concrete is the major construction material used all over the globe and its use is second only to water. The construction industry has become the largest consumer of global natural resources. According to the United Nations, it is expected that by the year 2050 more than 65% of the population will dwell in cities [1]. The growth of the developing countries means implementing the use of electricity and building infrastructures and houses, in other words, electricity and concrete. Ordinary Portland Cement (OPC) is the binding material used for concrete production. It is estimated that the production of 1 ton of cement will release a similar amount of carbon dioxide (CO_2) into the atmosphere. Generally, the cement industry accounts for 5-8% of the global CO_2 emissions [2]. A technological turnaround should be adopted in order to reduce the CO_2 emissions of this industry. Reduction of the CO_2 emissions is a global need and we should understand that the current emission rate if it continues, will form a real threat to the future generations.

Geopolymer binders have been proved to be green building materials that can totally replace the OPC in the concrete industry. Geopolymer technology can utilise many by-product materials, such as FA, granulated blast furnace slag (GBFS), palm oil fuel ash, rice husk ash, and mining wastes. Other

* Civil and Environmental Engineering Department, Universiti Teknologi PETRONAS, 32610 Seri Iskandar, Perak, Malaysia, Tel.: +60-133641734, fax: +60 5-368 4088. E-mail: fadhilnuruddin@petronas.com.my

IOP Conf. Series: Materials Science and Engineering **133** (2016) 012021 doi:10.1088/1757-899X/133/1/012021

resources may include natural reactive aluminosilicate powders or thermally activated aluminosilicates, which can provide a wide range of geopolymer binder sustainability and availability worldwide.

Geopolymer synthesis is based on the inorganic alkali activation chemistry, where aluminosilicate rich material is activated using a strong alkali solution to produce a three-dimensional aluminosilicate gel that has a characterisation that can compete with OPC. Geopolymer binders have shown the potential to be alternatives to OPC and currently attract more attention due to their superior features, including higher early strength, dimensional stability, durability, fire resistance, superior bond to reinforcement and aggregates etc. [3]. Currently, many researchers all over the world are focused on this topic. The geopolymer institute reported the exponential increase in the number of scientific publications in peer-reviewed journals with the keyword geopolymer; more than 400 publications were recorded in 2013. The current research trend is shifting from geopolymer chemistry to the engineering applications and commercialisation.

2. Structural Behaviour of Reinforced Geopolymer Concrete

Geopolymer binders showed increasing potential for use as a cement replacing materials in the concrete industry. These binders can produce concrete with a wide range of physical and mechanical properties. In order to prepare geopolymer concrete (GPC) with desired properties, Si/Al, M/Si (M denotes any alkali cations), and water/binder ratios need to be controlled. Hence, it is required to carefully characterise the alkali activator and the source material to determine these ratios.

GPC has a different strength development mechanism compared to cement concrete mixtures; hence, the available methods for analysis and design of cement concrete members should be evaluated and verified before use in GPC members. Available literature on GPC dwelled much on production, physical, mechanical, and durability properties. However, less attention was given to reinforced geopolymer concrete (RGPC) behaviour and structural applications.

2.1. Flexural behaviour of RGPC beams

Kumaravel et al. [4] studied the flexural behaviour of RGPC beams cast using the FFA+GBFS-based GPC and cured at 60°C for 24 hours. The beams were cast with grade 40 concrete with dimensions of 125x250x3000 mm and tested under the 4 points loading flexural test. The results were compared with the reference reinforced cement concrete (RCC) beams of the same concrete grade. The RGPC beams showed similar load-deflection curves for that obtained for the reference RCC beams; however, the RGPC beams displayed higher load capacity in terms of the first crack appearance, and ultimate and service loads. Compared to the RCC reference beams, the RGPC beams showed higher yield load, ultimate load, and maximum load by 3.57%, 2.7%, and 11.25%, respectively. Figure 1 shows the similarity of the cracking patterns for both RCC beams and RGPC beams under flexural testing. All the beams failed by yielding of the reinforcement in the tension zone followed by crushing of the concrete in the compression zone. The results for all the beams were predicted using ANSYS software and showed close agreement with the experimental ones.

A similar study conducted by Dattatreya et al. [5]; however, the beams were cured under ambient conditions. The beam dimensions were 100x150x1500 mm and the tensile reinforcement ratio varied between 82-110% of the balanced reinforcement ratio. The first crack load was 9-11% for the RGPC beams and 13-16% for the RCC beams as a percentage ratio of their ultimate loads. The average service loads were reduced by 12% for the RGPC beams compared to the RCC beams. The cracking patterns including the crack number, spacing, and width, and the failure modes were almost the same. Using the ACI 318 code [6] equations to predict the cracking moment, ultimate moment, and maximum deflection gave good correlation and fair agreement with the experimental results; however, the agreement degree was not the same for all cases. It was suggested to include an additional reduction factor. Further researches were recommended to investigate the stress block shape in the RGPC members.

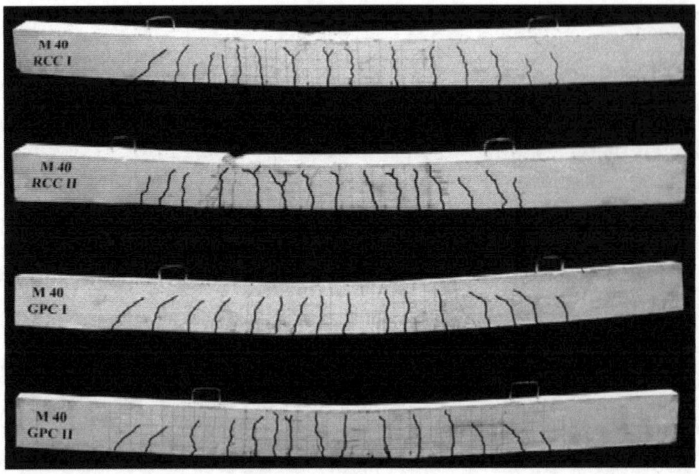

Figure 1. Cracking patterns of RCC (RCC I&II) and RGPC (GPC I&II) beams [4].

Another interesting study by Yost et al. [7, 8] has investigated the flexural behaviour of full dimension RGPC beams. The beams were cast using the FFA-based GPC of the dimensions 305x152x3200 mm and cured at 60°C for 24 hours using their newly developed system which was called "environmental curing chamber". This system used the halogen lights as the source of the temperature increase. The compressive strength was in the range of 52-57 MPa. The beams were reinforced with three different arrangements representing the cases of the under-reinforced (U), over-reinforced (O), and shear critical reinforcement (S). The beams were tested under the 4-point loading system and the results compared with the reference RCC beams. The U-beams from both the RCC and RGPC showed nearly identical load-deflection curves. The reinforcement yielding loads were slightly higher than the predicted values for both the RCC and RGPC beams. A fully ductile behaviour was observed after which the concrete in the compression zone crushed. However, the RGPC beams showed a more brittle manner where the concrete disintegrated and the load fell directly at the point of failure whilst the RCC beams sustained awhile after failure. The RGPC beams recorded higher deflection values at the maximum loads. The side strain gauges showed that the neutral axis position could be described in 3 phases: transition, elastic, and inelastic. The theoretical and experimental results for the neutral axis location during different testing stages were fairly identical, which implies that the ACI 318 code [6] equations of the Whitney rectangular stress block can be used for the RGPC beam design. In the case of the O beams, the beams showed a linear load-deflection diagram up to failure while the RCC beams displayed a slight curvature prior to failure. The beams' ultimate load was 8% higher than the predicted values. For the beams, the failure mode was by the diagonal shear mode. Identical cracking patterns were observed for both the RCC beams and RGPC beams, which suggests the similarity of the shear transfer mechanism for both types of concrete.

2.2. Shear behaviour of RGPC beams

The shear behaviour of the RGPC beams was investigated by Chang [9]. Nine beams of the dimensions of 200x300x1680 mm were cast using the FFA-based GPC. It was noted that the failure mode and the cracking patterns were similar for both the RGPC and RCC beams. The shear capacity of the beams was also found to be dependent on the longitudinal reinforcement ratios. The shear cracking load and the shear strength of the RGPC beams were predicted using the ACI 318 code [6] provision for shear calculations and gave conservative results. The VecTor2 programme was used to simulate the cracking patterns and the failure modes and to predict the shear strength. The predicted results indicated a very good correlation with the experimental data.

Madheswaran et al. [10] investigated the shear behaviour of thin webbed T-beams produced using the FFA+GBFS-based GPC. The results showed that both the RGPC and the RCC beams had similar shear behaviour, where the beams' shear capacity was affected by the stirrups' spacing and the shear span to effective depth ratio. The beams failed by the typical diagonal tension shear failure mode, in which one of the flexural cracks in the shear span developed an inclined crack that extended toward the loading and supporting points after which a sudden brittle shear failure occurred. The load-deflection diagrams showed a linear relationship until cracking and even the post cracking portion also showed a linear relationship. The experimental values for the deflections were higher than the calculated values using the available ACI 318 code equations for the RCC beams. This may attribute to the lower modulus of elasticity of the GPC. It was reported that the available ACI 318 code provision for the shear design of the RCC beams is valid and can be safely used for the RGPC beam design; the upper limit provided by the code is also conservative. The same researchers has conducted similar study; however, on rectangular beams [11]. Twelve beams were cast with a span of 1600 mm and designed as shear deficient beams. The strain compatibility method was used to determine the ultimate moment capacity of the beams and the non-linear stress-strain relationship was used to estimate the complete load-deflection diagram. These methods showed an excellent correlation with the experimental results for both the RGPC the OPC beams.

Laskar et al. [12] highlighted the enhanced capacity of the RGPC beams against cyclic loading effects. It was reported that the RGPC beams had an increased capacity by almost 30% over the RCC beams. Moreover, the RGPC beams displayed a lower degradation in stiffness over time. Also, the RGPC beams showed higher capability for energy dissipation by 45% over the RCC beams, which promoted the higher ability of the RGPC beams to sustain earthquakes.

2.3. Behaviour of RGPC columns

Sarker [13] analysed the RGPC short slender columns under the combined stress of compression and uniaxial bending. They found that the analytical methods available for the RCC column analysis could be conservatively used for the analysis of the RGPC columns. However, some modifications were suggested for a proper stress-strain relationship. The predicted values of the ultimate loads, mid-height deflection, load deflection curves, and deflected shapes correlated well with the experimental data.

Rahman et al. [14] investigated the behaviour of the slender columns under axial compression and biaxial bending stresses. The columns were cast using the FFA-based GPC. The tested parameters included: concrete grade, reinforcement ratio, and eccentricity distance. The failure mode for all the columns was by the spalling of the concrete cover followed by the concrete crushing. For the small eccentricity distance, the failure was in a more brittle manner with a shorter post-peak on the load-deflection diagram. Increasing the eccentricity distance increased the measured deflection at the mid-height of the columns; also, the deflection increased for a higher reinforcement ratio and higher compressive strength. Generally, the failure modes and load-deflection behaviour are similar to that observed for the RCC columns. Bresler's reciprocal load formula and the stress block formula provided by the Australian standards were used to estimate the column load capacity. The results were found to be conservative and close enough to the experimental results. However, these formulas give higher accuracy in the case of columns with smaller eccentricities. A similar research by Sumajouw et al. [15] also confirmed the conservative results predicted by the Australian standards AS3600 and the ACI 318 code provision. The average experimental to estimated ratio was 1.15. Another study by Sujatha et al. [16] compared the loading behaviour of the RGPC short slender columns with reference to the RCC columns. The columns were prepared with a circular cross-section and 1800 mm height and tested for axial compression loading. The results showed that the RGPC columns had higher load capacity by 30% over the RCC columns. On the other hand, lower mid-height deflections were observed for the RGPC columns.

2.4. Behaviour of RGPC slabs

The Flexural behaviour of the reinforced GBFS+FA-based GPC solid slabs reported to be similar to the behaviour of OPC reinforced concrete slabs [17]. Slab strips were prepared with the dimensions of 1300x650x75mm and reinforced 8mm bars. The slabs were tested under different supporting conditions and different loading types. The measured deflections at failure ranged from 4-30mm at mid span. Comparing the experimental results with the results calculated using the available OPC concrete equations showed good agreement for the maximum deflections; however, the calculated deflections at firs cracks loadings gave higher values than the measured ones. Madheswaran et al. [18] investigated the behaviour of the reinforced GBFS+FA-based GPC solid slabs under impact loading. The results were compared with similar OPC concrete slabs. It was reported that the energy absorption of GPC slabs was higher than that for the OPC concrete slabs at both cracking and failure stages, and this may be attributed to the lower modulus of elasticity which will reduce the GPC stiffness. On the other hand, it was observed that introducing steel fibres will increase the absorption of the OPC concrete slabs more than that for the GPC slabs and the performance of both types became similar.

Ganesan et al [19] investigated the behaviour of the reinforced FA-based GPC panels in one-way action in compare to the OPC reinforced concrete panels. Similar cracking patterns and mode of failure were observed in both types of them. The failure mode was by crushing of the concrete near the edges associated with large lateral deflections with maximum values at mid span. The load-deflection curves showed linear response until the appearance of the first cracks, after which it showed a nonlinear behaviour. The GPC concrete panels showed steeper curves, which indicated the higher ductility of the OPC panels. This was referred to the softening behaviour due to the existence of more fine particles in GPC mixes, which resulted in a less ductile behaviour. The current available ACI 318 code model found to give conservative results in predicting the ultimate load of the GPC panels. However, the calculated results were always lower that the experimental results. The error ranged from 20-35% depending on the aspect ratio and slenderness ratio. This may introduce the need to apply additional safety margins for predicting of the ultimate load of the GPC panels.

3. Stress-strain Relationship

The stress-strain behaviour of the FA-based GPC is not much different from that for normal strength OPC concrete [20, 21]. Using the available OPC concrete equations to draw the stress-strain relationship or to predict the strain at peak stresses for the GPC gave acceptable results [7, 22]. Compared to the OPC concrete, the FA-based GPC showed a similar behaviour up to the ultimate strength, after which a rapid decline in stress occurred during the post-peak strain softening; whilst, the OPC concrete showed a gradual strain softening after the ultimate strength was reached. The GPC displayed a more brittle behaviour compared to the OPC concrete [20, 21]. This may be related to the predominance of microcracks spread over the geopolymer microstructure [23, 24].

Sarker [13] suggested a modified model to properly predict the post-crack portion by introducing a new expression for the curve fitting value (n) using the same equations proposed for the high strength concrete by Thorenfeldt et al. [25]. The new model gave good predictions correlated well with the measured values. The peak strain for the different mixtures of GPC was recorded in the range of 0.0015-0.0026 [7, 22, 26, 27]. This is less than the 0.003 strain that is usually used in the OPC concrete design equations. Similar results [28] displayed that for 25 different geopolymer concrete mixes based on different FA sources, the ultimate strain was less than 0.002 whilst the stress ranged between 20-55 MPa. It was reported that the GPC was largely affected by the Si/Al ratio, and it was possible to produce a ductile GPC that failed in a ductile manner by deformations, and this was achievable by using a Si/Al ratio higher than 24 molars [29].

Prachasaree et al. [30] proposed a simplified empirical stress-strain model and stress-strain block parameter for the GPC design. They used the Thorenfeldt et al. [25] model for the OPC concrete as a base model. They suggested that the maximum strain at the peak stresses should be used rather than the maximum compressive strength to find the appropriate factor in terms of the elasticity modulus. The proposed model displayed a good correlation with the experimental data; however, for the post-peak portion, a small deviation appeared. The proposed model was used to find the equivalent rectangular stress block parameters and new expressions were proposed. The flexural capacity was calculated using the new model and based on the ACI 318 provision for flexural analysis. The proposed model showed very good correlation with the experimental results with an error range from

9-16% compared to 34% when using the currently available ACI 318 code models. And, this was validated over the FA-based GPC of strength below 75 MPa. Ganesan et al. [21] noted that for the same value of strain on the stress-strain diagram, the stress values for the FA-based GPC were greater than that of the OPC concrete. In addition, the initial deformation increased at a slower rate until the stresses reached 80% of its ultimate values; then, the deformation increased rapidly at a higher rate compared to the OPC concrete.

4. Limitations and Challenges

Despite the huge researches that have been carried out, the geopolymer-based concrete faces many challenges that need to be addressed. One of the main challenges is the absence of the standards which should be created by a global committee. The adaptation of such new materials will be limited by the institutional issues and lack of specifications' or standards' flexibility, rather than the obstructions of the technical issues. Appropriate standards that consider the performance as a base for concrete evaluation may be the most suitable solution for the adaptation of such new materials. On the other hand, the term "geopolymers" covers a wide range of source materials, which can confuse the designers or specifiers; hence, a correct category of the source materials should be selected for use in the concrete industry.

The creation of new codes will be very expensive so it requires the collaboration of governments, industries, and researchers. The main motivation for the adoption of such new materials is sustainability and the environmental issues. However, these issues are not enough to motivate the unwilling markets; the demand for new material should be required by the market itself. This can be achieved by increasing the global awareness and the adaptation of new regulations for greenhouse gas emissions' taxes and awards. End users are more willing to deal with the enhanced substantial properties like strength, cost, and durability rather than the environmental issues. In terms of the relatively low cost of the by-product raw materials, GPC was expected to reduce the cost by 10-30% compared to the conventional concrete [29]. In reality, most of these raw materials are controlled by long period contracts supported by the cement companies. It requires the development of new effective methods to reduce the activating solution requirements and produce cheaper activating solutions taking into account the efficiency in terms of greenhouse gas emissions. In light of the currently expected prices, GPC commercialisation is likely to be limited to the high-performance applications, including chemical, heat and fire resistance and hazardous waste management.

The long-term performance creates another main question for the acceptance of GPC. Most of the design engineers request at least 20-30 years of real world verification before such new materials are adapted for the construction industries. Lack of such data makes it unsuitable when the safety of the user is a critical concern. The available durability tests give indications about the expected performance; however, a definitive approach is urgently needed. Other hurdles to be overcome are more specifically related to the variation in the mechanical and physical properties which are related to the variability of the quality and compositions of the source materials [31]. This makes it difficult to directly compare the results of the available research. Adaptation of any new material requires the predictability and reproducibility of the fresh and hardened properties. GPC may require a new chemical and engineering point of view in which the chemical composition and rheological properties form the base for the evaluation of the product. The requirement for elevated temperature curing of the GPC makes another challenge that limits the application of GPC to the precast applications. Ambient cured GPC and one part geopolymer binder are highly required for the wide acceptance of GPC products. Some researchers have tried to increase the reactivity of the FA by increasing the fineness [32] or the addition of a calcium source material [33, 34]; however, this may affect the cost and durability of the GPC. The location of calcium in the GPC structure is needed to be clarified [3]. It is required for a deeper understanding of the physicochemical nature of the geopolymerisation reactions. Also, this may contribute to understanding the role of the other oxides that may exist in the raw materials, such as Fa, Mg, etc.

Heidrich et al. [35] performed a survey in Australia regarding the barriers that are facing the usage and acceptance of GPC. The survey included a wide range of the concrete industry stakeholders. More than 60% of the respondents found that the absence of standards made the first of the main challenges for GPC acceptance, in the second position came the concerns about the long-term properties, and productivity and safety issues came in the third position; however, cost and liability were found to be lower concerns.

5. Conclusions

The present work has reviewed the structural behaviour of the RGPC beams, columns, and slabs in comparison to normal OPC reinforced concrete members. The FA and GBFS-based GPC members can provide physical and mechanical properties which are comparable with the conventional cement concrete. The existing design provision available in the ACI 318 code and the AS3600 code standards are reported to be applicable for the analysis and design of the RGPC structures and in most cases will give conservative results. However, it is recommended to apply an additional safety factor to adjust for the unexpected long-term behaviour. Compared to the OPC concrete, the stress-strain curves of the FA-based GPC show similar behaviour up to the ultimate strength, after which a rapid decline in stress occurs during the post-peak strain softening. The GPC displays a more brittle behaviour comparing to the OPC concrete members. The peak strain for the different mixtures of GPC was recorded in the range of 0.0015-0.0026, which is less than 0.003 known for OPC concrete. Regardless of the wide body of available literature geopolymer concrete, there is still a significant gap regarding the engineering properties and the structural behaviour of the RGPC. It is required to clearly determine the relationships between the different properties of the GPC including elastic modulus, Poisson's ratio, tensile strength, flexure strength, compressive strength, shear strength, and bond strength. More research is required regarding these areas. The unavailability of the standards makes the major challenge for the acceptance of the GPC; evaluation of the GPC based on its performance characteristics seems to be the best way for acceptance of such a new material.

References

[1] U. N. Department of Economic and Social Affairs, 2014 *World urbanization prospects, the 2014 revision*, New York: United Nations.

[2] G. Habert, J. B. d'Espinose de Lacaillerie, and N. Roussel, 2011 *Journal of cleaner production*, vol. 19, pp. 1229-1238.

[3] P. Duxson, A. Fernández-Jiménez, J. L. Provis, G. C. Lukey, A. Palomo, and J. S. J. van Deventer, 2007 *Journal of materials science*, vol. 42, pp. 2917-2933.

[4] S. Kumaravel, S. Thirugnanasambandam, and C. A. Jeyasehar, 2014 IUP *Journal of structural engineering*, vol. 7, pp. 45-54.

[5] J. K. Dattatreya, N. P. Rajamane, D. Sabitha, P. S. Ambily, and M. C. Nataraja, 2011 *International journal of civil & structural engineering*, vol. 2, pp. 138-159.

[6] ACI-318-Committee, 2011 Building code requirements for structural concrete (aci 318-11) and commentary, *Farmington Hills: American concrete institute.*

[7] J. R. Yost, A. Radlińska, S. Ernst, and M. Salera, 2013 *Materials and structures*, vol. 46, pp. 435-447.

[8] J. R. Yost, A. Radlińska, S. Ernst, M. Salera, and N. J. Martignetti, 2013 *Materials and structures*, vol. 46, pp. 449-462.

[9] E. H. Chang, 2009 Shear and Bond Behaviour of Reinforced Fly Ash-Based Geopolymer Concrete Beams, PhD, *Department of civil engineering, Curtin university of technology*,

[10] C. K. Madheswaran, P. S. Ambily, N. Lakshmanan, J. K. Dattatreya, and S. J. Sathik, 2014 *ACI materials journal*, vol. 111, pp. 89-97.

[11] C. K. Madheswaran, P. S. Ambily, J. K. Dattatreya, and G. Ramesh, 2015 *Journal of the institution of engineers (India): Series A*, vol. 96, pp. 139-149.

[12] S. M. Laskar, R. A. Mozumder, and B. Roy, 2015 *Advances in Structural Engineering*, ed:

Springer, pp. 1643-1653.

[13] P. K. Sarker, 2009 *Materials and structures*, vol. 42, pp. 715-724.

[14] M. Rahman and P. K. Sarker, 2011 Geopolymer concrete columns under combined axial load and biaxial bending, *Concrete 2011 Conf.*, Australia, Western, pp. 12-14.

[15] D. M. J. Sumajouw, D. Hardjito, S. E. Wallah, and B. V. Rangan, 2007 *Journal of materials science*, vol. 42, pp. 3124-3130.

[16] T. Sujatha, K. Kannapiran, and S. Nagan, 2012 *Asian journal of civil engineering*, vol. 13, pp. 635-646.

[17] R. N. Laxminarayan, N. B. Mahantesh, and K. Amarnath, 2015 *International journal of engineering research & technology*, vol. 4, pp. 523-528.

[18] C. K. Madheswaran, J. K. Dattatreya, P. S. Ambily, and P. R. Karansingh, 2014 *International journal of innovative research in science, engineering and technology*, vol. 3, pp. 10775-10786.

[19] N. Ganesan, P. V. Indira, and A. Santhakumar, 2013 *Construction and building materials*, vol. 48, pp. 91-97.

[20] G. M. Haider, J. G. Sanjayan, and P. G. Ranjith, 2014 *Construction and building materials*, vol. 69, pp. 196-202.

[21] N. Ganesan, R. Abraham, S. Deepa Raj, and D. Sasi, 2014 *Construction and building materials*, vol. 73, pp. 326-331.

[22] A. I. Laskar and R. Bhattacharjee, 2013 *ACI materials journal*, vol. 110, pp. 513-518.

[23] R. J. Thomas and S. Peethamparan, 2015 *Construction and building materials*, vol. 93, pp. 49-56.

[24] C. D. Atiş, C. Bilim, Ö. Çelik, and O. Karahan, 2009 *Construction and building materials*, vol. 23, pp. 548-555.

[25] E. Thorenfeldt, A. Tomaszewicz, and J. Jensen, 1987 "Mechanical properties of high-strength concrete and application in design, *Proc. of the Symp. "Utilization of High Strength Concrete*, pp. 149-159.

[26] G. S. Manjunatha, Radhakrishna, K. Venugopal, and S. V. Maruthi, 2014 *Transactions of the indian ceramic society*, vol. 73, pp. 149-156.

[27] D. Hardjito and B. V. Rangan, 2005 Development and Properties of Low-Calcium Fly Ash-Based Geopolymer Concrete, PhD, *Curtin university of technology*, Perth, Australia

[28] E. I. Diaz-Loya, E. N. Allouche, and S. Vaidya, 2011 *ACI materials journal*, vol. 108, pp. 300-306.

[29] K. Komnitsas and D. Zaharaki, 2007 Minerals engineering, vol. 20, pp. 1261-1277.

[30] W. Prachasaree, S. Limkatanyu, A. Hawa, and A. Samakrattakit, 2014 *Arabian journal for science and engineering*, vol. 39, pp. 8549-8558.

[31] C. Gunasekara, D. W. Law, S. Setunge, and J. G. Sanjayan, 2015 *Construction and building materials*, vol. 95, pp. 592-599.

[32] K. Somna, C. Jaturapitakkul, P. Kajitvichyanukul, and P. Chindaprasirt, 2011 *Fuel*, vol. 90, pp. 2118-2124.

[33] J. Temuujin, A. Van Riessen, and R. Williams, 2009 *Journal of hazardous materials*, vol. 167, pp. 82-88.

[34] A. M. Rashad, 2014 *Materials & Design*, vol. 53, pp. 1005-1025.

[35] C. Heidrich, J. Sanjayan, M. L. Berndt, S. Foster, and K. Sagoe-Crentsil, 2015 Pathways and Barriers for Acceptance and Usage of Geopolymer Concrete in Mainstream Construction, 2015 *World of Coal Ash (WOCA) Conference*, Nashville, Tennessee.

Mixed Consolidation Solution for a Reinforced Concrete Structure

M Lute[1]

[1] CCI Department, Politehnica University Timisoara, 2 T. Lalescu Str., Timisoara, Romania

E-mail: marina.lute@upt.ro

Abstract. During the last years, reinforced concrete structures become subject for rehabilitation due to two factors: their long life span and large change in norms that leaded to a large increase of seismic loads in Eastern Europe. These lead to a necessity for rehabilitation of existing building stock in order to use them during their entire life span at the maximum potential. The present paper proposes a solution for rehabilitation for three reinforced concrete building of a hospital, that consumed a half of their life span and do not correspond anymore to present norms. The chosen solution is a combination between CFRP rehabilitation and increase of structural elements cross section in order to achieve the stiffness balance in the structure nodes that is required by present norms. As a further matter, correction in stiffness of local elements diminished the lateral drifts of the structure and improved the global seismic response of the building.

1. Introduction

Structural rehabilitation of reinforced concrete structures become a matter of sustainable development of modern cities [1,2] because of both economic and technical reasons. The technical side of the problem lies in the fact that the design norms [2-13] for seismic structural response suffered a large increase of design values, making that a large part of building stock designed before 1990 to became unsafe for leaving from seismic point of view, especially for Romania. From the economical side, demolishing all these buildings is not an option, so reliable and easy to use rehabilitation solution should be provided by Engineers.

The buildings that are subject to this debate, have structural flows only because the change in design values, their behaviour at vertical loads still being accepted by actual requirements. Moreover, the development of non-destructive testing methods [5-8] shown that the materials integrity was preserved during the years. Also, tests regarding the materials behaviour at polluted environments were conducted [6] for areas where acids were used for laboratory testing. These considerations contributed to the rehabilitation decision concerning the existing modern buildings.

The three buildings descripted in the present paper are part of a large hospital complex that fulfil all the above hypothesis.

2. Historical overview. Existing situation

The considered buildings are part of the Timis County Hospital, the largest hospital in the Western Romania. The hospital was created as a mixture of fourteen different buildings, containing all medical specialties and provide education act at the same time for medicine students. Eight buildings are arranged to form an interior courtyard, four lateral buildings and one building for heating plant.

IOP Conf. Series: Materials Science and Engineering **133** (2016) 012022 doi:10.1088/1757-899X/133/1/012022

The buildings are different both in terms of architecture, with different height levels, ranging between one and ten levels, different functionalities, and different strength structures adapted to support each of these requirements. The assembly and was designed in 1970 and built in 1974.

From the total assembly, the buildings subject of this paper are marked in figure 1. The buildings F and G are four levels high and building H is 3 levels high, one level being basement for all buildings.

The surgical department for the entire hospital is placed at the second floor of all these three buildings, and has the functional arrangement given in figure 2. Given its importance, the general rehabilitation of the hospital complex started with this department.

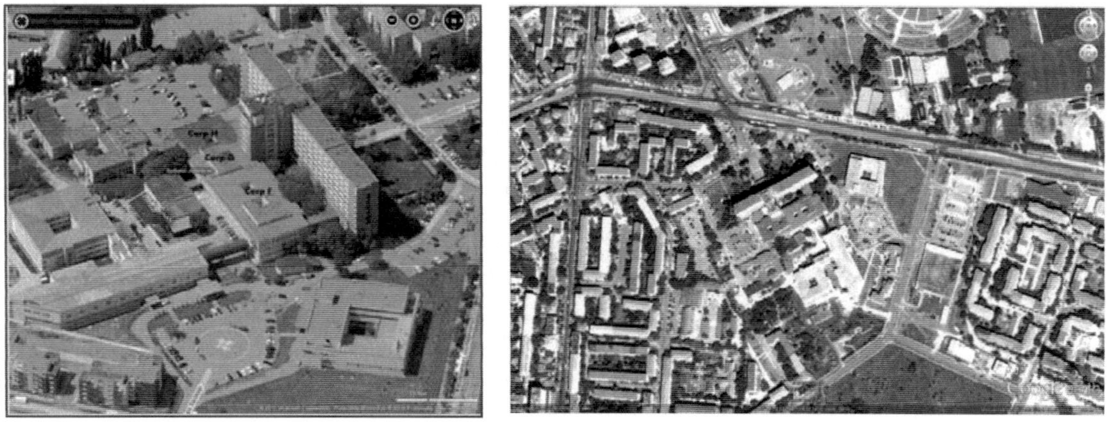

Figure 1. Timis County Hospital – aerial view

Made in 1974, the surgery department functioned without interruption, a situation that led to justified degradation both of interior finishing, construction of specific facilities, medical facilities and medical devices supplied. Above these considerations, the current situation was also characterized by the fact that the operating procedures worked without the specific standards and regulations occurred mainly after 2000 when Romania joined the EU. Moreover, no integrated modernization or upgrading works were carried out during this time. A general upgrade of the functional spaces, finishes and instrumentations was then required.

Structural description

Al the three buildings are spatial frames reinforced concrete structures, and the structure integrity is almost perfectly preserved. Only interior and exterior finishing were damaged, as seen in figure 2.

Figure 2. State of finishing degradations.

The only problem appeared concerning structure integrity is the change in norms, so that the location that has been an area free of earthquakes [1,3] after the norms valid in 1970 [10], changed in 2006 into an area strongly affected by seismic loads [11].

3. Improvement measures proposed at structure

Architectural interventions considered

In order to implement the measures for good function of surgical block, the existing space was proposed to change its functions both as dimensions and interior traffic management. With the reconfiguration, resizing and redistribution of surgical area, it requires rethinking traffic of clean / dirty items both on horizontal and vertical directions. In the current situation the traffic flow for clean / dirty items is overlapping horizontally and intersected in vertical transport, thing not accepted by European regulations.

The measures proposed can be summarized as follows:

- Overall increase of total area designed for surgical department by adding cantilever slab around the floor perimeter. The space is added both at operational floor as seen in figures 2 and 3, and at terrace (figure 4 and 5).
- Creating a technical space on terrace, by placing there all the necessary technological equipment.
- Create auxiliary closed space on extended terrace with role for technical staff and equipment.
- Add supplementary elevators in order to increase the vertical flux for patients.
- Add two new staircases for vertical access
- Increase all the existing stair with one additional floor

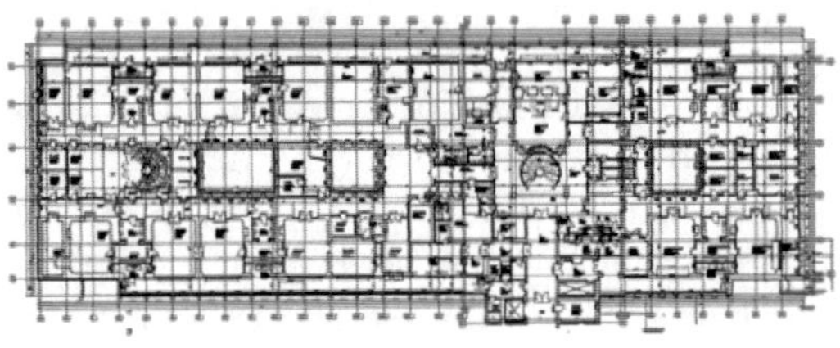

Figure 2. Surgery floor: existing situation.

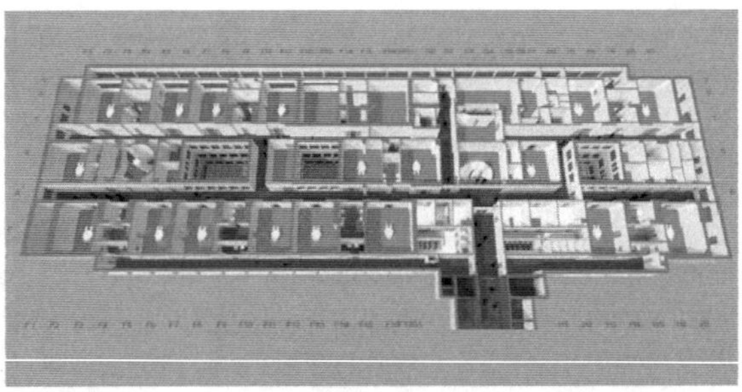

Figure 3. Surgery floor: proposed solution.

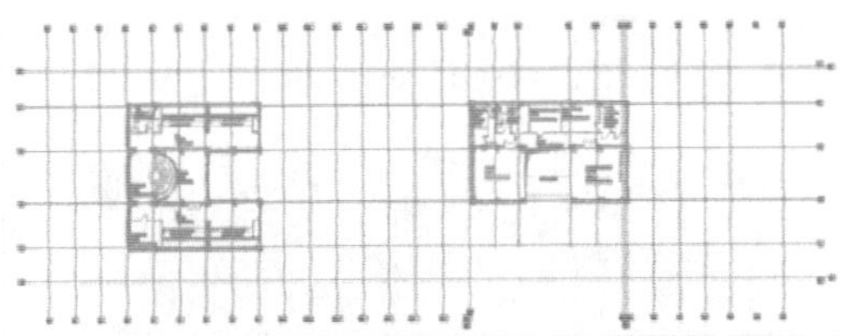

Figure 4. Terrace existing situation.

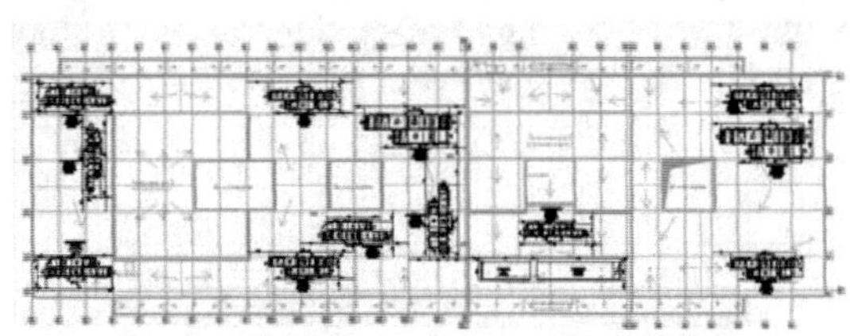

Figure 5. Terrace proposed solution.

The proposed modifications increased the existing surface of surgery from 3276 m^2 to 4854 m^2. The supplementary added space was used in order to obtain:

- Increase in dimension of all surgical rooms, and arrange them upon medical specialties;
- Separate wards for sick – before and after surgery – could be arranged;
- Auxiliary spaces for medical personnel could be added;
- Continuity in flows of "clean" aseptic with the "dirty" septic routs and total separation of them, so no accidental overlapping of the two routes could be possible. This was possible by

reserving the mid lobbies to septic routes and perimeter lobbies to aseptic routes. In this way, the central core of the floor is kept completely aseptic;

- The new configuration insures more transversal connection between longitudinal lobbies;
- The vertical transport could be divided between sterile and non-sterile routes;
- All supporting technical and technological equipment could be moved on the extended terrace (figure 5).

Structural interventions considered

The structural interventions were generated by architectural requirements and can be summarized as follows:

- Slab floors enlarged with cantilevers in composite metallic-reinforced concrete structure (metallic cantilever beams with connector at top side that support a concrete slab)
- Close the slab openings (where needed) using metallic solutions (as network of metallic girders)
- Heightening existing stairwells with metal spiral staircase
- Create new opening in floor slabs for new elevator emplacement
- Crate new storey for technical purpose using spatial metallic frame solution.
- New independent structures for staircases and elevator shaft in reinforced concrete solution: shears walls and spatial frames

As it can be seen the solutions for spatial interventions were chosen to add the minimum additional load possible, taking into account the fact that the seismic behaviour will be further diminished by any additional load. The heavy elements that had to be added – new staircases and elevator shaft – were created as new independent structures that do not interfere with the existing one. Great care was considered at placement of the new foundations between the existing ones to that the foundation system not to be disturbed.

4. Structural assessment

The structural assessment for these buildings had to take into account several distinct aspects as loads change due to norms and new added structural and technological loads together with the characteristic seismic behaviour of the existing structure due to their design particularities.

Changes in load levels

The buildings' design was made in 1971, when Timisoara city was included by seismic zoning map in the area that did not require structural calculation of seismic loads, dimensioning being made for gravitational loads only. The seismic norms weren't the only more permissive norms. The levels of design loads were also lower than that provided by current regulations, and partial safety coefficients considered in evaluating the loads were much lower than that required by Eurocodes.

Table 1 bellow shows the comparison of these modifications. It also shows some spotlights regarding the existing reinforcements toward the new technological requirements and change in seismic design requirements.

Table 1. Changes in load, intensity safety factors and technological requirements.

1971 – year of building design	2012
SEISMIC ZONING	
- according P13-70 was not necessary to calculate the seismic loads	- P100/1-2006 Normative : ground acceleration for seismic design 0.16g
PARTIAL SAFETY COEFFICIENTS	
- dead load of structural elements	
1.1	1.35

Table 1. Changes in load, intensity safety factors and technological requirements.

1971 – year of building design	**2012**
- dead load of finishing elements	
1.3	1.35
- live load offices	
200 x 1.3 = 260 daN/m^2	300 x 1.5 = 450 daN/m^2
EXISTING SITUATION ACCORDING TO DESIGN	**MINIMAL CONDITIONS IMPOSED BY NORMATIVE P100/1-2006**
* *Columns reinforcement*	
Total amount of longitudinal reinforcement - A_s = 11.12 cm^2	Minimum acceptable level: - for the minimum percentage of reinforcement p_{min}=1%$\Rightarrow$ A_s =15.75 cm^2
Transversal reinforcement: - stirrups ϕ6/12.5cm $\Rightarrow$ p=0.11%	Minimum acceptable level: - stirrups ϕ8/10cm - p_{min} = 0.5%
Concrete quality: - class C16/20	Minimum acceptable level: - minimum admitted class C25/30 (H ductility class)
- strength rebars placed in cross section corners only, no matter the distance between them	- distance required between strength rebars: maximum 25 cm

From table 1 it can easily be seen that, besides the loads change, the structures do not respect a large part of constructive conditions imposed by seismic norms for concrete quality, quantity and position of reinforcement, things that must be corrected by the rehabilitation solution.

Seismic behaviour of existing structure

Even in the conditions shown above, the buildings' conception meet some of the issues specified in P100/1-2006. The particularities of structure behaviour relative to seismic standards consist in:

- Structural simplicity: the structure has been divided into 3 separate structure with expansion joints in-between to provide better structural regularity of the structural assembly. The structural system is mostly a continuous one, and provides a clear path, a direct and uninterrupted transmission of seismic forces, regardless of their direction towards the foundation soil
- Seismic forces that arise in all building elements are sent to the 20 cm thick floor slabs that works as horizontal diaphragms. For thin slabs with 8-10 cm height, the horizontal forces are transmitted through the longitudinal and frame beams and sent further to vertical columns.
- There are no important discontinuities in the path of seismic loads transmission: the slabs are continuous, without notable openings, the beams have horizontal, continuous paths after two perpendicular directions, and the columns show a vertical continuity.
- Structural redundancy: from the geometric viewpoint, one can consider that a plastic mechanism was achieved and have enough potential plastic hinges, enabling the exploitation of strength reserves of the structure and have to show an advantageous seismic energy dissipation pattern.
- Structure configuration: structural regularity in plan and elevation of all three buildings is partially achieved; thus, all the buildings show some geometrical discontinuities in limited areas and present short columns.
- The cross section sizes for structural columns and beams do not meet the condition of rigidity required by P100-2006. The beams' stiffness is in many cases greater than the columns', so the structure nodes are compromised from the point of seismic behaviour.

– The concrete used for structure do not meet the basic requirement for structures in ductility class H (C25/30), or even for M class (C20/25)

5. Rehabilitation solutions

The MEP seismic spatial design was conducted first for the existing buildings without structural changes, these being added at a later stage. At all stages, the loads were considered with their real value and amended with partial factor of safety according to the regulations currently in force. The levels of lateral stiffness of the structure are non-compliant, none of lateral drifts meet the accepted level of P100. Moreover, all three buildings are subject of second order effect of torsion. The bearing capacity of structural elements – beams, columns and part of slabs – does not cover the efforts that has to be supported by the building, and the ratio between the stiffness of beams and columns do not meet P100 regulations for directing the plastic hinges in beams.

Consequently, the superstructure must be strengthened so that can meet the current requirements of strength and stability. On the other hand, the infrastructure does not require structural interventions because the foundation blocks have the appropriate dimensions so the pressure transmitted to the foundation soil not to exceed the permitted values. In addition, the geotechnical study shows that the admitted soil pressures can be 10% increased, value that covers the supplementary loads brought by the new metallic light storey.

The solution choice was made by combining the qualities of traditional strengthening solutions with modern solution of bearing capacity increase in order to match the goals of the entire building. Thus the advantages of beams and columns strengthening by reinforced concrete cross section increase:

- increase of structure lateral stiffness and limitation of lateral drift
- gives the possibility to add additional reinforcement to reach the minimum required quantity
- reliable technological solutions for node anchorage of longitudinal rebars from beams
- eliminate the second order effects that introduced additional efforts in structure
- solution economically advantageous

With the advantages of carbon fiber use [4] for bearing capacity increase:

- short execution time;
- execution can be done on limited areas one by one, so that the medical activity can be stopped on small areas;
- increase both flexural ad shear force bearing capacity of structural elements;
- economic disadvantage – 75% more expensive than the classic rehabilitation solution

The strengthening measurers agreed to cover all these aspects were the following:

a. Increase the lateral stiffness of the structure through the increase of reinforced concrete cross sections and add new reinforcement for vertical structural elements. This measure insures also the correction of the ratio between stiffens of beams and columns in node. In this way, the bearing capacity of the column is corrected and brought to the required bearing capacity to support all the structural loads. The transversal reinforcement could also be corrected with this solution.
b. Increase the bearing capacity of horizontal strength elements that do not have enough bearing capacity in regard with norms and actual load level. There has been adopted several measures, taking into account the actual behaviour for each structural element:
   - Ground level: carbon fiber strengthening at top and bottom side of the slab
   - First and second level: consolidation of all beams with 6 cm reinforced concrete at top side, with additional reinforcement connected to the old one, together the bottom side consolidation of beams using carbon fibers.
   - Second level: additional consolidation for cantilever beams using fiber carbon on top side
   - Third level: general rehabilitation for beams and slab with 5 cm thick reinforced concrete at top side, together with new cuts for openings required by the new technological circuits.

6. Conclusions

By strengthening the structure in the presented manner, the following results were achieved:

- The lateral drifts are within the permissible limits, both at the ultimate limit state and service limit state;
- Bearing capacity of beams and columns matches the efforts level in the structure
- The effects of torsion do not affect the structure, the second order effect being cancelled
- The reinforcement layout both for longitudinal and transversal rebars, complies with the normative requests for all structural elements
- No unnecessary interventions were made. The blocks of foundation that have appropriate dimensions did not suffer any structural interventions.
- The overall structure will work as new structure, designed according to P100/1-2006.

These goals were achieved only by a smart combination of strengthening solutions with carbon fibers and reinforced concrete and by considering all supplementary elements added to the structure in light metallic solutions.

References

[1] D. Anastasescu, 2012, Aspecte ale evoluţiei reglementărilor tehnice privind protecţia antiseismică a construcţiilor din municipiul Timişoara, *Buletinul AGIR*, vol.**3**, pp. 9-15.

[2] C. Bob and T. Dencsak, 2012, RC structures a solution for sustainable development, *Global Thinking in Structural Engineering: Recent Achievements*

[3] D. Cretu and S. Demetriu, 2006, Metode pentru calculul răspunsului seismic în codurile româneşti de proiectare. Comparaţii şi comentarii, *Revista AICPS*, pp. 1-9.

[4] A.C., Dăescu, T., Nagy-György, V., Stoian and D. Dan, 2009, Experimental study on the retrofitting procedures for RC columns, *Proceedings of the Twin Coimbra International Conferences on Civil Engineering - Towards a Better Environment, CE 2009 and the Concrete Future, CF 2009*, pp. CF119-CF128

[5] F. Iacob, 2010, Relativistic pseudo-Gaussian oscillators, *Physics Letters, Section A: General, Atomic and Solid State Physics*, vol. **374 (11-12)**, pp. 1332-1335

[6] A.M. Izzat, A.M.M. Al Bakri, H. Kamarudin, L.M. Moga, G.C.M. Ruzaidi, M.T.M. Faheem and A.V. Sandu, 2013, Microstructural analysis of geopolymer and ordinary Portland cement mortar exposed to sulfuric acid, *Materiale plastice,* Vol. **52(2)**, pp.171-174

[7] M. Lute, 2015, Non-destructive vibration tests on reinforced concrete structures, *Key Engineering Materials*, vol. **660**, pp. 186-191

[8] S.L., Matthews, T., Ueda, T. and A., Bigaj-van Vliet, 2012, Conservation of concrete structures in fib model code 2010, Concrete Repair, Rehabilitation and Retrofitting III - *Proceedings of the 3rd International Conference on Concrete Repair, Rehabilitation and Retrofitting, ICCRRR 2012*, pp. 549-554

[9] J., Valença, C., Almeida and E.S., Júlio, 2012, Concrete heritage: Tentative guidelines for the 'Patch Restoration Method', Concrete Repair, Rehabilitation and Retrofitting III - *Proceedings of the 3rd International Conference on Concrete Repair, Rehabilitation and Retrofitting, ICCRRR 2012*, pp. 1023-1028

[10] P13-1970, Seismic design code

[11] P100-1/2006, Seismic design code

[12] SR EN 1992-1, Design of concrete structures. General rules and rules for buildings

[13] M. Boştenaru Dan, The economic impact of seismic retrofit on heritage reinforced concrete buildings, Urbanism.Architecture.Constructions, vol 5(2), pp. 71-82

IOP Conf. Series: Materials Science and Engineering **133** (2016) 012023 doi:10.1088/1757-899X/133/1/012023

Properties and Microstructural Characteristic of Kaolin Geopolymer Ceramics with Addition of Ultra High Molecular Weight Polyethylene

Romisuhani Ahmad[1*], Mohd Mustafa Al Bakri Abdullah[1,2] , Kamarudin Hussin[1,2], Andrei Victor Sandu[3], Mohammed Binhussain[4], Nur Ain Jaya[2]

[1] Faculty of Engineering Technology, Universiti Malaysia Perlis (UniMAP), P.O Box 77, D/A Pejabat Pos Besar, 01000 Kangar, Perlis, Malaysia.
[2]Center of Excellence Geopolymer and Green Technology (CEGeoGTech), School of Materials Engineering, Universiti Malaysia Perlis (UniMAP), P.O Box 77, D/A Pejabat Pos Besar, 01000 Kangar, Perlis, Malaysia.
[3] Gheorghe Asachi Technical University of lasi, Faculty of Materials Science and Engineering, Blvd. D. Mangeron 71, 700050, lasi, Romania
[4] King Abdul Aziz City Science & Technology (KACST), P.O. Box Riyadh 11442, Kingdom of Saudi Arabia

[1*]romisuhani@studentmail.unimap.edu.my, [1,2]mustafa_albakri@unimap.edu.my

Abstract. In this paper, the mechanical properties and microstructure of kaolin geopolymer ceramics with addition of Ultra High Molecular Weight Polyethylene were studied. Inorganic polymers based on alumina and silica polysialate units were synthesized at room temperature from kaolin and sodium silicate in a highly alkaline medium, followed by curing and drying at 80 °C. Alkaline activator was formed by mixing the 12 M NaOH solution with sodium silicate at a ratio of 0.24. Addition of Ultra High Molecular Weight Polyethylene to the kaolin geopolymer are fabricated with Ultra High Molecular Weight Polyethylene content of 2, 4, 6 and 8 (wt. %) by using powder metallurgy method. The samples were heated at 1200 °C and the strength and morphological were tested. It was found that the flexural strength for the kaolin geopolymer ceramics with addition of UHMWPE were improved and generally increased with the increasing of UHMWPE loading. The result revealed that the optimum flexural strength was obtained at UHMWPE loading of 4 wt. % (92.1 MPa) and the flexural strength started to decrease. Microstructural analysis showed the samples appeared to have more number of pores and connected of pores increased with the increasing of UHMWPE content.

[1] Faculty of Engineering Technology, Universiti Malaysia Perlis (UniMAP), P.O Box 77, D/A Pejabat Pos Besar, 01000 Kangar, Perlis, Malaysia : romisuhani@studentmail.unimap.edu.my

1. Introduction

Since the concept of geopolymer was introduced by Joseph Davidovits in 1978, geopolymer have become familiar and prevalent materials all over the world due to their specific performance of high temperature resistance, high compressive strength, low density, low permeability and so on [1]. With the many advantages offered, geopolymers are considered as fine engineering materials with a broad application [2 – 5]. It can be used as fire and heat resistant fibre composites, sealants, concretes, ceramics, etc., depending on the chemical composition of the source material and the activators. It also can be used as waste encapsulation to immobilize toxic metals. Geopolymer are normally made by mixing materials that contain aluminosilicate, such as kaolin, metakaolin and fly ash, into alkaline solution and curing the mixture at a certain temperature [6 – 7]. Theoretically, any alkali and alkali earth cation can be used as an alkali element in this exothermic process, geopolymerization. Besides, any material composed of alumina and silica can be alkali-activated. In this study, kaolin was used as aluminosilicate material to prepare geopolymer.

The strength of geopolymers influenced by the nature of the source materials. Geopolymers that are synthesized using calcined source materials, e.g., fly ash, slag, and metakaolin, have higher compressive strength than those synthesized using non-calcined materials, such as kaolinite, and naturally-occurring minerals [8 – 10]. Several studies have focused on the use of metakaolin [11 – 13] because of its higher reactivity, which contributes to optimal performance, slightly than the less reactive of kaolin [14]. Besides, the use of metakaolin instead of other materials has been found to produce more homogeneous geopolymers [15]. Thus, there is a lack of study on green polymeric composites by using kaolin. In the development of geopolymer ceramic, kaolin is a potential material to be used since it can be calcined at high temperature [7].

The amorphous to semi-crystalline behaviour of geopolymer will change into crystalline ceramic phases upon heating [11, 16, 17]. Therefore, geopolymer technology offers a new route to produce advance ceramics of profitably low cost and near net-shape forming. As a result of their excellent physical and mechanical properties, geopolymer ceramic composites are expected to emerge as potentially stronger and tougher materials that can meet the demanding requirements of material applications. Jonathan et al. 2009 [11] also proved that it was possible to produce pollucite ceramic from the natural matakaolin-based geopolymer in the formation of pollucite ($Cs_2O \cdot Al_2O_3 \cdot 4SiO_2$). A geopolymer that consist of a mixture of metakaolin and cesium silicate solution were prepared using an IKA overhead mixer. Then the resultant slurry was cured at 50 °C for 24 h. The pollucite crystallization occurred gradually during heating at temperatures in the range of 900 - 1250 °C. While He et al. 2010 [12] studied the influence of temperature on microstructure and mechanical properties of carbon fibre reinforced geopolymer composites that heat treated those materials between 1000 °C and 1400 °C. They found that the mechanical properties of these composites can be greatly improved by heat treated at 1400 °C and tend to fracture in a very brittle manner owing to the seriously degraded carbon fibres together with the matrix melting and crystal phases dissolve.

A new approach of addition of polymer (UHMWPE) into geopolymer ceramic materials prepared by using powder metallurgy method and sintered at 1200 °C. Since the melting point of UHMWPE is low and expected to decompose when heated at high temperature, the effect of various UHMWPE loadings on flexural strength and microstructure were studied.

2. Experimental

2.1 Materials

Kaolin used in this study was supplied by Associated Kaolin Industries Malaysia as Si-Al sources materials. Table 1 summarised the chemical composition of kaolin analysed by X-ray fluorescence (XRF). The particle size distribution of aluminosilicate materials was obtained using Malvern particle size analyser. Results showed that kaolin consists of very fine particles with the most abundant particle size in the range of 90 - 120μ.

Sodium hydroxide (NaOH) powder used was of caustic soda micropearls, 99% purity with brand name of Formosoda-P made in Taiwan. Sodium silicate (Na_2SiO_3) chemical composition are 30.1% of

SiO_2, 9.4% of Na_2O and 60.5% of H_2O (modulus, SiO_2/Na_2O = 3.2), specific gravity at 20 °C = 1.4 g/cm^3 and viscosity at 20 °C = 0.4 Pa s.

The powdered formed UHMWPE, with a molecular weight of 5 X 106 g/mol and a density of 0.94 g/ml was used as a binder in kaolin geopolymer ceramics and purchased from Ticona Engineering Polymer, China.

Table 1: Chemical composition of kaolin.

Composition	SiO_2	Al_2O_3	Fe_2O_3	K_2O	TiO_2	MnO_2	ZrO_2	LOI
Percentage (%)	54.0	31.7	4.89	6.05	1.41	0.11	0.10	1.74

2.2 Samples preparation

Alkaline activator was formed by mixing the 12M of NaOH solution with Na_2SiO_3 at a ratio of 0.24. The solution was then allowed to mature for 24 hours in order to dissolve the silica completely. The geopolymer paste was prepared by mixing kaolin powder with alkaline activator with the solid-to-liquid ratio of 1.0 and were stirred well for a few minutes by using a mechanical mixer. The details of the mixtures involved were presented in Table 2. This ratio was chosen based on our previous work, as it allowed for optimal mechanical and thermal properties. The samples were then cured at 80 °C for 24 h.

To obtained fine powder, kaolin geopolymer were crushed using mechanical crusher and sieved manually by using 150µ siever. The kaolin geopolymer powder were then mixed with UHMWPE by using planetary mill in dry condition method. The mixing process was done for 4 minutes at 100 rpm with reverse direction to get the homogeneous mixture. The mixing formulae for UHMWPE and kaolin geopolymer are shown in Table 3.

The mixture was then compacted using a 12-mm-diameter cylindrical stainless steel die at 5 ton for 2 minutes. The final green body was sintered in a high temperature Carbolite furnace 1200 °C with 3 h soaking time and a cooling and heating rate of 5 °C/ min. All samples were sintered on an alumina plate as it can withstand high temperature firing.

Table 2: Mix design details.

S/ L	NaOH molarity	Na_2SiO_3/ NaOH	Total molar ratio		
			Na_2O/ SiO_2	SiO_2/ Al_2O_3	H_2O/ Na_2O
1	12	0.24	0.3950	3.2022	11.5730

Table 3: Mixing formulation of UHMWPE/kaolin geopolymer at different content.

Materials	**Formulation**
Kaolin geopolymer (wt. %)	100
UHMWPE (wt. %)	0, 2, 4, 6, 8

2.3 Testing

2.3.1 Flexural Strength

Flexural strength measurements of clay geopolymer ceramics were conducted on specimens (7 mm x 5 mm x 52 mm) using a three-point-bending fixture on an instron-500 tester, with a span length of 30 mm at a crosshead speed of 0.5 mm/min. A minimum of five samples for each group were subjected to flexural strength testing in accordance with ASTM C1161.

2.3.2 Morphology Analysis

JSM-6460LA model Scanning electron microscope (JEOL) was performed to reveal the microstructure of kaolin geopolymers at S/L of 1.0. The specimens were prepared and coated by using Auto Fine Coater; model JEOL JFC 1600 prior to examination.

3. Results and Discussion

3.1 Flexural Strength

The effect of UHMWPE content on the kaolin geopolymer ceramics heated at 1200 °C is shown in Figure 1. The flexural strength of kaolin geopolymer ceramics has been improved after incorporation of UHMWPE. As compared to the kaolin geopolymer ceramics without the addition of UHMWPE, the flexural strength increased at 2 wt. % of UHMWPE content and started to decrease when the UHMWPE content exceeded 4 wt. %. Kaolin geopolymer ceramics with addition of UHMWPE at 4 wt. % having the highest strength of 90 MPa. This is due to the decomposition of UHMWPE at high temperature attributed to increasing of carbon in the geopolymer system. As reported from previous researchers the fine distributed excess of carbon can be advantageous for the resistance against crystallization up to high temperatures and at the same time generates a complex microstructure for carbon-rich of geopolymer ceramics hence improve in flexural strength [18]. However, the flexural strength decreased at the optimum value were due to the porosity increased to a large extent hence reduced the strength [19]. Xu et al., (2015) observed the similar findings in their research on the effect of dolomite and spodumene for the performances of andalusite composite ceramics. They found the addition of dolomite has improved the mechanical properties of the andalusite composite ceramics and the excessive of dolomite decreased the bending strength [20].

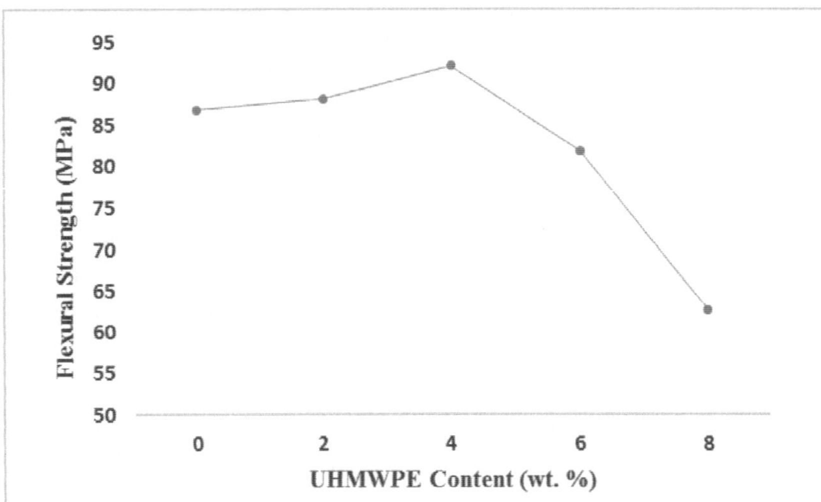

Figure 1: Flexural Strength of Kaolin Geopolymer Ceramics with various UHMWPE loading sintered at 1200 °C.

3.2 Morphology Analysis

Figure 2 shows SEM micrograph for raw kaolin while figure 3 kaolin geopolymer ceramics sintered at 1200 °C. Raw kaolin (Figure 2) has plate-like structure which contributes to the low reactivity compared to fly ash or any other aluminosilicate materials [21]. The appearance of small pores throughout the sample of kaolin geopolymer ceramics sintered at 1200°C denotes the transformation of amorphous to nepheline crystalline ceramic [22]. Figure 4 shows SEM micrographs for kaolin geopolymer ceramics with various UHMWPE content sintered at 1200 °C. The microstructure of fractured surfaces of kaolin geopolymer ceramics at various UHMWPE content showed that the microstructure of the investigated materials was considerably affected by the content of the UHMWPE. The number of pores and connected of pores increased with the increasing of UHMWPE content. As mention above that the more UHMWPE addition would result in higher porosity, the flexural strength decreases with the increasing of UHMWPE content. The strength is highly related to the size, quantity and distribution of pores, as well as the morphology and size of crystals [23].

Figure 2: SEM micrograph of raw kaolin

Figure 3: SEM micrograph of kaolin geopolymer ceramic sintered at 1200 °C.

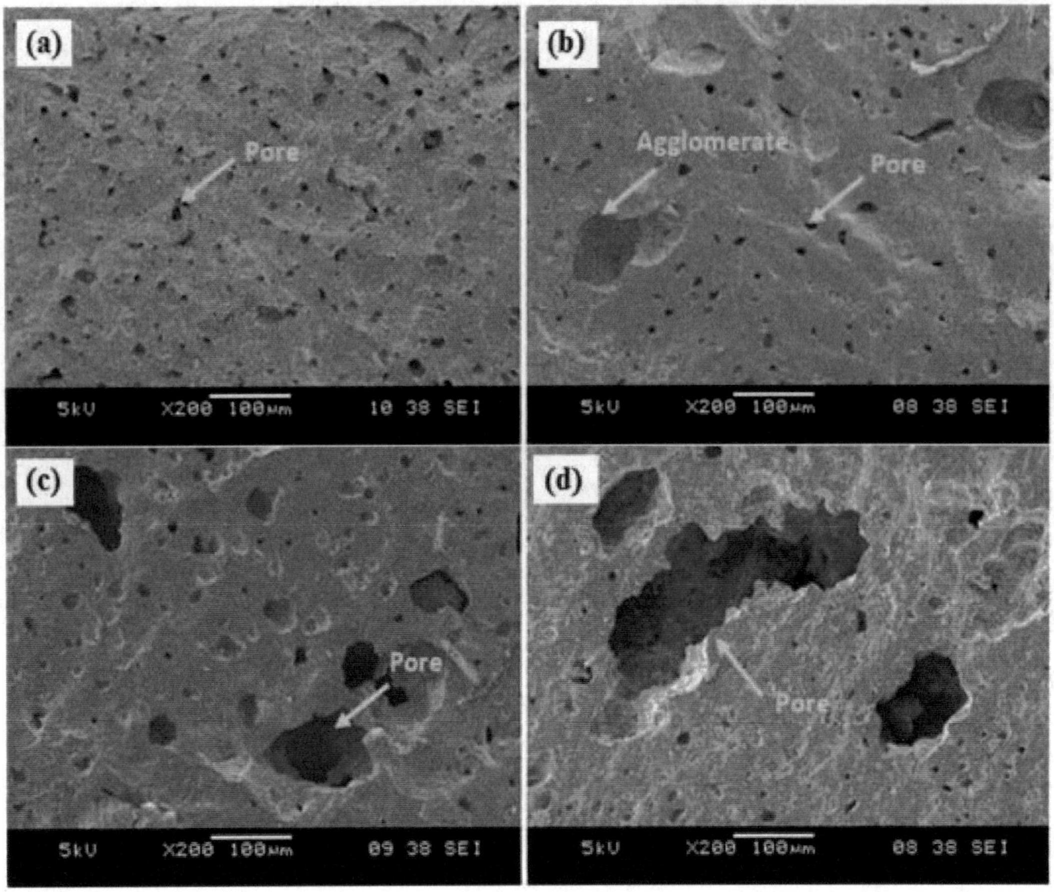

Figure 4: SEM micrograph of kaolin geopolymer ceramics at various UHMWPE loading (a) 2 wt.%, (b) 4 wt.%, (c) 6 wt.%, and (d) 8 wt.%.

4. Conclusion

The aims of this paper is to study the effect of addition of Ultra High Molecular Weight Polyethylene (UHMWPE) on kaolin geopolymer ceramics. The results concluded that the addition of UHMWPE affected the strength of the kaolin geopolymer ceramics. The flexural strength of kaolin geopolymer ceramics with addition of UHMWPE increased compared to the kaolin geopolymer ceramics without addition of UHMWPE. Kaolin geopolymer ceramics with addition of 4 wt. % UHMWPE give the highest strength which is 92.1 Mpa. However, the strength started to decrease due to the excessive content of UHMWPE lead to the increasing of pores size appeared in microstructure of kaolin geopolymer ceramics.

Acknowledgments

The authors of the present work wish to dedicate their great thanks to the Centre of Excellence Geopolymer System Research @UniMAP.

References

[1] Davidovits J 1991 *J. Therm. Anal.* **37**(8) 1633
[2] Duxson P, Lukey G C, and van Deventer J S J 2006 *J. Non-Crystal. Sol.*, **352** 5541

[3] Swanepoel J C, and Strydom C A 2002 *Appl. Geochem.* **17**(8) 1143
[4] Roviello G, Ricciotti L, Ferone C, Colangelo F, and Tarallo O 2015 *Cem. Conc. Compo.* **59** 89
[5] Cheng T W, Lee M L, Ko M S, Ueng T H, and Yang S F 2012 *Appl. Clay Sci.* **56** 90
[6] Al Bakri, A M M, Kamarudin H, Bnhussain M, Nizar I K, Rafiza A R, and Izzat A M 2011. *J. Appl. Sci. Res.* **7** 1199
[7] Liew Y M, Kamarudin H, AlBakri A M M , Binhussain M, Luqman M, Nizar I K, Heah C Y 2011 *Phy. Procedia* **22** 312
[8] Xu H and Van Daventer J S J 2000 *Int. J. Miner. Pro.* **59** 247
[9] Barbosa V F F, Mackenzie K J D and Thaumaturgo C 2000 *Int J. Inorg. Mater.* **2** 309
[10] A Palomo, M T Blanco-Varela, M L Granizo, F Puertas, T Vazquez and M W Grutzeck 1999 *Cem. Concr. Res.* **29** 997
[11] Bell J L, Driemeyer P E, and Kriven W M 2009 *J. Amer. Ceram. Soc.* **92**(3) 607
[12] He P, Jia D, Lin T, Wang M, and Zhou Y 2010 *Ceram. Inter.* **36**(4) 1447
[13] He P, Jia D, and Shengjin W 2013 *J. Euro. Ceram. Soc.* **33** 689.
[14] Heah C Y, Kamarudin H, Al Bakri A M M, Binhussain M, Luqman M, Khairul Nizar I, Liew Y M 2011 *Aus. J. Basic and Appl. Sci.* **5** 1026.
[15] Xie N, Bell L J, and Gordon M 2010 *J. Am. Ceram. Soc.* **93** 2644.
[16] Barbosa V F F, and MacKenzie K J D 2003 *Mater. Res. Bull.* **38**(2) 319
[17] Duxson P, Fernandez-Jimenez A, Provis J L, Lukey G C, Palomo A, and Van Deventer J S J 2007 *J. Mater. Sci.* **42**(9) 2917
[18] Mera G, Riedel R, Poli F, and Müller K 2009 *J. Europ. Ceram. Soc.* **29**(13) 2873
[19] Lu Z, Lu J, Li X, and Shao G 2016 *Ceram. Int.* **42**(2) 3452
[20] Xu X, Xu X, Wu J, Zhang Y, Lao X, Li K, and Xu Y 2015 *Ceram. Inter.* **41**(9) 11861
[21] Heah C Y, Kamarudin H, Al Bakri A M M, Binhussain M, Luqman M, Khairul Nizar I, Liew Y M 2012 *Adv. Mater. Res.* **479** 357
[22] Kuenzel C, Grover L M, Vandeperre L, Boccaccini A R, and Cheeseman C R 2013 *J. Europ. Ceram. Soc.* **33**(2) 251
[23] Guan Z D 1992 Proceedings of Performance of Inorganic Nonmetallic Materials, Tsinghua University Press, Peking 83.

IOP Conf. Series: Materials Science and Engineering **133** (2016) 012024 doi:10.1088/1757-899X/133/1/012024

Effect of Fluxes on 60Sn-40Bi Solder Alloy on Copper Substrate

M. N. Ervina Efzan [1, 2*], W.L. Ng[1], Mohd Mustafa Al Bakri Abdullah[3, 2]
[1]Faculty of Engineering and Technology, Multimedia University, 75450 Ayer Keroh, Malacca, Malaysia.
[2]Center of Excellence Geopolymer & Green Technology (CeGeoGTech), School of Material Engineering, Universiti Malaysia Perlis (UniMAP), Perlis, Malaysia
[3]Faculty of Engineering Technology, Universiti Malaysia Perlis (UniMAP), Perlis, Malaysia

E-mail: ervina.noor@mmu.edu.my

Abstract. This paper investigated the effect of different types of fluxes on the wettability of a type of low temperature lead-free solder, 60Sn-40Bi alloy. The purpose of this paper is to investigate the effect of different types of fluxes on the wettability of 60Sn-40Bi solder (T_m : 138 -170 °C), so that the most compatible flux to be used with low temperature alloy can be determined.The results of this paper showed that the water soluble flux sample has the highest spread area and lowest contact angle. This meant that the solder has the highest wettability when water soluble flux is used, followed by RMA flux and low solids flux. Therefore, it was determined that water soluble flux is the most compatible to be used with the low temperature 60Sn-40Bi solder. The characteristic of this type of flux enables it to function well even at a low working temperature.

1. Introduction

Flux is a very important part of soldering as it is needed to prevent oxides forming when hot metals are put into contact with air (Dongkai, 2005; Alvin, 1989). In the electronic manufacturing industry, many types of flux are currently being used in the assembly process. However, the decision to choose which type of flux to be used depends on the product's end use environment (Dongkai, 2005). Some products require that its printed circuit board (PCB) have low flux residues while some may need it to be able to withstand high working temperature. Therefore, the type of flux chosen must satisfy the required conditions or else the end product may be defective (Ervina & Amares, 2015).

Nowadays, there are various types of flux used in soldering which is a major part of the electronic manufacturing process, namely water soluble flux, rosin mildly activated flux (RMA) and RMA/low solids flux (Dongkai, 2005; Giles and David, 2004). Sometimes, it is a problem for manufacturers to choose the most suitable type of flux to be used in a certain process as each type of flux inhibits different characteristic and caters to different needs. If the wrong type of flux is used, the soldered PCB might have defects and this will directly affect the performance of the electronic products.

Other than that, the wettability of the solder is also a main consideration (Ervina and Amares, 2012). Wettability is the ability of a solid surface to reduce the surface tension of a liquid in contact with it such that it can spread over the solid surface and wet it (Ervina et. al, 2008). In this case, the solid surface is the PCB board that is about to be soldered while the liquid is the molten solder. The wettability of a solder is dependent on the type of flux used thus it is important to use the suitable type of flux during a soldering process.

Previously in the soldering process, there are two types of solder used: Sn-Pb (Tin-Lead) solder or SAC (Tin-Silver-Copper) solder. For eutectic SAC solder, the melting point is 223°C (Ervina & Faziera, 2016) while for eutectic Sn-Pb solders the melting point is 183°C. At these temperatures, the fluxes are compatible with the solder and are able to work well. However, recently there are various types of low temperature solders being used in the industry which has melting points below 180°C. The main concern of the industry is whether the existing fluxes are compatible to be used with the low temperature solders or not. In this paper, testing will be done to study the effect of different types of fluxes on the wettability of 60Sn-40Bi solder, which is a type of low temperature solder.

2. Materials and Methods

(A) Preparation of Solder in Required Composition

Tin (Sn) of 6g and bismuth (Bi) of 4g were weighed using an electronic balance. Both types of metals were put into a crucible which was then placed on the hot plate. After that, the hot plate was switched on and the temperature inside the crucible was measured using a multimeter. The crucible was heated to a temperature of 271°C where both metals melted and mixed together (T_m tin: 232°C and T_m bismuth: 271°C). After both metals have mixed together, the hot plate was then switched off and the crucible was allowed to cool down so that the alloy can solidify. Next, the hot plate was switched on again and the crucible was heated to a temperature of 170°C so that the alloy becomes molten form again(Tm of 60Sn-40Bi alloy: 138 -170°C). After the alloy has become molten, the hot plate was then switched off and the alloy was allowed to cool down and solidify. The heating and cooling process were repeated two more times to ensure that the mixture of alloy is homogenous throughout the crucible. Next, the alloy was heated one more time to a temperature of 170°C where it became molten form and immediately it was poured out onto a pot pedestal. After the solder has cool down and solidified, it was removed from the pot pedestal. The solid solder was then flattened to a uniform thickness of 1mm using a hydraulic hot molding machine. Using a puncher, the solder was punched into multiple billets of 7mm diameter.

(B) Applying Flux and Melting the Samples onto Copper Sheets

Prior to the experiment, multiple copper sheets of size 15 x 15mm were prepared. The billets from Part (A) were taken to be used here. A total of nine billets were needed, three billets for each type of flux. Using a cotton bud, the RMA flux were applied on the top and bottom surface of three billets. Each billet was then placed on the center of a copper sheet. After that, the hot plate was switched on again and heated until it reached a temperature of 150°C. Using a tweezers, the copper plate containing the billet were placed on top of the hot plate for 10 seconds and then removed from the hot plate. These 10 seconds are the reflow time for the chemical reaction between the solder, flux and copper sheet to occur. For a consistent result, the reflow time should be the same for all samples. The previous steps were repeated twice but using a reflow temperature of 160°C and 170°C instead. After that, the same steps were repeated again by using the other two types of flux: low solids flux and water soluble flux. All samples were left to be cooled down. Before moving on to Part (C), the spread area of the solder on the copper plates for all samples were observed and recorded for comparison and analysis.

IOP Conf. Series: Materials Science and Engineering **133** (2016) 012024 doi:10.1088/1757-899X/133/1/012024

(C) Mounting the Sample and Observing using Optical Microscope
One side of the short PVC pipe was sealed using a cellophane tape with the sticky side facing the inside of the PVC pipe. Then, two small pieces of plasticine were sticks onto two edges of a soldered copper sheet from Part (B), and then the copper sheet was placed inside the PVC pipe. The side with plasticine was faced inside so that it would stick onto the cellophane tape. This is to ensure that the copper sheet will remain straight upright and not be slanted. After that, an epoxy mixture was prepared by mixing resin and hardener with a ratio of 10:1. For each 10ml of resin used, we need to mix in 1ml of hardener. The mixture was stirred until it become homogenous and then it was carefully poured into the PVC pipe containing the copper sheet until it completely fill up the PVC pipe. Care must be taken so that when pouring the epoxy mixture into the pipe, the copper sheet remains straight upright and not be slanted. The previous steps were repeated for all the samples prepared in Part (B). After that, all the samples were left to be hardened for one day. Each PVC pipe was labeled for easy identification later on. On the next day, each pieces of PVC pipe was cut about halfway through the length of the pipe. The cut surface was then polished until the surface is flat, clean and shiny so that the contact angle can be clearly observed under the microscope. Each sample was put under the microscope, and the contact angles between the solder and the copper sheet were measured for each sample. For each sample, there are two contact angles, one on the left and one on the right. Both angles were measured and the angle with the smaller value was chosen. The contact angles for solder using different type of flux and also different reflux temperatures were recorded and tabulated for analysis.

3. Results and discussions

(a) Spread Area Comparison

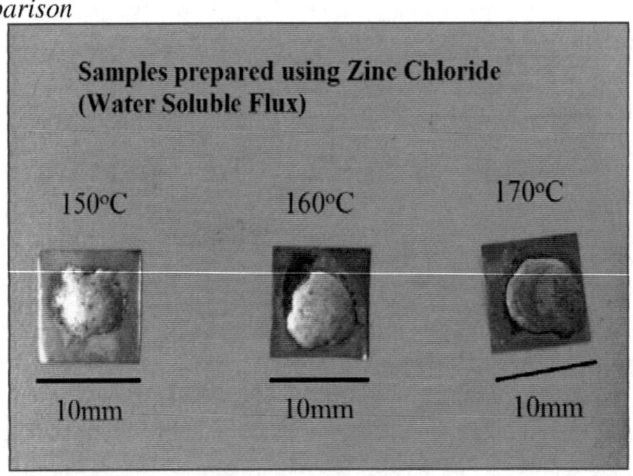

Figure 1: Spread Area for Samples prepared using Water Soluble Flux

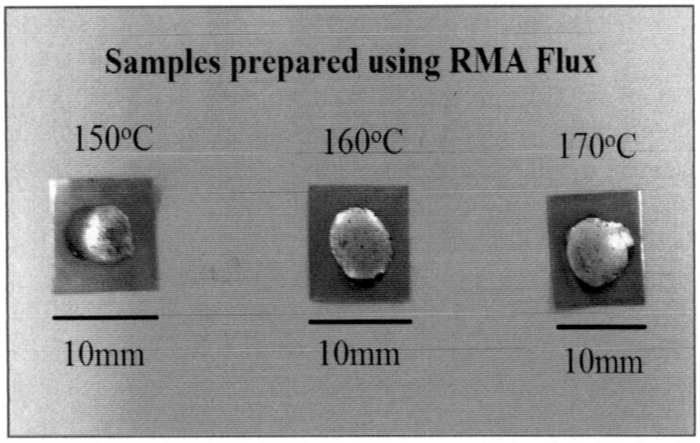

Figure 2: Spread Area for Samples prepared using RMA Flux

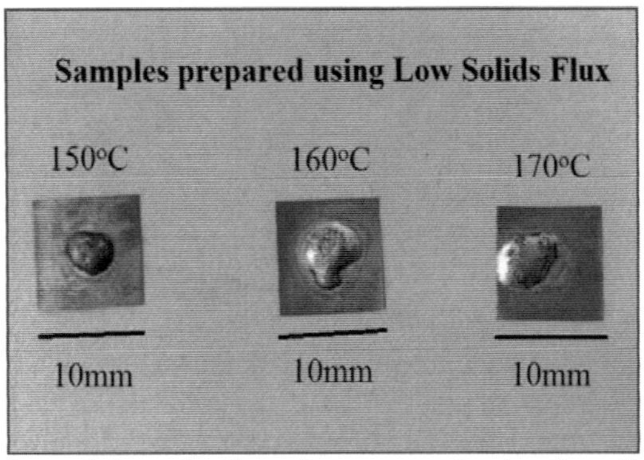

Figure 3: Spread Area for Samples prepared using Low Solids Flux

Figure 1, 2 and 3 show the spread area of the solder for all the samples. By comparing the spread area between samples using different type of flux, the type of flux that provides the most wettability to the solder can be determined. Solder with a higher wettability will be able to spread more easily therefore resulting in a larger spread area. From the figures, it can clearly be seen that the solder that spread the most was from the samples using water soluble flux, followed by samples using RMA flux and finally samples using low solids flux. From this we can deduce that water soluble flux provides the most wettability to the solder. This result is similar to what Lawrence et al. had obtained during their testing on the wetting of Sn-Bi alloys. In their paper, they found that solder of 60Sn-40Bi have a higher spread area when water soluble flux was used during the soldering process instead of a no-clean flux (Lawrence et al., 1993). Besides that, it is also important to note that for each type of flux, the spread area increases when the reflow temperature increases. When the reflow temperature is higher, the flux becomes more reactive and was able to function better in cleaning the copper sheet and providing solder wetting. Therefore, the wettability increases when the reflow temperature is higher which in turn results in a larger spread area.

The contact angles between the copper sheet and the solder could be measured accurately using an optical microscope. Figure 4 to 12 shows the contact angle for all the samples. The result of the contact angle must be an inverse of the spread area, which means that the solder with best wettability have the highest spread area and lowest contact angle (Ervina and Tan, 2013). As mentioned earlier, a solder with a higher wettability will be able to spread across the copper sheet earlier, thus resulting in the high spread area and low contact angle. From Figure 4 to 12, it can be clearly seen that the contact angle is lowest when water soluble flux is used, followed by RMA flux and finally low solids flux. The result shows that for Sn-Bi solder, water soluble flux is the best to be used as it gives the highest wettability to the solder. This is due to the compatibility between the solder and flux. Besides that, the water soluble flux contains a lot of carboxylic acid which acts as activators thus it is very active. Dongkai said that due to this property, the flux is able to perform well and increases solder wetting even at low temperature (Dongkai, 2005).

For each specimen, there are two contact angles (left and right), both angles were recorded and their average was calculated. Then, all the data were tabulated into Table 1 for analysis. After that, a graph of contact angle versus reflow temperature was drawn as shown in Figure 13. From this figure, it can be seen that for samples using water soluble and RMA flux, the contact angle decreases as the reflow temperature increases. This is due to the fact that the higher the reflow temperature, the more reactive the solder becomes thus its wettability increases. A higher wettability will then result in a lower contact angle between the solder and the copper plate. For the samples using low solids flux, the contact angle decreases a lot when the reflow temperature was increased from 150 to 160°C. However, when the reflow temperature was again increased to 170°C, the contact angle does not have any signification reduction. This shows that for low solids flux, the wettability was already optimum at 160°C. Therefore it does not improve anymore even if the reflow temperature was increased. As what Ervina and Tan have found, there exist some cases where the wettability does not improve anymore even if the reflow temperature is increased (Ervina and Tan, 2013).

Table 1: List of Contact Angle

Type of Flux	Reflow Temperature (°C)	Contact Angle (°) Left	Right	Average
Water Soluble	150	32.00	41.14	36.57
	160	35.59	24.47	30.03
	170	19.60	15.38	17.49
RMA	150	52.40	47.62	50.01
	160	42.45	34.67	38.56
	170	19.50	21.69	20.595
Low Solids	150	127.63	91.09	109.36
	160	48.40	43.98	46.19
	170	31.77	59.38	45.575

(b) Contact Angle Comparison

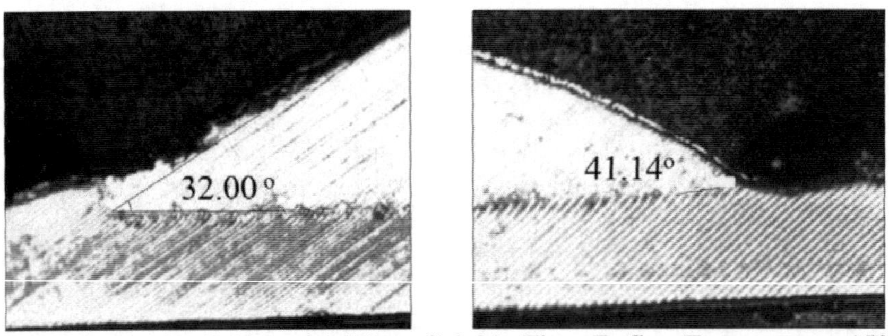

Figure 4: Contact Angle for Solder with Water Soluble Flux (Reflux Temperature: 150 °C) at 10X magnification

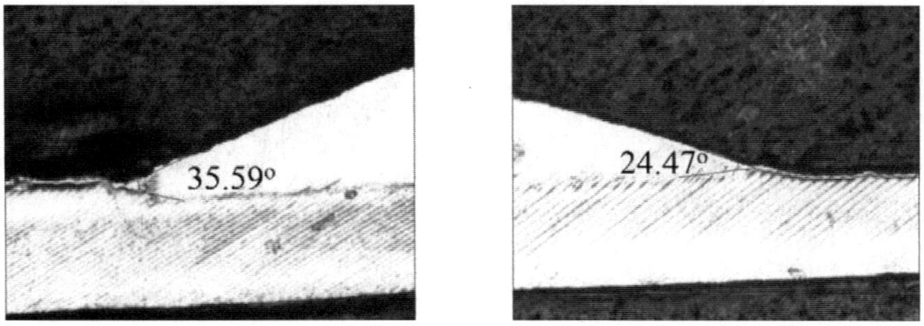

Figure 5: Contact Angle for Solder with Water Soluble Flux (Reflux Temperature: 160 °C) at 10X magnification

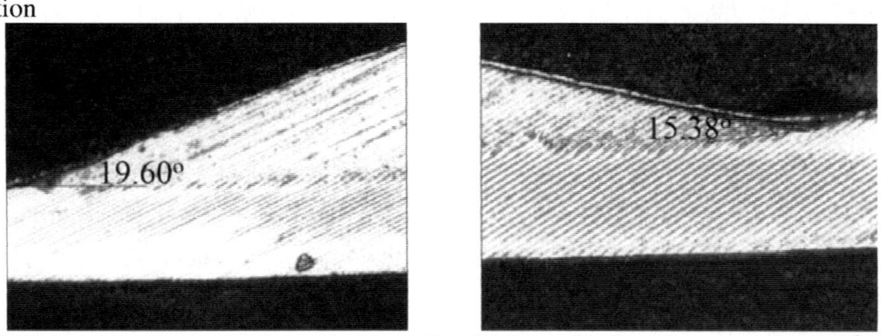

Figure 6: Contact Angle for Solder with Water Soluble Flux (Reflux Temperature: 170 °C) at 10X magnification.

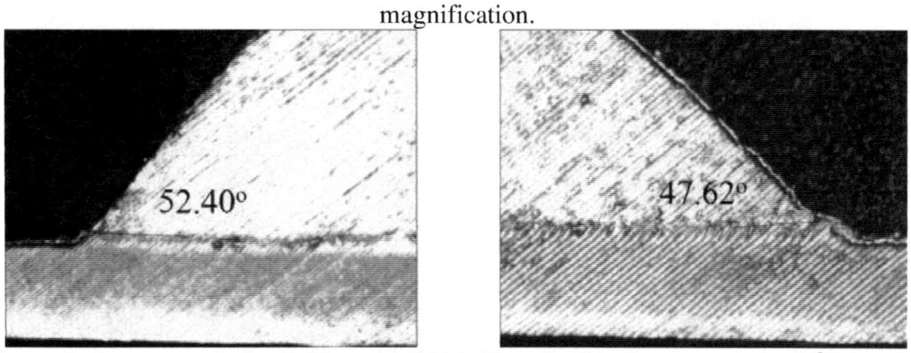

Figure 7: Contact angle for Solder with RMA Flux (Reflux Temperature: 150 °C) at 10X magnification

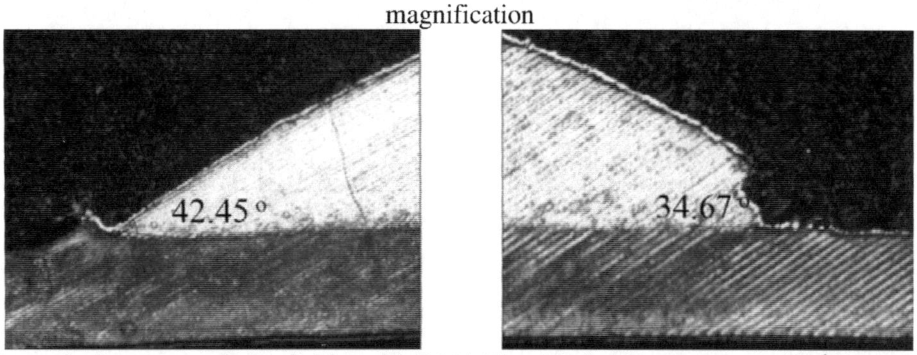

Figure 8: Contact Angle for Solder with RMA Flux (Reflux Temperature: 160 °C) at 10X magnification

IOP Conf. Series: Materials Science and Engineering **133** (2016) 012024 doi:10.1088/1757-899X/133/1/012024

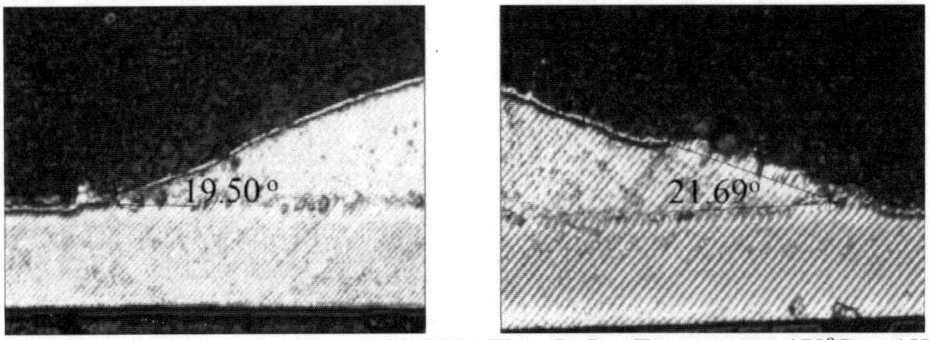

Figure 9: Contact Angle for Solder with RMA Flux (Reflux Temperature: 170°C) at 10X magnification

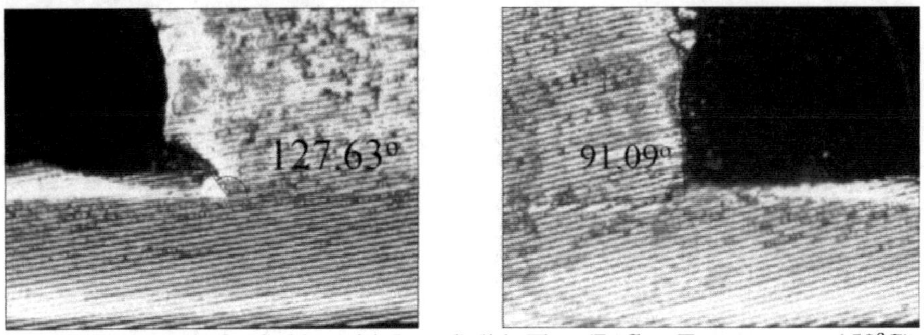

Figure 10: Contact Angle for Solder with Low Solids Flux (Reflux Temperature: 150°C) at 10X magnification

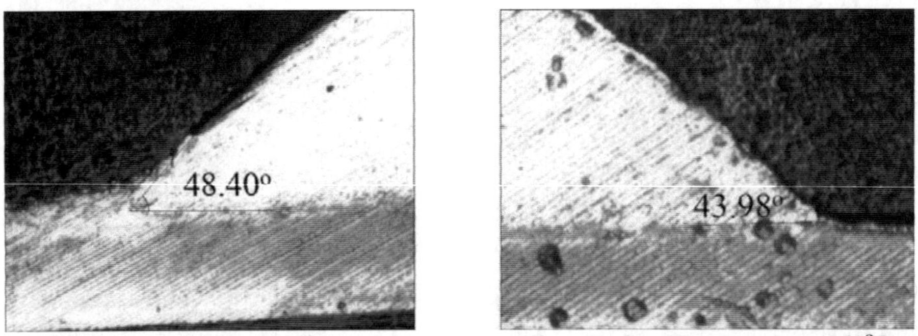

Figure 11: Contact Angle for Solder with Low Solids Flux (Reflux Temperature: 160°C) at 10X magnification

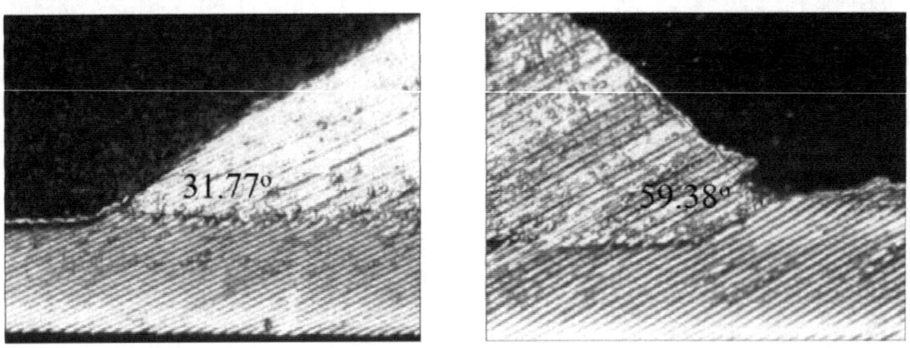

Figure 12: Contact Angle for Solder with Low Solids Flux (Reflux Temperature: 170 °C) at 10X magnification

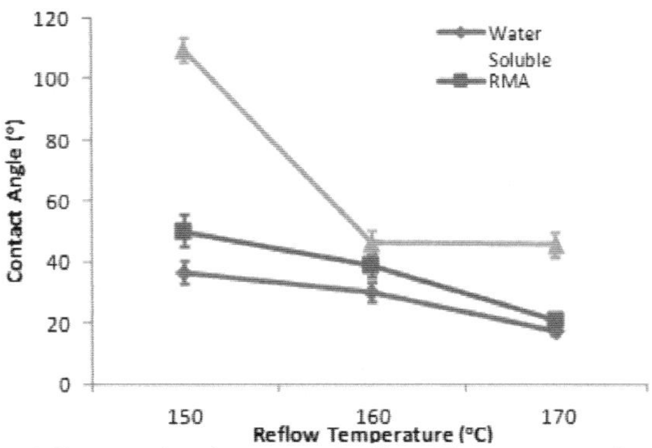

Figure 13: Graph of Contact Angle versus Reflow Temperature for Each Type of Flux

To further study the effect of different fluxes on the solder, a higher magnification image of the interaction at the edge of the solder for water soluble flux and low solids flux were taken. The results are as shown in Figure 14 and 15 below.

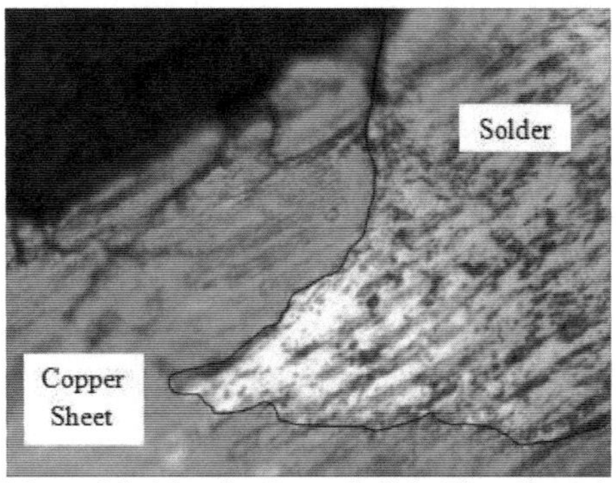

Figure 14: Interaction at the Edge of Solder for Low Solids Flux Sample using Optical Microscope at 50X magnification

Figure 15: Interaction at the Edge of the Solder for Water Soluble Flux Sample using Optical Microscope at 50X magnification

Figure 14 shows the interaction at the edge of the solder for a low solids flux sample which has a low wettability. The orange part is the copper sheet while the yellowish part is the Sn-Bi solder. From this figure, it can be seen that part of the copper plate is above the solder and blocking its spreading. This is due to the less active low solid flux which was unable to fully dissolve and clean the absorption layer of the copper sheet prior to the soldering process (Mulugeta and Selvaduray, 2000). The absorption and reaction layer of the copper sheet that was not dissolved and cleaned by the flux obstructs the Sn-Bi solder from flowing and therefore resulting in a poor wettability of the solder.

Figure 15 on the other hand, shows the interaction for the high wettability sample using water soluble flux. The orange part in this figure is the copper sheet while the darker brown part is the solder. In this figure, the solder is seen to be able to flow freely as there are no obstructions from the copper sheet. This is because the highly active water soluble flux has fully dissolved and cleaned the absorption and reaction layer of the copper sheet. The only layer left is the pure copper layer which the solder is able to spread over easily thus resulting in a high wettability (Mulugeta and Selvaduray, 2000). The downside of using water soluble flux however is it leaves behind a residue that needs to be cleaned. This is why in Figure 15 the solder appears to be darker as there are residues left on the edge.

4. Conclusions

By comparing and analyzing the results obtained from the wettability test, it was found that the samples soldered using water soluble flux have the highest spread area while the samples soldered using low solids flux have the lowest spread area. Moreover, it was found that at any reflow temperature, the water soluble flux samples yield the lowest contact angles, followed by RMA flux samples and finally the low solids flux samples. For example, at the reflow temperature of 150 C, the water soluble flux sample has a contact angle of 36.57°C and the RMA flux sample has a contact angle of 50.01°C while the low solids flux sample has a contact angle of 109.36°C. Therefore, it can be concluded that the wettability of Sn-Bi solder is highest when it is used with water soluble flux, followed by RMA flux and finally low solids flux. This result also proved that water soluble flux is indeed the most compatible with the low temperature solder used in this experiment which is the Sn-Bi solder. Furthermore, the water soluble flux sample soldered at a reflow temperature of 170°C has the lowest contact angle of 17.49°C. Thus the optimum reflow temperature for water soluble flux was determined to be 170°C.

The results presented in this paper are indeed useful and may contribute a lot to the electronics assembly industry. For any Sn-Bi solders to be used in future assembly lines, the company will be able to decide which type of flux to be used by just referring to this paper. This will directly save a lot of time and cost for the company. However, it is important to take note that the results obtained in this test are only limited to Sn-Bi solders. In a case where another type low temperature solder alloy was to be used, further testing needs to be done to check whether the results still stand true or not.

5. Acknowledgments

Special thanks to Multimedia University for financial support through ERGS Research Grant No. MMUE/110133 and MMU Internal Grant No. MMUI/140082.

References

[1] Dongkai, S. (2005), *Lead-free Solder Interconnect Reliability,* ASM International, Materials Park, Ohio.

[2] Alvin, F. S. (1989), "*Fluxes*", Electronic Materials Handbook: Packaging, Vol. 1, pp. 643-650.

[3] Giles, H. and David, M. J. (2004), *Principles of Soldering*, ASM International, Materials Park, Ohio.

[4] Felton, L. E., Raeder, C. H. and Knorr, D. B. (1993), "*The Properties of Tin-Bismuth Alloy Solder*", The Journal of Minerals, Metals and Materials Society, Vol. 45, Issue 7, pp. 28-32.

[5] Ervina, M. N. and Tan, S. Y. (2013), "*Wettability of Molten Sn-Zn-Bi Solder on Cu Substrate*", Applied Mechanics and Materials, Vol. 315, pp. 675-680.

[6] Mulugeta, A. and Selvaduray, G. (2000), "*Lead-free Solders in Microelectronics*", Material Science and Engineering, 27th Edition, pp. 95-141.

[7] Ervina, E. M. N., Amares, S. "Characterization of mechanical testing on lead free solder on electronic application" Electronic Manufacturing Technology Symposium (IEMT), 2012 35th IEEE/CPMT International , vol., no., pp.1,9, 6-8 Nov. 2012 doi: 10.1109/IEMT.2012.6521780

[8] Ervina, E. M. N. & Nur Faziera, M. N. Effects of Wetting Behavior on Gallium Addition in In-Zn Solder Alloy. (2016) Materials Science Forum, Vol. 857, pp 22-25

[9] Ervina, E. M. N., Sharif, N. M., Yew, C. K., Ariga, T., Ismail, A. B., & Hussain, Z. (2008). Characteristic of low temperature of Bi-In-Sn solder alloy. Electronic Manufacturing Technology Symposium (IEMT), 2008 33rd IEEE/CPMT International

[10] Ervina, E. M. N., Zuhailawati, H. & Radzali, O. "Low temperature In–Bi–Zn solder alloy on copper substrate", (2015) Journal of Materials Science: Materials in Electronics, 1-8

[11] Ervina, E. M. N & Amares, S. "A Study of Temperature, Microstructure and Hardness Properties of Sn-3.8 Ag-0.7 Cu (SAC) Solder Alloy", (2015) MATEC Web of Conferences 27, 2003

[12] Ervina, E. M. N. & Siti, N. "Effect of Different Aging Times on Sn-Ag-Cu Solder Alloy" (2015) TRANSACTIONS ON ELECTRICAL AND ELECTRONIC MATERIALS 16 (3), 112-116

IOP Conf. Series: Materials Science and Engineering **133** (2016) 012025 doi:10.1088/1757-899X/133/1/012025

ESD morphology deposition with WZr8 electrode on austenitic stainless steel support

M.C. Perju[1,2], C.A. Ţugui[1], C. Nejneru[1], M. Axinte[1], P. Vizureanu[1,2]

[1] Technical University "Gheorghe Asachi" of Iasi-Romania, Department of Technologies and Equipments for Materials Processing, Blvd. Mangeron, No. 51, 700050, Iasi, Romania
[2] Centre of Excellence Geopolymer& Green Technology (CEGeoTech), School of Materials Engineering, Universiti Malaysia Perlis, Kompleks Pengajian Jejawi 2, 02600 Arau, Perlis[3] Department, University, Address

E-mail: cryss_ela@yahoo.com

Abstract. Stainless steels are used to obtain mechanical parts, working in severe conditions with high dynamic loads in wet, chemically active environments. For this reason, these materials have good corrosion resistance in acidic or basic chemical agents. The main drawback is the relatively low wear and resistance to mechanical stress. This paper proposes a remedy by deposition of the hard thin films of tungsten electrode by spark electro-deposition method (ESD). Tungsten is an alfagen element and causes an increase for the mechanical properties at high and low temperatures for the austenitic stainless steels. Tungsten does not alter the corrosion resistance of stainless steels. The morphology for the obtained layers was analyzed using SEM, in 3D images, and profilographs.

1. Introduction

In industry, pumps, compressors, cutting machines, etc., works in difficult operating conditions. The surfaces of the moving parts are subject to static and dynamic mechanical loads, at various temperatures, in abrasive wearing and corrosive environment conditions. For this reason it is necessary to use protective coatings that improve the surface qualities of the parts by various deposition methods [1-3].

Stainless steels are used in almost all industrial domains and due to corrosion resistance characteristics, combined with its characteristics and surface processing possibilities, have led to increased interest in this class of materials. It presents a good corrosion resistance, so it can operate successfully in chemically aggressive environments, but do not withstand severe wearing. To remedy this situation, the paper proposes the use of additive materials based on tungsten, which are deposited by ESD method and increases the surface resistance to dynamic loads and severe wearing.

Electro-spark deposition is a microwelding process that is capable of depositing wear and corrosion resistance coatings to repair, improve, and extend the service life of the components [4].

In the electric discharge pulse deposition process, besides the great importance of the the mass transfer, the working process parameters and the physico-chemical characteristics of the whole surface, the deposition method presents a significant importance to the development of appropriate structures, especially the adherence and microallying of the metallic "drop" [5].

2. Materials and Methods

2.1. Base Material

For the experiment was used as the base material, the austenitic stainless steel X2CrNiMo18 14 3, DIN 1.4435. This austenitic steel deforms very easily and is not appropriate for subsequent heat treatments, [6]. The chemical composition of this material was determined using the spectrometer equipment Foundry Master from the Department of Materials Processing Technology and Equipment at the Faculty of Materials Science and Engineering of Iasi. The percentage for each chemical element is shown in table 1.

Table 1. Chemical analysis of the base material.

Chemical element	Fe	C	Si	Mn	Cr	Mo	Ni
%	balance	0,028	0,99	2,12	17,24	2,9	13,41

The samples from the X2CrNiMo18 14 3 were processed on a cutting machine obtaining the samples with the following sizes: 20mm long, 35mm wide and 4mm height. After cutting, these samples were prepared by successive grinding on metallographic paper to the 600-grain size. By grinding there is obtained a decrease of the surface peaks. If the deposition would be done on rough samples, a local burning of the peaks would occure. This local burning will cause several defects on the surface.

2.2. Deposition Material

The electrode used for deposition is WZr8, ISO 6848 AWS A5.12 and contains ZrO2 (0,70...0,90%), impurities ≤0,20%, and W for balance. The electrode is white marking type. This type of electrode shows good weldability properties, acts as low melted tip contamination and has good anti-corrosion properties, creating a relatively smooth surface (with small roughness). It also produces a stable electric arch and presents good resistance at pulling the soft areas of the electrode.

2.3. Deposition Material

ESD deposition method is used to cover the installations components, which operate in severe conditions as abrasive wearing in wet or dry environments. In wet environments beside the mechanic loads, the parts must work in high chemically active environments. The electrical discharge method, in gas, as a method to obtain a surface layer, is based on the polarization effect and the material transfer from the anode (electrode) to the cathode (part). Installation used for depositions is Elitron 22A, [7].

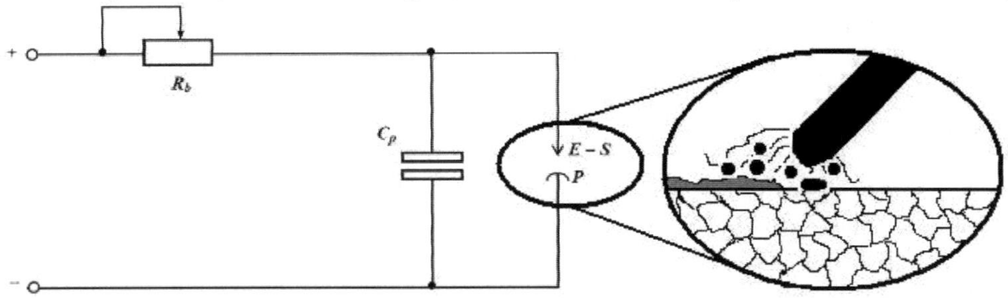

a) b)

Figure 1. Processing of metal surfaces by electrical discharge impulses: a) Electrical diagram of the processing device: Cp = capacitor , Rb = variable resistance , E-S = electrode coupled to the anode, P = part coupled to the cathode ; b) Surface processing diagram.

The two schemes, electrical diagram and working principle of processing by the electric discharge pulse are shown in Figure 1. Electrical diagram represents a generator, a part, which is coupled as cathode, and the anode as the deposition electrode.

The processing of the surface begins by approaching the electrode to the sample. At the critical distance, piercing triggers the electrical discharge pulse. In most of the cases, the discharge is continuous and is finished when the two electrodes are touching each other.

After piercing the interval between the electrodes, due to the energy accumulated in the condenser Cp, at the contact surface of the electrodes, strong heated areas appear (local melting and evaporation points), causing electrical errosion of electrodes (part and electrode itself).

Different depositions were made on a support of austenitic stainless steel. Depositions with one and two layers were studied, pointing to the morphology and surface characteristics obtained after coating.

3. Results and Discussions

Electrode substrate and parts surfaces make physical contact in small areas, involving just a small number of points from the surface. In these contact points electrical current is extremely intense, a large amount of energy causes the metal to melt and even quickly vaporize, [8-10].

It will increase heat and will ionize the gas between the electrode and the substrate. At the end of this period, many charged particles are generated, such as free electrons, ions and other particles, which fulfills the criteria of a pulsed discharge at low voltage. When the metal melts at contact points, it will occur a pitting (indentation) between the electrode and the substrate. Furthermore, a non-uniform electric field is created between the anode (electrode) and the cathode (substrate). The resistance of air between the two electrodes is pierced and the discharge occurs.

W-ZR 8 electrode creates thin hard layers on the steel surface. Moving the electrode holder in semicircular patterns melts locally the surface, including in the formed microbath the tungsten and nitrogen atoms absorbed from surrounding atmosphere and the plasma arc.

Tungsten sublimation temperature is lower than that obtained in the electric discharge ($\approx$10000°C), obtaining a plasma cloud where chemically active W, easily absorbed is present, forming on the surface of molten metal complex carbonitrides Ni, Mo and Cr, which are tough and stable.

Another feature that gives extra hardness for the outer layer is the solidification ultraspeed, which forms on the outside a vitrified amorphous structure well anchored in the surface layer.

Morphology analysis shows the presence of the solidified wavy areas formed from droplets. Full investigation of the samples coated by the ESD method was made on a scanning electronic microscope SEM, model VEGA II LSH, [11].

3.1. One Layer Deposition

The ignition of the arc between the two electrodes causes the appearance of a thermal flux effect that leads to the melting of the electrode tip. As the temperature varies by a paraboloid shape, initial electrode material will melt so that the electrode tip will gain an inner shape, which is exactly as the shape of the temperature field.

The molten area is deposited on the material with a relatively high speed in the form of droplets pulse, which due to the deposition rate, presents concave meniscus in the middle and microdroplets fast cooled on the outer areas.

One-layer depositions shows a relatively high roughness due to the deposition variations like the gas absorbed the edge drops, meniscus inner droplet area due to deposition dynamics, adherent droplets, material overlapping, etc.

An important role has a coefficient of expansion of W, which is very small compared to that of the substrate causing microcracks. Heat transfer coefficient and mass transfer coefficient of the electrode and the substrate material influence the surface quality.

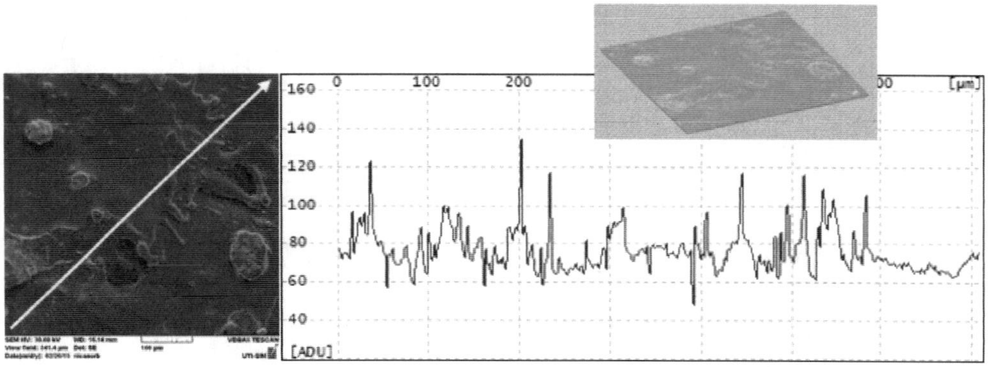

Figure 2. SEM image, and the 3D profilograph; 100μm.

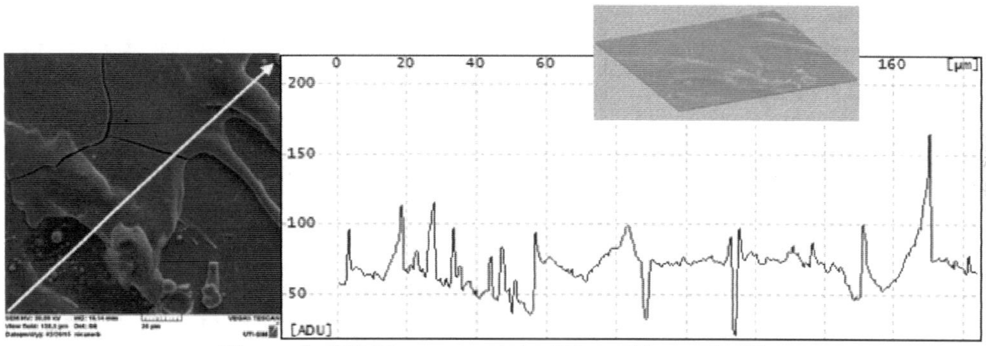

Figure 3. SEM image, and the 3D profilograph; 20μm.

One-layer depositions shows a relatively high roughness due to the deposition variations like the gas absorbed the edge drops (Figure 2 and Figure 3), meniscus inner droplet area due to deposition dynamics, adherent droplets, material overlapping, etc.

An important role has a coefficient of expansion of W, which is very small compared to that of the substrate causing microcracks. Heat transfer coefficient and mass transfer coefficient of the electrode and the substrate material influence the surface quality. Zirconium facilitates the dissolution of tungsten in the metal resulting in fluidization of drops and the compaction of the deposited layer.

3.2. Two Layers Deposition

When submitting two layers, some of the drawbacks of coating with a single layer diminishes.

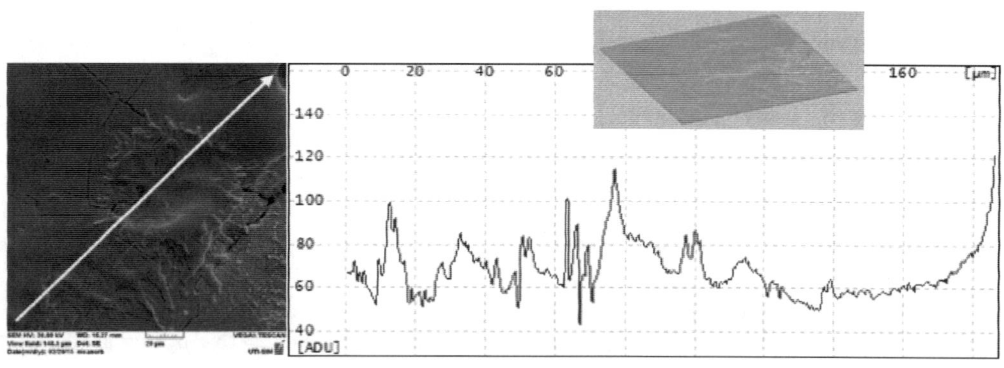

Figure 4. SEM image, and the 3D profilograph; 100μm.

Thus splashing drops made of the first layer are remelted and the bumps due to dynamics droplets forming attenuates. The surface microcracks are closed by remelting.

The concentrations of the surface layers elements modifies due to the increasing of the amount of deposited mass.

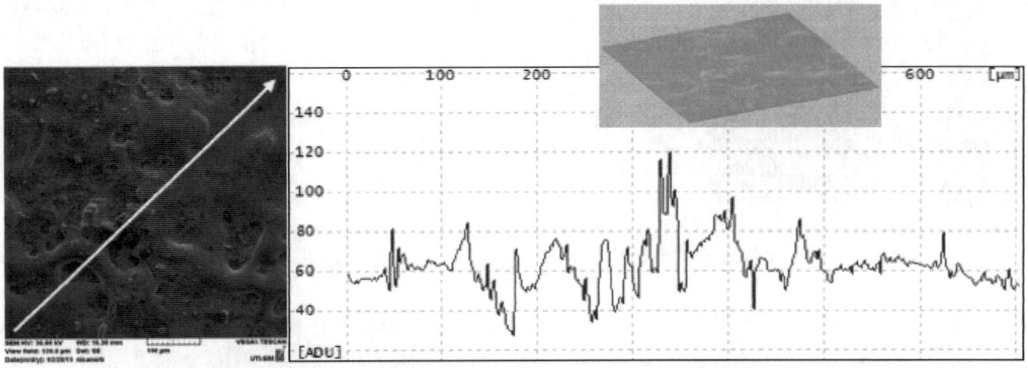

Figure 5. SEM image, and the 3D profilograph; 20μm.

Comparing the two scales magnification profilographs for coatings with one and two layers can be observed an accentuated surface smoothing in the two layers case, (Figure 4 and Figure 5). This can be observed also with 3D images.

4. Conclusions

Depositions on stainless steel with WZr8 by vibrating electrode method forms hard layers, increasing the reliability of the parts used in aggressive corrosive environments for wet and dry friction mechanical drives.

Due to the thermo-physic characteristics of tungsten, it hardens the surface of the piece mainly by melting and solidifying with the vitrification structure, the layer obtained presents an amorphous structure and high hardness.

Single layer deposition presents problems related to the unevenness of the surface, showing areas with microcracks, material overlapping, droplets and partially dissolved gases.

The second layer deposition creates a noticeable smoothing of the surface, achieving an uniform appearance, without microcracks, by reshaping areas uncovered or partially covered from first layer.

References

[1] A.E. Reitera, B. Brunnerb, M. Antea, J. Rechbergerb, 2006 *Surface and Coatings Technology*, vol. 200, Issues 18–19, pp. 5532–5541.

[2] I. Ciftci, 2006 *Tribology International*, vol. 39, Issue 6, pp. 565–569.

[3] N. Shaigana, W. Qua, D. G. Iveyb, W. Chenb, 2010 *Journal of Power Sources*, vol. 195, Issue 6, pp. 1529–1542.

[4] S. K. Tang, T. C. Nguyen and Y. Zhou, 2010 *Welding Journal*, vol. 89, pp. 172-180.

[5] M. C. Perju, G. D. Găluşcă, C. Nejneru, M. Agop, *Thin films: discharge pulse*, Publisher Ars Longa, Iaşi, 2010.

[6] ASM International, Austenitic Stainless Steels, 2008.

[7] *** Instalaţie Elitron 22, Academia de Ştiinţe, Republica Moldova, Chişinău, 1992.

[8] S. Durdu, S. Levent Aktuğ, K. Korkmaz, 2013 *Surface and Coatings Technology*, vol. 236, pp. 303-308.

[9] J. Gould, 2011 *Welding Journal*, vol. 90, pp.191-197.

[10] C. A. Tutui, P. Vizureanu, C. Nejneru, M. C. Perju, M. Axinte, 2015 *Applied Mechanics and Materials*, Vols. 809-810, pp 417-422.

[11] D. Mareci, N. Cimpoesu, M.I. Popa, 2012 *Materials and Corrosion*, No.63, pp. 176-180.

IOP Conf. Series: Materials Science and Engineering **133** (2016) 012026 doi:10.1088/1757-899X/133/1/012026

Self-cleaning geopolymer concrete - A review

Siti Norsaffirah Zailan[1], Norsuria Mahmed[1], Mohd Mustafa Al Bakri Abdullah[1,2], Andrei Victor Sandu[3]

[1]Centre of Excellence Geopolymer and Green Technology, School of Materials Engineering, Universiti Malaysia Perlis, Pusat Pengajian Jejawi 2, 02600 Jejawi Perlis, Malaysia
[2]Faculty of Engineering Technology, Unicity Alam Campus, Universiti Malaysia Perlis, 02100 Sungai Chuchuh, Padang Besar, Perlis, Malaysia
[3]Gheorghe Asachi Technical University of Iasi, Faculty of Materials Science and Engineering, D.61A Mangeron Str., 700050 Iasi, Romania.

E-mail: sitinorsaffirah@yahoo.com

Abstract. Concrete is the most widely used construction materials for building technology. However, cement production releases high amounts of carbon dioxide (CO_2) to the atmosphere that leads to increasing the global warming. Thus, an alternative, environmental friendly construction material such as geopolymer concrete has been developed. Geopolymer concrete applies greener alternative binder, which is an innovative construction material that replaces the Portland cement. This technology introduced nano-particles such as nanoclay into the cement paste in order to improve their mechanical properties. The concrete materials also have been developed to be functioned as self-cleaning construction materials. The self-cleaning properties of the concrete are induced by introducing the photocatalytic materials such as titania (TiO_2) and zinc oxide (ZnO). Self-cleaning concrete that contains those photocatalysts will be energized by ultraviolet (UV) radiation and accelerates the decomposition of organic particulates. Thus, the cleanliness of the building surfaces can be maintained and the air surrounding air pollution can be reduced. This paper briefly reviews about self-cleaning concrete.

1. Introduction

Concrete is the most widely used construction materials in the world. Concrete is the second most consumed substances on earth after water. Concrete is used everywhere such as office buildings, residential apartment, bungalow, any public places, etc [1]. Unfortunately, the main components contain in the cement contributed to the greenhouse gas emissions. Dust emissions during the cement manufacturing are one of the main issues that have been faced by the industries. A major source of air pollution is caused by concrete dust released by building demolitions and natural disasters [2]. Thus, the building surfaces will become dull and dirty. In addition, concrete also faces the problem of tending to become dirty when exposed to polluted area. These situations have pressured the industries to develop an innovative technology and find alternative materials that are environmentally friendly in order to support the sustainable development. Geopolymer concretes with self-cleaning behavior are green alternative to overcome this problem. This is because they offer tremendous environmental benefits such as decreasing the CO_2 emissions, economical, durable and environmentally friendly.

Geopolymer cement has their own unique properties which are able to withstand at high temperature and also resistant to salt and acid [3]. Self-cleaning concrete or photocatalytic concrete will keep the concrete surface free of dirt and keeps its colour, even in polluted condition and industrial areas. This concrete also called as green concrete due to its self-cleaning properties [1].

Self-cleaning technology is commercialized for indoor and outdoor applications. There are various applications of self-cleaning technology, such as in paints, glasses, textiles, tiles, etc. Glasses windows or tiles are modified with the addition of TiO_2 as photocatalytic thin films for their self-cleaning functions, which have been applied in the glass surface of the National Opera Hall, China. In Japan, eco-friendly, self-cleaning windows and roof tiles are widely used. Japan and Europe country were using this technology to make their polluted cities keep-clean and reduce the effect of air pollution [1]. Nanoparticles of TiO_2 are also applied on the construction materials especially in concrete in order to enhance the self-cleaning properties. Addition of titania in the concrete materials can contribute to the air-purifying and self-cleaning surfaces due to the photocatalytic process that has been activated by the UV light radiation. Thus, the building can maintain their aesthetical appearance and improve the quality of air by the degradation of airborne pollutants [4]. Self-cleaning is an automatic cleaning process on buildings or concrete [5]. This can be seen through the example of the white cement containing TiO_2 nanoparticles has been coated on the roof of Dubai Sports City's Cricket Stadium and also used for designing church at Rome, Italy. From the result, it was proved that the buildings could keep their original colour over the years. While in automotive industries, self-cleaning coatings with antifogging properties have been used in order to develop the clean and glare free windows [5]. In latest research, ethanol suspension of perfluorooctyltriethoxysilane ($C_{14}H_{19}F_{13}O_3Si$) coated with anatase TiO_2 nanoparticle is used to produce paint that can create the self-cleaning surfaces [5]. There are several reviews found on TiO_2 coating, where they can improve the mechanical properties of cotton fibers and also high stain removal of bacteria [5]. Thus, the purposes of applying the self-cleaning materials such as titania onto the geopolymer concrete are, to enhance the aesthetic durability and depollution effect of the products.

2. Approaches to self-cleaning

In 1973, the principle of self-cleaning was discovered by the botanist Wilhelm Barthlott. The products from the reaction of self-cleaning are easily removed by rain or simple rinsing due to the phocatalytic actions with the addition of TiO_2 due to the hydrophilic surface as shown in Figure 1[5]. The application of titania as a photocatalyst to the concrete are focusing on the self-cleaning effect and air purifications [6]. Self-cleaning concrete involved photocatalytic reaction, which contribute to the elimination of volatile organic compounds (VOCs) such as formaldehyde. Photocatalytic reaction occurs with the presence of UV radiation from the sun. When light and heat strike the surface of concrete, TiO_2 use the energy to break the dirt into the basic oxygen, water, CO_2, nitrate and sulphate molecules. Photocatalyst convert air pollutants that contain toxic to lesser toxic forms in the presence of UV light [7]. Thus, depollution removed contaminants and impurities from the surrounding environment [8].

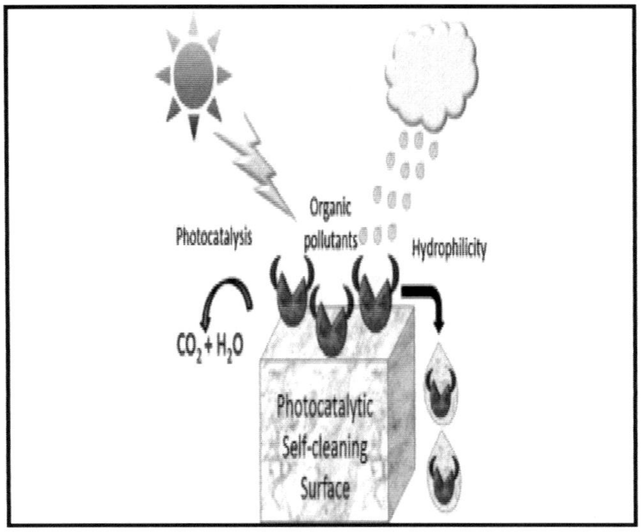

Figure 1. Photocatalytic reactions on self-cleaning surface [5]

2.1. Photocatalyst

There are a lot of materials that act as photocatalyst. It can be seen through a few researches that have been carried out in order to investigate the influence of photocatalyst in cementitious materials [9]. Nanoparticles are used as photocatalyst in order to improve the performance of the cement properties. Nanoparticles that being added to cement can function as either an admixture in cement paste or to replace part of cement. These nanoparticles act as fillers in the empty space, well distributed, increasing rate of hydration and tendency to agglomerate during mixing. Nanoparticles that most common used in cement products are silicon oxide (SiO_2), zinc oxide (ZnO), titania or titanium dioxide (TiO_2), aluminium oxide (Al_2O_3) and carbon nanotubes [9]. Research reported that the addition of nano-silica into the cement paste influenced the mechanical properties of the products [9]. A few studies showed that by adding nanoparticles of TiO_2 are mostly performed with photocatalytic activity due to its chemical stability and high catalytic activity. In concretes that contain nano-TiO_2, the resistance to penetration of chloride is higher than concretes that contain the same amounts of nano-SiO_2. This is due to the refinement of pore structure increases while the chloride penetration decreases with the content of nano-particles [3]. The addition of nano-TiO_2 shows a positive result with fewer cracks in compact microstructure and enhances the formation of geopolymer [3]. The compressive strength can be improved by increasing the amount of nano-TiO_2 in the modified cement [10].

2.1.1. Role of nanoparticles TiO_2 in self-cleaning.

Titania is widely used to produce concrete pavements by mixing the nano-particles within the pavement surface. TiO_2 has three different crystal structures, which are anatase, rutile and brookite. The efficiency of photocatalytic reaction was influence by TiO_2 structure where anatase nanoparticles shows more efficiency in crystal condition and transparent when applied on the substrate. The cleanibility of the substrate can be enhanced by nanostructured titania [11]. The main reason TiO_2 was loaded with cementitious materials because only water, oxygen and solar light required for photocatalysis reaction. In addition, TiO_2 is a photocatalyst that can break down almost any organic compound with the presence of sunlight. If any dirt present on the surface, the water will penetrates under the dirt and removes it as illustrated in Figure 2 [12]. TiO_2 photocatalyst are usually are coated on a surface as thin films. Most photocatalytic coatings are prepared by post-deposition of the catalyst, and then apply onto substrate. They're various method of coating TiO_2 with the cementitious materials such as sputtering, spray coating, sol-gel dip coating, electrodeposition and etc. These techniques of coating are used to incorporate the photocatalysts into the building materials [7]. When the properties of nanoparticles are transferred to the surface of substrates, it will become functionalization for self-cleaning application. TiO_2 are widely applied in different field such as air purification, anti-bacterial, food industry, production of paper, cosmetics, building materials and etc. TiO_2 are widely used because of its features, which is high catalysis efficiency, good chemical and thermal stability, low toxicity and low cost materials [11].

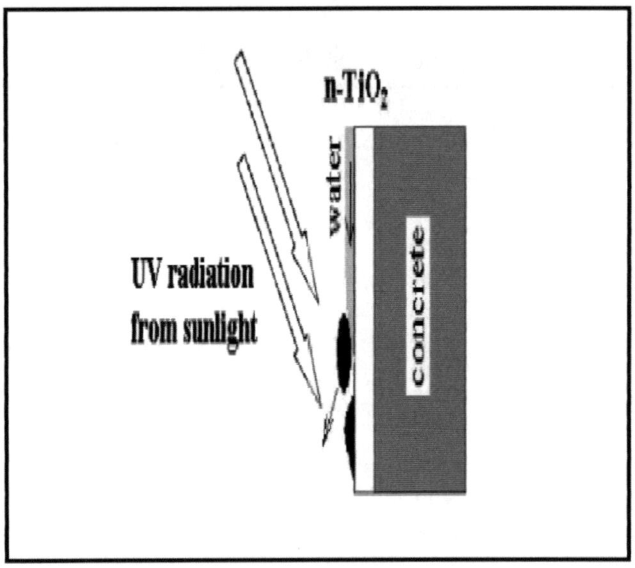

Figure 2. Self-cleaning effect of n-TiO_2 at concrete surface [12]

2.2. *Methods of sample preparations*

There are a lot of methods to prepare the sample by adding nano-TiO_2 onto the cement paste. There is a study about nano-TiO_2 with several of particles size (ground and sieved for 60 minutes in the sieve shaker), which have aims to investigate the effect of addition of nano-TiO_2 on fluidized bed fly ash based geopolymer. [3]. There is a recommendation to use TiO_2 as a solution by mixing water for better dispersion [3]. In another approach, the TiO_2 nanoparticles were soaked in water before mixing with cement, then, vibrated ultrasonically in 30 minutes for well distribution. After 3 minutes, the water was mixed with cement under rate of stirring 800 rph [14]. For sample preparation, ball are milling is used to mixed cement with various TiO_2 nanoparticles content (0%, 2.5%, 5.0% and 7.5%). Before mixing process, the TiO_2 powder was added to dry cement and subsequently water was added [15-16].

3. Self-cleaning testing

The self-cleaning test is conducted in order to evaluate self-cleaning abilities and depollution effects of the material. Photocatalytic performance is measured by photo degradation of methylene blue (MB) solution or rhodamine B (RhB) in the concrete surface in order to evaluate self-cleaning activities. The photo-induced bleaching of MB dye to evaluate photocatalytic materials can be measured by using the UV/visible spectrophotometric. There are a rapid and inexpensive method to test photocatalytic activity of a film, which is using digital scanner and indicator ink such as resazurin (Rz), basic blue 66 (BB66) and acid violet 7 (AV7) [5]. RhB is extensively used in the laboratory test method of photocatalytic degradation because it can simulate particulate pollutants due to its similar structure as the airborne particulate compounds. RhB also water soluble with reddish colour, which are stable in alkaline environments such as mortars and make it good for testing [5]. The experimental study on the ability of silicate coating material, which contains various dosages of TiO_2 to clean the surface by itself and for degradation of particulate pollutants. Based on the result of the experiment, an optimum dosage of TiO_2 in the silicate coating is 15% TiO_2 by mass of solid silicate are most effective in photocatalytic degradation of RhB [5]. The investigation proved that building materials which is coated with TiO_2 have an ability in soot removal. These are based on changes in sample colour in RhB and photocatalytic degradation was tested by closed chamber experiment [4]. Cement plates that are prepared using Portland cement are co-modified with TiO_2-N, C are tested by Multi N/C carbon analyzer in order to measured total carbon and nitrogen content. The degradation rate of Monoazo dye Reactive Red is determined by UV/vis spectrophotometer. The result shows that, the cement plates containing TiO_2-N,C photocatalyst exhibit higher photocatalytic efficiency than unmodified cement plates [5-6].

In order to evaluate the depolluting effect, degradation of nitrogen oxide was determined by using continuous gas flow reactor under UV exposure (as shown in Figure 3).The study was designed a stirred flow reactor in order to measure photodegradation potential at the surface of photocatalytic cementitious materials. The result of the study shows that, the oxidation rate increased with the order of benzene, toulene, ethyl-benzene and o-xylene. Moreover, another study tested the photocatalytic activity of TiO_2 by using a 2-propanol as a reactant of VOCs in the mortar. Different amount of 2-propanol are mineralized to acetone after a certain time under UV irradiation, which are analyzed using gas chromatography (GC) [7].

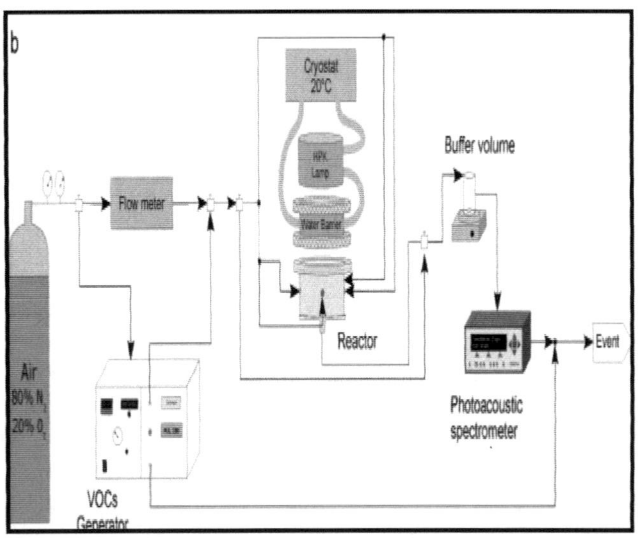

Figure 3. Continuous gas flow reactor [16]

4. Importance of self-cleaning

Air pollution has given negative implication for human health. Major pollutants that are produces from human activity are nitrogen oxide (NOx), sulphur dioxide (SO_2), volatile organic compound (VOCs), etc. Therefore, when photocatalytic reactions absorb UV from the sun, the hydroxyl radicals and superoxide anions are formed which have the ability to react with pollutant molecules such as SOx, NOx, and etc. Nowadays, atmospheric pollution also causes the external decay of buildings due to the deposition of organic matter and contaminants. Self-cleaning concrete has a potential to keep the city clean by reducing the air pollutants. Due to its properties, self-cleaning can save maintenance cost, thus extent the lifetime of the products. Moreover, self-cleaning also can reduce the heat that builds up on building with reflect more light, therefore, help in keeping city become cooler during hot season. Photocatalysts will remove odours and indoor air quality, so that, the city will become clean and beautiful [17]. The photocatalyst that involved in self-cleaning are also being used to purify water and fight the disease by reducing the spread of germs. The developments of photocatalytic cements have given an impact to depollution effect [5].

Self-cleaning are widely used in cementitious materials because it improve the mechanical properties and performance due to its properties. There are a study reported that mechanical properties of TiO_2 nanoparticles that added partially into cement are increase in tensile and flexural strength of the cement mortar [10]. There are a review about efficiencies of self-cleaning when TiO_2 are added to mortars or concretes. Most of TiO_2 mixed in mortars and concretes are remain unutilized because TiO_2 only active when exposed to light irradiation. There are also a few research studied about the addition of TiO_2 as coatings because coating layer are direct contact and exposed to pollutant. The result shows the concrete or mortar, which is covered by TiO_2 are less affected by cement hydration products. TiO_2 mixing with nano-particles are widely utilized to produce concrete pavements, thus contribute to mitigate urban air pollution [7]. White cement with addition of TiO_2 is used at the surface of buildings that can attribute to the durability of the visual aspect of the building. The whiteness of building will remain and maintained due to photocatalytic action.

5. Conclusions

This review aims to give an overview on self-cleaning concrete. This applications is promising to be utilized as a self-cleaning building materials. With these development of technology, air pollutions can be reduced because self-cleaning contribute to the elimination of VOC's that present in indoor and outdoor atmosphere. Nowadays, air pollution is a serious problem in urban environment that can affect human health. Thus, self-cleaning concrete is an effective alternative to provide cleaner environment and maintain building appearance. Through the development of cementitious materials, UV radiation from the sun may represent one of the solutions to the pollution issues. In conclusion, self-cleaning can be used in various application, especially cementitious materials which contribute to sustainability and towards green environment. Further research need to be focused on producing a functional geopolymer concrete with self-cleaning behavior.

IOP Conf. Series: Materials Science and Engineering **133** (2016) 012026 doi:10.1088/1757-899X/133/1/012026

References

[1] A. Khitab, M. Alam, H. Riaz, and S. Rauf, 2014 *International Journal of Advances in Life Science and Technology*, vol. 4, pp. 47-53.

[2] S. H. Saani and R. B. Khadiranaikar, 2013 *International Journal of Research in Engineering and Technology*, pp. 366-371.

[3] P. Duan, C. Yan, W. Luo and W. Zhou, 2016 *Construction and Building Materials*, vol. 106, pp. 115-125.

[4] M. Smits, C. K. Chan, T. Tygat, B. Craeye, N. Costarramone, S. Lacombe and S. Lenaerts, 2013 *Chemical Engineering Journal*, vol. 222, pp. 411-418.

[5] S. Banerjee, D. D. Doinysiou and S. C. Pillai, 2015 *Applied Catalysis B : Environmental*, vol. 176, pp. 396-428.

[6] M. Janus, J. Zatorska, A. Czyzewski, K. Bubacz, E. K. Nejman and A. W. Morawski, 2015 *Applied Surface Science*, vol. 330, pp. 200-206.

[7] J. Chen, S. C. Kou, and C. S. Poon, 2011 *Building and Environment*, vol. 46, pp. 1827-1833.

[8] G. M. Sabnis 2012 *Green Building with Concrete : Sustainable Design and Construction* (United states, CRC Press)

[9] M. Stefanidou and I. Papayianni, 2012 *Composites : Part B*, vol.43, pp. 2706-2710.

[10] R. Khataee, V. Heydari, L. Moradkhannejhad, M. Safarpour and S. W. Joo, 2013 *Journal of Nanoscience and Nanotechnology*, vol. 13, pp. 1-6.

[11] E. Quagliarini, F. Bondioli, G. B. Goffredo and C. Cordoni, 2012 *Construction and Building Materials*, vol. 37, pp. 51-57.

[12] H. S. Muller, M. Haist and F. Acosta 2012 *Proceedings of the 9th fib International PhD Symposium in Civil Engineering* (Germany, KIT Scientific Publishing)

[13] E. J. Relinque, J. R. R. Garcia, A. Castillo and M. Castellote, 2015 *Cement and Concrete Research*, vol. 71, pp. 124-131.

[14] D. Feng, N. Xie, C. Gong, Z. Leng, H. Xiao, H. Lui and X. Shi, 2013 *Industrial and Engineering Chemistry Research*, vol. 52, pp. 11575-11582.

[15] W. Shen, C. Zhang, Q. Li, W. Zhang, L. Cao and J. Ye, 2015 *Journal of Cleaner Production*, vol. 87, pp. 762-765.

[16] A. H. Aissa, E. Puzenat, A. Plassais, J. M. Herrman, C. Haehnal and C. Guillard, 2011 *Applied Catalysis B : Environmental*, vol. 107, pp. 1-8.

[17] M. Chusid, 2005 *Concrete DECOR*, vol.5.

IOP Conf. Series: Materials Science and Engineering **133** (2016) 012027 doi:10.1088/1757-899X/133/1/012027

Influence of different treatment condition on biopolymer yield production for coagulation-flocculation process

I S Aisyah[1], M F Murshed[2] and I Norli[3]

[1, 2]School of Civil Engineering, Engineering Campus, Universiti Sains Malaysia, 14300 Nibong Tebal, Seberang Perai Selatan, Pulau Pinang, Malaysia.
[3]School of Industrial Technology, Universiti Sains Malaysia, 11800 Minden, Penang, Malaysia.

E-mail: cefaredmurshed@usm.my

Abstract. Two different agro wastes (banana pseudostem and rice straw) were utilized in order to extract biopolymer (pectin) known as coagulant aid in water and wastewater treatment. Factors such as pH, temperature and time were chosen due to the critical role in hot acid extraction process. The yield of biopolymer extraction from banana pseudostem was found to be higher at 28 % meanwhile only 18 % from rice straw was manage to produce from the dry weight 10 g, respectively. It was found that extraction temperature and extraction time were the most important factors influencing the biopolymer yield which increased with temperature and time or decreasing pH. Based on two level factorial design, the same condition of pH 1.5, temperature 90 °C and 4 hours extraction time can produce high amount of extracted biopolymer. Fourier Transform Infrared Spectroscopy (FTIR) was used to detect the existence of functional group which helps in the coagulation-flocculation process. Result indicates a similar functional group of biopolymer were detected for both difference agro wastes.

1. Introduction

In 2012, Americans produced about 251 million tons of garbage, from which 87 million tons were recycled and composted [13]. Garbage should not only be reduced by recycling the trash but also by minimizing the waste generated globally. Likewise, agricultural waste production too has the potential to be recycled and produce new material. In conjunction, banana pseudostem and rice straw are agricultural wastes that have the capability to treat effluent from wastewater through extracted biopolymer. Annually, about 17 million tons of bananas are produced worldwide whilst global paddy production is more than 700 million tons [4]. It has been estimated that a few tons per hectare of banana pseudo-stem are produced every year [2]. Waste generated from the banana plantation was disposed on land and the effluent flows into water bodies, which result in serious ecological hazard. Similarly, a report by the International Rice Research Institute (IRRI) stated that in 2008 approximately 620 million tons of rice straws were produced in Asia alone. These agro wastes are disposed off in various ways and have been reported to be increasing every year. Thus, it can be concluded that inefficient solid waste disposal methods will create series of problems, namely air and water resource pollution, accident hazard, increase of rodents and insect vector disease which leads to serious risk towards the communities [3].

Nowadays, agro wastes are found to be interesting organic material due to the availability value in the market which can be used in wastewater treatment as biopolymer. Many researchers studied the effect of pH, temperature, extracted time, salting out time and solvent to solid ratio in order to get the optimal extracting biopolymer. However, only three of the parameters seem to influence considerably on the quantity of pectin; which is pH, temperature and extracted time [14]. The extraction rate of biopolymer increases at higher temperature of the extraction process [12]. While previous study discovered that pH and time is the most significant parameter on the biopolymer yield [13].Biopolymer known as pectin, has an important functional group to react as bioflocculating agent, which is efficient in wastewater treatment process. It proves to be the best biopolymer since it contain a huge number of carboxyl group (COOH) and able to supply more effective sites for ferric cations to form the bridge [10]. The electrolyte group can be exposed from the longer chain of biopolymer [6]. The functional group of carboxyl and hydroxyl with the negative charges are able to stretch out the biopolymeric chain due to the electrostatic repulsion [15].

The first part of this study is to identify the highest biopolymer yield from banana pseudostem and rice straw at various treatment conditions (pH, temperature and time) by using 2^k factorial design in triplicate. The influence factors that affecting the percentage of biopolymer yield (%) was investigated. The functional group for both extracted biopolymer also been analysed.

2. Methodology

The laboratory work of this study involves two phases which is extraction of biopolymer from banana pseudostem and rice straw. In order to study the effects of factors involved towards the yield, three effects were chosen include pH, temperature and time using two level factorial designs. Design Expert 6.0.2 software was used to evaluate normal probabilities plot in order to see the influent factors. The Analysis of Variance (ANOVA) was carried out with regards to validate the model with p-value at 95 % confidence interval.

2.1 Sampling of banana pseudostem and rice straw

The agro wastes from banana plantation site and paddy fields were obtained after they were harvested. Samples of banana pseudostem and rice straw were then clean, cut into smaller pieces and then the samples were washed with Sodium Metabisulphate 0.02% for preservation purpose before they were heated at 90°C in water bath for 15 minutes [8]. Later the samples were dried at 60°C for 48 hours before being ground into the size 500 μm.

2.2 Hot acid extraction process

Approximately 10 g of banana pseudostem and rice straw powders were mixed with distilled water at 1:10 ratio. Extraction of biopolymer known as pectin was carried out at different pH, temperature and time (Table 1). The extraction process was slightly modified from previous researcher [5]. The pH of the mixture was adjusted to the desired pH using sulphuric acid, H_2SO_4. The mixture was then extracted in a water bath stirrer at 80°C/ 90°C for 0.5/ 4 h, while being stirred. The mixture was left to cool to room temperature before the pH of mixture was increased to pH 3.5 by adding KOH 0.2M to promote pectin extraction [11]. Next it was filtered through 125 mm Whatman No.1 filter paper. 70% of Isopropyl Alcohol (IPA) was added for 4 hours to precipitate the biopolymer and freeze dried for 2 days. The collected biopolymer was ground to powder form and weighed for yield assessment. Finally, the homogenous biopolymer powder was stored at room temperature. Yield of biopolymer extracted was calculated as follow:

$$Yield\ of\ biopolymer(\%) = \frac{w_o}{w} x100 \quad (1)$$

#

Where:
W_0 (g): the dried product weight after freeze dried
W (g): the dried raw material weight.

The factors were analyzed through two level factorial designs in order to get the analysis of variance and determine the interaction of the factors involved. Biopolymer yield from banana pseudostem and rice straw went through the various treatment conditions (pH, temperature and time). The comparison of biopolymer yield (%) was studied between banana pseudostem and rice straw. The condition of biopolymer extraction for both agro wastes was tabulated in Table 1 as followed:

Table 1. Condition for biopolymer extraction

Experimental Runs	**A: pH**	**B: Temperature/ °C**	**C: Time/ h**
1	1.5	90	4.0
2	2.0	80	4.0
3	1.5	80	0.5
4	2.0	90	0.5
5	1.5	80	4.0
6	1.5	90	0.5
7	2.0	80	0.5
8	2.0	90	4.0

2.3 Determination of biopolymer functional group
Fourier Transform Infrared (FTIR) Spectroscopy Attenuated Total Reflectance (ATR) was used to determine the functional group of biopolymer at mid infrared region (4000-600cm^{-1}). Biopolymer powder was placed in the spechtrophotometer and scanned in the FTIR. Ethanol was used intentionally for all the cleaning part in the FTIR.

3. Result and Discussion

3.1 Yield of biopolymer extracted from banana pseudostem and rice straw
As shown Figure 1, biopolymer powder was observed in the beaker after the extraction process. The colour of the extracted biopolymer obtained was light brown powder. There are many factors that contribute to the colour discrepancy, such as surface contamination, amount of isopropyl alcohol used during precipitation or environmental factors [7].

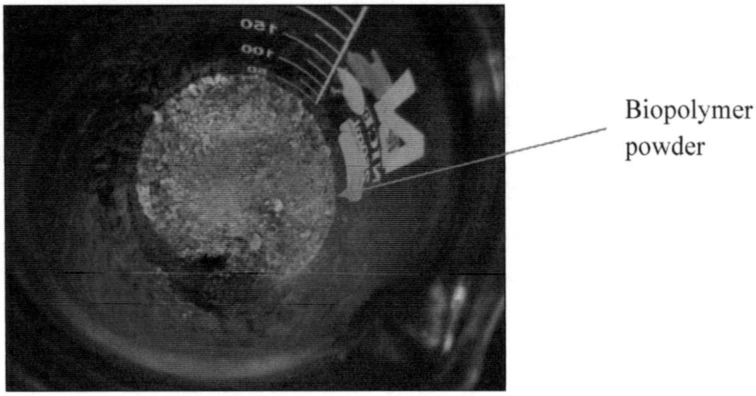

Figure 1. Biopolymer in powder form after the extraction process.

The result for both biopolymer yields derived from banana pseudostem and rice straw after the extraction process is illustrated in Figure 2. Three selected factors were involved: pH (1.5 and 2.0); Temperature (80°C and 90°C) and Time (0.5 and 4.0) as shown in Table 1.

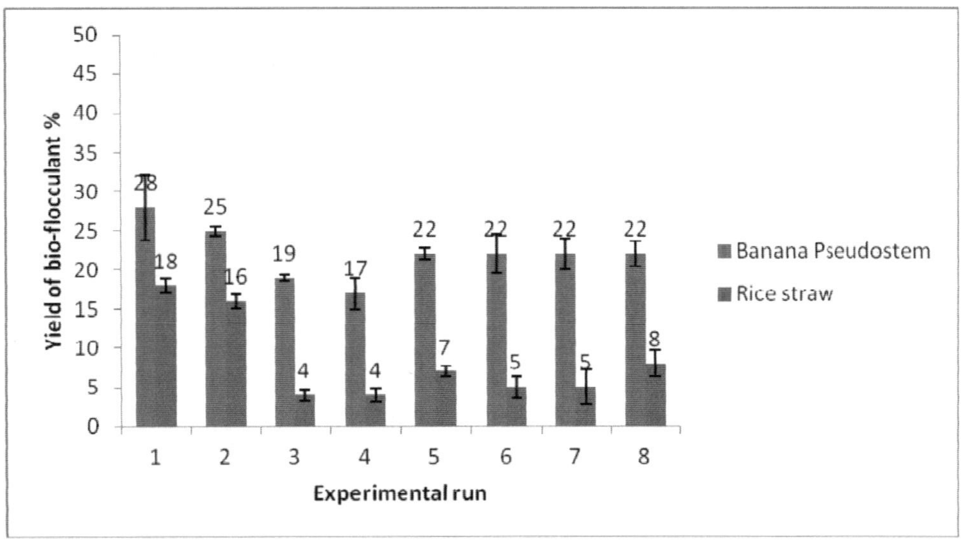

Figure 2. Yield of Extracted Biopolymer from banana pseudostem and rice straw.

The yield of extracted biopolymer from banana pseudostem ranged from 17 % to 28 % of the dry weight. The extraction for 4 hours at 90 °C, with pH 1.5, produced the highest yield among others. The variance analysis from Table 2 showed small interactive effects between pH and temperature with p-value at 0.0348. As for individual effects, temperature and time were the most influential factors at p-value 0.0246 and 0.0001 respectively. It is believed that by increasing the temperature in a harsh condition, it could increase the extraction rate of the extracted biopolymer [12]. In addition, favourable acidic conditions contribute to increase of biopolymer recovery by hydrolysing the insoluble biopolymer constituents into soluble biopolymer. Besides, more galacturonic acid content can be achieved at pH 1.5 rather than pH 2 [14]. However, galacturonic acid content did not influence the extraction rate of biopolymer.

Normal probabilities plot was used to analyze the factors influenced in the study. Figure 3 shows a normal probabilities plot for biopolymer extraction from banana pseudostem. At Point A, B, C, AB and BC, the distance from the red line due to the significance effects with P value < 0.05. In contrary, point AC and ABC were on the red line which means the effect was not significant in the treatment. It is seen that the larger significant effect showed a great distance from the line compare to non significant effects.

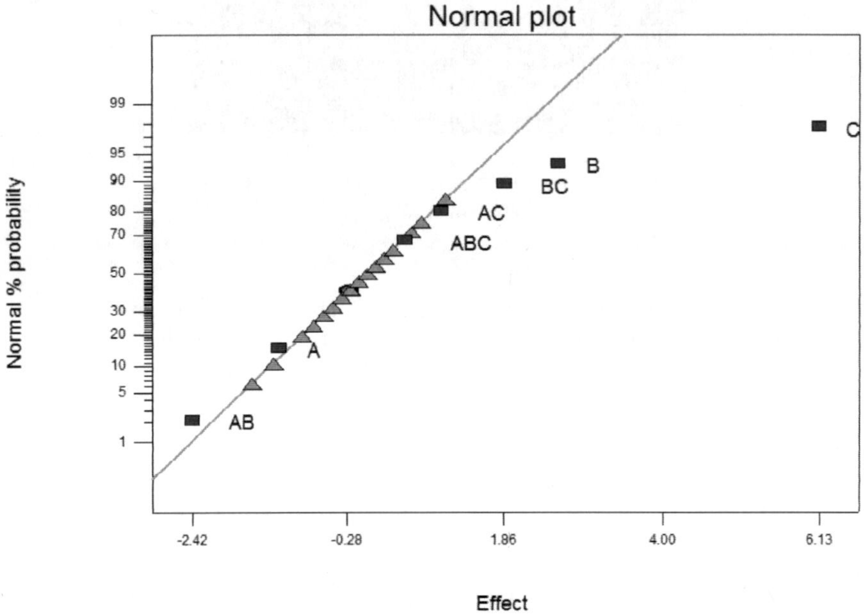

Figure 3. Percentage of normal probabilities plot of extracted biopolymer from banana pseudostem

Table 2. Analysis of variance (ANOVA) for extracted biopolymer from banana pseudostem

Extraction of Pectin from Banana Pseudostem					
Source of variance	**Sum of Squares**	**DF**	**Mean square**	**F-value**	**p-value**
Model	338.9	7	48.41	7.35	0.0005
A	8.88	1	8.88	1.35	0.2627
B	40.56	1	40.56	6.15	0.0246
C	225.71	1	225.71	34.25	0.0001
AB	35.04	1	35.04	5.32	0.0348
AC	6.20	1	6.20	0.94	0.3464
BC	20.91	1	20.91	3.17	0.0939
ABC	1.60	1	1.60	0.24	0.6287
Pure error	105.44	16	6.59		
Cor total	444.34	23			

Meanwhile, the yield obtained from the biopolymer extracted rice straw varied from 4 % to 18 % of the dry weight. As expected from the extraction process from banana pseudostem, the extraction of the highest yield for extracted biopolymer from rice straw were also produced at the same condition of pH 1.5, temperature 90 °C and 4 hours. The variance analysis from Table 3 shows no interactive effects between pH and temperature and between pH and time at p-value 0.2178 and 0.2105, respectively except for interaction effects between temperature and time at p-value 0.0015. Whereas, all factors were significant for individual effects of pH at p-value 0.0397 and for temperature and time was at p-value 0.001. At constant pH and temperature, the yield of biopolymer produced for 4 h was higher than those at 0.5 h. The yield of biopolymer has been reported to increase with the increase of temperature and time [13, 14].

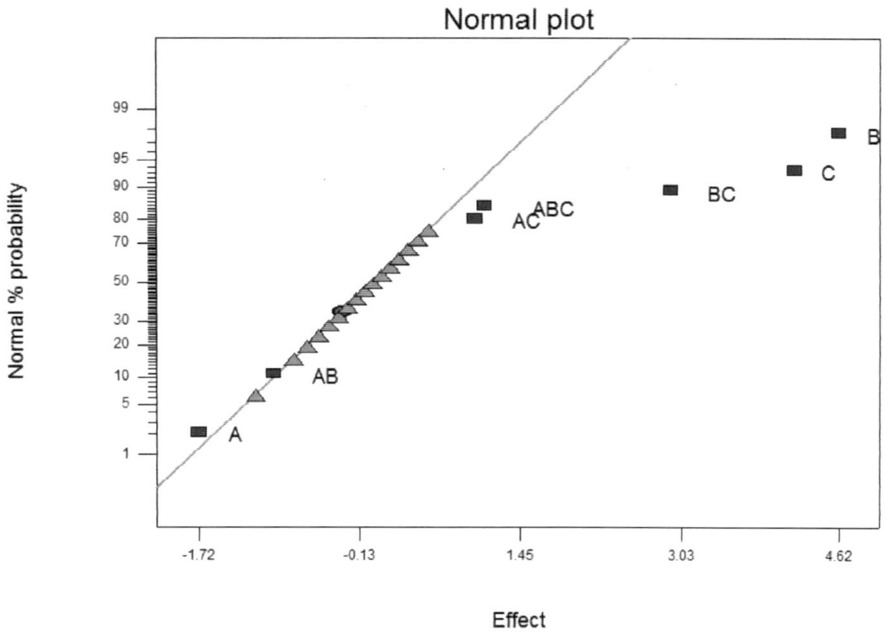

Figure 4. Percentage of normal probabilities plot of extracted biopolymer from rice straw

Normal probabilities plot was used to analyze the factors influenced in the study. Figure 4 show a normal probabilities plot for biopolymer extraction from rice straw. Point A, B, C, AC, BC and ABC showed a distance from the red line due to the significance effects with P value < 0.05. In contrary, point AB was on the red line which means the effect was not significant in the treatment. It is seen that the larger significant effect showed a great distance from the line compare to non significant effects.

Table 3. Analysis of variance (ANOVA) for extracted biopolymer from rice straw

Extraction of Biopolymer from Rice Straw					
Source of variance	**Sum of Squares**	**DF**	**Mean square**	**F-value**	**p-value**
Model	320.42	7	45.77	12.98	0.0001
A	17.68	1	17.68	5.02	0.0397
B	127.88	1	127.88	36.27	0.0001
C	104.17	1	104.17	29.55	0.0001
AB	5.80	1	5.80	1.65	0.2178
AC	6.00	1	6.00	1.70	0.2105
BC	51.63	1	51.63	14.64	0.0015
ABC	7.26	1	7.26	2.06	0.1705

Pure error	56.41	16	3.53
Cor total	376.82	23	

This indicated that the biopolymer was difficult to be extracted from the plant cell wall when the condition was too minimal in temperature and time. According to other study, degradation of biopolymer chain molecules occurred during longer extraction time and the higher extraction temperature which affected biopolymer extraction rate [12]. It was found that pH 1.5 with constant temperature and time can produce higher yield compared to pH 2. Similar results were obtained in previous study found that the increase amount of pectin extracted from agro wastes was affected by lower pH values [5].In order to see the influence of the factors involved in the extraction of biopolymer, the variance analysis was analysed and is shown in Table 4. The significant effects with P value < 0.05 indicate that the factors have an impact on the results obtained. On contrary, those factors with p value more than 0.05 were not significant in the treatment.

Table 4. Values of p-value for biopolymer yield from banana pseudostem and rice straw

Source of variance	Biopolymer Yield from Banana Pseudostem	Biopolymer Yield from Rice Straw
	P value (95% Confidence Level)	
Model	0.0005	0.0001
A	0.2627	0.0397
B	0.0246	0.0001
C	0.0001	0.0001
AB	0.0348	0.2178
AC	0.3464	0.2105
BC	0.0939	0.0015
ABC	0.6287	0.1705

3.2 Functional group of biopolymers using FTIR

FTIR was analysed for both biopolymer extracted from banana pseudo-stem and rice straw with regards to see the presence of effective functional group for flocculation. The FTIR spectrum of biopolymer was taken in the frequency range of 4,000–400 cm^{-1}. The functional group of extracted biopolymer from banana pseudo-stem and rice straw can be observed in Table 5.

Table 5. Intensity (cm^{-1}) of FTIR extracted biopolymer from banana pseudo-stem.

Biopolymer from banana pseudo-stem		Biopolymer from rice straw	
Intensity (cm^{-1})	**Functional Groups**	**Intensity (cm^{-1})**	**Functional Groups**
655.18, 1,031.43, 1,089.94, 1,120.28, 1,196 and 3,257	Hydroxyl (-OH)	656.10	Hydroxyl (-OH)
884.09	CH deformation (-CH)	753.5 – 778.25	Methylene groupl (-CH)
1,312	CH3 deformation (CH_3)	1,090.28 – 1,194.53	Hydroxyl (-OH)
1,424	Carboxyl group (COOH)	1,137.19	Alyphatic ether (C-O-C)
2,922	Methylene group ($-CH_2$)	1,313.75 – 1,609.96	Carboxyl group (COOH)

The FTIR spectra of biopolymer shows–OH stretching vibration peaks with clear absorption peaks at 655.18, 1,031.43, 1,089.94, 1,120.28, 1,196 and 3,257 cm^{-1}. The existence of symmetric C-H stretching vibrations of methylene groups at peak 2,922 cm^{-1} can be considered as the increment of C-H bond content after amidation reaction [9]. The peaks at 1,312 cm^{-1} and 884.09 could be assigned to CH_3 and CH deformation respectively. The FTIR spectra of biopolymer show that the spectrum also has an ester peak at 1196 cm^{-1}. There are carboxyl group detected at 1,424 cm^{-1}, which are strongly contributed to the flocculating mechanism [6].

Similarly, the functional group of biopolymer from rice straw is detected at the same range in FTIR spectra as observed in Table 5. Based on the result, hydroxyl group (-OH) which is one of the bio-flocculant's functional group is detected at peaks 656.10, 1,090.28, 1,122.01 and 1,194.53 cm^{-1} in an IR spectrum with the broad O-H absorption of alcohols. The peak at 753.85 and 778.25 cm-1 indicates C–H stretching vibration. The peak at 1090.28, 1,122.01 and 1194.53 cm-1 suggests –OH bending point out as the presence of secondary alcohol. The peak at 1137.19 and 1313.75 cm^{-1} could be assigned to aliphatic ethers C-O-C and carboxylic acid (-COOH) respectively. The peak at 1,609.96 also indicated the presence of carboxylate stretching. Nevertheless, the presences of hydroxyl and carboxyl group have a potential to increase electrostatic attraction in order to provide a strong bridging mechanism [1].

4. Conclusion

Three different factors were investigated to see their influence on biopolymer yield after hot acid extraction process. Two level factorial design indicates that, extraction temperature and extraction time were the major effects that contribute to higher amount of biopolymer. The range values of yield obtained from banana pseudostem were between 16 % up to 28 % whereas 4 % until 18 % were collected from rice straw. Thus, the functional group for both agro waste were similar which contained of hydroxyl group, methylene group and carboxyl group.

Acknowledgments

The authors would like to express their gratitude to Universiti Sains Malaysia, Pulau Pinang, Malaysia for the financial support granted under RUI grant no. 1001/PAWAM/814259.

References

[1] Bolto B and Gregory J. 2007 Organic polyelectrolytes in water treatment *Water research* **41** 2301-2324

[2] Cordeiro N, Belgacem M N, Torres I C and Moura, J C V P 2004 Chemical composition and pulping of banana pseudo-stems *Industrial Crops and Products* **19** 147-154

[3] Essien, J P, Akpan, E J and Essien, E P 2005 Studies on growth and biomass production using waste banana peel *Bioresource Technology* **96** 1451–1456

[4] FAO (Food and Agricultural Organization of the Unites Nations) 2013 FAOSTAT statistic data base, Agriculture, Rome, Italy

[5] Happi, E T, Robert, C, Ronkart, S N, Wathelet, B and Paquot, M 2008 Dietary fibre components and pectin chemical features of peels during ripening in banana and plantain varieties *Bioresource Technology* **99** 4346–4354

[6] Ho, Y C, Norli, I, Abbas Alkarkhi, F M and Morad, N 2010 Characterization of biopolymeric flocculant (pectin) and organic synthetic flocculant (PAM): A comparative study on treatment and optimization in kaolin suspension *Bioresource Technology* **101** 1166-1174

[7] Kanmani, P, Dhivya, E, Aravind, J and Kumaresan, K 2014 Extraction and Analysis of Pectin from Citrus Peels: Augmenting the Yield from Citrus limon Using Statistical Experimental Design *Iranica Journal of Energy and Environment* **5** 303-312.

[8] Kar, F and Arslan, N 1999 Effect of temperature and concentration on viscosity of orange peel pectin solutions and intrinsic viscosity-molecular weight relationship *Carbohyd Polym.* **40** 277–284.

[9] Mishra, A and Bajpai, M 2008 Flocculation behaviour of model textile wastewater treated with a food grade polysaccharide *Journal of Hazardous Materials* **118** 213–217.

[10] Piriyaprasarth, S and Sriamornsak, P 2011 Flocculating and suspending properties of commercial citrus pectin and pectin extracted from pomelo (Citrus maxima) peel. *Carbohydrate Polymers* **83** 561-568.

[11] Seggiani, M, Puccini, M, Pierini, M, Giovando, S and Forneris, C 2009 Effect of different extraction and precipitation methods on yield and quality of pectin *International journal of food science & technology* **44** 574-580.

[12] Qiu, L P, Zhao, G L, Wu, H, Jiang, L, Li, X F and Liu, J J 2010 Investigation of combined effects of independent variables on extraction of pectin from banana peel using response surface methodology *Carbohydrate Polymers* **80** 326-331.

[13] United States Environmental Protection Agency (USEPA), 2014.

[14] Wai, W W, Alkarkhi, A F and Easa, A M 2010 Effect of extraction conditions on yield and degree of esterification of durian rind pectin: An experimental design *Food and Bioproducts Processing* **88** 209-214.

[15] Yapo, B M, Robert, C, Etienne, I, Wathelet, B and Paquot, M 2007 Effect of extraction conditions on the yield, purity and surface properties of sugar beet pulp pectin extracts *Food Chemistry* **100** 1356-1364.

[16] Yokoi, H, Obita, T, Hirose, J, Hayashi, S and Takasaki, Y 2002 Flocculation properties of pectin in various suspensions *Bioresource Technology* **84** 287-290.

IOP Conf. Series: Materials Science and Engineering **133** (2016) 012028 doi:10.1088/1757-899X/133/1/012028

Self-Cleaning Technology in Fabric: A Review

Siti Rohani Saad[1], Norsuria Mahmed[1*], Mohd Mustafa Al Bakri Abdullah[1,2], Andrei Victor Sandu[3]

[1]Center of Excellence Geopolymer and Green Technology, School of Materials Engineering, Universiti Malaysia Perlis, Kompleks Pusat Pengajian Jejawi 2, 02600 Jejawi, Perlis, Malaysia.

[2]Faculty of Engineering Technology, Unicity Alam Campus, Universiti Malaysia Perlis, 02100 Sungai Chuchuh, Padang Besar, Perlis, Malaysia.

[3]Gheorge Asachi Technical University of Iasi, Faculty of Materials Science and Engineering, D.61A Mangeron Str., 700050, Iasi, Romania.

*E-mail: norsuria@unimap.edu.my

Abstract. This article gives an overview on photocatalytic self-cleaning technology on fabric resulting from titanium dioxide (TiO_2) and zinc oxide (ZnO) as photocatalyst which decompose the organic stain into water and carbon dioxide (CO_2) in presence of UV light source. The self-cleaning concept is useful in various application including the textiles materials which are normally used in daily life. This technology also can be developed in other application for instance medical textiles, athletic wear, and military uniform and also outdoor fabrics. Additionally, it is beneficial as it effectively conserves water and improves the appearance of the environment and in long term it will reduce energy, laundry cost and time as well.

1.0 Introduction

Self-cleaning concept has attained remarkable interest because of their distinctive features and wide range of possible applications in various fields. As well as increasing the demand for sanitary, self-disinfecting and contamination free surfaces, interest in self-cleaning protective materials and surfaces has developed rapidly throughout the years since it has high potential as commercial product, which is able to meet the market demands globally. There are numerous materials that utilized the self-cleaning technology including interior applications such as fabrics, furnishing materials, window glasses, and outdoor construction materials such as roof tiles, car mirrors, and solar panels [1].

The self-cleaning theory was instigated from nature phenomenon which can be noticed on leaves of lotus plant, rice plant, butterfly wings, fish scales, etc. [2-3]. For lotus leaf, it is a type of plant which grows in mud without let the mud affect the purity of the plant. The waxy surfaces of the lotus leaves combined with the presence of microscopic structures result in an extremely hydrophobic surface [4]. In 1997, the investigation of the self-cleaning ability of the plant leaves surface was done by Barthlott and Neihuis [2]. The surface characteristic was analysed by a high-resolution scanning electron

[1] To whom any correspondence should be addressed.

microscope (SEM) and the contact angle (CA) of leaves from 340 plant species. Majority of the investigated leaves have contact angle less than 110° without any prominent surface sculpturing especially epicuticular wax crystals. Contrary to water-repellent leaves, there are various different surface sculptured, mainly epicuticular wax crystals were exhibited in combination with papillose epidermal cells and their contact angle always exceeded 150°. They observed that on water-repellent surfaces, water contracted to form spherical droplets and came off the leaf very quickly, even at slight angles of inclination ($<5^{\circ}$), without leaving any residue.

Generally, all kinds of particles are always removed entirely from water-repellent leaves when exposed to natural or artificial rain, as long as the surface waxes are not destroyed. The dirt particles that deposited on the waxy surface of the leaves are generally larger than the microstructure of the surface of the leaf. Hence, deposited on the tips, which results in the minimizing of the interfacial area between both surfaces. In the case of a water droplet rolling over a particle, the surface area of the droplet exposed to air is reduced and energy through adsorption is gained. The particle is removed from the droplet's surface only if a stronger force overcomes the adhesion between particle and water droplets. Due to the very small interfacial area between particle and rough surface, adhesion is minimized. Therefore, the particle is 'captured' by the water droplet and removed from the surface. This phenomenon is called "Lotus effect".

There are two principal ways of self-cleaning materials, namely hydrophobicity and hydrophilicity. Both types of coating clean themselves with the action of water by rolling droplets for hydrophobic and sheeting water for hydrophilic that carries dirt away. Nevertheless, hydrophilic have an additional property, which can chemically break down the adsorbed dirt in sunlight through the help of photocatalyst which also known as hydrophilic photocatalytic coating [5].

2.0 Hydrophobic and Hydrophilic

Hydrophobic surface repels water with the properties of low wettability and contact angles more than 90°. The higher contact angle the lower the value of adhesion on the surface and increase the hydrophobicity. For contact angle more than 150° the surface is termed as superhydrophobic. Hydrophobicity also can be regulated by the roughness factor. With rougher surfaces, the contact angle of water on it will increase, and form bumps that trap air between water and the surface. The "Lotus Effect" was applied in this superhydrophobic mechanism [2]. The lotus plant cleans itself by having superhydrophobic leaf which consists of microscopic bumps all across the leaf's surface that play the important part to its water-repelling properties. A rough coating of nanoscopic wax crystals on these bumps further increases the effect. It allows the water droplets rolls across and removes dirt away. The water and the dirt have more affinity each other than the surface [2]. Hang Ji et al. [6] have created a superhydrophobic polymer structure by directly replicating the surface of a lotus leaf. Poly (dimethysiloxane) (PDMS) was used to replicate the lotus leaf structure. The leaf was used as a template to cast a complementary PDMS layer. An anti-stick layer was added to the PDMS, which was then used as negative template for a second PDMS casting step. The second PDMS layer was then a positive image of the lotus leaf. The complex lotus surface patterns are transferred with high fidelity. The artificial PDMS lotus leaf has the same water contact angles and very low water roll-off angle as the natural lotus.

Hydrophobic surfaces can be useful for other application such as anti-corrosive systems and anti-icing [7]. The superhydrophobic coating by using nanoparticle-polymer composite for anti-icing can prevent the ice formation by inhibiting the frost nucleation process. It was observed that the composites are able to prevent ice formation depends not only on their superhydrophobicity but also on the size particles exposed on the surface upon impact of supercooled water both in laboratory conditions and in naturally occurring environments. These results open up possibilities for rational design of anti-icing superhydrophobic surfaces by tuning surface textures in multiple length scales [7]. In addition, superhydrophobic coatings can enhance the fuel efficiency in maritime industry by reducing the skin friction drags occurring in ship hulls. Such a coating increases the ship speed and also acts as anti-corrosive systems, preventing any organic contaminant or marine microorganisms to come in contact with the ship hulls. In vehicles, superhydrophobic coatings are applied on the glasses to prevent rain droplets from clinging, thereby helping to clean the car itself.

Meanwhile, besides hydrophobic, researchers also seek more interest in hydrophilic surface which is advantageous in presence of water for self-cleaning application. The wettability is high and the contact angle is approximately 0°. Therefore, the water lies flat on the surface in sheet instead of droplets. It cleans the surface alongside the photocatalysis process, followed by sheeting of water which makes the surface superhydrophilic thus carry the dirt molecules away [5].

3.0 Photocatalytic Process

Photocatalytic process is the acceleration of photoreaction in the presence of catalyst. This process will decompose the dirt molecules by utilize the sunlight. By utilizing the photoreaction induced by photocatalyst, the organic contaminants will be degraded into air and water [8]. The mechanism of photocatalytic reaction begins when a photocatalyst is irradiated by light, usually ultraviolet light (UV). With the energy equal to or higher than its band gap, electrons on the photocatalyst surface are excited, and escaped from valence band to the conduction band, leading to the formation of electron-hole pairs in the surface-excited negative charged electrons (e^-) in the conduction band, and the positive charged holes (H^+) in the valence band [9-21]. The created pairs can recombine or get trapped and react with other material that absorbed on the photocatalyst. The pairs will cause redox reactions at the surface, the negative electrons (e^-) and oxygen will combine to form super oxide radical anions (O_2^-), whereas the positive electric holes and water will generate hydroxyl radicals (OH^-) [8]. Eventually, all the formed highly active oxygen species will oxidize organics compound to carbon dioxide (CO_2) and water (H_2O). Hence, photocatalyst can decompose common organic matters in the air such as odour molecules, bacteria and viruses [14].

4.0 Photocatalyst

The semiconductor photocatalytic process has shown a great potential as a low-cost, environmental friendly and sustainable treatment technology. Those semiconductor materials also have been used to functionalize onto the different textiles for self-cleaning properties. The functionalized fabrics have the ability to oxidize the coloring substances in the form of solutions and stains [22].

4.1 Titanium Dioxide (TiO_2)

Titanium dioxide (TiO_2) is a semiconductor material that acts as photocatalyst and has been proved to be an excellent catalyst in the photo degradation of colorants and other organic pollutants. It is widely used because of its various advantages, such as, non-toxicity, availability, cost effectiveness, chemical stability and favourable physical and chemical properties. Titanium dioxide exists in three crystalline phases; rutile, anatase and brookite. Among these three forms, rutile which is used in pigments as sun-blockers and paints is more sustainable than other two forms. The anatase has an open crystal structure, which makes it highly photocatalytic and always been used in semiconductor photochemistry, as this appears to be the most active and easiest to produce. Anatase and brookite, on the influence of heat, would change to rutile [1]. TiO_2 is used in paint and cosmetics as pigment and as a food-additive. It is also used in anti-pollution applications and for water purification. Currently, TiO_2 is used for self-cleaning surfaces and has now emerged into commercial products ranging from kitchen and bathroom ceramic tiles and fabrics, to indoor air filter and window glass section [1]. Pisitsak et al. [23] investigated the self-cleaning properties of cotton fabrics finished with nano-TiO_2 and nano-TiO_2 mixed with fumed silica. The self-cleaning effect was stronger for samples coated with higher TiO_2 concentrations. Mostly, the finished samples appeared clean after one wash whereas the untreated fabrics required repeated washing. However, with the addition of fumed silica, a reduction in self-cleaning ability was observed. Aside from photocatalytic ability, H. Yaghoubi et al. proved that polycarbonate substrate with a self-cleaning TiO_2 layer can improve its mechanical durability by exhibits better hardness and scratch resistance [24]. These results are good mechanical properties as the self-cleaning coatings are vital to make these coatings attractive in auto and construction industries.

4.2 Zinc Oxide (ZnO)

Another semiconductor material, which appears to be an alternative to TiO_2 is zinc oxide (ZnO) [11,15]. The application of ZnO as a degradation material for environment pollutants has also been extensively studied, because of its nontoxic nature, low cost and high photochemical reactivity [17]. ZnO semiconductor exhibits a better efficiency of photocatalytic reaction than anatase TiO_2 in photochemical degradation of reactive dyes [13]. ZnO in powder form has been extensively used for photocatalytic degradation of dyes present in effluent [9,13,16]. However, the ZnO powder has a recovery problem after degradation, because some of ZnO powder is lost while draining the solutions. To overcome the problem, ZnO particles can be attached to a substrate particularly textiles because they offer the highest surface area. These ZnO functionalized textiles can be used for effluent treatment as well as self-cleaning. A few attempts have also been made to functionalize textiles with ZnO to generate self-cleaning properties. Ashraf et al. [15] studied the photocatalytic solution discoloration and self-cleaning of functionalized polyester fabric with ZnO nanorods. The nanorods of ZnO were deposited on polyester fabric by hydrothermal method and degraded the stains and decolorize the solution of different types of dyes [15].

5.0 Self-Cleaning Fabric

Textile products play an important role in our basic need. They are commonly produced and used in clothing industry. Textiles or fabrics are also important in other industries such as agricultural industry, transportation, building material and also health industry. Therefore, the technological advances of textiles are improved in order to meet the need of various industries. The properties and performance of textiles fibres are essential to fabric manufacturing and utilization [25]. Textile finishing has always leaded to introduction of new technical properties, which is useful in diversified end uses. Technological diversification in finishing decides the performance domain of fabrics and rendered it special functional properties. There has been a continuous improvement and innovation in the area of chemistry related to textile finishing. Research interest in the use of nanotechnology in the textile industry has increased rapidly in recent times. It was demonstrated that nanotechnology can be used to enhance textile attributes, such as fabric softness, durability, breathability, water repellence, fire retardancy and anti-microbial property in fibers, yarns and fabrics [25]. Textile finishing involving the application of inorganic colloidal nanosuspension (nanosols) can give rise to new fabrics in which properties of inorganic nanoparticles are transferred to the textile's surface [26]. Ever since the introduction of nanoparticles in textiles, efforts were made to produce finished fabrics with multiple performances. This is mainly due to the fact that textiles fabrics constitute one of the best substrates for the application of nanotechnology because those consisting of cotton fibre which is normally known possess large surface area [27]. The tight binding of nanoparticles to the textile surfaces is a significant part in application of nanotechnology in the textile industry. This is to ensure the durability of desired properties is increased and also minimum released of nanoparticles into the environment. In order to apply the nanoparticles, the maximum chemical compatibility between nanoparticles and textile surface should be achieved. There is commonly used method for this purpose which is a method that uses covalent linking agents.

The dip-pad-dry-cure process is often used to create bonds between nanoparticles and a fabric. Chaudari et al. [28] examined the effect of nano TiO_2 pre-treatment on functional properties of cotton fabric. The cotton fabric was treated with nano TiO_2 colloid by pad-dry-cure method. The TiO_2 concentrations in the padding liquor were varied and the specimens were immersed in a fresh colloidal bath for 10 minutes and squeezed using a laboratory padder at constant pressure. The samples were dried for 30 minutes and cured at 140°C for 3 min. Similarly, Ortelli et al [29] also applied the TiO_2 nanosol directly on textiles by pad-dry-cure method. The fabric was dipped in TiO_2 nanosol and left for 3 min, then passed through a two-roller laboratory padder, oven dried and cured for 10 min at 130°C. SiO_2 was used as a binder since it prevents textile from degradation during the photocatalytic activity of TiO_2. Pakdel et al. [20] functionalized wool fabrics using TiO_2 and TiO_2/SiO_2 nanocomposite through a low temperature sol-gel method. The prepared TiO_2/SiO_2 composite were applied to the surface of wool fabrics using dip-pad-dry-cure method. In order to make the TiO_2-SiO_2 layer, Yuranova et al. [28] deposited it on the textile by soaking cotton in a 1:1 solution of colloidal TiO_2 and SiO_2 Ludox SM-30. The sample was then dried in air at 100°C for an hour. The results of high-resolution transmission electron microscopy (HRTEM) and infrared spectroscopy show that the corrosion of the upper layers of cotton would occur due to the effect of daylight on the TiO_2 particles, if there were no SiO_2 as a binder.

Besides SiO_2, researchers also mostly used acrylic binder in the deposition of nanoparticle to textile surface. Sivakumar et al [29] immersed the cotton fabric for 1 min in aqueous solution with 0.5% acrylic binder in padding mangle. The nanoparticle was uniformly coated using laboratory padding mangle at speed of 15 m/min with pressure of 15 kg/cm^2. After padding, the fabric was dried at 70°C for 5 min and cured at 120°C in curing chamber for 3 min. The SEM image showed that the nanoparticles were well dispersed on the fiber surfaces and were finely dispersed and embedded. Moreover, Kumar et al. [8] also used acrylic binder (1%) and added 0.5 % Lissapol-N as capping agent. Through the SEM analysis, TiO_2 nanoparticles were well dispersed and embedded on fiber surface. Apart from that, previous reports indicated that carboxylic acid groups can acts as surface modifier to anchor TiO_2 nanoparticle into fabrics. Therefore, Wijesena et al. [30] investigated a method to modify the cotton fiber surface with carboxylic acid groups through slight carbomethylation of cellulose using sodium monochloroacetate (MCAA) and NaOH. Then, the fabric was grafted with TiO_2 nanoparticles through a process similar to dyeing. Once complete the grafting process, the fabric sample was sonicated to remove unattached particles and dried at 60°C. The catalyst loading was determined by measuring the weight gained after the TiO_2 nanoparticles grafting. The results showed the TiO_2 nanoparticles loading was increased with increase of MCAA concentration used to treat the cotton fabric.

The self-cleaning fabrics are function by implementing the photocatalytic reaction on the fabric with photocatalyst such as titanium dioxide and zinc oxide. As reported by Samal et al. [14] the fabric is coated with a thin layer of titanium dioxide particles then when this semiconductor exposed to light, photons with energy equal to or greater than the band gap of the photocatalyst excite electrons up to the conduction band. The excited electrons within the crystal structure react with oxygen atoms in the air, creating free-radical oxygen. These oxygen atoms are powerful oxidizing agents, which can break down most carbon-based compounds through oxidation-reduction reactions. In these reactions the organic compounds (i.e. dirt, pollutants, and microorganisms) are broken down into substance such as carbon dioxide and water. The TiO_2-SiO_2 coated cotton textiles showed a high photocatalytic activity superior to TiO_2 coated cotton alone due to the high dispersion and the structural effects of the amorphous silica present [28]. Meilert et al [31] reported that it is possible to bind TiO_2 to cotton textiles through chemical spacers. The coating procedure was straightforward and used non-toxic reagents. The TiO_2 loaded cotton textiles presented stable self-cleaning properties and allowed to eliminate partially the chromophore(s) of the red wine under daylight irradiation presenting long-term stable performance. In addition, Mirjalili et al [1] claimed that degree of self-cleaning of cured fabrics with cross-link method is higher than cured fabrics with non cross-link method. The reason for this higher self-cleaning is due to the higher percentage and more uniform distribution of titania particles over surface of cotton fabrics in cross-link method than in non cross link approach.

Bedford et al. [32] concluded that the photocatalytic self-cleaning textile fibres fabricated by coaxial electrospinning are compatible to nonelectrospun photocatalytic self-cleaning textile fibers that use better photocatalytic materials and sub-band gap wavelength light sources. This illustrates the important role played by the high surface to volume ratio of the material fabricated by coaxial electrospinning. Sivakumar et al. [29] assessed the ultra-violet (UV) protection and self-cleaning action of cotton fabric coated with nano ZnO and nano TiO_2 with acrylic binder. In the case of self-cleaning action, the smaller nanoparticle size *in-situ* coating of ZnO derived using wet chemical technique showed better self-cleaning activity as compared to larger nanoparticles of TiO_2 treated fabric. They also found that the ZnO and TiO_2 nanoparticles finishing of cotton fabric not only imparts the UV protection and self-cleaning characteristics to the treated fabric but also added two more functional properties like antimicrobial activity and soil release action. Chaudhari et al [28] also reported that the self-cleaning action of the pre-treatment of cotton fabric with nano TiO_2 is higher at the higher concentration level. In the meantime, Cakir et al. [10] also investigated the self-cleaning properties of textile coated with ZnO nanoparticles which were synthesized within PS-b-PAA reverse micelles cores and the result showed that the ZnO nanoparticles coated fabric at ratio 40:1 exhibited fairly high photocatalytic activity on the degradation of methylene blue. Meanwhile, self-cleaning function and hydrophilicity of wool fabrics were successfully improved through the integration of silica in the TiO_2/SiO_2 nano composite [20]. By increasing the concentration of silica, the TiO_2/SiO_2 nanocomposite showed more capability in decomposing the stains. Oda et al. [19] used cotton fabric as carrier and stabilizer of ZnO/Ag^+ to test for the photocatalytic activity by using methylene blue dye and artificial UV light. The result reveals that ZnO and ZnO-Ag which stabilized on cotton fibre were efficient in the decolorization of dye.

Later, Ashraf et al. [15] studied the self-cleaning and solution discoloration by the immobilized ZnO nanorods which grown on polyester (PET) fabric by hydrothermal method. They found that the functionalised fabric degraded the stains and decolorised the solution due to photocatalytic effect of ZnO. The fabric sustained the repetitive usage for solution discoloration but its stain degradability decreases after each degradation. The functionalized fabric has excellent washing fastness but fairly good rubbing durability. Oda et al. [18] continued their research involves synthesis of a composite of Ag/ZnO with cotton texture by photodeposition method. Silver doped zinc oxide showed a higher activity in dye removal in comparison with un-doped oxide. Additionally, the composite of Ag/ZnO/texture showed a higher activity with respect to Ag/ZnO. The composite Ag/ZnO/texture showed an ability towards self-cleaning of dye under irradiation. It was found that the activity of zinc oxide was increased in pH range that are close to its point zero charge (PZC). According to Kumar et al. [8] who studied the self-cleaning action of TiO_2 nanoparticles coated cotton, the coated textile structure demonstrates the significant self-cleaning activity when exposing under UV light spectrum. He also confirmed that the increased of percentage of TiO_2 nanoparticle caused increased in self-cleaning activity and the high duration of exposing under UV light also accelerated the self-cleaning action. These functionalized fabrics are not only gained self-cleaning properties but also improved in water repellence, soil resistance, wrinkle resistance, anti-bacteria, anti-static and UV protection [14].

4.0 Self-cleaning Testing

Photocatalytic self-cleaning fabrics are able to clean themselves when exposed to light, but they also possess antibacterial and UV blocking functions. Therefore, several testing methods can be used for the evaluation of the photocatalytic activity of functionalized fabric. Photocatalytic efficiency of functionalized fabrics is typically assessed by degradation of organic pollutants such as natural colorants stain or synthetic dye which commonly used as model pollutant. The photodegradation of the colorant was determined by two type of colorant decomposition activities such as solution discoloration and stain degradation. As for solution discoloration, pieces of the functionalized fabrics were placed dye solution and exposed under UV light. The dye solutions were collected periodically after certain time and its concentration is measured by using UV-Vis spectrophotometer [12, 15–16, 19, 29–30].

Meanwhile, for stain degradation, the functionalized fabric is stained with dye, and irradiated to UV light source for certain time. The stained sample is assessed for color strength (K/S) values on color spectrometer. The self-cleaning action can quantify by comparing K/S values of the exposed and unexposed portions of the same stain as shown below [8, 24–25, 27]:

$$\% \text{ Decrease in K/S value} = \frac{\text{(K/S) unexposed} - \text{(K/S) exposed}}{\text{(K/S) unexposed}} \times 100$$

Where K is the absorption; and S, the scattering

The K/S represents the color strength on a surface and is directly proportional to the amount of dye present on it. The decrease in K/S values shows that the stains are disappearing [15]. Chaudhari et al. [27], Sivakumar et al. [24] and Kumar [8] evaluated the nano TiO_2 coated textile fabric by exposed the coffee stained samples to sunlight irradiation for 12-48 h and the percentage of decrease in K/S value was calculated. Ashraf et al. [15] stained the functionalized fabric samples with different types of dyes and exposed to UV light source. The K/S values as function of time were measured with color spectrometer.

Besides color strength value, the properties of self-cleaning fabric also can be determined based on highly oxidative intermediates generated at the stained cotton textiles surface, which can be measured by the gas chromatography. As reported by Meilert et al. [31], the released of carbon dioxide (CO_2) due to wine, coffee, make-up and perspiration stain were deposited on the treated textiles. These samples were irradiated for 24 h with Suntest solar light simulator with a light intensity of 50 mW/cm^2 and the release of CO_2 was followed by gas chromatography during 24 h. The higher amount of CO_2 measured, the better the self-cleaning properties.

The UV blocking function of textiles was usually evaluated by the determination of the ultraviolet protection factor values (UPF). The UPF is the ratio of UV radiation measured without the protection of the fabric compared to that with protection. UPF tests are conducted using spectrophotometer or a spectroradiometer and serve as a measure of the UV protection provided by clothing fabrics. A textile material must have a minimum UPF of 15 to be rated as UV protective [33]. Excellent UV-blocking ability of the ZnO-coated fabric, particularly in the region of the UVB (280–315 nm), was confirmed by Shateri-Khalilabad et al. [34]. Sivakumar et al. [29] assessed the ultra-violet (UV) protection of cotton fabric coated with nano ZnO and nano TiO_2 with acrylic binder. They found that the UV protection function of the fabrics treated with larger sized TiO_2 nanoparticles have better UV protection factor (UPF) value than the fabric treated with smaller sized ZnO nanoparticles.

Conclusion

Self-cleaning concept contributes a lot of benefits in various industries. Especially, self-cleaning fabric which has massive potential for improvement of products not only in clothing industry but also in health industry due to time, material, energy reduction and consequently cost-efficiency during production. Moreover, this technology embraces environmental friendly properties as it effectively decreases cleaning efforts and conserve a considerable amount of water and energy as well as saving time and laundering cost.

References

[1] Mirjalili M and Karimi L 2011 Photocatalytic degradation of synthesized colorant stains on cotton fabric coated with nano TiO_2, *JFBI, 3(4)*

[2] Barthlott W and Neinhuis C 1997 Self-cleaning of biological surfaces, *Planta*, 202 1-8

[3] Bixler G D and Bhushan B 2012 Bioinspired rice leaf and butterfly wing surface structures combining shark skin and lotus effects, *Soft Matter* 8 11271-11284

[4] Gulrajani M L 2006 Nano Finishes, *Indian J Fibre Text Res* 31 187-201

[5] Banerjee S, Dionysiou D D and Pillai S C 2015 Self-cleaning applications of TiO_2 by photo-induced hydrophilicity and photocatalysis, *Appl Catal B* 176 396–428

[6] Sun M, Luo C, Xu L, Ji H, Ouyang Q, YU D and Chen Y 2005 Artificial lotus leaf by nanocasting, *Langmuir* 21(19) 8978

[7] Cao L, Jones A K, Sikka V K, Wu J, and Gao D 2009 Anti-icing superhydrophobic coatings, *Langmuir* 25 (21) 12444–12448

[8] Kumar B S 2015 Self-cleaning finish on cotton textile using sol-gel derived TiO_2 nano finish, *IOSR-JPTE* 2(1) 2348-0181

[9] Pare B, Jonnalagadda S B, Tomar H, Singh P, Bhagwat V W 2008 ZnO assisted photocatalytic degradation of acridine orange in aqueos solution using visible irradiation, *Desalination* 232 80-90

[10] Cakir B A, Budama L, Topel O, Hoda N 2012 Synthesis of ZnO nanoparticles using PS-B-PAA reverse micelle cores for UV protective, self-cleaning and antibacterial textile applications, *Colloids Surf A* 414: 132– 139

[11] Soltaninezhad M and Aminifar A 2011 Study nanostructures of semiconductor zinc oxide (ZnO) as a photocatalyst for the degradation of organic pollutants, *Int J Nano Dim* 2(2) 137-145

[12] Meena V K and Meena RC 2010 Studies on photo degradation of methyl orange in aqueous solution using immobilized dowex-11photocatalyst, *J Ind Council Chem* 27(2) 180-184

[13] Gouvêa C A K, Wypych F, Moraes S G, Durán N, Nagata N and Peralta-Zamora P 2000 Semiconductor-assisted photocatalytic degradation of reactive dyes in aqueous solution, *Chemosphere* 40 433-440

[14 Samal S S, Jeyaraman P and Vishwakarma V 2010 Sonochemical coating of Ag-TiO_2 nanoparticles on textile fabrics for stain repellency and self-cleaning- The Indian scenario: A review, *JMMCE* 9(6) 519-525

[15] Ashraf M, Champagne P, Perwuelz A, Campagne C and Leriche A 2014 Photocatalytic solution discoloration and self-cleaning by polyester fabric functionalized with ZnO nanorods, *J Ind Text* 1–15

[16] Welderfael T, Yadav O P, Taddesse A M and Kaushal J 2013 Synthesis, characterization and photocatalytic activities of Ag-N-Co doped ZnO nanoparticles for degradation of methyl red, *Bull Chem Soc Ethiop* 27(2) 221-232

[17] Elamin N and Elsanousi A 2013 Synthesis of ZnO nanostructures and their photocatalytic activity, *JAIS* 1(1) 32-35

[18 Oda A M, Ali H H, Lafta A J, Esmael H A, Jameel A A, Mohammed A M and Mubarak I J 2015 Study self-cleaning of congo red from cotton fabric loaded by ZnO-Ag, *Int J Chem* 7(2)

[19] Oda A M, Kadhum S H, Farhood A S and Alkadhum H A 2013 Photocatalytic activity of silvered zinc oxide stabilized on cotton fiber for methylene blue dye decolorization, *Int J Environ Water*, 2(2)
[20] Pakdel E, Daoud W A and Wang X 2013 Self-cleaning and superhydrophilic wool by TiO_2/SiO_2 nanocomposite, *Appl Surf Sci* 275 397– 402
[21] Jašková V, Hochmannová L and VytLasová J 2013 TiO_2 and ZnO nanoparticles in photocatalytic and hygienic coatings, *Int J Photoenergy* 1-6
[22] Chaudhari S B, Mandot A A and Patel B H 2012 Effect of nano TiO_2 pretreatment on functional properties of cotton fabric, *IJERD* 1(9) 24-29
[23] Pisitsak P, Samootsoot A and Chokpanich N 2013 Investigation of self-cleaning properties of cotton fabrics finished with nano-TiO_2 and nano-TiO_2 mixed with fumed silica, *KKU Res J* 18(2) 200-211
[24] Yaghoubi H, Taghavinia N, Alamdari E K 2010 Self-cleaning TiO_2 coating on polycarbonate: surface treatment, photocatalytic and nanomechanical properties, *Surf Coat Technol* 204 1562–1568
[25] Sawhney A P S, Condon B, Singh K V, Pang S S and Li G 2008 Modern applications of nanotechnology in textiles, *Text Res J* 78(8) 731–739
[26] Ortelli S, Costa A L and Dondi M 2015 TiO_2 nanosols applied directly on textiles using different purification treatments, *Materials* 8 7988-7996
[27] Anita S, Ramachandran T, Koushik C V, Rajendran R and Mahalakshmi M 2010 Preparation and characterization of zinc oxide nanoparticles and a study of the anti-microbial property of cotton fabric treated with the particles, *JTATM* 6 (4)
[28] Yuranova T, Mosteo R, Bandra J, Laub D, Kiwi J 2005 Self-cleaning cotton textiles surfaces modified by photoactive SiO_2/TiO_2 coating, *J Mol Catal A Chem* 244 160-167
[23] R. D. Cramer, E. A. Ponomarenko, and J.C.T.R B.S. Laurent, (2003) Method of Applying Nanoparticles, U.S. Pat. No: 6,645,569
[29] Sivakumar A, Murugan R, Sundaresan K and Periyasamy S 2013 UV protection and self-cleaning finish for cotton fabric using metal oxide nanoparticle, *IJFTR* 38 285-292
[30] Wijesena RN, Tissera N, perera R and de Silva K M 2014 Side selective surface modification of chition nanofibers on anionically modified cotton fabrics, *Carbohydr Polym* 109 56-63
[31] Meilert K, Laun D, Kiwi J 2005 Photocatalytic self-cleaning of modified cotton textiles by TiO_2 clusters attached by chemical spacers, *J Mol Catal A Chem* 237 101-108
[32] Bedford N M and Stecki A J. 2010 Photocatalytic self-cleaning textile fibers by coaxial electrospinning. *ACS Appl Mater Interfaces* 2(8) 2448-2455
[33] Senic Z, Bauk S, Vitorovic-Todorovic M, Pajic N, Samolov A andRajic D 2011 Application of TiO_2 nanopqrticles for obtaining self-decontaminating smart textiles, *Sci Tech Rev* 61(3-4):63-72
[34] Shateri-Khalilabad M and Yazdanshenas M E 2013 Bifunctionalization of cotton textiles by ZnO nanostructures: Antimicrobial activity and ultraviolet protection, *Text Res J*,83(10)

Morphological and physical - chemical issues of metal nanostructures used in medical field

L. D. Duceac[1,2], **N. Velenciuc**[3], **E. C. Dobre**[1,2]

[1]*Apollonia* University of Iasi, Romania.
[2]*Academician Ioan Haulica* Institute of Researches, Apollonia University of Iasi
[3] Oncology Regional Institute of Iasi, Romania.

E-mail: vilinat2001@yahoo.com

Abstract. In recent years applications of nanotechnology integrated into nanomedicine and bio-nanotechnology have attracted the attention of many researchers from different fields. Processes from chemical engineering especially nanostructured materials play an important role in medical and pharmaceutical development. Fundamental researches focused on finding simple, easily accomplished synthesis methods, morphological aspects and physico-chemical advanced characterization of nanomaterials. More over, by controlling synthesis conditions textural characteristics and physicochemical properties such as particle size, shape, surface, porosity, aggregation degree and composition can be tailored. Low cytotoxicity and antimicrobial effects of these nanostructured materials makes them be applied in medicine field. The major advantage of metal based nanoparticles is the use either for their antimicrobial properties or as drug-carriers having the potential to be active at low concentrations against infectious agents.

1. Introduction

In our century, microbes type virulent strains of bacteria have developed resistance to antibiotics that lead to a tremendous health impact and need to deliver solutions to this worldwide problem. The use of nanotechnology in medicine field, particularly nanomedicine, has already proved its major action on the pharmaceutical industry, especially in the antimicrobial field [1]. Over the last century antibiotics and vaccines arguably reduced human morbidity and mortality being considered the most important medical descoveries against infectious diseases [2,3].

Antimicrobial agents development represent a necessary help for classic medicine considering high resistant of pathogen micro-organisms. Researchers identified several inorganic, organic, and hybrid materials with possible uses in prevention and treatment of infections reporting the current information on the action mechanism of nano/microparticles based antimicrobial agents [4].

2. Materials and methods

Nowadays, a well-established branch of science called micro/nanomedicine, manages with tailoring small size devices, i.e. micro/nanoparticles, possessing unique properties for human health. Figure 1 shows a schematic exemplification of nano/ microparticles frequently used for drug delivery and therapeutic systems. Their biological properties are correlated to structural and functional characteristics.

There are studies specifying the relationship between cells and micro/ nanoparticles correlated to their structural and textural characteristics like size, shape, charge and functional groups having a considerable impact on their internalization.

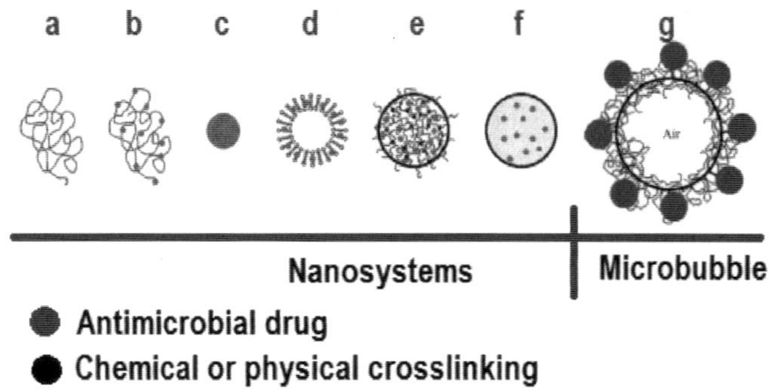

Figure 1. Exemple of nano/microsystems: a) antimicrobial polymers, b) polymer chains, c) inorganic nanoparticles, d) liposomes, e) polymeric nanoparticles, f) solid lipid nanoparticles loaded with antimicrobial drug, g) microbubbles loaded with nanoparticles.

Negative surface charge versus particles positive surface charge shows a significant reduction of particle association with cells [5].

3. Experimental part

Inorganic and hybrid materials were identified exhibiting antimicrobial activity depending on their physico- chemical and morpho-functional properties. Ionic form of metal based nanoparticles express antimicrobial ability compared to the metal, exhibiting an equivalent or superior antimicrobial effect [6-8]. Size, shape, porosity and surface properties are of great importance features inducing the antimicrobial efficacy of metal base nanoparticles. For example figure 2 shows TEM images of Ag nanoparticles revealing particles wide distribution ranging from 5 to 25 nm in diameter. Smaller particles are spherical in shape where as the larger ones are in an elongated form and shape that could be the result of the aggregation degree of the two or more particles together. Silver size and shape nanoparticles are also influenced by the concentration silver salt solution [9]. A major increase in the surface area consists in a stronger interaction of nanoparticles with the surrounding materials. Disadvantages of these material synthesis methods are represented by the toxicity of the reagents and the agglomeration and precipitation of nanoparticles [10].

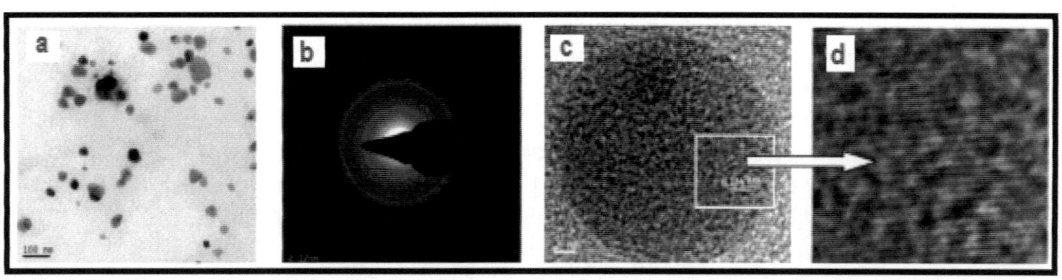

Figure 2. TEM and HRTEM images of silver nanoparticles on the GrO sheets at different salt concentrations: a) 8×10^{-3} moldm^{-3} $AgNO_3$, b) 1×10^{-3} moldm^{-3} $AgNO_3$, c) HRTEM with fringe spacing, d) enlarge image of fringe spacing.

IOP Conf. Series: Materials Science and Engineering **133** (2016) 012029 doi:10.1088/1757-899X/133/1/012029

Other uses of nanocomposite materials refer to coating medical devices and production of antimicrobial cotton textiles already being demonstrated to reduce postoperative infections [11]. Bionanotechnology use plants and micro-organisms such as bacteria, yeasts, fungi, and actinomycetes for fabrication of inorganic nanoparticles [12].

Multidisciplinary nano-scaled molecules used in diagnosis and treatment of different diseases are represented by metal oxide nanoparticles already applied as effective antimicrobial agents. Metal oxide nanoparticles are well known to target some cellular signaling pathways. Combination of metal oxides with other therapeutic agents enhances their bioavailability and efficacy. These remarkable nanomaterials have been used due to their various action mechanisms as antioxidant, anticancer, antimicrobial, and anti- inflammatory agents [13].

A serious threat to public health worldwide is considered to be the emergence of antibiotic-resistant microbial strains. Advanced nano-biotechnology enable the researchers to synthesize specific nanoparticulated metal oxide with well establish properties as antimicrobial activity. Zinc oxide nanoparticles were proven to inhibit the growth of wide spectrum bacterial strains such as gram-negative E. coli and Pseudomonas aeruginosa as well as gram-positive Staphylococcus aureus [14].

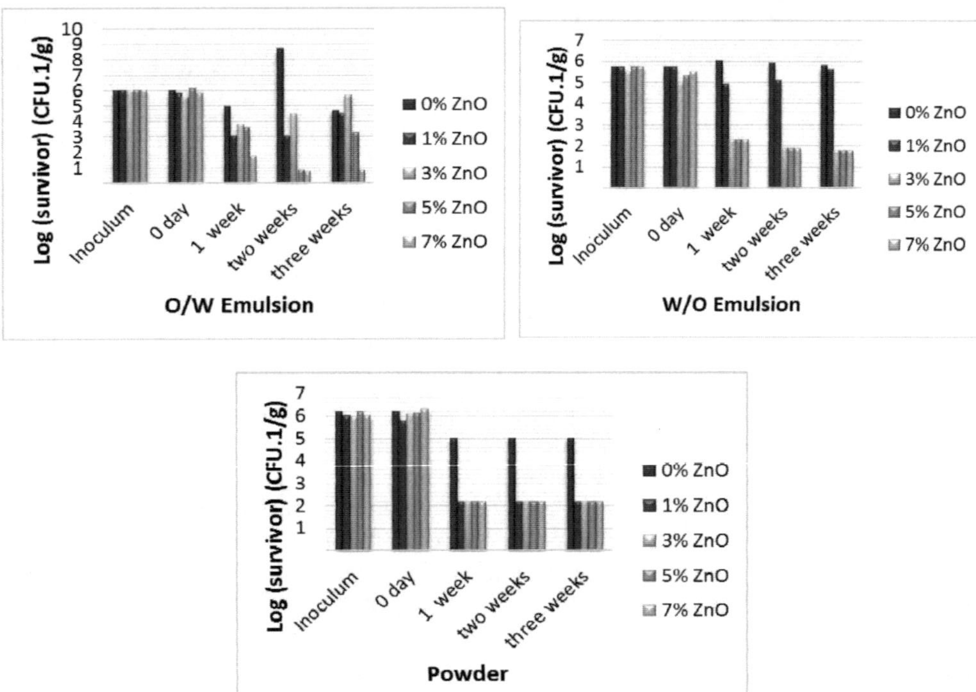

Figure 3. Test results of zinc oxides effect on P. aeruginosa in the oil-in-water emulsion, in the water-in-oil emulsion and in the dry powder.

Decreasing metal oxide nanoparticles size led to distortion and damage of bacterial cell membrane. More over, leakage of intracellular contents and possible death of bacteria demonstrate the antibacterial effects of these outstanding materials [15]. Different concentrations of ZnO nanoparticles in the anhydrous product were tested concerning P. aeruginosa as shown in figure 3.

IOP Conf. Series: Materials Science and Engineering **133** (2016) 012029 doi:10.1088/1757-899X/133/1/012029

4. Results and discussions

Results concerning P. aeruginosa bacterial strain revealed that bacterial population finally exceeded the control without zinc oxide nanoparticles at 28 days and decreased over 7 days, but then proliferated again during storage after 32 days [16].

Metal oxide nanoparticles known to possess a unique electronic structure provides a surface for direct free radical scavenging activity promising to act as an antioxidant therapeutic agent (Fig. 4).

Researchers found in a study of nerve cells (HT22), that CeO2/or Y2O3 nanoparticles, protect nerve cells suggesting that small size metal oxides could be the modulator of the oxidative stress in the biological systems [17].

Recent studies revealed that synthesized copper oxide nanoparticles using a thermal plasma technique have showed a minimum inhibitory concentration in the range 100 - 5000 µg/ml against drug resistant bacterial strains [18]. Other researches found that photo-activation of TiO2 nanoparticles using the ultra violet (UV) radiations, support the bactericidal effect which might be due to loss of respiratory activity caused by peroxidation of the polyunsaturated phospholipids of cell membrane [19].

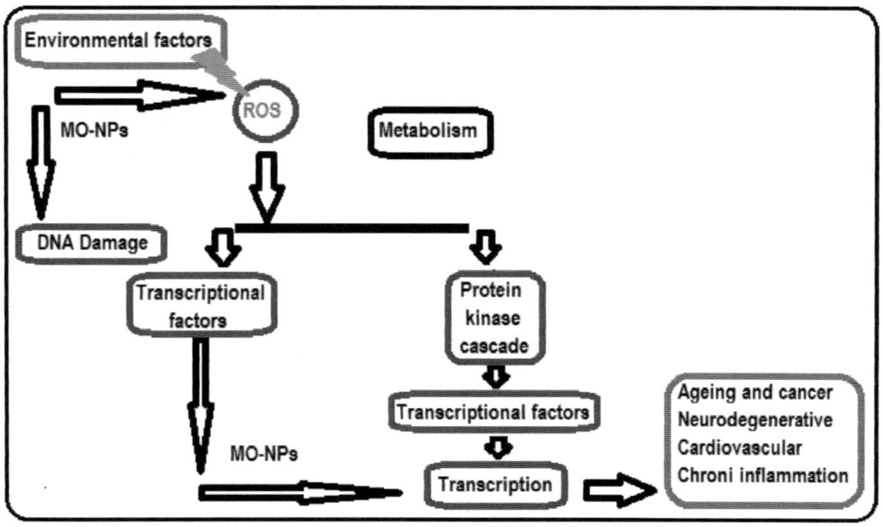

Figure 4. Brief representation of the direct free radical scavenging potential of metal oxide nanoparticles.

Among aluminum, silicon, titanium and zinc oxide nanoparticles against bacterial strains of Bacillus subtilis, E. coli and Pseudomonas fluorescens, ZnO nanoparticles were found to be more potent with 100% motility followed by SiO2 and Al2O3 nanoaprticles [20].

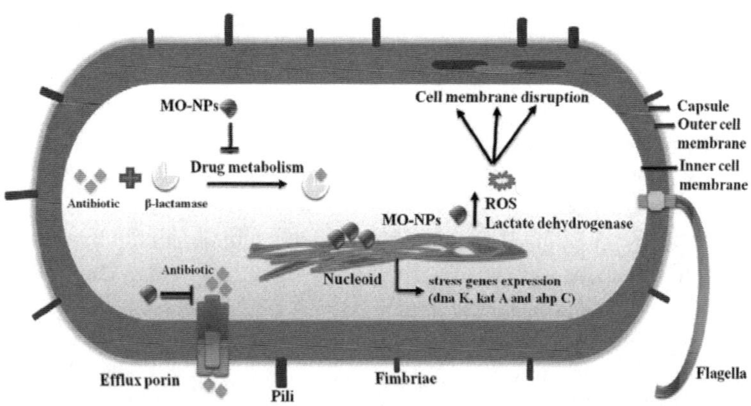

Figure 5. Representation of possible antibacterial mechanism of metal oxide nanoparticles activity.

The mechanism of antimicrobial action of metal oxide nanoparticles is not completely understood, it may depends on chemical composition, surface changes and intrinsic properties of nanomaterials along with the preponderant bacterial species. Nanoparticles were proved to be assigned to the microbial cell with electrostatic interactions consisting of cell wall damage concomitant with the increase in permeability of the cell [21, 22]. Figure 5 shows electrostatic interactions of metal oxide nanoparticles followed by reactive oxygen species (ROS) generation and activation of oxidative stress genes and the inhibitory effect of these nanoparticulate materials on drug resistance mechanisms mediated by β-lactamases and antibiotic efflux pumps.

Mutagenic potential of ZnO, CuO, TiO2, Al2O3, and Co3O4, and type metal oxides was investigated and reported significant colony inhibition activity of CuO nanoparticles. Generally, metal and metal oxide nanoparticles are known to destroy cell wall due to the interaction with the thiol (–SH) group of membrane proteins. Further more, metal oxide nanoparticles can deactivate the β-lactamases preventing the evolution of antibiotic resistance [23].

5. Conclusions

Remarkable morphology properties, physical-chemical characteristics, biological applications of metal nanoparticles significantly improve classic medicine. Metal and metal oxide nanoparticles incorporated into matrices act as active components in nanocomposites being biocompatible and suited for biomedical applications. In order to perform controlled release of drugs, by incorporating specific functional groups into matrices chemical engineering research has a major impact on nanomedicine research. Over the last few years nanosized materials have considerable evolved bringing to nanomedicine unprecedented advances in the diagnosis, imaging and treatment of a variety of hard to treat diseases. Nowadays, around 250 nanoproducts exist in various stages of development, including different compositions, nanostructures, physicochemical characteristics, surface functionality and morphology. Thus, nanomedicine researches will have to exceed chemical engineering and consider new medical aspects including in vivo studies and clinical trials.

References:

[1] Zhu X, Radovic-Moreno A F, Wu J, Langer R and Shi J 2014 Nanomedicine in the management of microbial infection—Overview and perspectives *Nano Today* **9** pp 478-498

[2] Cohen M L 2000 Changing patterns of infectious disease *Nature* **406** pp 762-767

[3] Armstrong G L, Conn L A and Pinner R W 1999 Trends in infectious disease mortality in the United States during the 20th century *J. Am. Med. Assoc.* **281** pp 61-66

[4] Cavalieri F, Tortora M, Stringaro A, Colone M, Baldassarri L 2014 Nanomedicines for antimicrobial interventions *Journal of Hospital Infection* **88** pp 183-190

[5] Gratton S E A, Ropp P A, Pohlhaus P D et al. 2008 The effect of particle design on cellular internalization pathways *Proc. Natl. Acad. Sci. USA* **105** pp 11613-11618

[6] Dallas P, Sharma V K and Zboril R 2011 Silver polymeric nanocomposites as advanced antimicrobial agents: classification, synthetic paths, applications, and perspectives *Adv Colloid Interface Sci* **166** pp 119-135

[7] Park H, Park S, Roh J et al. 2013 Biofilm-inactivating activity of silver nanoparticles: a comparison with silver ions *J Ind Engng Chem* **19** pp 614-619

[8] Nie M, Sun K, Meng D D 2009 Formation of metal nanoparticles by short distance sputter deposition in a reactive ion etching chamber *J Appl Phys* 2009 **106** 054314

[9] Das M R, Sarma R K, Saikia R, Kale V S, Shelke M V and Sengupta P 2011 Synthesis of silver nanoparticles in an aqueous suspension of graphene oxide *Colloids and Surfaces B: Biointerfaces* **83** pp 16–22

[10] Morones J R, Elechiguerra J L, Camacho A et al. 2005 The bactericidal effect of silver nanoparticles *Nanotechnology* **16** pp 2346-2353
[11] Seo S Y, Lee G H, Lee S G, Jung S Y, Lim J O, Choi J H 2012 Alginate-based composite sponge containing silver nanoparticles synthesized in situ *Carbohydr Polym* **90** pp 109-115
[12] Li X, Xu H, Chen Z and Chen G 2011 Biosynthesis of nanoparticles by microorganisms and their applications *J Nanomat* **3** pp 270974-270990
[13] Hardeep S T, Dharambir K, Simranjeet K B, Pardeep K, Gaurav K and Sardul S S 2015 Molecular aspects of metal oxide nanoparticle (MO-NPs) mediated pharmacological effects *Life Sciences* **143** pp 71–79
[14] Premanathan M, Karthikeyan K, Jeyasubramanian K and Manivannan G 2011 Selective toxicity of ZnO nanoparticles toward gram-positive bacteria and cancer cells by apoptosis through lipid peroxidation *Nanomedicine* **7** pp 184–192
[15] Sharma D, Rajput J, Kaith B S, Kaur M and Sharma S 2010 Synthesis of ZnO nanoparticles and study of their antibacterial and antifungal properties *Thin Solid Films* **519** pp 1224-1229
[16] Pasquet J, Chevalier Y, Couval E, Bouvier D and Bolzinger M A 2015 Zinc oxide as a new antimicrobial preservative of topical products: Interactions with common formulation ingredients *International Journal of Pharmaceutics* **479** pp 88–95
[17] Schubert D, Dargusch R, Raitano J and Chan S W 2006 Ceriumand yttriumoxide nanoparticles are neuroprotective *Biochem. Biophys. Res. Commun.* **342** pp 86–91
[18] Rena G, Hub D, Cheng E W C, Reusc M A V, Reipd P, Allakerc R P 2009 Characterisation of copper oxide nanoparticles for antimicrobial applications *Int. J. Antimicrob. Agents* **33** pp 587–590
[19] Tsuang Y H, Sun J S, Huang Y C, Lu C H, Chang W H S and Wang C C 2008 Studies of photokilling of bacteria using titanium dioxide nanoparticles *Artif. Organs* **32** pp 167–174
[20] Jiang W, Mashayekhi H and Xing B 2009 Bacterial toxicity comparison between nano and micro scaled oxide particles *Environ. Pollut.* **157** pp 1619–1625
[21] Feris K, Otto C, Tinker J, Wingett D, Punnoose A, Thurber A, Kongara M, Sabetian M, Quinn B, Hanna C and Pink D 2010 Electrostatic interactions affect nanoparticle mediated toxicity to gram-negative bacterium Pseudomonas aeruginosa PAO1 *Langmuir 26* pp 4429–4436
[22] Sinha R, Karan R, Sinha A, Khare S K 2011 Interaction and nanotoxic effect of ZnO and Ag nanoparticles on mesophilic and halophilic bacterial cells *Bioresour. Technol.* **102** pp 1516–1520
[23] Jayaraman R 2009 Antibiotic resistance: an overview of mechanism and a paradigm shift *Curr. Sci.* **96** pp 1475–1484

IOP Conf. Series: Materials Science and Engineering **133** (2016) 012030 doi:10.1088/1757-899X/133/1/012030

Development of tailored ceramic microstructures using recycled marble processing residue as pore-former

A Domopoulou[1], X Spiliotis[2], G Charalampides[1], A Baklavaridis[1], G Papapolymerou[2], V Karayannis[1*]

[1]Department of Environmental Engineering, Technological Education Institute of Western Macedonia, Kila, 50100, Kozani, Greece
[2]Department of Civil Engineering, Technological Education Institute of Thessaly, 41100, Larissa, Greece

[*]Corresponding author e-mail: vkarayan@teiwm.gr

Abstract. Recycling of marble processing residue is significant since marble processing constitutes an important industrial sector. Therefore, the sustainable management and the valorisation, in an economically profitable manner, of this industrial by-product should be considered. In this work, the potential use of marble residue as pore-former into clayey mixtures for the production of lightweight, porous and thermal insulating ceramics is investigated. Four samples consisting of clayey ceramic body incorporating up to 50 wt.% fine marble residue powder were produced. The final ceramic products were produced upon firing (sintering) at 950°C. Porosity and thermal conductivity measurements were carried out in order to assess the thermal insulating behavior of the produced sintered ceramics. The porosity of the sintered ceramics increases substantially by increasing the marble residue admixture loading. This, in turn, leads to a decrease in thermal conductivity. Consequently, the marble residue can be successfully employed as pore-forming agent, in order to improve the insulating behavior of the ceramic materials.

1. Introduction

Marble processing is currently regarded as one of the most important industrial sectors not only in Greece but also worldwide. In Greece the marble mining sites are mainly situated in the northern part of the country, including the Western Macedonia, Kozani region. The processing of marble ore includes various cutting, grinding and polishing stages that involve water recirculation, so as the cutting and polishing equipment to cool down. In this way, the fine marble dust particles are removed. In the end of the marble processing procedure, substantial quantities of solid marble residue are recovered from the resulting sludge stream.

On the other hand, the production and testing of ceramics for thermal insulating applications has attracted significant interest recently [1-4]. The success, in terms of applications, of an insulating material relies on its reduced heat conduction ability under a specific temperature gradient. The heat flow throughout a material is analogous to the applied temperature gradient. Thermal conductivity represents the proportionality factor (constant) and it is a material dependent property. Hence, in composite or multiphase materials, thermal conductivity can be modified by the modification of the materials constituents.

The addition of various constituents in the ceramic mixture that reduce the thermal conductivity, and consequently improve the thermal insulating behaviour, in fired clay ceramic production, is extremely important for the ceramic manufacturing industry. In practice, the decrease in thermal conductivity is usually achieved by the pore formation in the ceramic body. The incorporation of various pore-formers in standard clayey mixtures has been reported. The most important research efforts are related to the partial replacement of clay with bauxite in the ceramic material, the consolidation of used earth from biodiesel filtration and glycerine into the final ceramic material and the valorisation of paper processing residues, due to recycling, towards the production of novel ceramics [5-11].

Moreover, an alternative way for the porosity formation in the ceramic clay body is the incorporation of marble waste. Marble waste is mainly consisted of calcium carbonate ($CaCO_3$). The use of recycled marble residues (marble waste) for the engineering ceramics production could lead to significant benefits both from the environmental and economical point of view. Most works regarding marble wastes valorisation focused on rather conventional applications such as cement and concrete production, the production of asphalt aggregates used in road construction, and other building materials. However, not many works report the valorisation of marble residues to efficiently manipulate the properties of compacted ceramics [10,12-15]. Therefore, the investigation of the effect of various marble residue loadings, which are incorporated in the raw ceramic mixture, on the properties of the final ceramic product is important. Furthermore, the cautious examination of the properties of the final ceramic that incorporates waste assists to the correct evaluation of the ceramic material considering the standard requirements for commercial use.

In this work, the role of the marble processing residue as a pore-forming agent is investigated, by its incorporation in clayey raw materials as an admixture. Additionally, the possibility of employing powder metallurgy procedures for the production of innovative engineering ceramics is assessed. Moreover, the effect of the marble processing residue on thermal insulation behaviour (thermal conductivity) of the produced ceramics is also investigated. Furthermore, this work is expected to contribute to the sustainable management of this abundant industrial residue and also accentuate potential economic benefits regarding the production of novel ceramic materials.

2. Materials and Methods

2.1. *Raw materials*

In the present study, three different clays A, B and G were used. These clays were collected from Central Greece. Moreover, their composition is shown in table 1. The clay mixture ("Viokeral" mixture) used in this study consisted of 50% A, 33% B and 17 % G clay type. The Viokeral mixture was utilized for manufacturing prototypes (Pr – blank samples, without marble residue) and samples incorporating various marble residue loadings.

Table 1. Chemical analysis of clays

	Composition (%)										
CLAY	Loss on Ignition (LOI)	SiO_2	Al_2O_3	Fe_2O_3	CaO	MgO	K_2O	Na_2O	TiO_2	$CaCO_3$ (eq)	CO_2
A	11.9	49.4	12.9	7.1	8.6	4.9	2.9	1.6	0.8	14.7	6.5
B	9.9	52.8	13.5	7.6	6.3	4.3	3.2	1.6	0.9	-	-
G	16.5	51.0	8.5	4.7	11.6	3.9	1.5	1.4	0.6	24.5	10.7

The marble residue used in in this work resulted from the treatment of marble processing wastewater, from the Western Macedonia region. This region is one of the largest sources of marble in Greece. The Western-Macedonian marble is characterized by a large variety of white and coloured species of marble, and other stones as well, which are nowadays used in the marble industry [16]. The mineralogical composition of marbles from Tranovalto, Kozani area is: Calcite (98%), Dolomite (1%), Quartz (0.5%) and Muscovite (0.5%). The chemical composition of these marbles is given in table 2. Prior to the sample preparation the clay mixture and the marble residue were crushed and sieved so as fine aggregates (with particle size <0.63mm). is received.

Table 2. Chemical composition of marbles

Component	CaO	MgO	SiO_2	Fe_2O_3	Al_2O_3	K_2O	Na_2O	MnO	CO_2
Composition (%)	53.50	1.52	0.25	0.18	0.25	0.02	0.04	0.01	44.4

2.2. Fabrication of compacted ceramic specimens.
The ceramic samples consisting of marble residue and clay samples were produced by mixing 0, 12.5, 25 and 50 % wt. marble residue in the aforementioned clays, and adding 5-6% wt. water in the mixture. A full list of the samples fabricated in this work is given in table 3.

Table 3. The marble residue/clay ceramics

sample name	clay (%wt)	marble residue (%wt)
Pr	95	0
SA	72.5	12.5
SB	70	25
SG	45	50

The constituents shown in the table 3 were initially formed in wet mixtures. Subsequently, the wet mixture was uniaxially cold pressed in a stainless steel die using a hydraulic press. In that way a series of disc-shaped, with 5 cm diameter, green compacted samples were produced. All ceramic compacts were obtained by applying a 20 tn load and they had a reasonable initial strength for the subsequent fabrication step. Thereafter, the green disks were weighed in order to determine the moisture. Then the disks were dried naturally for 12h and then forced dried, at 110°C, until no additional weight loss is observed. Dried sample pieces were fired, following a gradual temperature increase protocol up to the temperature of 950°C. This sintering temperature typically applied in standard ceramic manufacturing. The sample pieces remained at 950°C for 2h. The whole firing procedure took place in a programmable electric chamber furnace. A comprehensive schematic diagram of the proposed production method is shown in figure 1.

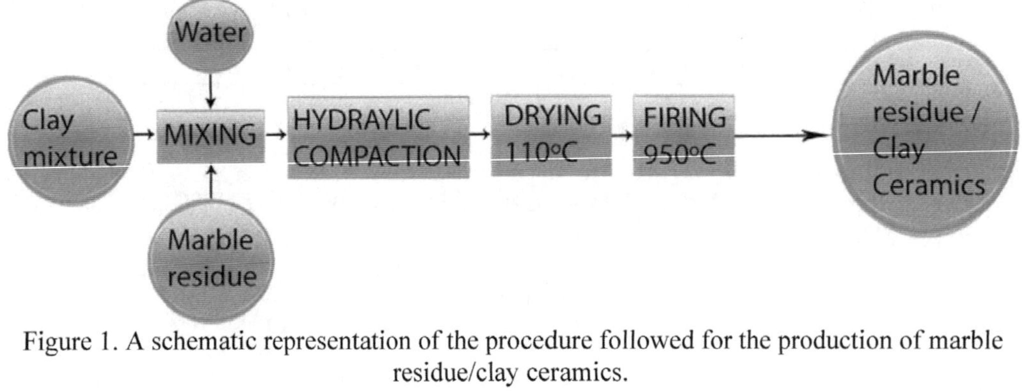

Figure 1. A schematic representation of the procedure followed for the production of marble residue/clay ceramics.

2.3. Ceramic specimen characterization.

The fired ceramic sample's shrinkage, weight loss upon sintering (LOS), density, water absorption capacity and open porosity was determined according to ASTM C67. The thermal conductivity coefficient (k) was measured at 25°C by using an Anter Unitherm device (Model 2022), based on the guarded heat flow meter method. The thermal conductivity measurements were performed according to ASTM E1530. All the measurements presented hereunder are the mean value of ten specimen measurements.

Morphological examination and elemental analysis of the ceramic sample surfaces was carried by means of Scanning Electron Microscopy (SEM) and Energy Dispersive Spectroscopy (EDS) using a JEOL 6610LV microscope carrying a large area ($80mm^2$) silicon drift detector (X-Max 80, Oxford Instruments). All ceramic sample surfaces were coated with carbon prior observation, using a Quorum 150R E device. The elemental maps shown in this work were acquired and analysed using AZtech Nanoanalysis software (Oxford Instruments).

In order to gain further insights into the porosity of the ceramic samples produced, a well-established methodology, based on SEM and image analysis, was used. In brief this method includes pixel binarization and counting [17]. The porosity was calculated from the ratio of the total pore area to the net area of the image. This image analysis method has been already successfully used in food and polymer science [17,18]. The SEM images, used for the porosity measurements, were recorded using the backscattered electron (BSE) detector of the microscope in order to achieve higher pixel contrast [19]. At least five images of each sample were recorded and analysed for the porosity determination.

3. Results and discussions

3.1. Ceramic specimen appearance and weight loss on sintering.

Photographs of the fired ceramic samples produced are shown in figure 2. The apparent change in the appearance and the color of the samples is attributed to the incorporation of marble residue into the clay mixture, which was sintered via firing at 950°C. Specifically, the sample color changes from red (Pr sample) to brownish (SA sample) and subsequently to pale brownish (SB and SG samples) by increasing the marble residue loading up to 50%. Moreover, by visually inspecting the samples we noticed that the disk geometry is not affected by the marble residue incorporation and that disk volume slightly increases only for the ceramic samples that incorporate higher waste loadings (50%).

The effect of the marble residue incorporation on the weight loss (LOS) of the fired samples is presented in figure 3. The weight loss substantially increases, almost proportionally to the marble residue addition. This increase in the weight loss should be ascribed to the calcium carbonate ($CaCO_3$) decomposition during firing.

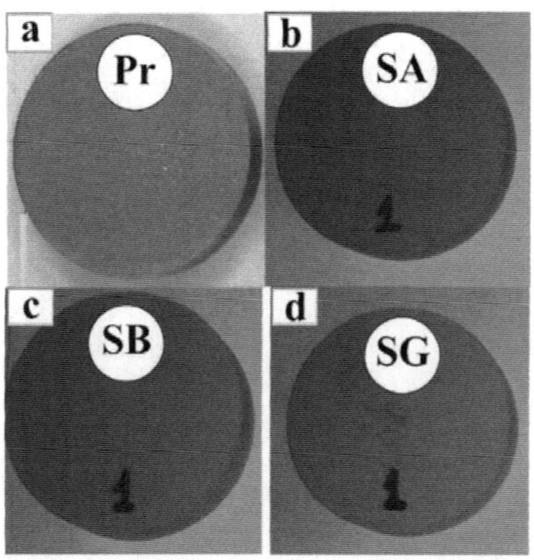

Figure 2. Appearance of the fired ceramic samples, with various marble residue loadings: (a) 0% (Pr-Prototype), (b) 12.5% (SA), (c) 25% (SB), (d) 50% (SG).

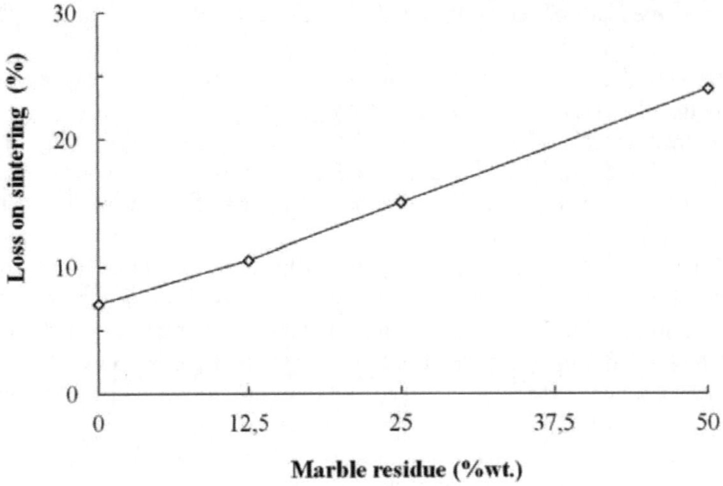

Figure 3. The effect of the marble residue addition on the % weight loss upon sintering (LOS).

3.2. SEM-EDS Mapping

In order to assess the dispersion of the marble residue inside the ceramic body, as well as to observe the morphology of the samples, SEM-EDS analysis was used. In figure 4, SEM images with EDS maps of prototype (without marble residue) ceramic samples (figures 4a and 4b) and also ceramic samples with 50% marble residue (figures 4c and 4d) are presented. The addition of marble residue fine particles increases the surface roughness of the ceramic sample. Calcium (magenta colored) appears to be evenly spread on the elemental map (figure 4d), indicating a satisfactory dispersion of marble residue into the clay matrix upon specimen preparation.

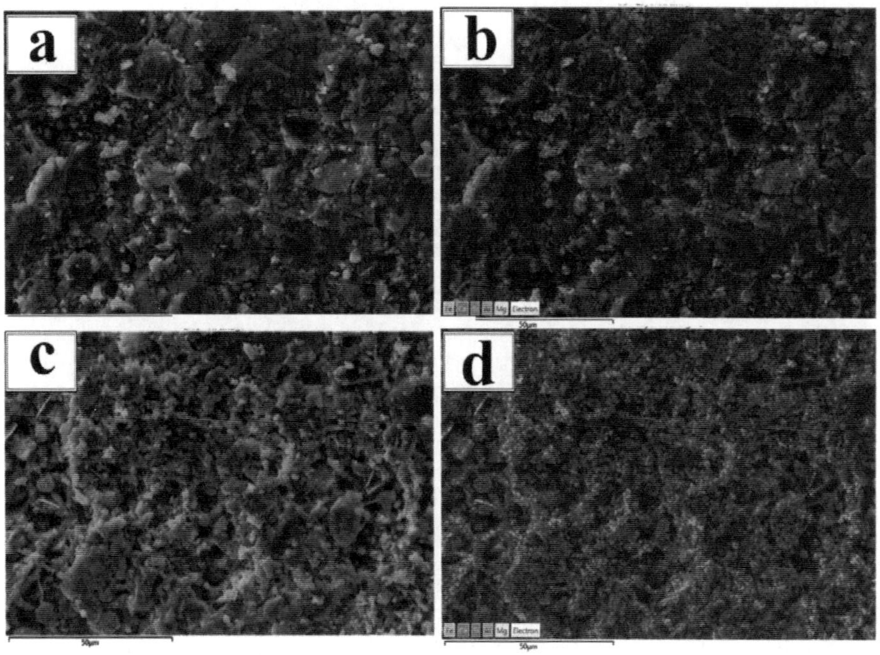

Figure 4. SEM images and the corresponding elemental maps of the Pr sample (a and b), and the SG4 sample (c and d). The scale bar is set to 50 μm.

3.3. Bulk density/ Open porosity

The effect of the marble residue addition into the clayey mixture on the bulk density (d) and open porosity (OP) of sample disks sintered at 950°C was also investigated (figure 5). The experimental data in Figure 5 show that the open porosity increases significantly while the bulk density decreases respectively with the marble residue incorporation into the clayey raw materials (by 17% and 83% respectively for 50% residue percentage). This finding suggests that, marble residue acts as pore forming agent when it is incorporated inside bulk ceramic bodies, which are subjected to firing in high temperatures (950°C).

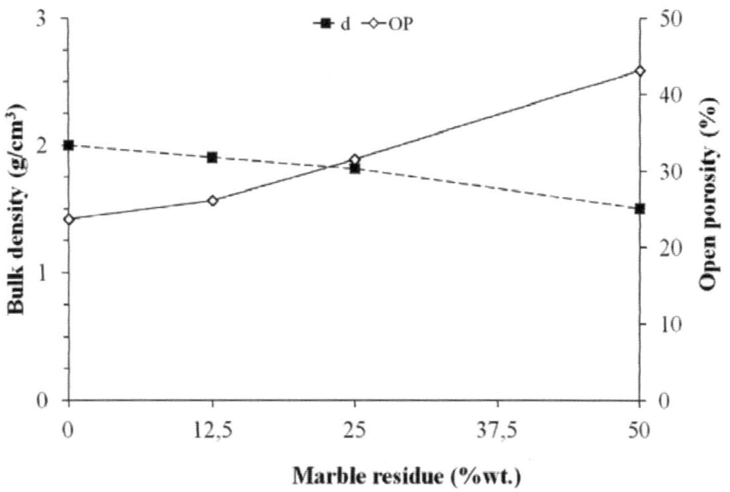

Figure 5. The effect of the marble residue addition on the bulk density (d) and open porosity (OP).

3.4. Porosity calculation from SEM micrographs.
SEM images of the ceramic samples surfaces are shown below (figure 6).

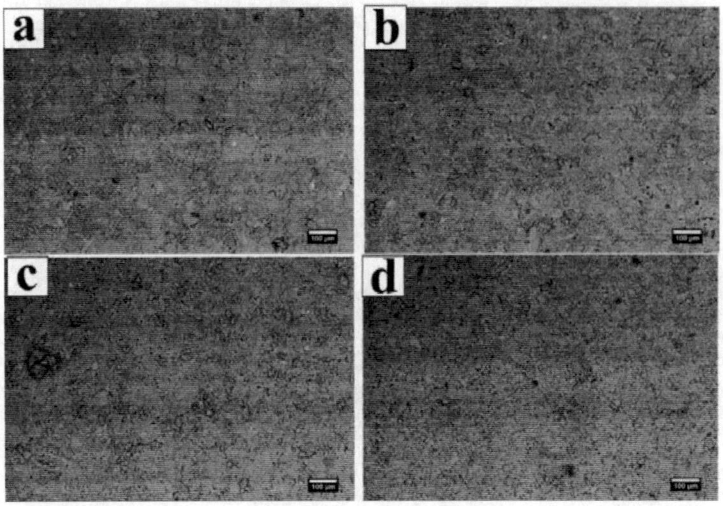

Figure 6. SEM micrographs of four representative samples studied in this work. (a) Pr (0% marble residue), (b) SA3 (12.5% marble residue), (c) SB4 (25% marble residue) and (d) SG4 (50% marble residue). Scale bar is set to 100 μm.

The addition of the marble residue increases the (dark colored) voids observed on the surface of the ceramic samples. In figure 7, the % porosity (as calculated from the SEM micrographs) versus the marble residue (%wt.) content is shown. The error bars in the graph correspond to the standard deviation from the average value. The addition of the marble residue significantly increases the porosity of the ceramic samples. This also confirms the pore-forming capabilities of the marble residue. Moreover, the % porosity values calculated from the SEM micrographs appear to be in good agreement with the open porosity (OP) values.

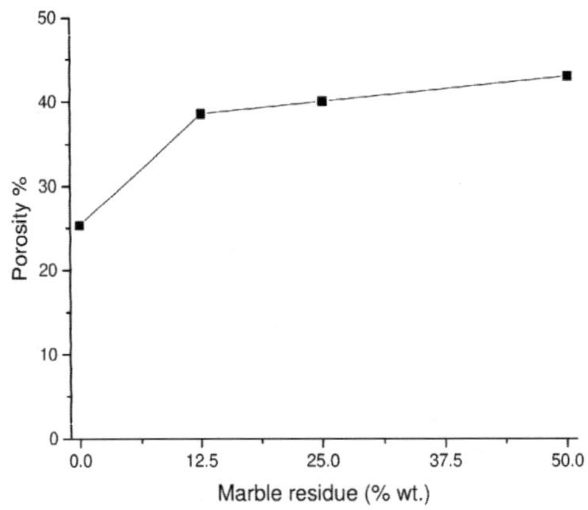

Figure 7. Porosity % (calculated from the SEM micrographs) versus marble residue (% wt.).

3.5. Thermal conductivity
Thermal conductivity measurements were conducted in order to verify the ability of marble residue to improve the insulating behaviour of the produced ceramics.

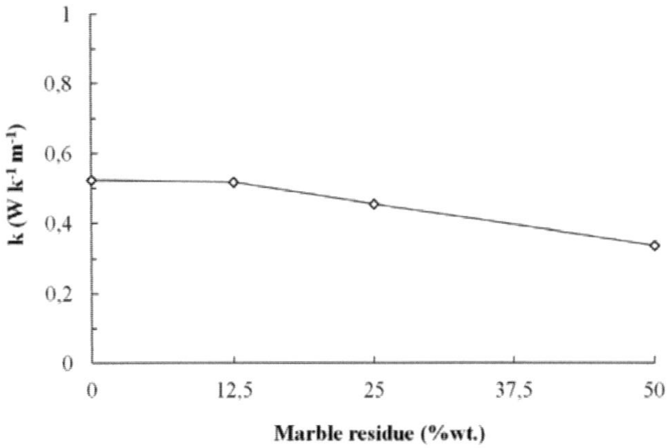

Figure 8. The effect of marble residue on the thermal conductivity coefficient (k).

The thermal conductivity coefficient (k) of compacted specimens fired at 950°C as a function of the marble residue (% wt.) content is shown in figure 8. The incorporation of high loadings (up to 50%) of marble residue inside the clayey body substantially reduces (up to 35 %) the thermal conductivity. In that sense the insulating behaviour of the ceramics is improved. This finding is in agreement with the porosity measurements. The presence of pores inside the ceramic body is expected to reduce the thermal conductivity, due to the very low thermal conductivity values of air which exists in the pore volume. Therefore, the porosity determination of the ceramic materials destined to thermal insulation application is rather significant.

4. Conclusions

Ceramics loaded with different % marble processing residue content were successfully produced upon firing at 950°C. Up to 50 wt.% residue content was successfully consolidated in the clayey ceramic body, by employing an established powder metallurgy procedure. SEM-EDS results suggest that the fine particles of marble residue were evenly incorporated and dispersed into the clayey ceramic body, even for higher admixture percentages. The characterization of the ceramic samples by means of porosity and thermal conductivity measurements reveals that, by increasing the marble residue loading, the porosity of sintered ceramic products increases, which, in turn, leads to a decrease in thermal conductivity. Consequently, the marble processing residue can be employed as an efficient pore former, in order to improve the insulating behavior of the ceramic materials.

References

[1] P. Meukam, Y. Jannot, A. Noumowe and T. C. Kofane, 2004 *Construction and Building Materials*, 18, pp. 437-43.
[2] W. D. Kingery, H. K. Bowen and D. R. Uhlmann, 1976 *Introduction to Ceramics*: Wiley).
[3] L. Hu, C. A. Wang and Y. Huang, 2010 *Journal of Materials Science*, 45, pp. 3242-6.
[4] M. M. Akiyoshi, A. P. Da Silva, M. G. Da Silva and V. C. Pandolfelli, 2002 *American Ceramic Society Bulletin*, 81, pp. 39-43.
[5] M. Sutcu and S. Akkurt, 2009 *Ceramics International*, 35, pp. 2625-31.
[6] S. P. Raut, R. V. Ralegaonkar and S. A. Mandavgane, 2011 *Construction and Building Materials*, 25, pp. 4037-4042.
[7] V. Karayannis, G. Papapolymerou, S. Zaoutsos, S. Lamprakopoulos, K. Ntampegliotis and X. Spiliotis, 2015 *Key Engineering Materials*, 660, pp. 268-272.
[8] V. Karayannis, X. Spiliotis, A. Domopoulou, K. Ntampegliotis and G. Papapolymerou, 2015 *Revista Romana de Materiale / Romanian Journal of Materials*, 45(4), pp. 358-363.
[9] D. Eliche-Quesada, S. Martínez-Martínez, L. Pérez-Villarejo, F. J. Iglesias-Godino, C.

IOP Conf. Series: Materials Science and Engineering **133** (2016) 012030 doi:10.1088/1757-899X/133/1/012030

Martínez-García and F. A. Corpas-Iglesias, 2012 *Fuel Processing Technology*, 103, pp. 166-73.
[10] D. Eliche-Quesada, F. A. Corpas-Iglesias, L. Pérez-Villarejo and F. J. Iglesias-Godino, 2012 *Construction and Building Materials*, 34, pp. 275-84.
[11] M. Al-Amaireh, 2006 *Journal of Applied Sciences*, 6, pp. 2605-2610.
[12] Jr. F. Saboya, G. C. Xavier and J. Alexandre, 2007 *Construction and Building Materials*, 21, pp. 1950-60.
[13] M. A. Montero, M. M. Jordán, M. S. Hernández-Crespo and T. Sanfeliu, 2009 *Appl. Clay Sci.*, 46, pp. 404-408.
[14] M. A. Montero, M. M. Jordán, M. B. Almendro-Candel, T. Sanfeliu and M. S. Hernández-Crespo, 2009 *Appl. Clay Sci.*, 43, pp. 186-189.
[15] N. Bilgin, H. A. Yeprem, S. Arslan, A. Bilgin, E. Günay and M. Maroglu, 2012 *Construction and Building Materials*, 29, pp. 449-457.
[16] A. Domopoulou, K. Gudulas, E. Papastergiadis and V. Karayannis, 2015 *XV Modern Applied Science*, 9(6), pp. 137-144.
[17] G. Impoco, S. Carrato, M. Caccamo, L. Tuminello and G. Licitra, 2007 *Communications to SIMAI Congress* ISSN 1827-9015, Vol. 2, pp. 1-10.
[18] A. Baklavaridis, I. Zuburtikudis and C. Panayiotou 2015 *Polymer Engineering and Science*, 55 pp. 1856-63.
[19] M. Von Bradke, F. Gitzhofer and R. Henne, 2005 *Scanning*, 27, pp. 132-135.

The aluminizing in powder technology of AISI 304 steel

D B Băitanu[1], D G Găluşcă[1], D C Achiţei[1,2], M G Minciună[1,2], Mohd Mustafa Al Bakri Abdullah[2]

[1]Department Technologies and Equipments for Materials Processing, Gheorghe Asachi Technical University of Iasi, D Mangeron Street, no 41A, Iasi
[2]Center of Excellence Geopolymer & Green Technology (CEGeoGTech), School of Materials Engineering, Universiti Malaysia Perlis, 01000 Kangar, Perlis, Malaysia

E-mail: doru_baitanu@yahoo.ro, dangalusca@yahoo.com, dragos_adc@tuiasi.ro, mirabela.minciuna@yahoo.ro

Abstract. The paper presents a study about the aluminizing treatments applied to AISI 304 stainless steel, with the purpose to improve the corrosion resistance. The aluminizing is realized in a powder medium, composed by aluminium powder (with 99.95 % purity), aluminium oxide Al_2O_3 and ammonium chloride NH_4Cl. The structural characterization was made by scanning electronic microscopy to highlight the structure of layer after aluminizing, at different magnitudes.

1. Introduction

The aluminizing represents a superficial saturation process of steel or cast iron, with aluminium, which imprint to these a high resistance to oxidation and atmospheric corrosion. Steel and cast iron, after aluminizing, are used like substitutes of steels or alloyed materials, which resist to oxides formation at 800–900°C, and in some cases up to 950–1000°C.

Can be subjected to aluminizing the cylinders of high pressure of turbines made by pearlite steels which work at 560-585°C, gases turbine blades, guide disks of gas turbines and shafts of plant by gases turbine made from austenite refractory steels. There are cases by aluminizing of alloys with complex composition and on metals (like Ti, Cu, Co). Aluminizing is made at 700–1100°C and can obtain a thickness of aluminized layer between 0.02-0.8 mm limits.

The superficial zone of aluminized layer has a high hardness (up to 500HV). However, the aluminized layer has a small usage resistance. The aluminized layer is distinguished by fragility, if the aluminium content is high. To decrease the fragility of this layer, can apply a diffusion annealing. If in superficial zone of aluminized layer, the content of aluminium is 20-30%, the layer has a good tenacity.

The aluminized layer presents noncorrosive properties, with high values for some mediums, like air. A good corrosion resistance in air is possessed by aluminized steels in melting by aluminium and iron and it is used for protection of wires, bands, household items, parts for automotives. The aluminized steel is much resistant against atmospheric corrosion and corrosion in sea water, than galvanized steels in melting zinc. After aluminizing, the parts must be cleaned and oiled.

The minimum content in aluminium from ferro-aluminum alloy, which ensures a high resistance at oxide formation, is by 8%. But in the aluminizing case, in the superficial layers of steel it is realizing a higher content of aluminium. The heating of aluminized layer in oxidizing medium leads to formation of a thin film on steel surface, but resistant, by aluminium oxide, which protect the steel against oxidation.

The resistance of parts in work conditions it is conditioned by the thickness of aluminized layer and his content in aluminium. In the exploitation process of thin parts at high temperature, the aluminium has time to diffuse until their mid and from this reason, the decrease of aluminium content in the superficial layer is slowed.

The advantages of aluminizing parts at high temperature is bigger at working in hydrogen sulphide medium, which favourite discarding of steel parts, inclusive the parts made from steels resistant to oxides formation.

2. Experiments

2.1. Determination of chemical composition for base material

The analysis of chemical composition was made on Foundry Masters Spectrometer, 01J0013 model, manufactured by WAS Worldwide Analytical Systems AG. With WASLAB software and the expandable calibration soft, were obtained the analysis bulletins, which present the determined values.

The analyzed samples was cut, and after, polished on abrasive papers, using cooling water, for not overheat the samples in the processing time.

The average values determined for alloying elements, are present in table 1 and fall the material in the stainless steel class.

Table 1. Chemical composition determined for the analyzed samples

Fe	C	Si	Mn	P	S	Cr	Mo
64.60	0.16	2.28	1.01	< 0.01	< 0.01	18.90	0.33
Ni	Al	Co	Cu	Nb	Ti	V	W
12.10	0.02	0.12	0.15	0.02	0.01	0.07	0.04

From the industrial applications of AISI 304 steel can remember: chemical equipment, food processing equipment, equipments for beer processing, cryogenic dishes, gutters, and downpipes.

2.2. Structural modifications induced by thermo-chemical treatment

The aluminizing consists by the pieces heating in boxes which contain a mixture by aluminium powder (figure 2.a), ammonium chloride (NH_4Cl) (figure 2.b), aluminium oxide (figure 2.c). At the temperature between 900-1100^{0}C take place the reactions: $NH_4Cl \rightarrow NH_3 + HCl$; $6HCl + 2Al \rightarrow 2AlCl_3 + 3H_2$; $AlCl_3 + Fe \rightarrow FeCl_3 + Al_{activ}$; $2NH_3 \rightarrow 3H_2 + 2\,N_{activ}$

The aluminium diffuse in steel, and nitrogen and hydrogen form an unoxidized atmosphere. The aluminium oxide is designed to not form the powder agglomerations of aluminium. The thickness of alloyed layer, which contain 40-50 % aluminium is 0.1-1 mm and depends by temperature and maintaining time.

The aluminizing was made at a maintaining temperature T = 1000°C, conformable the diagrams from figure 1.

Before aluminizing treatment, the samples were polished on abrasive papers, to obtain a good surface quality and remove any impurities.

The aluminizing was made in a steel recipient, with thick walls (figure 2.d). The samples are placed inside recipient, ensuring a mixture aluminizing layer around these. The precinct was sealed with a metallic cover, in which are provided some exhausted holes for reaction gases.

The aluminizing was made in iron or chrome-nickel boxes. After casting the first layer of powder, this is slightly beaten. Between the pieces is placed the control piece.

The ammonium chloride NH_4Cl (figure 2.b) is used like reaction accelerator of deposition process. The aluminium oxide doesn't participate at the process, but prevent the particles melting and implicit, the adherence to the pieces surface or recipient.

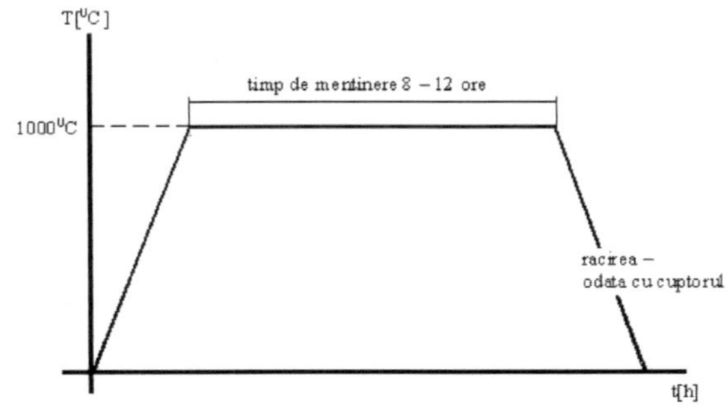

Figure 1. Aluminizing diagram.

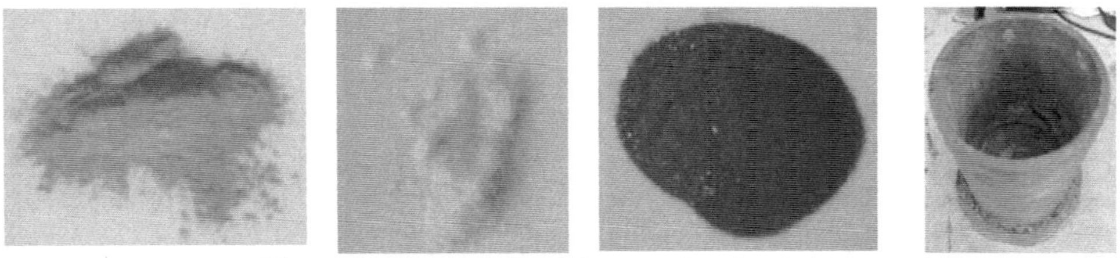

Figure 2. Materials used for powder aluminizing:
a) Al powder; b) NH_4Cl ammonium chloride; c) Al_2O_3 aluminium oxide; d) steel recipient.

After aluminizing, it is recommend a diffuse annealing for aluminium, at 900–1050^{0}C, for 4-5 hours, when the layer thickness increase with 20-40% and layer fragility (due to the presence of Al_4C_3 carbide) decrease. The structure of layer, after aluminizing, consist by a solid solution by alloyed ferrite with aluminium, rich in Fe_3Al, Fe_2Al, Fe_2Al_5, Al_4C_3.

2.3. Structural analysis by scanning electronic microscopy
The analysis is made on SEM model VEGA II LSH, manufacturing by TESCAN Cehia, coupled by a EDX detector, type QUANTAC QX2, manufactured by ROENTEC Germany.

The chemical composition of steel resistant at corrosion is established in function of medium in which the pieces will work. The principal mode to increase the corrosion resistance is to adding the elements which form on surface thin layers which protect the base material, preventing the contact between the piece and the aggressive medium.

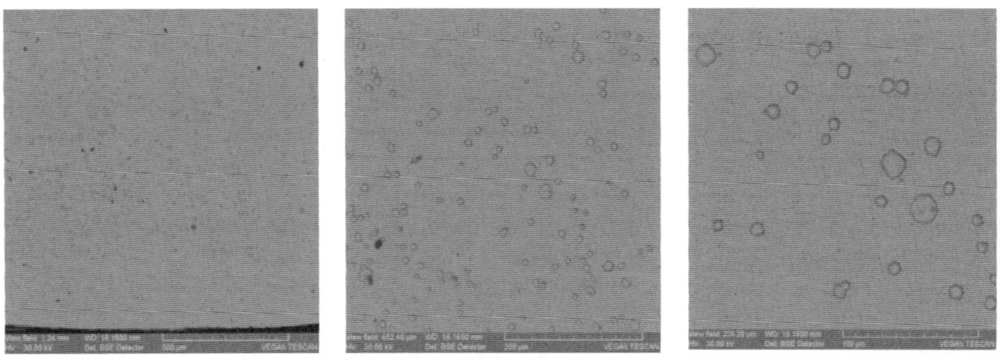

a. b. c.

Figure 3. SEM microstructures of AISI 304 steel: a) 200X BSE; b) 500X BSE; c) 1000X BSE; attack withFeCl$_3$.

The experiments were made on samples with small dimensions. The alloyed steel is part of austenite stainless steels.

The increase of oxidation resistance is obtained especially by adding Cr, Al and Si, elements which are dissolved in Fe and which form in time of heating the protector layers of oxides $(Cr, Fe)_2O_3$; $(Al, Fe)_2O_3$.

In the equilibrium state, the microstructure of these steels is formed by austenite, ferrite and carbides (figure 3). These steels are characterized by an acceptable breaking resistance, small yield strength, and good characteristics of ductility, tenacity and high corrosion resistance in oxidant mediums.

After immersion and maintaining of steel samples in aluminizing recipient, it is developed the formation of a aluminium layer, with 1 mm thickness, at the exterior of piece (figure 4 a,b). The aluminium layer which is adherent to the piece, present a small roughness (figure 4 c).

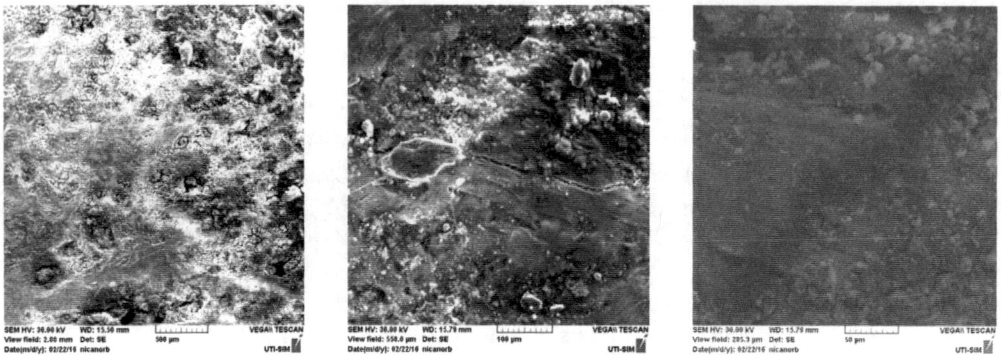

a. b. c.

Figure 4. SEM microstructures of aluminizing layer:
a) 100X SE; b) 500X SE; c) 1000X SE.

By applying a diffusion heat treatment, in the superficial layer of sample are placed an increase of aluminium content, fact that provided that the thermo-chemical treatment of aluminizing is reached the goal.

3. Conclusions

The principal mode to increase the corrosion resistance is the addition of elements which form on surface the thin films which protect the base material, preventing the contact between the piece and the aggressive medium.

The aluminizing thermo-chemical treatment is a solution to improve the corrosion resistance of pieces, which work in automotive industry.

The aluminium layers obtained by powder aluminizing that adheres to the pieces, is thin and slightly rough and ensure a good corrosion resistance in different aggressive mediums.

References

[1] A. S. Podder, A. Bhanja, 2013 *Advanced Materials Research*, Vol. 794, pp. 731-740.
[2] D. Moore, 2001 *Heat Treating Progress*, pp. 29-33.
[3] D. Scott MacKenzie, 2001 *Heat Treating Progress*, pp. 37-43.
[4] G. E. Toten, H. M. Tensi, 2002 *Heat Treating Progress*, pp. 01-04.
[5] N.I. Kobasko, G.E. Totten, G.M. Webster and C.E. Bates, 1998, *18th Heat Treating Society Conference Proceedings*, ASM International, Materials Park, OH, pp. 559-567.
[6] G. Negrea, A. Molinari, 1998, *Procedee avansate în ingineria suprafețelor*, Editura U.T.Pres, Cluj-Napoca.
[7] L. Shuqing, S. Daxim, C. Yihing, C. Zhuyou, 2000, *Chemical Journal of Chinese Universities*, vol. 21, (3), pp. 339.
[8] K. Leclere, C. Briens, T. Gauthie, J. Bayle, P. Guigon, M. Bergougnou, 2004 *Chemical Engineering and processing*, vol. 43, pp. 693.
[9] S. Bruhns, J. Werther, 2005 *AIChE Journal*, vol. 51 (3), pp. 766.

IOP Conf. Series: Materials Science and Engineering **133** (2016) 012032 doi:10.1088/1757-899X/133/1/012032

Special features of structural changes of amorphizing nickel alloys in liquid and heterogeneous state

G B Tyagunov[1], E E Baryshev[1] and K Yu Shmakova[1]

[1] Ural Federal University, Mira,19, Yekaterinburg, Russia

E-mail: e.e.barysev@urfu.ru

Abstract. Polytherms of kinematic viscosity of nickel-based alloys with Cr, Si, Fe and B additives have been studied. It was found out that the first liquid phase volumes appear when temperatures are well below the tabular values of the solidus temperature ts. The critical temperature of the alloy was defined. The ways of obtaining materials of higher amorphization ability were suggested.

1. Introduction

In the process of structure fabrication different types of welding are used. However, technically it is not always appropriate and possible. A very promising alternative to welding is to use solders to connect details made of fundamentally different materials. Soldering has a complex of advantages over welding and its capabilities expand significantly because of the use of solder in metastable state either amorphous or nanocrystalline.

The uniqueness of such solder is provided by the following properties: low melting temperature, homogeneous composition of the whole material volume, increased capillary and adhesion activity, ease of fabricating embedded elements by cutting or by punching.

2. Main part

In the National Research Nuclear University "MEPhI" a complex of works aimed at the development of solders was conducted. This resulted in solving several complex technical tasks [1]. In particular, nickel-based alloys are widely used in aviation technology for soldering of heat exchangers and gas turbine engine blades [2-3], in rocket-and-space technology for soldering rocket engine nozzles, in atomic engineering for soldering stubs, fuel elements [4] and so on. The use of nanocrystalline materials in these fields also greatly simplifies assembly operations and a soldered joint has high reliability at elevated temperatures and non-uniform loads.

The degree of volume amorphization and the structure of nanopowders are essentially determined by the elemental composition. However, an analysis of the reasons for the low quality of products demonstrates the essential role of different methods of obtaining materials by direct synthesis [5-11].

The easiest way to make nanocrystalline powders is to evaporate material at a proper temperature either under inert gas with subsequent condensation of steam or in the reaction zone with the formation of spherical particles, or on surfaces, where particles then will have faceting.

Among all the chemical methods the most widely used one is plasma-chemical method, in which the formation reaction of new phase germs takes place under non-equilibrium conditions at a high speed, but under a low intensity of growth.

Nanoparticle deposition in a colloidal solution involves a process of synthesis, formation of a disperse system and its transition from a liquid state to a dispersed solid one. The method has high selectivity and allows to produce nanopowders with very narrow size distributions.

The most productive method is a mechanical abrasion of components. During this process grinding and a plastic deformation of substances takes place, the mass transfer is accelerated, the mixing of components is performed at an atomic level, and their chemical interaction is activated.

The amorphous or nanocristalline structure of metals and alloys forms at fast (the speed $\geq 10^6$ deg/sec) cooling of the melt on the surface of the spinning disc or drum which is called the process of spinning. When required, an amorphous ribbon is annealed which leads to the formation of the nanostructure.

The melt is analogous to the amorphous material. According to the authors of work [12] the presence of nuclear local nanodomains (clusters) in the structure of amorphous materials and melts is as important for the theory of these condensed states of matter as an elementary cell for the theory of crystals.

Molten materials and alloys are special states of matter, caused by the lack of degenerated interfaces of clusters, a distortion of the translational symmetry, more electronic density in clusters as compared to the inter-cluster space, and an increased dynamic excitation of atoms in an outer atomic layer of clusters. The part of the last-mentioned ones in total volume of the matter is greater than 15%.

The role of the outer atomic layers of clusters for the formation of the structure and properties of materials becomes even more significant, given their different electronic configurations due to the change of valence and a decrease of electron density. The result of these changes is weakening of the interatomic interaction with other atoms of the cluster and a higher amplitude of oscillations.

The multiformity of a short-range order structure of burden materials determines a non-equilibrium and heterogeneity of the structure, achieved in the melting process of melts. However, the non-equilibrium is stipulated by an uneven distribution of atoms in liquid metal within the nanovolume space, and the heterogeneity is stipulated by an uneven distribution of clusters differing in structure within the nanovolume of the matter.

The speed at which the equilibrium is reached in metallic melts is much slower than changes of external environments. This determines the prolonged retention of the non-equilibrium states. As a result, the melts of the identical composition can be distinguished by short-range order parameters, the nature of inter-particle interactions and the distribution of elements and, therefore, the conditions of diffusion processes, crystallization, and others [5].

Numerous experimental data testify to the existence of the following patterns: the higher is the melt equilibrium and the homogeneity before the solidification, the better are the quality parameters of the solid metal [5-7].

The equilibrium can be achieved in different ways, but long-term experience shows that the maximum efficiency can be achieved by the temperature-time processing (TTP) of the melt during the melting process [5-11]. TTP processing modes are set by analyzing the results of the research of physical and chemical properties of the melt, as well as the micro-structure and properties of the solid metal.

Investigation of properties of melts, in particular, of kinematic viscosity (ν) is of a great interest to understand the processes occurring in the heterogeneous field, i.e. in the solidus-liquidus temperature range. Soldering technology provides for the formation of a liquid phase with further transition of the forming joint into the liquid-solid and solid-liquid, and then into the solid state. Because of the inability to accurately determine the relation of these phases due to the lack of many-component state diagrams, the methods to study the melt properties are used. In particular, the results of the study of viscosity for the first time allowed authors to determine the interface between the solid-liquid and liquid-solid states of iron-carbon alloys.

Thus, the study of the concentration and temperature dependences of melt properties allows to scientifically justify optimum modes of melting and to study the effect of liquid metal on the processes of the structure formation in the heterogeneous field.

Studies on temperature dependences of kinematic viscosity ν of nickel-based solder samples STEMET 1301, with additives of Cr, Si, B and Fe are given in figure 1. According to the developers the solidus temperature of the studied compound is 980°C and the liquidus temperature is 1010°C. The alloy is suitable for soldering of stainless steels, dispersion-strengthened steels, nickel-based heat-resistant melts and nichrome at temperatures of 1010-1170°C.

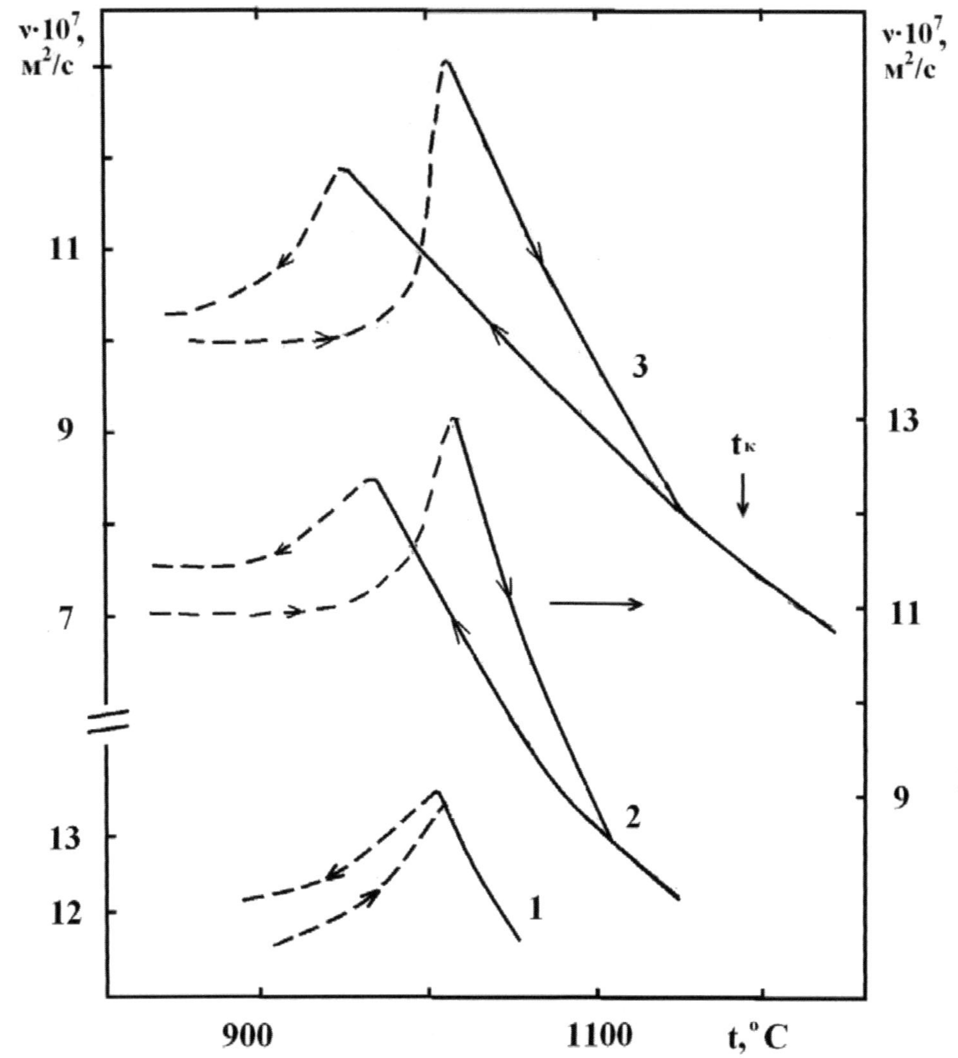

Figure 1. Temperature dependences of the kinematic viscosity (ν) of STEMET 1301 alloy during heating and cooling.

Amorphization of materials is carried out when relationship between amorphization temperature (t_a) and liquidus temperature (t_L) is t_a/ t_L=0.45. The required temperature for a given alloy t_a=440°C is assured by the sharp supercooling of the melt at a high cooling rate, the presence of silicon and boron in the alloy composition, which together with the basic elements forms complex low-melting eutectic molecular entities, and by other factors. The process is accompanied by an increase in viscosity by 10-15 times.

Viscosity polytherms which are represented on figure 1 consist of two parts. The left picture shows the temperature dependence on the attenuation decrement (δ) of combined oscillations of the crucible with the melt which is suspended on an elastic metal thread (broken line). This sensor is for the determination of the absolute values of kinematic viscosity by the method of Shvidkovskiy E.G. A conversion of the value δ into values ν will not allow to show dependency ν (t) in the studied temperature range, since "melt-solid" state in heterogeneous field exceeds the viscosity value of molten metal in tens of times.

The processes occurring in a heterogeneous field are important for solders, since the formation of the joint structure and of its properties takes place below the liquidus temperature.

When the solder is heated, the first volumes of liquid metal appear long before the solidus temperature is reached and the attenuation decrement is slowly increasing. Reaching of solidus temperature known from the work [12], is accompanied by a dramatic increase in value δ, i.e. the amount of the molten metal increases.

Further heating above t_L is accompanied by the reduction of the ν melt in accordance with the well-known exponent ν(t). When cooled at the temperature higher than 1150°C the polytherm does not coincide with the dependency obtained by heating. There is branching or a hysteresis of polytherms. In the heterogeneous field the value of δ is higher when cooled than when heated.

When cooled at the temperature higher than 1230° C the hysteresis value is increasing and the critical temperature is close to 1180°C.

3. Conclusions

Thus, the study of the liquid alloy STEMET 1301 led to the following conclusions:

- the first liquid phase volumes appear when temperatures are well below the tabular values of solidus temperature t_s;
- the intensive growth of the liquid phase volume in heterogeneous field is observed, when the temperature is close to t_s, and it ends at t_L;
-when the temperature increases further, the viscosity polytherms have a classic exponential type, typical of metallic liquids;
- polytherms ν in the heating-cooling mode coincide only when the super-heating temperature is not a 100°C higher, than t_L. A further increase in heating temperature leads to the hysteresis of polytherms, where its value remains constant only when the heating temperature of the melt reaches its critical point. When the melt is cooled polytherms ν are located below ν (t) when heated. In the heterophase field the viscosity value or the attenuation decrement of oscillations is higher than when cooled;
- the higher is the heating temperature of the melt in relation to $t_{(L)}$, the greater is the degree of supercooling during the solidification process;
- setting of exposure time of the melt at a temperature of 1150°C for 75 minutes allows to reduce the critical temperature by 100°C;
-considerable supercooling of the melt after the high-temperature heating and elevated values of attenuation decrement in the heterogeneous field contribute to obtaining of the highly amorphized material in the process of rapid hardening. This material is characterized by less surface roughness of the film, better ductility properties and good manufacturability in the process of fabricating of embedded elements;
- the nanocrystalline material obtained by the crystallization of the amorphous films, is qualitatively different from other nanopowders, obtained under other technologies by the distribution of elements and dimensional uniformity.

Acknowledgments

This work was supported by the Scientific researches of higher education institutions within the State task of the Russian Federation No. 2014/236.

References

[1] V. T. Fedotov , A. N. Suchkov , B. A. Kalin et. al, 2014 Tsvetnye Metally, 12, pp. 32.
[2] S. V. Maksimov, 2007 *Adhesion of melts and soldering materials*, 40, pp. 70.

[3] O. N. Sevryukov , A. N. Suchkov , E. V. Guseva, 2015 *Non-ferrous Metals*, 1, pp. 45.
[4] I. S. Logvenchev , A. A. Ivannikov , A. A. Volkov et. al, 2014 *Inorganic Materials: Applied Research* , 5, pp. 263.
[5] B. A. Baum and G. A. Khasin et.al, 1984 *Molten steel* (Moscow: Metallurgy), pp. 206.
[6] E. E. Baryshev , G. V. Tyagunov and N. A. Stepanova, 2010 *Influence of melt structure on the properties of superalloys in the solid state* (Yekaterinburg: UrBr RAC), pp. 198.
[7] V. S. Tsepelev, V. V. Konashkov et.al, 2008 *Properties of molten metals* (Yekaterinburg: USTU-UPI), p. 357.
[8] G. V. Tyagunov and E. E. Baryshev et.al, 2014 *Liquid metal. Powders* (Yekaterinburg: UMC-UPI), pp. 192.
[9] A. I. Gusev. 1988 *Nanocrystalline materials: methods for their preparation and properties* (Yekaterinburg: UrBr RAC), pp. 115.
[10] S. Sulaimankulova, E. Omurzak and J. Jasnakunov, 2009 *Journal of Cluster Science*, 20, pp. 37.
[11] B. Gunther, H. Weiss, 1993 *International Journal of Materials and Product Technology*, 8, pp. 424.
[12] V. K. Mashkovskii and V. N. Novikov et.al, 1993 *Physics-Uspekhi*, 5, pp. 119.

IOP Conf. Series: Materials Science and Engineering **133** (2016) 012033 doi:10.1088/1757-899X/133/1/012033

Classic and New Materials Used for Structural Rehabilitation. Case Study

M Lute[1]

[1] CCI Department, Politehnica University Timisoara, 2 T. Lalescu Str., Timisoara, Romania

E-mail: marina.lute@upt.ro

Abstract. New materials development with different combination of properties were always a challenge in terms of their adequate use in civil engineering. Introduction of carbon fibres as strength material for structures was a beginning of a new approach in structural rehabilitation, and sometimes meant the end of classic rehabilitation solution use. The present paper gives an example of a building rehabilitation that use a melt of both new and old solutions in order to achieve the optimum result for building itself. The problem was even more challenging, because the structure considered is only 22 years old, but having some design faults in terms of seismic behaviour and, in addition, one floor was added to existing structure. The chosen solution was a compromise between the use of old and new materials in places where their qualities were best suitable and their minuses could be compensated by the other material.

1. Introduction

Buildings rehabilitation is a challenge in terms of calculus, solution choice and technological possibility to apply the designed solutions. In addition, use of new materials become a "fashionable design way", used everywhere needed or not, and the old ones use was diminished even when they were the best choice for a certain purpose.

In this line, the rehabilitation solutions for concrete structures shown in the last years by the scientific literature, were divided between new materials use [1-6] and new innovative technological solutions [4-7], the use of classic rehabilitation solutions being avoided. The only thing in common of these trends was the preservation of the materials of the original structure [1-4].

The building rehabilitation solution presented here combines this new trend with the old school solutions, in order to achieve the best possible result.

2. Existing situation

The considered building was designed in 1994 and built in the next year, as orthopedic clinic of a regional hospital (figure 1). The construction as a whole, is a building consisting of two independent L-shape sections forming an interior light yard. The original blue print for the ground floor is shown in figure 2. Resistance structure for both sections is a spatial frame – diaphragm structure of reinforced concrete, monolithic solution. Currently it was considered the necessity of increasing the operational space by adding an additional floor to the existing building.

IOP Conf. Series: Materials Science and Engineering **133** (2016) 012033 doi:10.1088/1757-899X/133/1/012033

Figure 1. Existing building.

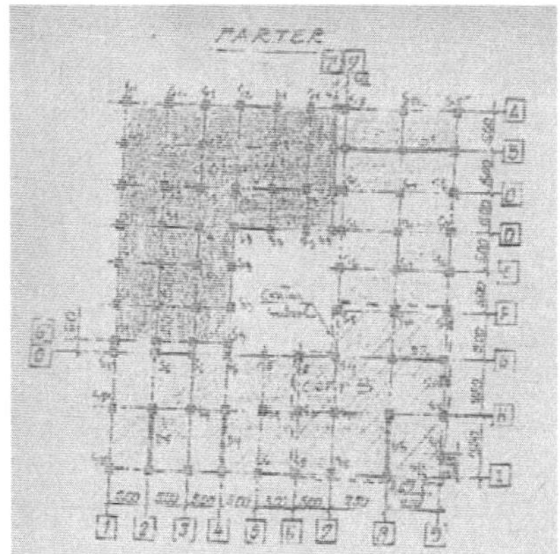

Figure 2. Original design – ground floor.

2.1. *Strength structure*
The two sections named A and B, have the following characteristics. The A section, an L-shaped building with dimensions of 30.32 x 30.32 m x 16.30m is 2 floors high, ground floor and 1 withdrawn floor. The structure is made of spatial frames reinforced concrete monolith with spans of 5.0x5.0 m except in areas of operating rooms where the spans are 10.0x 10.0 m. The B section, also an L-shaped building, is the largest building, having the dimensions 48.00 x 48.00 m x 15.98m. Its height is also larger, being a 4 floor building with one basement. The structure is made of monolithic reinforced concrete frame – diaphragm solution. Frames are developed after two perpendicular directions with 5.0 x 7.5 m spans respectively 7.5 x 7.5 m on margins. The diaphragms are placed inside the structure in axes 2, 4 and 8 between I and H axes, and in axis G between 8 and 9 and B between 7 and 9. The frames and diaphragms are connected and work as a continuous structure. The slabs for both sections are monolithic reinforced concrete slabs with constant thickness of 15cm supported on frame beams, secondary beams and transverse diaphragms for intermediate floors. In the basement slab is monolithic reinforced concrete slab with constant thickness of 30 cm frame that stands on beams and reinforced concrete diaphragms.

Due to the difference in level of buildings infrastructure, the foundation system is slightly different for the two wings. The foundation system of B wing consists of rigid isolated foundations under the columns and continuous foundations under diaphragms masonry walls that makes the building envelope. The basement has perimeter concrete diaphragms with role of forming a rigid box. The wing A foundation system consists of rigid bearing isolated foundations under reinforced concrete columns and continuous foundations under masonry perimeter closing walls.

2.2. *Architectural and structural interventions considered*
After the owner requirements, at Building A, it is required a secondary floor over the existing terrace considered in a solution of light steel structure, and increase the technical areas for HVAC equipment. In this way the hospitalization capacity is increased by a third, and the new HVAC equipment could fulfill the requirements for health operating rooms. To achieve vertical access in these conditions is necessary to create a new lift access in the interior yard with the structure of reinforced concrete diaphragms designed independently of the existing structure. Given the nature of the interventions described above, it follows that only the section A is affected by structural interventions, section B not being affected in structural terms.

At Building A, the changes in functionality led to drastic changes compared to the original solution, as follows:

- Arrangement of an operating room and its annexes, in place of former lobbies and terrace. Creating the necessary space for the operating room required the demolition of columns and changing the mode of transmission of the loads.

- Replacing protocol space with the technical area for installations with high increase in value of live loads, namely from 2.10 kN/m^2 to 11.25 kN/m^2 (78% increase)
- Increasing the area for equipment installation space from 100 m^{2}to 168 m^2
- Change of circulation areas with room for residents and safety airlocks.
- Increase width all lobbies from 2.20m to 2.40m, taking over space from areas with destination hospital room.
- Moreover, the architectural rearrangement from existing floors is doubled by addition of an entire new floor.

3. Technical factors to be considered for structural assessment

Due to the conditions presented, the assessment of the structure needed to consider four factors, besides the assessment of real condition of the structure, namely: change of norms provisions [9-11] for seismic loads, change in loads value due to norms and change in loads positon due to functional change, in addition to new supplementary loads that could not be considered by the original design and geometrical alteration due to functional change.

3.1. Geometrical alteration of the structure

The last two factors led to the following changes in general geometry of the structure:

Demolition of a column at the first floor leads to a modified static scheme for the entire structural strength with the consequent redistribution of efforts in structure and increase of efforts and strains in the elements that take on the role of the demolished column.

The new floor is proposed as metal structure with transversal and longitudinal bracing frames. The elements are designed to be made of laminated profiles. The columns are considered with hinged supports, the connection with the existing reinforced concrete structure being achieved with two chemical anchors. The beam - columns joint was considered in embedded solution. All these modifications altered the height of the building differently, as shown in table 1.

Table 1. Changes in height due to functional change.

Built area	**Levels**	
[m^2]	**Designed**	**Proposed**
391.15	1	2
67.14	1	3
280.98	2	2

3.2. Change in loads intensity and position

The geometrical modifications changed both the total of dead and live loads. Besides that, the intensity of live loads were changed for certain areas of the building [11], as shown in table 2.

Table 2. Changes in live load intensity due to functional change.

Design loads	**Loads after functional change**	**Change type**
Areas designed as hospital rooms turn into lobbies		
200 x 1.5 = 300 daN/m^2	500 x 1.5 = 750 daN/m^2	Increase with 150%
Areas designed as offices turn into technical areas		
200 x 1.5 = 300 daN/m^2	750 x 1.5 = 1125 daN/m^2	Increase with 275%
Areas designed as lobbies turn into hospital rooms		
500 x 1.5 = 750 daN/m^2	200 x 1.5 = 300 daN/m^2	Decrease with 60%

In the same time, the change of loads intensity and safety coefficients values were also altered, as shown in table 3.

Table 3. Changes in live load intensity and safety factors due to norms change.

Design values	Assessment values
SAFETY PARTIAL COEFFICIENTS	
- for dead loads – structural elements	
1.1	1.35
- for dead loads – finishing	
1.3	1.35
- for wall loads	
1.2	1.35
LOADS INTENSITY	
-live loads – hospital rooms	
150 x 1.4 = 210 daN/m^2	200 x 1.5 = 300 daN/m^2
- live loads - lobbies	
300 x 1.4 = 420 daN/m^2	500 x 1.5 = 750 daN/m^2
- live loads - terrace	
75 x 1.4 = 105 daN/m^2	Turn into functional space with different live loads
MINIMUM REQUIREMENTS FOR MATERIALS	
- concrete C12/15	- concrete C25/30 (H ductility class)
LIVE LOADS FOR INSTALLATIONS	
1. Heating system: 300 daN/m^2 x1.4 = 420 daN/m^2 Not provided by design	1. Heating system: 750 daN/m^2 x1.5 = 1125 daN/m^2 2. Ventilation and cooling system 750 daN/m^2 x1.5 = 1125 daN/m^2

In conclusion, the modifications changed the live load as follows: in about 10% of the structure, the live loads increased by 78%, in 26% the live loads has increased by 62.66%, and for 64% of the area the live loads remain unchanged. Considering that in addition, the structure has the shape of L, it follows that increasing load by about 70% on 36% of the area leads to change in the position of gravity center away from rigidity center, and affecting the efforts path and leading to their concentration on the sides of the building.

4. Structural assessment

Because the placement of the structure is in a seismic area of 7.5 degrees on Richter scale, the seismic response of the structure is the first thing to be considered. The local normative [10] for seismic design impose certain regulations that are partially fulfilled by the considered structure:

Structural simplicity is a requirement that is not totally respected because a building is divided into two L-shaped wings is not the most rational decision for a building placed in a seismic area. The stiffness center differs from center of gravity so that the first vibration mode will be a torsional mode – a real disadvantage for a seismic response.

The structural system is mostly continuous and provide a clear, direct and uninterrupted path for seismic forces transmission, regardless of the force direction, directly to the foundation, except columns SB2 and SB4 which stand directly on a beam – situation that is to be avoided as seismic design – but can be accepted for the last two levels of a building as it is the present situation.

All the seismic forces that can appear in the structure will be supported in all the floors of the building that can work as horizontal diaphragms due to their thickness – 15 cm.

Structural redundancy: from the geometric viewpoint one can consider that a plastic mechanism with sufficient plastic hinges is provided, allowing the use of structure strength reserves and a good seismic energy dissipation. At the same time, reinforcement of beams and columns (figure 3) compromise this system, by not taking into account the provisions of [10].

Structure configuration: structural regularity in plan and elevation sections is not satisfied.

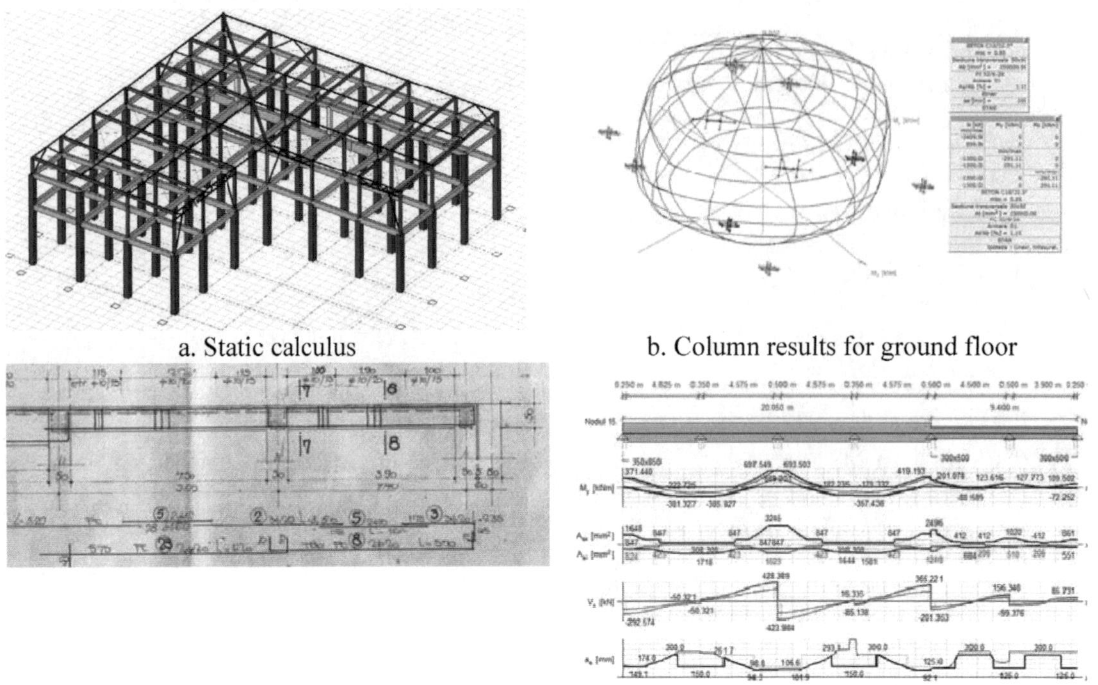

a. Static calculus b. Column results for ground floor

c. Beam reinforcement – original design d. Beam results

Figure 3. Static calculus.

Cross sections selected for strength elements satisfy the condition of rigidity required by [10], except beams section 35x80cm and 35x85 cm, where the beams rigidity is higher than the columns', the structure nodes being then compromised.

The materials used do not meet the minimum requirement for ductility class H (C25 / 30) being used concrete C18 / 22.5.

5. Rehabilitation solutions

Choosing rehabilitation solutions [2] was imposed by two things: combining the results obtained with the possibility of applying the chosen solutions in a space that could not be partially dismantled.

The solutions had to meet the following technical findings.

Increasing the lateral stiffness of the structure for seismic reasons: vertical elements must have increased cross sections. By adopting this solution the main scope is reached and, besides this, a correct stiffness ratio between beams and columns in junction will be achieved, and the stiffness of the column to exceed the sum of beams rigidities for each node. At the same time, supplemented longitudinal reinforcement leads to the increase of columns bearing capacity of columns up to the level required for supporting the efforts of structure. The increase of cross section was made symmetrically for intermediate columns and only on the interior sides for envelope columns (figure 4). For all cases, the longitudinal reinforcement was supplemented. A particular case appeared for the envelope columns, because the thickness of the new concrete layer. In this particular case, a strengthened corner was created by adding a supplementary stirrup in double L shape together with double longitudinal reinforcement (figure 4).

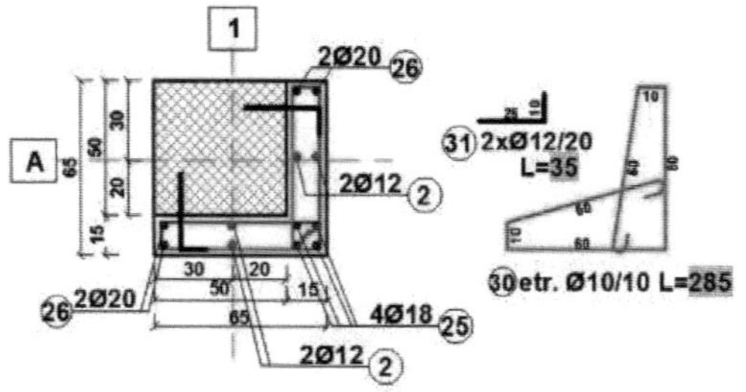

Figure 4. Column rehabilitation.

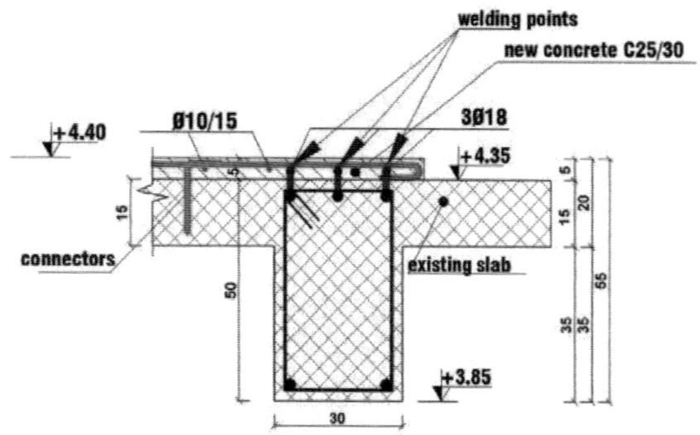

Figure 5. Slab rehabilitation.

Increasing the bearing capacity of the horizontal resistance strength elements: the structural calculus revealed that structural elements are not adequately reinforced for the loads required by current standards and architectural changes. The problem appeared on beams and slab over the ground floor.

The slab rehabilitation (figure 5) was done by adding supplementary reinforcement on the top side. The connection was made by welding between old and new rebars, through connection Z shape reinforcement. The floor top side being mainly free (terrace and technical areas), the solution was easy to apply. In addition, it increased the slab thickness, contributing also to the limitation of lateral displacements.

The same idea (figure 6) was applied to the upper side of beams consolidation. The result was not enough for the efforts supported by beams, but an increase of section on the lower side of the beam was not possible for two reasons: the ground floor was fully working during rehabilitation process the walls below beams could not be demolished, and another increase of beam stiffness could not be supported by the stiffness balance in the structure nodes, even with the designed change in section for the columns.

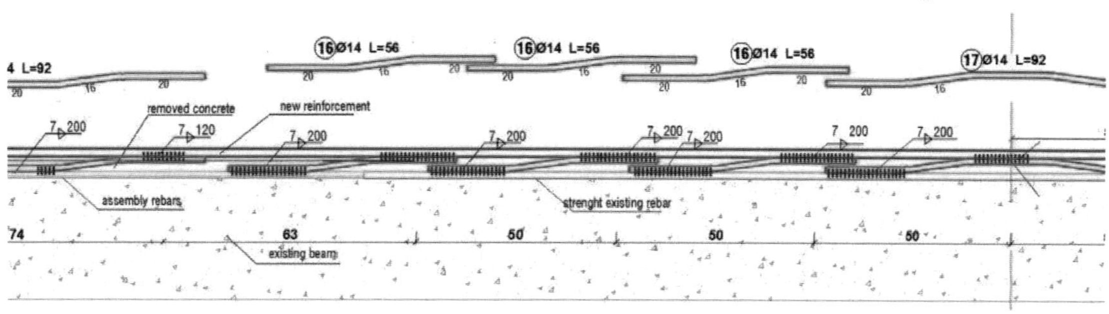

Figure 6. Beam rehabilitation on upper side using classic solution.

The carbon fiber solution appeared to be the most suitable for that purpose. A mix of fiber strips and fiber net applied on the lateral sides (figure 7) and bottom of the beams created the perfect solution for this particular case: increase of bearing capacity and keeping the same stiffness [3].

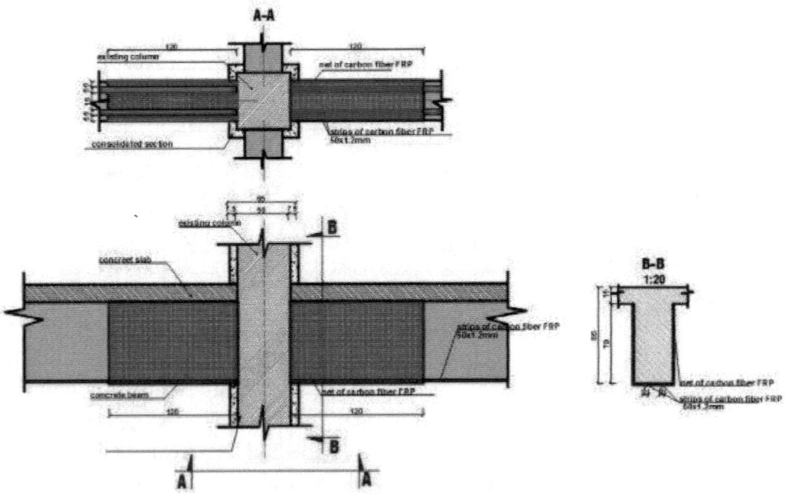

Figure 7. Beam rehabilitation using carbon fiber.

6. Conclusions

Structural rehabilitation became a normal fact for fair new building that did not finished their designed life span. This appeared due to a quick advance in technical development followed by change in norms.

The adopted solution implied using of a melt of new and classic rehabilitation solutions, each of them contributing to achievement of a certain task: the classic increase of concrete cross section improved the seismic behavior of the structure from the point of lateral displacements together with an increase of bearing capacity for columns. The use of FRP increased the bearing capacity of beams without affecting their stiffness. In this way, the two solutions together rebalanced the stiffness balance in nodes and insured the correct strength level for all structural elements.

In this way it was once more demonstrated that all technical solutions – new or old – have certain advantages that cannot be undermined by a subjective preference for one of them. One must keep an open mind to both directions and use each solution wherever is best suited, and a calculated blend between them may sometimes lead to the optimum result.

References

[1] F. Anamaria, G. Aurelian, and E. Felicia, 2015, *Sustainable old steel structure at corrosion protection, International Multidisciplinary Scientific GeoConference Surveying Geology and Mining Ecology Management, SGEM*, vol. **3(5)**, pp. 607-612.

[2] S., Dan, C., Bob, and A., Gruin, 2008 Analysis of Reinforced Concrete Existing Structures in Seismic Regions, *WSEAS International Conference: Engineering Mechanics, Structures, Engineering Geology EMESEG'08*, Heraklion, Greece, pp 96-103

[3] A.C., Dăescu, T., Nagy-György, V., Stoian and D. Dan, 2009, Experimental study on the retrofitting procedures for RC columns, Proceedings of the Twin Coimbra International Conferences on Civil Engineering - Towards a Better Environment, CE 2009 and the Concrete Future, CF 2009, pp. CF119-CF128

[4] D. Dubina, I. Marginean, F. Dinu, D. Dan, 2014, Assessment of an existing composite steel-concrete multisorey building for the purpose of its rehabilitation, *Proceedings of the 2nd International Conference on Protection of Historical Constructions*, pp. 851-856.

[5] M. Fofiu, A. Bindean, , and V. Stoian, 2015, Carbon fiber strips retrofitting system for precast reinforced concrete wall panel, *Key Engineering Materials*, vol. **660**, pp. 208-212.

[6] P. Lazer, C. Bejinariu, A.V. Sandu, A.M. Cazac, and I.G. Sandu, 2015, Chemical deposition on thin layers on reinforcing steel, *Key Engineering Materials*, vol. **660**, pp. 213-218.

[7] M. Mosoarca, and G. Victor, 2013, Structural safety of historical buildings made of reinforced

concrete, from Banat region-Romania, *Journal of Cultural Heritage*, vol. **14(3SUPPL)**, pp. e29-e34.
[8] T. Nagy-Gyorgy, I. Demeter, and D. Dan, 2014, Retrofitting strategies of RC wall panels with cut-out openings using CFRP composites, *Proceedings of the 12th International Conference on Steel, Space and Composite Structures*, pp. 267-273.
[9] A.M., Vaysburd, P.H., Emmons and B., Bissonnette, 2012, Concrete Repair, Rehabilitation and Retrofitting III - *Proceedings of the 3rd International Conference on Concrete Repair, Rehabilitation and Retrofitting, ICCRRR 2012,* pp. 862-869
[10] P100-1/2013, Seismic design code
[11] SR EN 1992-1, Design of concrete strucures. General rules and rules for buildings

IOP Conf. Series: Materials Science and Engineering **133** (2016) 012034 doi:10.1088/1757-899X/133/1/012034

Physical and Mechanical Properties of Bodymill Sludge (BS) Incorporated Into Fired Clay Brick

Aeslina Abdul Kadir[1,2], Ahmad Shayuti Abdul Rahim[1], Mohd Ikhmal Haqeem Hassan[1] & Mohd Mustafa Al Bakri Abdullah[3,2]

[1]Faculty of Civil and Environmental Engineering, Universiti Tun Hussein Onn Malaysia, 86400 Parit Raja, Batu Pahat Johor, Malaysia

[2]Center of Excellence Geopolymer & Green Technology (CeGeoGTech), School of Material Engineering, Universiti Malaysia Perlis (UniMAP), Perlis Malaysia

[3]Faculty of Engineering Technology, Universiti Malaysia Perlis (UniMAP), Perlis, Malaysia

Email: aeslina@uthm.edu.my, mad_sha89@yahoo.com.my, mohd.ikhmal.haqeem@gmail.com & mustafa_albakri@unimap.edu.my

Abstract. The huge volume of mosaic sludge that has been produced and the effect towards the environment had lead to the investigation of incorporating mosaic sludge into fired clay brick. In this study, the research attempt to reuse bodymill sludge (BS) from mosaic manufacturing process. The mosaic sludge is used to replace the raw material of clay up to 30%. In this investigation, the composition and concentration of heavy metal were determined by using X-Ray Fluorescence Spectrometer (XRF). Physical and mechanical properties test were also conducted such as compressive strength, shrinkage, density and initial rate of suction. Scanning Electron microscope was carried out to determine surface changes of the manufactured sludge brick. From the results, it shows that brick with 5% of BS sludge obtained the highest compressive strength and lower total shrinkage compared to other percentages. Nevertheless, all the other properties for all bricks incorporated with different percentages of mosaic sludge were complied with the standard (BS 3291:1985). Thus, mosaic sludge could be an alternative low cost material for brick and at the same time provide an environmental friendly disposal method for the waste.

Keywords: Bricks, Mosaic sludge, Compressive strength, Density, Shrinkage, Initial rate of suction, SEM.

Introduction

In general, sludge is one of the waste that has been produced in liquid and semisolid. Few different industrial sludge such as textile laundry (Herek et al, 2012; Jahagirdar et, 2013; Mary and Sreeja, 2014; Swarna and Venkatakrishnaiah, 2014; Baskar et al, 2006), waste water treatment (Victoria, 2013; Hegazy et al, 2012; Saijun, 2011; Ramadan et al, 2008; Weng et al, 2003), petroleum treatment plant (Sengupta et al, 2002) were successful incorporated into clay brick. Industrial sludge could be in organic or inorganic form. Inorganic content of industrial sludge such as heavy metals should get the specific treatment to prevent environmental pollution (Abdul Kadir *et al*, 2014). Furthermore, heavy metal has effected in the environment appealed concerned amongst ecologist around the world (Lokhande et al, 2011; Zhoa et al, 2015), also huge amount of waste was disposed to the landfill and the limited availability of land is a major concern (Abdul Latif, et al, 2012). Moreover, huge quantities of industrial sludge been disposed to landfill for a long time may release hazardous substances and may infiltrate into soil and groundwater (Abdul Rahim & Kadir, 2014; Swierk et al, 2007). Mosaic is pieces of hard material that has made and known as stone, tile or glass. Mosaic or tiles was made from quarts, feldspar, clay, sand and water, it was used for decoration walls and pavement, also provides a durable surface. These materials are mixed homogenously to form body slip of the mosaic. Mosaic body slip is used to be a base for topping. Drying and heating process was carried out on body slip until it turns into powder and dry dust and then followed by dry pressing process. Higher temperature use will produce stronger tiles and can allowing chemical bonds between the flakes (Jones, 2008). Maximum temperature is 2500 degree Fahrenheit (Molly, 2008). In mosaic fabrication process, it produces several of waste that may constitute of sludge, dust and solid waste. These wastes will further increasing the environmental pollutants. The majority of traditional ceramic products will produce oxides such as silica, alumina, lime, alkaline oxides and magnesium oxides (Segadaes et al, 2005). Mosaic sludge content chemicals from mosaic colouring substances and need to be managed properly. This is also supported by Vallette (2014) in his research on a healthy choice for ceramic tiles. Recently, many types of sludge waste have become alternative low cost raw material in brick and this is due to high demand and flexibility of brick. The utilization of these wastes in clay bricks usually has positive effects on the properties such as lightweight bricks with improved shrinkage, porosity, thermal properties, and strength (Agrawal *et al.* 2014; Kadir *et al*, 2014; Abdul Kadir and Mohajerani, 2011). The lightweight bricks will reduce the transportation and manufactured cost. Moreover, this incorporation will reduce clay content in the fired clay brick that eventually reduce the manufacturing cost (Abdul Kadir and Sarani, 2015; Abdul Kadir and Mohajerani, 2015). This motivates many researches to investigate more on the potential of different sludge waste to be incorporated into the brick. Therefore, this research is to investigate the potential to manufacture fired clay brick incorporated with mosaic sludge.

Material and Method

Material. Mosaic sludge waste was collected in semisolid form at one of the local mosaic manufacture in Kluang, Johor. Clay soil was collected from brick manufacturer at Sedenak, Johor. Both of the materials, clay soil and sludge mosaic was kept properly in a closed container and been storage individually.

Method. Pellet shape was made before being tested with X- ray fluorescence (XRF) machine by using model S4-Pioneer Bruker-AXS (Germany). Crush samples for every brick percentage were scanned for scanning electron microscope (SEM) by using model ZEISS EVO LS10 (Germany). All raw materials (mosaic sludge and clay soil) were dried in the oven for 24 hours. After that the mosaic sludge and clay soil were crushed with suitable tools and being weighed with suitable ratio.In the brick moulding process, different percentages of sludge were mixed with clay soil using a mixer machine. Water content is an important factor affecting the quality of the brick, therefore, compaction test was conducted to determine the optimum moisture content (OMC). Using this OMC, the mixtures with various proportions of mosaic sludge and clay soil were prepared in brick manufacturing. The mixtures were then moulded into a brick mould. Control brick with 0% of sludge were also made for comparison purposes. After 24 hours drying in room temperature followed by another 24 hours at 105°C oven-dry period, the moulded mixtures were fired into a furnace at temperature 1050°C for another 24 hours. As a requirement by BS 3921:1985 standard, the brick properties including firing shrinkage, density, compressive strength and initial rate of suction were determined.

Result and Discussion

X-ray Fluorescence (XRF). The composition of clay soil and mosaic sludge produced from BS process. The properties vary considerably depending on the type of raw material and the process that have been done during the manufacturing. The results shown the highest percentage of clay was silica dioxide (SiO_2) and aluminium oxide (Al_2O_3) which is between 57.6% to 58.10% and 31.5% to 32.0% respectively. The mosaic sludge also shows that the highest percentage was demonstrated by silica dioxide (SiO_2) and aluminium oxide (Al_2O_3) between 65.5% to 68% and 21.6% to 23.8% respectively.

The concentration of heavy metals for the mosaic sludge and clay soil shows that the clay soil and mosaic sludge (BS) contain high value of titanium (Ti), manganese oxide (Mn), iron (Fe), zinc (Zn), zirconium (Zr) and barium (Ba) and all these heavy metals were more than 100ppm. This was due to the process of the sludge that the amount of heavy metals was higher because of the processing of colouring and glazing of the mosaic.

Scanning Electron Microscope (SEM). Scanning Electron microscope (SEM) test was conducted at Analytical environment laboratory (Faculty of Civil and Environmental Engineering). Samples were analysed by using SEM to determine the shape and sizes of pores development in the bricks as well as to examine the brick texture (Cultrone et al, 2004).

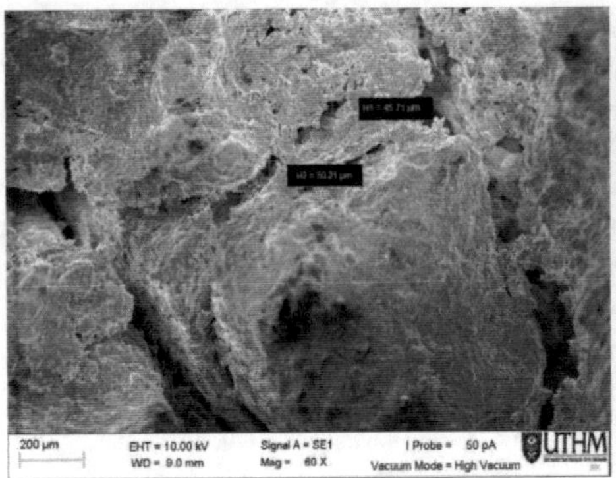

Figure 1a: Control Brick

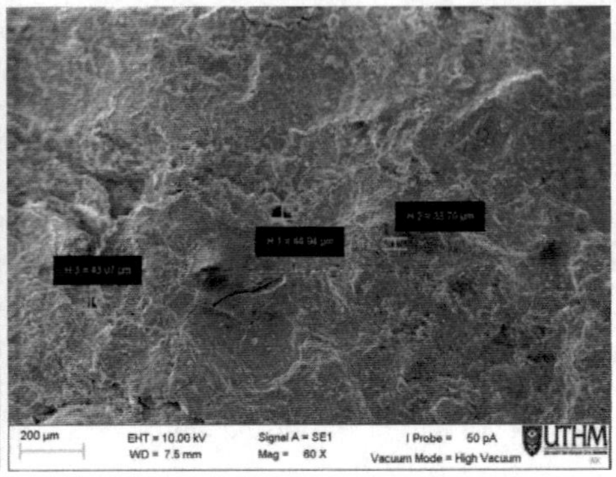

Figure 1b: BS Brick 1%

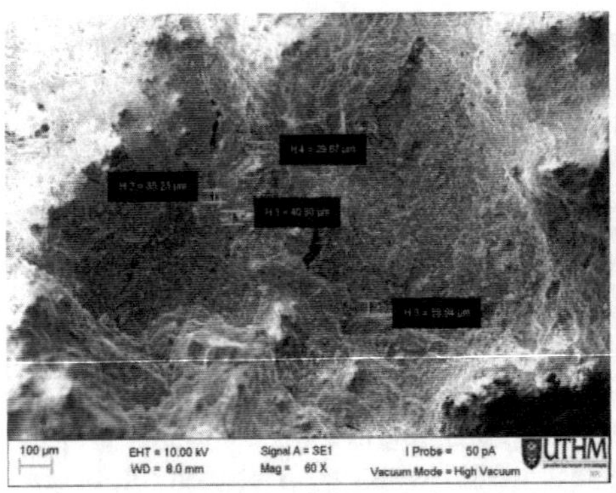

Figure 1c: BS Brick 5%

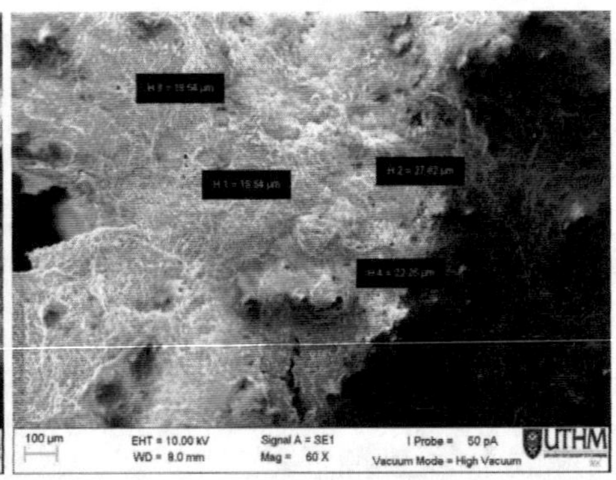

Figure 1d: BS Brick 10%

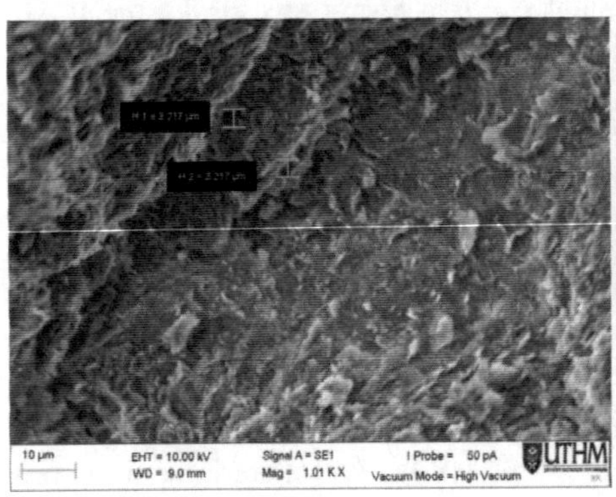

Figure 1e: BS Brick 20%

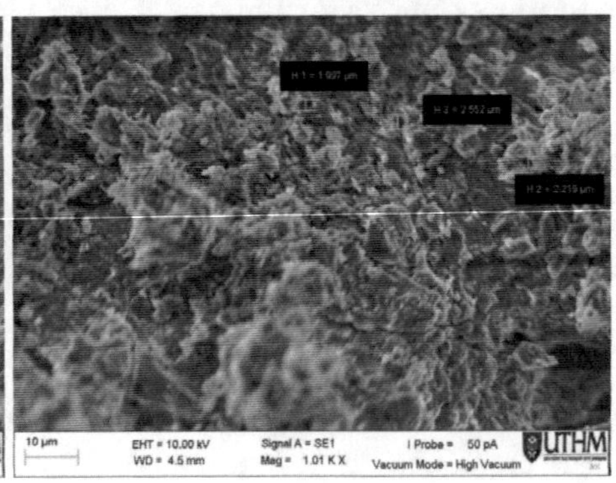

Figure 1f: BS Brick 30%

SEM were conducted in order to get a better understanding about microstructure in terms of porosity, Figure 1a to Figure 1f show SEM images of control brick and BS brick. The figures show that the surface structure appearances of the brick incorporated with mosaic sludge are finer compared to control brick. Furthermore, the images illustrate more course surface of brick with 0%, 1%, 5% and 10% compared to brick with 20% and 30%. The control brick shows the pore size was between 45μm to 50μm. The BS brick shown 1% to 10% pore sizes samples were between 18μm to 44μm. Meanwhile, for 20% and 30% it shows the pore sizes were between 1.9 μm to 3 μm. This was due to the texture of mosaic sludge powder which is more fine compared to clay soil and this enable the mosaic sludge to cover and fill the gaps between the clay soil in the mixture. At this state, mosaic sludge brick (20% and 30%) gave smooth surface but the texture is more and less cohesion between the soils. According to Babu and Ramana (2013), addition of sludge may be attributed to changes in plasticity characteristic and filler action of sludge. Furthermore, BS up to 40% was visibly affected and it was supported by Ingunza et al. (2011) he also reported that the brick manufactured sludge up to 35% were very brittle and fractured as they were removed.

Physical and Mechanical Properties

Total Shrinkage. Firing shrinkage is defined as the contracting of hardened mixture due to the loss of capillary water (Koratich, 2009) Shrinkage depends upon several factors. These factors include the properties of the components, the proportion of the components, mixing manner, the amount of moisture, and dry environment. Furthermore, Rouf and Hossain (2003) claimed that drying and firing of the brick influences shrinkage behaviors. From Table 1 and Figure 2, it shows that the 5% of BS brick obtained the lowest shrinkage with 0.34% followed by 1%, 20%, 30% and 0% of BS brick with 0.43%, 0.56%, 0.57% and 0.59% respectively. On the other hand, BS brick (10%) showed the highest values with 0.69%. Nevertheless, all manufactured bricks do not exceed 1% of shrinkage and far from the good quality requirement by previous researcher that should less than 8% (ASTM, 1998; Weng *et al,* 2003; Baskar *et al,* 2006).

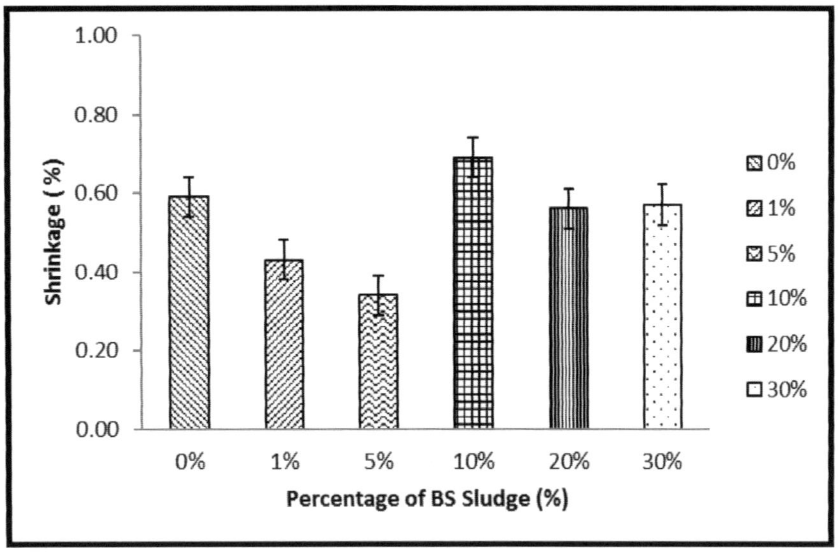

Figure 2. Shrinkage between different percentages of BS brick

Table 1. Total shrinkage of the manufactured bricks

	Control	BS				
Percentage of mosaic sludge (%)	0%	1%	5%	10%	20%	30%
Drying Shrinkage (%)	0.15	0.15	0.26	0.26	0.22	0.22
Firing Shrinkage (%)	0.69	0.46	0.31	0.56	0.52	0.67
Total Shrinkage (%)	0.59	0.43	0.34	0.69	0.56	0.57

Density. Table 2 and Figure 3 show the density of BS brick. BS brick (30%) shows the lowest density with 1665.24kg/m^3 and the value slightly increase from 20%, 10%, 5%, 1% and 0% with 1677.43kg/m^3, 1679.89kg/m^3, 1683.28kg/m^3, 1688.06kg/m^3 and 1698.24 kg/m^3 respectively. Furthermore, BS brick shows trend with increasing the percentages of sludge into fired clay brick was decreased the density value. In general, based on the other resources (Riza et al, 2010), density values are generally between 1500kg/m^3 to 2000kg/m^3. Therefore, the bricks are in the range and complied with common brick weight. According to Abdul Kadir et al. (2011), lighter bricks are feasible and the lightweight bricks will reduce the transportation and manufactured cost.

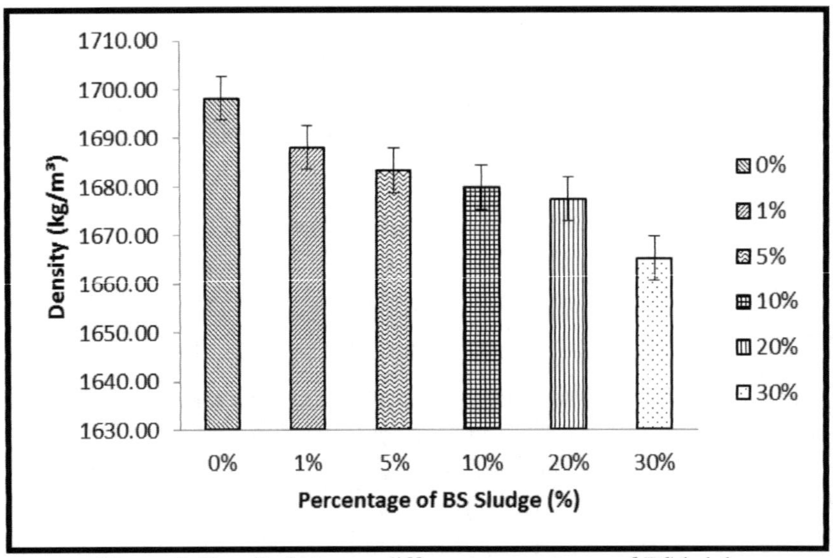

Figure 3. Density between different percentages of BS brick

Table 2. Density of the manufactured bricks

	Control	BS				
Percentage of mosaic sludge (%)	0%	1%	5%	10%	20%	30%
Density (kg/m^3)	1698.24	1688.06	1683.28	1679.89	1677.43	1665.24

Compressive Strength. Table 3 and Figure 4 show the result of the compressive strength of BS brick. The highest compressive strength obtained was BS brick (30%) with 26.58 N/mm^2 and followed by 20%, 10%, 5% and 1% BS brick with 26.17N/mm^2, 25.81N/mm^2, 24.80N/mm^2 and 17.91N/mm^2 correspondingly. The results determined control brick was the lowest among the other samples with 15.32N/mm^2. The reason for these was, clay soil was a more organic matter that burns and weak inter-particulate (Jahagirdar et al., 2013; Johari et al., 2010). Apparently, all results obtained and all the tested bricks are comply with the standard (BS 3921:1985) which is between 7N/mm^2 to 100N/mm^2 but not satisfied to be classified as engineering brick.

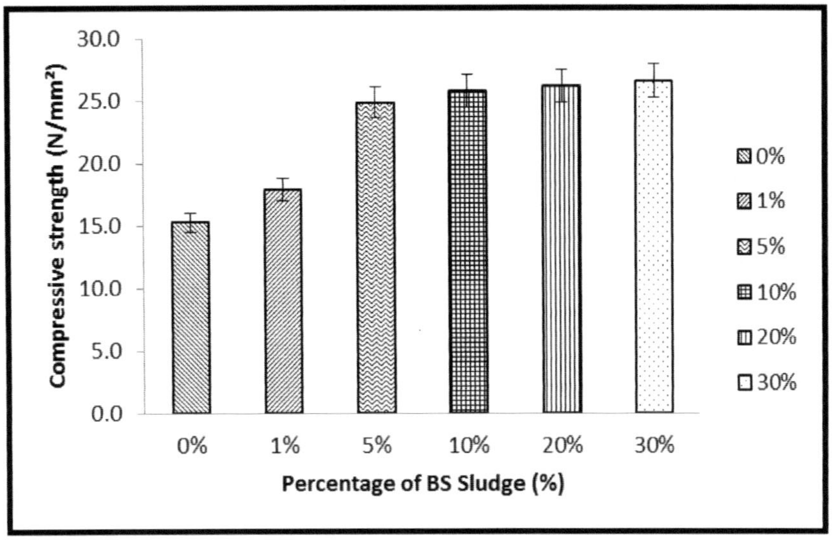

Figure 4. Compressive strength between different percentages of BS brick

Table 3. Compressive strength of the manufactured bricks

	Control	BS				
Percentage of mosaic sludge (%)	0%	1%	5%	10%	20%	30%
Compressive strength (N/mm^2)	15.32	17.91	24.80	25.81	26.17	26.58

Initial Rate of Suction. Based on Table 4 and Figure 5, its shows the amount of the initial rate of suction for BS brick slightly decreased from the 0% to 30% of sludge wastes. The highest was determined by BS brick (1%) with 12.65g/mm^2 and the lowest was BS brick (30%) with average 3.94g/mm^2. The results obtained for control brick shows the highest with 12.96g/mm^2. Based on the classification of brick in terms of the initial rate by BS 3921:1985. The water initial rate for control brick and BS brick for 1% to 10% was unsatisfied to be classified as damp-proof course brick but can be categories as all other bricks (no limits). Otherwise, for BS brick 20% and 30% can be classified as damp-proof course brick course because do not exceed more than 5g/mm^2.

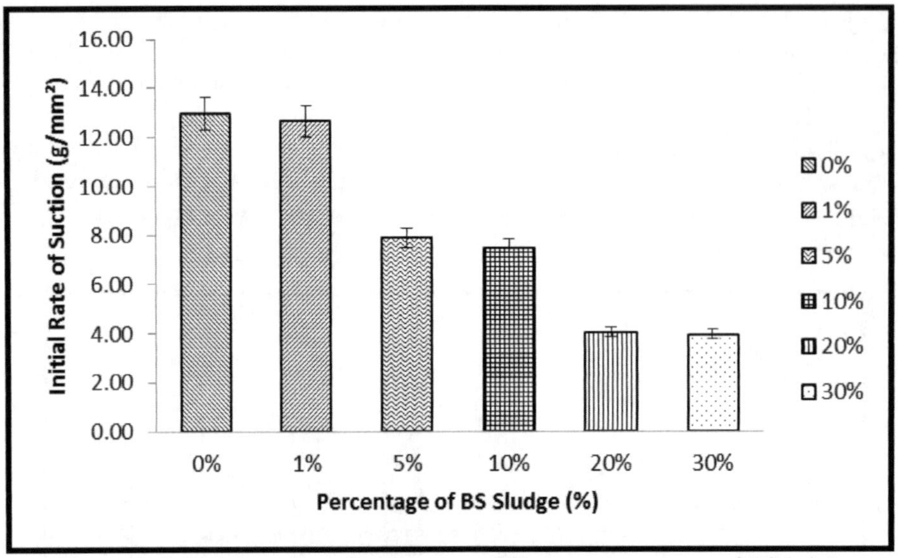

Figure 5. Initial rate of suction between different percentages of BS brick

Table 4. Initial rate of suction of the manufactured bricks

	Control	BS				
Percentage of mosaic sludge (%)	0%	1%	5%	10%	20%	30%
Initial rate of suction (g/mm^2)	12.96	12.65	7.90	7.47	4.03	3.94

Conclusion

As a conclusion, physical and mechanical properties incorporated with bodymill mosaic sludge waste were determined. The characteristic that was obtained by XRF shows the chemical composition of raw material of clay soil and mosaic sludge were high with silica dioxide (SiO_2) and aluminium oxide (Al_2O_3). Therefore, with the same characteristic, mosaic sludge is suitable to replace clay soil as a raw material. The properties of fired clay brick were significantly affected by different percentages of incorporated BS sludge. The results showed fully utilization of mosaic sludge brick obtained the highest compressive strength up to $25N/mm^2$ and low initial rate of suction under the limit of $5g/mm^2$. Increasing the sludge ratio in general will decreases the macropores develop and thus increases the compressive strength value of the samples. On the other hand, the brick with higher pore size will absorb more water compared to the brick with low porosity. SEM result shows the surface structure appearances of the brick incorporated with mosaic sludge are finer compared to control brick. This was due to the texture of mosaic sludge powder which is finer compared to clay soil and this enable the mosaic sludge to cover and fill the gaps between the clay soils in the mixture and decreases the IRS. Furthermore, it could be happen due to the cohesion between the soil and mosaic sludge that has been incorporated with suitable percentages, as the mixture can be mix properly and balance with the suitability on the bond. Nevertheless, all the mosaic sludge brick comply with the standard. It can be concluded that mosaic sludge is suitable an alternative raw material in brick manufacturing while providing a suitable waste disposed method.

References

C. S. L. Herek, C. E. Hori, M. H. M. Reis, N. D. Mora, C. R. G. Tavares, R. Bergamasco, Characterization of ceramic bricks incorporated with textile laundry sludge. *Ceramics International*. 38 (2012), pp.951-959.

S. Jahagirdar, S. Shrihari, B. Manu. Reuse of Textile Mill Sludge in Burnt Clay Bricks. *International Journal of Advanced Technology in Civil Engineering*. 2013. 2(1): 96-99.

P. N. L. Mary and M. S. Sreeja, Utilization of Sludge in Manufacturing Energy Efficient Bricks. *Journal of Mechanical and civil engineering (IOSR-JMCE)*. (2014). Vol. 11(4), pp. 70-73.

M. A. Swarna and R. Venkatakrishnaiah, Manufacturing of Bricks Using Tannery Effluent Sludge. *International Journal of Recent Development in Engineering and Technology*. (2014) Vol. 3 (4), pp. 33-36.

R. Baskar, S. B. K. M. Meera, S. Sundaram,. Characterization and Reuse of Textile Effluent Treatment Plant Waste Sludge In Clay Bricks, *Journal of the University of Chemical Technology and Metallurgy,* vol. 41(4), pp. 473-478, 2006.

A. N. Victoria,. Characterisation and Performance Evaluation of Water Works Sludge As Brick Material. *International Journal of Engineering and Applied Science*. 3(3) (2013), pp.69-79.

B. E. Hegazy, , A. F. Hanan, and M. H. Ahmed,. Brick Manufacturing from Water Treatment Sludge and Rise Husk Ash. *Australian Journal of Basic and Applied Sciences*. 2012. 6(3): 453-461.

Z. Saijun. "Experiment on Making Autoclaved Sludge Bricks with Water-Treatment Plant Sludge", *Scientific Research Fund of Human Provincial Education Department. IEEE*. 2011. 6568-6571.

M. O. Ramadan, A. F. Hanan, and M. H. Ahmed. Reuse of Water Treatment Plant Sludge in Brick Manufacturing. *Journal of Applied Sciences Research*. 2008. 4(10):1223-1229.

C. H. Weng, D. F. Lin, P. C. Chiang, Utilization of Sludge as Brick Materials. *Advances in Environment Research*. 7(2003), pp.769-685.

P. Sengupta, N. Saikia, P. C. Borthakur. Bricks from Petroleum Effluent treatmant plant Sludge: Properties and Environmental Characteristics. *Journal of Environmental Engineering*. 2002. 28(11): 1090-1094.

A. A. Kadir, A. S. A. Rahim, H. H. Jamil. Utilization of Mosaic Sludge Waste into Fired Clay Brick: Properties and Leachability. *Advanced Materials Research*. Vol 1025-1026. (2014) pp 117-121.

R. S. Lokhande, P. U. Singare, D. S. Pimple. Toxicity Study of Heavy Metals Pollutants in Waste Water Effluent Samples Collected from Taloja Industrial Estate of Mumbai, India. *Resources and Environment*. Vol 1, No 1.(2011). pp 13-19.

K. Zhoa, W. Fu, Z. ye, C. Zhang. Contamination and Spatial Variation of Heavy Metals in the Soil-Rice System in Nanxun Country, Southeastern China. *International Journal of Environmental Research and Public Health*. Vol 12. (2015). pp 1577-1594.

A. S. Abdul Rahim & A. A. Kadir. Physical and Mechanical Properties of Fired Clay brick Incorporating with Mosaic Sludge Waste. *Materials Science Forum*. Vol 803. (2015). pp. 203-208.

A. B Aziz Abdul Latif, A. Tarmizi Abd.Karim, A. Shukri Ahmad, M. Baharudin Ridzuan and Y.-Tse Hung, Phytoremediation of Metals in Industrial Sludge by Cyperus Kyllingia-Rasiga, Asystassia Intrusa and Scindapsus Pictus Var Argyaeus Plant Species. *International Journal of Integrated Engineering*. 4(2) (2012), pp. 1-8.

K. Swierk, A. Brielicka, I. Bojanowska, Z. mackiewicz. Investigation of heavy Metals Leaching from Industrial Wasterwater Sludge. *Polish J. of Environ*. Stud. Vol. 16, No. 3 (2007). Pp447-451.

L. Loretta Jones (2008). Chemistry Explained, Foundations and Application. Retrieved on 9 February 2015, Information on http://www.chemistryexplained.com/Bo-Ce/Ceramics.html

Edmonds Molly (2008). How is ceramic Tile Made?. Retrieved on 14 April, 2013 Information on http://home.howstuffworks.com/home-improvement/construction/materials/how-is-ceramic-tile-made.htm.

A. M. Segadaes, M. A. Carvalho , W. Acchar, Using Marble and granite rejects to enhance the processing of clay product. *Applied Clay Science*. 30(1) (2005), pp.42-52.

J. vallette (2014). Pharos, made in the USA: A healthy Choice for Ceramic Tiles. Retrieved on 27 may 2015, Information on https://www.pharosproject.net/blog/show/184/ceramic-migration.

A. Agrawal, M. I. M. Loya, A. M. Rawani. An Empirical Study on Influence of Quality on Adoption of Fly Ash Bricks. *International proceedings of Economics Development and Research (IPEDR)*. vol. 75(8). pp 33-35.

A. A. Kadir, H. H. Jamil, A. S. Abdul Rahim. Leachability of Fired Clay Brick Incorporating with Sludge Waste from Mosaic Industry. *Materials Science Forum*. Vol. 803. (2015). pp. 233-238.

A. Abdul Kadir & A. Mohajerani, Recycling cigarette butts in lightweight fired clay bricks *Proceedings of Institution of Civil Engineers: Construction Materials*. 164(5) (2011), pp. 219-229.

A. Abdul Kadir, N. A. Sarani, & A. M. Leman, Testing on Building Material Using Waste Material in Fired Clay Brick. *Material Science Forum*, Vol. 803 (2015), pp. 330-336.

A. Abdul Kadir & A. Mohajerani, Effect of Heating Rate on Gas Emission and Properties of Fired Clay Bricks and Fired Clay Bricks Incorporated with Cigarette Butts. *Applied Clay Science*, Vol. 104 (2015), pp. 269-276.

G. Cultrone, E. Sebastian, K. Elert, M. J. D. L. Torre, O. Cazalla, C. R. Navarro. Influence of Mineralogy and Firing Temperature on Porosity of Bricks. *Journal of the European Ceramic Society*. (2004). Vol 24. pp 547-564.

R. G. Babu & V. N. Ramana. Durability of Bricks Cast with Industrial Sludge. *IOSR Journal of Machanical and Civil Engineering (IOSR-JMCE*. 2013. 6(4), 43-46.

D. M. P.Ingunza, , C. L. Anazsandra. Duarte and Nascimento R. M., "Use of Sewage Sludge as Raw Material in the Manufacture of Soft-Mud Bricks", *Journal of Materials in Civil Engineering.* 2011. 852-856.

D. Koratich (2009). Dry shrinkage. Retrieved on 20 November, 2013 information on http://www.engr.psu.edu/ce/courses/ce584/concrete/library/cracking/dryshrinkage/dryingshrinkage.html

M. A. Rouf & M. D. Hossain. Effects of using arsenic iron sludge in brick making, Fate of Arsenic in the Environment, *Proceedings of the BUET-UNU International Symposium*, 5–6 February, Dhaka, Bangladesh, 2003, 193-208.

ASTM (American Society of Testing Materials), ASTM C67 Standard Test Method for Sampling and Test Brick and Structural Clay Tile, Annual Book of ASTM Standards, Section 4, Construction, 04. 08, 04. 09, Soil and rock (I) and (II), ASTM, West Conshohocken, Pa, USA, 1998.

F. V. Riza, I. A Rahman,. A. A. M. Zaidi. A Brief Review of Compressed Stabilized Earth Brick (CSEB). *International Conference on Science and Social Research (CSSR 2010)*, 2010. pp 1011- 1016.

A. Abdul Kadir & A. Mohajerani, Properties improvement of fired clay bricks incorporating with cigarette butts. *Advanced Materials Research*, vol 535-537 (2012) , pp.1723-1730.

I. Johari, S. Said, B. Hisham, A. Bakar, Z. A. Ahmad. Effect of the Change of Firing Temperature on Microstructure and Physical Properties of Clay Bricks from Beruas (Malaysia). *Science of Sintering*, 2010. vol 42. pp 245-254.

British Standard Institution. *Specification for Clay Bricks.* London, BS3921. 1985.

Leakage current measurement of protective equipment insulating materials used in electrical installations

G Buică, R. Dobra [2*], D. Păsculescu [3], A. Tătar[4]

[1] Department of Explosion-Proofnes, National Research and Dev. Institute for Labor Protection "Al. Darabont", G-ral Vasile Milea str., no. 32-34, Bucharest ROMANIA
[2] Department of Computers and Power Engineering, University of Petrosani, 20 Universitatii str, Petrosani, România
[3] Department of Computers and Power Engineering, University of Petrosani, 20 Universitatii str, Petrosani, România
[4]Department of Energy, Environment and Sustainable Development, Faculty of Engineering and Sustainable Development, University "Constantin Brâncuşi" of Tg-Jiu, România

E-mail: remusdobra@upet.ro

Abstract. This research describes the behaviour of equipment and safety devices during use in extreme environmental conditions, in order to establish the technical conditions and additional health and safety requirements during operation, to ensure the health and safety of users, regardless of conditions and working environment in which they are use. The studies have been conducted both on new equipment and means of protection used in electrical installations. There has been evaluated protective equipment made of insulating rubber, reinforced fiberglass or PVC. They have been followed the technical characteristics and protection against electric shock by measuring the leakage current of different insulating materials.

1. Introduction

Following the studies in the power system regarding the risk assessment in transport, distribution and use of electricity, it was observed that a significant proportion of the overall risk factors generated by work equipment (means of production) is given by the equipment and means of protection. If this risk factor is detailed, it can be found that there is a large percentage of equipment and means of protection in use that are not checked regularly, and technically damaged and some of the workers do not know how to use them, including the technical characteristics and environmental conditions in which they can be used. A significant proportion is occupied by the protective equipment and means of protection which have not been diagnosed or assessed and certified from technical point of view to meet the essential safety and health requirements, in accordance with national and European legislation in force.

The total leakage current is obtained by adding the partial currents, namely, multiply leakage current circuit corresponds to the number of circuits and leakage currents are added intensity, then because of connecting motors and appliances. In figure 1 is presented a diagram regarding the weight of risk factors identified in the use of equipment and protective devices.

* University of Petrosani, remusdobra@upet.ro

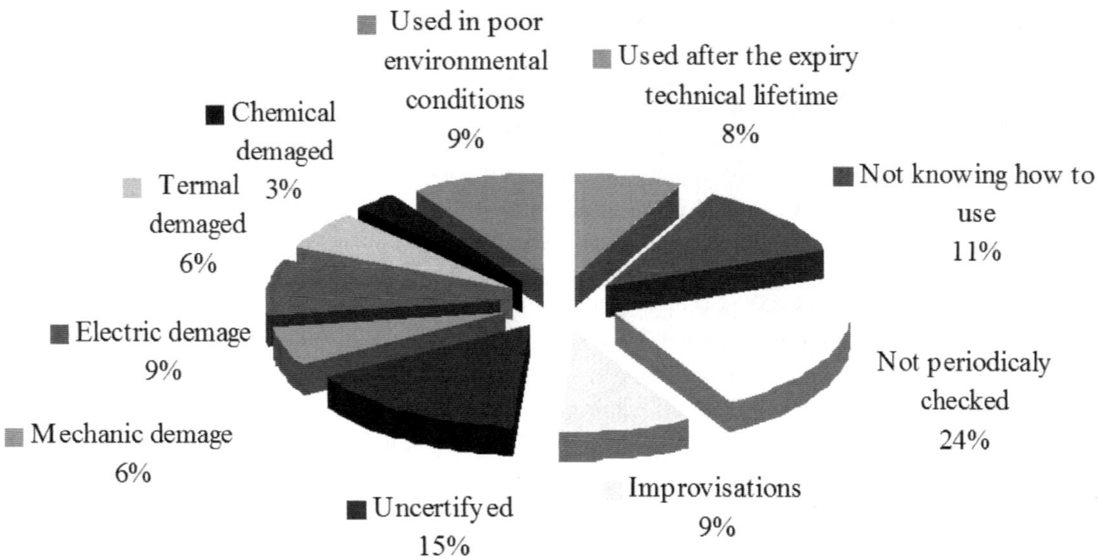

Figure 1. Weight of risk factors identified in the use of equipment and protective devices.

2. Safety requirements for equipment and means of protection used in electrical installations

Protective equipment and means of protection, used in electrical installations, are designed to protect workers and unskilled people in work areas, and near electrical installations, by the dangers of electricity.

In this regard, to ensure the safety and health of workers, new or in use protective equipment and means of protection must be evaluated, in terms of compliance with the minimum safety and health requirements, before being put on the market:

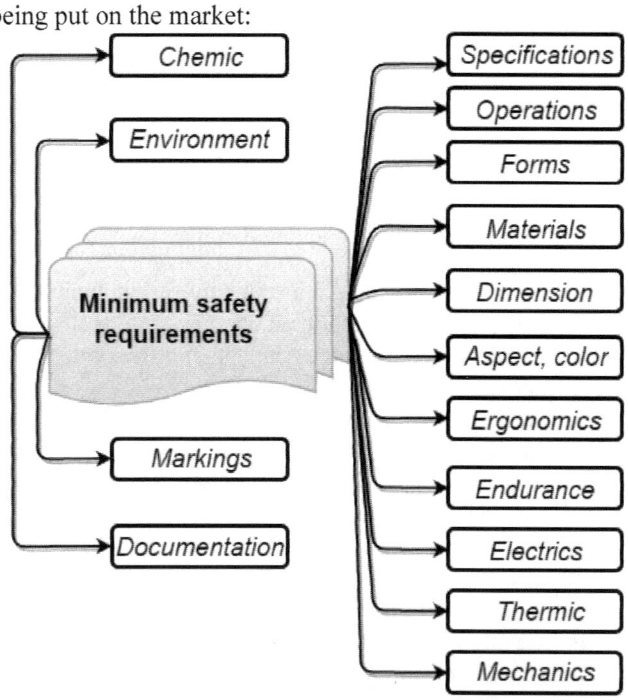

Figure 2. Minimum safety requirements for equipment and means of protection.

Consequently they must ensure the safety function for which they were made, so that failure or damage during use or malfunction does not endanger exposed persons. Protective equipment and means of protection used in electrical installations have to: (GD 457, 2004), (GD no. 1146, 2006)

Protect against dangers of electricity, so that:

- People should be protected from the danger of injury that can be caused by electric shock;
- Insulation of protective equipment and means of protection are appropriate for the provided conditions (e.g. rated voltage, work environment).
- Do not produce temperatures, arcs, radiation or fire;
- The construction of the protective equipment and means of protection is ergonomic and appropriate for working environment;

Provide safety against the dangers arising from external influence, so that:

- to meet requirements for physical and chemical conditions;
- to provide immunity to the effects resulting from environmental conditions;
- to provide immunity to electromagnetic interference and do not disrupt the operation of other electrical and electronic equipment.

Meeting these minimum requirements, are the main criteria for assessing the quality of safety of equipment and means of protection whether they are new or already in use.

2.1. Study on protective equipment and means of protection that are in use in electrical installations

In electrical installations there are possible many defects, complex as progress and the effects that they can produce. From the multitude of defects that can occur in electrical installation, it is of interest, for work equipment, equipment and means of protection, in terms of accidental voltage, the failures consisting in damage, breakthrough and creepage of insulation.

In this sense, the studies have been conducted both on new equipment and means of protection and in use in electrical installations. There has been evaluated protective equipment made of insulating rubber, reinforced fiber glass or PVC. They have been followed the technical characteristics and protection against electric shock.

There have been evaluated new means of protection which have been immersed in water, conditioned at extreme temperatures and oil. It was followed for three years the behavior of the technical and protection characteristics of equipment and means of protection for use in wet environments, environments with extreme temperatures, and environments with dust, static electricity and oil.

2.2. Results and discussions

From the tests conducted on various rubber insulating protective equipment and means of protection, it has been found, after verifying dielectric strength after conditioning, that the colour has an influence over insulation strength. It is found that for the same product with the same rubber recipe to which was added different color, the leakage current is different.

In figure 3, are presented the leakage current values of a group of 30 insulated mats having blue, black and red colour. Test voltage was 20 kV for Class 1 electrical insulating mats. The electrical insulating mats were tested in normal environment (ASRO, SR EN 61111:2010).

In order to validate this conclusion, it was tested and other protective equipment and means of protection, made of insulated rubber. A batch of insulated platforms of different colours (brown, grey, dark-grey, black), from the same producer, made of the same rubber, was evaluated and tested in normal dielectric environment, and conditioned by immersion in water and extreme temperatures.

In figures No. 4 ÷ 8 is presented the leakage current value for a group of 12 rubber insulated platforms, having brown, light grey, dark-grey and black colour. Test voltage was 40 kV. There is no national or European standard applicable for insulated platforms.

In figure 3, are presented the leakage current values of a group of 30 insulated mats having blue, black and red colour.

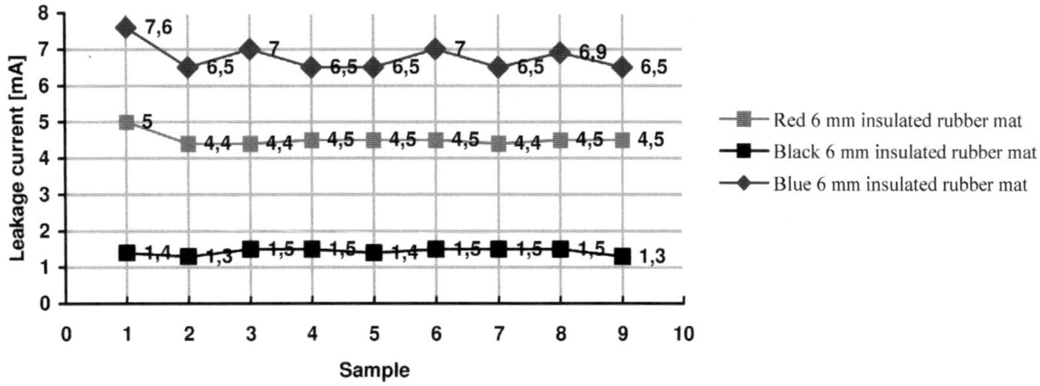

Figure 3. Leakage current – Electrical insulating matting.

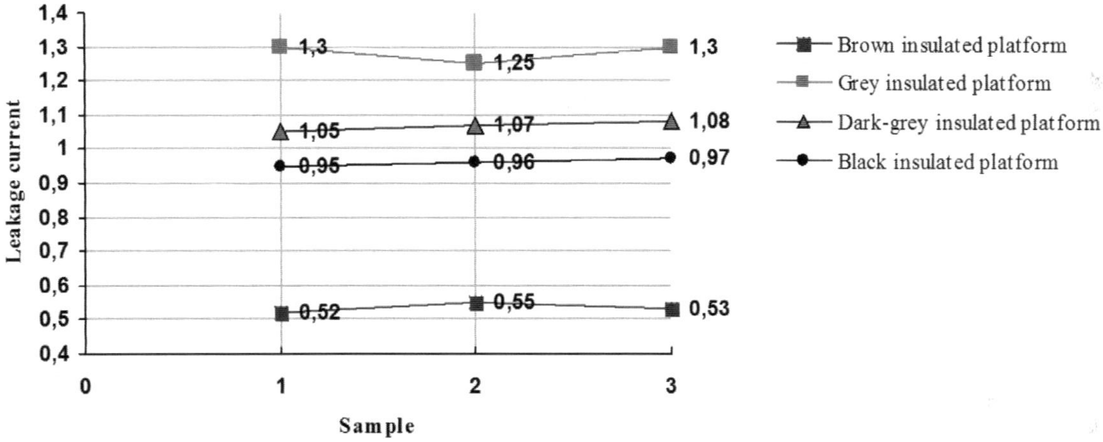

Figure 4. Leakage-current values - in normal environmental conditions (20 ÷23^0 C).

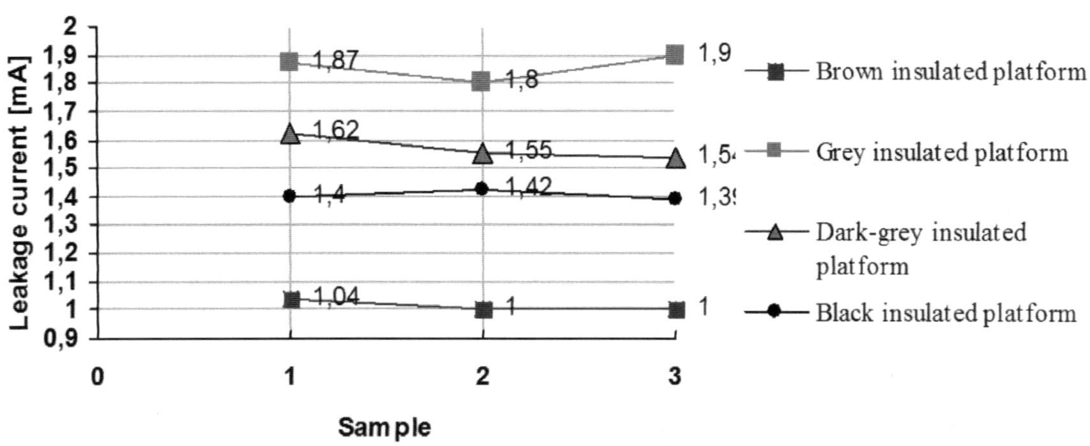

Figure 5. Leakage-current values – after immersion in water.

To establish the test voltage of 40 kV, it was taken into consideration the technical norm 3RE-I2-83 - "Instructions for electrical testing of the means of labour protection", "Test procedures on insulated

IOP Conf. Series: Materials Science and Engineering **133** (2016) 012035 doi:10.1088/1757-899X/133/1/012035

platforms / culverts" developed after following studies and research by the Electric Risks Laboratory through ICSPM-CS Certification Body within INCDPM "Al Darabont" institute and the Instructions issued by the transmission and distribution electricity facilities.

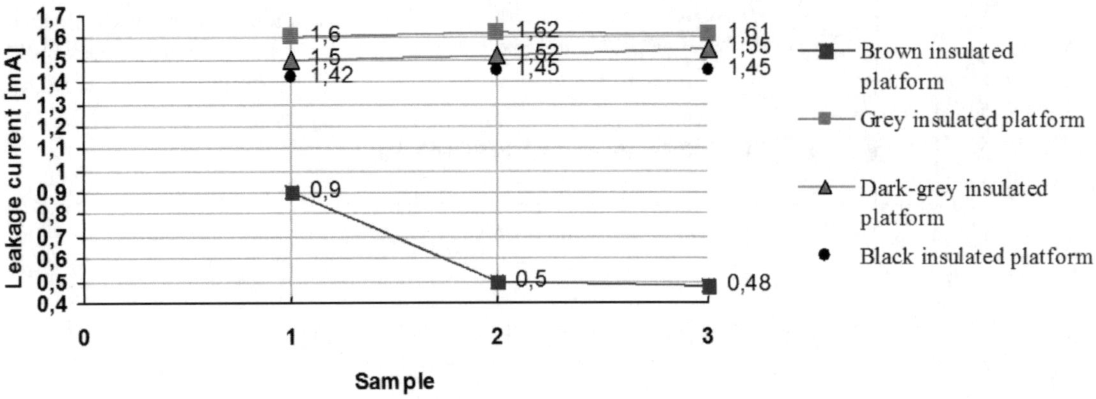

Figure 6. Leakage current values –low temperature conditioning (-30^{0}C, 24 hours).

In concordance with the legal provisions on ensuring safety and health at work have established safety requirements for the following equipment and means of protection: insulated rods, carpets electrical, couplers fixed / mobile binding phase devices used in electrical installations with nominal voltage up to 1000 V ac 1500 V D.C.

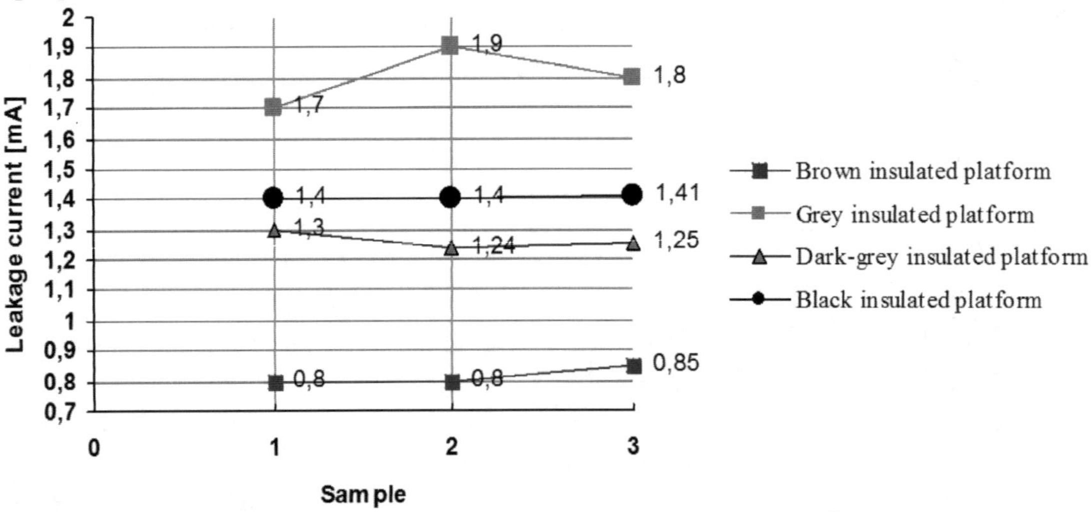

Figure 7. Leakage current values – high temperature conditioning (+60^{0}C, 24 hours).

The results have led to same conclusion that for the same product with the same rubber recipe for which was added different colour, the leakage current is different.

In this sense the producers of protective equipment and means of protection made from insulating rubber must take into consideration that the products on the market must be identical with the product submitted to the certification body for the certification procedure, respectively the safety assessment. Brown insulated platforms that have met health and safety requirements, have been certified and have been given in use for three years. It was followed during this time, technical and electrical insulation characteristics. Also it was monitored the protective behaviour in humid environments with extreme temperatures, with static electricity and steam edible oil, dust and lubricants for machines.

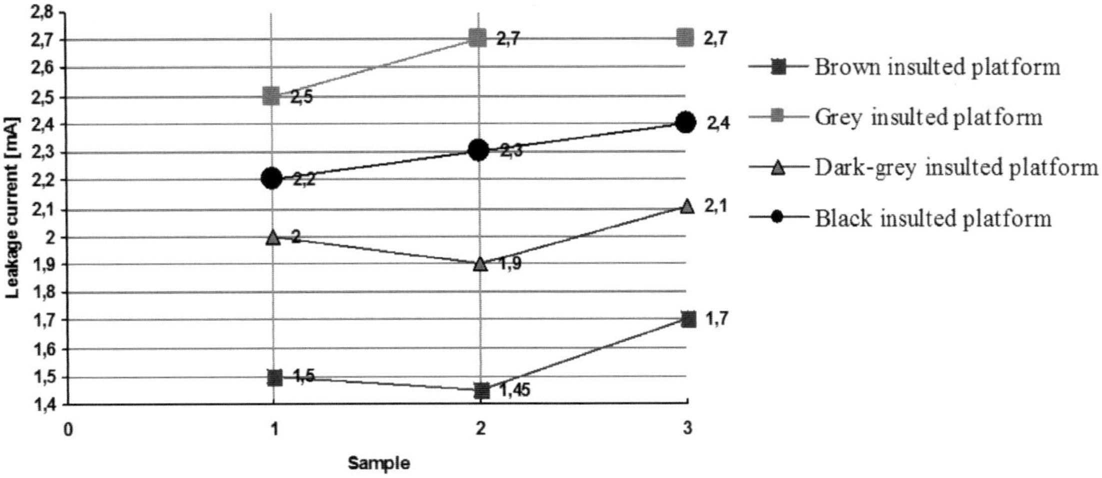

Figure 8. Leakage current values after conditioning in oil.

They must fully respect the characteristics (technical, functional, constructive) declared, including the colour, the material or the recipe. The results of the electrical characteristics respectively leakage current are shown in figure 9.

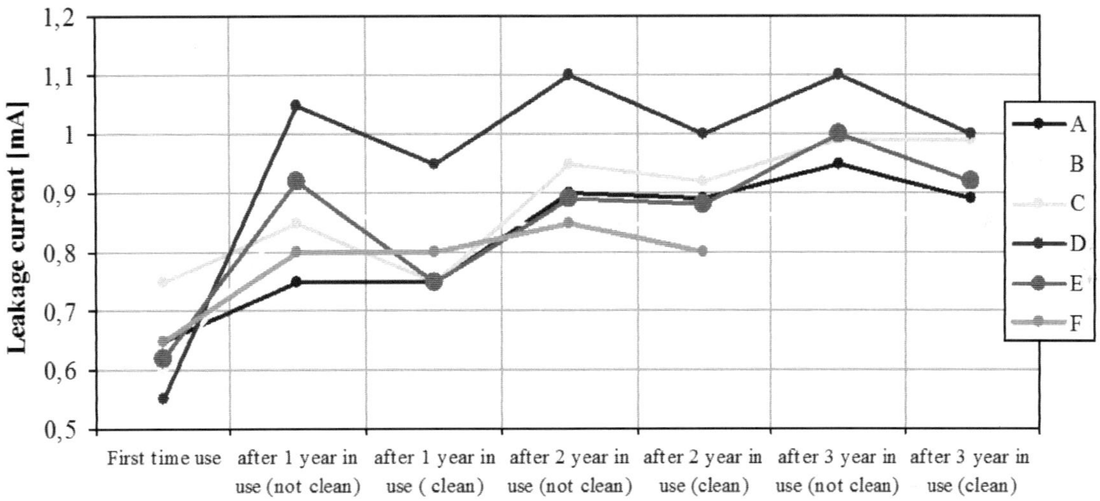

Figure 9. Leakage current values – insulated platforms.

The insulated platforms have been installed in front or behind the electrical board and electrical panels in production facilities. At the first time use and annually they were tested. The test method used was the "plate" method with the test voltage of 40 kV.

It was found that in time insulated material loses the electrical insulation properties due to ageing, coming into contact with oily solutions, with dust and so on, by the increasing leakage current, unable to keep the equipment clean.

Evaluating the results, it is recommended that the protective equipment and means of protection to be periodically tested for insulation characteristics, even if the producer or distributor do not specify in the technical documentation, and before every use, each worker should check visually the equipment for physical condition, including cleaning and if there are no faults (missing parts /components, cracks, nicks, burrs, foreign inclusions, etc.).

National legislation provides that electrical protective equipment and means of protection to be tested in accordance with the regulations and technical rules applicable before use, at the first time in use, after repairs or modifications and then periodically (in use) (GD no. 1146, 2006), (GD no 1091, 2006), but health and safety legislation in force does not specify which equipment and tools are subject to be periodically tested and their periodicity.

Periodically test of the protective equipment and means of protection used in electrical installations is a safety requirement. It must be required by the producers, meaning that it must be included in the technical specifications or in the instructions for use and to be imposed to users.

Health and safety policy at work with the integrated with environmental protection policy made by the management at the highest level and appropriated by all employees must include objectives concerning the protective equipment and means of protection in use in electrical installations.

The results of periodic safety tests and visual examinations are important information in risk management in an efficient enterprise for a better selection of protective equipment for the most appropriate control measures. Besides, management of risks generated by the protective equipment and means of protection in use, allow targeting the studies and research to develop better assessment procedures, diagnosis and testing.

3. Conclusions

When running the protective equipment suffers degradations and damage that alter the technical and safety characteristics and in this sense it should be periodically tested. Regular safety verifications guarantee efficiency of maintenance applied to protective equipment in use. Safety and health policy at work integrated with the environmental protection policy made by management at the highest level must include objectives concerning the protective equipment and means of protection. The results of periodic safety tests and examinations are important information in risk management in an efficient enterprise both for a better selection of protective equipment and for improving OSH measures.

For certain categories of equipment and protection devices there are not covered conditions related to design and production and there are no specific requirements established in the national or European standards, so in this papers was presented some technical criteria about security and methods evaluation.

References

[1] G. Buică, 2010 *Contributions of diagnostic methods of safty and health at work in electrical installations*, PhD Thesis, Universitas, Petroşani, pp. 70 – 149.

[2] S. Kurihara, et al., 2003 *Construction of remote monitoring system for separative measurement of leakage current of outdoor insulators*, Properties and Applications of Dielectric Materials,. Proceedings of the 7th International Conference on. Vol. 1. IEEE.

[3] K. Alveera, A. Mohammad, J. Shirish, & V. Lyashenko, 2015 *Elixir Crystal*, 87, pp. 35801-35805.

[4] S. Pece, 2010 *Risk assessment in work system*, Rubin, Bucharest, pp. 170-226.

[5] D. C. Petrilean, et al., 2014*Environmental Engineering and Management Journal*, pp. 1383-1388.

[6] SR EN 61111:2010 – Live working - Electrical insulating matting.

[7] SR ISO 1817:2000- Rubber, vulcanized. Determination of the effect of liquids.

[8] SR EN 61477:2009 - Live working - Minimum requirements for the utilization of tools, devices and equipment.

[9] 3RE-I2-83 (1995)- Technical instruction- Instructions for electrical testing of the means of labor protection.

IOP Conf. Series: Materials Science and Engineering **133** (2016) 012036 doi:10.1088/1757-899X/133/1/012036

Investigation on Leaching Behaviour of Fly Ash and Bottom Ash Replacement in Self-Compacting Concrete

Aeslina Abdul Kadir[1,2], Mohd Ikhmal Haqeem Hassan[1] & Mohd Mustafa Al Bakri Abdullah[3,2]

[1]Faculty of Civil and Environmental Engineering, Universiti Tun Hussein Onn Malaysia, 86400 Parit Raja, Batu Pahat Johor, Malaysia
[2]Center of Excellence Geopolymer & Green Technology (CeGeoGTech), School of Material Engineering, Universiti Malaysia Perlis (UniMAP), Perlis Malaysia
[3]Faculty of Engineering Technology, Universiti Malaysia Perlis (UniMAP), Perlis, Malaysia

Email: aeslina@uthm.edu.my, mohd.ikhmal.haqeem@gmail.com & mustafa_albakri@unimap.edu.my

Abstract. Fly ash and bottom ash are some of the waste generated by coal-fired power plants, which contains large quantities of toxic and heavy metals. In recent years, many researchers have been interested in studying on the properties of self-compacting concrete incorporated with fly ash and bottom ash but there was very limited research from the combination of fly ash and bottom ash towards the environmental needs. Therefore, this research was focused on investigating the leachability of heavy metals of SCC incorporated with fly ash and bottom ash by using Toxicity Characteristic Leaching Procedure, Synthetic Precipitation Leaching Procedure and Static Leaching Test. The samples obtained from the coal-fired power plant located at Peninsula, Malaysia. In this study, the potential heavy metals leached out from SCC that is produced with fly ash as a replacement for Ordinary Portland Cement and bottom ash as a substitute for sand with the ratios from 10% to 30% respectively were designated and cast. There are eight heavy metals of concern such as As, Cr, Pb, Zn, Cu, Ni, Mn and Fe. The results indicated that most of the heavy metals leached below the permissible limits from the United States Environmental Protection Agency and World Health Organization limit for drinking water. As a conclusion, the minimum leaching of the heavy metals from the incorporation of fly ash and bottom ash in self-compacting concrete was found in 20% of fly ash and 20% of bottom ash replacement. The results also indicate that this incorporation could minimize the potential of environmental problems.

1. Introduction

Malaysia is one of the thriving economies and a well-developing country in the world. Because of that, economic growth has contributed to the increase in the demand for electricity primarily that was produced by coal-fired plants. It will lead to the increment in coal utilization by the power plants thus will generate waste such as fly ash (FA) and bottom ash (BA) to landfills [1,2]. These wastes needed to manage appropriately and disposed of without causing any harmful environmental effects. There are about 10% total weights of the coal burned produces ash [3] with very high environmental risk. FA can be recognized as a fine grey powder consisting mostly of spherical in shape, and their surfaces appeared to be smooth [4]. On the other hand, BA is coarse, with grain sizes spanning from fine sand to fine gravel ranging between 10 mm to 0.075mm and typically grey to black in color [5, 6]. At present, the utilization of FA and BA from power stations includes in road construction, brick making and the production of cement and concrete. Fly ash is the famous type of cement replacement material to replace a certain amount of cement content in the concrete and increased the workability properties of SCC mix. The cementitious compound in FA may produce when siliceous material reacts with calcium hydroxide in the presence of water at ordinary temperature. Meanwhile, many investigations found that the BA has some cementitious properties in which may increase the strength and long-term than concrete with natural sand [7, 8].Even though the utilization of FA and BA have been practiced in worldwide with positive results, but weathering and erosion over time will ultimately cause the heavy metal contents have a tendency to leached out, and a diversely would affect human and environmental health [9]. Therefore, this research was conducted to determine the leachability of SCC incorporated with fly ash and bottom ash by using Toxicity Characteristic Leaching Procedure (TCLP), Synthetic Precipitation Leaching Procedure (SPLP) and Static Leaching Test (SLT).

2. Materials and Methods

2.1. Materials

The fly ash and bottom ash were obtained from a power plant located at Peninsula, Malaysia. The material characterization of FA, BA and OPC were carried out through X-ray fluorescence (XRF) test, performed using Bruker AXS S4 Pioneer. Fly ash and bottom ash samples used for the test were prepared in the forms of pellets, abiding a sample to a wax ratio of 8:2 using Pressed Pellet Technique. The chemical compositions of the materials are shown in Table 1. A total of 26 chemical compounds or elements were detected in FA, BA, and OPC. By considering the constraint of laboratory apparatus, only selected parameter were considered to be tested by using Atomic Absorption Spectrometry (AAS). Henceforth, the analysis of the leachability of SCC incorporated with FA and BA is narrowed down to these few specific heavy metals. The heavy metals selected are As, Cr, Pb, Zn, Cu, Ni, Mn, and Fe.

Table 1: Heavy metals of FA, BA, and OPC

Heavy metals	**Formula**	**Concentration (mg/L)**		
		FA	***BA***	***OPC***
Chromium	Cr	217	145	69
Manganese Oxide	MnO	900	700	700
Iron (III) Oxide	Fe_2O_3	38800	48700	29600
Nickel	Ni	99	74	17
Copper	Cu	94	42	20
Zinc	Zn	50	30	132
Arsenic	As	35	12	33
Lead	Pb	54	14	41

2.1.1. Mix Proportions of SCC Incorporated with fly ash and bottom ash

The mix design of SCC incorporating with FA and BA are presented. Ten mixtures were designed for control sample with 0% FA and BA and mixture with various percentages of FA and BA (10%, 20%, 30% of FA and BA). The different percentages of FA and BA designed by considering in minimizing heavy metals leachability that can leach out from leaching testing under the regulatory level. Total binder for all the samples was between 530 kg/m^3 to 550 kg/m^3 respectively. Fly ash was replacing the cement in the percentages of 10%, 20% and 30% meanwhile bottom ash was replacing the sand with the same percentages.The details of the mix design for this research as tabulated in Table 2. Note that the proportion is for one batch casting (0.0475 m^3).

Table 2: Mix proportions off SCC mixes incorporated with FA and BA

Mix Design	Total Binder (kg/m^3)	Cement (kg/m^3)	FA (kg/m^3)	CA (kg/m^3)	Sand (kg/m^3)	BA (kg/m^3)	Water (kg/m^3)	SP (kg/m^3)
FA0BA0	550	550	0	593	914	0	228	5
FA10BA0	550	495	55	593	914	0	220	4
FA20BA0	540	432	108	593	914	0	216	4
FA30BA0	530	371	159	593	914	0	212	4
FA0BA10	550	550	0	593	822	67	228	5
FA0BA20	550	550	0	593	731	133	228	5
FA0BA30	550	550	0	593	640	200	228	5
FA10BA10	550	495	55	593	822	67	220	4
FA20BA20	540	432	108	593	731	133	216	4
FA30BA30	530	371	159	593	640	200	212	4

*CA: Coarse aggregate; SP: super plasticizer

2.2. Methods

2.2.1. Toxicity Characteristic Leaching Procedure (TCLP)
The Toxicity Characteristic Leaching Procedure (TCLP) was adopted by the USEPA in 1990 to replace the Extraction Procedure (EP) as the USEPA regulatory method for classifying wastes as hazardous based on toxicity [10]. This method is designed to determine the mobility of both organic and inorganic contaminants present in liquid, solid and multiphase wastes. In this research, the cube concrete sample was crushed into four portions finer than 9.5 mm for the analysis. The extraction fluid used for the extraction depends on the alkalinity of the waste material. Very alkaline waste materials are leached with a fixed amount which of glacial acetic acid (CH_3CH_2OOH) with distilled water while other waste materials are leached with glacial acetic acid, distilled water and sodium hydroxide (NaOH). There were two types of extraction fluid as described in TCLP namely Extraction Fluid No.1 and Extraction Fluid No.2. After the determination of the extraction fluid has been done, the samples were prepared in screw-capped polyethylene bottles which were filled with crushed concrete samples and leaching fluid at the ratio of 1:20.The bottles were agitated at 30 rpm for 18 hours in an end-over-end manner (Figure 1). The leachate collected was then filtered using 0.7 μm glass fiber filters, preserved the sample before being analyzed using Atomic Absorption Spectrometry to determine the dissolved metals.

2.2.2. Synthetic Precipitation Leaching Procedure (SPLP)
The Synthetic Precipitation Leaching Procedure (SPLP) is an agitated extraction that is used to provide information on the mobility (leachability) of organic and inorganic constituents from liquids, soils, and wastes [11]. This procedure is similar to the TCLP but instead of the acetic acid mixture used with the TCLP to simulate landfill leachate, nitric and sulfuric acids are utilized to simulate the acid rain resulting from airborne nitric and sulfuric oxides. For sites east of the Mississippi River, the leaching fluid is a solution of 60% of sulfuric acid and 40% of nitric acid in distilled water until the pH 4.2. For sites west of the Mississippi River, a solution of sulfuric acid and nitric acid in distilled water, pH 5.0, is used as the leaching fluid. If the sample is a waste or wastewater, the extraction fluid employed is the pH 4.2 solution. For this research, extraction fluids with pH 4.2 solution were used. The next procedure is same with TCLP that has been described earlier.

Figure 1: Rotary agitation apparatus for TCLP and SPLP

2.2.3. Static Leaching Test (SLT)

Another leaching testing used in this research was Static Leaching Test (SLT). A modified and simplified method from American Nuclear Society for solidified a low-level radioactive waste which is conducted for long term duration was considered in this research [12]. Static leaching methods are used to evaluate the release of an element from a material that typically exists as a massive solid and are frequently used to characterize the release of pollutants from stabilized waste materials [13]. The release element is a function of the exposed surface area as opposed to the mass. In SLT, a regular particle geometry and known surface area are immersed in a volume of leaching solution. The same leaching solution is sampled with no renewal of the solution to ensure leachate concentration is not interrupted by leachate renewal process. Also, the main objective of this test is to study long-term leaching behavior and to determine optimum quantity of heavy metals that leached out from FA and BA. In this test, ten whole solid concrete samples, containing 0% (control), 10%, 20%, 30% FA and BA was prepared. The whole concrete samples were suspended in a closed container for up to 200 days. The static leaching test was conducted as shown in Figure 2.

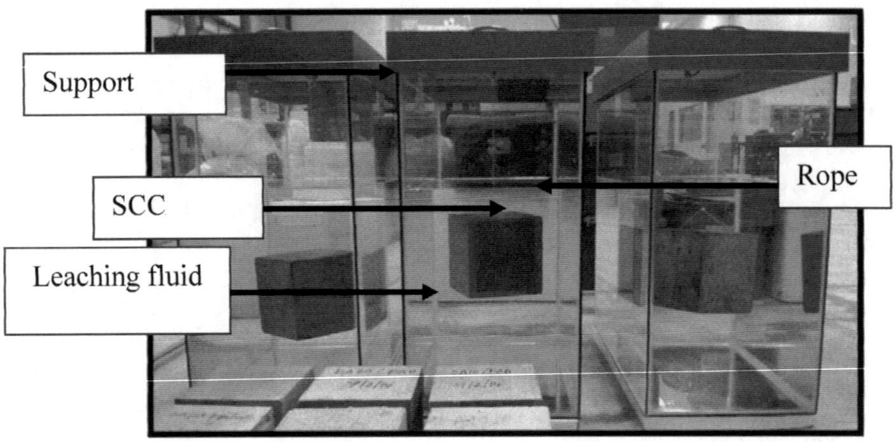

Figure 2: Static Leaching Test (SLT) sample

3. Result and Discussion

3.1. Toxicity Characteristic Leaching Procedure (TCLP) result

From the results of the TCLP test in Table 3, it shows that there are insignificant levels of heavy metals leached out from the samples. The results were then compared to United States Environmental

Protection Agency regulatory levels. Explanations of the results were discussed further according to Figure 3a, 3b, and 3c.

Table 3: Leachability of heavy metals in SCC samples in TCLP

Heavy metals	Concentration level (mg/L)*	FA0BA0	FA10BA0	FA20BA0	FA30BA0	FA0BA10	FA0BA20	FA0BA30	FA10BA10	FA20BA20	FA30BA30
		Concentration (mg/L)									
As	5	12.060	14.705	14.895	15.370	13.640	13.675	14.640	16.510	18.200	18.576
Cr	5	0.112	0.128	0.132	0.140	0.143	0.154	0.170	0.174	0.184	0.207
Pb	5	0.447	0.783	0.847	0.909	0.578	0.646	0.695	0.940	1.024	1.028
Zn	NA	0.173	0.145	0.202	0.222	0.145	0.144	0.160	0.172	0.177	0.181
Cu	100	0.026	0.030	0.030	0.031	0.029	0.026	0.031	0.033	0.034	0.035
Ni	NA	0.167	0.065	0.066	0.069	0.065	0.071	0.075	0.062	0.063	0.077
Fe	NA	0.095	0.074	0.083	0.083	0.053	0.034	0.017	0.136	0.037	0.000
Mn	NA	ND	ND	ND	ND	ND	ND	ND	ND	ND	ND

Where,
As = Arsenic Pb = Lead Cr = Chromium Zn = Zinc Cu = Copper Ni = Nickel
Fe = Ferum Mn = Manganese

* US Environmental Protection Agency, USEPA (1996); NA – Not Available; ND – Not Detected

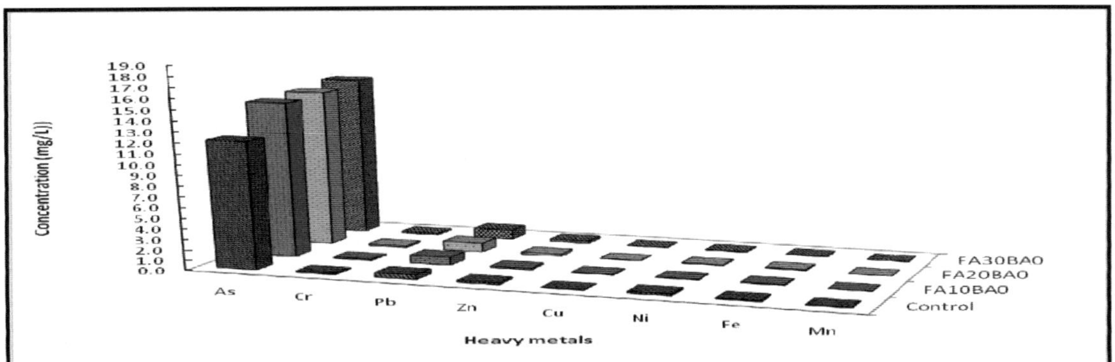

Figure3a: Comparison of heavy metal concentrations for control sample (FA0BA0) with samples incorporated with FA (FA10BA0, FA20BA0, and FA30BA0) using TCLP

According to Figure 3a, the results showed the comparison of heavy metals concentrations for control sample which is FA0BA0 sample with samples incorporated with different percentages of FA. The As concentrations in FA0BA0, FA10BA0, FA20BA0 and FA30BA0 (12.060 mg/L, 14.705 mg/L, 14.895 mg/L, 15.370 mg/L respectively) was the highest leaching among other heavy metals. In fact, only Arsenic leached higher than the regulatory level set by USEPA (5 mg/L). Several factor which affect leaching potentials is type of extraction fluid, solid to liquid ratio, the number of extraction and the length of test duration. This is also explained by [14] in his research, whereby the metals leachate concentration is directly affected by the pH of the solution. The Cr, and Pb, concentrations in FA30BA0 (0.170 mg/L, 0.695 mg/L, and 0.202 mg/L respectively) were slightly higher than control sample (FA0BA0) (0.112 mg/L, 0.447 mg/L and 0.173 mg/L respectively). In comparison to USEPA, these heavy metals were far below the regulatory level. The Cu and Ni concentrations were also recorded with similar trends of leaching. The highest leached of Cu and Ni concentration value were from the FA30BA0 sample with 0.031 mg/L and the lowest was from control sample (0.026 mg/L).The Ni concentration was recorded highest from the control sample with 0.167 mg/L and the lowest was from FA10BA0 sample with 0.065 mg/L while Fe and Mn concentrations were marginally low throughout the TCLP test.

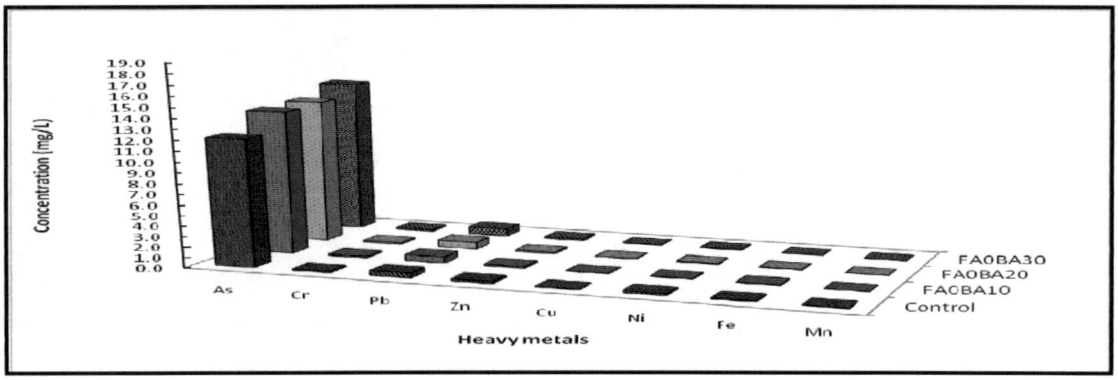

Figure 3b: Comparison of heavy metal concentrations for control sample (FA0BA0) with samples incorporated with BA (FA0BA10, FA0BA20, and FA0BA30) using TCLP

Figure 3b shows the comparison of heavy metal concentrations for the control sample (FA0BA0) with samples incorporated with BA only (FA0BA10, FA0BA20, and FA0BA30) by using TCLP method. From the observation at different percentage of BA, it shows that the As leaching are the highest even in control (FA0BA0) samples. The highest peak of As leaching was recorded from FA0BA30 sample with 14.640 mg/L followed by FA0BA20 (13.675 mg/L), FA0BA10 (13.640 mg/L) and the lowest from the control sample (12.060 mg/L). Zn and Ni concentrations were recorded the highest value in the control sample (0.173 mg/L and 0.167 mg/L respectively) without the mixture of BA in it.. Other heavy metals such as Cr, Pb, Cu, Fe and Mn concentrations were slightly low in all the samples and still in allowable value when compared to USEPA except Arsenic. Arsenic releases from acidic solution increase with pH, whereas in alkaline solution this trend is reversed [15].

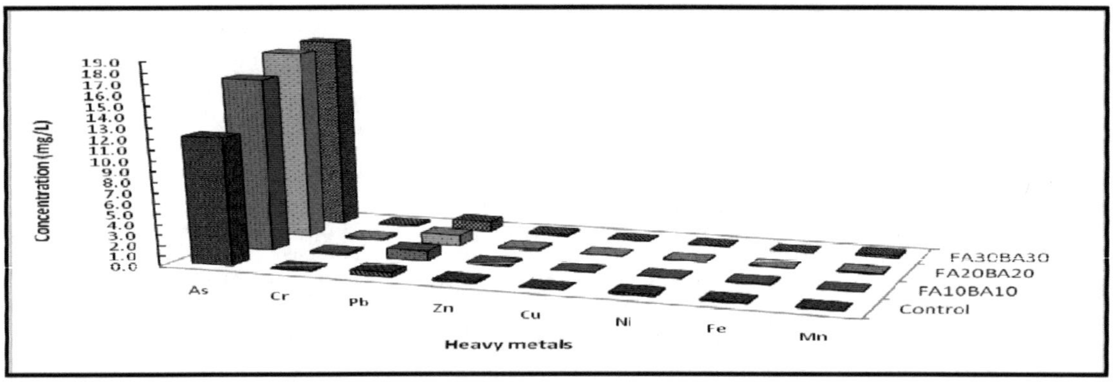

Figure 3c: Comparison of heavy metal concentrations for control sample (FA0BA0) with samples incorporated with FA and BA (FA10BA10, FA20BA20 and FA30BA30) using TCLP

As can be seen in Figure 3c, Arsenic leaching was the major heavy metals leached from the samples. As such, the highest value for As was recorded from FA30BA30 sample with 18.576 mg/L. It is because the percentage of FA and BA in this sample was the highest to be compared to other samples.

heavy metal such as Cr, Pb, Zn and Cu concentrations in the 30% replacement of FA and BA in SCC (0.207 mg/L, 1.024 mg/L, 0.181 mg/L, and 0.035 mg/L respectively) were slightly higher than the control sample (FA0BA0) with (0.112 mg/L, 0.447 mg/L, 0.173 mg/L and 0.026 mg/L respectively). It shows that the results recorded higher concentration with the increasing of FA and BA replacement in the samples.

3.2 Synthetic Precipitation Leaching Procedure (SPLP) result

The leaching fluid for this test is a solution of 60% sulfuric acid and 40% of nitric acid in distilled water until the pH is 4.2. Table 4 summarizes the results determined for heavy metals by using the SPLP method. Further results were then explained according to Figure 4a, 4b, and 4c.

Table 4: Leachability of heavy metals in SCC samples in SPLP

Heavy metals	Concentration level (mg/L)*	Concentration level (mg/L)**	Samples									
			FA0BA0	FA10BA0	FA20BA0	FA30BA0	FA0BA10	FA0BA20	FA0BA30	FA10BA10	FA20BA20	FA30BA30
			Concentration (mg/L)									
As	5	0.01	12.060	14.705	14.895	15.370	13.640	13.675	14.640	16.510	18.200	18.576
Cr	5	0.05	0.112	0.128	0.132	0.140	0.143	0.154	0.170	0.174	0.184	0.207
Pb	5	0.01	0.447	0.783	0.847	0.909	0.578	0.646	0.695	0.940	1.024	1.028
Zn	NA	NA	0.173	0.145	0.202	0.222	0.145	0.144	0.160	0.172	0.177	0.181
Cu	100	2	0.026	0.030	0.030	0.031	0.029	0.026	0.031	0.033	0.034	0.035
Ni	NA	0.07	0.167	0.065	0.066	0.069	0.065	0.071	0.075	0.062	0.063	0.077
Fe	NA	NA	0.095	0.074	0.083	0.083	0.053	0.034	0.017	0.136	0.037	0.000
Mn	NA	0.4	ND	ND	ND	ND	ND	ND	ND	ND	ND	ND

As = Arsenic Pb = Lead Cr = Chromium Zn = Zinc Cu = Copper
Ni = Nickel Fe = Ferum Mn = Manganese
* US Environmental Protection Agency, USEPA (1996); NA – Not Available; ND – Not Detected
**Guideline values for chemicals that are of health significance in drinking-water, WHO (2006) [16]

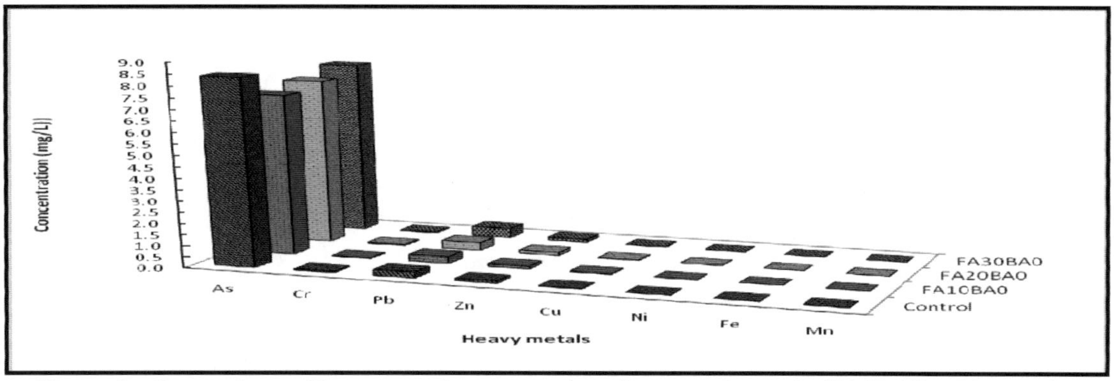

Figure 4a: Comparison of heavy metal concentrations for control sample (FA0BA0) with samples incorporated with FA (FA10BA0, FA20BA0, and FA30BA0) using SPLP

Referring to Figure 4a, the results demonstrated the comparison of heavy metals concentrations for control sample which is FA0BA0 sample with samples incorporated with different percentages of FA using SPLP. The As concentrations in FA0BA0, FA10BA0, FA20BA0 and FA30BA0 with 8.349 mg/L, 7.216 mg/L, 7.477 mg/L, and 7.868 mg/L respectively were the highest leaching concentrations among other heavy metals. In fact, only the As reached higher than the regulatory level set by USEPA (5 mg/L) and WHO (0.01 mg/L). The second greatest heavy metal in the SPLP samples was Pb ranging from 0.266 mg/L to 0.436 mg/L. Meanwhile, heavy metals such as Ni, Fe, and Mn concentrations were not detected in the samples used in SPLP method. Other than that, Pb, Zn, Cr, and Cu were leached in very low concentrations. In comparison to USEPA (1996), all the heavy metals were recorded below the concentrations level except for As. The concentration of As and Pb were higher than WHO level for all the samples.

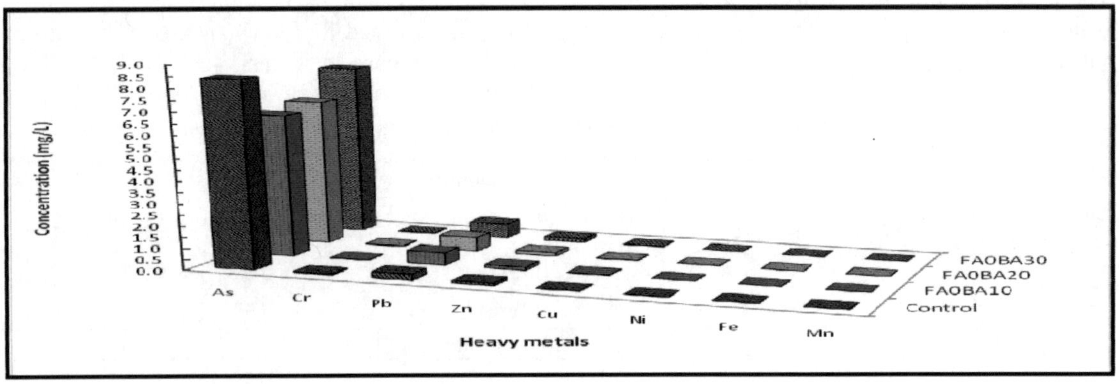

Figure 4b: Comparison of heavy metal concentrations for control sample (FA0BA0) with samples incorporated with BA (FA0BA10, FA0BA20, and FA0BA30) using SPLP

From Figure 4b, the results had shown different leaching level with different percentages of BA in the samples. From the observation, it shows that As is the highest heavy metals that leached out followed by Pb, Zn, Cr, and Cu. Meanwhile, for Ni, Fe, and Mn, the results were not detected in the samples incorporated with BA. As was recorded the highest with 8.349 mg/L and the lowest for As was 6.376 mg/L from FA0BA10 sample. Meanwhile Pb is the second largest heavy metals that leached out was ranging from 0.266 mg/L to 0.667 mg/L. From the observation, only As was exceeded the permissible limit given by USEPA. On the other hand, the concentration of As and Pb were slightly higher than WHO regulatory level but still considered as low concentration as well.

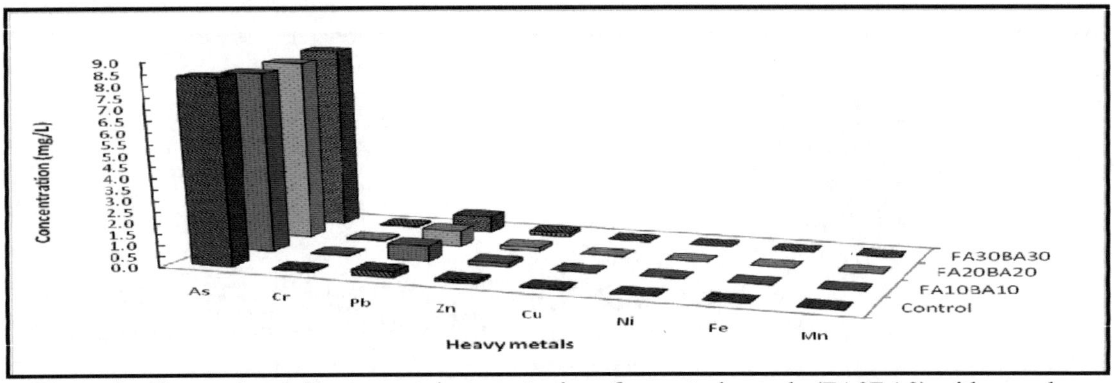

Figure 4c: Comparison of heavy metal concentrations for control sample (FA0BA0) with samples incorporated with FA and BA (FA10BA10, FA20BA20, and FA30BA30) using SPLP

Figure 4c shows heavy metal concentrations for control samples with samples incorporated with FA and BA using SPLP. As can be seen, the trend was similar to samples incorporated with FA only and samples incorporated with BA only. It shows that highest heavy metal concentration was Arsenic. The Arsenic concentration for control sample is 8.349 mg/L. Meanwhile, for samples incorporated with FA and BA which is FA10BA10 (8.133 mg/L), FA20BA20 (8.213 mg/L) and FA30BA30 (8.415

IOP Conf. Series: Materials Science and Engineering **133** (2016) 012036 doi:10.1088/1757-899X/133/1/012036

mg/L) of As concentration. It shows that higher percentages of FA and BA in the samples produce higher heavy metal concentrations. Another observation has been made for heavy metals such as Ni, Fe, and Mn. The result shows that these three heavy metals were not detected in SPLP samples. Meanwhile for Cr, Pb, Zn and Cu, the results were leached in low concentration. However, As and Pb concentrations were once again over the WHO limit and did not fulfill the requirements.

3.3 Static Leaching Test (SLT)) result

Static Leaching Test (SLT) test was tested to simulate weathering for continuous exposure to acid rain. Hence, the leaching fluid used in this SLT test used the same leaching fluid for SPLP. The concentrations of selected heavy metals element such as As, Mn, Cu, Cr, Zn, Ni, Fe, and Pb along the period of 4, 8, 12, 36, 40, 60, 80, 100, 120, 140, 160,180 and 200 days were measured using AAS. Figures 5a to 5j summarized the results of heavy metal concentrations at the specified sampling periods.

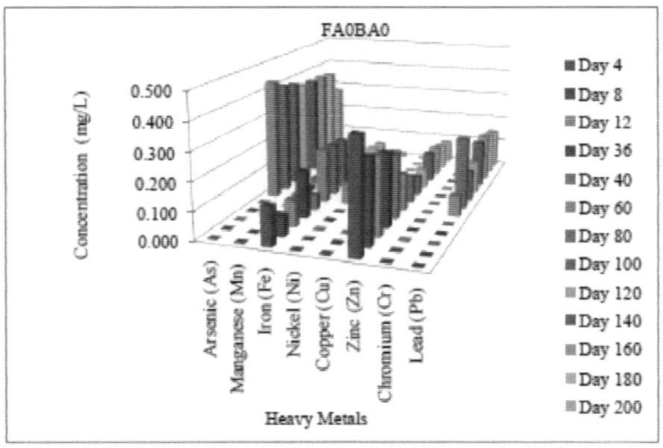

Figure 5a: Comparison of heavy metal concentrations using SLT (FA0BA0)

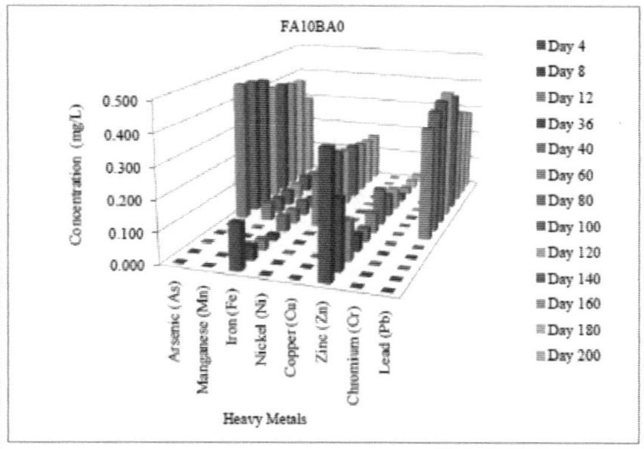

Figure 5b: Comparison of heavy metal concentrations using SLT (FA10BA0)

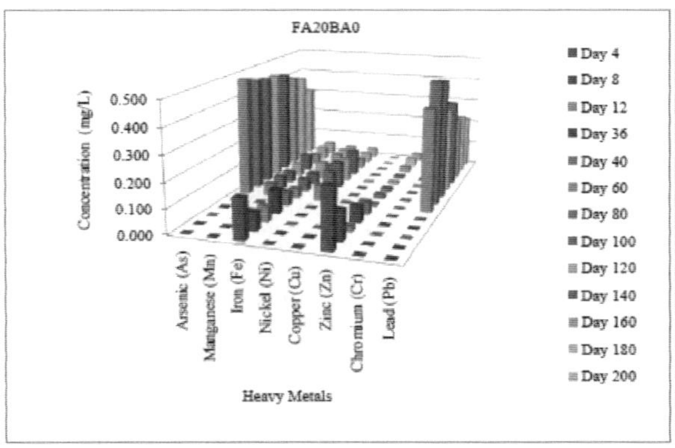

Figure 5c: Comparison of heavy metal concentrations Using SLT (FA20BA0)

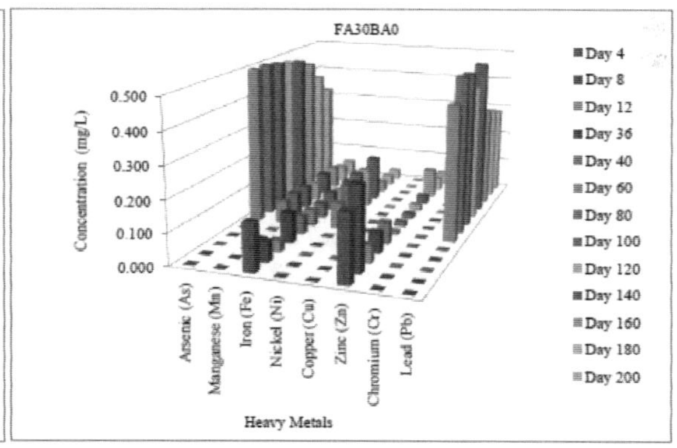

Figure 5d: Comparison of heavy metal concentrations using SLT (FA30BA0)

IOP Conf. Series: Materials Science and Engineering **133** (2016) 012036 doi:10.1088/1757-899X/133/1/012036

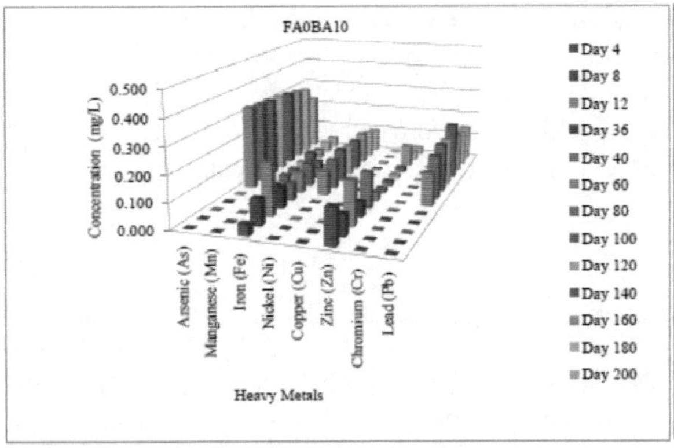

Figure 5e: Comparison of heavy metal concentrations using SLT (FA0BA10)

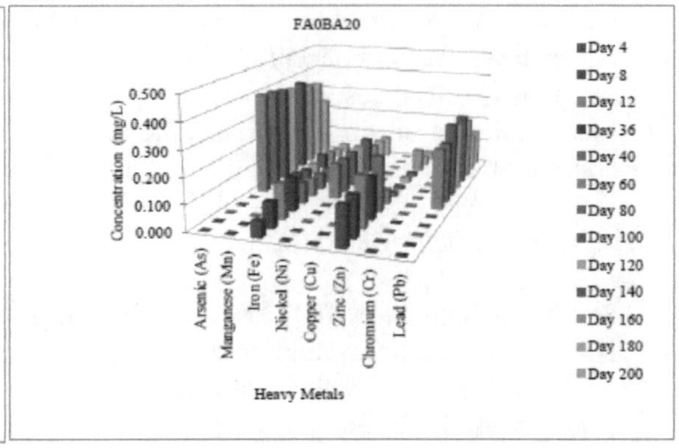

Figure 5f: Comparison of heavy metal concentrations using SLT (FA0BA20)

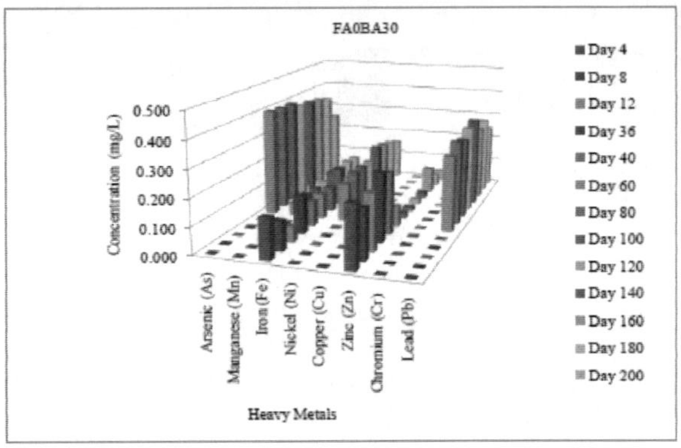

Figure 5g: Comparison of heavy metal concentrations using SLT (FA0BA30)

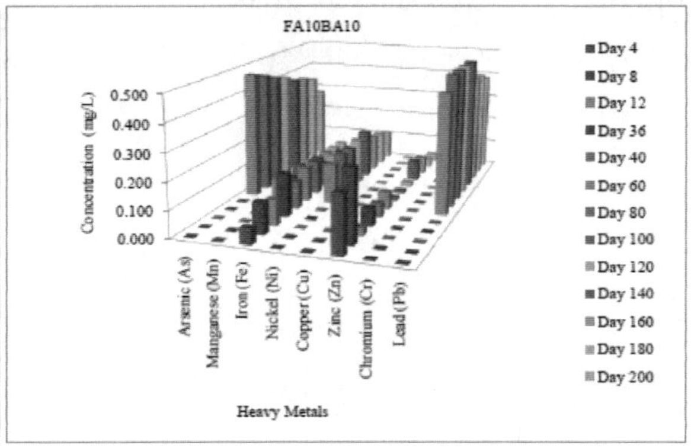

Figure 5h: Comparison of heavy metal concentrations using SLT (FA10BA10)

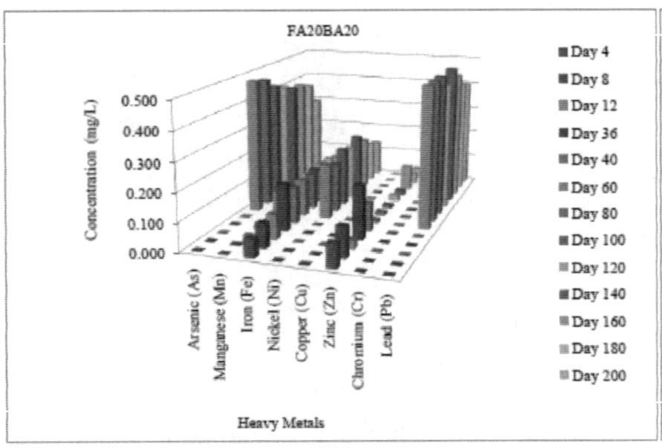

Figure 5i: Comparison of heavy metal concentrations using SLT (FA20BA20)

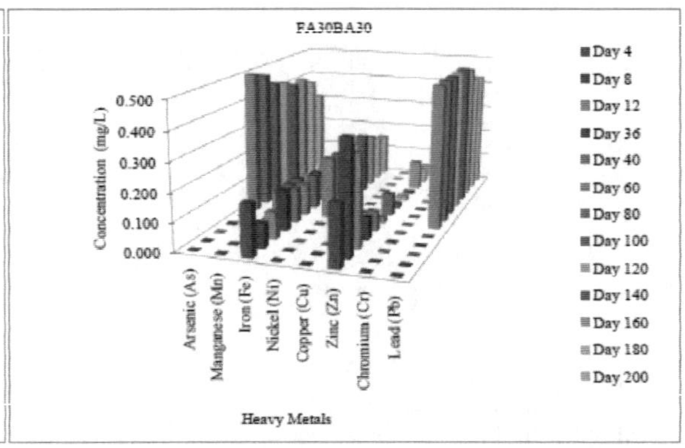

Figure 5j: Comparison of heavy metal concentrations using SLT (FA30BA30)

The concentration of selected heavy metals (As, Mn, Cu, Cr, Zn, Ni, Fe, and Pb) in the synthetic acid rain leachate collected from the SCC specimens were significantly lower than the limit specified by the USEPA (1996) [17]. The results produced slightly lower values than TCLP and SPLP due to the differences in the types of samples and the leachant used. Even though SLT were carried out after 200 days, the results revealed that all the elements were leached in small value compared to the TCLP and SPLP results. As a whole samples (without crushing), these observations indicate that the cement chemistry of the hydrated cement paste in each concrete controls the leaching concentration of most materials at alkaline condition [15]. As for a comparison with WHO guidelines, elements such as As, Pb, and Ni in most cases have exceeded the guidelines. However, these elements were detected in low concentration in SLT test. Much of this variability can be explained by differences in the compositions of the coal combustion products which is in this case FA and BA [18].

The materials of FA and BA maybe bound up in the SCC due to the lack of heavy metals concentration leach from the specimens. The concentrations of heavy metals in the leachate were lower than the detection limit is due to the solidification/stabilization process that takes place when to apply cement binders. Besides that, the immobilization of heavy metals is believed to be caused by chemical changes like pozzolanic reactions incorporate into a solid matrix during the process of solidification take place [19].

4. Conclusion

The concentration of heavy metals elements such as Cr, Mn, Cu, Zn, As and Pb showed that FA contained the highest concentration value compared to OPC and BA. From TCLP, As leaching was the only heavy metals that leached out from the samples and exceeded the limit set by USEPA. Meanwhile for SPLP, once again the results show that that the highest heavy metal concentration that leached out from the samples was As. The Arsenic concentrations for control sample are 8.349 mg/L. For samples incorporated with FA and BA, which is FA10BA10 (8.133 mg/L), FA20BA20 (8.213 mg/L) and FA30BA30 (8.415 mg/L) of As concentration. The important factor that effects the leaching of heavy metals concentrations is the pH value of the leachant. TCLP results show higher value compared to SPLP result because the leaching is more acidic compared to SPLP. Meanwhile for SLT results, it revealed that heavy metals concentrations in all the samples were leached in low concentration even the test has been conducted up to 200 days. From these three results, it indicated that 20% of FA and BA (FA20BA20) in SCC are recommended since it leached less heavy metal. As environmental concerns is a top priority, it was concluded that FA20BA20 sample will serve better in concrete technology regarding economic-viability as it also provides a good strength. All in all, FA and BA are both man-generated wastes that are steadily accumulating every year. Hence, a proper sustainable and safe way of reusing such waste materials as in this research is useful towards the environment and sustainability for construction purposes.

5. References

[1] Naganathan, S., (2012). Development of Brick Using Thermal Power Plant Bottom Ash and Fly Ash. *Asian Journal of Civil Engineering (Building and Housing)*, 13(1), pp.275–287.

[2] Kadir A. A., Hassan M. I. H.,(2014) "An Overview of Fly Ash and Bottom Ash Replacement in Self Compaction Concrete" *Key Engineering Materials* Vols. 594-595 pp 465-470.

[3] Kim, B, Prezzi, M., and Salgado R.,(2005), "Geotechnical properties of fly and bottom ash mixtures for use in highway embankments." *Journal of Geotechnical Geo-environment Engineering*. 131(7), 914–924,

[4] Prajapati K, Chandak R. and Dubey S.K., (2012) "Development and Properties of Self Compacting Concrete Mixed with Fly Ash", *Research Journal of Engineering Sciences Vol.* (3), 11-14.

[5] Muhardi, A., (2010). Engineering characteristics of Tanjung Bin coal ash. *Electronic Journal of Geotechnical Engineering*, Vol. 15, pp.1117–1129.

[6] USEPA, (United State Environmental Protection Agency). Wastes – Resource Conservation – Reduce, Reuse, Recycle – Industrial Materials Recycling. Available online http://www.epa.gov/epawaste/conserve/rrr/imr/ccps/flyash.htm (accessed on 12th April 2010).

[7] Syahrul, M., Muftah F., Muda Zulkifli (2010). The Properties of Special Concrete Using Washed Bottom Ash (WBA) as Partial Sand Replacement. *International Journal of Sustainable Construction Engineering & Technology*, 1(December 2010), pp.65–76.

[8] Sua-Iam, G. & Makul, N., (2015). Utilization of coal- and biomass-fired ash in the production of self-consolidating concrete: A literature review. *Journal of Cleaner Production*, 100, pp.59–76.

[9] Kadir, A and Mohajerani, A (2012), 'Leachability of heavy metals from fired clay bricks incorporated with cigarette butts',*Proceedings from the 8th International Symposium on Lowland Technology*, pp. 1-7.

[10] TCLP. Toxicity Characteristic Leaching Procedure. (1992). US Environmental Protection Agency, Method 1311.

[11] SPLP. Synthetic Precipitation Leaching Procedure. (1994). US Environmental Protection Agency, Method 1312.

[12] American Nuclear Society (2003). ANSI/ANS.16.1:2003. Measurement of the Leachability of Solidified Low-Level Radioactive Wastes by a Short-Term Test Procedure.

[13] Kim B., Prezzi M., Salgado R. & Lee J. (2005). Mechanical properties of Class F fly and bottom ash mixtures for embankment application. *Indian Geotechnical Conference.* pp. 239 – 242.

[14] Garrabrants, A.C., Kosson D., DeLappa, R., van der Sloot H., (2014). Effect of coal combustion fly ash use in concrete on the mass transport release of constituents of potential concern. *Chemosphere*, *Vol.* 103, pp.131–9.

[15] Kosson, D.S. (2014). "PH dependent leaching of constituents potential concern from concrete materials containing coal combustion fly ash". *Chemosphere*, *Vol.* 103, pp.140–7.

[16] World Health Organization (WHO) (2006). Guidelines for Drinking water Quality.

[17] USEPA Environmental Protection Agency, United States (1996). Hazardous wastes characteristics (scoping study).

[18] Jones, K, Leslie F, and Sharon M. (2012). "Leaching of Elements from Bottom Ash , Economizer Fly Ash , and Fly Ash from Two Coal- Fired Power Plants" *International Journal of Coal Geology*, Vol 94 pp 337–348.

[19] Gines, O., Chimenos, J. M., Vizcarro, A., Formosa, J. &Rosell, J. R., (2009). Combined use of MSWI bottom ash and fly ash as aggregate in concrete formulation: Environmental and mechanical considerations. *Journal of Hazardous Materials*, *Vol.* 169, pp. 643 – 650.

IOP Conf. Series: Materials Science and Engineering **133** (2016) 012037 doi:10.1088/1757-899X/133/1/012037

Investigation of the photovoltaic cell/ thermoelectric element hybrid system performance

D T Cotfas[1*], P A Cotfas[1], O M Machidon[1], D Ciobanu[2]

[1] Electronics and Computers Department, Transilvania University of Brasov, Eroilor 29 Brasov

[2] Product Design, Mechatronics and Environment Department, Transilvania University of Brasov, Eroilor 29 Brasov

E-mail: dtcotfas@unitv.ro

Abstract. The PV/TEG hybrid system, consisting of the photovoltaic cells and thermoelectric element, is presented in the paper. The dependence of the PV/TEG hybrid system parameters on the illumination levels and the temperature is analysed. The maxim power values of the photovoltaic cell, of the thermoelectric element and of the PV/TEG system are calculated and a comparison between them is presented and analysed. An economic analysis is also presented.

1. Introduction

The solar energy can be converted in electrical energy using photovoltaic cells and panels (PV) or in thermal energy using solar collector. These two energy sources are very important in the actual climatic context and the global installed capacity is 177 GW for solar PV and 406 GWth for solar hot water [1].

The researchers and producers work has been translated into a variety of photovoltaic cells and panels. Variety consists in: the types of materials and structures used to produce photovoltaic cells, type and number of junctions, the indoors or the outdoor applications, levels of illumination normal light up one sun (1 sun = 1000W/ m^2) or concentrated light more than one sun (until 1000 suns). The silicon is the most commonly used material in residential and industrial applications. The monocrystalline and polycrystalline silicon photovoltaic cells are used in large applications and the amorphous silicon solar cells are used in small applications such as garden illumination lamps. The multijunction photovoltaic cells are used in concentrated light.

The efficiency of the photovoltaic cells for the representative type at standard test condition STC [2]: the illumination level 1 sun, the temperature of photovoltaic cell 25°C and air mass 1.5 is given in table 1. The efficiency of the photovoltaic cells is influenced by the levels of illumination, the temperature, the inclination angle between light and cell, the wind speed, the degradation rate and the material parameters [3]. The temperature of the photovoltaic cells is a critical factor for their efficiency. The efficiency of monocrystalline photovoltaic cell decreases with 0.4%/^{o}C if the temperature increases. The time life of the photovoltaic cells substantially decreases, if they work at high temperature, over 80°C [4]. The irradiance is another critical factor for the efficiency of the photovoltaic cells. The spectral distribution of the natural sunlight contains the infrared part, thus the

[*] Address and email of corresponding author

temperature of the photovoltaic cells increases when the sunlight falls on them and the irradiance increases, reaching in some cases a temperature of 83°C [5].

Table 1. Efficiency for different types of photovoltaic cells at STC.

Type	**Area**	**Efficiency**
Monocrystalline Si	143.7 cm^2	25.6 ± 0.5 %
Polycrystalline Si	243.9 cm^2	20.8 ± 0.5 %
Amorphous Si	1.001 cm^2	10.2 ± 0.3 %
GaAs (thin film)	0.99 cm^2	28.8 ± 0.9 %
InGaP/GaAs/InGaAs	1.047 cm^2	37.9 ± 1.2 %
GaInP/GaAs, GaInAsP/GaIn[a]	0.052 cm^2	46.0 ± 2.2 %

[a]The light intensity is 508 suns

The efficiency of the photovoltaic cells can be increased if the irradiance increases and the cells temperature decreases. The decreasing of the photovoltaic cells' temperature has to be realized without energy consumption. This can be made by using natural cooling, in this case the efficiency being lower, by using the solar collector - realizing the PVT system or by using the thermoelectric element.

The thermoelectric element can be used for heating or cooling if DC current is applied based on the Peltier effect, or it can generate electrical power if there is a temperature gradient between the two parts of the thermoelectric element based on the Seebeck effect.

The figure of merit, ZT where T is the absolute temperature and Z is given by the equation (1) dictated the efficiency of the thermoelectric element. The power factor of the thermoelectric element is defined as $\alpha^2\sigma$.

$$ZT = \frac{\alpha^2\sigma}{k}T \tag{1}$$

where α is the Seebeck coefficient, k represents the thermal conductivity and σ is the electrical conductivity.

The figure of merit for the first generation of the thermoelectric elements, before the years 1990, was under 1 [7] and their efficiencies were very low, thus their applications were limited. After this period, the number of the applications quickly increases because the figure of merit becomes double or more than doubles [8]. The figure of merit for the important thermoelectric elements is given in table 2.

Table 2. Figure of merit for different types of thermoelectric materials.

Materials	**ZT**	**Temperature [K]**
Bi_2Te_3/Sb2Te_3 [9]	2.4	300
Cu–BiTeSe [10]	1.1	373
PbTe/Ag_2Te [6]	1.6	773
SnSe [11]	2.6	923
$AgPb_mSbTe_{m+2}$ [6]	1.8	800

2. Materials and methods

The temperature of the solar cell increases when the natural or artificial light falls on it. The generated power of the photovoltaic cell can be increased using a thermoelectric element for cooling it, while at the same time generating additional power.

The main goals of the paper are to design and study the behavior of the PV/TEG hybrid system which consists of the photovoltaic cells, PV, and the thermoelectric element TEG, see figure 1.

The experimental set-up consists of, see figure 2:

- The cooling system based on an aluminium block cooled with water. The water flowing is assured using a water pump. The inlet water temperature is quasi constant due to using a water tank.
- The thermoelectric element used is Stonecold, made from Bi_2Te_3, the ceramic material used is alumina, Al_2O_3, with dimensions 6.2 cm/6.2 cm/4.8 cm. The TEG is mounted on the aluminium block. The thermic contact between the TEG and the aluminium block is realized with thermal adhesive tape with 0.1 mm thickness and its thermal conductivity is 2 W/m·K.
- The photovoltaic cells, with dimensions of 6 cm/6 cm. Three types of photovoltaic cells were used for this study: monocrystalline silicon, mSi, polycrystalline silicon, pSi, and amorphous silicon aSi. The photovoltaic cell is thermally connected with the hot part of the thermoelectric element using a thermal tape with 1 mm thickness and its thermal conductivity is 6 W/m·K.
- The illumination system based on nine halogen bulbs powered with a Keithely programmable dc power source.
- The linear actuator - it is used to vary the illumination levels of the photovoltaic cells.
- The temperature monitoring system-it consists of nine k thermocouples and a NI cRIO module for temperature measurement. The nine thermocouples are distributed as follows: three for monitoring the temperature of the cold part of TEG, three for monitoring the temperature of the hot part of TEG and the temperature of the back of the photovoltaic cell and three for monitoring the water temperature: inlet, outlet and for the water tank.
- The I-V characteristic system – two electronic loads and NI cRio integrated system are used to measure the I-V characteristic of the photovoltaic cell and the thermoelectric element simultaneously.
- The PC with a proper software for control, measurements and analysis of the data, developed LabVIEW graphical programming language.

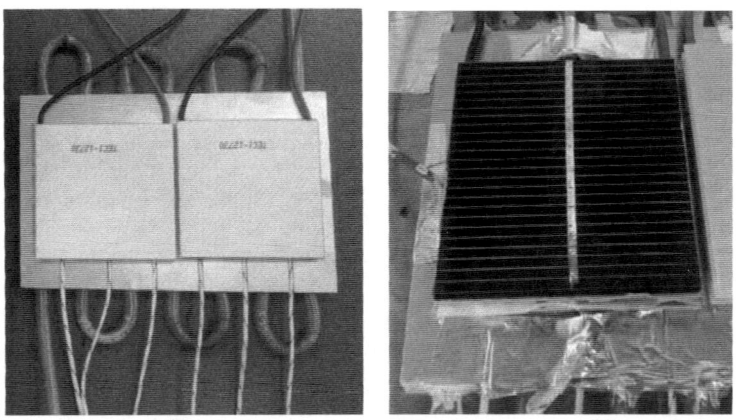

a) b)

Figure 1. The PV/TEG hybrid system a) the thermoelectric element mounted on the aluminum block; b) the monocrystalline silicon photovoltaic cell mounted on the thermoelectric element.

Figure 2. The experimental set-up.

2.1. *Photovoltaic cells*

The I-V characteristic is a very important tool for the characterization of photovoltaic cells. The analysis of the photovoltaic cell I-V characteristic can be done using the equivalent circuit and the mathematical model [12]. The choice of the equivalent circuit and the mathematical model is in function of the conduction mechanism. The most commonly used equivalent circuit is the one diode model, see figure 3.

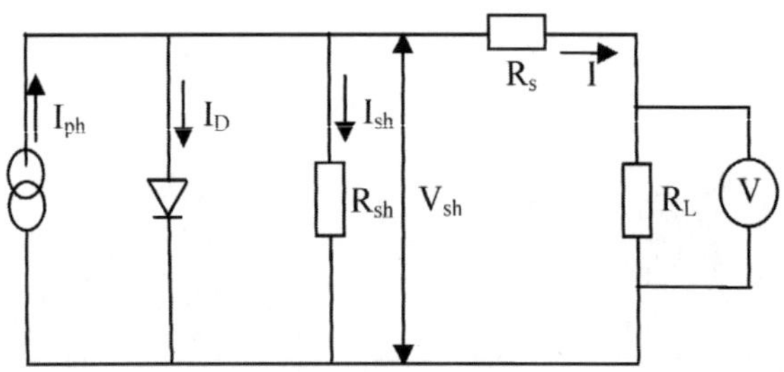

Figure 3. The equivalent circuit of the photovoltaic cell-one diode model.

The mathematical model which corresponds to the one diode model equivalent circuit is given by equation (2)..

$$I = I_{sc} - I_o\left(e^{\frac{V+IR_s}{nV_T}} - 1\right) - \frac{V + IR_s}{R_{sh}} \tag{2}$$

where I_{sc} is the short circuit current, n represents the ideality factor of diode, R_s is the series resistance, R_{sh} is the shunt resistance, I_o represents the reverse saturation current, V_T represents the thermal voltage, VT =kT/q, k denotes the Boltzmann constant, T is the temperature of the solar cell and q is the electronic charge.

The analytical five point method developed by Chan et al. can be used to determine the important parameters of the photovoltaic cells [12].

2.2. *Thermoelectric element*

The equivalent electric circuit of the thermoelectric element is very simple. It consists of a voltage source and the internal resistance [13]. The thermal-electrical equivalent circuit is presented in figure 4 [14,15]. The thermoelectric element is characterised by the thermal conductance K, the internal resistance and the Seebeck coefficient [14].

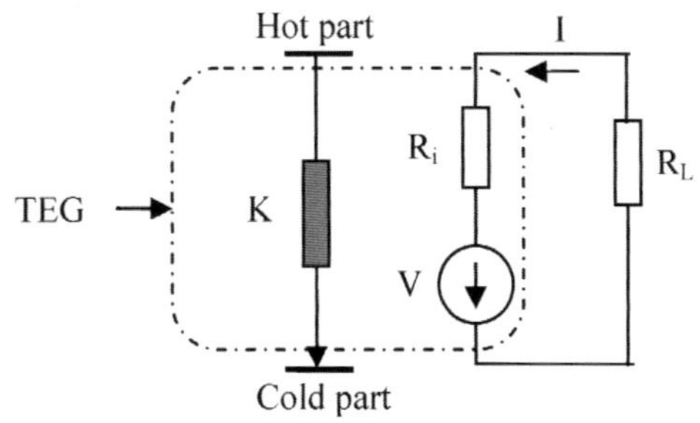

Figure 4. The thermal-electrical equivalent circuit of the thermoelectric element.

The I-V characteristic is used to characterize the thermoelectric element. It is, in this case, a linear dependence.

Using the equation (3), the output current can be calculated. The internal resistance, R_i, of the thermoelectric element is given by equation (4). The efficiency, η, of the thermoelectric element can be calculated using the equation (5).

$$I = \frac{\alpha N \Delta T - V}{R_i} = \frac{\alpha N \Delta T}{R_i + R_L} \tag{3}$$

where N is the number of p-n junctions, ΔT represents the temperature difference between the hot part and cold part of the thermoelectric generator and R_L is the load resistance.

$$R_i = \frac{2NL}{a\sigma} \tag{4}$$

where L is the length, a represents the area of the thermoelectric element and σ is the effective conductivity of the material.

$$\eta = \frac{(T_h - T_c)}{T_h} \frac{\sqrt{1+ZT_m} - 1}{\sqrt{1+ZT_m} + \frac{T_c}{T_h}} \tag{5}$$

where T_h is the temperature of the thermoelectric element hot part, T_c is the temperature of the thermoelectric element cold part, T_m is the average temperature between the T_h and T_c.

2.3. *Methods*

The important parameters of the photovoltaic cells, such as the short circuit current, the open circuit voltage V_{oc}, the maximum power P_{max}, the fill factor FF, the efficiency and the parameters of the thermoelectric element, such as the short circuit current, the open circuit voltage, the maximum power and the internal resistance are calculated.

The experimental set-up allows the determination of the important parameters of the two components of the PV/TEG hybrid system in function of the illumination levels. The spectral distribution of the halogen light has an important infrared part and the temperature of the photovoltaic cell increases. The performance of photovoltaic cell decreases with the temperature growth. The thermoelectric element was introduced in the system to assure the temperature decrease. The temperature difference between the temperature of the hot part of the thermoelectric element, which is bonded on the back of the photovoltaic cell and the temperature of the cold part of the thermoelectric element, which is bonded on the aluminum block, is dependent on the illumination levels. The aluminum block is cooled with water. Its goal is to maintain the cold part temperature of the thermoelectric element quasi constant.

The I-V characteristics of the PV/TEG hybrid system components are measured for each illumination level every two minutes. The measurements begin when the light is turned on and the temperature of the components is equal, and they are stopped when the temperature of the hot part of the thermoelectric element remains quasi constant.

The temperature of the PV/TEG hybrid system components is measured in three points using the k type thermocouples.

The system control, the parameters monitoring and the data acquisition is realized using the cRIO system and the adequate software built in the LabVIEW programming language.

3. Results and discussions

3.1. Effects of illumination levels

The PV/TEG hybrid system is characterized at eight illumination levels from 720 W/m^2 to 1080 W/m^2.

The I-V characteristics of the photovoltaic cells in function of the irradiance are presented in figure 5. The short circuit current increases proportionally with irradiance for both the photovoltaic cells. The open circuit voltage remains quasi constant because the decreasing caused by the temperature increasing is compensated by the increase in irradiance.

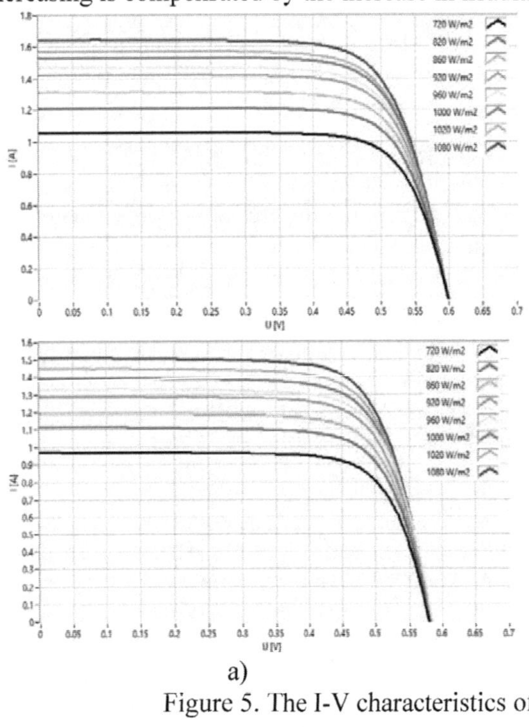

a) b)

Figure 5. The I-V characteristics of the photovoltaic cells: a) mSi; b) pSi.

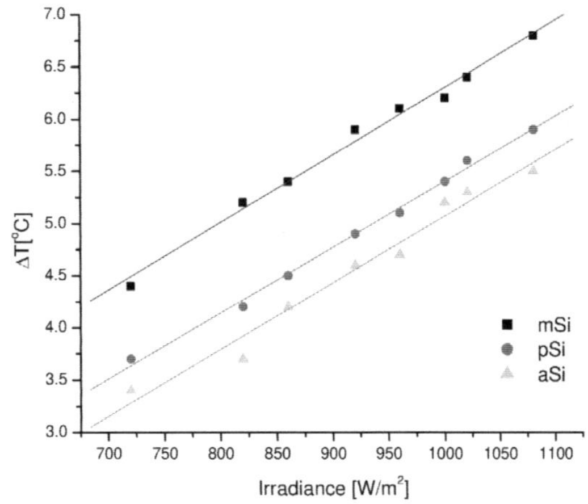

Figure 6. The difference temperature of the thermoelectric element vs irradiance.

The temperature difference between the hot and cold parts of the thermoelectric element is dependent on the irradiance, see figure 6. The temperature difference increases proportionally with the irradiance grow. The highest temperature difference is obtained for mSi/Bi_2Te_3 hybrid system and the lowest for mSi/Bi_2Te_3 hybrid system. This can be explained by the fact that the amorphous silicon photovoltaic cell is encapsulated and the heat transfer is lower than for the other photovoltaic cells.

The important parameters of the photovoltaic cells and the thermoelectric element were determined and they are given in Table 3. The maximum power generated by the photovoltaic cells and by the thermoelectric element, as well as the efficiency of the photovoltaic cell are higher for the mSi than for the pSi and aSi. This result makes the mSi/ Bi_2Te_3 hybrid system the most suitable one to be used in real applications.

Table 3. The parameters of the PV/TEG hybrid system components.

Materials	**Photovoltaic cell**						**Thermoelectric element**					
	I [W/m²]	**V_{oc} [V]**	**I_{sc} [A]**	**P_{max} [W]**	**FF [%]**	**η [%]**	**T_{PV} [°C]**	**V_{oc} [V]**	**I_{sc} [A]**	**P_{max} [mW]**	**R_i [Ω]**	**ΔT [°C]**
mSi/ Bi_2Te_3	720	0.599	1.057	0.474	74.90	18.30	28.6	0.064	0.107	1.69	0.597	4.4
	820	0.599	1.214	0.542	74.44	18.35	30.0	0.078	0.133	2.59	0.591	5.2
	860	0.600	1.316	0.583	73.90	18.84	30.8	0.082	0.137	2.79	0.600	5.4
	920	0.599	1.424	0.627	73.46	18.93	31.8	0.089	0.155	3.40	0.577	5.9
	960	0.599	1.470	0.645	73.25	18.66	32.1	0.093	0.156	3.62	0.598	6.1
	1000	0.599	1.533	0.671	73.11	18.65	32.5	0.095	0.159	3.77	0.600	6.2
	1020	0.599	1.574	0.687	72.89	18.70	33.0	0.099	0.165	4.07	0.602	6.4
	1080	0.599	1.646	0.716	72.62	18.41	33.2	0.101	0.167	4.19	0.603	6.8
pSi/ Bi_2Te_3	720	0.580	0.973	0.418	74.12	16.14	29.9	0.057	0.094	1.34	0.607	3.7
	820	0.581	1.114	0.476	73.56	16.12	30.8	0.067	0.110	1.84	0.607	4.2
	860	0.582	1.192	0.511	73.69	16.50	31.2	0.072	0.123	2.19	0.585	4.5
	920	0.582	1.291	0.551	73.24	16.62	31.8	0.079	0.130	2.55	0.605	4.9
	960	0.585	1.333	0.570	73.14	16.50	31.4	0.081	0.139	2.87	0.604	5.1

	1000	0.582	1.394	0.592	72.92	16.43	32.5	0.087	0.144	3.14	0.607	5.4
	1020	0.582	1.451	0.614	72.64	16.71	32.9	0.092	0.151	3.44	0.609	5.6
	1080	0.583	1.512	0.639	72.56	16.44	33.1	0.094	0.155	3.63	0.606	5.9
aSi/ Bi_2Te_3	720	3.95	0.062	0.075	30.27	2.88	30.8	0.073	0.062	1.12	1.2	3.4
	820	3.87	0.070	0.078	28.58	2.63	32.4	0.073	0.062	1.15	1.2	3.7
	860	3.89	0.075	0.082	28.32	2.66	32.9	0.086	0.072	1.54	1.2	4.2
	920	3.81	0.080	0.082	26.78	2.48	33.9	0.101	0.088	2.17	1.1	4.6
	960	3.83	0.083	0.085	26.98	2.47	33.5	0.097	0.081	1.96	1.2	5
	1000	3.81	0.086	0.086	26.31	2.40	33.7	0.106	0.086	2.36	1.2	5.2
	1080	3.74	0.092	0.087	25.12	2.23	35.7	0.118	0.103	2.98	1.1	5.5

3.2. Temperature effect
The temperature of the photovoltaic cell can vary in general, under 1000 W/m^2 natural sunlight conditions, from 50°C to 80 °C depending on the climatic zone and on the type of the materials being used.

Using the PV/TEG hybrid system, the temperature of the photovoltaic cell at 1000 W/m^2 irradiance is for mSi 32.5°C, pSi 33.1°C and aSi 35.7°C. A substantial difference can be observed between the working temperatures. The gain in maximum power for the mSi and pSi is 7%, and for the aSi is 5%, because the temperature of the photovoltaic cell can reach 53°C for the irradiance by 1000 W/m^2 without the TEG element.

The evolution of the I-V characteristic's thermoelectric element in time when the mSi photovoltaic cell is considered, is presented in figure 6. The temperature difference between the hot and cold part of the TEG becomes quasi constant after 8 min for the mSi and pSi and after 14 min for the aSi. The time difference can be accounted for due to the photovoltaic structure: mSi and pSi are unencapsulated and aSi is encapsulated.

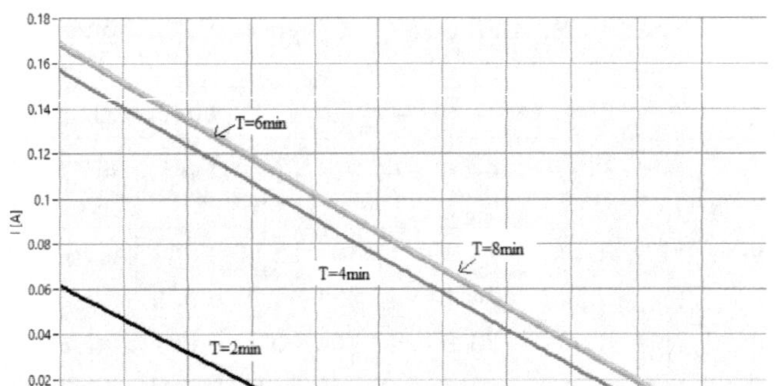

Figure 7. The evolution in time of the I-V characteristics of the thermoelectric element.

3.3. Economic analysis
The cost of the PV/TEG hybrid system is calculated using the available price of the photovoltaic cells and of the thermoelectric element. The price of the W_p (Watt peak – is the generated power by the photovoltaic cell under 1000W/m^2 irradiance) photovoltaic cell depends on the used material. The number of the p-n pairs, the material and the internal resistance of the thermoelectric element give the effective cost.

The comparative cost of the PV/TEG hybrid system components in function of the maximum power is presented in Table 3. The PV/TEG hybrid system is not cost effective yet, because the price of the thermoelectric element is high. The system becomes attractive when the price of the thermoelectric element substantial decreases and its efficiency increases.

Table 4. The comparative cost of the PV/TEG hybrid system.

Materials	PV [\$]	TEG [\$]	PV/TEG [\$]	P_{max}[W]	\$/ W_p
mSi (nonencapulated)	1.4	40	41.4	0.675	61.3
pSi (nonencapulated)	1.2	40	41.2	0.595	69.2
aSi (encapsulated)	1	40	41	0.088	465

4. Conclusions

Three types of the PV/TEG hybrid systems, built using mSi, pSi, aSi photovoltaic cells and Bi_2Te_3 thermoelectric element, were realized and characterized. The important parameters, such as the short circuit current, the open circuit voltage, the maximum power, were determined at different illumination levels and temperatures for both of the PV/TEG hybrid system components. The fill factor and the efficiency of the photovoltaic cell and the internal resistance of the thermoelectric element were also determined.

Analysing the three types of the PV/TEG hybrid systems, the mSi/ Bi_2Te_3 hybrid system is the best solution. Also, a comparison of the price required to obtain 1 W_p with the three systems shows that the mSi/ Bi_2Te_3 has a lower price. Due to the high price of the thermoelectric element there are few practical applications. The number of applications will increase if the researcher's effort will be materialized in new types of materials with higher efficiency and optimized structure. Using the graphene and nanomaterials, whose efficiency can reach 20%, could prove to be a promising solution for the future.

The temperatures of the photovoltaic cells decrease if the thermoelectric element is used. This has two positive effects: the photovoltaic cells produce more energy and the life time of the photovoltaic cells increases, because the temperature is a critical ageing factor.

Acknowledgments
This work was supported by a grant of the Romanian National Authority for Scientific Research and Innovation, CNCS - UEFISCDI, project number PN-II-RU-TE-2014-4-1083.

References

[1] J. L. Sawin and et al., 2015, Renewables 2015 Global Status Report, REN21 Secretariat, Paris, France.
[2] M. A. Green, K. Emery, Y. Hishikawa, W. Warta and E. D. Dunlop, 2015 *Progress in Photovoltaics: Research and Applications*, vol 23, pp.805–812.
[3] A.Makki, S. Omer, and H. Sabir, 2015 *Renewable and Sustainable Energy Reviews*, vol. 41, pp. 658–684.
[4] D. T. Cotfas, P. A. Cotfas, D. Floroian, L. Floroian, and M. Cernat, 2015 *Intl Aegean Conference on Electrical Machines & Power Electronics (ACEMP), 2015 Intl Conference on Optimization of Electrical & Electronic Equipment (OPTIM) & 2015 Intl Symposium on Advanced Electromechanical Motion Systems (ELECTROMOTION)*, pp. 599 – 604.
[5] M. Benghanem, A.A. Al-Mashraqi, and K. O. Daffallah, 2016 *Renewable Energy*, vol. 89, pp. 51-59.
[6] L. D. Zhao, V. P. Dravid, and M. G. Kanatzidis, 2014 *Energy Environmental Science*, vol. 7, pp. 251-268.

IOP Conf. Series: Materials Science and Engineering **133** (2016) 012037 doi:10.1088/1757-899X/133/1/012037

[7] R. Ahıska and H. Mamur, 2014 A review: Thermoelectric generators in renewable energy, *International Journal Of Renewable Energy Research*, vol. 4, pp. 128-136.

[8] C. A. Gould, N. Y. A. Shammas, S. Grainger, I. Taylor, 2008 *In: Proceedings of the 26th International Conference on Microelectronics*, 11–14 May 2008; NIS, Serbia. pp. 329–332.

[9] R. Venkatasubramanian, E. Siivola, T. Colpitts, and B. O'Quinn, 2001 *Nature*, vol. 413, pp. 597-602.

[10] W.-S. Liu, Q. Zhang, Y. Lan, S. Chen, X. Yan, Q. Zhang, H. Wang, D. Wang, G. Chen, and Z. Ren, 2011 *Advanced Energy Materials*, vol. 1, pp. 577–587.

[11] H. Zhang, D. V. Talapin, 2014 *Angewandte Chemie International Edition*, vol. 53, pp. 9126–9127.

[12] D. T. Cotfas, P. A. Cotfas, S. Kaplanis, 2013 *Renewable and Sustainable Energy Reviews*, vol. 28, pp. 588–596.

[13] P. Vizureanu, C. Samoilă, D. Cotfas, Materials Processing using Solar Energy, Environmental Engineering and Management Journal, March/April 2009, Vol.8, No.2, pg. 301-306

[14] H. Chen, N. Wang, and H. He, 2014 *Advances in Condensed Matter Physics*, vol 2014, pp. 1-6.

[15] P. Camacho-Medina, M. A. Olivares-Robles, A. Vargas-Almeida, and F. Solorio-Ordaz, 2014 *Entropy*, vol. 16, pp. 2890-2903.

[16] B. S. Yilbas, A. Z. Sahin, 2010 *Energy*, vol. 35, pp. 5380-5384.

Effects of the restoration mortar on chalk stone buildings

R M Ion[a,b, 1], S Teodorescu[c], R M Ştirbescu[c], I D Dulamă[c], I R Şuică-Bunghez[b], I A Bucurică[c], R C Fierăscu[b], I Fierscu[b], M L Ion[d]

[a]Valahia University, Materials Engineering Department, 13 Aleea Sinaia, Targoviste, Romania.
[b]ICECHIM, 202 Splaiul Independentei, Bucharest-060021, Romania.
[c]Multidisciplinary Scientific and Technologic Research Institute, 13 Aleea Sinaia, Targoviste, Romania.
[d]Valahia University of Targoviste, History Department, 13 Aleea Sinaia, Targoviste, Romania.

E-mail: rodica_ion2000@yahoo.co.uk

Abstract. The monument buildings as components of cultural heritage are exposed to degradation of surfaces and chemical and mechanical degradation, often associated to soiling and irreversible deterioration of the building. In many conservative and restorative works, a cement-based mortar was used without knowing all the adverse effects of this material on the building. This paper deals with the study of the effects of natural cement used in restorative works in the particular case of the Basarabi-Murfatlar Churches Ensemble. Cement-based materials exposed to sulfate present in the chalk stone - gypsum ($CaSO_4.2H_2O$), can induce signs of deterioration, due to ettringite ($[Ca_3Al\ (OH)_6 \cdot 12H_2O]_2 \cdot (SO_4)_3 \cdot 2H_2O$) or thaumasite ($Ca_3[Si(OH)_6 \cdot 12H_2O] \cdot (CO_3) \cdot SO_4$) formation. These phases contribute to strain within the material, inducing expansion, strength loss, spalling and severe degradation. Several combined techniques (XRD, EDXRF, ICP-AES, SEM, EDS, sulphates content, FT-IR and Raman analysis were carried out to put into evidence the effects of them on the building walls.

1. Introduction

For understanding deterioration and conservation of stone, especally those from arhitectural buildings heritage is absolutely necessary to know the structural, chemical and mineralogical composition of the roc and the mechanisms of stone degradation due to the external factors [1].

Carbonate rocks have been widely used as building stones since ancient times, because they are soft, easy to process and cut, mainly due to their composition: various amounts of matrix, biogenic and nonbiogenic components. But, the most important negative aspect of these stones are their damages susceptibility, generally due to the increasing atmospheric pollution by acidic agents [2,3]. Also, the carbonate rocks are characterized by various porosity and water absorption that affect their resistance to frost and salt-crystallization damage [4,5].

Limestone is known as a sedimentary carbonate rock, fine-grained and massive, without visible pores and lack of distinct bedding. Many of limestone buildings have been restored with concrete

[1] To whom any correspondence should be addressed.

mortars, and because after a short time after restoration they show visible damages, is imperious necessary to find the causes responsible for such processes.

Concrete, is a versatile and robust building material, but due to its porosity is susceptible to potentially harmful agents, such as chlorides, sulphates, carbon dioxide, and other pollutants that determine the high decay rate and deterioration of concrete [6]. It has been used in many conservative and restorative works, mostly as cement-based material, but without knowing all the adverse effects of this material on the building. Cement-based materials exposed to sulfate present in the carbonateous stones (limestone) as gypsum ($CaSO_4.2H_2O$), can induce signs of deterioration, due to ettringite ($[Ca_3Al(OH)_6\cdot 12H_2O]_2\cdot(SO_4)_3\cdot 2H_2O$) or thaumasite ($Ca_3[Si(OH)_6\cdot 12H_2O]\cdot(CO_3)\cdot SO_4$) presence. These newly solid phases can lead to strain within the material, inducing expansion, strength loss, spalling and severe degradation of the wall building [7-9].

Our study aimed to determine the influence in time of the cement-based mortars used to restoration procedures, and to show how these mortars are a continuous source for a deterioration process for the building walls, due to thaumasite and ettrungite formation. As a particular case, will be evaluated the effects of these cement-based mortars used in restorative works in the Basarabi-Murfatlar Churches Ensemble. Several analytic techniques (XRD, EDXRF, ICP-AES, SEM, sulphates content, FT-IR and Raman analysis) were carried out to put into evidence the effects of degradation on the building walls.

2. Materials and methods

As materials, has been used a small piese of chalk (B4) and a small piese of restoration mortar (BMR) both of them detached from the church and without patrimonial value.

X-ray diffraction analyses were performed using a Rigaku SmartLab equipment, operating at 45 kV and 200 mA, using Cu Kα radiation (1.54059 Å), in parallel beam configuration, from 3 to 90 2θ degrees; the components were identified using the Rigaku Data Analysis Software PDXL 2 and available literature data.

The X-ray fluorescence was used to identify the elements in certain pigments, fillers. The determinations were carried out using an energy-dispersive spectrometer, EDXRF PW4025, Minipal 2 - PANalytical, with a Si-PIN detector.

Trace elements were quantified using **ICP-AES** (inductively coupled plasma-atomic emission) spectrometer Varian Liberty 110 Series. The samples were cut off from the original shreds and were finely powdered in an agate mortar. Multielement, matrix matched standards were used for the quantitative determinations. Microwave assisted digestions were performed in a Berghof microwave oven with the use of high-pressure closed Teflon PFA vessels and online pressure and temperature control.

Raman spectra have been obtained with a portable dual wavelength Raman analyzer IR - XANTUS 2 –RIGAKU, with the following parameters: Dual wavelength source 1064nm & 785nm, spectral range (cm-1) 200 – 4000, Spectral Resolution (cm-1) 7 – 10, Laser Output Power (mW) 400 – 490, Laser Output Power (mW) 30 – 490, cooled detectors – CCD and InGaAs. The paint cross sections were analyzed using Raman spectroscopy in order to identify pigments and fillers. The laser beam was focused on a diameter of about 25, 5 and 1.5–2 μm, respectively. Spectra were acquired using 10 s of signal collection time and five accumulations.

Light Optical Microscopy (LOM) has been recorded by using NOVEX microscope with a Leica EC3 camera under a magnification of 40x to 600x, to determine the matrix heterogeneity, particle size, color, shape and transparency.

Scanning electron microscopy (SEM) has been achieved with a SU-70 (manufacturer Hitachi, Japan) COUPLED ENERGY dispersive spectrometer (EDS) spectrometer wavelength dispersive (WDS). SEM magnification range of 30X-800.000X and the resolution accelerating voltage of 15 KV is 1 nm. The spectrometer EDS attached allows qualitative and quantitative analysis (from Be (Z = 4) to Pu (Z = 94)) on point, rectangle, circle or the freedom of choice, analysis and line type grid mapping X-ray solutions for overlapping peaks. WDS spectrometer attached allows qualitative and quantitative analysis (B (Z = 5) to Pu (Z = 94)).

The sulphates attack tests carried out on brushed samples were made according to the SR EN ISO 787-13:2003 "General methods of test fot pigments Part 13: Determination of water soluble sulphates, chlorides and nitrates" and SR EN ISO787-8:2003: "General methods of test for pigments and extenders – Part 8: Determination of matter soluble in water – Cold extraction method".

3. Results and discussions

As a first step of the samples identification procedures, the two samples were subjected to X-ray fluorescence analysis (figure 1). The two samples are dominated by calcium [10]. The main difference regarding the trace elements composition is the presence of aluminium and iron in higher amount in the wall filler than in the wall itself. Those results are confirmed by the ICP-AES analyses: for the wall samples, 350 ppm Al, and, respectively, 242 ppm Fe; for the wall filler samples, 575 ppm Al, and, respectively, 358 ppm Fe.

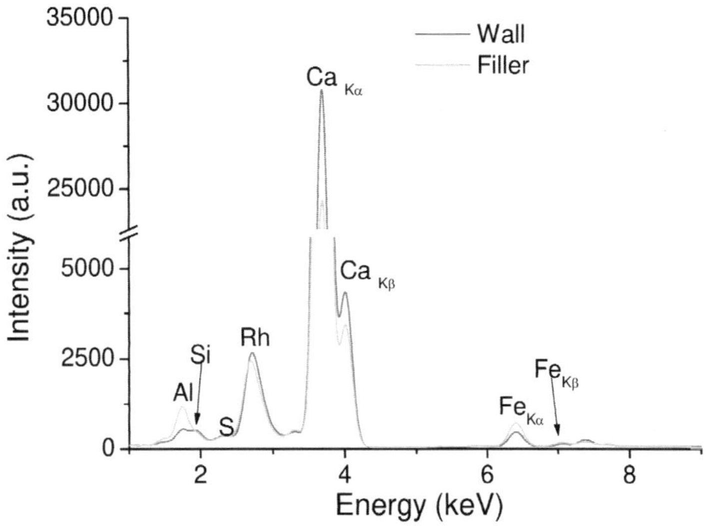

Figure 1. EDXRF analyses of the two samples.

WDS spectrometer attached to SEM equipment allows qualitative and quantitative analysis both of chalk sample and of restoration mortar, Figure 2. There are visible the difference between both samples and the changes in the Al/Si ratio, which is resposible for thaumasite and ettringite identification, too.

The X-ray diffraction patterns (figure 3) reveals that the mineralogical composition of the two samples is dominated by calcite (CC) and, in lesser extension, quartz (Q). A possible explanation for the differences observed by EDXRF and ICP-AES is the presence of traces of ettringite (E) and thaumasite (T) in the wall filler sample, concomitent with a decrease of calcite peaks. Thaumasite is a phase newly recognised in some sulphate-attacked building constructions. XRD giving direct qualitative evidence is generally accepted.

Thaumasite, $[Ca_3Si(OH)_6{\cdot}12H_2O](SO_4)(CO_3)$ or $CaSiO_3{\cdot}CaCO_3{\cdot}CaSO_4{\cdot}15H_2O$, or $Ca_6[Si(OH)_6{\cdot}12H_2O]_2 (SO_4)_2 (CO_3)_2$, is formed from external sulfate attack of cement concrete at low temperatures [11]. Its crystal structure is most unusual in that it contains silicon atoms that are surrounded by an octahedron of OH^-# ions. This structure is closely related to that of ettringite, $Ca_6[Al(OH)_6{\cdot}12H_2O]2(SO_4)_3(H_2O)_2$, and the two minerals form a series of solid solutions. Thaumasite tends to form at low temperatures (4 ^OC - 10 ^OC), in the presence of water. The presence of sulfate induces the formation of ettringite that uses aluminium provided by the cement. In the presence of sulfate and carbonate, thaumasite can continue to form until the calcium silicate hydrate is completely decomposed. Sulfate can be supplied from the chalk damaged surfaces, and carbonate can be supplied from atmospheric CO_2 or from limestone present in the concrete or mortar.

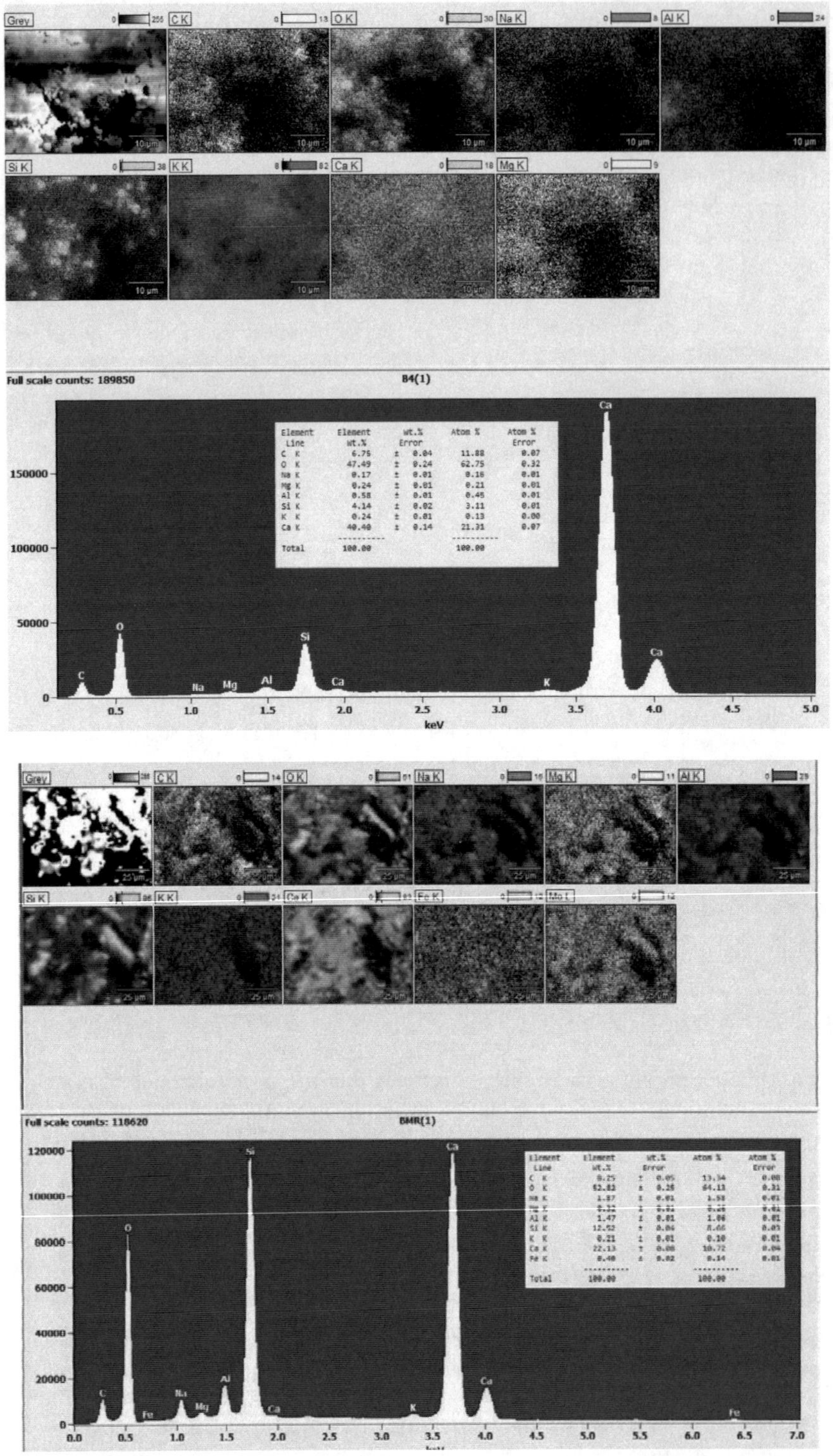

Figure 2. EDS results for chalk sample (up) and mortar sample (down).

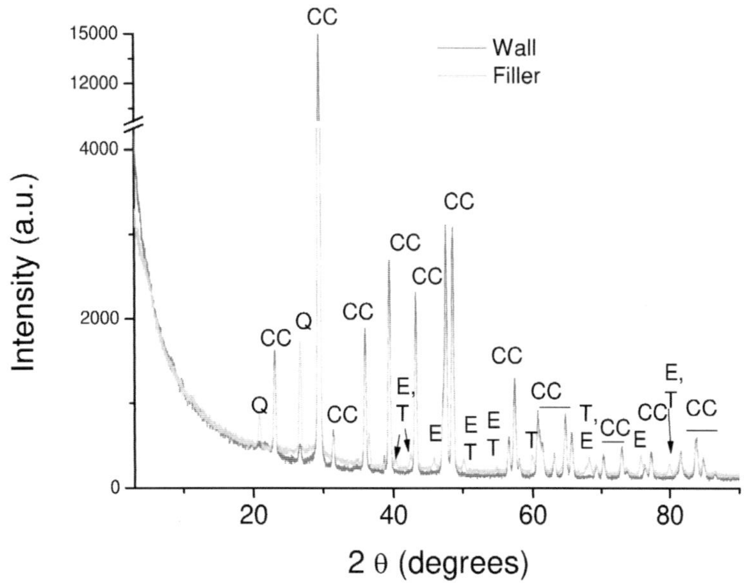

Figure 3. The XRD diagramms for chalk sample and mortar sample used for restoration.

The sulphate determination for chalk and mortar samples showed us higher concentration in mortar than in chalk sample, as follows: 0.87 % in chalk and 1.275 % in mortar. Both samples are affected by sulphate but in mortar the concentration is higher, this mortar becoming a dangereous damage factor for chalk.

By degradation, on the chalk surface no visible evidence could occurs, except effluorescence and small cracks, but on mortar large brown area could be observed most probably due to pyrite formation during degradation. Also, large smoothy deposits are present, with different structure, easy recognizible as thaumasite and probably ettringite, figure 4.

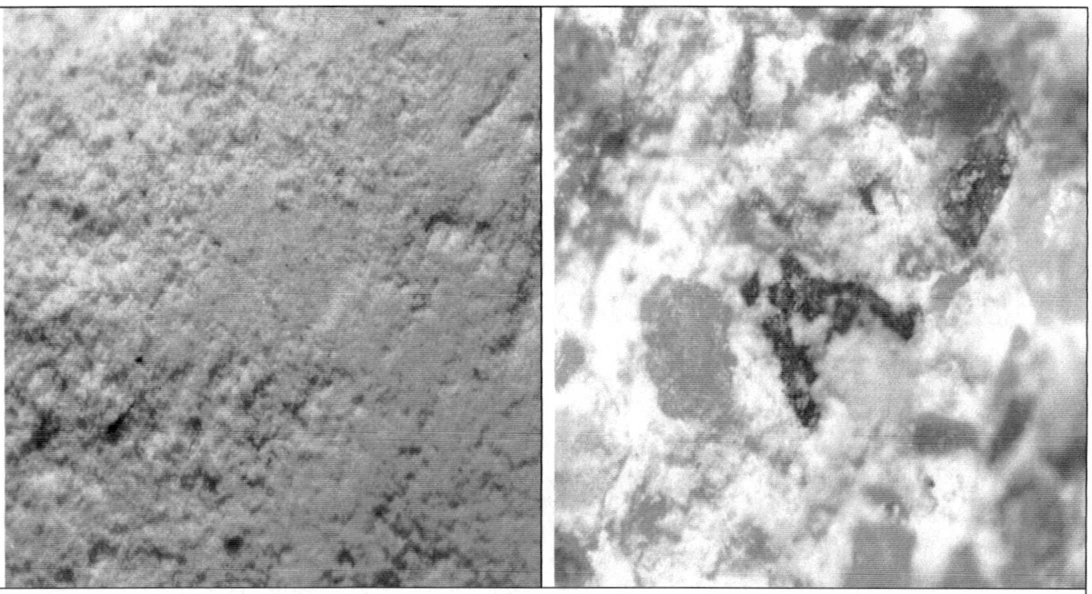

Figure 4.	
The optical microscopy of chalk sample	The optical microscopy of restoration mortar

The chalk was found to be extensively deteriorated in both out-door and indoor environments in the studied historical monument, showing flaking, subflorescence, efflorescences, crumbling and black and white crusts as a result of the deterioration phenomena [12]. The infrared spectroscopic analysis carried out on the sample prelevated from Basarabi Church allowed the identification of teh following products, as follows (figure 6):

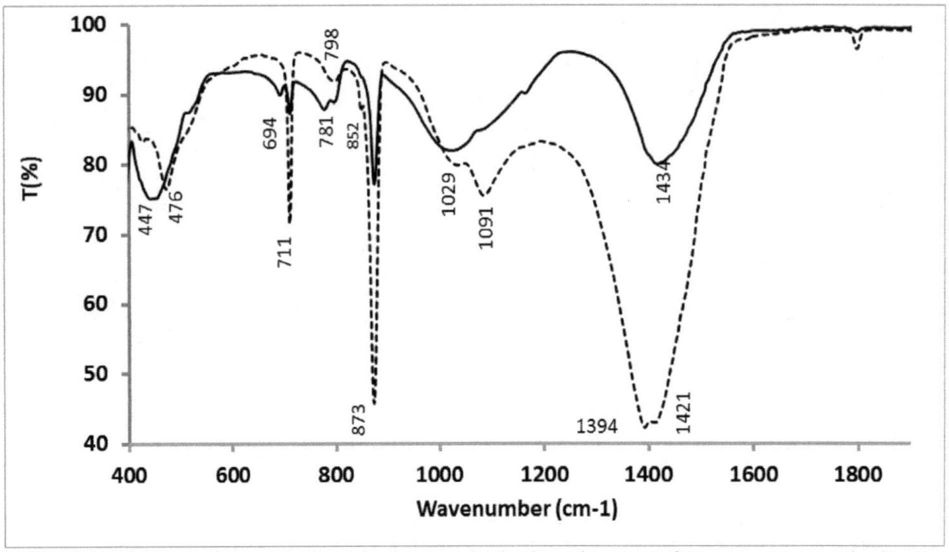

Figure 5. The FTIR spectra of chalk sample (---) and restoration mortar sample (___).

The spectrum shows specific bands for CO_3^{2-} stretching at 1409-1426, 873 SO_4^{2-}stretching at 1109 and 1029 cm-1, band for Al-OH at 873 cm-1, SiO6 stretching at 798 and 476 cm-1, Ca-OH bending at 611 cm-1. These results are in good agreement with Perkins [13], Alvarez-Ayuso et al. [14], Myneni et al. [15] and Ciliberto et al.[16].

Also, Raman spectrum of the same sample (Fig. 6) displays characteristic bands of calcium carbonate that correspond to the vibration modes of the free CO_3^{2-} ion, located at approximately 1078 cm^{-1}, recognized as the strongest feature in this spectrum [*16, 17*]. *Thaumasite* has three major peaks at 658, 990, 1076 cm^{-1} and three minor peaks at 417, 453, 479 cm^{-1}. *Ettringite* has major peaks at 990, 1088 cm^{-1}.

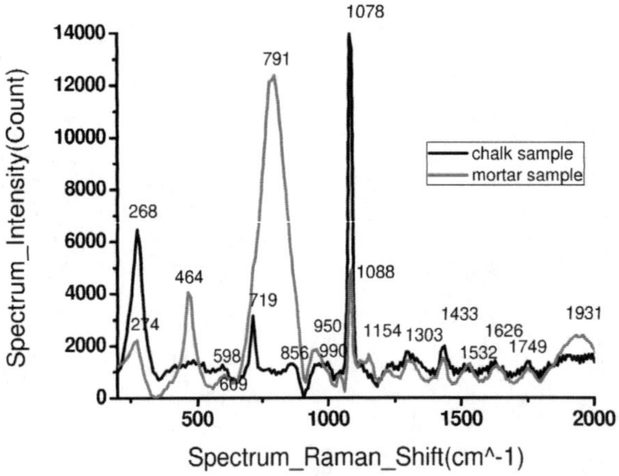

Figure 6. The Raman spectra of chalk (black) and restoration mortar (red line) samples.

Raman bands and assignments of ettringite are: 988 cm^{-1}(v1 symmetric stretching of (SO_4), 448 cm^{-1} (v2 symmetric bending of (SO_4), 1119 cm^{-1} (v3 asymmetric stretching of (SO_4), 612 cm^{-1} (v4 asymmetric bending of (SO_4), 547 cm^{-1} (Al-OH stretching) in very good agreement with other literature data [18]. Is important to mention that the above Raman frequencies are not of gypsum (1009, 414, 493, 1136, 618, 670 cm^{-1}). The presence of thaumasite and ettringite could be observed by SEM microscopy, these forms having shape of plaques (thaumasite-left image) and with some fluorish decorations (in mortar) for ettringite (right image), figure 7.

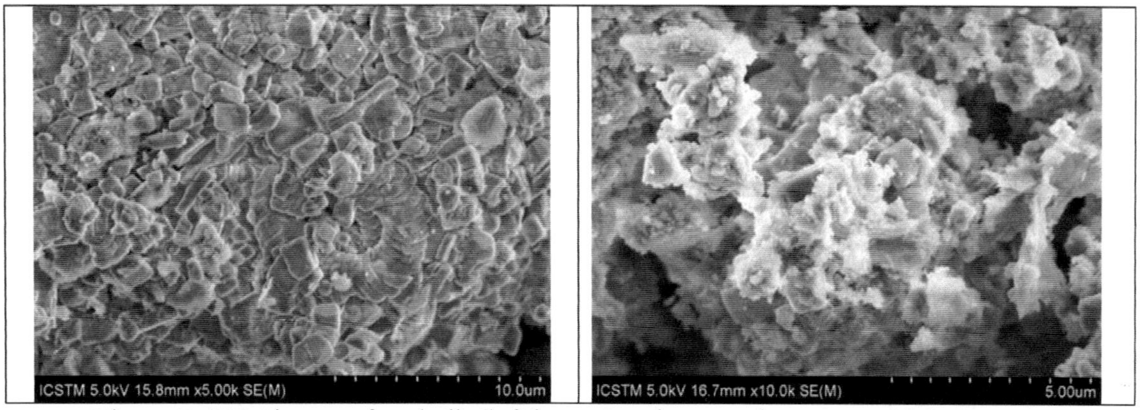

Figure 7. SEM images for chalk (left image) and restoration mortar (right image).

4. Conclusions

Cement-based materials exposed to sulfate present in the chalk stone - gypsum ($CaSO_4.2H_2O$), can induce signs of deterioration, due to ettringite ($[Ca_3Al\ (OH)_6\cdot 12H_2O]_2\cdot (SO_4)_3\cdot 2H_2O$) or thaumasite ($Ca_3[Si(OH)_6\cdot 12H_2O]\cdot (CO_3)\cdot SO_4$) formation. These phases contribute to strain within the material, inducing severe degradation. Several combined techniques (XRD, EDXRF, ICP-AES, SEM, EDS, sulphates content, FT-IR and Raman) were carried out to put into evidence the effects of them on the building walls.

5. Acknowledgements

This paper has been prepared with the financial support of the projects PN II 261/2014 and PNII 222/2012.

References

[1] P. Ferreira Pinto and J. Delgado Rodrigues, 2008 J. Cult. Herit., 9, pp. 38.

[2] A. E. Charola and R. Ware, 2002 Acid deposition and the deterioration of stone: a brief review of a broad topic, In: (eds) S. Siegesmund, T. Weiss and A. Vollbrecht, *Natural stone, weathering phenomena, conservation strategies and case studies* (Geological Society) Special Publication 205, pp. 393-406.

[3] G. A. Pope, J. M. Stavash and J. C. Walker, 2002 Correlation of acid wet deposition with trends in stone deterioration at the local scale, *R.* Prikryl and H. A.Viles *Understanding and managing of stone decay ed* , pp. 297–316.

[4] E. Doehne, 2002 Salt weathering: a selective review, (eds) S. Siegesmund, T. Weiss and A. Vollbrecht, *Natural stone, weathering phenomena, conservation strategies and case studies* (Geological Society) Special Publication 205, pp. 51-64.

[5] J. Ondrasina, D. Kirchner and S. Siegesmund, 2002 Freeze - thaw cycles and their influence on marble deterioration: a long - term experiment, (eds) S. Siegesmund, T. Weiss and A. Vollbrecht, *Natural stone, weathering phenomena, conservation strategies and case studies* (Geological Society) Special Publication, 205, pp. 9-18.

[6] T. Yates, 2003 Mechanism of air pollution damage to bricks, concrete and mortar, ed P. Brimblecombe, *The Effects of Air Pollution on the Built Environment*, pp. 107-122.
[7] E. F. Irassar, V. L. Bonavetti, M. A. Trezza and M. A. González, 2005 *Cement and Concrete Composites*, 27(1), pp. 77.
[8] M. Collepardi, 2003 *Cem. Conc. Comp.*, 25, pp. 401.
[9] G. Cultrone, E. Sebastián and M. Ortega Huertas, 2005 *Cement and Concrete Research*, 35, pp. 2278.
[10] R. M. Ion, D. Turcanu-Carutiu, R. C. Fierascu, I. Fierascu, I. R. Bunghez, M. L. Ion, S. Teodorescu, G. Vasilievici and V. Raditoiu, 2015 *Applied Surface Science*, 358B, pp. 612-618.
[11] V. Gosselin, A. Verges-Belmin, V. Royer and G. Martinet, 2009 *Materials and Structures*, 42, pp. 749.
[12] R. M. Ion, 2013 *Journal of Optoelectronics and Advanced Materials*, 15, pp. 888.
[13] R. B. Perkins and C. D. Palmer, 1999 *Geochim. Cosmochim.Acta*, 63, pp. 1969.
[14] E. Alvarez-Ayuso and H. W. Nugteren, 2005 *Water Res.*, 39, pp. 65.
[15] S. C. B. Myneni, S. J. Traina, G. A. Waychunas and T. J. Logan, 1998 *Geochim. Cosmochim. Acta*, 62, pp. 3499.
[16] E. Ciliberto, S. Ioppolo and F. Manuella, 2008 *J.Cult.Herit.*, 9, pp. 30.
[17] Y. Yue, Y. Bai, P. A. Muhammed Basheer, J. J. Boland and J. J. Wang, 2013 *Optical Engineering*, 52(10), pp. 104107.
[18] S. Kramar, M. Urosevic, H. Pristacz and H. Mirtic, 2010 *J. Raman Spectrosc.*, 11, pp. 1441.

International Conference on Innovative Research 2016 - ICIR Euroinvent 2016 IOP Publishing
IOP Conf. Series: Materials Science and Engineering **133** (2016) 012039 doi:10.1088/1757-899X/133/1/012039

Properties and Leachability of Self-Compacting Concrete Incorporated with Fly Ash and Bottom Ash

Aeslina Abdul Kadir[1], Mohd Ikhmal Haqeem Hassan[2], Norwati Jamaluddin[1] & Mohd Mustafa Al Bakri Abdullah[3, 2]

[1]Faculty of Civil and Environmental Engineering, Universiti Tun Hussein Onn Malaysia, 86400 Parit Raja, Batu Pahat Johor, Malaysia
[2]Center of Excellence Geopolymer & Green Technology (CeGeoGTech), School of Material Engineering, Universiti Malaysia Perlis (UniMAP), Perlis Malaysia
[3]Faculty of Engineering Technology, Universiti Malaysia Perlis (UniMAP), Perlis, Malaysia

Email: aeslina@uthm.edu.my, mohd.ikhmal.haqeem@gmail.com & mustafa_albakri@unimap.edu.my

Abstract. The process of combustion in coal-fired power plant generates ashes, namely fly ash and bottom ash. Besides, coal ash produced from coal combustion contains heavy metals within their compositions. These metals are toxic to the environment as well as to human health. Fortunately, treatment methods are available for these ashes, and the use of fly ash and bottom ash in the concrete mix is one of the few. Therefore, an experimental program was carried out to study the properties and determine the leachability of self-compacting concrete incorporated with fly ash and bottom ash. For experimental study, self-compacting concrete was produced with fly ash as a replacement for Ordinary Portland Cement and bottom ash as a replacement for sand with the ratios of 10%, 20%, and 30% respectively. The fresh properties tests conducted were slump flow, t_{500}, sieve segregation and J-ring. Meanwhile for the hardened properties, density, compressive strength and water absorption test were performed. The samples were then crushed to be extracted using Toxicity Characteristic Leaching Procedure and heavy metals content within the samples were identified accordingly using Atomic Absorption Spectrometry. The results demonstrated that both fresh and hardened properties were qualified to categorize as self-compacting concrete. Improvements in compressive strength were observed, and densities for all the samples were identified as a normal weight concrete with ranges between 2000 kg/m^3 to 2600 kg/m^3. Other than that, it was found that incorporation up to 30% of the ashes was safe as the leached heavy metals concentration did not exceed the regulatory levels, except for arsenic. In conclusion, this study will serve as a reference which suggests that fly ash and bottom ash are widely applicable in concrete technology, and its incorporation in self-compacting concrete constitutes a potential means of adding value to appropriate mix and design.

1. Introduction

In a developed country like Malaysia, the growth in production and manufactured sector had lead to increased industrial by-product waste. According to [1], Malaysia had generated approximately 1.1 million metric tons of scheduled waste in the year 2007. Every year, the coal-fired power plants produced large amounts of FA and BA worldwide. Malaysia was not left behind in created of fly ash and bottom ash. Although there was no reported about the production of coal ash every year in

Malaysia, about 10% of the total weight of the coal ash has been contributed [2].The schedule industrial waste generated in Malaysia in the year 2009 was 1,705,308 metric tons, and the major components of the waste were from dross, slag, clinker, ash, gypsum, oil and hydrocarbon [2, 3, 4].

Both FA and BA disposed of as waste. The landfill has become the primary method of disposal of waste materials. However, dumped the BA and FA in the landfill was not the best sustainable solution [5]. These wastes must adequately manage without caused any harmful environment effects. Alternatively, a better approach in solved and handled of coal combustion by-product should be explored. The FA and BA were effectively used in many applications [6]. The problems that occurred to disposed of coal ash with limited landfill area can be reduced by reuse of FA and BA in the construction industry technology. The used of FA from coal-fired power plants and ground granulated slag from the blast-furnace industry in blended Portland cement was an excellence example of industrial that reduced environmental impact [7, 8].

Previous studies proved that FA and BA could potentially be used as an alternative replacement material for cement and sand in the concrete mix. This research was conducted based on the awareness of the need to utilize the waste product to preserve the environment [9,10] It was also to promote the use of environmentally friendly technology and sources in construction. Therefore, this research was conducted to determine the properties as well as leachability of self-compacting concrete (SCC) incorporating with FA and BA and to formulate the suitable ratio for this waste replacement.

2. Materials and Methods

2.1. Materials

Fly ash and bottom ash were collected from one of the coal-fired power plant located in Peninsula Malaysia. FA and BA were become part of the SCC mix composition by replacing cement and sand in the percentages of 0%, 10%, 20% and 30% respectively. ADVA 181 superplasticizer (SP) that specially formulated to produce high strength concrete with good slump retention was used. According to Table 1, ten design mix proportions were made, which had total binder content of 530 kg/m^3 to 550 kg/m^3 at water to binder ratio (w/b) of 0.41. The workability of the mix proportions was determined by referring to specification and guidelines for SCC prepared by EFNARC [11].The size of specimens used throughout this research was 150 mm cubes.Total binder for all the samples was between 530 kg/m^3 to 550 kg/m^3 respectively. Note that the proportion is for one batch casting (0.0475 m^3).

Table 1: Mix proportions off SCC mixes incorporated with FA and BA

Mix Design	Total Binder (kg/m^3)	Cement (kg/m^3)	FA (kg/m^3)	CA (kg/m^3)	Sand (kg/m^3)	BA (kg/m^3)	Water (kg/m^3)	SP (kg/m^3)
FA0BA0	550	550	0	593	914	0	228	5
FA10BA0	550	495	55	593	914	0	220	4
FA20BA0	540	432	108	593	914	0	216	4
FA30BA0	530	371	159	593	914	0	212	4
FA0BA10	550	550	0	593	822	67	228	5
FA0BA20	550	550	0	593	731	133	228	5
FA0BA30	550	550	0	593	640	200	228	5
FA10BA10	550	495	55	593	822	67	220	4
FA20BA20	540	432	108	593	731	133	216	4
FA30BA30	530	371	159	593	640	200	212	4

*CA: Coarse aggregate; SP: super plasticizer

2.2. Methods

Each mixed proportions were tested for their fresh and hardened properties of SCC to classify it at SCC. These tests are the filling ability (slump flow and T_{500} test), passing ability (J-ring test) and segregation resistance (sieve segregation test) according to BS EN 12350 respectively [12, 13, and 14]. Curing was done in accordance with BS EN 12390-2 and density, compressive strength and water absorption at 28 days were determined by referring BS EN 12390 and BS 1881: part 122:1983 [15, 16].

After compressive strength test, the samples were then crushed further using the Aggregate Impact Value (AIV) equipment so as to reduce the solid particles sizes to less than 9.5 mm. Using a sieve with the corresponding size, the crushed fragments were sieved and collected as samples for Toxicity Characteristic Leaching Procedure (TCLP) [17].

TCLP was used as a set of instructions to prepare extracted samples of concrete samples for leaching analysis to be done. As SCC cubes were solid samples, the procedure of extraction had to be done under the category for sample containing more than 0.5% dry solid. The pH of each solid sample was checked in order to determine the suitable extraction fluid to be used. It was found that the pH of all samples exceeded 5.0; hence extraction fluid No.2 was prepared and used to extract the samples.

Extraction fluid No.2 was prepared by diluting 5.7 mL glacial acetic acids (CH_3CH_2OOH) with distilled water to a volume of 1 litre. A total of 50 g sample was prepared and placed into a 2 liter extraction bottle, and the extraction fluid was poured in after. Fixing the extraction bottle to a rotary agitation apparatus, it was left to rotate from end to end for 18 hours. The solutions within the extraction bottle were then filtered to dispel the solid particles. The fluid portion of the sample was preserved to a pH less than 2.0 and stored in a refrigerator at 5 °C for leachate determination analysis using AAS, which was performed using Perkin Elmer AAnalyst 800.

3. Result and Discussion

4. *3.1. Fresh properties of SCC.* Figure 1 showed the result of filling ability by using slump flow test of ten different percentages FA and BA in SCC. Generally, diameters of the spread for all the mixes were recorded higher than the minimum requirement of the SCC, which is 500 mm. 4 mixes come out with result below 600 mm which is control, FA10BA0, FA20BA0, and FA30BA0 of SCC samples with diameter of spread 565 mm, 553 mm, 580 mm and 518 mm respectively. The other six mixes were recorded higher than 600mm diameter of the spread. A slump flow value ranging from 500 to 700 mm for a concrete to be self-compacting was suggested by Liu [20]. At slump flow higher than 700 mm, the concrete might segregate, and at lower than 500 mm, the concrete might have insufficient flow to pass through highly congested reinforcement.The time spread of t_{500} for the samples was recorded as shown in Figure 2. The red line in the figures indicates the permissible line to recognized the concrete as SCC.

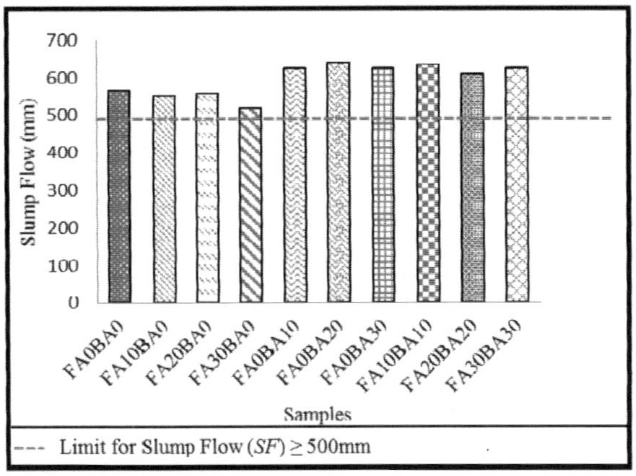

Figure 1: Slump flow against type of SCC samples

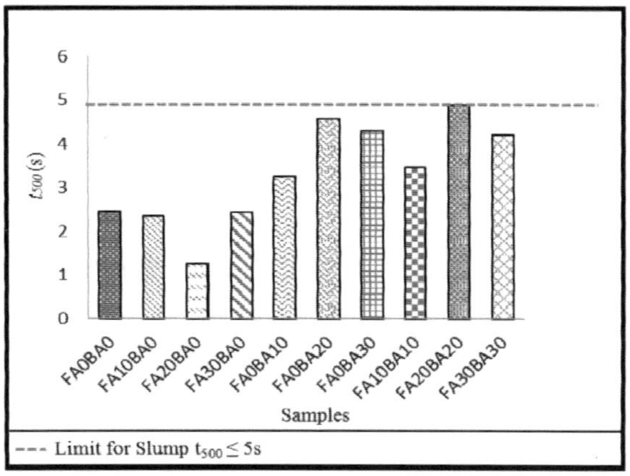

Figure 2: t_{500} (s) against type of SCC samples

Meanwhile, Figure 3 showed the result of passing ability by using J-ring test. According to the result, all the design mix proportions have fulfilled the requirement of the J-ring test. Those samples showed the ability to pass the height of spread $\leq$ 2cm and distance from spread end to $t_{500} \leq 10$ cm. The highest reading for the height of spread for J-ring test was recorded by the sample FA30BA30 with 1.025cm while the sample FA30BA0 recorded the height of spread for J-ring test with 2.08cm.

Based on the sieve segregation test result from Figure 4, all the design mix proportions have achieved the target of the test that was conducted to determine the ability to remain homogeneous in composition during transport and placing of the SCC. The segregated portion ≤ 20% was the requirement need to be obtained by the samples. The lowest value of segregated portion, 3.33% was recorded by the sample FA0BA30 and the highest values of segregated portion are from the control (FA0BA0) sample with 15.85%.

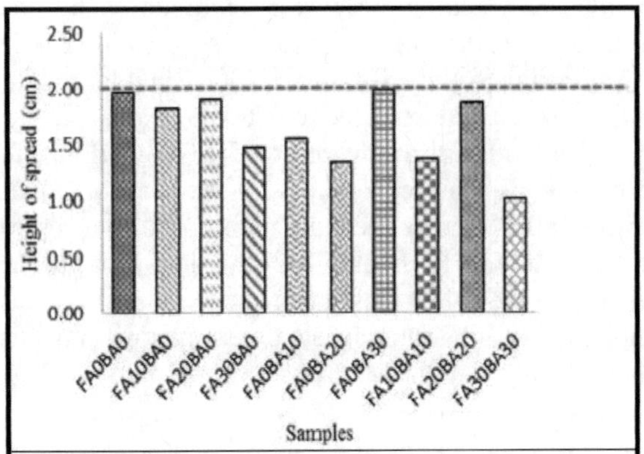

Figure 3: Average height (h) of spread (cm) against types of SCC samples

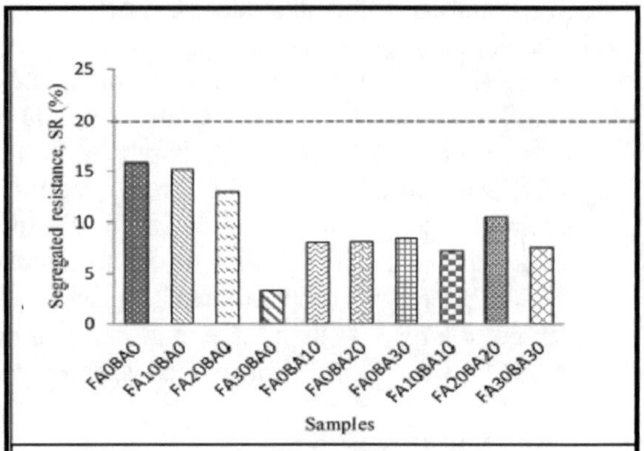

Figure 4: Segregated resistance, SR (%) against types of SCC samples

3.2 Hardened properties of SCC

3.2.1 Density

The density of SCC incorporated with FA and BA was determined by weighing the cube samples before conducting the compressive strength test. For accuracy, the average density is calculated for each concrete mix design and the results as tabulated in Table 2. According to Table 2, FA30BA30 samples achieved the lowest density which is 2170 kg/m^3. Meanwhile the highest density of SCC samples was recorded by FA10BA0 and FA20BA0 samples with 2304 kg/m^3. It shows that with the replacement of FA up to 20% in the samples, the density was recorded higher compared to the samples with BA alone. Nevertheless, all the samples were comparable with BS EN 206: 2013 as a normal weight concrete with the range between 2000 kg/m^3 and not exceed 2600 kg/m^3. Other than that, [18] suggested that density of hardened mortar or concrete varies, depending on the amount and density of the aggregate, the amount of air voids that is entrained or entrapped and the water and cement contents.

Table 2: Density of SCC samples

Sample	Average (kg/m^3)
FA0BA0	2189
FA10BA0	2304
FA20BA0	2304
FA30BA0	2274
FA0BA10	2244
FA0BA20	2289
FA0BA30	2244
FA10BA10	2296
FA20BA20	2207
FA30BA30	2170

3.2.2 Water absorption

Water absorption is used to determine the amount of water absorbed under specified conditions which indicate the degree of porosity of material [19]. This test was conducted according to BS 1881: 2011 on 28 days to determine the water absorption of SCC incorporated FA and BA. The results for water absorption of SCC samples were tabulated in Table 3. According to Table 3; it shows that with the increase of BA only to the samples, the water absorption decreased slightly for FA0BA10 sample with 1.08 %, FA0BA20 sample with 1.02% and FA0BA30 sample with 0.67%. It is suggested by Siddique [19] that incorporating of BA in concrete resulted in low absorption characteristic that is less than 10% of absorption. The same patterns are also demonstrated for the incorporation of FA and combination of FA and BA. For samples with FA only, the results are 1.10%, 0.66% and 0.53% with the addition of 10%, 20% and 30% of FA respectively. For the samples incorporated with FA and BA, the results are 0.79%, 0.55% and 0.46% for FA10BA10, FA20BA20 and FA30BA30 respectively. According to the result, it showed that the more percentage of FA and BA added to the samples, the lower the percentages of water absorption recorded. It was supported by Liu [20] that this could because the FA is finer than the cement .Therefore it fills the voids in the samples and leading to lower porosity. It is noticeable that all SCC mixes samples had low absorption characteristic (less than 10%).

Table 3: Water absorption results

Sample	Water Absorption (%)
FA0BA0	0.97
FA10BA0	1.10
FA20BA0	0.66
FA30BA0	0.53
FA0BA10	1.08
FA0BA20	1.02
FA0BA30	0.67
FA10BA10	0.79
FA20BA20	0.55
FA30BA30	0.46

3.2.3 Compressive Strength

In terms of mechanical properties, the compressive strength of the cube samples with a size of (150x150x150) mm was tested at the age of 7 days, 14 days and 28 days. The compressive strength was determined accordingly to BS EN 12390-3:2009. All the SCC incorporated with FA and BA obtained compressive strengths with similar range but higher than normal concrete. By referring to BS EN 206: 2013, compressive strength classes for the samples were range from class C45 to C70 at 28 days and meets the requirement to be categorized as normal-weight and heavy-weight concrete. Figure 5a, 5b, and 5c show the comparison of compressive strengths between control sample (FA0BA0) and samples with various percentages of FA and BA replacement in SCC.

Figure 5a shows the compressive strength at various FA contents in SCC samples. At various replacement of OPC with 0%, 10%, 20%, and 30% of FA, strengths were observed to be in the range of 30 MPa to 42 MPa at seven days, 41 MPa to 59 MPa at 14 days and 49 MPa to 69 MPa respectively. All the samples incorporated with FA were higher than control sample which is FA0BA0. It shows that replacement of FA has improved the compressive strength of the SCC. The highest strength was recorded from FA10BA0 sample with 68.79 MPa at 28 days. The increase in strength of fly ash concrete may be attributed to continuous hydration and the filling of pores with Calcium Silicate Hydrated gel formed due to pozzolanic action of coal fly ash [21].

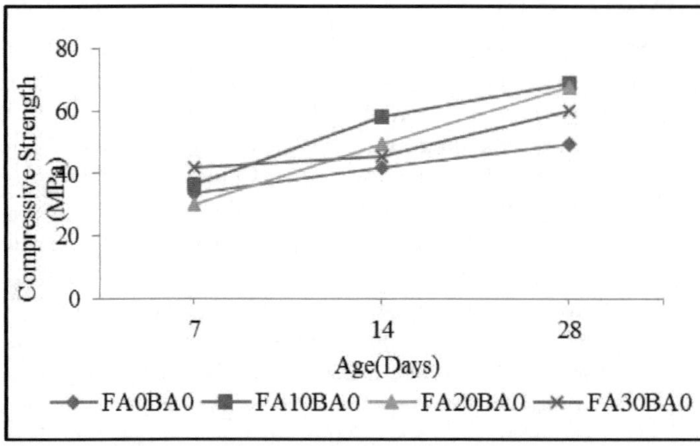

Figure 5a: Compressive strength at various FA contents (0%, 10%, 20% and 30% of FA) in SCC samples

Meanwhile, for compressive strength with a various replacement of sand with 10%, 20%, and 30% of BA, the results are shown in Figure 5b. It shows that strengths were observed to be in the range of 33 MPa to 50 MPa, 41 MPa to 57 MPa, and 49 MPa to 59 MPa at 7, 14 and 28 days respectively. SCC incorporated with BA gained higher strength at early ages of the SCC ranging from 47 MPa to 49 MPa but slowly increased at 14 days and 28 days with the highest value recorded was from FA0BA10 sample with 58.01 MPa.

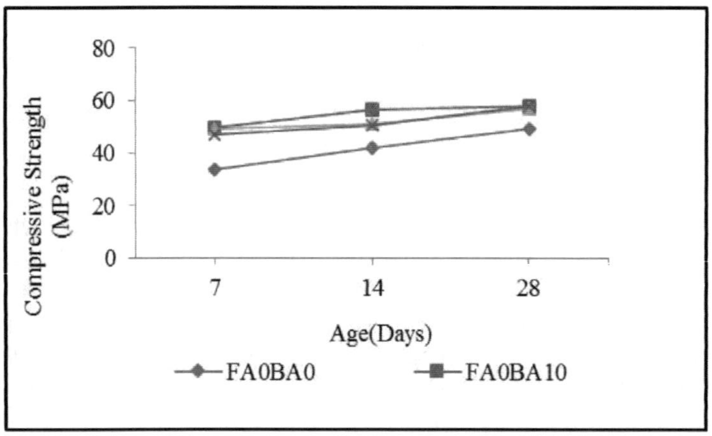

Figure 5b: Compressive strength at various BA contents (0%, 10%, 20% and 30% of BA) in SCC samples

Figure 5c shows the compressive strength with various replacements of FA and BA (10%, 20%, and 30%) replacing cement and sand in SCC. Again, the result showed that SCC incorporated with FA and BA were higher than control samples with a range of 33 MPa to 46 MPa, 38 MPa to 53 MPa, and 46 MPa to 58 MPa at 7, 14 and 28 days respectively. However, only FA30BA30 recorded lower value than the control sample. Therefore, the reduction in strength was observed up to 30% replacement of the fine aggregate of FA and BA. It was supported by [22] that increasing in percentages of BA will decrease the compressive strength of SCC.

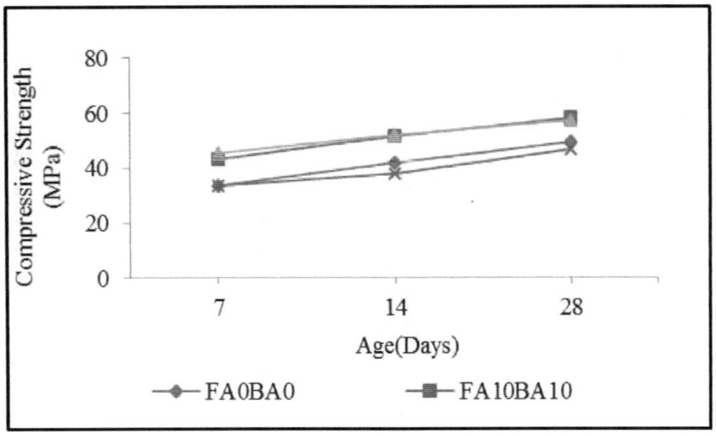

Figure 5c: Compressive strength at various FA and BA contents (0%, 10%, 20% and 30% of FA and BA) in SCC samples

3.2.4 Morphology Analysis

Morphology analysis using Scanning Electron Microscope (SEM) was done in order to determine to investigate the surface changes in the SCC samples and also its porosity with 1000x magnification as illustrated in Figure 6a to Figure 6j. In addition, SEM is one of the most versatile instruments available for the examination and analysis of the microstructure characteristics of solid objects SEM observation was carried out on the surface of control SCC sample, samples incorporated with FA only, samples incorporated with BA only and samples incorporated with both FA and BA samples. As can be seen, the majority of the FA and BA particles are spherical. This ash consists of a series of spherical vitreous particles of different sizes (diameters ranging from 200 μm to 10 μm). The morphology of the particles was similar to the control sample. The spheres seem to be almost intact or appear within other spheres, depending on the percentages of the FA in the samples. Finer particles with higher density, which have the appearance of metallic lustre in SEM images, can be seen on larger particles. Observation also revealed that there are significant changes in the growth of pore size with the incorporation of BA compared to control (FA0BA0) sample. This is because BA has a larger size to be compared to FA. The mechanical performance of the concrete can be related to morphological changes on the surfaces. Apart from that, the uses of bottom ash as replacement of natural sand in self-compacting concrete, the number of fines and irregular shaped, rough textured and porous particles increases and thereby increasing the inter-particles friction. The increased inter-particle friction hinders the flow characteristics of fresh concrete [23].

Figure 6a to 6c: (a) FA0BA0 sample (control); (b) FA10BA0 sample; (c) FA20BA0 sample

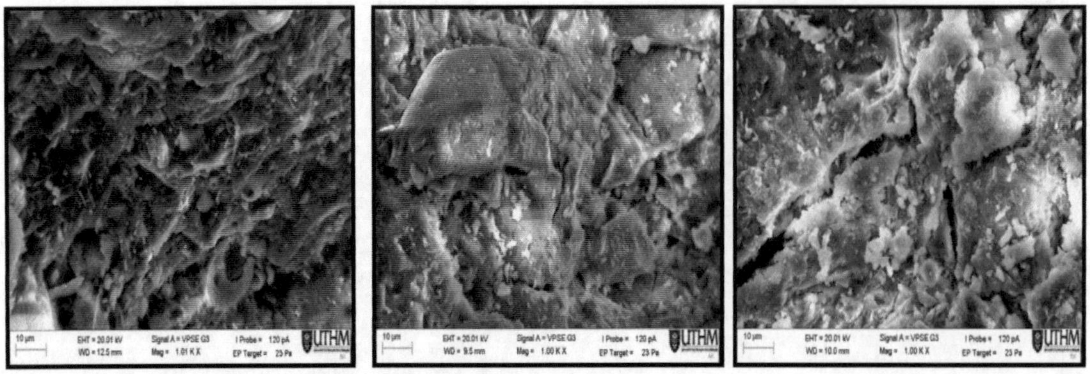

Figure 6d to 6f: (d) FA30BA0 sample; (e) FA0BA10 sample; (f) FA0BA20 sample

Figure 6g to 6i: (d) FA0BA30 sample; (e) FA10BA10 sample; (f) FA20BA20 sample

Figure 6j: FA30BA30 sample

3.3 Toxicity Characteristic Leaching Procedure (TCLP) result

From the results of the TCLP test in Table 4, it shows that there are insignificant levels of heavy metals leached out from the samples. The results were then compared to United States Environmental Protection Agency regulatory levels.

Element concentrations in all the samples in TCLP were below the permissible limit set by USEPA except for arsenic. In addition, samples with replacement of FA alone and samples with a combination of FA and BA resulted in the higher leaching of As to be compared with samples with BA alone. The highest leaching of As was recorded by sample FA30BA30 with 18.576 mg/L. Arsenic has gained considerable attention due to the fact that it is mobile throughout a wide pH range. Arsenic releases from acidic solution increase with pH, whereas in alkaline solution, this trend is reversed [24]. The difference in leaching is caused by the Ph dependent of most heavy metals elements [25]. Another study by [24] showed that the types of FA are also a factor in leaching of As. It is because FA consists aluminum oxide and iron hydroxide that are common sorbents for As adsorption.

Table 4: Leachability of heavy metals in SCC samples in TCLP

Heavy metals	Concentration level (mg/L)*	Samples									
		FA0BA0	FA10BA0	FA20BA0	FA30BA0	FA0BA10	FA0BA20	FA0BA30	FA10BA10	FA20BA20	FA30BA30
		Concentration (mg/L)									
As	5	12.060	14.705	14.895	15.370	13.640	13.675	14.640	16.510	18.200	18.576
Cr	5	0.112	0.128	0.132	0.140	0.143	0.154	0.170	0.174	0.184	0.207
Pb	5	0.447	0.783	0.847	0.909	0.578	0.646	0.695	0.940	1.024	1.028
Zn	NA	0.173	0.145	0.202	0.222	0.145	0.144	0.160	0.172	0.177	0.181
Cu	100	0.026	0.030	0.030	0.031	0.029	0.026	0.031	0.033	0.034	0.035
Ni	NA	0.167	0.065	0.066	0.069	0.065	0.071	0.075	0.062	0.063	0.077
Fe	NA	0.095	0.074	0.083	0.083	0.053	0.034	0.017	0.136	0.037	0.000
Mn	NA	ND	ND	ND	ND	ND	ND	ND	ND	ND	ND

Where,
As = Arsenic Pb = Lead Cr = Chromium Zn = Zinc Cu = Copper Ni = Nickel
Fe = Ferum Mn = Manganese

* US Environmental Protection Agency, USEPA (1996); NA – Not Available; ND – Not Detected

4. Conclusion

As a conclusion, it was observed that fly ash and bottom ash have potential to been used as replacement in SCC. All the samples incorporated with these waste materials showed the fresh properties within the ranges specified to be categorized as SCC. As for compressive strength, it shows that compressive strength increased from day 7 up to day 28. By referring to BS EN 206: 2013, compressive strength classes for the samples were range from class C45 to C70 at 28 days and met the requirement to be categorized as normal-weight and heavy-weight concrete. From water absorption, it shows that with the increase of BA only to the samples, the water absorption decreased slightly for FA0BA10 sample with 1.08%, FA0BA20 sample with 1.02% and FA0BA30 sample with 0.67%. SEM images for SCC incorporated with FA shows the morphology of the particles were similar to the control sample. The spheres seem to be almost intact or appear within other spheres, depending on the percentages of the FA in the samples. Leachability of heavy metals of SCC incorporated with FA and BA was determined using TCLP. As environmental concerns is a top priority, the uses of FA and BA are not only could reduce the amount of these wastes materials sent to the landfill, but it is also will serve better in concrete technology in terms of economic-viability as a low-cost replacement materials.

5. References

[1] Ahmaruzzaman, M., (2010). A review on the utilization of fly ash. *Progress in Energy and Combustion Science, Vol.* 36(3), pp.327–363.

[2] Kadir A. A., Hassan M. I. H.,(2014) "An Overview of Fly Ash and Bottom Ash Replacement in Self Compaction Concrete" *Key Engineering Materials* Vols. 594-595 pp 465-470.

[3] Naganathan, S., (2012). Development of Brick Using Thermal Power Plant Bottom Ash and Fly Ash. *Asian Journal of Civil Engineering (Building and Housing)*, 13(1), pp.275–287.

[4] B.Kim, M. Prezzi, & R. Salgado(2005), "Geotechnical properties of fly and bottom ash mixtures for use in highway embankments." *Journal of Geotechnical, Geo-environment Engineering*. 131(7), 914–924,

[5] Siddique, R., Paratibha, A., Yogesh, A., (2012) "Influence of water/powder ratio on strength properties of self-compacting concrete containing coal fly ash and bottom ash", *Construction and Building Materials* 29 pp. 73–81.

[6] Syahrul, M., Muftah F., Muda Zulkifli (2010). The Properties of Special Concrete Using Washed Bottom Ash (WBA) as Partial Sand Replacement. *International Journal of Sustainable Construction Engineering & Technology*, 1(December 2010), pp.65–76.

[7] Prajapati K, Chandak R. and Dubey S.K., (2012) "Development and Properties of Self Compacting Concrete Mixed with Fly Ash", *Research Journal of Engineering Sciences Vol.* (3), 11-14.

[8] Muhardi, A., (2010). Engineering characteristics of Tanjung Bin coal ash. *Electronic Journal of Geotechnical Engineering*, *Vol.* 15, pp.1117–1129.

[9] Sarode, D.B.,Jadhav R. N.,Khatik V. A., Ingle S. T., Attarde S.B., (2010). Extraction and leaching of heavy metals from thermal power plant fly ash and its admixtures. *Polish Journal of Environmental Studies*, 19(6), pp.1325–1330.

[10] Kadir, A and Mohajerani, A (2012), 'Leachability of heavy metals from fired clay bricks incorporated with cigarette butts',*Proceedings from the 8th International Symposium on Lowland Technology*, pp. 1-7.

[11] EFNARC, (2002), "Specifications and Guidelines for Self-Compacting Concrete", (www.efnarc.org), pp 1-32.

[12] British Standards Institution (2010). Testing fresh concrete – Part 8: Self-compacting concrete – Slump-flow test. London. BS EN 12350-8.

[13] British Standards Institution (2010). Testing fresh concrete – Part 11: Self-compacting concrete – Sieve segregation test. London. BS EN 12350-11.

[14] British Standards Institution (2010). Testing fresh concrete – Part 12: Self-compacting concrete – J-ring test. London. BS EN 12350-12.

[15] British Standards Institution (2000). "Testing hardened concrete – part 2: making and curing specimens for strength tests, BS EN 12390-3: 2000,

[16] BS 1881-122:1983**,** Testing concrete. Method for determination of water absorption.

[17] TCLP. Toxicity Characteristic Leaching Procedure. (1992). US Environmental Protection Agency, Method 1311.

[18] Wongkeo, W., Thongsanitgarn P., Ngamjarurojana A., Chaipanich A., (2014) Compressive strength and chloride resistance of self-compacting concrete containing high level fly ash and silica fume. *Materials & Design*, *Vol.* 64, pp.261–269.

[19] Siddique, R., (2013)Properties of self-compacting concrete containing class F fly ash, Material and Design 32 1501–1507.

[20] Liu, M., (2010). Self-compacting concrete with different levels of pulverized fuel ash. *Construction and Building Materials*, *Vol.* 24 (7), pp.1245–1252.

[21] Siddique, R., (2010) "Utilization of coal combustion by-products in Sustainable Construction materials, resources, conservation and recycling, Vol. 54, *Construction and Building Materials*, pp. 1060-1066.

[22] Kasemchaisiri, R. & Tangtermsirikul, S., (2008). Properties of Self-Compacting Concrete in Corporating Bottom Ash as a Partial Replacement of Fine Aggregate. *ScienceAsia* 34, pp.87–95.

[23] Singh, M. & Siddique, R., (2015). Properties of concrete containing high volumes of coal bottom ash as fine aggregate. *Journal of Cleaner Production*, 91, pp.269–278.

[24] Kosson, D.S. (2014). pH-dependent leaching of constituents of potential concern from concrete materials containing coal combustion fly ash. *Chemosphere*, *Vol.* 103, pp.140–7.

[25] Kim B., Prezzi M., Salgado R. & Lee J. (2005). Mechanical properties of Class F fly and bottom ash mixtures for embankment application. *Indian Geotechnical Conference*. pp. 239 – 242.

ZnO for photocatalytic air purification applications

I V. Tudose and M. Suchea*

Chemistry Faculty, "Al.I.Cuza" University of Iasi, 11 Bulevard Carol I, Iasi, 700506, Romania, Romania and Center of Materials Technology and Laser, School of Applied Technology, Electrical Engineering Department, Technological Educational Institute of Crete, Heraklion, Greece

E-mail: *mirela.suchea@uaic.ro; mirasuchea@staff.teicrete.gr

Abstract. Nano and micro-structured ZnO coatings onto various substrates were grown by chemical methods and optimized with respect to their photocatalytic activity against in-doors common air pollutants. Excellent quality coatings with high stability and photocatalytic efficiency were obtained with the scope to be integrated in a novel air-purification system.

1. Introduction

The photocatalytic materials have various applications such as self-cleaning [1], antifogging [2], antibacterial actions [3-5], deodorization [6] or decomposition and removal of pollutant [7-13]. These applications are responsible for the development of a broaden range of so called smart materials used constructions. TiO_2 and ZnO are most widely used semiconductor photocatalysts because of their high photosensitivity, photochemical stability, large band gap, strong oxidizing power and non-toxic nature. Particularly, anatase TiO_2 has been reported as the most extensively used semiconductor photocatalyst for industrial applications and pollution clean-up since 1970s. Recent research reports ZnO to show an even better activity than TiO_2 in the photodegradation of some dyes in aqueous solutions [10-15], since it can have better light absorption efficiency. The successful exploitation of such catalysts requires the development of techniques for controlling their size, morphology, structural and surface characteristics, as well as efforts to enhance their photochemical response to visible/solar illumination. Presently, there are quite few reports in the literature presenting state of art approaches of use of ZnO material onto textile substrates for several applications as antibacterial, deodorizing and UV protection, and none regarding any systematic approach of direct growth and optimization with respect the textile support. More than that, up to our knowledge, except our previously reported preliminary results [16, 17], there is no available study in the literature about ZnO coated textiles used as photocatalytic active support for gaseous compounds decomposition. The present presentation will bring to attention recent advances on photocatalytic ZnO onto textile supports for air purification applications. Nano and micro-structured ZnO coatings onto various textile substrates were grown by aqueous solution chemical growth from different precursors. Effect of the substrate was rigorously considered and optimized. Excellent quality coatings with high stability and photocatalytic efficiency were obtained onto mineral fibers substrates.

2. Materials and Methods

ZnO was chemically grown directly onto various different substrates from zinc acetate (ZnAc2) precursor with various concentrations in aqueous solutions as presented in Table1.

Table 1. ZnO onto various substrates samples table

Substrate	Precursor concentration	Growth conditions
PES+Viscose		
Organic Cotton		
Silk+Viscose	0.1 M ZnAC2 + 248 μL NH3 25% 20 mL	95°C/ 20hrs
Bambus		
Tencel		
Glass fiber		

All deposited materials were studied using scanning electron microscopy (SEM), energy dispersive X-ray spectroscopy (EDX) and Attenuated total reflectance (ATR) Fourier Transform Infrared Spectroscopy (FT-IR) and X-ray diffraction (XRD). SEM and EDX characterization were performed using a JEOL JSM 6362LV electron microscope equipped with an EDAX INCA X-act Oxford Instrument detector. SEM surface characterization was performed in low vacuum (3-12 Pa) without any sample preparation while EDX characterization and elemental mapping were performed in high vacuum. Elemental mapping of the sample surface was also collected at magnification x10000 and using large acquisition times (more than 80000 counts) for an accurate definition of surface points. The instrument used for FT-IR characterization is a spectrometer fitted with an ATR module manufactured by Shimadzu. For XRD characterization was used a D8 Discover XRD system from Bruker. Photocatalytic activity was preliminary tested on Methylene Blue (CI 52015) (MB) degradation under UV light exposure. Promising samples were tested as photocatalyst for formaldehyde degradation using a homemade gas chamber inserted in a Shimadzu FT-IR spectrometer.

3. Results and discussions

Crystalline ZnO was successfully deposited onto all kinds of textile substrates. Low magnification SEM characterization of the prepared samples is presented in Figure 1. As one can observe from Figure 1, all the organic substrates allowed deposition of ZnO onto them but they most of them were chemically decomposed during the deposition. ZnO deposited onto organic substrates showed less conductive behaviour than ZnO deposited onto glass fibre substrates.

Despite the chemical damage o the organic substrates, the textile fibres were quite homogeneously coated with ZnO nanostructures. Those proved to have a good adhesion to the fibre tested by successive washings. A larger magnification SEM analysis of ZnO nano-micro structures deposed onto various substrates shows that substrate has a very strong effect on ZnO structuring. Both morphology and size of ZnO particles are very different onto each substrate during exactly the same deposition procedure (all growth parameters identical except the substrate). This is shown in Figure 2.

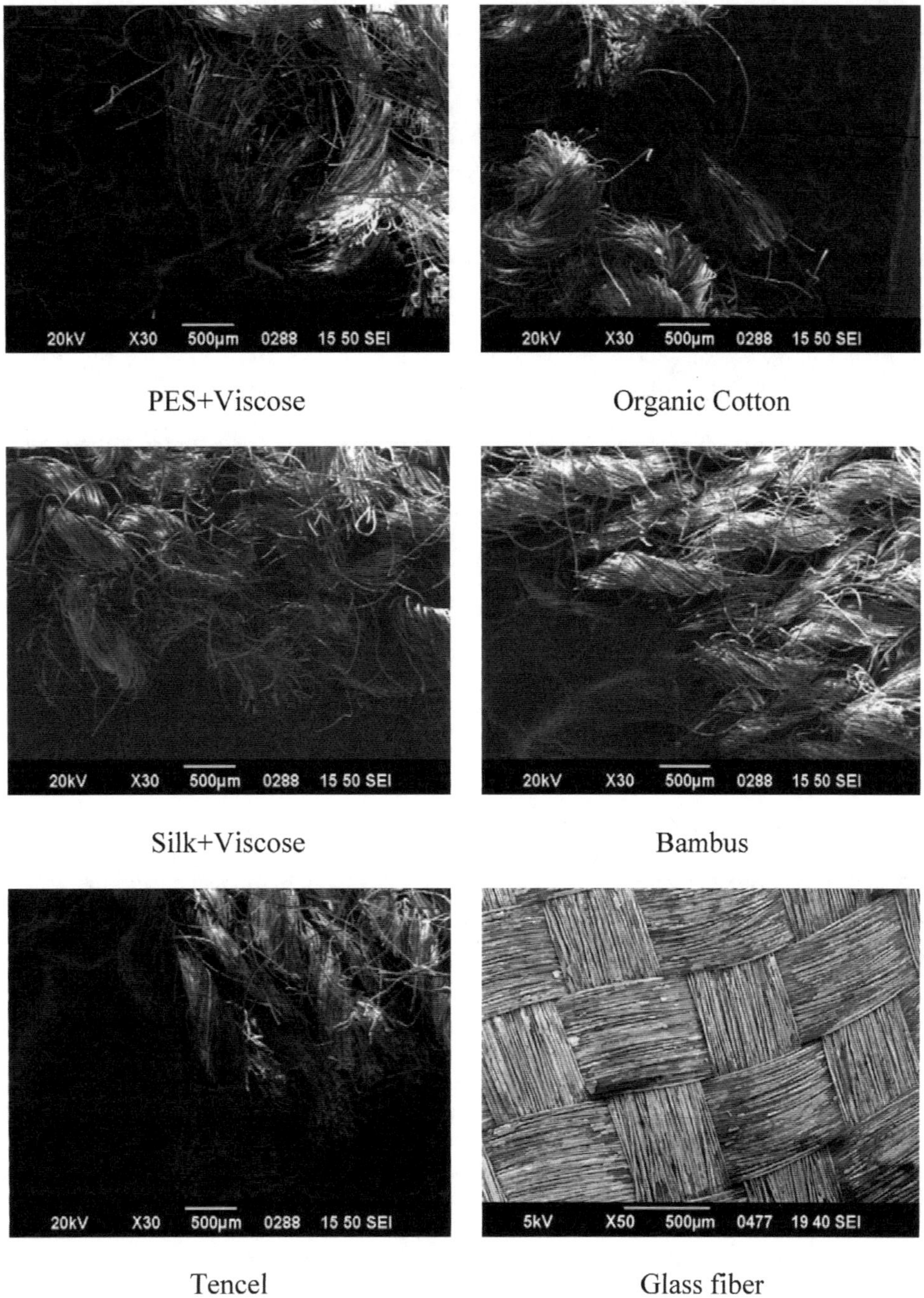

Figure 1. SEM images of ZnO onto various kinds of textile substrates

Figure 2 shows SEM images of ZnO deposited in similar conditions onto various kinds of organic substrates. PES substrates and Tencel leads to growth of ZnO assembled large nano-walls-like structures with large surface to volume ratio and good electrical conductivity while natural organic substrates (cotton, silk and bamboo) lead to smaller ZnO nonconductive nanostructures. High magnification SEM showed that onto the natural organic fibers ZnO forms hexagonal section nanorods and nanoflowers.

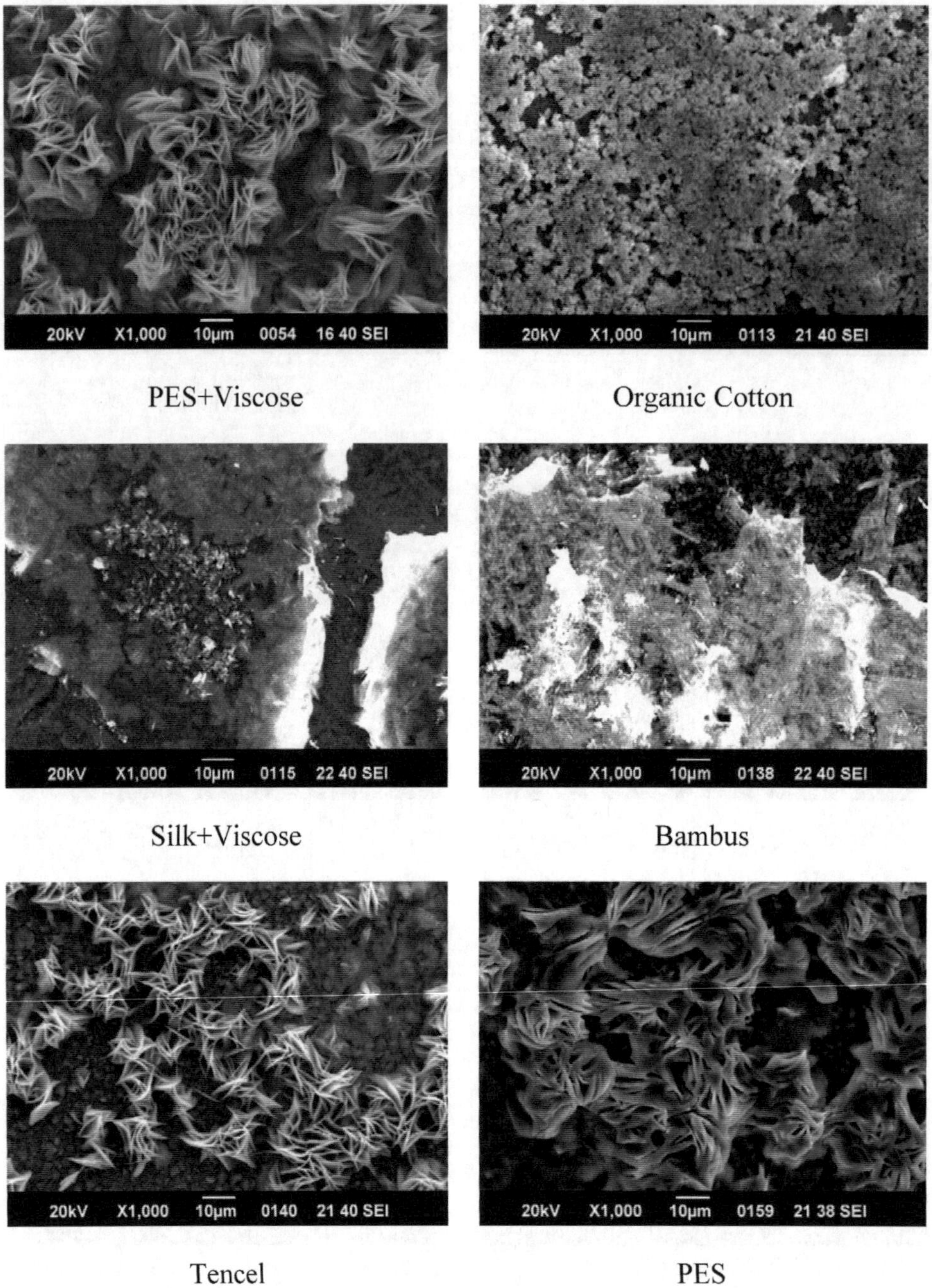

Figure 2. Larger magnification SEM images of ZnO onto various kinds of organic textile substrates

Unlike the organic substrates, the glass fiber substrate sustains the chemical growth and prove to be the best for further optimization of the ZnO growth for the photocatalytic air purification applications. Figure 3 shows large magnification images of glass fiber substrates ZnO coated. As one can see, ZnO deposited has granular dense structure and it shows a very good fiber coating as well as good electrical conductivity. Characterization was possible in high vacuum at 10kV and 20kV.

Figure 3. Excellent nanostructured ZnO coating onto glass fibre substrates SEM images.

As presented in figures 2 and 3 ZnO growth is strongly influenced by the substrate nature. Growth onto synthetic organic substrates from $ZnAC_2$ precursor solutions leads to ZnO structuring as walls-like structures while growth onto natural organic substrates promote the formation of smaller hexagonal ZnO micro and nano crystals onto textile fibers. XRD analysis, showed the presence of wurtzite ZnO (1 0 1), (1 0 0), (0 0 2) [JCPDS Card No. 36-1451] crystalline structure with highest peak intensity for (1 0 1) crystalline orientation and the best crystallinity for a PES-Viscose and Tencel substrates. Typical XRD diffractograms for coated and uncoated PES-Viscose substrate are presented in figure 4.

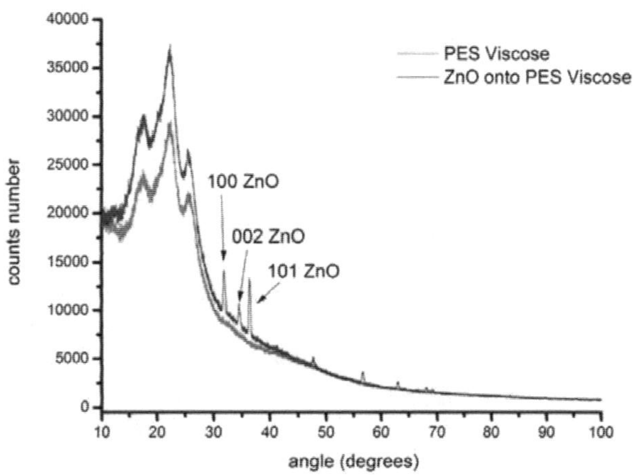

Figure 4. Typical XRD diffractograms for ZnO coated and uncoated PES-Viscose substrate.

XRD analysis of ZnO onto glass fiber substrates, showed the presence of wurtzite ZnO (1 0 1), (0 0 2), (1 0 0), [JCPDS Card No. 36-1451] crystalline structure with highest peak intensity for (0 0 2) crystalline orientation and high crystallinity. Typical diffractogram is presented in figure 5.

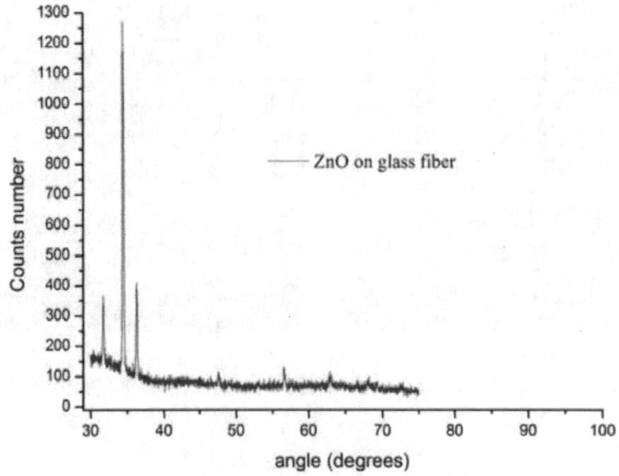

Figure 5. Typical XRD diffractogram of ZnO onto glass fibre substrate.

Elemental analysis EDX was performed in order to verify if except the observable ZnO structures, there is also an uniform coating of the textile fibers. EDX analysis confirmed the presence of ZnO onto the textile fibers but elemental maps couldn't be obtained with publishable resolution due to charging and damage under electron beam exposure. The EDX characterization had only a qualitative value. ATR FTIR characterization was also performed onto all deposited samples and the characteristic transmittance spectra analyzed. Characteristic picks for the organic substrates were observed : 3330 –OH stretching, 2910 –CH and CH2 stretching, 1010 and 1080 –CO and ring stretching, as well as specific peaks associated with each organic material substrate. Most of characteristic absorptions of ZnO were overlapping with the picks of the substrates and the only one that was sometime visible was the one at 2370 cm^{-1} but very faint and could not be used for quantitative estimations.

ZnO deposited onto the different substrates was tested successfully as photocatalyst for MB degradation under UV light exposure. An example of photocatalytic MB discoloration onto ZnO coated textile is presented in figure 6.

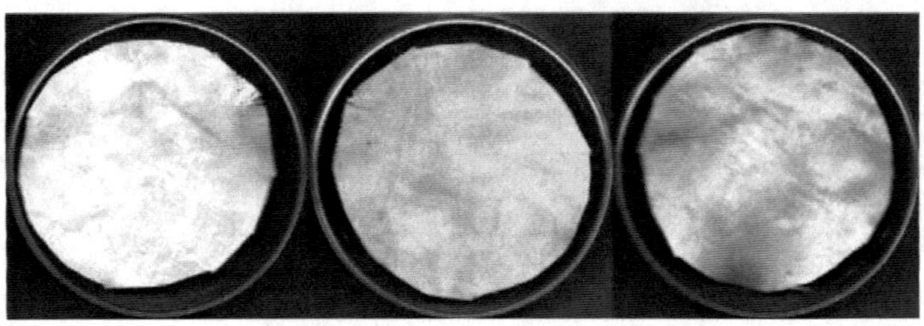

Initial	1 mL MB 10mg/mL	12 min UV

Figure 6. MB discoloration onto ZnO onto glass fibre substrate.

Promising samples will be further tested as photocatalyst for formaldehyde degradation (as one of the most common in-doors pollutants) and integrated in a device for indoors applications.

4. Conclusions

ZnO was chemically grown via a facile low cost method directly onto different substrates cheap precursor concentrations solutions. Characterization by SEM and XRD shows that the specific method leads to highly crystalline ZnO growth. SEM and EDX characterization proved that ZnO coat the fiber surface. ZnO deposition onto the textile fibers is different for each substrate. Nature of the substrate determine the growth of nano-walls like structures onto synthetic organic materials, nanostructures onto natural organic materials and nanocrystalline ZnO with (0 0 2) hexagonal wurtzite preferential orientation onto glass fibers. Due to low conductivity of textile coated samples, accurate elemental composition could not be achieved. ATR-FTIR confirmed that the textile organic substrates undergo chemical degradation during the deposition of ZnO. The characteristic absorption bands for ZnO are present in the spectra but are fairly week therefore no quantitative estimations can be made. ZnO deposited onto the different substrates was tested successfully as photocatalyst for MB degradation under UV light exposure. Promising samples will be further tested as photocatalyst for formaldehyde degradation (as one of the most common in-doors pollutants) and integrated in a device for indoors applications.

Acknowledgments
Part of the work was partially supported by a grant of the Romanian National Authority for Scientific Research, CNCS – UEFISCDI, project number PN-II-RU-TE-2012-3-0202.

References

[1] G. Kenanakis, D. Vernardou, N. Katsarakis, 2012 *Applied Catalysis A: General*, vol. 411-412, pp. 7–14.
[2] Wang, R.; Hashimoto, K.; Fujishima, A.; Chikuni, M.; Kojima, E.; Kitamura, A.; Shimohigoshi, M.; Watanabe, T.; 1997 *Nature*, vol. 388, pp. 431–432.
[3] A. Sirelkhatim, S. Mahmud, A. Seeni, N. H. M. Kaus, L. C. Ann, S. K. M. Bakhori, H. Hasan, D. Mohamad, 2015 *Nano-Micro Letters,* vol. 7, pp. 219-242.
[4] K. R Raghupathi, R. T. Koodali, and A. C. Manna, 2011 *Langmuir*, 27 (7), pp. 4020–4028.
[5] H. Zeng, W. Cai, P. Liu, X. Xu, H. Zhou, C. Klingshirn, H. Kalt, 2008 *ACS Nano*, 2, 1661.
[6] M. Saito, 1993 *Journal of Industrial Textiles,* vol. 23, pp. 150-164.
[7] E. Luévano-Hipólito, A. Martínez-de la Cruz, 2016 *Research on Chemical Intermediates*, vol. 42, pp 4879-4891.
[8] Y. Huang, S. Ho, Y. Lu, R. Niu, L. Xu, J. Cao and S. Lee, 2016 *Molecules*, , Vol. 21, pp. 56
[9] A. Rezaee , H. Rangkooy, A. Khavanin, A. J. Jafari, 2016, *Environmental Chemistry Letters,* vol.12, pp 353-357.
[10] Meng Z, Juan Z. , 2008 *Global Environmental Policy in Japan*, vol. 12, pp. 1-9.
[11] Adams L..K., Lyon D.Y., Alvarez P.J.J., 2006 *Water Research* , vol. 40(19), pp. 3527 – 3532.
[12] Khezrianjoo S, Revanasiddappa H., 2012 *Chemical Sciences Journal,* vol.(CSJ-85), pp. 1-7.
[13] C. Hariharan, 2006 *Applied Catalysis A: General* , vol. 304(1), pp. 55-61.
[14] Jin Y, Wang J, Sun B, Blakesley J.C., Greenham N.C., 2008 *Nano Letters*, vol. 8(6), pp. 1649-1653.
[15] Yang J.L., An S.J., Park W.I., Yi G.-C., Choi W., 2004 *Advanced Materials* , vol. 16(18), pp. 1661–1664.
[16] M. Suchea, I. V. Tudose, S. Ionita, I. Sandu, F. Iacomi, E. Koudoumas, 2015, *Revista de Chimie,* Vol 66 (12).
[17] M. Suchea, I.V. Tudose, N. Vrinceanu, B. Istrate, C. Munteanu, E. Koudoumas 2013 *Acta Chemica Iasi* , Vol. 21 (2), Pp. 107-118.

IOP Conf. Series: Materials Science and Engineering **133** (2016) 012041 doi:10.1088/1757-899X/133/1/012041

The effect on slurry water as a fresh water replacement in concrete properties

Aeslina Abdul Kadir[1,2], Shahiron Shahidan[1], Lau Hai Yee[1], Mohd Ikhmal Haqeem Hassan[1] & Mohd Mustafa Al Bakri Abdullah[3,2]

[1]Faculty of Civil and Environmental Engineering, Universiti Tun Hussein Onn Malaysia, 86400 Parit Raja, Batu Pahat Johor, Malaysia

[2]Center of Excellence Geopolymer & Green Technology (CeGeoGTech), School of Material Engineering, Universiti Malaysia Perlis (UniMAP), Perlis Malaysia

[3]Faculty of Engineering Technology, Universiti Malaysia Perlis (UniMAP), Perlis, Malaysia

Email: aeslina@uthm.edu.my, shahironshahidan@gmail.com, lhaiyee@gmail.com, mohd.ikhmal.haqeem@gmail.com & mustafa_albakri@unimap.edu.my

Abstract. Concrete is the most widely used engineering material in the world and one of the largest water consuming industries. Consequently, the concrete manufacturer, ready mixed concrete plant is increased dramatically due to high demand from urban development project. At the same time, slurry water was generated and leading to environmental problems. Thus, this paper is to investigate the effect of using slurry water on concrete properties in term of mechanical properties. The basic wastewater characterization was investigated according to USEPA (Method 150.1 & 300.0) while the mechanical property of concrete with slurry water was compared according to ASTM C1602 and BS EN 1008 standards. In this research, the compressive strength, modulus of elasticity and tensile strength were studied. The percentage of wastewater replaced in concrete mixing was ranging from 0% up to 50%. In addition, the resulted also suggested that the concrete with 20% replacement of slurry water was achieved the highest compressive strength and modulus of elasticity compared to other percentages. Moreover, the results also recommended that concrete with slurry water mix have better compressive strength compared to control mix concrete.

1. Introduction

According to the Construction Industry Development Board (CIDB) Malaysia (2015), the number of projects for building and civil engineering work in Malaysia recorded as the highest compared to electrical and mechanical work [1]. This is due to the huge invested in construction development projects under the 10th Malaysia plan from 2011 to 2015 [2]. Consequently, the ready mixed concrete (RMC) plant is increased dramatically duc to high demand of concrete production and unconsciously led to a higher consumption of fresh water [3,4]. However, at the end of each day, it is estimated that approximately 500 to 1500 litres of tap water needed for washing out one concrete mixer truck [5, 6, and 7]. At the same time, slurry water may produce from wash out the mixer truck which could lead to environmental problem [8].

According to the previous researchers, the typical pH value of slurry water is varied from pH 11 to 12.5 pH [5, 6, 7, and 9]. In line with that, slurry water is characterized as dangerous and could pollute surface bodies of water due to its alkali characteristics [6]. In addition, slurry water also considered as hazardous waste under the United States Environmental Protection Agency (USEPA) and European Environmental Agency Special Waste Regulation [6, 10, 11, and 12]. Apart from that, slurry water from ready mixed concrete plant is contains of dissolved solids such as sulphate from cement and chloride from calcium chloride as an admixture [5, 13].

Borger, J. et al [12] studied the recycled wash water in the production of fresh concrete. The experimental work includes compressive strength, sulphate resistance, workability and setting time [12]. In addition, the reused of slurry water to manufacture fresh concrete was studied by Sandrolini [14]. Their results suggested that at 28 days the compressive strength of concrete samples with slurry water was at least 96% of the control mix concrete.

Furthermore, Su [15] used slurry water, underground water and tap water to manufacture concrete and mortar. The results indicated that concrete mixed with slurry water and underground water satisfied the ASTM C94 standard and the compressive strength is as good as concrete with tap water [15]. Low [9] studied the use of slurry water from a concrete batching plant for making concrete. The study mainly focused on the effects of using slurry water in concrete mixes on the fresh and hardened properties of concrete. Their results showed that concrete with slurry water was encountered the specification from BS EN 1008 in term of compressive strength, setting time and drying shrinkage [9, 16]. Chatveera [17] studied manufacturing concrete using sludge water from RMC plant. The fresh and mechanical properties of concrete were carried out including slump, compressive strength and modulus of elasticity. Their results indicated that slurry water has a high alkaline content and the total solids content above the limit of ASTM C94 which may cause the concrete more porous and weaker matrix [17, 18]. The results also suggested that an increasing percentage of slurry water in mixing water caused the weight loss and drying shrinkage due to acid attacks [17].

Moreover, researchers from Greece studied the reused of slurry water from RMC plant [6]. The slurry water chemical properties which included chloride, sulphate and pH value were investigated. The result shows that slurry water has a high pH value which exceeding pH 11.5 and the chemical properties were complied with ASTM standard [6, 19]. In addition, the results also illustrated that the compressive strength of all concrete with slurry water was higher than control concrete specimens [6]. Asadollahfardi [5] used slurry water from RMC plant and concrete mixer truck in the production of fresh concrete. The compressive strength, tensile strength and flexural strength of concrete were carried out by an analysis of variance (ANOVA) technique. The results indicated the concrete mixed with slurry water was comparable to the control mix concrete [5]. In conclusion, there is lack research area on the characterization of slurry water in concrete application particularly in Malaysia. Therefore, this research is mainly focused on characterization of slurry water and the effect on concrete properties. The contaminated of chloride and sulphate in slurry water were analysed and compared with the ASTM standards [19]. However, the tests for concrete properties were performed according to the ASTM standards which include compressive strength, tensile strength and modulus of elasticity [20, 21, and 22].

2. Materials and Experimental Work

In this research, the replacement percentages of slurry water in concrete mixes are at 10%, 20%, 30%, 40% and 50% of tap water. The water cement ratio w/c is 0.45 and all the concrete specimens were designed to have a target mean strength 35 MPa with a slump value between 30mm to 60mm. The mix designed was according to the DOE method [23]. In addition, the properties of fresh concrete such as unit weight and slump test and the mechanical properties including compressive strength, tensile strength and modulus of elasticity (MOE) were carried out in this study. The concrete specimens testing for compressive strength was cast in the moulds with dimensions of (150 x 150 X 150) mm while tensile strength and MOE were cast in a 300 mm x 150mm diameter moulds. A triplicate concrete samples were made in each batch of concrete design mix.

2.1 Materials

2.1.1 Water

Slurry water was collected from three different RMC plant in Batu Pahat, Johor, Malaysia which are Hanson RMC plant, Chin Keng RMC plant and Renggam RMC plant. Before the slurry water was collected, the wash water from concrete mixer truck was discharged into a pond at RMC plant. The chemical properties of slurry water were analysed according to USEPA standard method and compared with the standards specification [16, 19].

2.1.2 Cement
A local Ordinary Portland cement type CEM 1 42.5N was used in this research. The cement comply with the requirements of MS EN 197-1 [24].

2.1.3 Aggregates
The fine aggregate was natural sand and meet the ASTM C33 requirements [25]. The coarse aggregate was crushed rocks and comply the ASTM C33 specification. ASTM C33 and BS 882 standards were used to control the aggregates sizes using a sieve analysis [25, 26]. The fine aggregate size was passing 5mm and retained on 75μm. The coarse aggregate size was passing 20mm and retained on 10mm. All the aggregates were air dried at 20+5^{0}C to obtain saturated surface dry condition.

2.2 Mix Proportion of Concrete
In this research, concrete mix design was according to the department of environmental, DOE method and all concrete specimens was design to have a target mean strength 35MPa. Table 1 shows the mix proportion of concrete.

Table 1: Mix proportion of concrete

Mix	Cement (kg/m^3)	Tap Water (kg/m^3)	SW (kg/m^3)	F. A (kg/m^3)	C.A (kg/m^3)
CM	414	180	0	535	1245
SW10%	414	162	18	535	1245
SW20%	414	144	36	535	1245
SW30%	414	126	54	535	1245
SW40%	414	108	72	535	1245
SW50%	414	90	90	535	1245

*F.A (Fine Aggregate); C.A (Coarse Aggregate); SW (Slurry Water)

2.3 Experimental Work

2.3.1 Wastewater Analysis
The slurry water pH value was measured by Hanna HI 8424 digital pH meter. The sulphate and chloride content in slurry water were analysed according to United States Environmental Protection Agency USEPA method 300.0. In addition, all the results of wastewater characterization were compared with the ASTM and BS standards [16] [19].

2.3.2 Concrete Testing
The manufacturing and curing process of concrete were according to ASTM standards [27]. However, the concrete slump tests were performed according to ASTM C 143 [28]. Besides, the tests for unit weight of concrete were conducted according to ASTM C138 [29]. On the other hand, the compressive strength and tensile strength of concrete at 7 and 28 days were carried out in accordance with ASTM C39 and ASTM C496 respectively [20, 21]. Meanwhile, the modulus of elasticity of concrete at 28 days was tested according to ASTM C469 [22].

3. Results and Discussion

3.1 Slurry Water Characterization

Table 2 and Table 3 showed the results of slurry water characterization and tolerance limit of water mixing in concrete respectively.

Table 2: Wastewater characteristics

Sources/Parameters	Parameters		
	pH	Chloride Cl^-, ppm	Sulphate SO_4^{2-}, ppm
	Slurry Water		
Hanson	10.0 – 11.3	7.5 – 46.9	108.6 – 454.9
Chin Keng	11.0 – 11.9	6.7 – 28.7	426. – 289.5
Renggam	11.6 – 12.1	19.9 – 33.7	69.4 – 475.8
	Tap Water		
Tap Water	7.6 – 8.1	19.4 – 19.7	13.3 – 13.4

Table 3: Tolerance limit of water in mixing concrete

Standards	Tolerance limit, ppm		
	Chloride Cl^-, ppm	Sulphate SO_4^{2-}, ppm	pH
ASTM C1602	<500	<3000	-
ASTM C685	<500	<3000	-
BS EN 1008	<500	<2000	>4
BS 3148	<500	<1000	-
	Previous Researchers		
[5, 6, 7, 9, 14, 15, 17 and 30]	20.8 - 355	12.7 - 1152	8.1 – 13..5

As referred to Table 2 and Table 3, the slurry water characteristics were completely complied with the standards as well as the typical values that suggested by previous researchers. Although the existing current standards do not clearly specify the used of slurry water in mixing water for concrete, but the results from Table 2 and Table 3 illustrated that the wastewater had a higher potential could be used in concrete mixes due to its characterization was in the recommended ranges of water for manufactured concrete.

3.2 Sieve Analysis

Figure 1 and Figure 2 demonstrated the results of sieve analysis for fine and coarse aggregate in this study.

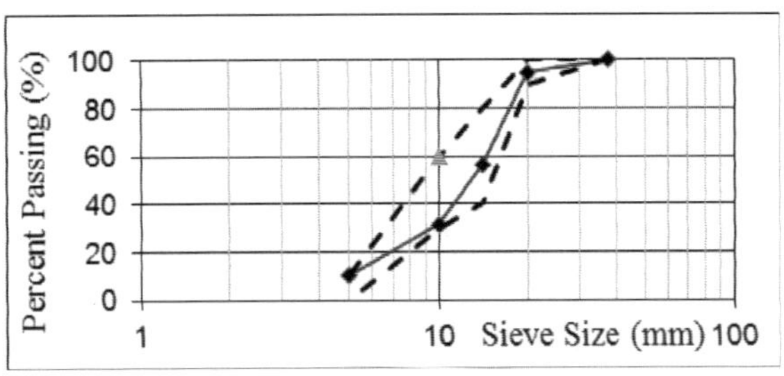

Figure 1: Coarse aggregate sieve analysis

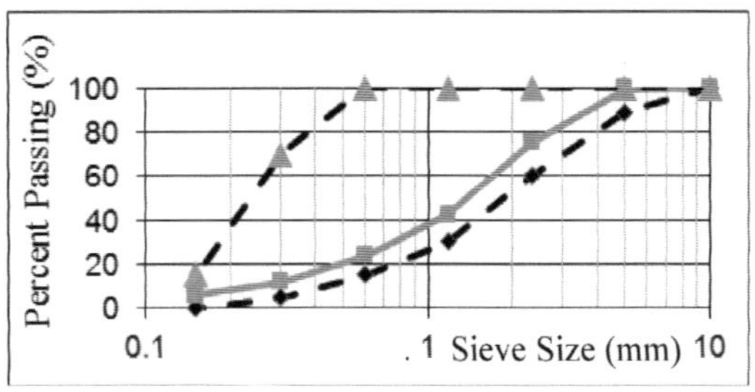

Figure 2: Fine aggregate sieve analysis

Based on Figure 1 and Figure 2, the coarse and fine aggregate was in the range that had been specified by a British Standard (BS 882: Table 3 and 4). So, it was a good correlation to show the aggregates was in ideal quality in order to enhance the good quality of concrete especially in term of strength.

3.3 Properties of Fresh Concrete

3.3.1 Slump Test

In this study, the concrete slump test was investigated according to the ASTM C143 and the results were listed in Table 4.

Table 4: Slump Test

Concrete Type	CM	SW 10%	SW 20%	SW 30%	SW 40%	SW 50%
Slump Value (mm)	32	43	39	40	42	48

As presented in Table 4, the slump value for both wastewater were comparable to the control mix. As mentioned in sub topic 2, the target mean slump value was varied from 30 mm to 60 mm. According to that matter, both slump value of concrete obtained with slurry water was in the range that had been designed. In addition, the results recommended the higher percentage of slurry water in concrete mixing the greater slump value was achieved.

3.4 Mechanical Properties of Concrete

3.4.1 Compressive Strength

Table 5 and Figure 3 illustrated the comparison of compressive strength between slurry water mix and control mix.

Table 5: Compressive strength of slurry water mix

Concrete Type	7 Days Strength (MPa)	28 Days Strength (MPa)	ASTM C1602 Minimum Wastewater Concrete At 7days (>90%) of Control Strength
CM	44.14	50.01	100.00%
SW 10%	35.73	50.78	80.95%
SW 20%	40.67	52.31	92.14%
SW 30%	41.91	49.07	94.95%
SW 40%	43.06	50.27	97.55%
SW 50%	41.20	44.68	93.34%

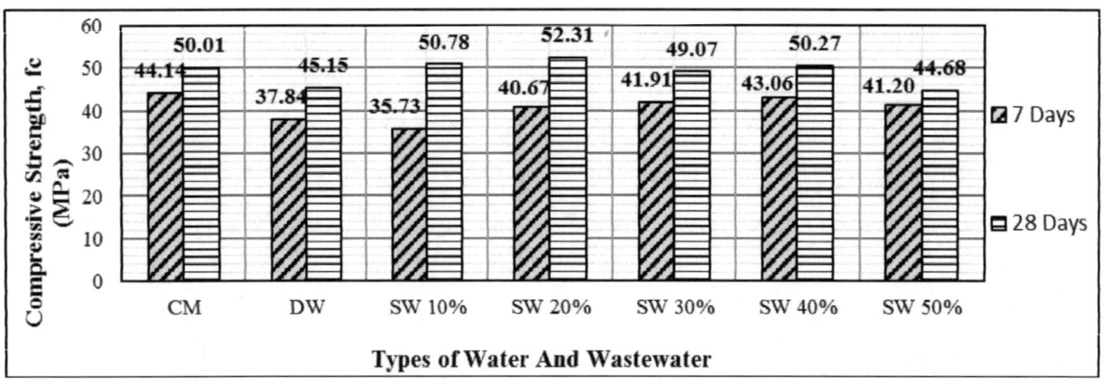

Figure 3: Comparison of compressive strength between slurry water mix and control mix

According to Table 5, all types of slurry water were higher than the minimum limit with 90% that specified in standards except slurry water 10% (SW10%) with only 80.95%. However, at 28 days the compressive strength of slurry water with 10% to 40 % replacement were above the limit but 50% replacement was apart from the minimum limit. This is because the mix proportion of water for concrete (SW50%) was mixed by 50% slurry water and 50% tap water, which would causes the pH value might higher than the others. This finding can be seen from Table 4.1, pH value for slurry water is between 10.8 – 11.9 compare to tap water 7.6 to 8.1. As mentioned by the Neville [31] and Kucche [32], the typical range of pH value for water used in concrete manufacturing is between pH 6.0 to 9.0 therefore the higher percentages of slurry water replacement may resulted in higher pH value. Consequently, whenever the water pH value above the standard limits it would results lower compressive strength of concrete. Therefore, it can be concluded that the suitable percentage of slurry water replacement in concrete mixes was 20% to 40% and the optimum percentage replacement was 20%.

3.4.2 Tensile Strength

Figure 4 demonstrated the comparison of tensile strength between slurry water mix and control mix.

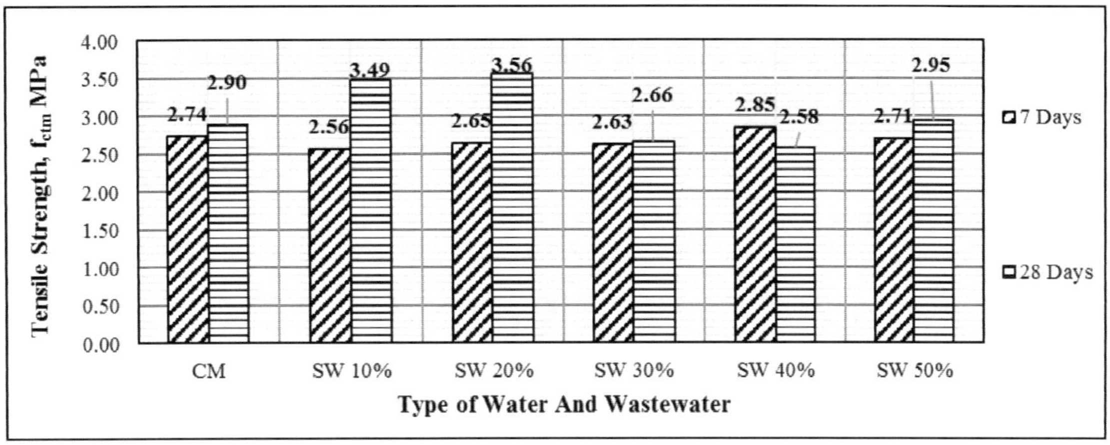

Figure 4: Comparison of tensile strength between slurry water mix and control mix

As mentioned in BS EN 1992, the mean tensile strength of concrete with mean target strength 35 MPa was about 2.8+1 MPa [33]. Based on the findings, it indicates that only concrete with 10% and 20% slurry water were in line with the requirement specified by BS EN 1992 standards. The highest tensile strength of concrete at 7 days was achieved by concrete with 40% slurry water with 2.85 MPa but later at 28 days it was demonstrated the lowest tensile strength with only 2.58 MPa. It means that 40% of slurry water replacement was good in early strength but not suggested used for long term. Accordingly, concrete with 20% slurry water achieved the highest tensile strength with approximately 3.56 MPa compared to others percentages replacement as well as tensile strength of control mix. This was a good correlation to show that the slurry water not only could be reused in concrete mixes but could also improve the tensile strength of concrete.

3.4.3 Modulus of Elasticity (MOE)

Figure 5 presented the comparison of modulus of elasticity between slurry water mix and control mix.

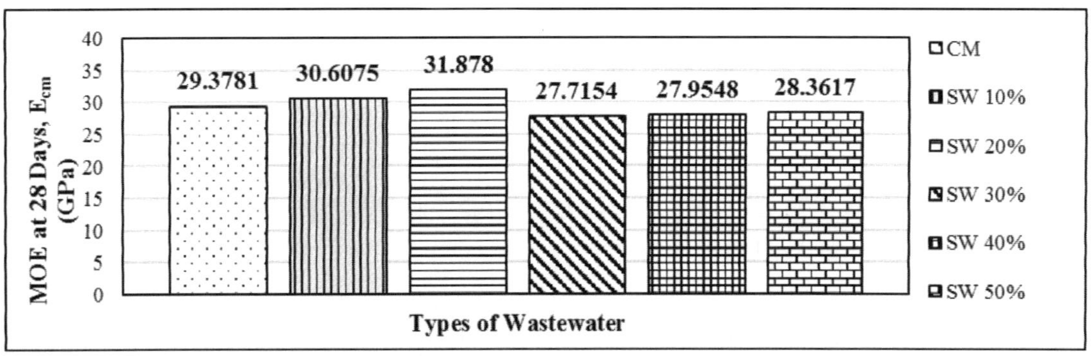

Figure 5: Comparison of MOE between slurry water mix and control mix

As stated in BS EN 1992, the typical range of secant modulus of elasticity of concrete (Ecm) with grade 35 was 32 + 6 GPa [33]. So, the results in Table 7 was completely achieved the typical range of MOE for concrete have a mean strength 35MPa. In addition, ACI 318-02 and Neville, (1995) also mentioned that the typical MOE value for concrete with compressive strength 35 was 30 GPa [31]. Therefore, as compared the data in Table 7 again the previous study, it was clearly shown the MOE value of slurry water were complied with the suggested ranges.

On the other hand, Table 7 indicated that the concrete with 20% of slurry water achieved the highest MOE value with 31.8780 GPa compared to others as well as control mix. This finding could be related to the highest compressive strength of concrete with slurry water as shown in Table 5. Therefore, it could summarize that the higher compressive strength the greater value of MOE. These findings were also similar with the point of view by previous researchers [31, 34]. The researchers mentioned that higher compressive strength value will result larger value of MOE. Overall, the MOE results as shown in Table 7 were in between the typical range that had been specified by past researchers. Thus, it can be concluded that the higher value of concrete compressive strength will lead to higher values of MOE.

3.5 Relative between Compressive Strength and MOE

Figure 6 shows the relationship between compressive strength and Modulus of Elasticity for concrete with slurry water mixed.

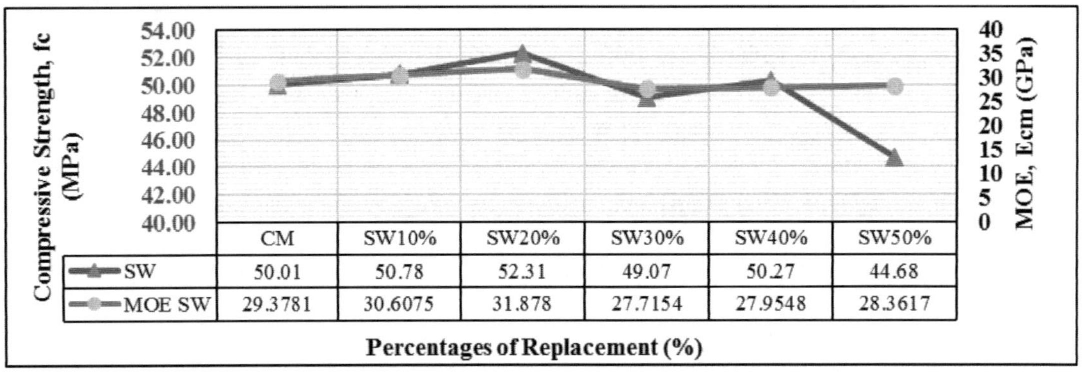

Figure 6: Relationship between compressive strength and MOE of slurry water replacement

According to Figure 6, the modulus of elasticity increased when the concrete compressive strength was increased. In addition, Figure 6 illustrated that concrete with 20% slurry water achieved highest compressive strength and modulus of elasticity. Therefore, it can be concluded that the optimum percentages of slurry water as a fresh water replacement in concrete mixes was 20%.

4. Conclusions

The pH value for slurry water from ready mixed concrete plant was in the range of 10 to 12 which is considerably higher than the tap water. However, the sulphate and chloride content in slurry water were complied with the ASTM and BS standards specification. The compressive strength, tensile strength and modulus of elasticity of concrete have a tendency to decrease with an increase in the percentages of slurry water replacement in concrete mixes. Generally, the compressive strength of concrete with slurry water is in the range of 81% to 98% of the control mix. The results recommended that the higher compressive strength value may lead to greater value of MOE. According to the results for compressive strength and modulus of elasticity, the optimum percentages of slurry water as a fresh water replacement in concrete mixes was 20%.

5. References

[1] CIDB, "Construction Statistics Quarterly Bulletin (Jun 2015)," Construction Industry Development Board Malaysia (CIDB), 2015.

[2] R. A. Khan, M. S. Liew and Z. Ghazali, "Malaysian Construction Sector and Malaysia Vision 2020: Developed Nation Status," Procedia - Social and Behavioral Sciences, vol. 109, pp. 507-513, 2014.

[3] A. H. Noruzman, B. Muhammad, M. Ismail and M. Z. Abdul, "Characteristics of treated effluents and their potential applications for producing concrete," Journal of Environmental Management, vol. 110, pp. 27-32, 2012.

[4] A. M. Wahid and H. Tanaka, "Potential Implementation of Wastewater Reclamation and Reuse in Malaysian Urban Area," International Sustainability and Civil Engineering Journal, vol. 1, no. 1, pp. 75-83, 2012.

[5] G. Asadollahfardi, M. Asadi, H. Jafari, A. Moradi and R. Asadollahfardi, "Experimental and statistical studies of using wash water from ready- mix concrete trucks and a batching plant in the production of fresh concrete," Construction and Building Materials, vol. 98, pp. 305-314, 2015.

[6] S. Tsimas and M. Zervaki, "Reuse of wastewater from ready-mixed concrete plants. Management of Environmental Quality," An International Journal, vol. 22, no. 1, pp. 7-17, 2011.

[7] S. Ekolu and A. Dawneerangen, "Evaluation of recycled water recovered from a ready-mix concrete plant for reuse in concrete," Journal of the South African Institution of Civil Engineering, vol. 52, no. 2, pp. 77-82, 2010.

[8] A. Sales, F. R. De-Souza and F. C. Almeida, "Mechanical properties of concrete produced with a composite of water treatment sludge and sawdust," Construction and Building Materials, vol. 25, no. 6, pp. 2793-2798, 2011.

[9] G. L. Low, W. L. Ng, C. T. Tam and R. B. Heng, "Use of recycled cement- based slurry water for making concrete," The Institution of Engineer, Malaysia, vol. 68, no. 4, pp. 47-55, 2007.

[10] B. J. Sealey, P. S. Phillips and G. J. Hill, "Waste Management Issues for the UK Ready-Mixed Concrete Industry," Resources Conservation and Recycling, vol. 32, pp. 321-331, 2001.

[11] M. Paolini and R. Khurana, "Conservation and Recycling," Cement and Concrete Composites, vol. 20, no. 2-3, pp. 221-229, 1998.

[12] J. Borger, R. L. Carrasquillo and D. Fowler, "Use of recycled wash water and return plastic concrete in the production of fresh concrete," Advance Cement Based Materials, vol. 1, no. 6, pp. 267-274, 1994.

[13] S. A. Chini and W. J. Mbwambo, "Environmentally Friendly Solution for the Disposal of Concrete Wash Water from Ready Mixed Concrete Operations," in Proceedings of CIB W89 Beijing International Conference, 1996.

[14] F. Sandrolini and E. Franzoni, "Waste wash water recycling in ready-mixed concrete plants," Cement and Concrete Research, vol. 31, no. 3, pp. 485-489, 2001.

[15] N. Su, B. Miao and F. S. Liu, "Effect of wash water and underground water on properties of concrete," Cement and Concrete Research, vol. 32, no. 5, pp. 777-782, 2002.

[16] BS EN 1008, Mixing Water for Concrete-Specification for Sampling, Testing and Assessing the Suitability of Water, Including Water Recovered from Processes in the Concrete Industry, as Mixing Water in Concrete, London: British Standards Institution, 2002.

[17] B. Chatveera, P. Lertwattanaruk and N. Makul, "Effect of sludge water from ready-mixed concrete plant on properties and durability of concrete," Cement and Concrete Composites, vol. 28, no. 5, pp. 441-450, 2006.

[18] ASTM C94, Specification for Ready-Mixed Concrete, Pennsylvania, United States: American Society for Testing and Materials, 2007.

[19] ASTM C1602, Specification for Mixing Water Used in the Production of Hydraulic Cement Concrete, Pennsylvania, United States: American Society for Testing and Materials, 2006.

[20] ASTM C39, Test Method for Compressive Strength of Cylindrical Concrete Specimens, Pennsylvania, United States: American Society for Testing and Materials, 2005.

[21] ASTM C496, Test Method for Splitting Tensile Strength of Cylindrical Concrete Specimens, Pennsylvania, United States: American Society for Testing and Materials, 2004.
[22] ASTM C469, Test Method for Static Modulus of Elasticity and Poisson Ratio of Concrete in Compression, Pennsylvania, United States: American Society for Testing and Materials, 2002.
[23] DOE 1988, Design of Normal Concrete Mixes. Building Research Establishment, Watford, United Kingdom: Department of the Environmental, 1988.
[24] MS EN 197-1, Composition, specifications and conformity criteria for common cements, Cyberjaya, Selangor: Department of Standards Malaysia, 2014.
[25] ASTM C33, Specification for Concrete aggregates, Pennsylvania, United States: American Society for Testing and Materials, 2003.
[26] BS 882, Specification for Aggregates from Natural Sources for Concrete, London: British Standards Institution, 1992.
[27] ASTM C192, Practice for Making and Curing Concrete Test Specimens in the Laboratory, Pennsylvania, United States: American Society for Testing and Materials, 2006.
[28] ASTM C143, Test Method for Slump of Hydraulic Cement Concrete, Pennsylvania, United States: American Society for Testing and Materials, 2005.
[29] ASTM C138, Test Method for Density Unit Weight, Yield and Air Content of concrete, Pennsylvania, United States: American Society for Testing and Materials, 2005.
[30] M. Elchalakani and E. Elgaali, "Sustainable Concrete made of Construction and Demolition Wastes using Recycled Wastewater in the UAE," Journal of Advanced Concrete Technology, vol. 10, pp. 110-125, 2012.
[31] A. M. Neville, Properties of Concrete, 4th ed., London: Pearson, 1995.
[32] K. J. Kucche, "Quality of Water for Making Concrete: A Review of Literature," Journal of Scientific and Research Publications, vol. 5, no. 1, pp. 201-205, 2015.
[33] BS EN 1992, Design of Concrete Structure, London: British Standards Institution, 2004.
[34] K. P. Mehta and P. J. Monteiro, Concrete: Microstructure, Properties, and Materials, 4th ed., New York: Mc Graw Hill, 2014.

IOP Conf. Series: Materials Science and Engineering **133** (2016) 012042 doi:10.1088/1757-899X/133/1/012042

Strength and Density of Geopolymer Mortar Cured at Ambient Temperature for Use as Repair Material

A.Z. Warid Wazien[2,1,a], Mohd Mustafa Al Bakri Abdullah[2,1,b], Rafiza Abd. Razak[2,1,c], M.A.Z. Mohd Remy Rozainy[3,d], and Muhammad Faheem Mohd Tahir[1,e]

[1] Center of Excellence Geopolymer and Green Technology (CEGeoGTech), School of Materials Engineering, Universiti Malaysia Perlis (UniMAP), P.O Box 77, D/A Pejabat Pos Besar, 01000 Kangar, Perlis, Malaysia.
[2] Faculty of Engineering Technology, Universiti Malaysia Perlis (UniMAP), P.O Box 77, D/A Pejabat Pos Besar, 01000 Kangar, Perlis, Malaysia.
[3] School of Civil Engineering, Universiti Sains Malaysia (USM), Engineering Campus, 14300 Nibong Tebal, P. Pinang, Malaysia

[a]warid.wazien92@yahoo.com, [b]mustafa_albakri@unimap.edu.my, [c]rafizarazak@unimap.edu.my, [d]ceremy@eng.usm.my, [e]faheem@unimap.edu.my

Abstract. Geopolymers produced by synthesizing aluminosilicate source materials with an alkaline activator solution promised an excellent properties akin to the existing construction material. This study focused on the effect of various binder to sand ratio on geopolymer mortar properties. Mix design of geopolymer mortar was produced using NaOH concentration of 12 molars, ratio of fly ash/alkaline activator and ratio Na_2SiO_3/NaOH of 2.0 and 2.5 respectively. Samples subsequently ware cured at ambient temperature. The properties of geopolymer mortar were analysed in term of compressive strength and density at different period which are on the 3rd and 7th day of curing. Experimental results revealed that the addition of sand slightly increase the compressive strength of geopolymer. The optimum compressive strength obtained was up to 31.39 MPa on the 7th day. The density of geopolymer mortar was in the range between 2.0 g/cm^3 to 2.23 g/cm^3. Based on this findings, the special properties promoted by geopolymer mortar display high potential to be implemented in the field of concrete patch repair.

1. Introduction

Geopolymer is an inorganic polymer produced from two main constituents which are the combination of raw materials known as aluminosilicate source containing alumina (Al) and silica (Si) with alkaline activator [1]. The combination of sodium silicate (Na_2SiO_3) and sodium hydroxide (NaOH) has been used widely as the alkaline activator to produce geopolymer.

Dissolution of solid aluminosilicate source by alkaline activator and followed with polycondensation to form 3D network of silico-aluminates structures described the geopolymerisation process [2]. A number of researchers reported excellent properties promoted by geopolymer such as high compressive strength, resistance to steel corrosion, low permeability and high durability [3,4]. Source materials which are rich in silica and alumina such as fly ash, slag, kaolin and metakaolin are commonly used to make geopolymers which promote excellent behaviour and properties. Fly ash is one of the major material approached as the source material of geopolymer due to its commercial and performance advantage. The amount of fly ash generated by electric power plant in Malaysia is increasing each year. Instead of their availability, silica and alumina content is mostly found in fly ash material thus making it the suitable raw material for geopolymer.

Curing temperature also affect the development of mechanical strength in most cementitious systems especially for geopolymer [5]. Geopolymerisation was suggested by previous researchers using a curing temperature less than 60 % for 4-48 hours which is one of the important conditions for the synthesis of geopolymer [3]. However, many researchers have tried to improve the strength development of fly ash geopolymers at ambient temperature [6-8]. The ability of geopolymer mortar to cure at ambient temperature is the advantage especially for concrete repair and rehabilitation [9-11]. Hu et al. [9] investigated the influence of steel slag added in geopolymer and achieved 40 MPa compressive strength and above when cured at ambient temperature and it displayed higher bond strength. Songpiriyakij et al. [10] also studied the rice hush bark ash geopolymer and discovered its high compressive and bonding strength properties. In addition, T. Phoo-ngernkham et al. [11] also studied the potential of geopolymer as repair material and reported that high calcium fly ash geopolymer containing Portland cement cured at ambient temperature exhibited higher compressive strength due to the reaction of calcium in the system. Therefore, the strength of geopolymers are sufficiently high and desirable for some application even when cured at ambient temperature and this can be improved.

Geopolymer mortar is the combination of fly ash, sand and the chosen activator. For the geopolymer formulation, Mustafa et al. [12] discovered that 12 M of NaOH is the optimum compressive strength achieved. In a previous research by Hardjito et al. [13], it was also found that Na_2SiO_3/NaOH ratio of 2.5 produced the best results. Geopolymer formulation determined by previous researchers was used in this study due to the same class and types of fly ash. However, the addition of sand in geopolymer improved the strength and exhibited different physical properties at different binder to sand ratio. The addition of sand as aggregates to geopolymer is not only economically favourable but also reduces pore density and promotes high strength [14]. C. Kuenzel et al. [15] studied the influence of sand on the mechanical properties of geopolymers and discovered that adding sand aggregate also means adding strong particles to the geopolymer matrix. Thus, sand acts as reinforcements in the geopolymers hence promoting it to the higher strength if properly formulated.
The objective of this paper is to determine the effect of various binder to sand ratio on the properties of geopolymer mortar. Determining the optimal binder to sand ratio is important in order to meet the standard and specified mechanical properties of mortars for used as repair material for concrete infrastructure rehabilitation.

2. Experimental

2.1 Material selection
The fly ash was obtained from Manjung Power Station in Lumut, Perak, Malaysia. It is an industrial by product which residue resulting from the combustion of coal. The combination of sodium silicate (Na_2SiO_3) and sodium hydroxide (NaOH) has been used in this study. The 12 M NaOH was prepared by mixing sodium hydroxide pellets of 97–99 % purity with distilled water. The river sand used in this study having water absorption of 1.6 % and specific gravity of 2.50.

2.2 Samples preparation

The ratio of fly ash/alkaline activator and Na_2SiO_3/NaOH was fixed at 2.0 and 2.5 for the entire experiment. The preparation of NaOH solution by diluting the NaOH pellets with distilled water in order to produce 12 M NaOH concentration. Meanwhile, the various binder to sand ratio was used in this study. The details of mix design was described in Table 1. The selection of ratio was based on the practicality of the previous researchers of geopolymer mortar and the acceptable workability to be used at various application.

Table 1. Mix design details of geopolymer mortar.

Sample	Binder:Sand, wt. ratio	Fly ash, wt. %	Sand, wt. %	12 M NaOH, wt. %	Sodium silicate, wt. %
Mortar A	0.25	13.33	80	1.90	4.76
Mortar B	0.33	16.54	75	2.36	5.91
Mortar C	0.5	22.22	67	3.17	7.94
Mortar D	1.0	33.33	50	4.76	11.90
Mortar E	2.0	44.45	33	6.35	15.87
Mortar F	3.0	50.00	25	7.14	17.86
Mortar G	4.0	53.33	20	7.62	19.05
Paste	-	66.66	-	9.52	23.81

Initially, the sand and fly ash were mixed together in the mixer for 3 minutes. After that, alkaline activator was added in the mixer and the mixing process was continue for another 5 minutes. When the mixture becomes homogeneous, the mixture was moulded using (50×50×50) mm^3 according to ASTM C109. The samples were demolded at the age of 1 day and immediately wrapped with vinyl sheet to protect moisture loss and kept in the room at ambient temperature for 3rd and 7th days. The compressive strength and density values were an average of testing of 3 samples for each ratio.

3. Results and Discussion

3.1 Chemical Composition with X-ray fluorescence (XRF)

The chemical composition of the fly ash was obtained by X-ray fluorescence (XRF) (Table 2). The main chemical composition of fly ash is silica oxide SiO_2, followed by Fe_2O_3, CaO, and Al_2O_3.

Table 2. Chemical composition of fly ash using XRF.

Chemical	SiO_2	Al_2O_3	Fe_2O_3	TiO_2	CaO	MnO	CuO	K_2O	P_2O_5	SO_3	SrO
wt.%	38.8	14.7	19.48	1.02	18.1	0.16	0.041	1.79	0.054	1.50	0.11

Based on the data, the fly ash is classified as Class F as referred to ASTM C618 where the calcium content (18.1 %) which is less than 20 % (Table 3). The summation of SiO_2 and Al_2O_3 content is 54 % which is the most required composition to produce a good geopolymer [16, 17]. In addition, the sufficient CaO in the fly ash geopolymer mixture is also reported as a suitable source to produce a geopolymer which able to be cured at room temperature [11].

Table 3. Standard classification.

Chemical	wt.%	ASTM C618 (%)
SiO_2+ Al_2O_3+ Fe_2O_3	72.98	70.00 min
CaO	18.10	20.0 max

3.2 Particle size distribution of sand
Grading curve described the particle-size distribution of an aggregate as determined by a sieve analysis. The particle size distribution (grading curve) of the sand is shown in Figure 1. The particle size distribution limitation was obtained according to ASTM C33. The standard size sieves for fine aggregate have openings range from 150 μm to 9.5 mm. Grading curve obtained indicates that the particle size distribution of sand is within the limit specified in the standard. The appropriate distribution of sand plays an important role which influence the mechanical properties of geopolymer and its workability.

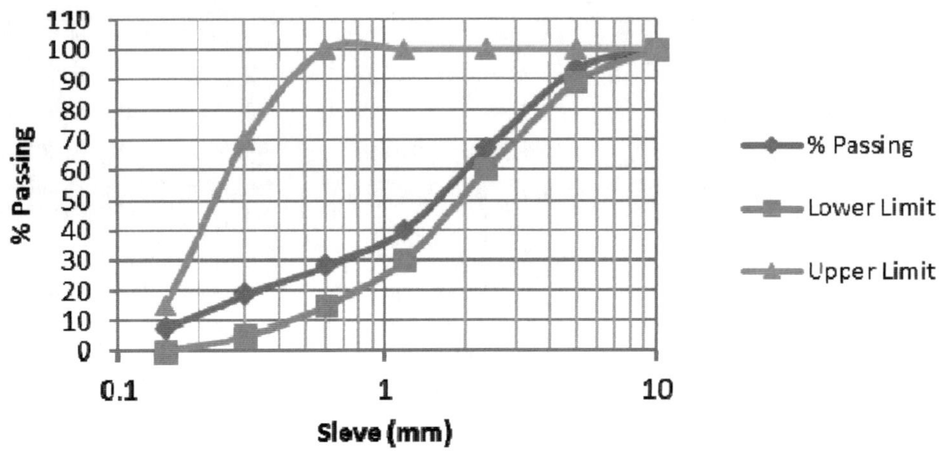

Figure 1. Particle size distribution of sand.

3.3 Characteristics of geopolymer mortar
Selection of the appropriate binder to sand ratio helps to ensure workability and performance expectations. Geopolymer mortar shows different behavior at different binder to sand ratio. Physically, each ratio with various binder to sand ratio has some basic characteristics (Table 4).
The composition of sand and its grading can influence the characteristics of mortars either in fresh or hardened state. Geopolymer mortar with 0.5 and 1.0 binder to sand ratio display acceptable workability as repair material for patching purposes while the rest were stiff with excessive percentage of sand and slurry due to small percentage of sand.

3.4 Density
Density of the geopolymer mortars are summarized in Table 5. The average value of geopolymer mortar density are in the range between 2.0 to 2.23 g/cm^3. As the level of aggregate was decreased the density decreased from Mortar A to paste. The density of a geopolymer paste is reported to be below 2 g/cm^3 depending on its composition [18].

Table 4. Physical characteristics of geopolymer mortar.

Sample	Binder : Sand, wt. ratio	Characteristics
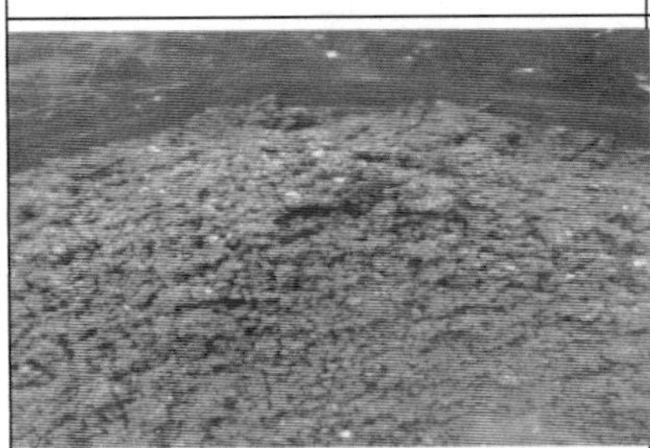 Mortar A and B	0.25 and 0.33 (75 - 80) % of sand	Difficult to mix and cast into the mould and not very workable
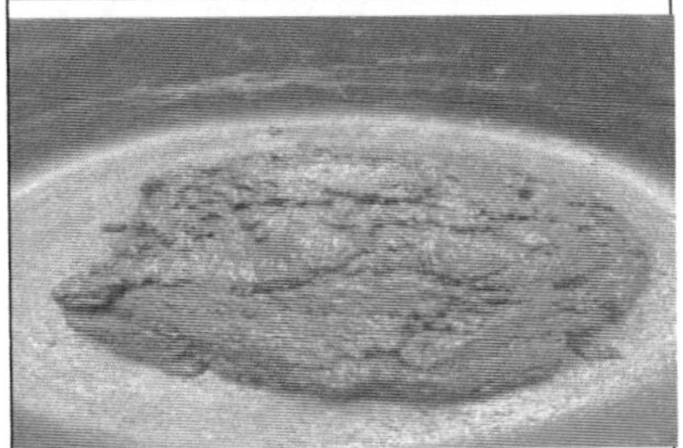 Mortar C and D	0.5 and 1.0 (50 - 67) % of sand	Easily placed into the mould. Suitable for restoration of concrete spalling.
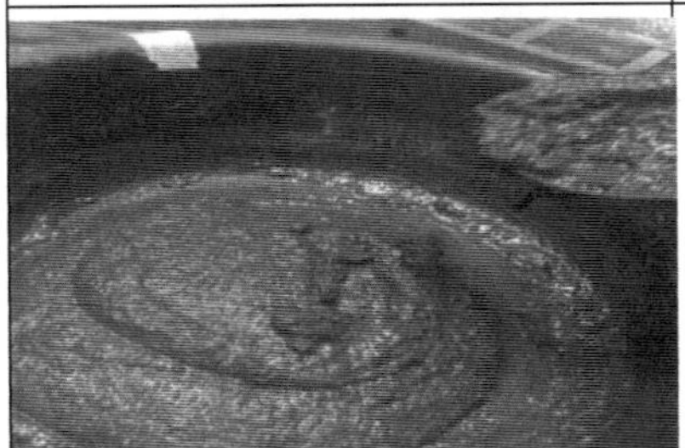 Mortar E and F	2.0 and 3.0 (25 - 33) % of sand	Easy to handle but slurry
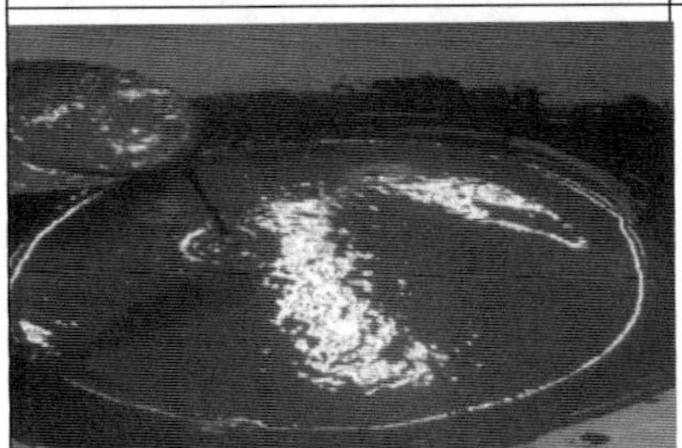 Mortar G	4.0 20 % of sand	Mixture is too wet, very high workability, not suitable for restoration of concrete spalling.

The main component of the sand aggregate is quartz which has a density of 2.65 g/cm^3 [19]. This factor contributes to the increase of geopolymer mortar from Mortar G containing 20 % sand content to Mortar A which is about 80 % sand content. The denser the sample the better the compressive strength is however, sufficient alkaline activator is required. This is due to the sand which help reduce porosity in the geopolymer mortar thus generally increasing the strength of geopolymer. However, the optimum binder to sand ratio from the compressive test result is the most accurate data to justify the best formulation of the mix design to avoid insufficient alkaline activator for geopolymerisation to take place.

IOP Conf. Series: Materials Science and Engineering **133** (2016) 012042 doi:10.1088/1757-899X/133/1/012042

Table 5. Density of geopolymer mortar.

Sample	Binder:Sand, (wt. ratio)	Sand content (wt. %)	Density (g/cm^3)
Mortar A	0.25	80	2.23
Mortar B	0.33	75	2.21
Mortar C	0.5	67	2.19
Mortar D	1.0	50	2.16
Mortar E	2.0	33	2.04
Mortar F	3.0	25	2.01
Mortar G	4.0	20	2.00
Paste	-	-	1.94

3.5 Compressive strength

The compressive strength values obtained from geopolymer mortar are compared with the compressive strength values obtained from geopolymer paste as shown in Figure 2.

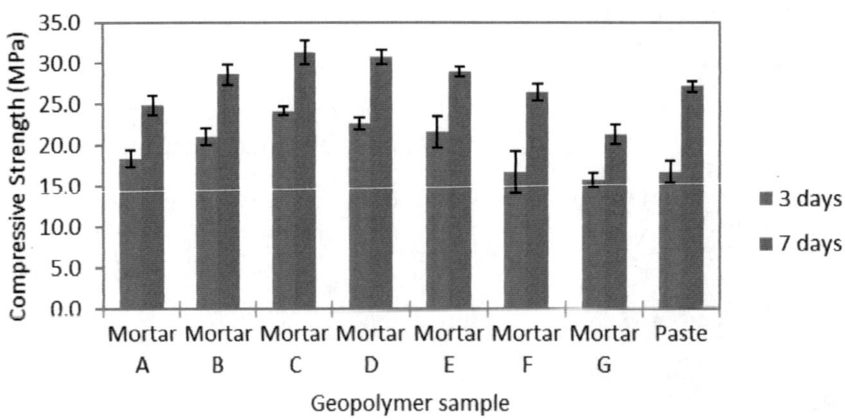

Figure 2. Compressive strength of various binder to sand ratio.

Compressive tests on geopolymer under different binder to sand ratio showed that fly ash based geopolymer mortar with a ratio 0.5 (Mortar C) had the highest strength which is 31.39 MPa at the 7th day of curing (Fig. 2). It also promotes higher compressive strength than geopolymer paste which is an increase of 13 %.

The compressive strength increased with further increase in the ratio of binder to sand until the ratio of 0.5 (Mortar C). However, the strength starts to decrease after the increase of the ratio. This shows that the compressive strength was affected by the binder to sand ratio. This result can be explained by the effect of sand to the geopolymer system. Sand is a material containing silica. An addition of sand beyond 50% causes the insufficient content of an activator for geopolymerisation process to take place. This causes the compressive strength of geopolymer mortar to decrease. Sand is responsible for the compressive strength of geopolymer mortar. Increasing sand content without increasing the amount of alkaline activator resulted in a decreasing level of geopolymerisation. It is proposed that compressive strength of geopolymer mortar with high levels of aggregate can be increased by optimizing the amount of alkali and properly distributed in the particle sand size [20].

Particle size distribution of the sand also plays an important role in the strength of the geopolymer mortar. The smaller particles will try to fit in between larger particles. Thus, the interlocking between the aggregate exist and this contributes to the strength of geopolymer mortar. The 0.5 binder to sand ratio was selected as an optimum compressive strength achieved. Therefore this ratio is the optimum compressive strength of geopolymer mortar.

4. Conclusion

The aim of this paper is to determine the best mix design of geopolymer mortar. Thus, the main purpose of this paper which to give an overview on the physical and mechanical properties of geopolymer is done. The various binder to sand ratio acts as an excellent repair material to be developed and applied to solve civil engineering problems. From the behavior of the geopolymer, it is clearly highlighted that the use of sand as fillers in geopolymer has the potential for use as repair material at optimal binder to sand ratio formulation. The geopolymer is compatible with the standard requirements and shows the potential to overcome damaged concrete structures. It is proposed that the compressive strength of geopolymer mortars with aggregate can be increased by optimising the binder to sand ratio and optimising the content of alkaline activator. Therefore, the suitable preparation of geopolymer mortar, 0.5 of binder to sand ratio is recommended on the basis of workability and compressive strength for concrete patch repair. However, the high compressive strength are not yet comfirmed to promise good bonding strength between repair materials to concrete substrate. Thus, further analysis is required.

5. References

[1] Davidovits, J., High-alkali cements for 21st century concretes, "In Concrete Technology, Past, Present and Future," Metha, P.K, American Concrete Institute, Farmington Hills, (1994).

[2] P. De Silva, K. Sagoe-Crenstil, and V. Sirivivatnanon, "Kinetics of geopolymerization: Role of Al_2O_3 and SiO_2," Cement and Concrete Research, 37 (2007) 512–518.

[3] M. M. A. Abdullah, H. Kamarudin, M. Bnhussain, I. Khairul Nizar, A. R. Rafiza, and Y. Zarina, "The Relationship of NaOH Molarity, Na_2SiO_3/NaOH Ratio, Fly Ash/Alkaline Activator Ratio, and Curing Temperature to the Strength of Fly Ash-Based Geopolymer," Advanced Materials Research, 328–330 (2011) 1475–1482.

[4] Z. F. Farhana, H. Kamarudin, A. Rahmat, and A. M. M. Al Bakri, "The Relationship between Water Absorption and Porosity for Geopolymer Paste," Materials Science Forum, 803 (2014) 166–172.

[5] M. M. Al Bakri Abdullah, M. F. Mohd Tahir, K. Hussin, M. Binhussain, and J. J. Ekaputri, "Effect of Microwave Curing to the Compressive Strength of Fly Ash Based Geopolymer Mortar," Materials Science Forum, 841 (2016) 193–199.

[6] J. Temuujin, R. P. Williams, and a. van Riessen, "Effect of mechanical activation of fly ash on the properties of geopolymer cured at ambient temperature," Journal of Materials Processing Technology, 209 (2009) 5276–5280.

[7] T. Phoo-ngernkham, P. Chindaprasirt, V. Sata, S. Hanjitsuwan, and S. Hatanaka, "The effect of adding nano-SiO_2 and nano-Al_2O_3 on properties of high calcium fly ash geopolymer cured at ambient temperature," Materials and Design, 55 (2014) 58–65.

[8] A. M. M. Al Bakri, H. Kamarudin, H. Mohamed, C. M. Ruzaidi, A. R. Rafiza, M. T. M. Faheem, and A. M. Izzat, "Properties And Microstructural Characteristics Of Geopolymers Using Fly Ash With Different Percentages Of Kaolin At Room Temperature Curing," Australian Journal of Basic and Applied Sciences, 5 (2011) 824–828.

[9] S. Hu, H. Wang, G. Zhang, and Q. Ding, "Bonding and abrasion resistance of geopolymeric repair material made with steel slag," Cement and Concrete Composites, 30 (2008) 239–244.
[10] S. Songpiriyakij, T. Pulngern, P. Pungpremtrakul, and C. Jaturapitakkul, "Anchorage of steel bars in concrete by geopolymer paste," Materials and Design, 32 (2011) 3021–3028.
[11] T. Phoo-ngernkham, V. Sata, S. Hanjitsuwan, C. Ridtirud, S. Hatanaka, and P. Chindaprasirt, "High calcium fly ash geopolymer mortar containing Portland cement for use as repair material," Construction and Building Materials, 98 (2015) 482–488.
[12] Mustafa Al Bakri A.M., H. Kamarudin, MBnhussain, I. Khairul Nizar, A.R. Rafiza, Y. Zarina, "Microstructure of different NaOH molarity of fly ash-based green polymeric cement," J Eng Tech Re, 3 (2011) 44-49.
[13] D. Hardjito, C. C. Cheak, C. Ho, L. Ing, and C. H. L. Ing, "Strength and Setting Times of Low Calcium Fly Ash-based Geopolymer Mortar," Modern Applied Science, 2 (2008) 3–11.
[14] Subaer. "Influence of aggregate on the microstructure of geopolymer," Dept. of Applied Physics, Curtin University of Technology, (2004).
[15] C. Kuenzel, L. Li, L. Vandeperre, A. R. Boccaccini, and C. R. Cheeseman, "Influence of sand on the mechanical properties of metakaolin geopolymers," Construction and Building Materials, 66 (2014) 442–446.
[16] J. Davidovits, First International Conference on Alkaline Cements and Concretes Kiev, (1994).
[17] I. Garcia Lodeiro, A. Fernandez-Jimenez, M. T. Blanco, and A. Palomo, "FTIR study of the sol-gel synthesis of cementitious gels: C-S-H and N-A-S-H," Journal of Sol-Gel Science and Technology, 45 (2008) 63–72.
[18] Duxson P, Provis JL, Lukey GC, Mallcoat SW, Kriven WM, van Deventer JSJ, "Understanding the relationship between geopolymer composition, microstructure and mechanical properties," Colloids Surf A: Physicochem Eng Aspects, 269 (2005) 47–58.
[19] M. Steinerova, "Mechanical properties of geopolymer mortars in relation to their porous structure," Ceramics - Silikaty, 55 (2011) 362–372.
[20] J. Temuujin, a. Van Riessen, and K. J. D. MacKenzie, "Preparation and characterisation of fly ash based geopolymer mortars," Construction and Building Materials, 24 (2010) 1906–1910.

Acknowledgments
The author would like to express his gratitude to the Center of Excellence Geopolymer & Green Technology (CEGeoGTech) for funding the research project.

IOP Conf. Series: Materials Science and Engineering **133** (2016) 012043 doi:10.1088/1757-899X/133/1/012043

Some Analysis of Major Impact of Geothermal Fluid Components in Power Plant Equipment

A Buzăianu[1*], I Csaki[2], P Moţoiu[3], K Leósson[4], S Serghiuţă[2], A Arnbjornsson[4], V Moţoiu[3], G Popescu[2], S Guðlaugsson[5], D Guðmundsson[6]

[1] Metav- R&D, Bucharest, 31 C.A.Rosetti St.,Code 020011, Romania.
[2] University "Politehnica" Bucharest, 313, Splaiul Independentei; Romania.
[3] Tehnoid Com Ltd.,48 Baritiu St,,Code 011295, Romania.
[4] Innovation Center Iceland-ICI, Arleyinir 2-8 Reykjavik, Iceland.
[5] Orka Náttúrunnar;110 Baejarhals , Iceland.
[6] Vélvík ehf, 110 Hofðabakka, Reykjavik, Iceland.

E-mail: buzaianu @ metav-cd.ro

Abstract. This paper presents the results from a some analysis and major impact of geothermal fluid composition on the equipment in use in geothermal power plant. The structural analysis of material deposition improve the direct influenced of chemical composition of stem and waters included CaO, MgO, Al_2O_3 and SiO_2 incorporated in the molten phase and the deposits in the scales formed due to equipment. The steam turbine corrosion damage, particularly of blades, discs and pomps, has long been recognized as a leading causes of reduced availability in the geothermal power plant. The corrosion process depends on temperature, pressure, chemisty and vaporous carryover by diversity of impurity. The experimental analysis procedure involves characterization of the fluid geothermal composition. Detailed information about surfaces morphological modification of the power plant components are obtained by electron microprobe analysis EDX and SEM investigation. References selection are obtaining by X-ray diffractometer patterns of the specimen.

1. Introduction

Geothermal steam contains various solid particles resulting in particle erosion and scale deposits. Geothermal steam contains hydrogen sulphide (H_2S) and carbon dioxide (CO_2). Both components are highly corrosive can promote heavily corrosion. The steam impurities could be easily transferred to the metal surface through this process may generate greater concentration of corrosive impurities and massive silicides and salts deposits. Aggressive corrosion in steam geothermal condition may be defined as an accelerated corrosion, resulting from the presence of many salt contaminants such as Na_2SO_4, NaCl, and silicates forming molten deposits, with a damaging impact on the protective surface of geothermal plant components [1,2]. Specific chemistry of geothermal steam include: high non-condensable gas contents, high content of solids salts dissolved in steam and corrosive chemical composition [3]. High H_2S content generally promotes metallurgical problems: stress corrosion, cracking and metal fatigue [4]. The high CO_2 content in geothermal fluid below 200°C, accelerate calcite scaling and promote active corrosion. Also one of the most commonly problems regarding high gas content in the geothermal steam are high non-condensable gas contents, sulphur deposition and corrosion due to low pH and oxygen into the geothermal power components steam [5]. For example the most commonly problems concerning high gas content in the geothermal steam are: high Cl^- 10.000 ppm as 1.8% NaCl and CO_2 300 ppm and SO_4 50 ppm as Na_2SO_4). This paper aims to present our current understanding, a new class of investigation and experimental data to provide a broader view on the effects of geothermal water and steam composition in Hellisheiði-Iceland geothermal power plant.

2. Materials and methods

The Hellisheiði geothermal field is located on the Hengill area. The Hengill geothermal system is in southwest part of Iceland on the Mid-Atlantic Ridge and comprises of Hellisheiði (Figure 1) geothermal power plant (there are four 40MWe/units and 45MWe/units steam *Mitsubishi Heavy Industry* turbines). The temperature in the Hellisheiði system change with drill depth; the maximum temperature is 180°C, found at 200-600m, but at 2 km, the temperature drops to 220-260°C. This suggests that a cooling process is taking place in the deepest part of the drill in Hellisheiði geothermal system. The geothermal steam that is produced from the Hellisheiði system contains dissolved CO_2 and H_2S gases. However, the concentration of non-condensable gases in the steam is quite low (<0.5%), when compared to other high temperature in Icelandic geothermal systems.

Figure 1.Hellisheiði power plants on a Hengill area in Iceland.

The concentration of CO_2 is similar or lower than usual, and we suppose that this is the case for insufficient supply of this gas to the fluid, to saturate it with calcite. The steam from the well then goes through the separator before entering the turbine. Table 1 present the chemical composition of the steam in turbines 1 to 6 in the Hellisheiði power plant.

Table 1. Chemical composition of the steam in the turbines at Hellisheiði power plant.

Turbine no.	H_2 [mg/kg]	N_2 [mg/kg]	CH_4 [mg/kg]	H_2S [mg/kg]	CO_2 [mg/kg]
1	31.14	28.85	4.43	804.27	1864.60
2	23.78	32.45	6.17	741.86	3087.17
3	29.35	22.71	5.04	928.96	2261.80
4	23.92	26.74	3.81	634.27	1725.63
5	15.92	82.93	7.58	472.50	4687.05
6	40.91	37.86	6.22	976.88	4551.85

Table 2 show examples of the chemical composition of the condensate/separator water. The temperature and pressure for some of the steam turbines that are operational at Helliheiði power plant is 170-175°C. The temperature and pressure is similar between turbines no.1–4 with the average temperature 170.8°C and pressure 7.17 bars. When the geothermal steam goes through the turbine, the steam temperature and pressure decrease. The steam pressure for the turbine is 7.21bar and temperature 170.8°C and after the turbine the pressure will be around 0.095 bars and temperature

45.5°C. The concentration of CO_2 is similar or lower than commonly, it is hypothesized that the cause is insufficient supply of this gas to the fluid to saturate it with calcite. The steam from the well then goes through the separator before entering the turbine. The concentration of H_2S and CO_2 in the steam are the highest. This aspect was expected due to the fact that they are common non-condensable gases dissolved in geothermal steam. Both H_2S and CO_2 are highly corrosive therefore are capable of aggressive corrosion in steel and alloys used in geothermal systems.

Table 2. Chemical composition ions of the steam and pH value of the separator and condensate water at Hellisheiði power plant.

Location [in water]	Na	K	Ca	Mg	Fe	Al	Cl	SO_4	F	pH
Separator [ppm]	208.8	36.1	1.054	0.025	0.05	1.851	179.03	54.59	1.37	9.23
Condensate [ppm]	0.15	0.2	0.01	0.001	0.011	0.003	1	0.2	0	6.32

It is important to get exact type of materials in the most relevant components of geothermal turbines power plants operated, as well as the most critical material damages of turbine components [6,7]. In a corrosive geothermal environment careful design and material selection is required to protect different parts of a turbine, particularly from: erosion, corrosion fatigue and Stress Cracking Corrosion (SCC). Surface damages such as corrosion and erosion depends not only on the steel type but also on the material condition as hardness of the steel and tempering process.

Turbine rotor at Hellisheiði power plant (figure 2) is manufactured by Mitsubishi Heavy Industry (MHI). Attention must be made to low-stress designs and applying rotor materials with low stress cracking corrosion susceptibility. The steel used for turbine components in power plant is a 12Cr steel and is a precipitation hardening martensitic stainless steel type 17-4 PH /AISI 630 (is the specification for martensitic stainless steel A17Cr-4Ni-3Cu; or Japan JIS G4303 (SUS 630). The producer MHI has developed two types of materials (Table 3) for the geothermal turbine rotors and chemical composition as shown in Table 4.The material 10325 MGB has a low SCC susceptibility because it contains extremely low levels of sulphur and impurity and GSR 1 is a 12% Cr-5 Ni material applied to turbine rotors exposed to highly corrosive geothermal steam. The steel used for turbine blades in Hellisheiði power plant is a 10325 MGB (low alloy ferrite steel of Cr, Mo,V type). In Fig.2 are shown examples of erosion and abrasion (index A), damages in blades (detailed index B) and rotor (detailed index C).

Table 3. Components turbine materials at Hellisheiði power plant and mechanical properties.

Components	Metallurgical characteristics	Yield strength $R_{0.2}$ [N/mm^2]	Tensile strength [N/mm^2]
Turbine blades 10325MGB	Forged low alloy ferrite steel	$\geq$635	$\geq$740
Turbine rotors GSR 1	12Cr precipitation hardening martensitic stainless steel	$\geq$635	$\geq$740

Table 4. Chemical composition of geothermal turbine materials.

Material	Chemical composition (% of mass)								
Turbine blades	**C**	**Mn**	**P**	**S**	**Si**	**Ni**	**Cr**	**Mo**	**V**
10325MGB	**0.23-0.3**	**0.7-1.0**	**$\leq$0.015**	**$\leq$0.005**	**0.2-0.4**	**$\leq$0.5**	**1.0-1.3**	**1.0-1.3**	**0.21-0.29**
Turbine rotors GSR 1	**$\leq$0.06**	**$\leq$1.0**	**$\leq$0.015**	**$\leq$0.005**	**<0.30**	**5.0-5.5**	**11.5-12.5**	**0.8-1.2**	**-**

Figure 2. Turbine MHI rotor at Hellisheiði power plant and examples of erosion (B) and abrasion (C) damages.

Method analyses of scales and deposits of the turbine components were examined using complex analyses. The samples were taken from different parts of the equipment used in Hellisheiði. The authors performed a series of tests on the samples to find the morphology and composition of scales/deposits. The method laboratory proposed three kinds of laboratory testing: a) Scanning Electron Microscopy (SEM) was used to evaluate the morphology of structure of the scale deposited materials; b) Energy Depressive Spectroscopy (EDS analysis) was used to analyse the composition of deposited materials; c) X-Ray Diffraction (XRD analysis) was employed to quantitative analyse of the compounds in the scales/deposited materials [8].

3. Results and discussions

High CO_2 content in geothermal fluid below about 140°C accelerates calcite scaling and promotes very active corrosion of geothermal plant components. The solids content and composition is varied, particularly regarding chloride and different trace elements, as: boron, fluoride etc. The effects of geothermal steam is providing nucleation seeds which may further develop into an acidifying early condensate and the formation of concentrated solutions generating impurity deposits with accentuated corrosion effects [9,10]. Experimental evidence shows that the deposit on the surface exposed to steam consists of many constituents with 0.5 microns to millimetre particles, small crystals and silicide particles with absorbed ionic impurities and their effect is to create strong adhesion and corrosion on the turbine components surface.

Figure 3. Massive chemical deposits in the geothermal power equipment caused by higth salts concentration in the geothermal steam.

Active corrosion occurs when metals are heated in the temperature range 200-280°C, in the presence of sulphate deposits formed as a result of the reaction between sodium chloride and sulphides compounds in the phase solutions (figure 3). Figure 4 reveals result of EDX analyse of Smart Phase Map collected at 30.00kV resolution, clearly showing the seven different phases in the samples and the spectra basis for the O, Al, Si, S, K, Ca and Fe elements analysis. Phase mapping uses spectra as the basis for characterizing the elemental as association and distribution by tracing variation in component peak intensity. This approach provides the fine correlation of the data between phases. The results are one single phase image describing the overall chemical map of the sample. Geothermal steam contains various solid particles which cause erosion and scale deposits. Special design and resistant material are needed to reduce the damage of Solid Particle Erosion (SPE). Wet steam can lead to droplet/moisture impact erosion. At Hellisheiði power plant erosion seems to be the main reason for damages of turbine parts, probably due to droplet impact erosion and solid particle erosion. In addition to silica the results of laboratory tests showed that there was also a small ion of Na, Ni and Fe particles and S and Si oxides.

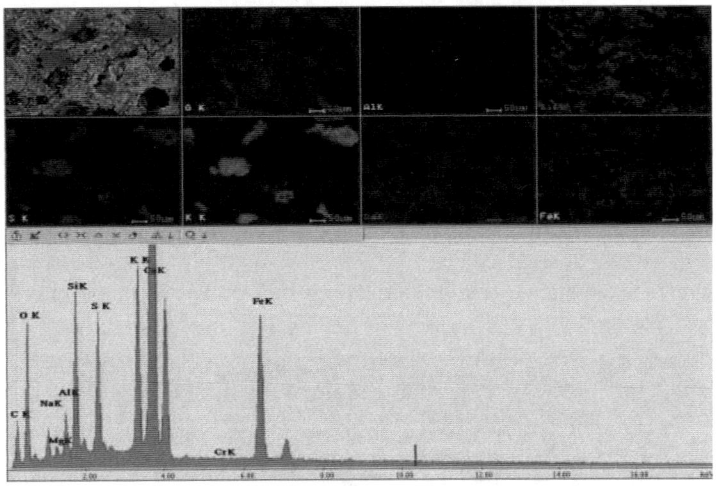

Figure 4. EDX micrography of the complex materials scaling deposition of the separator geothermal equipment.

Full knowledge of the droplet size and number spectrum is vital for assessment of the droplet impact turbine erosion process. The droplet impact pressure as well as the droplet impact angle is confirmed to be driving parameters affecting a material's volume loss, and volume loss increases exponentially with increasing perpendicular component of the droplets' impact velocity. The deposit also accumulated on the leading and trailing blades edges. Similar to the turbine diaphragm, the rotor and the blades experienced similarity problems (Figure 5 and Figure 6).

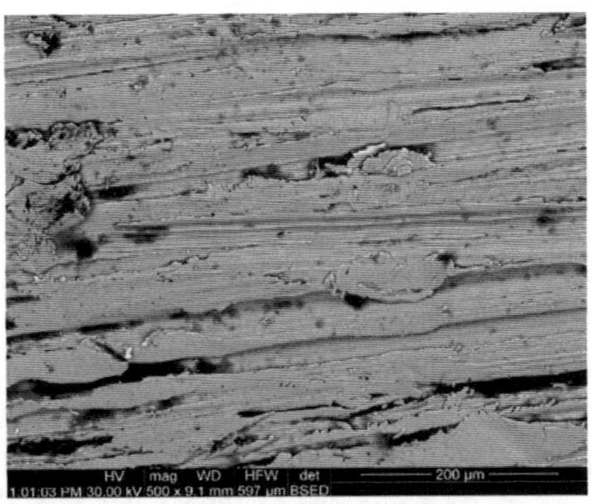

Figure 5. Electronic SEM micrography in surface erosion of blades (x 500).

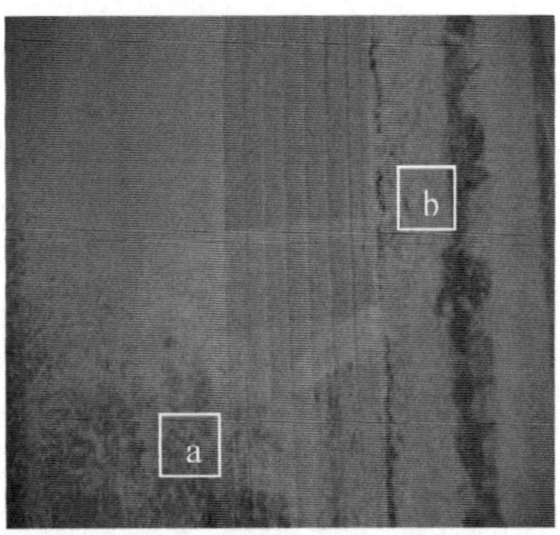

Figure 6. Optical image of the (a) corrosion and (b) erosion rothor surface.

The deposit thickness was reduced to 0.5-0.7 mm. Scaling and deposition occur also in the steamstainer (Figure 7), steam separator (Figure 8 and Figure 9), scrubber (Figure 10) and demister (Figure 11) elements. The demister's purpose is to capture water that is in the steam. Steam flows through the demister's elements and the resulting wet steam is depressed and water removed. The cleaning method of chisel chipping and fine grinding were used to remove the deposit from turbine blades.

Figure 7. Steamstainer with materials scaling deposition on its interior side.

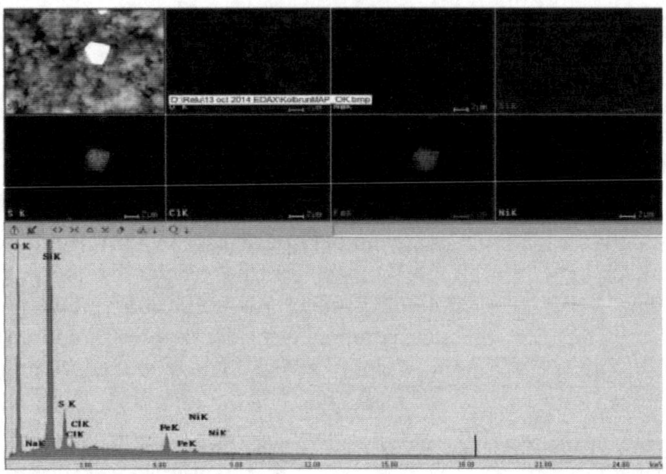

Figure 8. EDX mapping micrography of the complex materials scaling deposition of the separator geothermal equipment. Exampls of EDX showing seven different element phases of: O; Na; Si, S; Cl; Fe; and Ni in the samples.

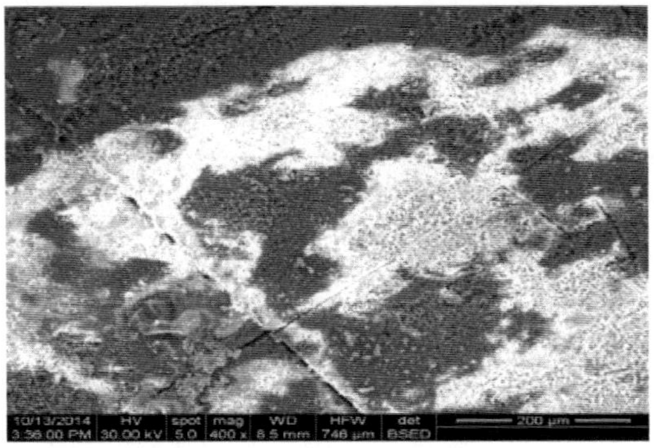

Figure 9. SEM image of the separator accumulated deposit (x 400).

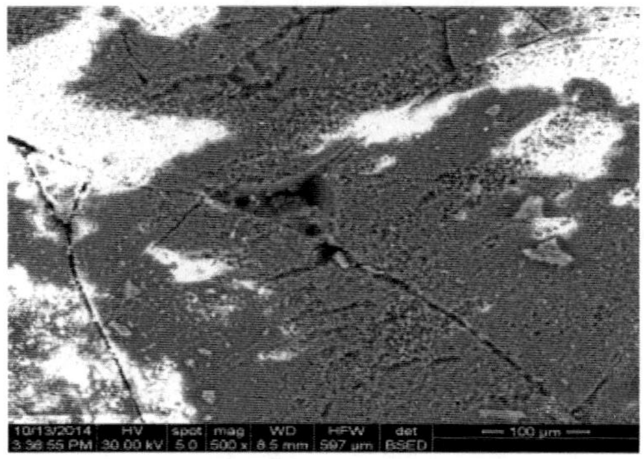

Figure 10. SEM image of the scrubber accumulated deposit (x 400).

Figure 11. SEM image of the demister equipment surface deposits (x 2000).

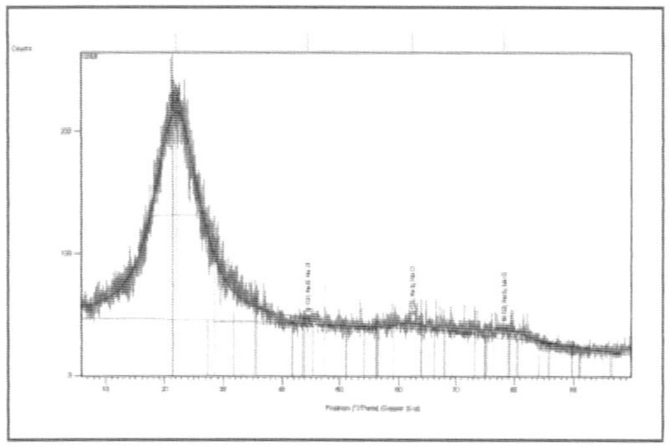

Figure 12. Scaling deposits X-ray diffraction pattern (Cu Kα).

Date: 19.02.2015 Time: 12:15:23 File: OXIZI User: user-1

No.	Visible	Ref. Code	Compound Name	Chemical Formula	Score	Scale ...	Display Color	Quality	Subfiles	Crystal System
1	☑	04-008-8251	Silicon Oxide	Si O2	55	0,863	Blue	B;N...	Alloy,...	Cubic
2	☑	00-023-1123	Iron Sulfide	Fe S	11	0,223	Lime	I	Alloy,...	Cubic
3	☑	00-001-0994	Sodium Chloride	Na Cl	11	0,038	Gray	B;D	Com...	Cubic

Figure 13. Identification and the colors code in the RDX scaling deposits difractogram for the major compounds distribution.

In this experiment the RDX analysis is a technique for the rapid determination of the ions in the samples. It displays the spectrum of composition existing in a given relative homogeneous phases from the shape of diffraction peak broadened (Figure 12 and Figure 13) by a range of lattice parameters in the phase.

5. Conclusions

Hellisheiði geothermal system is composed of basaltic hyaloclastites layers and other rock types, that is located within the Hengill central volcano where the volcanism is most intense. The maximum recorded temperature is around 280°C in the Hellisheiði system. Material analyses (SEM, EDX and XRD) showed that the problem of scaling deposit did not originate from corrosion. Supposed steam-carried particle were the main source of deposited material.The system contains dissolved CO_2 and H_2S gases. However, the concentration of non-condensable gases in the steam is quite low (<0.5%), in comparison to other geothermal systems in Iceland.Wet steam can lead to droplet/moisture impact erosion. At Hellisheiði power plant erosion seems to be the main reason for damages of turbine parts, probably due to droplet impact erosion and solid particle erosion.The deposits, consisting of manly SiO_2, FeS and NaCl were found in the main equipment used for steam passing, including the separator, scrubber, demister and on the turbine blades.

Acknowledgements
This work is supported by the Grant Romanian EEA Financial Mechanism Committee 2009-2014 and Iceland as Donor Stat. Bilateral Agreement Program Iceland-Romania „Research within priority sectors". Contract 16 SEE /30.06.2014.

References
[1] Cuevas-Artega C, Rodriguez J A, Clemente C M and Rodríguez J M 2014 Pitting Corrosion Damage for Prediction Useful Life of Geothermal Turbine Blade *American Journal of Mechanical Engineering* **2(6)** pp164-168
[2] Mazur Z, Garcia R, Aguirre J and Perez N 2008 Steam Turbine Blade Failure Analysis *Engineering Failure Analysis* **15(1-2)** pp 129-141
[3] Regenspurg S, Wiersberg T, Brandt W, Hueges E and Saadat A 2010 Geochemical Properties of Saline Geothermal Fluids from the in–Situ Geothermal Laboratory *Chemie der Erde-Geochemistry* **70(3)**
[4] Stefansson A, Arnórsson S, Gunnarsson I, Kaasalainen H and Gunnlaugsson E 2011 The geochemistry and sequestration of H2S into the geothermal system at Hellisheidi, Iceland *Journal of Volcanology and Geothermal Research* **202(3-4)** pp 179-188
[5] Ahned M and Sürken N 2009 Experimental assessment of droplet impact erosion resistance of steam turbine blade materials *Wear* **267(9-10)** pp 1605-1618
[6] Stapleton M 2002 Scaling and Corrosion in Geothermal Operation *Power ChemTechnology* pp15-18
[7] Corsi R 1886 Scaling and Corrosion in Geothermal Equipment: problems and preventive measures. *Geothermics* **15 (5–6)** pp 839–856
[8] Adiprana R, Yuniarto E and Salak G 2010 Geothermal Power Plant Experience of Scaling/Deposit: Analysis, Root Cause and Prevention *Proceedings World Geothermal Congress* Bali, Indonesia **25-29**
[9] Armannsson H, Gudmundsson A and Steingrímsson B S 1987 Exploration and development of the Krafla geothermal area *Jökull* **37** pp13-30
[10] Karlsdóttir S.N 2012 Comprehensive Renewable Energy *Elsevier Ltd* pp 239-257

Effect of Geopolymer filler in Glass Reinforced Epoxy (GRE) Pipe for Piping Application: Mechanical Properties

Mohammad Firdaus Abu Hashim[1,2, a], Mohd Mustafa Al Bakri Abdullah[1,2,b], Che Mohd Ruzaidi Ghazali[1,2, c], Kamarudin Hussin[1,2, d], Mohammed Binhussain[3,e]

[1]Faculty of Engineering Technology, Universiti Malaysia Perlis (UniMAP), Perlis, Malaysia.
[2]Center of Excellence Geopolymer & Green Technology (CeGeoGTech), School of Material Engineering, Universiti Malaysia Perlis (UniMAP), Perlis Malaysia
[3]King Abdul Aziz City Science & Technology (KACST), Riyadh, Kingdom of Saudi Arabia

[a]firdaushashim88@yahoo.com,[b]mustafa_albakri@unimap.edu.my,
[c]ruzaidi@unimap.edu.my, [d]vc@unimap.edu.my, [e]prof.binhussain@gmail.com

Abstract. The present work is aimed to carry out the effect of geopolymer material which is fly ash as filler in the glass reinforced epoxy pipe on the microstructure of fly ash geopolymer, compression properties, and bulk density using the filament winding method. Conventional glass reinforced epoxy pipes has its own disadvantages such as high corrosion resistance at acidic environment and low strength which can be replaced by the composite pipes. Geopolymer is a type of amorphous alumino-silicate and can be synthesized by geopolymerization process. A series of glass reinforced epoxy pipe and glass reinforced epoxy pipe filled with 10 - 40 weight percentage geopolymer filler which is fly ash with 4 Molarity were prepared. Morphology of the raw material fly ash and fly ash based-geopolymer surface was characterized using scanning electron microscopy. It was found that the additions of fly ash at the beginning with 10 wt% are showing higher compressive strength than glass reinforced epoxy pipe without fly ash geopolymer filler. The compressive test of these series of samples was determined using Instron Universal Testing under compression mode. It was found that compressive strength for samples fly ash based-geopolymer filler are higher as compared to glass reinforced epoxy pipe without geopolymer filler. However, the compressive strength of glass reinforced epoxy pipe with fly ash geopolymer filler continues to decline when added to 20 wt% - 40 wt% of geopolymer filler loading. The results showed that the mixing of geopolymer materials in epoxy system can be obtained in this study.

1. Introduction

Glass fiber reinforced epoxy (GRE) composite tubes have a great strength, lighter in weight, and lower risk of bursting which making them commonly preferred in the fields of aviation, structural engineering, and in oil and gas industries [1]. The increased usage of composite structures demands reliable testing methods to ensure the safety and predict their long-term performance. However, previous researches reported that GRE pipes have disadvantages such as lower strength compare to metal pipes, cannot stand at high temperature due to epoxy properties and have low corrosion resistance which can suffer strain corrosion at acidic environments [2-4]. This paper concerns these entire disadvantage includes the application in the offshore oil and gas industry, particularly composite pipelines for aqueous liquids and to produce higher strength of GRE pipe. GRE pipes are normally designed to withstand high pressure, lightweight, relatively thin-walled structure provides ease of handling and transportation, which results in reduced installation costs [5, 6]. Geopolymer filler which is fly ash is use in this research to overcome this problem.

Geopolymer is a type of amorphous alumino-silicate cementitious material and can be synthesized by polycondensation reaction of geopolymeric precursor and alkali polysilicates known as geopolymerization process [7]. Through a period of time, substantial research struggles have been focused towards the development of inorganic geopolymers, due to the wide range of potential applications for these materials. Several reports can be found in the literature on the synthesis, properties and applications of geopolymers [8]. Geopolymerization is an innovative technology that can transform several aluminosilicate materials into useful products called geopolymers or inorganic polymers. The remaining achievements in inorganic chemistry made through geopolymerization include mineral polymers which termed as polysiatate or geopolymers, making a great possibility to produce composite materials not only with excellent mechanical properties such as lightweight and high strength but also with ideal fire resistant, non-toxic fumes and smokes, and resisting all organic solvents [9-11]. For several years, researchers have been enthralled by the studies on geopolymers which offered low energy consumption and the absence of CO_2 emissions in the preparation process [12-14]. Composites based on clay have been more widely studied since clay materials are naturally available and contribute to excellent mechanical properties of either thermoplastic or thermosets matrices [15-17].

In this research, GRE pipe which is diglycidyl ether of bisphenol A (DGEBA) epoxy were mixed with fly ash geopolymer filler with 4 Molarity concentration of sodium hydroxide (NaOH) and ratio solid to liquid is 1. To produce a higher strength of geopolymer, the optimum sodium silicate to sodium hydroxide ratio should be in range of 0.67 to 1.00 [18]. The formulation design of geopolymer was referred to the composite system as stated from the previous researcher on composite system which is use the loading of flax fibers varied from 0 to 60 wt%. [19]. Epoxy resin DGEBA was mixed with geopolymer filler which one of the greatest current interests in thermoset-based clay composites since it offers low cost, ease of processing, fine adhesion to many substrates, and good chemical resistance for a wide range of applications [20]. Fly ash contains of finely divided ashes produced by pulverized coal in power stations. The spherical shape of fly ash improves the consolidation of pipe, which also reduces permeability [21]. Therefore, these materials (fly ash, kaolin, and white clay) has wide potential to be used as a source material to react with liquid alkaline activator within sodium silicate solution (Na_2SiO_3) and sodium hydroxide (NaOH) mixture [22].

2. Material and Experimental Details

2.1 Materials selection

The epoxy resin, Diglycidyl Ether of Bisphenol-A (DGEBA) supplied by Euro Pharma Sdn Bhd and hardener Isophorondiamine (IPDA) was obtained from Dr Rahmatullah Holdings was used in this study. The raw materials geopolymer filler was produced by Saudi local based materials which is Fly Ash was used in preparation of geopolymer. Geopolymer paste is formed by alkaline activator to induce the silicon and aluminum atoms in the material [23]. Alkaline activator liquid used in this research is combination of sodium hydroxide (NaOH) with 4 Molarity and sodium silicate (Na_2SiO_3).

Alkaline activator was prepared by mixing NaOH and sodium silicate before mixed with raw materials to increase the reactivity of solution with concentration 4 M of NaOH to determine the optimum concentration. The ratio of raw materials to alkaline activator and sodium silicate to NaOH is 1. The geopolymer pastes then were cured in oven for 24 hours to 48 hours at 60 °C – 80 °C and then were taken out to be crushed by using ring mill. Crushed geopolymer filler were sieved using sieve size of 150 μm.

2.2 Experimental procedure

The Epoxy Geopolymer was prepared according to the formulation in the table 1 by mechanical mixer using blade stirrer. The mixing epoxy geopolymer were cured with cycloaliphatic amine curing agent IPDA. Epoxy and geopolymer materials were mixed for about 2 hours to make it completely homogeneous, followed by curing agent/hardener for about 5 minutes. Then the resin was poured into the tank after mixing with epoxy hardener.

Continuous glass fibers (E Type) were impregnated ("wet-out") with geopolymeric resin by means of homemade "impregnation machine" as filament winding technique. The velocity of the fiber during the impregnation process was chosen based on the best penetration on geopolymer resin into the fibers. The winding speed was controlled to generate the desired winding angle patterns. The feeding velocity of the fiber into the resin tank will depends on the mandrel rotational speed. It was selected based on the best impregnation of geopolymer resin into the fibers.

After the appropriate number of layers has been applied according to fully wounded, filament winding samples were allowed to be cured at the mandrel in the room temperature for 24 hours. After the curing process completed, the sample is ready for testing. In order to determine the mechanical properties, several tests are performed on composite structures which are compressive test and elasticity modulus.

Table 1. Epoxy Geopolymer Formulation Design.

Geopolymer Material (4 M)	Epoxy + Hardener (%)	Geopolymer (%)
	100	0
Fly Ash	90	10
	80	20
	70	30

3. Result and discussion

3.1 Surface morphology of raw material fly ash

Figure 1 shows the surface morphology of the raw material fly ash. It can be seen that the microstructure of fly ash appeared to be a glassy, hollow and clearly shows the shape of fly ash as reported from previous researcher contains of spherical particles with smooth outer surfaces [24]. The smooth aluminosilicate spherical particles, also known as cenospheres are formed as a result of mineral particles during coal combustion process, where the minerals melt to form small droplets, which upon rapid cooling and action of surface tension forces adopt the spherical shape [25-27]. The raw material of fly ash shows that the spherical morphology of the raw fly ash was fully destroyed which mean there have some agglomeration of the fine grained particles due to mechanical milling processes.

Figure 1: SEM micrographs of the raw fly ash at 2000x and 5000x magnification

3.2 Bulk Density Analysis

Bulk density is one of the physical properties that can be measured by simple procedure and calculated manually with mass per unit volume of sample. From Figure 2, it can be seen that the fly ash geopolymer filler with 40 wt% of filler loading has the highest bulk density which is 1.585 g/cm3 as compared to other samples. GRE pipe with fly ash based-geopolymer with 10 wt% has the lowest bulk density which is 1.414 g/cm^3 due to low filler loading. As a result, the density of the sample will decrease or less dense as the porosity increase.

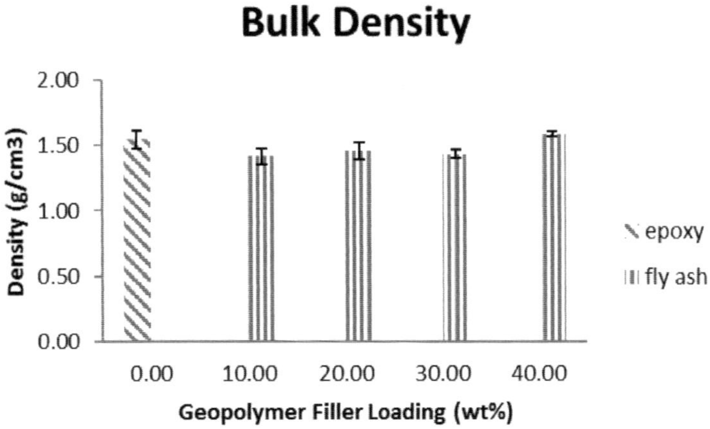

Figure 2: Comparison graph of bulk density between samples

3.3 Compression Analysis

Figure 3 shows the typical graph from the tests of compression strength tests for the glass reinforced epoxy (GRE) pipe without filler and glass reinforced epoxy (GRE) pipe filled with 10 wt% - 40 wt% of fly ash geopolymer filler. Compression test was performed according to the ASTM D3410 using Instron Universal Testing Machine [28]. The compression test consists of deforming a cylindrical hollow specimen to produce a thinner cylinder hollow of large diameter. It is a convenient method for determining stress strain response.

From Figure 3, it is clearly indicated that GRE pipe with 10 wt% of fly ash geopolymer filler shows the highest compressive strength compare to others composition. However, compressive strength of GRE pipe filled with geopolymer filler of fly ash are slightly continue to drop with the increase of the percentage of geopolymer filler loading and it was supported by the rule of mixture theorem which the addition of micro particles offer greater stiffness than epoxy matrix [29]. This may be due to the smaller agglomerated and disordered nature of clay and also the uses of lower molarity of NaOH. Even the compressive strength of GRE pipe with fly ash geopolymer filler drop when added with higher filler loading, the compressive strength is much higher than GRE pipe without geopolymer filler. Table 2 shows the values of compressive strength, compressive strain and compressive modulus of elasticity for the 0 wt% – 40 wt% fly ash geopolymer filler loading.

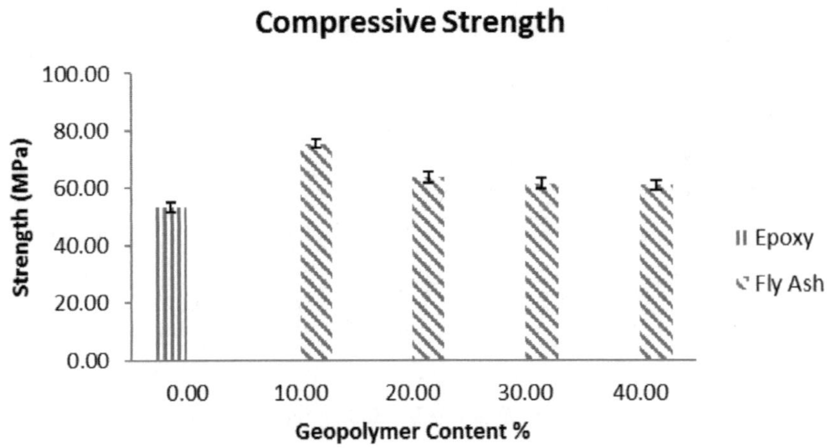

Figure 3: Compressive strength of GRE pipe with 0 wt% - 40 wt% geopolymer filler loading

The high compressive strength revealed a strong interaction between polymer chains which is epoxy hardener with the geopolymer filler itself. However, the decreasing of compressive strength in higher contain of fly ash which 20 wt% – 40 wt% may be controlling the capability and capacity of relocating load and plastic deformation between particles and matrix interface.

Table 2. Compression Properties of Epoxy Hardener and Epoxy Geopolymer

	Compressive Strength (MPa)				Compressive Strain (%)				Modulus Elasticity (MPa)			
	10	20	30	40	10	20	30	40	10	20	30	40
Epoxy	53.36				0.04				1681.28			
Fly Ash	75.49	63.75	61.55	60.89	0.07	0.07	0.04	0.06	2295.85	1755.33	2560.17	2450.45

3.4 Surface morphology of fly ash based-geopolymer

Figure 4 shows the fly ash based-geopolymer that undergoes 24 hours of curing time in the oven within 70 °C and has been crushed into small particles using a milling machine (ring mill). The changes on the surface morphological of raw material of fly ash and fly ash based-geopolymer observed show changes due to the reactive dissolution of SiO_2 and Al_2O_3 (contained in the cenospheres) in alkaline solution. This will leads to the formation of alumino silicate gel, which then acts as a precursor to geopolymer formation [30].

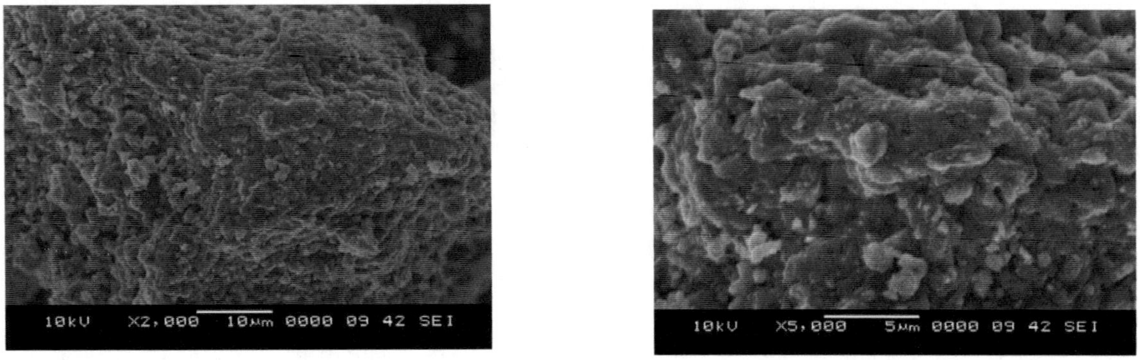

Figure 4: SEM micrographs of the fly ash based-geopolymer at 2000x and 5000x magnification

4. Conclusion

From this study, it can be concluded that glass reinforced epoxy (GRE) pipe filled with fly ash based-geopolymer was developed on epoxy resin with different weight percentage of the geopolymer material. The experimental result shows the performance of the product through the compression test, morphology, and bulk density test. Based on the results of the compression test, the product from GRE pipe with fly ash based-geopolymer shows the higher strength compare to the GRE pipe without geopolymer filler. GRE pipe with 10 wt% of fly ash based-geopolymer filler shows the highest compressive strength. Due to the good properties of the waste materials based geopolymer has bigger potential to be a matrix filler of composite with glass fiber in filament winding technique. Geopolymer composite can be as a filler in the piping system application through filament winding technique that not only environmentally but also reducing the production cost of the product.

Acknowledgement
We would like to extend our appreciation to the Center of Excellence Geopolymer & Green Technology (CEGeoGTech), King Abdulaziz City for Science and Technology (KACST) and all the people who helped to ensure that the successful completion of this study.

References

1. Faria, H. and R.M. Guedes 2010 *Long-term behaviour of GFRP pipes: Reducing the prediction test duration.* Polym. Test. **29** p 337-345.
2. Hahn, H. 1982 *Fatigue of composites- Environmental effects.* Fatigue and creep of composite materials 19-35.
3. Knox, E., M. Cowling, and S. Hashim 2000 *Fatigue performance of adhesively bonded connections in GRE pipes.* Int. J. Fatigue. **22** p 513-519.
4. Salibi, Z. 2001 *Performance of reinforced thermosetting resin pipe systems in desalination applications: a long-term solution to corrosion—the Arabian Gulf example.* Desalination. **138** p 379-384.
5. Gibson, A. 1989 *Composite materials in the offshore industry.* Met. Mater. (Inst. Mater.). **5** p 590-594.
6. Gibson, A. and D. Spagni 1991 *Recent developments in the use of composite materials offshore.* The Institute of Marine Engineers(UK) 3-9.

7. Dimas, D., I. Giannopoulou, and D. Panias 2009 *Polymerization in sodium silicate solutions: a fundamental process in geopolymerization technology.* J. Mater. Sci. **44** p 3719-3730.
8. De Silva, P., K. Sagoe-Crenstil, and V. Sirivivatnanon 2007 *Kinetics of geopolymerization: role of Al 2 O 3 and SiO 2.* Cem. Concr. Res. **37** p 512-518.
9. DAVIDOVITS, J. 2008 *Geopolymer: Chemistry & Applications. 2ª Edição.* França: Institut Géopolymère.
10. Davidovits, J. *Geopolymer chemistry and sustainable development. The poly (sialate) terminology: a very useful and simple model for the promotion and understanding of green-chemistry.* in *Proceedings of 2005 geopolymer conference.* 2005.
11. Sheppar, L. 2009 *Geopolymer Composites: A Ceramics Alternative to Polymer Matrices.* Proceedings of the 105th.
12. Wang, M., D. Jia, P. He, and Y. Zhou 2010 *Influence of calcination temperature of kaolin on the structure and properties of final geopolymer.* Mater. Lett. **64** p 2551-2554.
13. Muñiz-Villarreal, M., A. Manzano-Ramírez, S. Sampieri-Bulbarela, J.R. Gasca-Tirado, J. Reyes-Araiza, J. Rubio-Ávalos, J. Pérez-Bueno, L. Apatiga, A. Zaldivar-Cadena, and V. Amigó-Borrás 2011 *The effect of temperature on the geopolymerization process of a metakaolin-based geopolymer.* Mater. Lett. **65** p 995-998.
14. Elimbi, A., H. Tchakoute, and D. Njopwouo 2011 *Effects of calcination temperature of kaolinite clays on the properties of geopolymer cements.* Construction and Building Materials. **25** p 2805-2812.
15. Alexandre, M. and P. Dubois 2000 *Polymer-layered silicate nanocomposites: preparation, properties and uses of a new class of materials.* Mater. Sci. Eng.: R: Reports. **28** p 1-63.
16. Vaia, R.A., B.B. Sauer, O.K. Tse, and E.P. Giannelis 1997 *Relaxations of confined chains in polymer nanocomposites: glass transition properties of poly (ethylene oxide) intercalated in montmorillonite.* J. Polym. Sci. Part B: Polymer Physics. **35** p 59-67.
17. Messersmith, P.B. and E.P. Giannelis 1994 *Synthesis and characterization of layered silicate-epoxy nanocomposites.* Chem. Mater. **6** p 1719-1725.
18. Chindaprasirt, P., T. Chareerat, and V. Sirivivatnanon 2007 *Workability and strength of coarse high calcium fly ash geopolymer.* Cem. Concr. Compos. **29** p 224-229.
19. Arbelaiz, A., B. Fernandez, J. Ramos, A. Retegi, R. Llano-Ponte, and I. Mondragon 2005 *Mechanical properties of short flax fibre bundle/polypropylene composites: Influence of matrix/fibre modification, fibre content, water uptake and recycling.* Compos. Sci. Technol. **65** p 1582-1592.
20. Rattanasak, U. and P. Chindaprasirt 2009 *Influence of NaOH solution on the synthesis of fly ash geopolymer.* Miner. Eng. **22** p 1073-1078.
21. Pacheco-Torgal, F., J. Castro-Gomes, and S. Jalali 2008 *Alkali-activated binders: A review: Part 1. Historical background, terminology, reaction mechanisms and hydration products.* Construction and Building Materials. **22** p 1305-1314.
22. Temuujin, J., A. van Riessen, and K. MacKenzie 2010 *Preparation and characterisation of fly ash based geopolymer mortars.* Construction and Building Materials. **24** p 1906-1910.
23. Tran, D., D. Kroisová, P. Louda, O. Bortnovsky, and P. Bezucha 2009 *Effect of curing temperature on flexural properties of silica-based geopolymer-carbon reinforced composite.* Manuf. Eng. Mater. Process. **37** p 492-497.
24. Sukmak, P., S. Horpibulsuk, and S.-L. Shen 2013 *Strength development in clay–fly ash geopolymer.* Construction and Building Materials. **40** p 566-574.

25. Temuujin, J., R. Williams, and A. Van Riessen 2009 *Effect of mechanical activation of fly ash on the properties of geopolymer cured at ambient temperature.* J. Mater. Process. Technol. **209** p 5276-5280.
26. De Vargas, A.S., D.C. Dal Molin, A.C. Vilela, F.J. Da Silva, B. Pavão, and H. Veit 2011 *The effects of Na 2 O/SiO 2 molar ratio, curing temperature and age on compressive strength, morphology and microstructure of alkali-activated fly ash-based geopolymers.* Cem. Concr. Compos. **33** p 653-660.
27. Daud, Y.M., H. Kamarudin, C.M. Ruzaidi, A.F. Osman, and M. Al-Bakri 2014 *Epoxy layered-silicates filled with fly ash based geopolymer: Compressive properties.* in Mater. Sci. Forum Trans Tech Publ.
28. ASTM D3410 / D3410M-03, 2003, "*Standard Test Method for Compressive Properties of Polymer Matrix Composite Materials with Unsupported Gage Section by Shear Loading,*" ASTM International, West Conshohocken, PA, DOI:10.1520/D3410_D3410M-03, www.astm.org
29. Nuhiji, B., D. Attard, G. Thorogood, T. Hanley, K. Magniez, J. Bungur, and B. Fox 2013 *The effect of a rapid heating rate, mechanical vibration and surfactant chemistry on the structure–property relationships of epoxy/clay nanocomposites.* Materials. **6** p 3624-3640.
30. Mustafa Al Bakri, A., H. Kamarudin, M. Bnhussain, I. Khairul Nizar, A. Rafiza, and A. Izzat 2011 *Chemical reactions in the geopolymerisation process using fly ash-based geopolymer: a review.* Aust. J. Basic Appl. Sci. **5** p 1199-1203.

IOP Conf. Series: Materials Science and Engineering **133** (2016) 012045 doi:10.1088/1757-899X/133/1/012045

Performance of Straight Steel Fibres Reinforced Alkali Activated Concrete

Meor Ahmad Faris[1,a], Mohd Mustafa Al Bakri Abdullah[2,1,b], Khairul Nizar Ismail[3,c], Ratnasamy Muniandy[4,d] & Ramadhansyah Putra Jaya[5,1,]

[1] Center of Excellence Geopolymer and Green Technology (CEGeoGTech), School of Materials Engineering, Universiti Malaysia Perlis (UniMAP), Perlis, Malaysia.
[2] Faculty of Engineering Technology, Universiti Malaysia Perlis (UniMAP), Perlis, Malaysia.
[3] School of Environmental Engineering, University Malaysia Perlis, Perlis, Malaysia
[4] Department of Civil Engineering, University Putra Malaysia, Selangor, Malaysia
[5] Department of Geotechnics and Transportation, Faculty of Civil Engineering, Universiti Teknologi Malaysia, Johor, Malaysia

[a]meorfaris88@gmail.com, [b]mustafa_albakri@unimap.edu.my, [c]nizar@unimap.edu.my, [d]ratnas@upm.edu.my

Abstract. This paper focus on the performance of alkali activated concrete produced by using fly ash activated by sodium silicate and sodium hydroxide solutions. These alkali activated concrete were reinforced with straight steel fibres with different weight percentage starting from 0 % up to 5 %. Chemical composition of raw material in the production alkali activated concrete which is fly ash was first identified by using X-ray fluorescence. Results reveal there have an effect of straight steel fibres inclusion to the alkali activated concrete. Highest compressive strength of alkali activated concrete which is 67.72 MPa was obtained when 3 % of straight fibres were added. As well as flexural strength, highest flexural strength which is 6.78 MPa was obtained at 3 % of straight steel fibres inclusions.

1. Introduction

In most of country, the source of generation electricity is from the combustion of coal. This is proved by a report made by World coal Association where 41 % of global electricity is generated by the combustion of coal [1]. In details, there are approximately 2 million of coal will be fired for a capacity of 100 Mega-Watt coal fired station and produce 230 000 tons of coal ash annually [2]. This coal ash is a waste and may become harmful to the environment if not disposed properly. Nowadays, due to the most of coal ash is not consumed as alternative applications it become a waste product and normally disposed in landfill which can cause hazardous to the environmental [3]. So, the usage of fly ash as a raw material in the production of alkali activator material is the alternative solution for this environmental issue.

In 1990, there are about 22.7 billion tons of carbon dioxide was release to the environment over the world. The value then increase in 2012 where 33.9 billion tones of CO_2 was released and currently the emission of CO_2 reported is over than 34.5 billion tons [4]. Unfortunately, a main contributor to the emission of CO2 is claimed from the cement industry. In facts, the production of 1 tons of OPC will release about 1 tons of CO_2. Details of chemical reaction involved in the production of OPC where CO_2 was release to the environment as in equation (1):

$$5CaCO_3 + 2SiO_2 \rightarrow 3CaO\,(SiO_2) + 2CaO_2\,(SiO_2) + 5CO_2 \quad (1)$$

People are very depending on this material (OPC) due to the using of OPC in the production of buildings. This may lead the increment of emission of CO_2 to the environment and brings into a global warming issue. Regarding to this issue, alkali activated conc is the best alternative material to be used as a substitution of OPC in the production of cementitious binder with an excellent properties and proved to improves the greenness of OPC cement [5].

Steel fibres that normally used in OPC concrete have an ability to enhance the post cracking behavior and create a bridge between the crack spot with a sufficient amount of fibres [5]. The addition of steel fibres in OPC concrete proved to produce a better crack control [6-9] and improves tensile strength before and after cracking [7, 10]. Function of steel fibres in OPC concrete to influence the behavior of brittleness of normal concrete to ductile has been studied [7, 9, 11, 12]. Fatigue and dynamic resistance of OPC concrete also improved with high bond in steel fibres surface and cement paste that has been hydrated [13-16]. However, a drawback of steel fibres addition is same as any additions of other types fibres which is the workability of concrete is become lower [17]. Low workability leads to the increasing of bleed water at the top of concrete [18-20]. The effect of fibres additions have been extensively studied toward the OPC concrete. However there is still lack of data regarding to the using of steel fibres as reinforcement in alkali activated concrete. Therefore, the performance of alkali activated concrete with the additions of steel fibres need to be studied and discovered especially in terms of the mechanical properties.

2. Materials and Methodology

Alkali activated concrete was produced by using fly ash as raw material and mixed with alkaline activator. Fly ash was taken from electric power plant Manjung, Malaysia. Alkaline activator is composed of mixture between sodium silicate (Na_2SiO_3) and sodium hydroxide (NaOH). The Na_2SiO_3 is from South Pacific Chemicals Industry Sdn. Bhd. (SPCI), Malaysia with SiO/Na_2O ratio is 3.2 while the NaOH pellet is obtained from Formosa Plastic Corporation, Taiwan with concentration of 12 M after diluted with distilled water. Chemical composition of Fly ash was first characterized by using x-ray fluorescence (XRF). Ratio between fly ash to alkaline activator that has been used was 2.0. Meanwhile, ratio between Na_2SiO_3 to NaOH was 2.5.

River sand was used as fine aggregate with a maximum size of 4.75 mm. Meanwhile, limestone was used as coarse aggregate with a maximum size of 20 mm. Fibres used in this research was high quality of low carbon steel with type of straight end. Dimension of straight steel fibres used in this experiment was 0.13 mm in diameter and 5 mm in length. Amount of steel fibres additions were 0 %, 1 %, 3 %, and 5 % from the total weight of concrete.

Sample cube of dimension 100 x 100 x 100 mm were cast for compression test of alkali activated concrete. Meanwhile, beam with dimension of 100 x 100 x 500 mm were cast for flexural test. Every samples from compression and flexural were demoulded after 24 hours and cure at room temperature.

After 28 days, all samples were tested for compression and flexural by using universal testing machine. Four point bending was applied in flexural test as determined by ASTM C1018.

IOP Conf. Series: Materials Science and Engineering **133** (2016) 012045 doi:10.1088/1757-899X/133/1/012045

3. Results and Discussion

Table 1 shows XRF result for chemical analysis of fly ash. According to the result from Table 1, the total SiO_2 + Al_2O_3 + Fe_2O_3 value of the fly ash is 72.98 % while the total of CaO is 18.10. Since the chemical content of CaO is lower than 20 % therefore this fly ash is classified as Class F according to ASTM C618 [21]. The ratio between silica and alumina (SiO_2/Al_2O_3) in raw material of fly ash is 2.64.

Table 1. Chemical composition of fly ash.

Component	Percentage (%)
SiO_2	38.80
Al_2O_3	14.70
Fe_2O_3	19.48
TiO_2	1.02
CaO	18.10
MgO	3.30
SO_3	1.50
K_2O	1.79
MnO	0.16
BaO	0.27
SrO	0.11

The effect of steel fibres additions which are 0%, 1 %, 3 % and 5 % to the compressive strength of alkali activated concrete was plotted in a graph as in Figure 1. From the graph, 3 % of steel fibres addition shows the highest value of compressive strength which is 67.72 MPa, followed by 1 % of steel fibres (62.27 MPa), 5 % of steel fibres (60.21 MPa), and the lowest is with 0 % of steel fibres (57.92 MPa). This result showed the function of steel fibres addition in order to improve the compressive strength of alkali activated concrete. From the graph, the increasing of steel fibres additions will slightly increase the compressive strength of alkali activated concrete until at 3 % which is at the optimum value.

However, steel fibres addition more than 3 % shows a reduction in terms of compressive strength. The increasing of compressive strength is due to the contribution of steel fibres inclusion to enhanced post cracking behaviour. This is supported by the previous researchers where Cuenca et al. mentioned steel fibres functioned to bridge the crack spot if the total amount of steel fibres is sufficient [22]. The inclusion of steel fibres in alkali activated concrete give minor effect to the improvement of alkali activated concrete which is 16.91 % increment at 3 % steel fibres additions (optimum).

Meanwhile, Figure 2 shows the flexural strength (4 point bending) of alkali activated concrete on 28 days related to the steel fibres additions starting from 0 %, 1 %, 3%, and 5 %. From the graph, addition of 3 % steel fibres shows the highest flexural strength which is 6.78 MPa, followed by 5 % steel fibres content (6.15 MPa), 1 % steel fibres content (5.88 MPa), and the lowest value is 0 % steel fibres content which is 4.3 MPa. This graph result shows the importance of steel fibres in order to improve the flexural strength of alkali activated concrete. From the graph, the increasing of steel fibres content from 0 % to 3 % shows the increasing of flexural strength until maximum at 3 % steel fibres content.

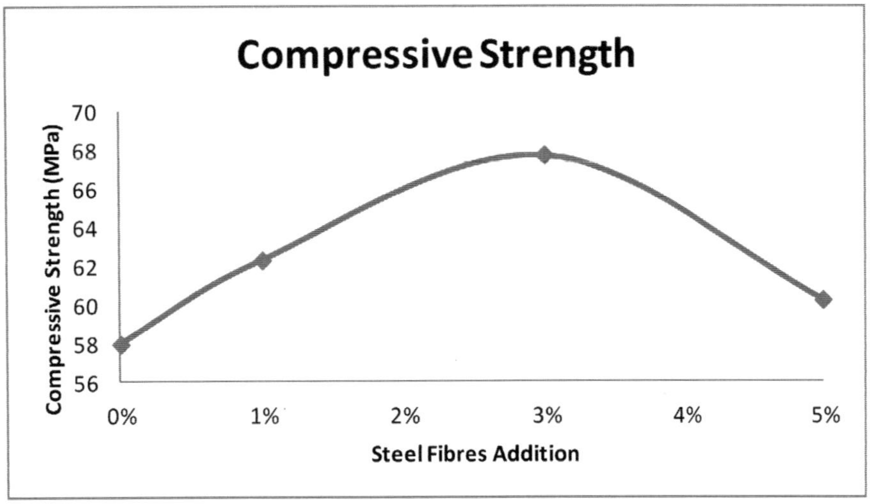

Figure 1. The effect of steel fibres addition on 28-day to the compressive strength of alkali activated concrete.

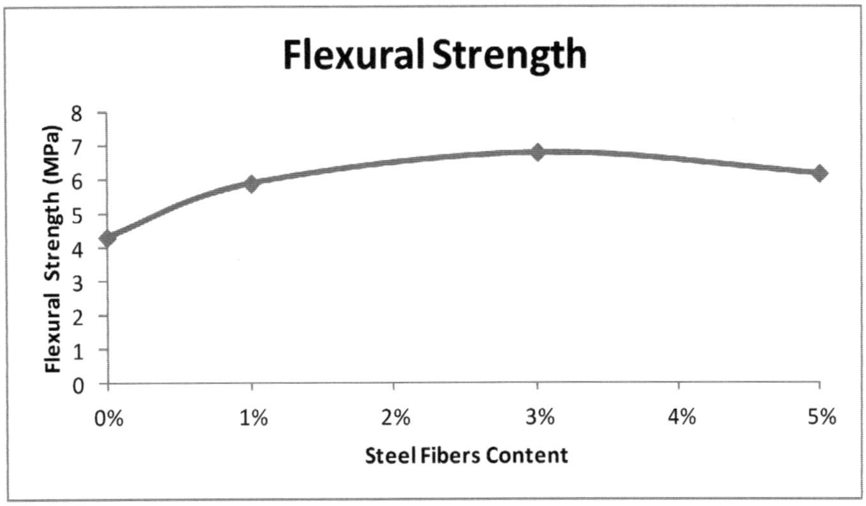

Figure 2. The effect of steel fibres addition on 28-day to the flexural strength of alkali activated concrete.

However, the flexural strength of alkali activated concrete slightly dropped when steel fibres additions over than 3 %. Increasing of flexural strength is due to the functions of steel fibres to change the properties of concrete from brittle to ductile behaviour. Generally, main function of steel fibres is excellent resistance to cracking and crack propagation. Hence, extensibility and tensile strength, both at first crack and at ultimate, specifically under flexural loading; and the fibres are able to hold the matrix together even after extensive cracking.

4. Conclusion

The additions of steel fibres in alkali activated concrete give impact to the mechanical properties which are compressive and flexural strength. The additions of steel fibres will enhance the compressive strength until at optimum content which is at 3 % and over than that, the value slightly getting lower. Meanwhile, the inclusions of steel fibres give significant effect to the flexural strength which is about 57.67% increase (maximum) at 3 % of fibres additions compare to the compressive strength which is 16.91 % (maximum) at 3 % of fibres additions. This is shown the functions of steel fibres additions to increase the compressive strength and mainly to increase the flexural strength.

5. References

[1] Association WC. Coal & Electricity. 2015 [cited; Available from: http://www.worldcoal.org/coal/uses-of-coal/coal-electricity/

[2] Agency IE. Case Studies in Sustainable Development in theCoal Industry. 2006 [cited; Available from: http://www.iea.org/publications/freepublications/publication/CIAB_Case_Studies_2006.pdf

[3] Hutagi A, Khadiranaikar RB, Shemblle R. A Study on Fibre Reinforced Geopolymer Concrete. Journal of Civil Engineering Technology and Research. 2014;2(1):15-22.

[4] Benhelal E, Zahedi G, Shamsaei E, Bahadori A. Global strategies and potentials to curb CO_2 emissions in cement industry. Journal of Cleaner Production. 2013;51(0):142-61.

[5] Rangan BV. Low calcium flyash based geopolymer concrete. New York: CRC Press 2007.

[6] Cuenca E, Serna P. Shear behavior of prestressed precast beams made of selfcompacting fibre reinforced concrete. Construction Building Material 2013;45:145-56.

[7] Kim K, Lee D, Hwang J, Kuchma D. Shear behavior model for steel fibrereinforced concrete members without transverse reinforcements. Composite Part B. 2012;43:2324-34.

[8] Tiberti G, Minelli F, Plizzari G, Vecchio F. Influence of concrete strength on crack development in SFRC members. Cemical Concrete Composite. 2014;45:176-85.

[9] Chalioris C. Analytical approach for the evaluation of minimum fibre factor required for steel fibrous concrete beams under combined shear and flexure. Constrion Building Material 2013;43:317-36.

[10] Soroushian P, Bayasi Z. Fibre-type effects on the performance of steel fibre reinforced concrete. ACI Material J 1991;88(2):129-34.

[11] Cuenca E, Serna P. Shear behavior of self-compacting concrete and fibrereinforced concrete push-off specimens. In: Khayat K, Feys D, eds. *Design, production and placement of self-consolidating concrete RILEM Bookseries*. Netherlands: Springer 2010:429-38.

[12] Cuenca E, Serna PF. Failure modes and shear design of prestressed hollow core slabs made of fibre-reinforced concrete. Composite Part B: Engineering. 2013;45(1):952-64.

[13] Uygunogˇlu T. Investigation of microstructure and flexural behavior of steel fibre reinforced concrete. Material Structure. 2008;41(8):1441-9.

[14] Jiuru T, Chaobin H, Kaijian Y, Y Y. Seismic behaviour and shear strength of framed joint using steel–fibre reinforced concrete. Journal Structuce Engineering. 1992;118(2):341-58.

[15] Wei S, Jianming G, Yun Y. Study of the fatigue performance and damage mechanism of steel fibre reinforced concrete. ACI Material J. 1996;93(3):206-2012.

[16] Yin W, Hsu T. Fatigue behaviour of steel fibre reinforced concrete in uniaxial and biaxial compression. ACI Material J. 1995;92(1):71-81.

[17] Balaguru P, Ramakrishnan V. Properties of fibre reinforced concrete: workability, behavior under long-term loading, and air-void characteristics. Material J. 1988;85(3):189-96.

[18] Wainwright P, Rey N. The influence of ground granulated blast furnace slag (GGBS) additions and time delay on the bleeding of concrete. Cem Concr Compos. 2000;22(4):253-7.

[19] Topçu I, Elgün V. Influence of concrete properties on bleeding and evaporation. Cem Concr Res. 2004;34(2):275-81.

[20] Ravindrarajah R. Bleeding of fresh concrete containing cement supplementary materials. The ninth east Asia-Pacific conference on structural engineering and construction; 2003 16–18 December; Bali, Indonesia; 2003.

[21] ASTM C 618. Standart Specification for Coal Fly Ash and Raw or Calcined Natural Pozzolan for Use as a Mineral Admixture in Concrete, Annual Book of ASTM Standard; 2000.

[22] Cuenca E, Echegaray-Oviedo J, Serna P. Influence of concrete matrix and type of fibre on the shear behavior of self-compacting fibre reinforced concrete beams. Composites Part B. 2015;75:135-47.

Acknowledgement
The authors would like to acknowledge the financial support and facilities in order to complete this research provided by University Malaysia Perlis (UniMAP) and Centre of Excellent Geopolymer & Green Technology (CEGeoGTech).

IOP Conf. Series: Materials Science and Engineering **133** (2016) 012046 doi:10.1088/1757-899X/133/1/012046

Microstructure and Mechanical Properties of Fly Ash Particulate Reinforced in LM6 for Energy Enhancement in Automotive Applications

M.N. Ervina Efzan *[1,2], N. Siti Syazwani[1], Mohd Mustafa Al Bakri Abdullah[3,2]

[1] Faculty of Engineering and Technology, Multimedia University, 75450 Ayer Keroh, Malacca, Malaysia.
[2] Center of Excellence Geopolymer & Green Technology (CeGeoGTech), School of Material Engineering, Universiti Malaysia Perlis (UniMAP), Perlis, Malaysia
[3]Faculty of Engineering Technology, Universiti Malaysia Perlis (UniMAP), Perlis, Malaysia

E-mail: ervina.noor@mmu.edu.my

Abstract. Fly ash has gathered widespread attention as a potential reinforcement for aluminium matrix composites (AMCs) to enhance the properties and reduce the cost of production. Aluminium alloy LM6 reinforced with three different amounts (0, 4, 5 and 6 wt. %) of fly ash particle that were prepared by compo-casting method. The fly ash particles were incorporated into semi-solid state of LM6 melt. In this study, the microstructure of prepared AMCs with the homogenous distribution of fly ash was analysed using optical microscope. The microstructure having refinement of structure with the decreasing of Si-needle structure and increasing the area of eutectic α-Al matrix as shown in figure. Besides, as the increasing amount of fly ash incorporated, there are more petal-like dark structure existed in the microstructure. The density of the AMCs decreased as the incorporation of fly ash increased. While the hardness and ultimate tensile strength of the AMCs increased with the incorporation of fly ash. The addition of fly ash particles improved the physical and mechanical properties of the AMCs. Thus lead to improve the energy consumption in automotive parts.

1. Introduction

Metal matrix composites (MMCs) are an advanced materials resulting from a combination of two or more materials in that the tailored properties (Kumar et.al,2013). MMCs possess significantly improved properties including high specific strength, better properties at elevated temperatures, lower coefficients of thermal expansion and good wear resistance compared to unreinforced alloys (Akio et.al,1999; Kulkarni et.al,2014). Meanwhile, aluminium based MMCs (AMCs) have gained significant attention in recent decades as engineering materials with most of them possessing the advantages of high strength, hardness and wear resistance (Anilkumar et.al,2011). Good physical and mechanical properties of AMCs have made attractive materials for automotive and aerospace applications (Kretschmer, 1988; Ibrahim et al., 1991; Surappa, 1997).

Cost is the key factor for the AMCs wider application in modern industry, while the potential benefits in weight reduction, increased composites life and improved recyclability should be taken into account (Klimowicz., 1994; Hashim et al., 2001). According to Shorowordi et.al, cost reductions can be achieved only by simpler fabrication methods, higher production volumes and cheaper reinforcements (Shorowordi et al., 2003). These composite materials represent a new product opportunity for automotive manufacturers to improve performance and thus reduce the cost. Currently, the particulate reinforced aluminium composites are gaining importance because of their low cost with better performance (Deepak and Mediratta,2013; Rohatgi,1994).

There are variety of methods for fabrication of AMCs including diffusion bonding, powder metallurgy, liquid casting and pressure infiltration have recently become available (Ebisawa,1991; Rohatgi,1990; Hunt et.al,1991). Even though there are various method in producing AMCs, the greatest attention has been on the liquid casting route, especially stir casting that appears to be most attractive (Rohatgi,1994). The conventional foundry process of making AMCs components readily lend themselves to the manufacture of high-volume, complex components at high production rates and low costs required by industry. In view of low cost requirements is imperative to industry, it appears that discontinuous composite incorporating inexpensive particles will find the widest applications in industry. Currently, many researchers works on the waste materials as the inexpensive reinforcement particles in AMCs such as rice husk ash, red mud, fly ash and coconut ash (Ervina et.al, 2016).

Among various discontinues disperoids used, fly ash is one of the most inexpensive and low density reinforcement available in large quantities as solid waste by product during combustion of coal in thermal power plant (Lokesh et.al,2013). Fly ash as a filler has been proved to increase the mechanical, physical and tribological properties including hardness, stiffness, wear and abrasion resistance and decrease density. It also improves the maintainability, damping capacity, and coefficient of friction which are which are desirable in automotive and industrial applications (Sukanya et.al,2014).

Previously, there were several researches on the characteristic of reinforcement materials that affected the properties of AMCs have been done. According to Rohatgi, mechanical properties of composites can be affected by the size, shape and volume fraction of the reinforcement, matrix material and reaction at the interface. These aspects have been discussed by many researchers (Rohatgi,1994). The same author reported that with the increase in volume percentages of fly ash, hardness value also increases in the Al–fly ash (precipitator type) composites. Besides, the tensile elastic modulus of the ash alloy increases with increase in volume percent (3–10) of fly ash. On the other hand, Aghajanian concluded that the interface between the matrix and reinforcement plays a critical role in determining the properties of MMCs.

Whereas, Anilkumar et al. investigated Al6061 with fly ash processed by stir casting and concluded that the tensile strength increased with an increase in the weight percentage of fly ash. This is due to the fly ash particles acted as barriers to the dislocations when taking up the load applied. The hard fly ash particles obstruct the advancing dislocation front, thus strengthening the matrix. However, as the size of the fly ash particles increased, there was a decrease in tensile strength (Saravindan,2015). Good bonding of smaller size fly ash particles with the matrix is the reason for this behaviour. Despite this, the decrease in the tensile strength of the samples with fly ash weight fraction beyond 15 % is due to the poor wettability of the reinforcement with the matrix (Anilkumar et.al,2011). In the present work, an attempt was made to fabricate aluminium hybrid composites with fly ash as reinforcements by using compo-casting method. Properties such as density, metallurgical, and mechanical behaviour were investigated and the related mechanisms have been discussed and presented.

2.Experimental Method

Aluminium alloy (LM6) is used as the matrix material while fly ash is used as the reinforce material. The aluminium alloy (LM6) was supplied from Foundary Equipment Supply with the elemental details as shown in Table 1. Fly ash powders (precipitator fly ash) were received from Jimah Power Plant, Port Dickson, Negeri Sembilan. The fly ash is sieved to 45μm to get smaller particle size. According to Saravindan, the smaller particle size of reinforcement the better properties of composite can be obtained (Saravindan,2015). The fly ash is then undergoing characterizations that are particle size analysis and chemical composition (XRF). Before incorporated into the aluminium slurry, the fly ash is pre-heated at 600°C for 3 hours to remove impurities and water content. The amount of fly ash used are 4, 5 and 6wt.%. The chemical composition of fly ash particle is shown Table 2.

Table 1. Chemical composition of LM6

Elements	Cu	Fe	Mg	Mn	Ni	Pb	Si	Al	Reference
Percentage (%)	0.10	0.60	0.10	0.50	0.10	0.10	10.00-1300	Remainder	Material Data Sheet FES Sdn.Bhd.

Table 2. Chemical composition of fly ash

Compound	Na_2O	MgO	Al_2O_3	SiO_2	P_2O_5	K_2O	CaO	TiO_2	Fe_2O_3	LOI
Weight Percent(%)	0.481	1.102	15.939	65.61	0.28	0.684	2.098	0.813	4.517	1.87

The LM6/fly ash composite (AMCs) was fabricated by using the compo-casting method where the reinforcement is added into aluminium slurry in semi-solid state. The LM6 matrix is melted in a graphite crucible in melting furnace at 800°C. Measured quantity of fly ash particles was added to the semi solid aluminium slurry. Simultaneous stirring of the semi solid aluminium slurry was carried out using a stainless steel mechanical stirrer driven by an electric motor with speed of 300rpm for 5 to 10 minutes. The stirring process was continued until all the fly ash particles were added to the semi solid aluminium alloy and then poured into the mould. The composite is then undergoing CNC machining process for tensile specimen preparation following the standard code of ASTM E8 which is suitable for automotive application.

The AMCs also being prepared for other testing and characterizations including density analysis, microstructural analysis, and hardness test. The sample for metallurgical analysis is prepared according to standard metallographic process that is ground using SiC paper and then undergo polishing process by using diamond paste before the etching process. The hardness was measured using a Rockwell hardness tester at 15kg load applied for 5s.

3.Result and Discussion

Aluminium alloy LM6 reinforced fly ash particulate composites were successfully fabricated using compo-casting process. The metallurgical and mechanical characterizations of the fabricated aluminium composite are discussed below.

The particles size of cenosphere fly ash is below than 45μm as shown in Figure 1. The fly ash is sieved to 45μm following the standard method ASTM C311-77. Fly ash particles are typically spherical, ranging in diameter from <1 μm to 150 μm. The type of dust collection equipment used largely determines the range of particle sizes in any given fly ash. The fly ash from boilers at some older plants using mechanical collectors alone is coarser than from plants using electrostatic precipitators (Malhotra et.al,1993). In this research, the used fly ash was taken from the plants using the electrostatic precipitators where the average particle size from <1μm to <100μm. According to Sudarshan and Surappa, the smaller particle size (<45μm) of fly ash will able to give superior properties of AMCs. It is the same way to the particle size of fly ash used in this research. The average particle size of fly ash after sieving process is 10μm which falls into the range of small size (<45μm) as shown in Figure 1. Thus the sieved fly ash was successfully incorporated into the aluminium slurry as the reinforcement.

Characterization of Fly ash

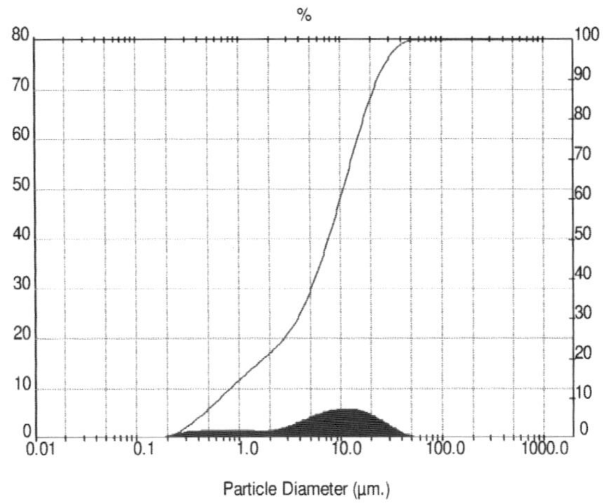

Figure 1. Average particle size of fly ash

Meanwhile, the shape of the powder particles was evaluated using scanning electron microscopy (SEM) micrograph as shown in Figure 2. The SEM micrograph shows the homogenous size distribution of fly ash particles (grey region) with only three particles in bigger size (100 and 75 μm). However, the average size distribution in the SEM micrograph is below than 50μm. Cenosphere particles were hollow spheres with an apparent density ranging from 0.4 - 0.6 g cm^{-3}(Rohatgi,1997). On the contrary, the shape of fly ash used in this research was spheres with no hollow structure. The micrograph in Figure 2 revealed the shape of fly ash is spherical with an average size below than 50μm.

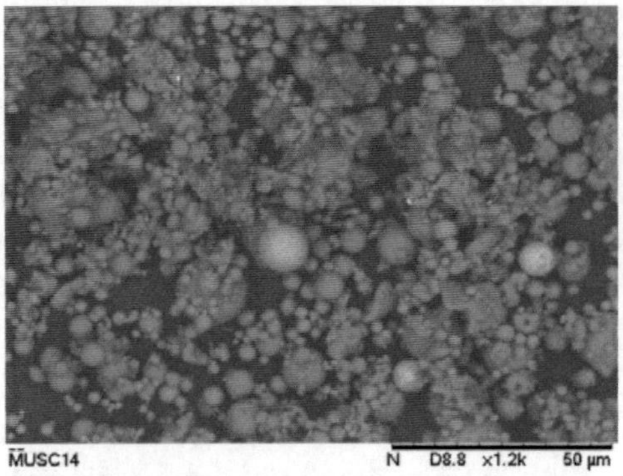

Figure 2. SEM micrograph of fly ash

The microstructure plays an important role in the overall performance of a composite. The physical properties depend on the microstructure, reinforcement particle size, shape and distribution in the alloy. The fabricated aluminium composites were examined using optical microscope to study the distribution pattern of fly ash in the LM6 matrix. Optical micrograph of cast aluminium alloy LM6 and the prepared aluminium fly ash composite (AMCs) with 4, 5 and 6wt% of fly ash compositions at 25x magnification are shown in Figure 3a, 3b and 3c respectively.

Before the addition of fly ash reinforcement, the microstructure of unreinforced LM6 alloy is dominated by Si-rich (dark region) and Al-rich (bright region) phases, as illustrated in Figure 3a. The Si-rich phase reveals itself in three different types of morphologies, the cuboid, the needle, and the irregular polygon. In some instances, the Si cuboids are surrounded by the primary α-Al phase. These Si-rich cuboids might be the primary Si phase. This result was confirmed by Pena and Lozano (Pena and Lozano,2006). According to them, the primary Si phase resembles cuboid-type of morphology. They serve as the nucleating sites of primary α-Al phase occasionally. Regarding the irregular polygons, they exhibit a similar morphology to the primary Si phase. The microstructure also reveals the formation of α-aluminium dendritic network structure which is formed due to the supercooling of composite during solidification.

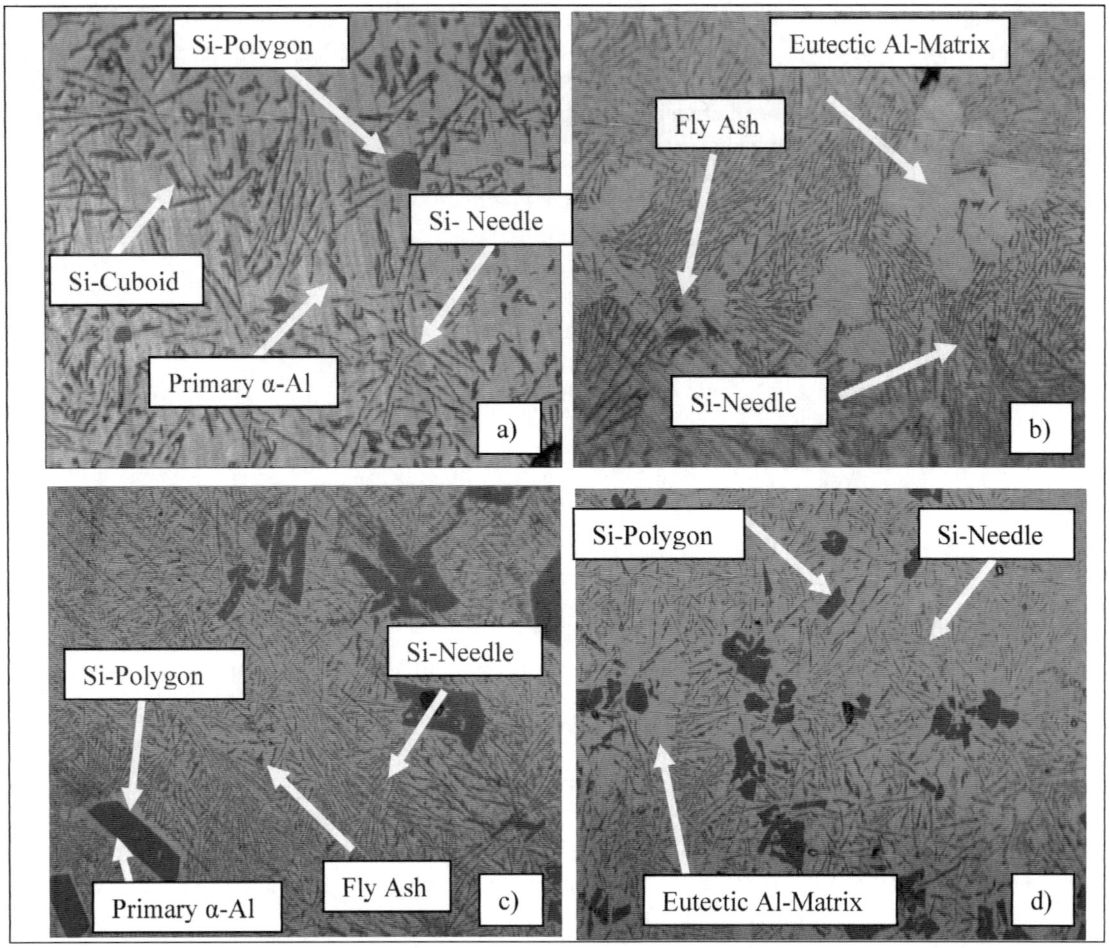

Figure 3. a) Microstructure of unreinforced LM6 alloy, b) microstructure of 4wt% of LM6/fly ash composites, c) microstructure of 5wt% of LM6/fly ash composite, d) microstructure of 6wt% of LM6/fly ash composites.

After the addition of 4wt.% fly ash reinforcement, the most distinguishable feature will be the increasing Si-rich (dark region) phase as demonstrates in Figure 3b. Si-needles and Si-cuboids have been gather in certain part in the fabricated AMCs, while a wide area of eutectic α-Al matrix has formed. The Si-needle become closely packed each other. However, upon the addition of 5w.% fly ash reinforcement, the microstructure of the composites still dominates by the Si-rich (dark region) and Al-rich (bright-region). There are also some petal-like dark structures and Si-cuboids existed as demonstrated in Figure 3c. The Si-needles were well distributed and dominated most of the structure whereas the eutectic α-Al matrix dominated minor region of the structure. Meanwhile, Figure 3d showed the microstructure of LM6 with 6wt.% of fly ash reinforcement. The microstructure shows the increasing region of petal-like dark structures and Si-polygon existed while the Si-needle structures are decreased. Conversely, the eutectic α-Al matrix (bright region) has increased. This is due to the homogenous distribution of fly ash particles in the fabricated AMCs. As for the needle-like structure, it is frequently referred as sharp eutectic Si embedded in the eutectic α-Al matrix that deteriotes the mechanical performances of Al-Si alloys(Pena and Lozano,2006; Sebaie et.al,2008; Ervina et.al,2013).

Thus, the decreasing amount of Si-needles in the structures as the increasing wt.% of fly ash incorporation is good as its can improves the properties of fabricated AMCs. It was shown that the addition of fly ash particle as the reinforcement can modify the microstructure of fabricated AMCs in better form. This is because, the increasing of Si in the composites is expected to improve the properties of the composite (Ervina et.al,2014).

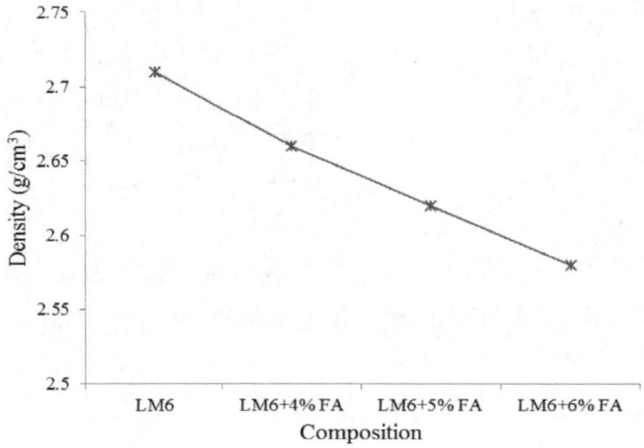

Figure 4. Density analysis of unreinforced LM6 and LM6/fly ash composites

The variation of density of LM6 and fabricated AMCs are shown in Figure 4. The density of the fabricated AMCs were decreased gradually as the wt.% of fly ash reinforcement added increased. Before the addition of fly ash particles as the reinforcement, the density of LM6 alloy is 2.71g/cm^3 which is nearly same to the theoretical density of Al 6xxx in the range of 2.69-2.72 g/cm^3 (Zhong et.al,2014) where is the current used aluminium materials in automotive parts. Although the density of Al 6xxx is relatively 50-60% lower than the conventional cast iron used, the automotive manufacturer still looking forward for the lower weight materials that can be used in the automotive parts. It is apparent that the lower weight parts of automotive potentially increasing the efficiency of the vehicles (Hadley et.al,1999). Efficiency gains can be used to lower the amount of energy needed for transportation, reducing the emissions while improving the energy security of the country. So that, in order to achieve the required density of materials for automotive part, the LM6 alloy is reinforced with a low density reinforcement materials that is fly ash.

The density of fly ash cenospheres particles is in the range of 0.4-0.6g/cm^3 where is far lower than the density of LM6. After the addition of 4wt.% fly ash particles as the reinforcement, the density of the fabricated AMCs was decreased to 2.66 g/cm^3. While the density of fabricated AMCs with 5wt.% of fly ash particles has decreased to 2.62 g/cm^3. Lastly, the density of fabricated AMCs with 6wt.% of reinforcement is decreased again to 2.58 g/cm^3. This is due to the homogenous distribution of fly ash in the composites as shown in the microstructure in Figure 3. The atomic structure of unreinforced LM6 is closed pack with some small space among the atom that leads to introduce the porosity which will deteriotes the properties of the materials. Upon the addition of fly ash particles, the particles distributed homogenously and filled up the space among the atom, thus replaced some of them. As the density of fly ash particles is lower than the LM6, it is able to reduce the density of the fabricated AMCs. According to the Figure 4, the addiction of fly ash particle is able to reduce the density of fabricated AMCs up to 13% reduction. There are some comparisons of density value of fabricated AMCs with others finding have been summarized in Table 3.

Table 3. Comparisons of density value of fabricated AMCs with other finding

Composition (wt.%)	4%	5%	6%	References
Density (g/cm^3)	2.66	2.58	2.64	(Kulkarni et.al,2014; Rohatgi, 1997; Lokesh et.al,2013)
	2.66	2.62	2.58	Current Study

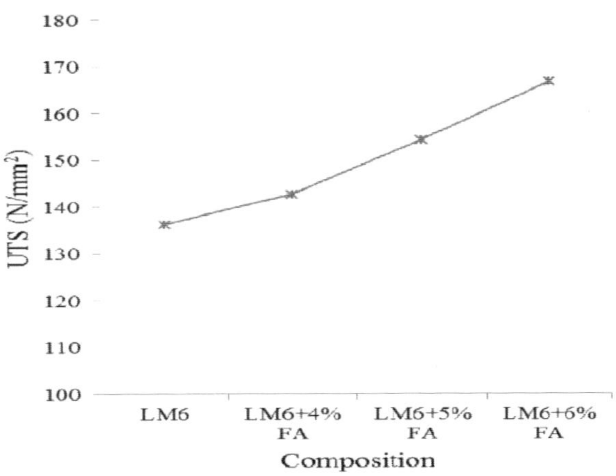

Figure 5. Tensile strength of unreinforced LM6 and LM6/fly ash composites

The effect of wt.% of fly ash particles on tensile strength of fabricated AMCs is respectively presented in Figure 5. Before the incorporation of fly ash particles, the ultimate tensile strength of the unreinforced LM6 is 136.3N/mm^2. While the current used materials for automotive application is cast iron alloy with ultimate tensile strength of 160N/mm^2. It is proved from the figures that the incorporation of fly ash particles significantly enhances the ultimate tensile strength of the composites. After the addition of 4 and 5wt% of fly as particles, the ultimate tensile strength of the fabricated AMCs increased from 136.3 N/mm^2 to 142.7 and 154.4 N/mm^2. Meanwhile, the fabricated AMCs with 6wt% fly ash particles exhibit 30.4% higher tensile strength that is 166.7 N/mm^2 compared to unreinforced LM6 alloy. The fracture surface of LM6 alloy exhibits brittle structure surface with small amount of elongation. However, after the incorporation of fly ash, it is able to slightly increase the elongation of the AMCs although still exhibits the brittle fracture surface. Similar results were reported by David et.al and also Ramachandra and Radhakrishna (Ramachandra and Radhakrishna, 2005; David et.al,2013). The grain refinement and reduction of ductile matrix content when the weight percentage of fly ash particles is increased can reduce the ductility of the AMCs.

During the solidification of the composite, strain fields are created around fly ash particles because of the difference in the thermal expansion coefficients while the strain fields piles up dislocations. The interaction between dislocations and fly ash particles offer resistance to the propagation of cracks during tensile loading. The grain refinement provided by the fly ash particles presents more area to resist the load. The uniform distribution of fly ash particles provides better strengthening (Zhang and Chen,2008). Therefore, the tensile strength of fabricated AMCs is enhanced by the fly ash particles. However, the tensile strength of the fabricated AMCs does not achieve the requirement of automotive spec since the required tensile strength of aluminium for automotive parts is 245 N/mm^2. Since the addition of fly ash particle able to increase the ultimate tensile strength of composite, the higher wt.% of fly ash need to use in order to obtain higher tensile strength, thus achieves the required tensile strength of automotive parts.

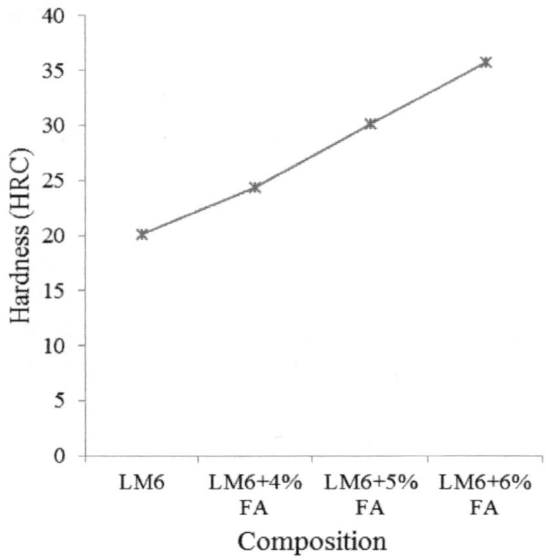

Figure 6. Rockwell Hardness of unreinforced LM6 and LM6/fly ash composites

Hardness provides a quick, simple and non-destructive means to access the mechanical behaviour of a material. It is defines as the resistance of a material towards plastic deformation. The Rockwell hardness testing is done with 15kg of load and 10s of dwell time. The test is done according to the standard test code ASTM E384 – 09. Hardness is necessary for automotive to ensure good machinability and for assembly of the component. Although the increasing of hardness will decrease the machinability of a material, however, higher hardness stills a vital property of an automotive component. This is because, hardness for automotive is basically the ability of a component to resist permanently deformed when a higher load is applied (Automotive Manual,2011). So that, higher hardness is important to avoid component failure as higher load is applied.

Figure 6 highlights the hardness of unreinforced LM6 and fabricated AMCs. It is showed the hardness of the fabricated AMCs increased as the increasing of wt.% of fly ash incorporated into LM6. Before the incorporation of fly ash reinforcement, the hardness of LM6 is 20.1 HRC. Upon the incorporation of 4wt.% of fly ash, the hardness of fabricated AMCs is increased to 24.4 HRC. And yet, the hardness of fabricated AMCs increased to 35.7 HRC with addition of 6wt.% fly ash reinforcement, equivalent to 46.31% of hardness increment. This hardness value already achieved the hardness requirement of automotive component as the required hardness is 34.0 HRC (American Foundary,2004). This significant rise in the hardness of fabricated AMCs is believed to be strongly attributed to the enhancing the microstructural features and hardness LM6.

4.Conclusion

Based on the optical micrograph, the fly ash particles were dispersed homogenously in the AMCs. The incorporation of fly ash particles into semi-solid aluminium alloy improved the wettability.

Meanwhile, the addition of fly ash particles enhance the hardness and tensile strength of fabricated AMCs. The fabricated AMCs with 6wt% of fly ash reinforcement exhibited almost 50% higher hardness and 30.4% higher tensile strength compared to unreinforced LM6 alloy. More the weight percentage of fly ash, more will be the effect of the above discussed factors which further increases the mechanical properties.

5. Acknowledgement
Special thanks to Multimedia University for financial support through ERGS Research Grant No MMUE/130133.

6.References

[1] Ibrahim A, Mohamed FA, Lavernia F.J. 1991. Particulate reinforced metal matrix composites- a review, Journal of Material Science, Vol. 26, pp.1137-1156.
[2] Kretschmer J. 1988. Overview: composites in automotive applications, state of the art and prospects. Material Science Technology, Vol. 4, No. 9, pp. 757-767
[3] Surappa M K., 1997. Microstructure evolution during solidification of DRMMCs: state of art, Journal of Material Processing Technology, Vol. 63, pp. 325-333
[4] Hashim J., Looney L., Hashmi M.S.J., 2001. The enhancement of wettability of SiC particles in cast aluminium matrix composites, Journal of Material Processing Technology, Vol. 119, pp. 329–335.
[5] Klimowicz T.F., 1994. The large scale commercialization of aluminium-matrix composites, Journal of the Minerals, Metals and Materials Society, Vol. 46, pp. 49–53.
[6] Shorowordi K.M., Laoui T., Haseeb A.S.M.A., Celis J.P., Froyen L., 2003. Microstructure and interface characteristics of B4C, SiC and Al2O3 reinforced Al matrix composites, Journal of Material Processing Technology, Vol. 142, pp. 738–743.
[7] Rohatgi, P.K (2006): "Applications of fly ash in synthesizing low cost Metal Matrix Composites for automotive and other engineering applications", JOM, vol. 58, issue no.11, pp.71-76, 2006.
[8] Rohatgi, P.K., 1994. Low-cost, fly ash-containing aluminium-matrix composites. JOM 46, 55-59.
[9] Sukanya M. S, Rachana K. and Sharma S.S. (2014)," Microstructure And Some Mechanical Properties of Fly Ash Particulate Reinforced Aluminium Alloy Composites Prepared By Different Casting Procedures With And Without Wetting Agents" International Journal Of Advances In Engineering And Applied Science (Ijaeas), Volume-1, Issue-1.
[10] Basavarajappa, S., Chandramohan.G., Dinesh, A. 2004, Mechanical properties of MMCs-An experimental investigation, Int. symposium of research on Materials and Engineering, IIT, Madras, December 20, 1-8.
[11] Seah,H.W, Sharma,S.C., Girish, B.M.1995, Mechanical properties of cast ZA-27/Graphite particulate composites, Materials and Design, 16, 271-275.
[12] Suresh,K.R., Niranjan, H.B., Jebraj, M.J. Chowdiah, M.P. 2003, Tensile and wear properties of aluminium composites, Wear, 255, 638-642.
[13] B. Pena and J. Lozano, "Microstructure and mechanical property developments in AL-12Si gravity die castings after Ti and/ or Sr additions," Materials Characterization, vol. 57, pp. 218-226, 2006.
[14] O. E. Sebaie, A. M. Samuel, F. H. Samuel, and H. W. Doty, "The effects of michmetal, cooling rate and heat treatment on the eutectic Si particle characteristics of A319.1, A356.2 and A413.1 Al-Si casting alloys," Materials Science & Engineering A, vol. 480, pp. 342-355, 2008.

[15] Zhong, H., Rometsch, P., & Estrin, Y. (2014). Effect of alloy composition and heat treatment on mechanical performance of 6xxx aluminium alloys. Transactions of Nonferrous Metals Society of China, 24(7), 2174–2178. doi:10.1016/S1003-6326(14)63329-X
[16] Zhang Q, Chen G, Wu G, Xiu Z, Luan B. Property characteristics of a AlN_p/Al composite fabricated by squeeze casting technology. Mater Lett 2003; 57:1453-58
[17] Anilkumar, H. C., Hebbar, H. S., & Ravishankar, K. S. (2011). Mechanical Properties of Fly Ash Reinforced Aluminium Alloy (Al6061) Composites, 6(1), 41–45.
[18] Aravindan, S., Rao, P. V., & Ponappa, K. (2015). Evaluation of physical and mechanical properties of AZ91D/SiC composites by two step stir casting process. Journal of Magnesium and Alloys, 3(1), 52–62. doi:10.1016/j.jma.2014.12.008
[19] Hadley, S. W., Das, S., & Miller, J. W. (2000). Aluminium R & D for Automotive Uses and the Department of Energy's Role. Energy.
[20] David, R. S. J., & Robinson. S. D. S. D. (2013). Synthesis and characterization of Al6061-Fly Ash p -SiC p composites by stir casting and compo casting methods. Energy Procedia, 34, 637–646. doi:10.1016/j.egypro.2013.06.795
[21] Kumar, A., Lal, S., & Kumar, S. (2013). Original article Fabrication and characterization of A359 / Al 2 O 3 metal matrix composite using electromagnetic stir casting method. Integrative Medicine Research, 2(3), 3–7. doi:10.1016/j.jmrt.2013.03.015
[22] Lokesh, G. N., Ramachandra, M., Mahendra, K. V, & Sreenith, T. (2013). Characterization of Al-Cu alloy reinforced fly ash metal matrix composites by squeeze casting method, 5(4), 71–79.
[23] Ramezanianpour, A. A. (2014). Cement Replacement Materials (p. 336). doi:10.1007/978-3-642-36721-2
[24] Singla, D., & Mediratta, S. R. (2013). Evaluation of Mechanical Properties of Al 7075-Fly Ash, 2(4), 951–959.
[28] Mohd Noor. E. E., Kong. H. J., Kok .C. K. (2013). Review: Effect ofAlloying Element on Al-Si Alloys. Advanced Materials Research 845, pp.355-359.
[29] Ervina Efzan. M. N., Kong. H. J., Kok. C. K. (2014). The Properties of Na2CO3 and Ti Hybrid Modified LM 6 Alloy using Ladle Letallurgy. International Science Index, Mechanical and Mechatronics Engineering 2 (4), pp. 356.
[30] Ervina Efzan. M. N., Siti Syazwani. N., Emerson J. (2016). Properties of Aluminum Matrix Composites (AMCs) for Electronic Application. Materials Science Forum. Vol 857, pp. 18-21.

IOP Conf. Series: Materials Science and Engineering **133** (2016) 012047 doi:10.1088/1757-899X/133/1/012047

Morphological characterisation of complex powder used for protective coatings for geothermal plant components

I. Csaki[1], S. N. Karlsdottir[2], A. Buzaianu[3], S. Serghiuta[1], G. Popescu[1], V.A. Motoiu[4], K.R. Ragnarstottir[5], S. Guðlaugsson[6]

[1]Materials Science and Engineering Faculty, University Politehnica Bucharest, Splaiul Independentei 313, 060042
[2]University of Iceland, Sæmundargata 2, 101 Reykjavík
[3]METAV – R&D, CA Rosetti 31, Bucharest, 020011
[4]TEHNOIND com Ltd, Baritiu 48, Bucharest, 011295
[5]Inovation Center Iceland, ICI, Arleyinir 2 – 8 Reykjavik, Iceland
[6]Orka Natturunnar, 110 Baejarhals, Iceland,

E-mail: ioana.apostolescu@upb.ro

Abstract. This paper aims to review the morphological characteristics, microstructures, physical and chemical properties of two complex composite powders: Ni18Cr5Si2B and Ni21Cr11Al2.5Y. These powders will be used as an option for coating geothermal turbine blades to prevent corrosion. The corrosion process in the steam turbine results in damages being recognized as the leading cause of reduced availability in geothermal power plants and is depends on temperature, mechanical and vaporous carryover of impurities and water treatment. Thermal spraying is a suitable technique for coating layers with wear and corrosion resistance. Therefore this technique could be successfully used in geothermal applications for obtaining coatings layers from new complex composite powders protecting the turbine blades from corrosions and good control of steam chemistry. The composite powders were investigated using X-ray diffraction and electronic microscopy to provide detailed information about composites morphological modifications. The results obtained after morphological evaluation are encouraging for using these composite powders as an option for coating geothermal components using thermal spraying technique.

1. Introduction

The corrosion process in the geothermal turbine depends on temperature, pressure, chemistry, mechanical and vaporous carryover of impurities and water treatment (distribution between the vapours, the surface film and rotor blades material, heat transfer properties etc.). Corrosion is considered to be the most severe problem in geothermal power plants. The extreme high temperature and pressure conditions present in geothermal systems, as well as the existence of almost an entire periodic system of elements as corrosive salts [1,2] result in a high corrosion rate in geothermal power plants. In order to increase the competitiveness of the geothermal energy against the conventional energy resources, there is a great necessity for reducing the costs of geothermal power plant construction and providing safer and continuous energy exploitation. One of the major obstacles that affect a stable and continuous energy production is the corrosion of constructional materials and equipment due to their interaction with an aggressive environment [3-6].

However, high-quality materials are inevitable for some process components when conventional materials have failed or are expected to lack performance. This includes equipment that is required to have a high reliability and near-zero corrosion allowance for reasons of performance, maintenance, and/or safety [7,8].

Multi-composite materials technology provides the construction of the main components exposed to very corrosive environment. Rating the principal technological measures already applied to current geothermal turbines is studied and discussed. The geothermal steam inlet pressure and temperature are relatively variable with time and a decreasing in geothermal steam initial parameters may appear. Even in the situation when the fluid reinjection is ensured, usually the source loses pressure, being necessary a period of time for renewing. The renewing time depends on the reservoir nature and its permeability [8].

A very important issue in the thermal spraying processes is the application of appropriate coating materials. The method of producing coating materials has a determining effect on the possibilities of their use for different thermal spraying technologies.

Producers of materials apply advanced technologies which make it possible to produce materials with the assumed and complex chemical and phase composition as well as of the specified particle morphology. The structure of such materials is optimized for the given thermal spraying method and it includes the specificity of the transformations taking place in the material during the process of thermal spraying, formation and use of the coating [9-15].

The corrosion process can be stopped with new materials, with a good steam chemistry control and new blades surfaces structure. These surfaces can be obtained using ceramic or metallo-ceramic materials as turbine blades coating layers, preventing corrosion and providing good reliability during the life geothermal turbines.

2. Materials and Methods

Elemental powders and composite structures used for plasma spray coatings are currently investigated due to their promising results regarding obtaining gradual structures and life extension of geothermal turbines.

The stainless steel represents an important segment of the powder metallurgy (P/M) industry, the cost of stainless steel and the potential application being very promising. More complex powder compositions have been investigated to determine the final composition and microstructure responsible for adequate geothermal turbine applications. In this context master alloyed metallic compounds with controlled particle morphology were used to determine the composite homogenization and densification. Commercial Sulzer Metco metal powders were used to design two compositions that probably will behave good regarding the corrosion protection: Cr (7-12 μm-99.5% pure); Ni (4-8μm-99.5% pure); Al (4.5μm-97.5% pure); B (2-4μm-99.5% pure).

The powder components were weighted according to the designed composition. The powders then placed into a sealed container and filled with argon gas to prevent oxidation of powders. The powders particles were mixed in a high velocity ball milling RETSCH type, using stainless stell vials and balls, shaped and bonded together. Samples were tapped for analysis for scanning electron microscopy (SEM) and X-ray diffraction (XRD).

The composite powders obtained are NiCrBSi and NiCrAlCoY.

3. Results and discussions

Aggressive corrosion in geothermal steam condition may be defined as an accelerated corrosion, resulting from the presence of many salt contaminants such as Na_2SO_4, NaCl, and silicates forming molten deposits, with a damaging impact on the protective surface of plant geothermal components.

Active corrosion occurs when metals are heated in the temperature range 250-300°C in the presence of sulphate deposits formed as a result of the reaction between sodium chloride and sulphides compounds in the solutions phase.

Geothermal steam specifics chemistry include: high non-condensable gas contents, high total dissolved solids salts in steam and corrosive chemical composition. High H_2S content generally promotes metallurgical problems: stress corrosion, cracking and metal fatigue. The high CO_2 content in geothermal fluid below about 200 ^{0}C accelerate calcite scaling and promote active corrosion. Also one of the most commonly problems regarding high gas content in the geothermal steam are high non condensable gas contents, sulphur deposition and corrosion due to low PH and oxygen into the geothermal power components steam.

The most common problems concerning high gas content in the geothermal steam are: high Cl^- (10.000ppm as 1.8% NaCl); CO_2 (300ppm) and SO_4 (50ppm as Na_2SO_4). High CO_2 content in geothermal fluid below about 140°C accelerates calcite scaling and promotes very active corrosion of geothermal plant components. The solids content and composition is diversified, particularly as regards chloride and different trace elements, as: boron, fluoride etc. Geothermal fluids contains about 130g/kg (130,000ppm), mostly sodium chloride.

Formation of the protective coating oxide layer with plasma deposition is use to decrease their oxidation and corrosion rate. Thermal plasma deposition of Ni, Cr, B, Si or Al coatings form the stable oxides with low diffusion coefficients for oxygen. In spite of Cr, Al and Si-B having affinity for oxygen, after the formation of the thin oxide layer, resisted oxygen diffusion and limits further oxidation of the alloy.

Some powder metals components such as Ni, Co, W, and Mo do not form protective oxide layers. Ti has a high affinity for oxygen and oxygen is highly soluble in Ti, thus the oxidation rate is not limited. The W and Cr, Mo, C can form volatile oxides which may accelerate the oxidation rate due to the absence of a protective oxide layer and the metal could lose weight during oxidation process.

We investigated by SEM microscopy all the aspects regarding particles morphology, including size, shape, surface topography, coating or thin film characteristics, oxides and inclusions. In different manufacturing PM processes the powder results in different: morphology, density, initial phases, particle size distribution and grain size.

Powder with a almost spherical morphology has excellent flow and feed ability through the system. The powder particles shape influences the flow rate. As a result, very low deposition rates for angular powders are commonly observed.

The effect of energy ball milling on the experiment is obtaining complex matrix composite powders, and these powders will be compared with the result reaction in the as-mixed powder. Diffusion and adhesion couple were formed during ball milling processing. The sizes of the composites, the conglomerate and the diffusion metallic element were reduced by increasing the ball milling time.

For powder mixtures analyse we used a sizing instrument type Retsch AS 200 with sieves with dimensions starting from 125μm till 25μm. Sizing method used was volumetric dry sieving. The results are shown in figure 1.

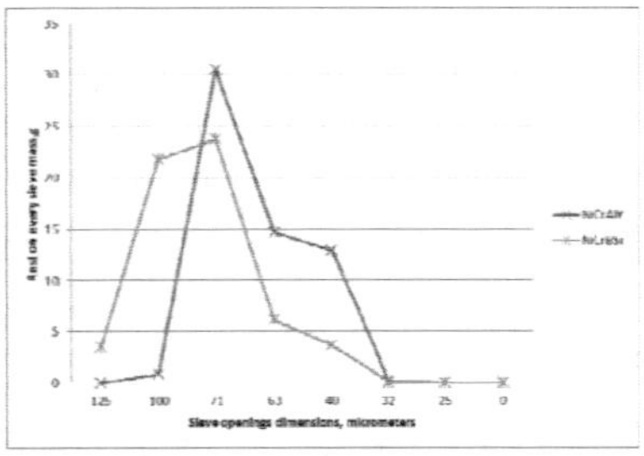

Figure 1. Size distribution of NiCrBSi and NiCrCoAlY complex powders.

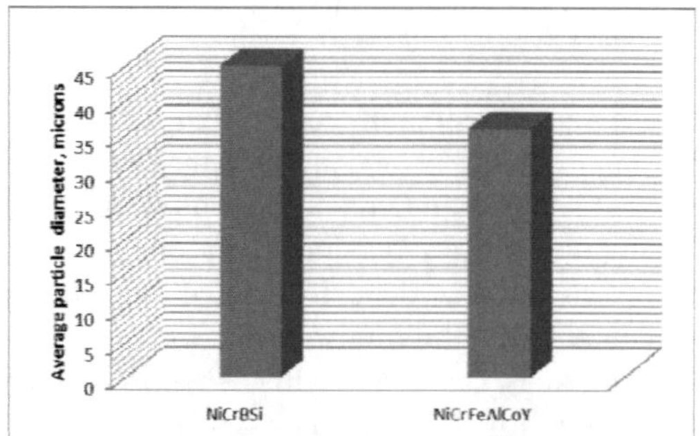

Figure 2. Average particle diameter claculated after sieving analyses for NiCrAlCoY and NiCrBSi.

The apparent density decreased when the mean diameter for the powders was low and the small particles implies a large specific surface for the powder used. This phenomenon increases the friction between particles and subsequently and decreases the apparent density. Complex powder Ni based composites coating shows good high-temperature wear and corrosion resistance. They have good wear resistance after adding percent of Mo elements to the powders [10,14]. Ni based coatings are used in applications when wear resistance is combined with oxidation or hot corrosion resistance. The primary reason for selecting a stainless steel composition usually is its improved corrosion resistance.

Ni based powders are widely used. We selected NiBSi powders system and additions of alloying elements are strongly recommended. Addition of chromium promotes the oxidation and corrosion resistance at elevated temperatures and increases the hardness of the coating by formation of the hard phases. The mean diameter of the complex powder was calculated. In figure 2 the results of average diameter of the complex composite powders is shown. The average diameter for the powder to be sprayed using plasma spraying technique or high velocity oxygen fuels technique is around 45 microns; figure 2 showing that the characterized powders are in the appropriate range.

Flowing rate and density were calculated for the studied powders and the results are shown in figure 3 and 4.

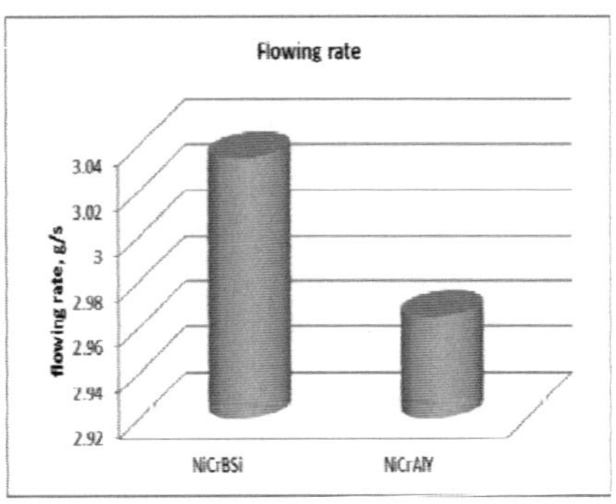

Figure 3. Flowing rate of the complex composite powders.

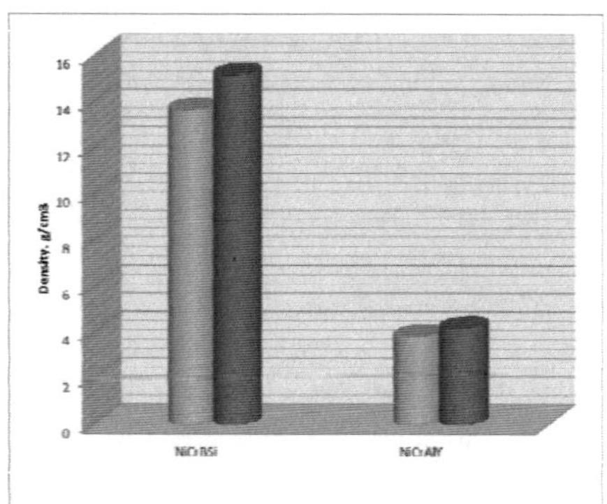

Figure 4. Apparent density and tap density measured for complex composite powders.

The complex composite powder NiCrAlCoY had a flowing rate lower than NiCrBSi powder and the apparent density and tap density was lower as well. The composite coatings will have different properties that will be assessed in future papers. Silicon is added to increases the self-fluxing properties, increasing the flowing rate of the NiCrBSi complex powder. The average diameter, flowing rate and density values for both powders are in the range of HVOF coating process.

Figure 5 reveals SEM micrograph displaying typical rounded shape of the composites powder. The finer grades of powder require higher pressures and small molten metal streams complex mixture based on complex Ni-Cr-B-Si powders. In SEM imagens, the particle shape and surface topography can be observed. A significant increase of equivalent diameter could be observed (Figure 6).

The chemical composition of the NiCrBSi composite powder is shown in table 1.

Table 1. Chemical composition of the complex powders.

Elements	Ni	Cr	Si	B	Al	Fe	O
Composition (wt %)	67.78	18.09	5.52	2.02	0.52	4.47	1.60

In figure 6 the electronic microscope image of the surface morphology is investigated. On the spherical particles in zone 1 a small coating from the same powder could be observe. The powder homogeneity is very good and the chemical composition in zone 1 is shown in figure 7, on the EDS analyses results. In the EDS analyses all the designed species were identified and the coating on the particle placed in zone 1 are from the same powder composition.

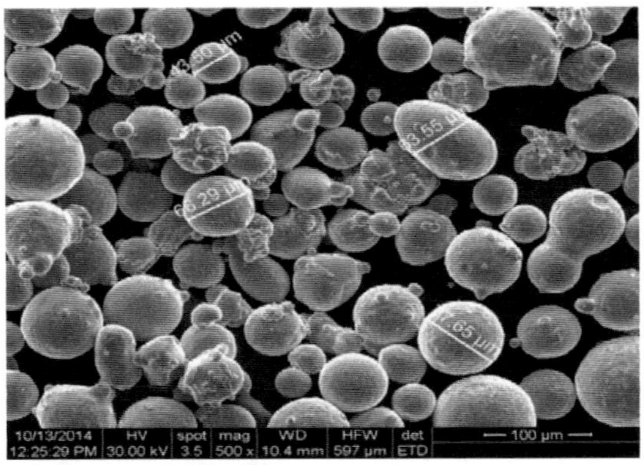

Figure 5. SEM image of the complex composite powder NiCrBSi having rounded to spherical in shape with relative smooth surfaces.

Figure 6. SEM image of the surface morphology revealing the relative occurrence of the forms depending on the distributed of elemental powder on the particle surfaces.

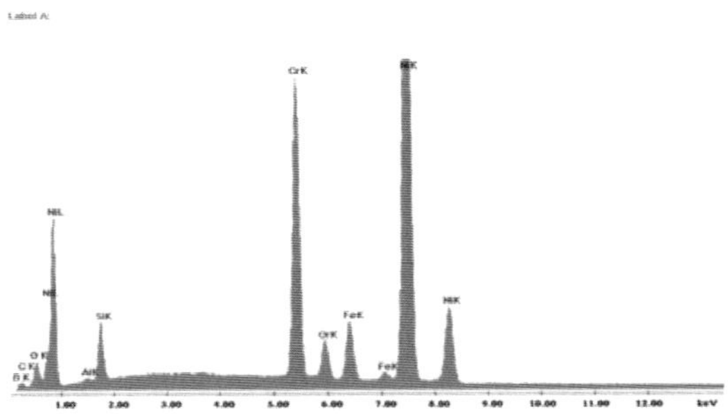

Figure 7. EDS analysis results of the complex matrix Ni-Cr-B-Si mixture.

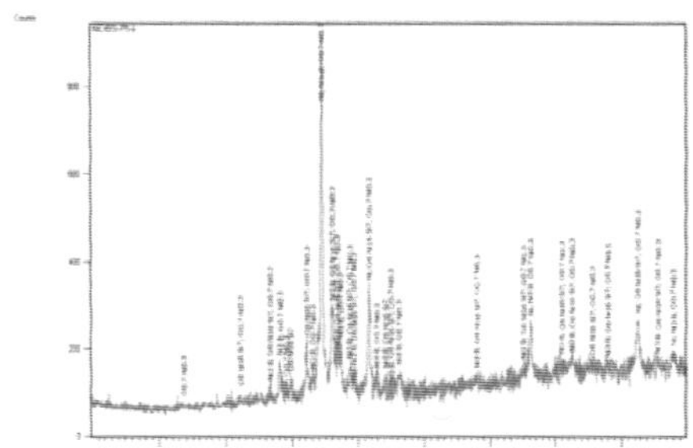

Figure 8. X rays diffraction pattern for NiCrBSi complex powder.

The presence of boron lowers the melting temperature and contributes to the formation of hard phases. The alloyed elements such as: Ni, Cr and Fe were relatively uniform distributed over the composite (zone1 and zone2 are similar in the EDX - quantitative examinations). Electron diffraction

IOP Conf. Series: Materials Science and Engineering **133** (2016) 012047 doi:10.1088/1757-899X/133/1/012047

powder patterns (XRD) are difficult to analyse and experimental data must be carefully select (Figure 8). In the samples were observed the major cubic $Cr_6Ni_{16}Si_7$ and tetragonal phases $Cr_{0.7}Ni_{0.3}$ with the effects in carburizing and chlorination resistance. The minor orthorhombic phases Ni3B were also observed. The presence of complex Ni_3B peak shows the formation of intermetallic compounds. Ni_3B with similar structure of cementite and similar properties of stable and inert chemically silicides, the Si in excess cause the formation of Ni_3Si.

Complex matrix of Ni-Cr-Al-Co-Y powders could be use as alternative composite coatings. In the presence of Al, NiAl compounds appear and coatings exhibit system performance and excellent carburization resistance at high temperature. The Y addition for this powder may act as a corrosion resistance increaser and Co addition could affect positively the coatings adhesion properties. Also Al added to this complex powder could promote oxide formation and Al2O3 layer could protect the coating at high temperature. In figure 9, SEM microstructure images are shown and the homogeneity of the sample could be observed.

In this case, the EDS analysis (Figure 10) is a technique for the rapid determination for samples homogeneity degree. This analyse displays the spectrum of composition existing in a given inhomogeneous phases from the shape of diffraction peak broadened by a range of lattice phase parameters. In table 2 the complex powder chemical composition is shown.

Table 2. Chemical composition of powders.

Elements	Ni	Cr	Al	Y	Co	O
Composition (wt %)	58.42	21.47	11.71	2.36	2.02	3.02

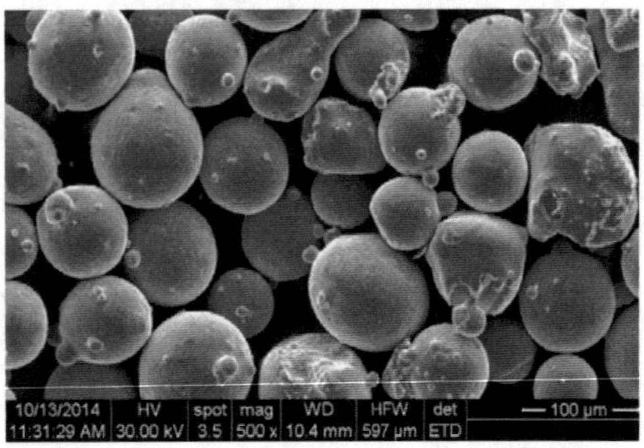

Figure 9. SEM of gas atomized complex Ni-Cr-Al-Co-Y powders.

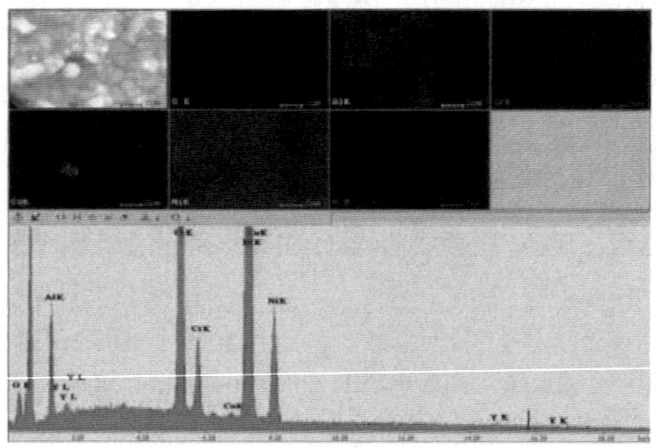

Figure 10. EDS mapping micrography of the matrix complex Ni-Cr-Al-Co-Y powders.

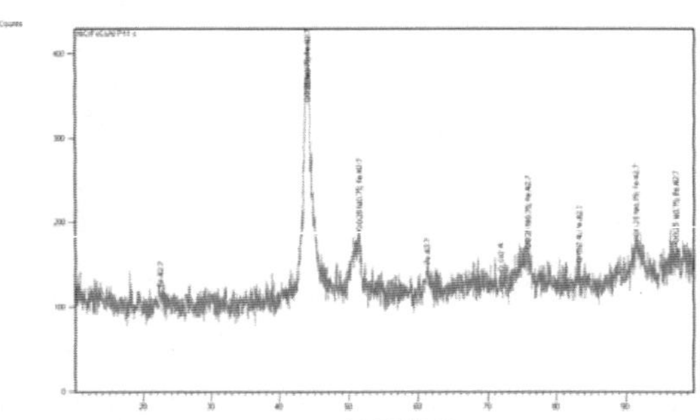

Figure 11. X-ray diffraction pattern (Cu Kα) of: a) matrix Ni21Cr11Al2.5Y_2O_3.

The major cubic $Cr_{0.25}Ni_{0.75}$ and orthorhombic $FeAl_{2.7}$ phases and the minor cubic phase $Cr_{0.25}Co_2Al$ were observed in the samples. In the XRD spectrum (Figure 11) was observed with low intensity and broadened width, indicating that Y_2O_3 particles into the complex-powders samples. The XRD pattern showed the brooded multiple diffraction peaks with low intensity, confirming the crystalline and fine size of the NiAl and Y_2O_3 components in the composites. The Y_2O_3 improves adherence and appalling resistance of oxide layer, and hence improves oxidation, sulfidation and carburization resistance of surfaces composites.

4. Conclusions

Several complex composite powders were studied to determine the final composition and microstructure responsible for adequate applications in geothermal power plants. Two sets of complex alloying powders were prepared by the P/M process. The new complex composites syntheses have been correlated with the elemental powders of B, Si, Co, Y and Al with different additions of NiCr powder.

The analysis and the results suggested that for the NiCrBSi complex powder were observed the major cubic $Cr_6Ni_{16}Si_7$ and tetragonal phases $Cr_{0.7}Ni_{0.3}$ and the minor orthorhombic phase Ni_3B were observed.

The complex powder matrix increases the corrosion and oxidation resistance and maintains the hardness and wear resistance of the materials.

The major cubic $Cr_{0.25}Ni_{0.75}$ and the minor cubic phase $Cr_{0.25}Co_2$ were observed in the samples $Ni21Cr11Al2.5Y_2O_3$.

The existence of the ternary compounds and the mutual crystallographic orientation relationship of this ternary phase with the matrix were also reported.

The morphological study of these complex powders suggested that these powders are suitable for high temperature corrosion resistance coatings realised by HVOF process.

Acknowledgements:
This work is supported by the Grant Romanian EEA Financial Mechanism Committee 2009-2014 and Iceland as Donor Stat. Bilateral Agreement Program Iceland-Romania „Research within priority sectors"- Contract 16 SEE /30.06.2014.

References

[1] Stapleton, M., 2002 Scaling and Corrosion in Geothermal Operation *PowerChemTechnology*, Nevada, USA
[2] Matthews S, James B and Hyland M 2008 Erosion of oxide scales formed on Cr3C2-NiCr thermal spray coatings *Corrosion Science* **50** pp 3087-3094
[3] Corsi R 1986 Scaling and corrosion in geothermal equipment: problems and preventive measures *Geothermics* **15(5–6)** pp 839–856
[4] Elders A W, Fridleifsson O G and Palsson B 2014 Iceland Deep Drilling Project: The first well, IDDP-1, drilled into Magma *Geothermics* **49** pp 1-128
[5] Ouyang H and S. Sasaki S 2007 Microstructure and tribological properties of low-pressure plasma- sprayed ZrO_2-CaF_2-Ag_2O composite coating at elevated temperature *Wear* **249** pp 440-451
[6] Clay A and Tansley G D 2011 Exploration of a simple, low cost, micro gas turbine recuperator solution for a domestic combined heat and power unit *Applied Thermal Engineering* **31(14-15)** pp 2676
[7] Thomas R 2003 Titanium in the geothermal industry *Geothermics* 32 pp 679–687
[8] Buzăianu A, Petra Moţoiu P, Csaki I, Popescu G, Ragnarstottir K, Guðlaugsson S, Guðmundsson D and Arnbjornsson A 2015 Experiments on Life Cycle Extensions of Geothermal Turbines By Multi Composite Technology *Geothermics* **57** pp 1-7

[9] Szymanski K, Hernas A, Moskal G and Myalska H 2014 Thermally sprayed coatings resistant to erosion and corrosion for power plant boilers - a review *Surface & Coatings Technology* **268** pp 153-164

[10] Kamal S, Jayaganthan R and Prakash S 2009 Evaluation of cyclic hot corrosion behaviour of detonation gun sprayed Cr3C2–25%NiCr coatings on nickel- and iron-based superalloys *Surface and Coatings Technology* **203(8)** pp 1004-1013

[11] Ossai C I 2012 Advances in asset management techniques: an overview of corrosion mechanisms and mitigation strategies for oil and gas pipelines *ISRN Corrosion* (Article ID 570143)

[12] Kumar S, Mudgal D and Singh S 2013 Cyclic oxidation of bare and Cr_3C_2 -25(NiCr) coated super alloy at elevated temperature *Advanced Materials Letters* **4 (10)** pp 754-761

[13] Kosman W 2010 Thermal Analysis of Cooled Supercritical Steam Turbine Components *Energy* **35(2)** pp 1181-1187

[14] Yoshihiro S, Yoshiki O and Hideo K 2011 The Latest Geothermal Steam Turbines *Fuji Electric Review* **55(3)** pp 87-92

[15] Stefănoiu R, Geantă V, Voiculescu I, Csaki I and Ghiban N 2014 Researches regarding the influence of chemical composition on the properties of AlxCrFeCoNi alloys *Revista de Chimie* **65(7)** pp 819-821

Mechanical and Physical Properties of In-Zn-Ga Lead-Free Solder Alloy for Low Energy Consumption

M.N. Ervina Efzan[1, 2*], M.N. Nur Faziera[1], Mohd Mustafa Al Bakri Abdullah[3, 2]

[1]Faculty of Engineering and Technology, Multimedia University, 75450 Ayer Keroh, Malacca, Malaysia.
[2]Center of Excellence Geopolymer & Green Technology (CeGeoGTech), School of Material Engineering, Universiti Malaysia Perlis (UniMAP), Perlis, Malaysia
[3]Faculty of Engineering Technology, Universiti Malaysia Perlis (UniMAP), Perlis, Malaysia

E-mail: ervina.noor@mmu.edu.my

Abstract. Due to the demand in the use of electronics devices in industry, the usage of solder connections has increased. Concerning with the toxicity of lead in Sn-37Pb solder alloy, developing lead free solder alloy with low melting temperature is one of the most important issues in electronic industry. Previously, researchers found out that the most promising candidate of lead free solder alloy is Sn-3.0Ag-0.5Cu (SAC). However, the melting temperature of this solder alloy is 217°C, 34°C higher than Sn-37Pb. This can lead to high energy consumption in electronic industry. In this paper, In-Zn-Ga solder alloy was investigated as a potential candidate replacing SAC. This study covers on the physical and mechanical properties of the solder alloy. Differential Scanning Calorimetry (DSC) testing shows that this solder alloy gave low melting temperature as low as 141.31°C. The addition of Ga in In-Zn solder alloy lowered the melting temperature compared to SAC and Sn-37Pb. From coefficient of thermal expansion (CTE) analysis, the In-Zn-Ga solder alloy gives good expansion properties and able to avoid the mismatch between the solder and copper substrates. The density of In-Zn-Ga solder alloy is 6.801g/cm^3, lower than SAC and Sn-37Pb. For the strength, single lap shear testing was conducted on the In-Zn-Ga solder alloy and the results is near to the strength of SAC.

1. Introduction

Conventional soldering technology has become indispensable for the interconnection and packaging of virtually all electronic devices and circuits. The traditional tin-lead (Sn-Pb) solders have been widely employed as electrical interconnects in electronics industries (Zhang et. al, 2008). However, the toxic in Pb can cause harmful influence on the environment and health (Zhang et. al, 2008; Shen & Chan, 2009; Lin et. al, 2009; Islam et. al, 2005). The Europian Union directive on waste electrical and electronic equipment (WEEE) announced the prohibition of the use of Pb in the consumer electronics market after January 2006 (Tsao, 2011). Thus, it is urgent task to develop the lead-free solders with excellent properties.

Recently, a great amount of lead free solders have been reported and promising candidates such as Sn-Ag, Sn-Cu, Sn-Zn, Sn-Bi, Sn-Sb, Au-Sn, and Sn-Ag-Cu systems have been of interest (Goh et. al, 2013; Zhang et. al, 2008). Among the lead free candidates' solders, the Sn-Ag-Cu (SAC) family is believed to be the most promising (Tsao, 2011; Abtew and Selvaduray, 2000; Laurila et. al, 2005; Hung et. al, 2006; Yoon & Jung, 2008; Islam et. al, 2005). SAC alloys with melting temperature 217°C is extensively used as solder materials in the electronic industry (Alena et. al, 2010). It has been proposed as the most promising substitute for lead containing solders because of its relatively low melting temperature, its superior mechanical properties, and its relatively good wettability (Anderson, 2007; Glazer, 1994; Yoon et. al, 2009).

Despite this, its properties still not good enough for to be employed in temperature sensitive components, optoelectronics modules, step soldering processes and thin printed wiring boards that require a relatively low melting temperature (Ervina et. al, 2010). A higher melting point leads the increase in the peak reflow temperature required for assembly compared to that required for Sn-Pb eutectic solder (Whitney & Corbin, 2006). An increased temperature is detrimental to many microelectronic components, making assembly very difficult to be applied. In the other hand, high melting temperature may leads to high energy consumption in the industry.

A low processing temperature is desirable for preventing heat damage to electronic devices during soldering (Wong, 2006; Li et. el, 2006: Kim et. al, 2007). Low temperature soldering also reduce the risk of thermal shock induced by the thermal expansion mismatch among different materials in an electronic package (Ervina et. al, 2010). The alloying element employed to lower the melting temperature of solder alloys must be conform to the Restriction of Hazardous Substances (RoHS) directive, and possesses a low melting point, such as bismuth (Bi), gallium (Ga), and indium (In) (Kim & Suganuma, 2003; Fukuda et. al, 2003). Besides reducing the risk of thermal shock, step soldering is another application for low temperature solders. Step soldering is commonly used when soldering a device that requires more than one step. According to Mei et. al, (1996) step soldering is the availability of solders with lower melting points will make possible multiple reflow processes on a single board (Mei et. al, 1996). In this application, the solder used for subsequent steps should have a lower melting point than that used for preceding step (Ervina et. al, 2010).

The main elements of low temperature alloys are Sn, Pb, Bi, and In, (the Cd-bearing alloys are not considered because of their extreme toxicity) (Mei et. al, 1997; Mei et. al, 1996). Various compositions of these elements produce alloys that melt at any given temperatures between 50°C and 183°C (Mei et. al, 1997). Indium (In) is a kind of soft metallic element. Kanlayasari et. al, 2009) reported that In has the influence of reducing the solidus and liquidus temperatures. From the process point of view, Indium (In) has a lower melting point of 156°C compared to Sn, Pb and Bi (Liu et. al, 2003). Thus, it can reduce the soldering temperature. According to Kumar et. al the currently available low temperature solders with high indium-content have the requisite wetting properties, meet the desired reflow temperature requirements (Kumar et. al, 2008).

Nevertheless, the studies on the In-based lead free solder alloys are scarce. This paper discussed In-Zn-Ga as a good candidate for a low temperature lead free solder alloys. Since this solder alloy has low liquidus temperature, it may be used for low temperature soldering and lower the energy consumption in the production. The results of the investigation of the physical and mechanical properties of In-Zn-Ga will be reported in this paper.

2. Materials and Methods

In-Zn-Ga solder alloy was prepared from indium (99.9% pure, Alfa Aesar), zinc (99.9% pure, Alfa Aesar), and gallium (99.99% pure, Alfa Aesar) to obtain low melting temperature solder alloy. These elements were first weighed by digital electronic balance based on each nominal and then melted in a furnace by temperature of 450°C for 1 hour. In order to obtain a homogeneous composition, the ingot was re-melted for three times.

Then, the ingots were compressed to 1mm thickness by using Hydraulic Compression Machine and punched into the shape of billets by using puncher. The diameter of this billet is 5 mm with the thickness of 1 mm each as in Figure 1. The billets was ground with abrasive paper and polished with alumina powder into disk-like billets.

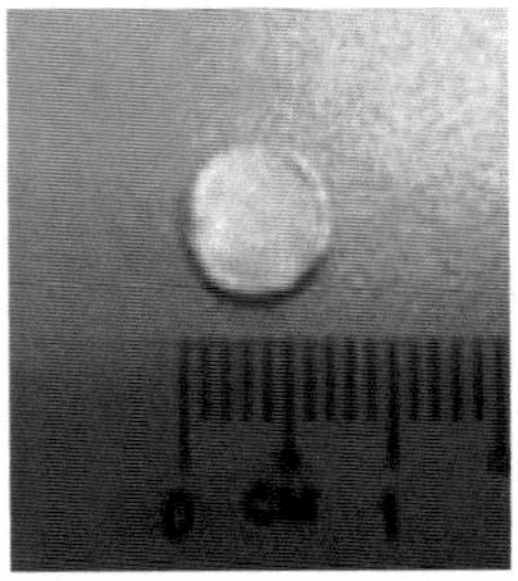

Figure 1. The diameter of the billets.

The billets were used for density testing. The density of In-Zn-Ga solder alloy was measured using an electronic densimeter machine for 3 billets to make sure the value was precise by calculate the average density. For thermal analysis, the melting temperature was analyzed using Differential Scanning Electron (DSC) from TA Instruments. The analysis was carried out at 10°C/min of heating rate with the temperature from 0 to 500°C under nitrogen atmosphere. The weight of the billets was approximately 10g. The coefficient of thermal expansion (CTE) of In-Zn-Ga was investigated using the L75 Platinum Series machine from Leiseis. The solder billet were prepared cylindrically with the diameter of 5mm and the height of 10mm. Shear testing were performed to test the joint strength. The maximum shear force was calculated with the crosshead speed of 0.01mm/s and then being converted to the shear strength using the Eq. 1. The soldered area on the substrate was 10×10mm and the substrate's dimension is 30×10×0.01mm.

$$\tau = F/A \text{.............. (Eq. 1)}$$

Where τ = shear strength (MPa), F = shear force (N), and A = soldered area

3. Results and discussions

In order to serve as an alternative solder alloy, the solder should consume less mass and space as well as possess a density near to or lower than the Sn-37Pb solder (Humpston & Jacobson, 2004). In this investigation, the density value for In-Zn-Ga, SAC and Sn-37Pb are 6.801, 7.440 and 8.790 g/cm^3 respectively as tabulated in Table 3.1. The table also shows the density value of each elements used in In-Zn-Ga, SAC and Sn-37Pb.

Table 3.1. The density values for the compositions and elements.

Composition	Density (g/cm^3)	References
In-Zn-Ga	6.90	This research
SAC	7.44	Humpston & Jacobson, 2004
Sn-37Pb	8.36	Humpston & Jacobson, 2004
In	7.31	Callister & Rethwisch, 2010
Zn	7.14	Callister & Rethwisch, 2010

Ag	10.49	Callister & Rethwisch, 2010
Sn	7.31	Callister & Rethwisch, 2010

The density of In-Zn-Ga is 8.59% lower than most promising lead free solder, SAC and 22.63% lower than the traditional Sn-37Pb solder. This is due to the density value of the elements in the composition of solder alloys. According to Chandra et. al, the density property of a solder is directly related to the atomic structure and the volume of the elements mixed (Chandra et. al, 2010). An addition of element with lower density value tends to lower the overall density of the solder alloys. This is due to the atomic structure of the lower density material has more space than the higher density material and will affect the density as the mass and volume of the composition also affected. According to (El-Daly et. al, 2014), the solder alloys should consume lower density value as the size of the pitch interconnections is becoming smaller. This requirement will be able to satisfy the need of miniaturization in the electronic packaging industry nowadays. A solder alloys should be lighter and less dense compared to lead solder (Sn-Pb) that is 8.79g/cm3 to accommodate the small space in the devices as the electronic devices are getting smaller.

The melting temperature is a critical solder characteristic because it determines the maximum operating temperature of the system and the minimum processing temperature that components must survive (El-Daly et. al, 2008). The melting curves of differential heat flow against temperature are shown in Figure 3.1. The melting point of In-Zn-Ga was determined from the single sharp endothermic peak of DSC heating curve within the temperature of 0 to 500°C. Onset temperature is the temperature when the solder starts to melt and termed as solidus temperature, while end temperature is the temperature when the solder is fully solidifies and termed as liquidus temperature. Meanwhile, peak temperature is the melting temperature.

Table 3.2. The analysis value of the DCS curve.

Composition	Onset (T_{onset})	End (T_{end})	Peak temperature (T_m)	References
In-Zn-Ga	119.60°C	148.45°C	141.31°C	This research
SAC	212.8°C	217.1°C	216.9°C	Sung et. al, 2004

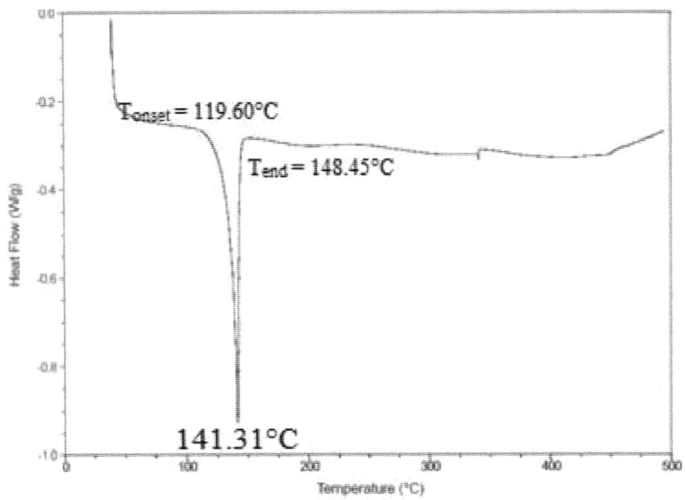

Figure 3.2. DSC curve showing the heating process of the In-Zn-Ga.

The DSC results in Figure 3.2 are summarized in Table 3.2. The melting temperature (T_m) of the In-Zn-Ga was measured as 148.45°C, 34.85% lower than the eutectic temperature of the SAC that is 217°C and 18.88% lower than the melting temperature of Sn-37Pb (T_m=183°C). Ga has the lowest melting point of all the metals, and it has been reported that Ga can enhance the flow ability of some lead-free solders (Zhang et. al, 2004; Chen et. al, 2010). Qiao et. al, 2008 investigated the effects of Ga alloying on melting behaviors. They founded that addition of Ga element on Sn-Ag-Cu solder can obviously decrease the melting point. Based on the experimental results and discussions, each percentage of Ga addition can decrease the melting point of the Sn–Ag–Cu eutectic solder alloy by about 4°C.

Global mismatch will predominate when dissimilar base materials are joined together. The coefficient of thermal expansion is an important parameter for solder alloys. If the thermal expansion coefficient of a solder alloy does not match with the base material, thermal stress will be formed at solder joints and affect the soldering quality.

IOP Conf. Series: Materials Science and Engineering **133** (2016) 012048 doi:10.1088/1757-899X/133/1/012048

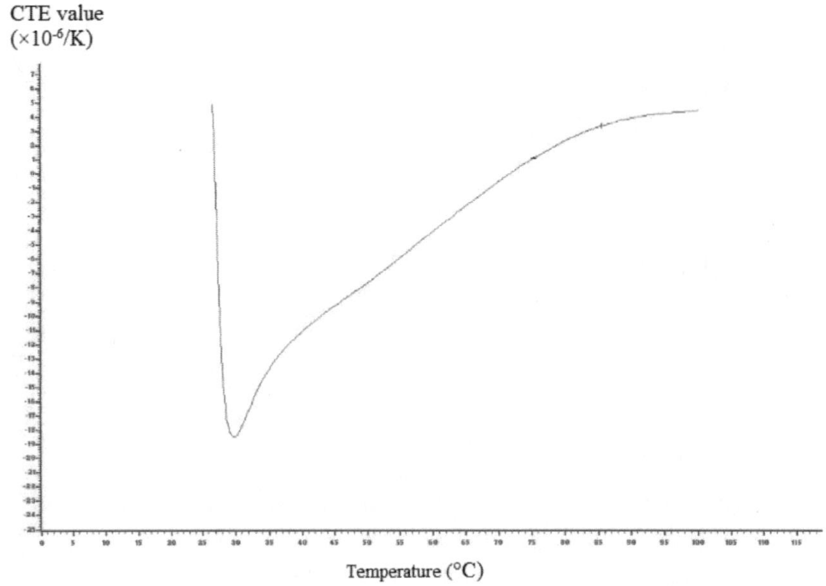

Figure 3.3. Thermal expansion for In-Zn-Ga.

From Figure 3.2, the result was decreasing from 0 to 30°C. This mean the shrinkage occur for the solder at the beginning. This shrink is happened because in the beginning, the atom have more energy to vibrate and the atom with smaller size have high energy to enter the space among the In-Zn-Ga solder alloy. However, it was increase at the temperature of 30 until 100°C. This showed that In-Zn-Ga solder alloy was expand when the temperature increase. When the temperature increase, the atom gain more energy and the vibration will lead to a big gap between the atom rather than fill the space among itself. It has been reported that the CTE value for Cu substrate is 17.0×10^{-6}/K while the CTE value for SAC is 40×10^{-6}/K. It is necessary to decrease the difference in the CTE value between the substrate and the solder alloy for a joint to be effective (Fallahi et.al, 2012). The CTE value of In-Zn-Ga is 1.1×10^{-6}/K. Hence, the difference of the Cu substrate and In-Zn-Ga is acceptable than the CTE value of SAC. If all of the components within the device have identical CTE value, then they will expand and contract at the same rate.

To investigate the shear strength of In-Zn-Ga, single lap shear test was conducted. The shear strength will acknowledge the quality of the solder alloy in terms of the strength in application when it endures shaking (Jeon et. al, 2008). Figure 3.4 show the stress versus train curve of In-Zn-Ga. The shear strength of this solder alloy is 190.188MPa and the shear strength of SAC and Sn-37Pb are 276 and 51.7MPa respectively. The shear strength is higher than Sn-37Pb but lower than SAC. According to Luo et. al (2012), the shear strength of a solder alloy joint depends on their microstructural. Hence, from this result the microstructure of In-Zn-Ga can be predict as less lamellar shape and nearer to each other as the strength is strong. In fact, In-Zn-Ga has an element with smaller atomic size. Hence, the smaller atomic size will fill up the blank space among the atoms making the atoms close to each other and higher the strength at once.

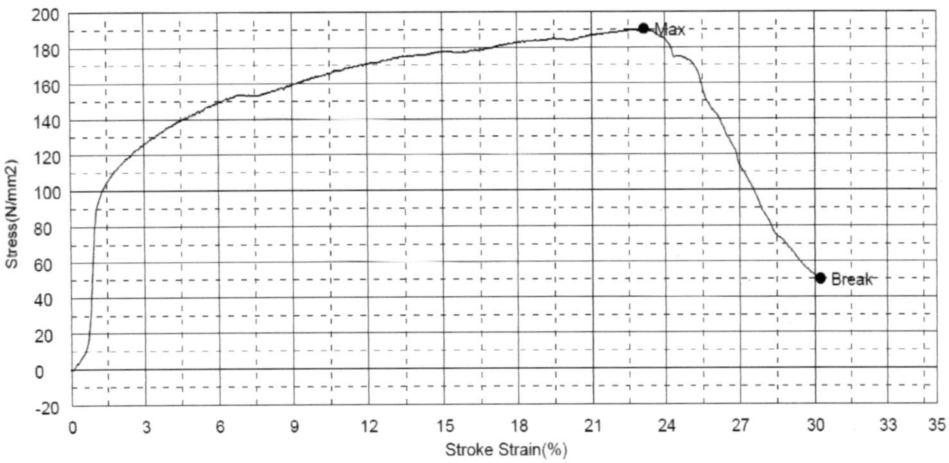

Figure 3.4. The stress strain curves for In-Zn-Ga.

4. Conclusions

The melting temperature of In-Zn-Ga is 148.45°C since the gallium have the capability to reduce the melting temperature effectively. The density of In-Zn-Ga is 6.90g/cm3 that is lower than SAC and Sn-63Pb. The CTE of In-Zn-Ga able to cover the mismatch between the solder and the Cu substrate. The CTE value is 1.1×10-6/K. As the shear strength of In-Zn-Ga is higher than Sn-37Pb, it is acceptable even though lower than SAC. The In-Zn-Ga is a new developed lead free solder alloy that able to replace the SAC. The low melting temperature of In-Zn-Ga can lower the energy consumption in the industry due to less heat need in soldering process.

Acknowledgments

Special thanks to Multimedia University for financial support through ERGS Research Grant No. MMUE/110133 and MMU Internal Grant No. MMUI/140082.

References

[1] Kim, K. S., Suganuma, K. (2003). Third International Symposium on Environmentally Conscious Design and Inverse Manufacturing, pp. 414–415.

[2] Fukuda, Y., Casey, P., Pecht, M. (2003). IEEE Transactions on Electronics Packaging Manufacturing, 26 305–312.

[3] Zhang, L., Cui, J. H., Han, J. G., Guo, Y. H., He, C. W. (2012). Microstructures and properties of SnZn-xEr lead-free solders [J]. Journal of Rare Earths, 2012, 30(8):790-793.

[4] Subedi, D. (2011). Contact Angle Measurement for the surface Characterization of solids. Himalayan Physics, (1), 1–4. Retrieved from http://www.nepjol.info/index.php/HP/article/view/5201

[5] Sharif, A., Chan, Y. C. (2004). Dissolution Kinetics of BGA Sn–Pb and Sn–Ag Solders with Cu Substrate During Reflow. Materials Science and Engineering, pp. 126–131.

[6] Krishnakumar, P. (2010). Wetting and spreading phenomena, 1–12.

[7] Liu, C. C., Lin, Y. K., Houng, M. P., & Wang, Y. H. (2003). The Microstructure Investigation of Flip-Chip Laser Diode Bonding on Silicon Substrate by Using Indium-Gold Solder. IEEE Transactions on Components and Packaging Technologies, 26(3), 635–641. doi:10.1109/TCAPT.2003.817655

[8] Mei, Z., Hua, F., Glazer, J., & Key, C. C. (1997). Low temperature soldering. Twenty First IEEE/CPMT International Electronics Manufacturing Technology Symposium Proceedings 1997 IEMT Symposium, (d). doi:10.1109/IEMT.1997.626966

[9] Ervina, E. M. N., Sharif, N. M., Yew, C. K., Ariga, T., Ismail, A. B., & Hussain, Z. (2010). Wettability and strength of In-Bi-Sn lead-free solder alloy on copper substrate. Journal of Alloys and Compounds, 507(1), 290–296. doi:10.1016/j.jallcom.2010.07.182

[10] Mei, Z., Holder, H., & Plas, H. Vander. (1996). Low-temperature solders. Hewlett Packard Journal, (August), 1–10. Retrieved from http://www.hpl.hp.com/hpjournal/96aug/aug96a10.pdf

[11] Goh, Y., Haseeb, A. S. M. A., & Sabri, M. F. M. (2013). Effects of hydroquinone and gelatin on the electrodeposition of Sn-Bi low temperature Pb-free solder. Electrochimica Acta, 90, 265–273. doi:10.1016/j.electacta.2012.12.036

[12] Kumar, P., Dutta, I., Raj, R., Renavikar, M., & Wakharkar, V. (2008). Novel Liquid Phase Sintered Solders With Indium As Minority Phase For Next Generation Thermal Interface Material Applications School of Mechanical and Materials Engineering Department of Mechanical Engineering University of Colorado , Boulder , CO 80309-0. Thermal Issues in Emerging Technologies, ThETA 2, Cairo, Egypt, 325–332.

[13] Lin, C.Y., Chou, J.H., Lee, Y.G., Mohanty. U.S. (2009). J. Alloys Compd. 470, 328–331.

[14] Islam, R. A., Wu, B. Y., Alam, M. O., Chan, Y. C.,Jillek, W. (2005). J. Alloys Compd. 392, 149–158.

[15] Abtew, M., Selvaduray, G. (2000). Mater. Sci. Eng. R27, 95–141.

[16] Laurila, T., Vuorinen, V., Kivilahti, J. K. (2005). Mater. Sci. Eng. R49, 1–60.

[17] Hung, F. Y., Lin, H. M., Chen, P. S., Liu, T. S., Chen, L. H. (2006). J. Alloys Compd. 415, 85–92.

[18] Yoon, J. W., Jung, S. B. (2008). J. Alloys Compd. 458, 200–207.

[19] Yu, D. Q., Wang, L. (2008). J. Alloys Compd. 458, 542–547.

[20] Islam, M. N., Chan, Y. C., Rizvi, M. J., Jillek, W. (2005). J. Alloys Compd. 400, 136–144.

[21] El-Daly, A. A., Hammad, A. E. (2010). Mater. Sci. Eng. A, doi:10.1016/j.msea. 2010.04.078.

[22] Shen, J., Chan, Y. C. (2009). J. Alloys Compd. 477, 552–559.

[23] Anderson, I. (2007). J. Mater. Sci.: Mater. Electron. 18, 55–76.

[24] Glazer, J. (1994). J. Electron. Mater. 23, 693–700.

[25] Yoon, J. W., Noh, B. I., Kim, B. K., Shur, C. C., Jung, S. B. (2009). J. Alloys Compd. 486, 142–147.

[26] Li, Y., Wong, C. P. (2006). Mater. Sci. Eng. R 51, 1-35

[27] Li, J. F., Mannan, S. H., Clode, M. P., Whalley, D. C., Hut, D. A. (2006). Acta Mater. 54, 2907-2922

[28] K. S. Kim, T. Imanishi, K. Suganuma, M. Ueshima, R. Kato, Microelectron Reliab 47 (2007) 1113-1119

[29] El-Daly, a. a., & Hammad, a. E. (2010). Elastic properties and thermal behavior of Sn-Zn based lead-free solder alloys. Journal of Alloys and Compounds, 505(2), 793–800. doi:10.1016/j.jallcom.2010.06.142

[30] Sun, L., & Zhang, L. (2015). Properties and Microstructures of Sn-Ag-Cu-X Lead-Free Solder, 2015.

[31] Swilem, Y., & Hammad, A. E. (2009). Creep properties of Sn – Sb based lead-free solder alloys, 471, 98–104. doi:10.1016/j.jallcom.2008.03.097

[32] Zhang, G. S., Jing, H. Y., Xu, L. Y., Wei, J., & Han, Y. D. (2009). Creep behavior of eutectic 80Au / 20Sn solder alloy, 476, 138–141. doi:10.1016/j.jallcom.2008.09.009

[33] El-Daly, A. A., Swilem, Y., Makled, M. H., El-shaarawy, M. G., & Abdraboh, A. M. (2009). Thermal and mechanical properties of Sn – Zn – Bi lead-free solder alloys, 484, 134–142. doi:10.1016/j.jallcom.2009.04.108

[34] Alena, P., Bednarcik, J., & Durisin, J. (2011). In situ investigation of SnAgCu solder alloy microstructure ˇ, 509, 1550–1553. doi:10.1016/j.jallcom.2010.09.153

[35] Tsao, L. C. (2011). Solder alloy on the intermetallic growth with Cu substrate during isothermal aging. Journal of Alloys and Compounds, 509(33), 8441–8448. doi:10.1016/j.jallcom.2011.05.116

[36] Smith, D. R., & Madeni, J. C. (2002). Properties of Lead-Free Solders Properties of Lead-Free Solders

[37] Chandran Rao, B., Weng, J., Shen, L., Lee, T., & Zeng, K. (2010). Morphology and mechanical properties of intermetallic compounds in SnAgCu solder joints. Microelectronic Engineering, 2416-2422.

[38] Humpston, G., & Jacobson, D. (2004) . principles of Soldering. ASM International

[39] Kotadia, H., Mokhtari, O., Clode, M., Green, M., & Mannan, S. (2012). Intermetallic compound growth suppression at high temperature in SAC solders with Zn addition on Cu and Ni-P substrates. Journal of Alloys and Compounds, 229-237

[40] Kanlayasiri, K. Mongkolwongrojn, M., & Ariga, T. (2009). Influence of indium addition on characteristics of Sn-0.3Ag-0.7Cu solder alloy. Journal of Alloys and Compounds, 225-230

[41] Shalaby, R. (2010). Effect of rapid solidification on mechanical properties of a lead free Sn-3.5Ag solder. Journal of Alloys and Compounds, 113-117.

[42] Satyanarayan, & Prabhu, K. (2011). Reactive wetting, evolution of interfacial and bulk IMCs and their effect on mechanical properties of eutectic Sn-Cu solder alloy. Advances in Colloid and Interface Science, 87-118

[43] Smedskjaer, M., Jensen, M., & Yue, Y. Z. (2008). Theoretical calculation and measurement of the hardness of diopside. Journal of the American Ceramic Society, 514-518

[44] Whitney, M., & Corbin, S. F. (2006). Lead contamination of a transient liquid-phase-processed Sn-Bi lead-free solder paste. Journal of Electronic Materials, Vol. 35, No. 2.

[45] Ervina, E. M. N. & Aisyah, M. (2012). A review of solder evolution in electronic application. Europe, 1(1), 1–10.

[46] Kang, S. K., Lauro, P., Shih, D.-Y., Henderson, D. W., Bartelo, J., Gosselin, T., … Choi, W. K. (2004). The Microstructure, Thermal Fatigue, and Failure Analysis of Near-Ternary Eutectic Sn-Ag-Cu Solder Joints. Materials Transactions, 45(3), 695–702. doi:10.2320/matertrans.45.695

[47] Manko, H. H, Solder and Soldering, 4th Edition, McGraw-Hill, New York, 2001.

[48] Callister, W. D., & Rethwisch, D. G, Materials Science and Engineering, 8th Edition, John Wiley & Sons, Hoboken, US, 2011.

[49] Ervina, E. M. N., Amares, S. "Characterization of mechanical testing on lead free solder on electronic application" Electronic Manufacturing Technology Symposium (IEMT), 2012 35th IEEE/CPMT International , vol., no., pp.1,9, 6-8 Nov. 2012 doi: 10.1109/IEMT.2012.6521780

[50] Ervina, E. M. N. & Nur Faziera, M. N. Effects of Wetting Behavior on Gallium Addition in In-Zn Solder Alloy. (2016) Materials Science Forum, Vol. 857, pp 22-25

[51] Ervina, E. M. N., Sharif, N. M., Yew, C. K., Ariga, T., Ismail, A. B., & Hussain, Z. (2008). Characteristic of low temperature of Bi-In-Sn solder alloy. Electronic Manufacturing Technology Symposium (IEMT), 2008 33rd IEEE/CPMT International

[52] Ervina, E. M. N., Zuhailawati, H. & Radzali, O. "Low temperature In–Bi–Zn solder alloy on copper substrate", (2015) Journal of Materials Science: Materials in Electronics, 1-8

[53] Ervina, E. M. N & Amares, S. "A Study of Temperature, Microstructure and Hardness

Properties of Sn-3.8 Ag-0.7 Cu (SAC) Solder Alloy", (2015) MATEC Web of Conferences 27, 2003

[54] Ervina, E. M. N. & Siti, N. "Effect of Different Aging Times on Sn-Ag-Cu Solder Alloy" (2015) TRANSACTIONS ON ELECTRICAL AND ELECTRONIC MATERIALS 16 (3), 112-116

IOP Conf. Series: Materials Science and Engineering **133** (2016) 012049 doi:10.1088/1757-899X/133/1/012049

Methods for elimination of dampness in Building walls

Cristina Campian[1], **Maria Pop**[1]

[1]Technical University of Cluj Napoca, Faculty of Civil Engineering, Cluj-Napoca, Romania

E-mail: cristina.campian@dst.utcluj.ro, maria.pop@dst.utcluj.ro

Abstract. Dampness elimination in building walls is a very sensitive problem, with high costs. Many methods are used, as: chemical method, electro osmotic method or physical method. The RECON method is a representative and a sustainable method in Romania. Italy has the most radical method from all methods. The technology consists in cutting the brick walls, insertion of a special plastic sheeting and injection of a pre-mixed anti-shrinking mortar.

1. Introduction

Dampness in buildings gives rise to processes of change in building materials usually with very evident symptoms. In our profession, we often come across chipped plaster, crumbling bricks, corroded metal components and wooden beams which, because of their hydroscopic nature, swell up, crack and rot.

Let us consider the various causes of dampness in walls:

- Consideration dampness:

This is difficult to evaluate if the materials used in the construction and seasonal in changes in conditions are not known (winter maximum, summer minimum). It is to be found the inside of buildings at certain critical points (corners) or widespread (e.g. north facing walls) with mould in greater or lesser proportions.

In order to eliminate this dampness it is necessary to insulate the wall and ventilate the rooms both naturally and by using forced air.

In order to eliminate the mould it is necessary to use disinfectant products mixed in paint or wash the walls with vinegar or bleach.

- Descending damp:

This generally results from the breakage of drainage piping or defective or poorly made roofing with imperfect sealing of expanding joints.

Careful repair and successive regular maintenance should solve all these problems.

- Rising damp:

This is the worst type of damp as it always implies much more radical and expensive treatment compared to the others.

This type of damp may occur because of water rising from water tables in the ground, from water dispersed in the ground from a faulty or defective water supply or sewerage systems.

[1] Maria.pop@dst.utcluj.ro.

This problem may be solved by finding the fault and carrying out repairs with care.

Regarding this problem, there are several procedures which vary in their effectiveness and duration but which can be carried out to prevent this problem.

These procedures can be divided into three approaches: chemical, electro osmotic and physical.

2. Chemical method

The first stage of this procedure consists of drilling holes in the wall, usually only on one side and up to a certain depth of the wall. Depending on which company carries out the work, either injectors or simple transfusion tubes are used.

Resins are then fed into these tubes either under pressure using a suitable pump or simply by gravitational force so that a chemical barrier is created against the capillary rise of dampness and to saturate the treated part.

This de-humidification process may meet difficulties if the injected product does not penetrate the affected surfaced uniformly; it must be added that very often the walls to be treated are very old and may have vertical cracks, internal cavities and air spaces which impede the resins injected either by gravity or under pressure, from spreading in a capillary and correctly horizontal way which could jeopardize the entire treatment.

Even very few untreated points allow dampness to pass beyond the level of the line of holes.

A second chemical method consists of using special plasters which dry out the wall without stopping the capillary rise of damp through the wall but encourages evaporation of the dampness in the wall.

This effect is produced by including progenies in the mixture which create intercommunicating macrospores in the mortar and increase the total evaporation surface area.

In the presence of a high rate of rising damp it is possible that an insufficient exchange between the capillary action of the rising damp and the evaporation in walls treated in this wall can occur.

This type of plaster could be used together with the de-humidifying chemical barrier technique with holes.

We believe that these methods are too recent development to judge their effectiveness over a significant length of time. These procedures must be reapplied after few years (2-5 years), because this is not a definitive solution and the moisture reappears.

3. Electro osmotic system

One of the first procedures of this type consisted of introducing triangular, trapezoidal or round elements of brick, called siphons, into the wall which contained a small copper plate in an internal cavity whose function was to attract the dampness contained in the wall inside the element; water resulting from this process was transferred to the outside because of the inclination of the element itself.

After a period of enormous popularity (several public buildings and churches in Europe still bear the signs of this method) it was completely abandoned as the copper plate corroded and was no longer able to fulfil its function.

Another method using this system, mostly used in Switzerland, consists of encircling the entire building with cooper conductor wire connected to probes, also made of cooper, which are inserted at predetermined intervals in the wall to be treated.

The line of connection is then tied to iron or cooper pegs which are inserted in the ground so that a difference of potential is created which brings about de-humidification.

4. Physical Method

It must be started that this is a system which provides a radical solution to the problem of rising damp.

Whichever method is chosen, an insulating barrier is created in the cut made at the base of the wall and provides an effective stop to rising damp.

There are two main methods used at the present:

4.1. Mechanical holes drilling method

In this method a series of holes are made in a line along the wall.

The diameter of the hole depends on the type of material to be treated and the type of insulating material to be inserted.

A first series of holes is created in a horizontal line at a constant height from ground level and in alternating sections. Silicone resin mortars are inserted into these cut sections.

When the resin has hardened, more holes are drilled as before in the parts of the wall which had not been touched.

This system is very costly because of the number of holes which have to be drilled.

4.2. Mechanical cutting system

This method is commonly known as the COMER-RECON Method from the name of the company in Italy, Recon being its representative in Romania, which manufactures patented cutting machines as well as the accompanying products and equipment.

It can be used on outside walls made of brick, tinstones, and concrete and expanded clay blocks.

It is necessary to use diamond tipped tools on hard rock or stone walls.

The first stage consists of removing a certain high of the wall plaster, of the wall to be cut.

If the remaining plastering is in very bad condition, it is a good idea to remove it before cutting the wall in order to encourage evaporation of water contained in the wall to be treated and the removal of soluble salts.

Cuts sections of lengths varying according to the existing wall structure are made and plastic sheeting is inserted into them.

These sheets act as a barrier to the rising damp (figure 1).

In the lagoon city of Venice, because of the very particular structure in which walls are supported by foundation piles, an insulating barrier is created by inserting polyethylene coated lead sheets into the cut.

This type of operation was carried out by Fiat engineering in 1985 during the restoration of Palazzo Grassi in Venice, property of Palazzo Grassi Spa.

The use of sheets creates a perfect insulating barrier with excellent dielectric properties, good chemical inertia, and excellent thermal resistance to compression.

The normal procedure involves applying mortar to the insulating material.

Figure 1. Introduction of anchoring wedges.

Then a number of patented plastic anchoring wedges are inserted in the gap for each wall section which, while they compress the mortar in the cut, secure the insulating material and assure the stability of the wall during the hardening of the saturating mortar.

Figure 2. Anchoring wedges.

The wedges are inserted in the gap first.
Pre-mixed anti-shrinking mortar is injected into the cut.

Figure 3. Wall in treatment (Arcalia Castle, Romania).

This penetrates not only the wedge grooves and the cut itself but also any existing cavities in the wall structure, reinforcing it at the base.

It is very important to wait until the whole wall has dried out before proceeding to re-plastering. Regarding modalities for this type of operation in seismic areas, in Germany, Austria, Great-Britain, France, Thailand, Czech Republic, Slovakia, Hungary, Spain and Belgium, wall cutting machinery is used regularly.

It should not be forgotten that even in new constructions a separating cut is made for the insertion of waterproofing insulating sheaths (so-called wall-cutter product once made of asphalt sheets) between the foundations and the base of the walls which are also used in multi-story buildings and are frequently simply laid without using any type of binder.

5. Procedure applied in Romania

In Romania, this procedure was applied with success at many buildings as old ones, historical monuments and even an old castle. These old buildings has become new buildings.

An interesting work was made at the Arcadia Castle in Bistrita-Nasaud in 2002. The castle was in a much degraded state, but now here it is the Centre of Biological Research oh Babes-Bolyay University, as it can be seen from the next pictures. Nothing has remain from the old state in which the building was before the intervention.

Figure 4. Wall before the treatment (Arcadia Castle, Romania).

Figure 5. Structural Wall the treatment.

Figure 6. Arcalia Castle today.

6. Conclusions

From the few methods presented here, some conclusions can be made.

The chemical method is too recent developed to judge their effectiveness over a significant length of time. This procedure must be reapplied after few years (2-5 years), because this is not a definitive solution and the moisture reappears.

The mechanical hole drilling method is costly because of the number of holes which have to be drilled.

The best method as we consider is the radical method, using the mechanical cutting system and the insertion of a special plastic sheeting with injection of anti-shrinking mortar, method successfully applied in Romania in many old buildings, common houses, and even an old castle.

References

[1] Agrement tehnic nr.007-03/051-1999 – S.C. RECON SRL, Cluj-Napoca, privind metoda de eliminare a umiditatii din pereti de caramida ai cladirilor, INCERC Cluj.

[2] T. Streza, Cr. Câmpian, 2002 *Tehnologia COMER de eliminare a umiditatii din ziduri*, 26-28 aprilie, „SIBSTIL 2002", Sibiu.

[3] C. Campian, 2005 *Methods for elimination of dampness in old masonry building walls*, Traditional Craft in 21 st Century Architecture", Conference Proceedings ISBN 83-921903-0-0, Szczecin University of Technology, Poland, pag 195-201.

[4] C. Campian, Z. Nagy, M. Pop, P. Pernes, 2015 *Sustainable Buildings* – 5th International U.A.B. – B.E.N.A. Conference, Alba Iulia 28-30 Mai, pp.46.

[5] N. Cobîrzan, S. Palacean, 2014 *Making existing building accessible*, Meditech, Springer Publishing Company, DOI 10.1007/978-3-319-07653-9_1, ISBN 10.1007/978-3-319-07653-9_1, Volume 4, pp 1-6.

[6] J. Carmody, B. Andreson, 2016 *Moisture in basements: causes and solutions*, University of Minnesota.

[7] C. Sharon Park, 1994 *Moisture in Historic Buildings and Preservation Guidance*, ASTM Manual Series, MNL 18 Febr., ASTM Publication Code Number 28-0018094-10.

[8] N. Moyer, D. Beal, D. Chasar, J. McIlvaine, *Moisture Problems in Manufactured Housing*, Florida Solar Energy Center, Publication Number FSEC-GP-212-01.

IOP Conf. Series: Materials Science and Engineering **133** (2016) 012050 doi:10.1088/1757-899X/133/1/012050

Applications of Radio Frequency Identification (RFID) in Mining Industries

Mohd Khairul Nizam Mahmad[1, 2, 4], Mohd Remy Rozainy M.A.Z[2, 3] and Norlia Baharun[1]

[1]School of Materials and Mineral Resources Engineering, Engineering Campus, Universiti Sains Malaysia, 14300 Nibong Tebal, Penang, Malaysia
[2]School of Civil Engineering, Engineering Campus, Universiti Sains Malaysia, 14300 Nibong Tebal, Penang, Malaysia
[3]Center of Excellence Geopolymer and Green Technology (CEGeoGTech), School of Materials Engineering, University Malaysia Perlis (UniMAP), P.O Box 77, d/a Pejabat Pos Besar, 01000 Kangar, Perlis, Malaysia
[4]Faculty of Business Management, Universiti Teknologi MARA (UiTM) Kedah, 08400 Merbok, Kedah, Malaysia

nizam.mahmad@gmail.com

Abstract. RFID technology has recently become a dream of many companies or organizations because of its strategic potential in transforming mining operations. Now is the perfect time, for RFID technology arise as the next revolution in mining industries. This paper will review regarding the application of RFID in mining industries and access knowledge regarding RFID technology and overseen the opportunity of this technology to become an importance element in mining industries. The application of Radio-Frequency Identification (RFID) in mining industries includes to control of Personal Protective Equipment (PPE), control of personnel to access mining sites and RFID solutions for tracking explosives.

1. Introduction

The mining industry is the oldest industries for worldwide and become a main economic contribution for many countries around the world. For example China, Australia, Indonesia, Malaysia and others. At this time, the mining industries is undergoing an economic boom that is causal to economic regaining and social development in various countries. In order to survive, it must encounter quite a lot of challenges related to initiate of new development in the projects [1].

Mining industries is categorized as high risk job that will reflect to occupational safety and health. The uncertainty event could happen anytime. The proactive action is become a must in control the management of risks associated with occupational safety and health. Relatively, related to systematic approach. Practically, once core decisions are being decided thru the life cycle of the mining business starting at the stage of exploration work concluded mine operations to mine rehabilitation. Consequently, the risk controlling is main impotence part subsequently the chance of opposing significances cannot be removed completely without causing economically non-viable or else mining operations technically [2].

This paper will review regarding the application of RFID in mining industries and access knowledge regarding RFID technology and overseen the opportunity of this technology to become an importance element in mining industries, Judgements regarding the choice of new technologies like RFID involve substantial investment and can have a strategic impact on the organization. We will discusses the practice of this technology from a management perspective, give some theoretical frameworks that are beneficial for the organization to implement the RFID technology in mining industries [3].

RFID technology has recently took the dreams of many company or organization because of its strategic potential and high operational in transforming mining operations. Now is the perfect time for RFID technology arise as the next revolution in mining industries. Truly, RFID technology is predicted to alter many of the mining operations including the actual time data collection and sharing amongst mining investors, the establishing of innovative business models and practices such as the total inventory management, products anti-fabricating strategies, build-to-order supply chain management and reverse logistics. Additionally, RFID technology permits the reshape of inter-organizational and intra-organizational business developments that could lead to the rise smart processes. It is a processes that can initiate business transactions without any human interferences and minimalism the integration of inter- organizational and intra-organizational information systems. The RFID technology principally should permit the tagged entity to turn into an intelligently communicating element of the organization's overall information internal frame and become a mobile based on the mobility of RFID technology [4].

2. Radio Frequency Identification RFID

Radio Frequency Identification (RFID) is the technology use of electromagnetic fields to transmit data. It also refer to the process of using the technology of wireless broadcast to collect and recovery data, [5]. RFID is an unique Automatic Identification and Data Capture (AIDC) technologies [6]. This technology is considered by many scholars and experts as an "Internet of Thing". It talk about to the prospect to find out information regarding the tagged object by surfing database entry or Internet address that matches to a particular RFID [4].

3. The Principle RFID

The usual Radio Frequency Identification technology is comprise of the electronic tags, reader, antenna and a computer with software to manage and control the RFID equipment [5]

Principally, an RFID system is comprise of four component:

i. An electronic tag consists of microprocessor chip (is an integrated circuit with memory). RFID tags can be passive (without batteries) or active (with batteries). [7]. This microprocessor chip is implanted in a physical object to be recognized [8]

ii. The reader is a device that can read data from RFID and respond without making any physical contact to the tags [8]. Communication between reader and tag allows the locality information of an items to be recorded and transmitted to a server via a computer network, thus permitting the movement of the items to be tracked and traced. [7].

iii. The antennae are the channels for the interaction of data between the reader and tag. It has a reading range both sideways and in front of the antenna. The location and design is vital in controlling the coverage area, efficiency and accurateness of communication between the tags [7]

iv. A computer with software to manage and control the RFID equipment, screening the data, and cooperates with enterprise applications [8]

4. Application of RFID in Mining sites
The following below is the applications of RFID in mining sites.

4.1. Control of Personal Protective Equipment (PPE)
In mining industries necessitate the employees to wear PPE. The usual PPE consist of safety vests, gloves, steel-toed, safety boot, clothing, ear plugs hardhats, eyewear, masks, and etc. Different work situation has different PPE requests that essential be uses referring to rules and regulations that must be compliance. In order to solve this problem, the RFID technology can be used, as a one step to create safety environment for the mining worker.

According to Kelm et al. (2013), to control the Personal Protective Equipment (PPE) required each PPE component must has attached with an RFID tags. The mechanism of RFID is when a person wearing PPE that equip with RFID tags passing through the gate, it will be detected by an RFID reader or an antenna system. The RFID tag may encompasses essential information of PPE that had been wearing by the worker. This technique guarantees that essential and compulsory PPE components that must be worn by the personnel before the employee starts to work. There are some important PPE information that can be stored on the chip of a passive RFID includes:

i. Serial number to differentiate from other equipment's `
ii. Operation manual to safely use the equipment's
iii. Maintenance records (lasting useful time or date of equipment's)
iv. Info or data log about individuals who use the equipment's
v. Disposal facility or recycle procedure (to ensure damaged or outdated equipment's is not reused over).
vi. Expiration date of PPE
vii. Latest use of PPE

4.2. Control of Personnel to Access Mining Sites
Main mission of logistics at mining places is to accomplish access control of personnel. On huge mining area, many labours are mostly employed and on site concurrently. It also involve several vendors, subcontractors and guests. They might enter frequently to the mining area daily. The illegal individuals should be not permit to entry to the mining site for the safety precaution, to prevent theft, and illegal employment in mining sites. In order to prevent it from happen, the mining area are secure usually by a barrier or buffer zone. Reliant on the how big of the mining site, the operated mining firm may employ extra access control, for example security guard that manage entry and exit to the mining site and also guarding the interior and borders of the mining site. [9]

To solve this problem, the implementation of RFID technology become main vital aspect. The use Identification (ID) cards for personnel that use RFID technology to operate may reduce the risk of safety precaution, theft avoidance and illegitimate employment in mining sites. Identity cards are distributed to control personnel from exiting and entering the mining area. The size of the ID cards usually similar to credit card. It will carry name and photo of the worker. The ID cards can be read by a barcode reader by touch the card. With RFID technology, it will ease the process of identifying personnel. Traditionally, before this the process of identifying personnel use a lot of labor and time-consuming. The security guards tend do not carefully check so that this technology will bring the future of new system in security aspect. It will measure when a person enters or leaves the mining site. The statistical data that get from the system, will become an important for business to measuring the workers performance for example attendance and punctuality. The systems currently for ID cards established on (passive) RFID and barcodes. The access control and time recording systems using automated-identification (ID) technologies joint with time and attendance recording competencies are suitable for mining site. Development advance RFID, for face detection system for individual to entering mining area not widely use the camera that function as biometric recognition be able to handle the problem encounter in mining site [9].

4.3. RFID Solutions for Tracking Explosives

RFID technology is proper for inventory management of boosters and detonators, tracing of detonators and explosives, security and recovery of the assembly from the explosion debris in the case of a misfire. This technology are be able to tracking and monitoring of detonators and boosters, repossession of the explosive materials or detonator in case of misfire, remote reading of detonators' IDs once detonators are in situation to blast are important for dependable inventory control also for safety features in mining work. Reading if possible at the time the detonator is being loaded in a blast hole can offer the trustworthy inventory control. This take account of the record keeping of depletion of detonators and permits bring up to date the inventory [10].

4.3.1. Tracking Of Detonator and Booster through Transit. To tracking of detonator and booster through transit, there are two possible method. The first is tags was attached with explosives and read by a mobile reader and second is by tags was attached to explosives and read by stationary readers. For the first method, the tags was attached with explosives and read by a mobile reader. The passive RFID tags may possibly be attached with explosives and so that it can be read by a mobile reader. During loading, the explosives can be traced and read by the mobile reader and data may possibly be linked to the computer inside the driver's compartment via mobile reader for upcoming indications. At the exit or entrance of the supplier or manufacturer, the explosives can over once more be traced and tracked by the reader positioned and contrary to the earlier data obtainable in computer. The driver can trace and track the explosives through the mobile reader in between the transportation for any mistake and ensure with the earlier data during transportation. The explosives can all over again be traced and rechecked with the data got at the time of loading at the entrance of shipment. The transferred power from tags to reader and then reader to tags would be inadequate to avoid from the lethal accidents due to impulsive explosion and to sustain the physical condition of the explosives available in the container [10].

For the second method, the tags was attached to explosives and read by stationary readers. Basically, the tags attached with explosives available in the vessel and readers should be located at considered place inside the truck. The readers will read the tags and eventually transmit the data to the computer that been placed in the truck. The main vital issues must be considered before attaching the tags with explosives and locating the readers at the considered places. For example location of the tags, ii) tag overcrowding, and the ability to handle coordination of readers and RFID tags. In order to avoid the risk of trigger of the explosives, the conveyed power from tags to reader and vice versa must be carefully control [10].

4.3.2. In Case of Misfire to Finding Detonators after Blast. Misfired might happen, if the external source trigger or initiate the explosive. Example electronic devices, flash, impact by a large piece from rock fragments and other. So Misfired explosives risk must be present in order to keep safety for the worker at mining site. During drilling into explosive also can trigger the blasting. In order to do second blasting, the blasting engineer must know the exact location the explosive that misfired. The existence of misfired explosives in the drilling zone may cause very bad injury and might cause death. So in order to prevent this, we must tracking, monitoring and retrieval of misfired explosives using RFID technology to prevent this accidents before happen. Misfired explosives can be trace or track using RFID tags that been attach or located on the explosives or detonators submerged inside the drilled bore hole of the surface of rock. When the explosion of explosive will destroyed the RFID tags during the event. Misfired explosives with the RFID tag will be able to sending and transfer signal to the reader. The data is expected by the reader right after the blasting, will give information regarding the explosives that are still not be blast completely and Might be dangerous for second blasting. So the action will be taken to proceed to trace and trace the remaining explosive. The challenges that must be encounter before implementation of RFID technology. The following is main importance aspect that must considered according Mishra et al. 2012:

i. Must be robust to mining environment
ii. Ability to resist vibration
iii. Economically such as low cost. Because the tags will destroyed by explosive

5. Conclusions
Mining industries is currently undergoing an economic boom that is causal to economic recovery and social development in various countries. Now is the perfect time for RFID technology arise as the next revolution in mining industries Truly, RFID technology is predicted to alter many of the mining operations including the real-time data collection and sharing among mining investors, the establishing of innovative business models and practices such as the total inventory management, products anti-fabricating strategies, build-to-order supply chain management and reverse logistics. The application of Radio-Frequency Identification (RFID) in mining industries includes to control of Personal Protective Equipment (PPE), control of personnel to access mining sites and RFID solutions for tracking explosives. This technology are able give an economic value to the industries and will become a new changes to mining commodity.

6. Acknowledgements
The authors thank to Universiti Teknologi MARA (UiTM) and Universiti Sains Malaysia (USM) for use of sources of information through the online journal database that been establish by the university.

References
[1] A. Badri, S. Nadeau, and A. Gbodossou, "A new practical approach to risk management for underground mining project in Quebec," *J. Loss Prev. Process Ind.*, vol. 26, no. 6, pp. 1145–1158, 2013.
[2] D. Komljenovic, W. a. Groves, and V. J. Kecojevic, "Injuries in U.S. mining operations - A preliminary risk analysis," *Saf. Sci.*, vol. 46, no. 5, pp. 792–801, 2008.
[3] G. Ferrer, N. Dew, and U. Apte, "When is RFID right for your service?," *Int. J. Prod. Econ.*, vol. 124, no. 2, pp. 414–425, 2010.
[4] S. Fosso Wamba, A. Anand, and L. Carter, "A literature review of RFID-enabled healthcare applications and issues," *Int. J. Inf. Manage.*, vol. 33, no. 5, pp. 875–891, 2013.
[5] Y. Y. Niu, Z. J. Zhu, and M. R. Fan, "The Application of RFID in Coal Mine Safety Management," *Adv. Mater. Res.*, vol. 710, pp. 720–722, 2013.
[6] M. K. Lim, W. Bahr, and S. C. H. Leung, "RFID in the warehouse: A literature analysis (1995-2010) of its applications, benefits, challenges and future trends," *Int. J. Prod. Econ.*, vol. 145, no. 1, pp. 409–430, 2013.
[7] E. W. T. Ngai, K. K. L. Moon, F. J. Riggins, and C. Y. Yi, "RFID research: An academic literature review (1995-2005) and future research directions," *Int. J. Prod. Econ.*, vol. 112, no. 2, pp. 510–520, 2008.
[8] S. Fosso Wamba, L. a. Lefebvre, Y. Bendavid, and É. Lefebvre, "Exploring the impact of RFID technology and the EPC network on mobile B2B eCommerce: A case study in the retail industry," *Int. J. Prod. Econ.*, vol. 112, no. 2, pp. 614–629, 2008.
[9] A. Kelm, L. Laußat, A. Meins-Becker, D. Platz, M. J. Khazaee, A. M. Costin, M. Helmus, and J. Teizer, "Mobile passive Radio Frequency Identification (RFID) portal for automated and rapid control of Personal Protective Equipment (PPE) on construction sites," *Autom. Constr.*, vol. 36, pp. 38–52, 2013.
[10] P. K. Mishra, M. Bolic, M. C. E. Yagoub, and R. F. Stewart, "RFID technology for tracking and tracing explosives and detonators in mining services applications," *J. Appl. Geophys.*, vol. 76, pp. 33–43, 2012.

IOP Conf. Series: Materials Science and Engineering **133** (2016) 012051 doi:10.1088/1757-899X/133/1/012051

Antioxidant activity and phytochemical compounds of snake fruit (Salacca Zalacca)

I R Suica-Bunghez[1], S Teodorescu[2], I D Dulama[2], O C Voinea[3], S Simionescu[4,5], R M Ion[1,6*]

[1]National R&D Institute for Chemistry and Petrochemistry – ICECHIM, 202 Splaiul Independentei Street, 060021- Bucharest Romania
[2]Multidisciplinary Science and Technology Research Institute of Valahia University of Targoviste - ICSTM-UVT, 13 Aleea Sinaia, 130082- Targoviste, Romania
[3]Carol Davila University of Medicine and Pharmacy, 37 Dionisie Lupu Street, Bucharest, Romania
[4]Polytechnica University of Bucharest, Faculty of Applied Chemistry and Materials Science, 1-7 Polizu Gheorghe street, Bucharest- 11061, Romania
[5]National Agency of Medicines and medical devices, 48 Av.Stefan Sanatescu Street, 011478-Bucharest, Romania
[6]Valahia University, Materials Engineering Department, 13 Aleea Sinaia, 130082 - Targoviste, Romania.

E-mail: rodica_ion2000@yahoo.co.uk

Abstract. Snake fruit *(Salacca zalacca)* is a palm tree species, which is found in Malaysia and Indonesia. This study was conducted to investigate and compare the composition, total phenolic, flavonoid, tanins and monoterpenoids contents in the core and shell fruits. Concentration values of extracts were obtained from standard curves obtained. Antioxidant activity was determined using DPPH method. For all methods it was used the UV-VIS Specord M40, using different wavelength. The infrared spectral analysis was carried out to caracterized the type of functional group existent in snake fruit parts (shell and core).

1. Introduction

Scientifically studies confirmed the fact that consumption of fruits and vegetables reduce incidence of different types of diseases, like cancer or cardiovascular disease [1, 2]. They contain a number of types of minerals and phytochemicals (phenols and flavonoids) that provide important health benefits [3-6].

Nowadays a variety of tropical fruits are found in Europe. The high nutritional value of the subtropical and tropical fruits led to significant increase in their consumption, but a lot of people are not familiar with the nutritional values of these fruits [7-9].

Snake fruit (Sallaca zallaca) is a specie of a palm tree, from family Arecaceae, is a good source of natural antioxidants, which containing many different radical scavenger [9, 10]. This fruit it is also called "The Future of Our Health" and "The Superheroes of Functionality". The skin fruit is brown and looks like a snake skin. The fruit has an astringent taste [11,12].

Leontowicz et al [3] discovered that snake fruit has the ability to prevent atherosclerosis in vivo and it has a significantly displayed higher antioxidant value than mangosteen.

The snake fruit resembling an egg like in shape and the skin of the mature fruit is brown. The fruit contain a kernel covered with white flesh. The aroma it is between pineapple, pear and banana. The weight is up to 70 g at the last maturation stage. Most of this fruit is freshly consumed and some are processed into fruit juice, canned fruit or jam [10-13].

This research presented the snake fruit characteristics and benefits for human body. The results indicate that snake fruit contain certain amounts of polyphenols, flavonoids, tannins and monoterpenoids, proving them to be perfect sources of antioxidants.

2. Materials and methods

2.1. Materials

Snake fruit was added from Bali Island (Indonesia). Standard substances used for curve calibration of methods are linalool (Merck), vanillin (Scharlau), gallic acid (Merck), catechin () and 2,2-diphenyl-1-picryl-hydrazyl-hydrate stable free radical (DPPH from Merck); methanol by Scharlau. For determinations of methods it was used also aluminium chloride ($AlCl_3$ from Sigma-Aldrich), $NaNO_2$, NaOH, H_2SO_4 (Merck), sodium carbonate (Na_2CO_3 from Merck), HCl and Folin–Ciocalteu reagent (Merck).

2.2. Preparation of plant sample

The Snake fruit (Sallacca zallacca) was peeled and cut into small pieces. The fruit (figure 1, a) core and b) shell) was weighed and exhaustively extracted in hidroalcoolic mixture (CH_3OH : H_2O dist.; 9:1), at room temperature for 24 hours. The macerate was kept in a brown volumetric dark flask in order to avoid degradation.

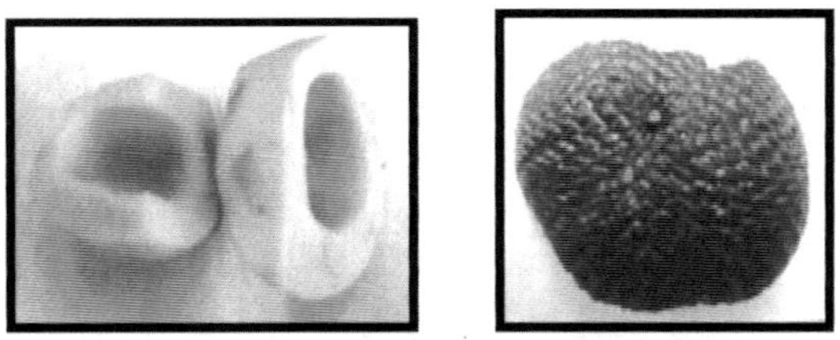

a) b)

Figure 1. Snake fruit: a) core and b) shell.

Figure 2. Colours of snake core and shell hidroalcoolic extracts.

2.2. Characterization Methods

2.2.1. UV-VIS Spectroscopy. The absorption spectra of the samples were recorded on a double beam M 400 Carl Zeiss Jena UV-VIS spectrophotometer from 400 to 800 nm, at the resolution of 1 nm, with 1 nm slit width and 0.3 nm/s scan rate.

2.2.2. Fourier transform infrared spectroscopy. The functional groups of the snake fruit (shell and core) were determinated using a Vertex 80 FTIR spectrometer with the Hyperion 3000 IR microscope.

2.2.3. Phytochemical Analyses. The phytochemical quantification procedures were used for the determination of total tannins, total flavonoids, total pholyphenols and total monoterpenoids in the snake fruit extracts. The assays are presented in table 1.

Table 1. Phytochemical assays performed on the Snake fruit extracts

No.	Assay	Reagents	Conditions	Monitoring system and standard curve	Ref.
1	Total Tanins	0.5 mL extract + 3 mL 4% vanillin-MeOH + 1,5 mL HCl	15 minutes of incubation at room temperature	Absorbance at 500 nm; Catechin curve calibration standard	[14,15]
2	Total flavonoids	1 mL extract + 4 mL distilled water + 0,3 mL $NaNO_2$ (5%); After 5 minutes, 0,3 mL $AlCl_3$ (10%); After 5 minutes, 2 mL 1M NaOH + 2,4 mL distilled water	30 minutes of incubation at room temperature	Absorbance at 510 nm; Catechin curve calibration standard	[15-17]
3	Total polyphenols	1 mL diluted extract + 5 mL Folin-Ciocalteau reagent. After 8 minutes, 4 mL Na_2CO_3	60 min of incubation at room temperature	Absorbance at 765 nm; Gallic acid curve calibration standard	[16-18]
4	Total mono-terpenoids	- 2 mL extract + 1 mL 2% vanillin-H_2SO_4	- heated at 60 °C/20 min, - cooled at 25 °C/5 min	Absorbance at 608 nm; Linalool curve calibration standard	[19]

2.2.4. Antioxidant activity determination. The principle of this method consists in reducing in the presence of an antioxidant molecule, giving rise to coloured methanol solutions. The utilization of DPPH method delivers an easy and rapid mode to estimate antioxidant activity against free radicals [20, 21]. DPPH solution is a methanolic solution prepared in lab (0.02 mg DPPH/mL MeOH), with violet colour. From each sample it was used 0,5 ml extract sample and it was mixed with 1 mL of DPPH solution. After it were agitated and it were left in the dark for an hour, the mixtures were tested by reading the absorbance λ=517 nm on a UV-VIS Specord M 400 spectrophotometer. As a blank, it was used a mixed solution prepared from 0.5 mL of MeOH with 1 mL of 0.02 mg/mL DPPH solution, which it was measured at the same wavelength. The antioxidant activity (AA%) percentage was calculated using the formula [20, 21]:

$$AA\% = [(A_{Control} - A_{Sample} / A_{Control}] \text{ x } 100$$

where: $A_{Control}$ = absorbance of a DPPH solution without sample; and A_{Sample} = absorbance of the sample mixed with 0.02 mg/mL DPPH solution.

3.1. UV-VIS results

UV-visible spectroscopy demonstrated to be very useful technique for the analysis of extracts. The UV-Vis analysis (figure 3) shows the UV-VIS spectra obtained for the snake fruit core and shell extracts, identified the maxima wavelengths specific flavonoids and other phenolic compounds at 280-300 nm and 300-350 nm. The peaks of hidroalcoolic extracts characterized by absorptions in the UV region 250-450 nm, corresponding to phenolic acids and their derivatives (flavones, flavonols, phenylpropenes, quinones). At 280 nm it is identified the specific wavelength for phenolics acids and between 300-350 are found flavonoids, quinines, coumarins [22-24].

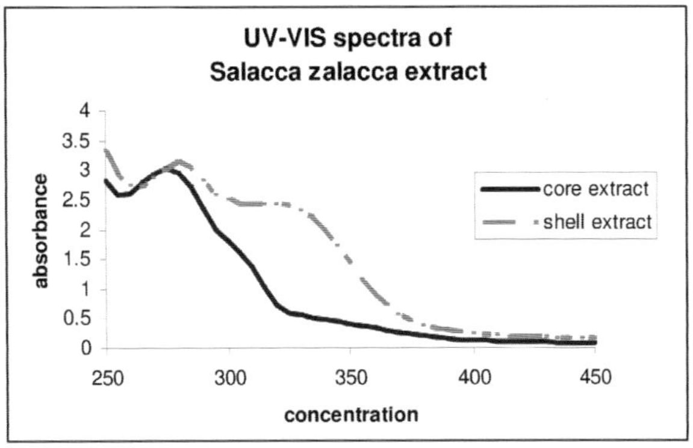

Figure 3. The UV–VIS spectra of Snake fruit extract.

3.2. FTIR results

The infrared spectral analysis was carried out to characterized the type of functional group existent in snake fruit parts (shell and core). FTIR spectra of components from snake fruit is used to observe the type of functional groups contained in this type of fruit. The spectrum was measured at a wavenumber range between 200 - 4000 cm^{-1} and the results are expressed in figure 4 (a) and b)).

FTIR spectrum from fig.5 a), presented shell snake fruit extract from outside (ext) and inside (int). Are observed strong bands at 3225 cm^{-1} attributed to hydroxyl groups and the band is more pronounced for exterior part of shell snake fruit. The band from 2950 cm^{-1} is characteristic to groups CH. The bands C = C and C = O are found at 1590-1458 cm^{-1} region. The aromatic group of the amide of type I and II are observed in the region between 1390 and 1320 cm^{-1}. The CO groups, esters, hidroxiflavones, catechins and amides of type III are visible at 1220-1150 cm^{-1} regions, especially at exterior part of shell snake fruit, and the aliphatic amine functional groups are predominant in 1000-700 cm^{-1} area. The shell snake extract expose weak IR bands between 1140-1075 cm-1 specific for C\N stretching vibrations of aliphatic amines or C\O stretching vibrations of alcohols or phenols, which are found due to different types of phytoingredients present in the fruit extract (polyphenols, polysaccharides and proteins) [25-26].

Also, FTIR results for core snake fruit (fig. 4 b) presented carboxyl groups -C=O at 1647 cm^{-1} band and OH absorption emerging appeared at 3465 cm^{-1}. Peak at 1321 cm^{-1} is specific for -C-O of the alcohol. -CH group was observed at 2875 cm^{-1}. Carboxyl groups shows uptake -C = O appeared at 1500 cm^{-1} and OH absorption emerging as a broad band at 3200 cm^{-1}. Peak at 1200 cm^{-1} was the peak -C-O of the alcohol and peak at 3200 cm^{-1} it is attributed to –CH group. The core snake fruit extract FTIR spectrum show a weak band at 1716 cm^{-1} specific to amide I, which result from stretch vibration of carbonyl C-O from proteins, and a strong band at 1350 cm^{-1} specific to C\O group of polyols: catechins or hydroxyflavones.

The short two bands observed in both parts (shell and core) of snake fruit extract, at approximate 2870 cm^{-1} and 2980 cm^{-1} are characteristic for C\H asymmetric and symmetric stretching vibrations of saturated C (sp^3) [25-26].

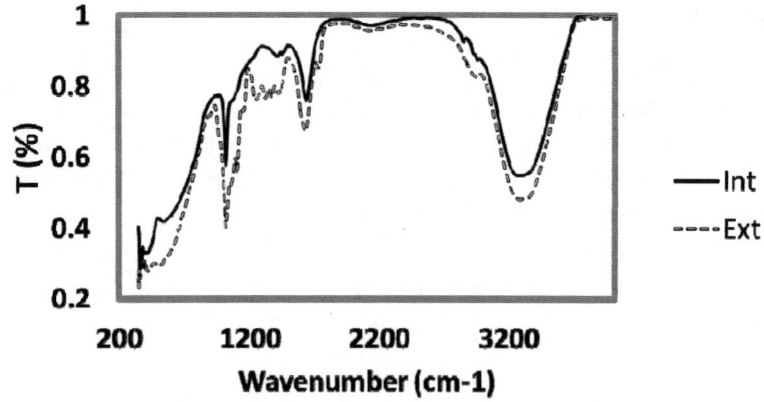

Figure 4.a). FTIR spectra of shell from Snake fruit.

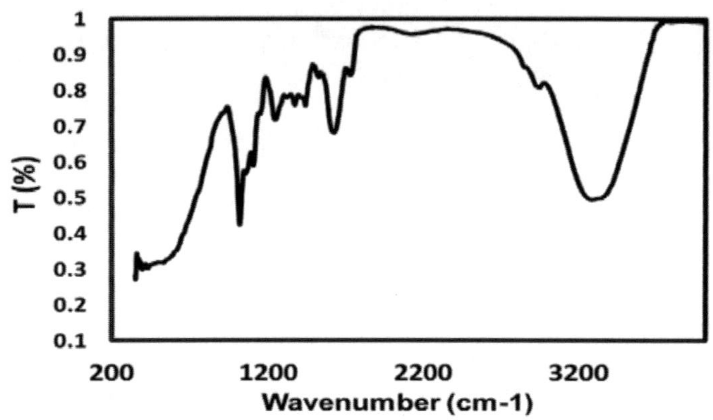

Figure 4.b). FTIR spectra of core from Snake fruit.

3.3. Antioxidant activity results

In table 2, are observed the results of antioxidant activity determination. Fruit represents a higher percentage of antioxidant activity against shell. Antioxidant activity of samples extracted from snake fruit, shows high values, which demonstrates a high content of vitamins, beneficial for body.

Table 2. Antioxidant activity of snake fruit

Sample	AA%
Snake fruit - core	82.675
Snake fruit - shell	73.135

3.4. Results of total flavonoids content

For preparation of the calibration curve, it was used Catechin as standard. It was prepared standard solutions of different concentrations: 20, 40, 60, 80 and 100 mg / g. The total flavonoids content (TFC) of the extracts was determined from the regression equation for the calibration curve ($y = 0.0003x + 0.0054$; $R^2 = 0.9988$).

The results of total flavonoid content are presented in table 3. It was observed a good flavonoid content in the both snake fruit samples (core and shell).

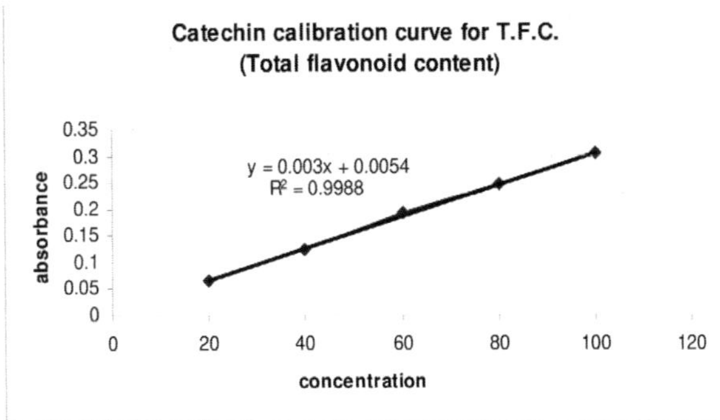

Figure 5. Catechin calibration curve.

Table 3. Total flavonoids content in snake fruit

Sample	T.F.C. mg/g Catechin
Snake fruit - core	129.8 ± 0.003
Snake fruit - shell	124.9 ± 0.004

3.5. Results of total tanins content

The amount of total condensed tannins is expressed as mg catechin/g DW. All samples were analyzed in triplicate and the results are presented in table 4. The total tanins content (TTC) of the extracts was determined from the regression equation for the calibration curve ($y = 0.0003x + 0.0054$; $R^2 = 0.9988$).

Table 4. Total tannins content in snake fruit

Sample	T.T.C. mg/g Catechin
Snake fruit - core	109.1 ± 0.0003
Snake fruit - shell	63 ± 0.0005

3.6. Results of total polyphenols content

For preparation of the calibration curve, it was used Gallic acid as standard. It was prepared standard solutions of different concentrations: 10, 20, 30, 40 and 50 mg / g.

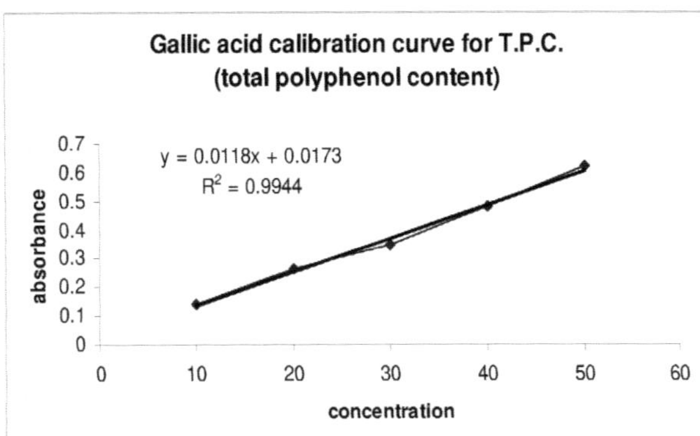

Figure 6. Gallic acid calibration curve.

The total polyphenols content (TPC) of the extracts was determined from the regression equation for the calibration curve ($y = 0.0118x + 0.0173$; $R^2 = 0.9944$).

The total phenolics content presented in table 5 and calculated from the calibration curve, were 1067,79 gallic acid equivalents/g and 946,61 gallic acid equivalents/g. Research articles detailed the fact that phenolic compounds from fruits have redox properties, which permit them to play as antioxidants [26].

Table 5. Total phenolics content in snake fruit

Sample	T.P.C. mg /g Gallic acid
Snake fruit - core	1067.796 ± 0.0002
Snake fruit - shell	946.610 ± 0.042

3.7. Results of total monoterpenoids content
For preparation of the calibration curve, it was used Linalool as standard. The total terpenoids content (TMtC) of the extracts was determined from the regression equation for the calibration curve ($y = 0.0139x +0.3939$; $R^2 = 0.9999$).

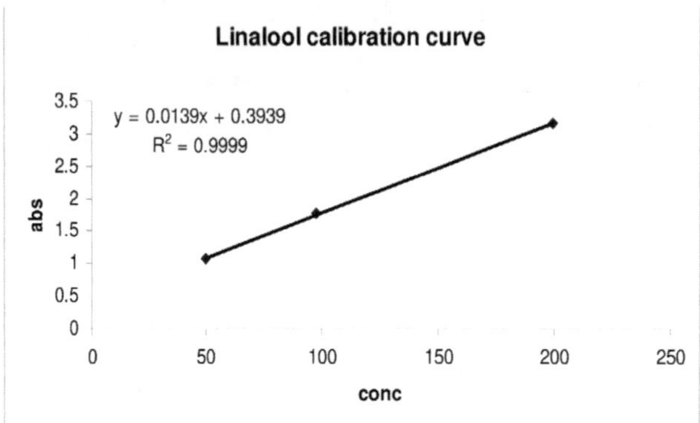

Figure 7. Linalool calibration curve.

The results of total terpenoids content are detaliated in table 6. The highest value was obtained for snake fruit core (223,89 mg/g), which is correct because in fruit are major terpenoids content, not in the core of this fruit.

Table 6. Total monoterpenoids content in snake fruit

Sample	T.Mt.C. mg /g Linalool
Snake fruit - core	223,892 ± 0.06
Snake fruit - shell	39,395 ± 0.0002

4. Conclusions

In the present study, it was determinated total flavonoids, polyphenols, monoterpenoids, tannins and antioxidant activity of a tropical fruit, snake fruit, brought from Indonesia. Also, the fruit compounds were characterizated using different types of analytical methods (FTIR, UV-VIS). The existence of phenolic compounds in the snake fruit was confirmed by the Folin-Ciocalteu method. ATR-FTIR results demonstrated the major amount of carbohydrates, aminoacids, proteins and of the phytoingredients and hydroxyl functional groups (polyphenols).

The antioxidant capacity was measured by the free radical scavenging methods DPPH. The methanolic solutions of the snake extract showed high antioxidant capacity (AA core = 82,67% and AA shell = 73,13 %). All results of phytochemical analyses were made in triplicate and calculated using results of calibration curves, with very good regression indices. In conclusion, the phytochemical estimation of the components of snake fruit, the results in this study indicate that snake fruit contain certain amounts of polyphenols, flavonoids, tannins and monoterpenoids, proving them to be perfect sources of antioxidants.

Acknowledgments
This work was supported by a grant of the Romanian National Authority for Scientific Research, CNDI-UEFISCDI, project number PN II 185/2014.

References

[1] S. Southon, 2000 *Food Res. Int.*, 33, pp. 211–217.
[2] P. Saiko, A. Szakmary, W. Jaeger and T. Szekeres, 2007 *Mutat. Res.*, 658, pp. 68-94.
[3] M. Leontowicz, H. Leontowicz, J. Drzewiecki, Z. Jastrzebski, R. Haruenkit, S. Poovarodom, Y. S. Park, S. T. Jung, S. G. Kang, S. Trakhtenberg and S. Gorinstein, 2007 *Food Chem.*, 102 pp. 192–200.
[4] M. Oktay, I. Gülçin and O. Küfrevioglu, 2003 *Lebensm.Wissen Technol.*, 36, pp.263-271.
[5] M. Karsheva, E. Kirova, S. Alexandrova and S. Georgieva, 2013 *J. Chem. Tech. Metall.*, 48, pp. 475-478
[6] A. E. Hegazy and M. I. Ibrahium, 2012 *World Appl, Sci, J*, 18, pp. 684-688.
[7] A. Duda-Chodak and T. Tarko, 2007 *Acta Sci. Pol. Technol. Aliment.*, 6, pp. 29-36.
[8] R. Gyawali and K. Su Kim. 2014 *J. Appl. Anim. Res.*, 4, pp. 85-95.
[9] S. Gorinstein, R. Haruenkit, S. Poovarodom, Y. S. Park, S. Vearasilp, M. Suhaj, K. S. Ham, B. G. Heo, J. Y. Cho and H. Gi Jang, 2009 *Food Chem.Toxicol*, 47, pp. 1884–1891.
[10] L. H. A. Priyatno, E. Y. Sukandar, S. Ibrahim and I. K. Adnyana, 2007 *J. App. Sci.*, 7, pp. 3127-3130.
[11] G. F. Deng, C. Shen, X. R. Xu, R. D. Kuang, Y. J. Guo, L. S. Zeng, L. L. Gao, X. Lin, J. F. Xie, E. Q. Xia, S. Li, S. Wu, F. Chen, W. H. Ling and H. B. Li, 2012 *Int. J. Mol. Sci.*, 13, pp. 8308-8323.
[12] E. Munaf, F. Hayuni, R. Zein and H. Suyani, 2014 *Res. J. Pharma. Biol. Chem. Sci.*, 5, pp. 1535-1543.
[13] T. Budiyanti, S. Hadiati, R. Prihatini and Sobir, 2015 *Int. J. Adv. Sci. Engin. Info. Technol.*, 5, pp. 4.
[14] A. Rebaya, S. I. Belghith, B. Baghdikian, V. Mahiou Leddet, F. Mabrouki, E. Olivier, J. K. Cherif and M. T. Ayadi, 2014 *J App Pharm Sci*, 5, pp. 052-057.
[15] F. Medini, H. Fellah, R. Ksouri and C. Abdelly, 2014 *J. Taibah. Univ. Sci.*, 8, pp. 216–224.
[16] S. A. Baba and S. A. Malik, 2015 *J. Taibah. Univ. Sci.*, 9, pp. 449–454.
[17] S. Kaur and P. Mondal, 2014 *J. Microbiol. Experiment,* 1, pp. 6.
[18] International Standard ISO 14502-1, Determination of substances characteristic of green and black tea — Part 1: Content of total polyphenols in tea — Colorimetric method using Folin-Ciocalteu reagent, 2005-03-01.
[19] C. Indumathi, G. Durgadevi, S. Nithyavani and P. K. Gayathri, 2014 *Int J ChemTech Res*, 6, pp. 4264-4267.
[20] I. R. Bunghez, O. Dumitrescu, R. Somoghi, I. Ionita and R. M. Ion, 2015 *Rev. Chim.*, 66, pp. 112-115.
[21] I. R. Bunghez, R. C. Fierascu, O. Dumitrescu, I. Fierascu and R. M. Ion, 2015 *Rev. Roum. Chim.*, 60, pp. 515-519.
[22] H. Zhu, Y. Wang, Y. Liu, T. Xia Yand Tang. 2010 *Food Anal. Method.*, 3, pp. 90–97.
[23] D. B. Rodriguez-Amaya, 1993 *Chemical, biological, physical and nutritional aspects* (Amsterdam, Elsevier) pp. 547-589.
[24] F. Bunghez, C. Socaciu, F. Zăgrean, R. M. Pop, F. Ranga and F. Romanciuc, 2013 *Bull UASVM Food Sci. Technol.*, 70, pp. 16-24.
[25] M. E. Barbinta-Patrascu, I. R. Bunghez, S. M. Iordache, N. Badea, R. C. Fierascu and R. M. Ion, 2013 *J. Nanosci. Nanotechnol.*, 13, pp. 2051-2060.
[26] I. R. Bunghez, M. E. Barbinta-Patrascu, N. Badea, S. M. Doncea, A. Popescu and R. M. Ion, 2012 *J. Optoel. Adv. Mat.*, 14, pp. 1016-1022.

Evolution of geopolymer binders: a review

M F Nuruddin[1*], A B Malkawi[1,2], A Fauzi[1,3], B S Mohammed[1], H M Almattarneh[2]

[1]Civil and Environmental Engineering Department, Universiti Teknologi PETRONAS, 32610 Seri Iskandar, Perak, Malaysia
[2]Civil Engineering Department, Najran University, 11001 Najran, Saudi Arabia
[3]Civil Engineering Department, Politeknik Negeri Lhokseumawe, 24301 Lhokseumawe, Aceh, Indonesia

E-mail: fadhilnuruddin@petronas.com.my

Abstract. This study aimed to present the current state of research about the terminology, chemical reactions, mechanisms, and microstructure modelling of geopolymer binders. Modelling the structure of the geopolymerization products is essential for controlling the product properties. The currently available models have shown some limitations in determining the rate of geopolymerization and setting time of the gel. There is a need for deeper knowledge regarding the physicochemical analysis of geopolymer binders. Most of the available models have used pure material like metakaolin; however, the less pure materials are expected to have different mechanisms. The FTIR and MAS-NMR analysis are considered as effective tools in providing information on the molecular deviations during geopolymerization. However, XRD analysis is not effective because most of the changes take place in amorphous phases. Also, the role of the iron oxides and some of the other impurities still not clear where none of the previous method of investigation can be used to detect the molecular changes of the iron compounds. This issue is very relevant hence the iron oxides are existed in substantial amounts in most of the waste materials that are suitable to be used as geopolymer source materials.

1. Introduction

The concrete industry is critical to the world economy with a share of 3.97US$ trillion of the global economy [1]. This industry, mainly based on the cement as a binding material with a worldwide production of 3.57 trillion tons [2]. Thus, concrete is next to water of the most used commodity worldwide. The huge consumption of concrete and cement have significant effects on the global environment; this is because of the enormous amount of energy needed for the quarrying and calcination of clinker, which is basically generated by fuel combustion. The cement industry is considered to be responsible for 5-8% of the global CO_2 emissions [3]. The global warming issues and the increasing demand in addition to the sustainability issues, create an urgent need to search for new materials.

Geopolymer binders are green materials that have the potential to replace cement in the construction industry. Geopolymer binders are based on the utilization of waste materials such as fly ash (FA), granulated blast furnace slag (GBFS), rice husk ash and red mud. In fact, any material that mainly consists of aluminosilicate can be used as a base material for geopolymer binder production.

* Tel.: +60-133641734; fax: +60 5-368 4088. E-mail: fadhilnuruddin@petronas.com.my

The use of such binder can reduce the carbon dioxide emissions by 80-90% compared to ordinary Portland cement (OPC) cement [4]. Geopolymer binders, display a wide range of properties that are superior to the cement and can be used to produce concrete with superior properties in term of strength and durability [5].

Implementation of geopolymers in the construction industry is still limited, currently, it is used by a few concrete companies mostly in Australia such as WAGNERS and Zeobond. The main hurdles facing the widespread of this material may include; the shortage of the information regarding its long-term behavior, in addition to the absence of the standards that control the use of such new technology. The markets will remain unwilling to take such a risk until the regulations of the carbon tax become actually implemented.

2. Geopolymer binders

Geopolymer binders are inorganic materials that can develop binding characteristics formed by the alkali activation of an aluminosilicate precursor. This binder buildup of a three-dimensional network, in which the aluminate and silicate species are covalently bonded by co-sharing of oxygen atoms. Elevated temperature sometimes is required to activate the aluminosilicate precursor particles; however, from a terminological point of view, this material should be hardened at the ambient temperature so we can call it geopolymer cement, else, it should be called geopolymer binders [6].

Geopolymer cement considered as a good alternative to OPC in the construction applications, pavement infrastructure applications, and offshore applications [7]. Geopolymer cement production requires mainly two components; the first is the source material, which can be any material that mainly composed of the alumina and silicate species. The second is the alkali-activating solution, which may be composed of any alkali silicate, hydroxide, carbonate, sulfate and thereof mixture. Sodium silicate or potassium silicate solutions are the most used solutions [8]. It is worth mentioned that the alkali solution should be user-friendly, and can be easily used in the field and mass applications. Using of strong alkali solution is unacceptable for geopolymer cement, and this can be user-hostile, and requires a special preparation for using in site and for the mass applications.

It should be differentiated between alkali-activated materials (AAM) and geopolymers. Many researchers got confused between them. Geopolymers are not alkali activated materials nor geopolymerization is the same as the alkali activation [9]. Geopolymer cement is based on the use of friendly alkali solution, which is usually sodium, or potassium silicates dissolved in water and can be produced at ambient temperatures. However, alkali-activated materials usually use strong activators and requires heat curing. Geopolymerization process should produce a stable durable product; however, alkali activation usually produces unstable products, where the alkali remains free and out of the structure, and can be easily leached out.

2.1. Geopolymer formation mechanism

Geopolymer binder has different formation mechanism compared to the known OPC reactions. The reaction processes in geopolymer binders are known as the geopolymerization reactions which mainly involve three processes; the first process is the dissolution processes where the alkali solution dissolves the aluminosilicate precursors to liberate the aluminum and silicon species in the mixture. These dissolved species, hydrolyze and deprotonated to produce the aluminate and silicate monomers $[Al(OH)_4]^-$ and $[Si(OH)_4]^-$, respectively. The second process is the condensation processes, where the aluminate and the silicate monomers adjoining together by co-sharing of the oxygen atoms to produce oligomers and form larger networks. During this process, the water that was consumed during the hydrolysis processes expelled out of the structure. The last stage occurs when the mixture become oversaturated with the aluminosilicate gel (which is initially rich with Al bonds). More silicates are dissolved to the mixture, where the rearrangement processes occur and more silicates are introduced to the aluminosilicate gel. This increases the connectivity of the network and the geopolymer gel starts to harden [10, 11]. These processes are described in Figure 1(a). Geopolymerization processes are controlled by several factors that mainly specify the dissolution rate. The main factors that affect this rate include; the amount of reactive phase in the source material, source material particle size, the alkalinity of the mixture, curing temperature, and the amount of the liquefied silicate that may be provided by the alkali solution [12, 13].

Fernández-Jiménez et al. [14] gave more details in their model for the FA geopolymerization. This model has assumed that most of the FA particles are composed of hollow spheres that have a crust housed additional smaller FA particles inside of it. They assumed that the alkali solution starts to dissolute the FA particle at one point, where the particle crust eroded forming a hole and then the solution enter inside the particle and the reactions start in the both directions from outside and inside also the inner particles start to react. The reaction products start filling the space and deposed on the surface of the unreacted FA particles, which prevents further reactions and the processes slowdown and become controlled by the diffusion process.

Provis [11] has proposed a detailed model for the geopolymerization of metakaolin (MK) and FA. This model includes more details regarding the silicate oligomers, which might be incorporated into the alkali solution. These oligomers, have been divided into two phases; amorphous phase and nanocrystalline phases, depending on the Si/Al ratio. Also, another reaction pathway was introduced for the direct formation of the aluminosilicate gel in addition to the zeolite phase. This sequence of geopolymerization reactions is shown in Figure 1(b).

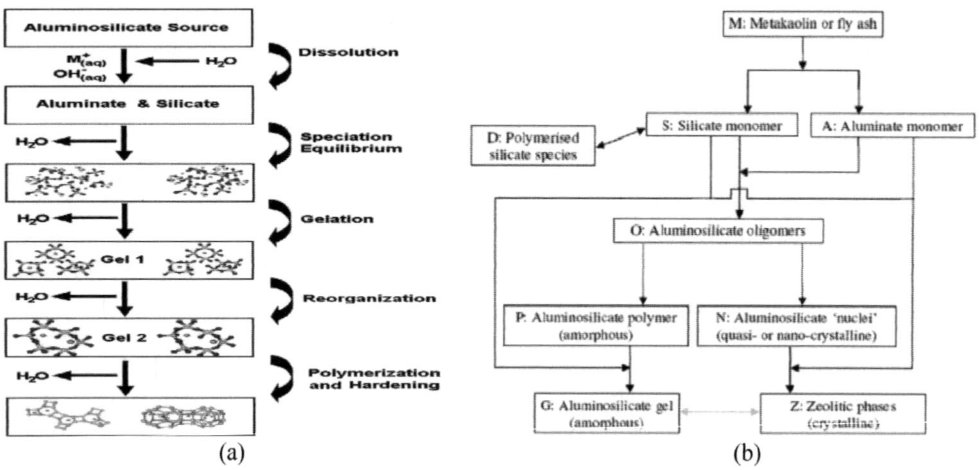

Figure 1. Geopolymerization reaction process (a) Duxson's model [10], (b) Provis's model [11].

2.2. Geopolymer chemical structure

The chemical structure of geopolymers is mainly described as a phase of aluminosilicate gel, which composed of covalently bonded aluminate and silicate 3D-tetrahedral units that sharing oxygen atoms. The negative charge of the aluminate groups is balanced by joining the alkali cations [15].

According to Davidovits [4], geopolymers has been described by the structural formula Mn[-$(SiO_2)_z$-$AlO_2)_n$.wH_2O where M represents any alkali cations, z is 1, 2, 3, w is the number of water molecules, and n represents the degree of polycondensation. Based on the Si/Al ratio, three different types of monomeric -Si-O-Al- structure are believed to define geopolymers [16]. The basic monomer unit is the poly-sialate unit, where Si/Al=1 [–Si–O–Al–O]. When the Si/Al=2 it is called poly(sialate-siloxo) which contains two Si atoms [–Si–O–Al–O–Si–O–], and when the Si/Al=3 it contains three Si atoms [–Si–O–Al–O–Si–O– Si–O–].

Many attempts were tried to draw the 3D geopolymer structure schematic diagrams. This type of efforts will probably be apprehensive with difficulty, because of the disordered nature of the geopolymer along with the difficulty involving perfectly representing a 3D framework in 2D representation. Figure 2(a) shows Davidovits's model [4] for geopolymer structure based on MK as a source material using the poly(sialate-siloxo) monomers as a base unit. This model has assumed a bulk polymer, which was similar to the organic polymers. However, this model did not account for the presence of water. Barbosa et al., [17] have proposed a new model. They have assumed that the geopolymers are similar to vitreous bodies like aluminosilicate glass. Both geopolymers and aluminosilicate glasses have similar 3D structure organization, nonetheless, aluminosilicate glasses structures do not contain any water and it exists without pores. A representative model of the proposed structure is shown in Figure 2(b).

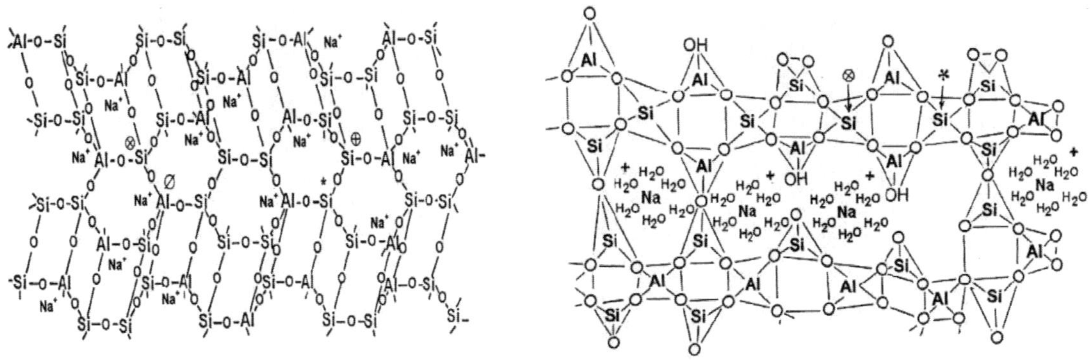

(a) (b)

Figure. 2. Geopolymer structure (a) Davitovits's model [4], (b) Barbosa et al.'s, model [17].

3. Geopolymers microstructure

Geopolymer microstructure is mainly composed of three phases that include; the geopolymer aluminosilicates gel phase, which is considered as the main phase of geopolymers, the unreacted particles and incompletely reacted particles phase, and the pore water phase [4, 11]. Figure 3 (A-D) shows the scanning electron microscopy (SEM) for FA based geopolymer. These micrographs show that the main composition of the FA based geopolymer is the aluminosilicate gel (points 4 and 5) also, we can see big amounts of unreacted and incompletely reacted rounded FA particles (points 2 and 3, respectively). The fly ash morphology seems to consist of spherical particles of different size in the range from 10 to 200 micrometer, most of these spheres are solid and also hollow spheres are existed (point 3 in Figure3a). This figure also shows the creation of new crystalline phase, which is similar to the zeolite crystals and this phase may be considered as the last stage of the geopolymerization processes. However, some of these crystals initially existed as a composition of the source materials, which may originally contain quartz and hematite as a crystalline phase; these crystals are represented by the points 6, 7, and 8 in Figure 3(B&D).

Figure 3(C) shows some of the FA particles that are incompletely covered by the reaction products, which indicates the incomplete or weaker geopolymerization processes. This figure also presents the variation in the FA reactivity, which suggests the variability in the properties of the FA particles or the partial or inadequate exposure of the FA particles to the geopolymerization reactions. Hence, it will be difficult to predict the degree of geopolymerization in the case of using the FA as a geopolymer source material.

Comparing to MK-based geopolymer, the MK-based geopolymer displays a more homogenous microstructure with less amount of the unreacted or incompletely reacted phase. This may refer to the higher reactivity of the MK material. In contrary, FA-based geopolymer displays a less homogenous microstructure, where a large amount of the unreacted particles are present. These particles are joining together by the reaction products [18]. Figure 3(E) and Figure 3(F) display the microstructure of MK and FA-based geopolymers, respectively. It should be stressed that the reaction degree is mainly reliant on the morphology of the source materials. The MK particles composed of alumina and silica in the form of layered sheets. Compared to the FA particles which are spherical in shape, the dissolution reaction of the MK particles will peel off the surface layers; while, for the rounded FA particles the dissolution reaction products will deposit on the outer surface. This will expose additional layers of the MK particles to the reactions and the reaction will continue; however, for the FA particles the precipitated products will encapsulate the surface of the particles and prevent further dissolution where the reaction will become diffusion controlled and this will slow down the reaction rates and more particles will remain unreacted or partially reacted.

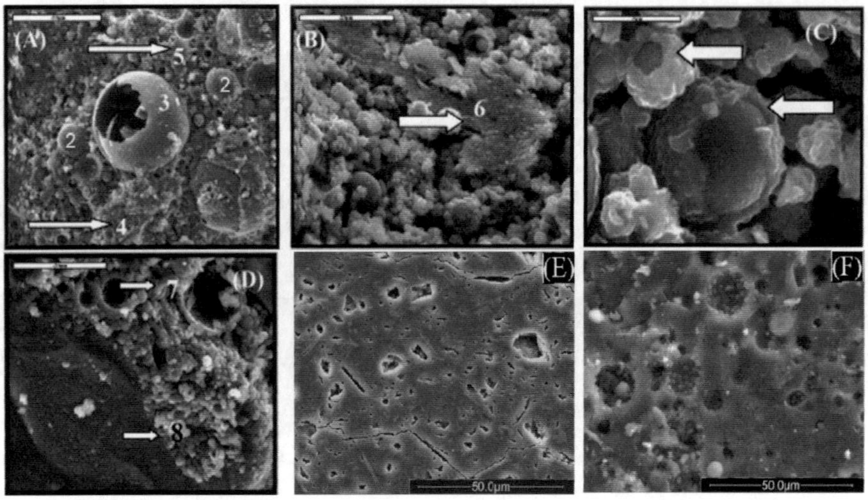

Figure 3: SEM micrographs of FA and MK-based geopolymers [11].

4. Geopolymer scientific means of investigation

The high resolution nuclear magnetic resonance (NMR) is considered as one of the powerful tools that may provide useful structural data. The NMR analysis of the geopolymer products ensures the Si(IV) and Al (IV) occurrence in geopolymers. The ^{27}Al MAS NMR analysis of MK-based geopolymers shows the presence of Al (IV) (around 60 ppm), the Al(VI) and Al(V) phases that was originally available in the MK particles has converted to the Al(IV) sites and the alkali cations has become connected to balance the charge. However, the Al(VI) (around 0 ppm) has detected in small amounts which indicate the presence of unreacted metakaolin particles but not the formation of new Al(VI) as a reaction product. The ^{29}Si MAS NMR analysis of MK shows a broad resonance in the range 80-100 ppm and this depends on the Si/Al ratio, however, this resonance mainly associated with the presence of Si(IV) and the presence of silicate in amorphous structure. Nonetheless, the use of the deconvolution method has allowed relating the featureless silicate resonance to the Q^4(mAl) centers that contain the aluminosilicate framework [19, 20]. Similar results were reported in [21] for the ^{29}Si MAS NMR of the FA-based geopolymers which showed main shifts of -97 ppm and -89 ppm that was related to the Si(IV)(2-3Al) and Al(IV)(4Si), respectively. The shift of -107 ppm which related to Si(IV)(0Al) that form the main shift for the unreacted FA powder was less represented in the FA-based geopolymer, which indicates the penetration of the Al into the Si(IV) skeleton.

Comparing the MAS NMR of the ^{27}Al and ^{29}Si spectra for MK powder and MK-based geopolymer with the ^{27}Al and ^{29}Si spectra for fly ash and fly ash based geopolymer that are shown in Figure 4, we can see that the main difference between the unreacted materials is that the fly ash contains higher amount of the Al(IV) however, after reaction this wide peak become sharper and have higher structural order. For the ^{29}Si spectra both metakaolin and fly ash geopolymers showed a broad resonance in the range 80-100 ppm, however, the sharper peaks were noted for the fly ash geopolymers which may be related to the presence of the crystalline originally provided by the unreacted materials like quartz [10, 23].

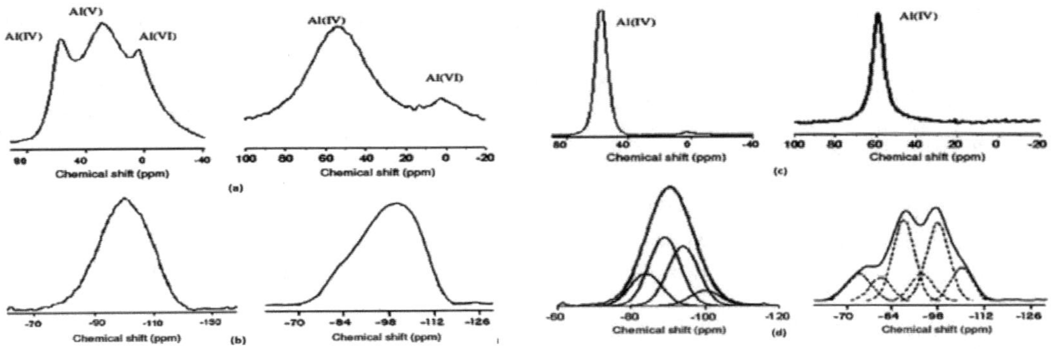

Figure 4. Spectra of (a) ^{27}Al NMR of MK (left) and MK-based geopolymer (right), (b) ^{29}Si NMR of MK (left) and MK-based geopolymer (right), (c) ^{27}Al NMR of FA (left) and FA-based geopolymer (right), (d) ^{29}Si NMR of FA (left) and FA-based geopolymer (right) [10].

Powder X-ray diffraction (XRD) analysis depends on the diffraction angle which gives an indication about the interlayer spacing of the crystal structure. Since the similarity of diffraction patterns indicates the similarity of the crystal structure. Many researchers reported that the main structure of geopolymers is described to be amorphous phase and their X-ray diffractograms have main hunched around 27-29° 2θ and this description looks to be true in spite of the type of the source material, curing conditions, and alkali activation solution type [23, 24].

Comparing the XRD diffractograms for geopolymer binder produced based on low calcium fly ash with that for the original fly ash materials, it can be seen that the geopolymerization processes have insignificant effects on the crystalline phase originally available in the fly ash (quartz, mullite, magnetite, etc.). However, the peaks observed in the geopolymer binder diffractograms seems to sharpen and slightly shifted to higher values over the original fly ash; also, the proportion of the amorphous phases have been increased [14, 25]. For shorter curing duration (4 hours) and in geopolymer binder cured at ambient temperature, the detection of the newly crystalline phase may refer to the formation of Thermonatrite ($Na_2CO_3.H_2O$) crystals [25]. However, this phase has disappeared when the samples cured for a longer duration (24 hours). This may attribute to the lower reaction rate which caused a higher concentration of remaining sodium hydroxide solution. The remaining sodium hydroxide will react with the carbon dioxide from the atmosphere which has resulted in forming of these new crystals.

A typical XRD micrograph is shown in Figure 5, this figure display the XRD patterns for unreacted fly ash and for geopolymer at different temperature, this figure displays several low and broad diffraction bands which indicate the presence of amorphous phases that may be combinations of the newly formed geopolymer gel and the remaining unreacted fly ash amorphous oxides.

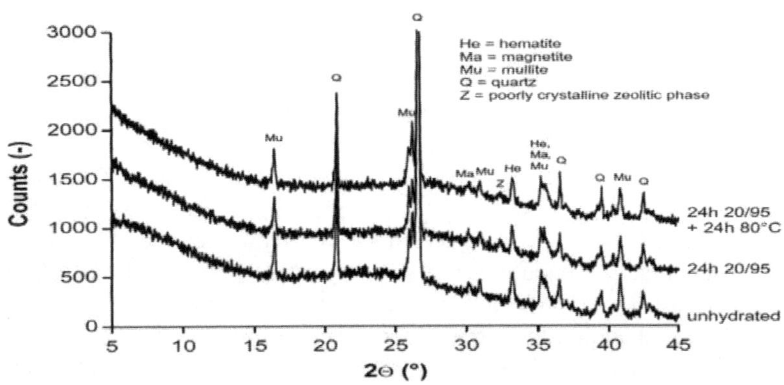

Figure 5: X-ray analysis of FA and FA-based geopolymer cured at different temperature [26].

IOP Conf. Series: Materials Science and Engineering **133** (2016) 012052 doi:10.1088/1757-899X/133/1/012052

The Fourier transform infrared spectroscopy (FTIR) analysis also provides a good evaluation of the geopolymerization processes and the reaction products formation. The infrared spectrum of the FA powder shows main bands in the range 1080-1090 cm^{-1} which related to the Si-O and Al-O bonds vibration. This range shifts to lower values after the geopolymerization reactions usually in the range at 900-1000 cm^{-1} [21, 27]. This shift is related to the penetration of Al in the network of the Si-O-Si. The higher shift indicates the higher penetration into the Si(IV) network. Other bands may include bands in the range 450-470 cm^{-1} corresponds to the Si-O bonds vibration and the band 797 cm^{-1} corresponds to the Al-O bonds vibration. After geopolymerization reactions, it was noted that the band at 797 cm^{-1} was almost disappeared; this may be related to the consumption of the Al(IV). From the other hand, some of the new bands were reported in the range 3440-3450 cm^{-1} corresponds to the formation of H-O bonds, 1650 cm^{-1} corresponds to the formation of H-OH bonds, and in the range of 1450-1500 cm^{-1} corresponds to the formation of C-O bond as reaction products. The H-O and H-OH bonds may indicate the formation of water and the C-O bonds may indicate the formation of alkali carbonates as a result of the reaction with the carbon dioxide from the atmosphere [28, 29]. A typical FTIR spectrum is shown in Figure 6.

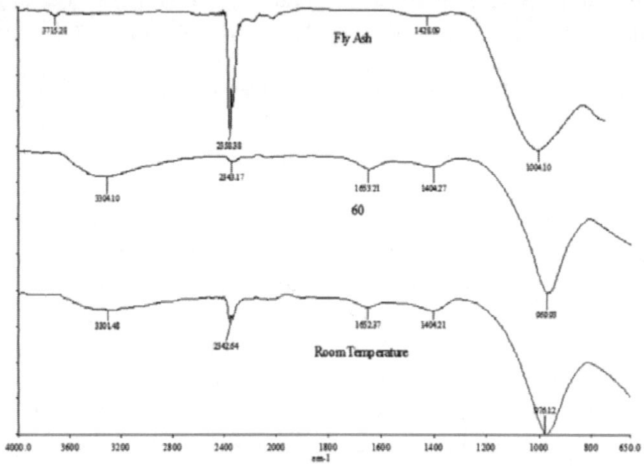

Figure 6: X-ray analysis of FA and FA-based geopolymer cured at different temperature [30].

5. Conclusions

Geopolymer cement is considered as a green building material, which can replace OPC in the concrete industry. Use of geopolymer binders can produce concretes with a wide range of properties. Until now, the geopolymerization model and mechanism are still not clearly understood. More research is needed to understand and control the properties of the produced geopolymers. Geopolymers and AAMs are different; geopolymerization reactions are known to produce a stable chemical structure, where the alkali cations are rigidly joined and not vulnerable to leaching. The geopolymer gel is composed of nanoparticles separated by nanopores, which produce a very tiny microporosity matrix that promotes the superior properties of geopolymers. The rounded shape of the FA particles plays an important role in increasing the mixture workability without additional amount of water, which reduce the product permeability. Also, the unreacted FA particles play the role of micro-aggregates, which increase the amounts of aggregates in the mixture and give higher stability. The ^{29}Si and ^{27}Al NMR spectroscopy is considered as an effective technique for rendering the data about the structural skeleton of geopolymer binders. Geopolymers are mainly structures of Al(IV)(4Si), Si(IV)(3Al), and Si(IV)(2-3Al) coordination. The Si(IV)(0Al) coordination that originally exist in the aluminosilicate precursor has a less representation after geopolymerization reactions, which confirms the penetration of the Al into the Si(IV) skeleton. The FTIR analysis shows the presence of water in the form of free water and it exists within the gel structure.

References

[1] United Nations: 2015 UN national accounts main aggregates database, http://unstats.un.org/unsd/snaama/dnllist.asp.
[2] U.S. Geological Survey: Mineral commodity summaries 2015: U.S. *geological gurvey*, pp. 196, http://dx.doi.org/10.3133/70140094.
[3] E. Benhelal, G. Zahedi, E. Shamsaei & A. Bahadori, 2013 *Journal of cleaner production*, vol. 51, pp. 142-161.
[4] J. Davidovits, 2011 *Geopolymer chemistry and applications*. Saint-Quentin, France.
[5] J. Davidovits, 2013 *Technical paper#8*, Geopolymer Institute, pp. 1-7.
[6] H. K. Tchakoute, C. H. Rüscher, J. N. Y. Djobo, B. B. D. Kenne & D. Njopwouo, 2015 *Applied clay science journal*, vol. 107, pp. 188-194.
[7] J. Davidovits, M. Izquierdo, X. Querol, D. Antennuci, H. Nugteren, V. Butselaar-Orthlieb, Y. Luna, 2014 *Technical paper # 22*. Geopolymer Institute, pp. 1-11.
[8] D. Hardjito, C. C. Cheak & C. H. L. Ing, 2008 *Modern applied science*, vol. 2(4), pp. 3.
[9] J. Davidovits, 2014 *geopolymer camp 2014*, Saint Quentin, France.
[10] P. Duxson, A. Fernández-Jiménez, J. L. Provis, G. C. Lukey, A. Palomo & J. S. J. van Deventer, 2007 Journal of materials and scince., vol. 42(9), pp. 917-933.
[11] J. L. Provis, 2006 *Modelling the formation of geopolymers. (PhD)*, University of Melbourne.
[12] A. A. Adam & X. X. X Horianto, 2014 *Procedia engineering*, vol. 95, pp. 410-414.
[13] J. L. Provis, P. Duxson, E. Kavalerova, P. V. Krivenko, Z. Pan, F. Puertas & J. S. J. van Deventer, 2014 *Alkali activated materials*, vol. 13, pp. 11-57.
[14] A. Fernández-Jiménez, A. Palomo & M. Criado, 2005 *Cement and concrete research*, vol. 35, pp. 1204-1209.
[15] M. Kanuchova, L. Kozakova, M. Drabova, M. Sisol, A. Estokova, J. Kanuch & J. Skvarla, 2015 *Environmental progress & sustainable energy*, vol. 34(3), pp. 841-849.
[16] F. Škvára, 2007 *International conference alkali activated materials–research, production and utilization*, Česká rozvojová agentura, Praha.
[17] V. F. Barbosa, K. J. MacKenzie & C. Thaumaturgo, 2000 *International journal of inorganic materials*, vol. 2(4), pp. 309-317.
[18] Sindhunata, 2006 *A conceptual model of geopolymerisation. (PhD)*, University of Melbourne.
[19] A. Favier, G. Habert, N. Roussel & J. B. d'Espinose de Lacaillerie, 2015 *Cement and concrete research*, vol. 75, pp. 104-109.
[20] M. Frías, S. Martínez-Ramírez, T. Blasco, M. F. Rodríguez & D. Viehland, 2013 *Journal of the american ceramic society*, vol. 96(7), pp. 2306-2310.
[21] F. Škvára, L. Kopecký, J. Nemecek & Z. Bittnar, 2006 *Ceramics-Silikaty*, vol. 50, pp. 208-215.
[22] P. S. Singh, T. Bastow & M. Trigg, 2005 *Journal of materials science*, vol. 40(15), pp. 3951-3961, 2005.
[23] S. A. Bernal, J. L. Provis, V. Rose, R. M. de Gutiérrez & J. Biernacki, 2013 *Journal of the american ceramic society*, vol. 96(6), pp. 1951-1957.
[24] T. Yang, X. Yao & Z. Zhang, 2014 *Construction and building materials*, vol. 69, pp. 109-115.
[25] J. Xie & O. Kayali, 2014 *Construction and building materials*, vol. 67, pp. 20-28.
[26] F. Winnefeld, A. Leemann, M. Lucuk, P. Svoboda & M. Neuroth, 2010 *Construction and building materials*, vol. 24(6), pp. 1086-1093.
[27] C. A. Casárez, S. P. Arredondo, J. M. Gómez, J. L. Alamaral, R. Corral, M. d. Chinchillas, et al., 2014 *International journal of advances in computer science and its applications*, vol. 4, pp. 221-226.
[28] I. Cătănescu, M. Georgescu & A. Melinescu, 2012 *U.P.B. Sci. Bull.*, Vol. 74(1), pp. 3-14.
[29] A. Allahverdi & E. N. Kani, 2009 *International journal of civil engineerng.*, vol. 7, pp. 154-160.
[30] M. M. Al Bakri, K. Hussin, M. Bnhussain, K. N. Ismail, Z. Yahya, & R. A. Razak, 2012 *International journal of molecular sciences*, vol. 13, pp. 7186-98.

IOP Conf. Series: Materials Science and Engineering **133** (2016) 012053 doi:10.1088/1757-899X/133/1/012053

Green bio-oil extraction for oil crops

H Zainab[1], N Nurfatirah[2], A Norfaezah[2] and H Othman[1]
[1]Faculty of Engineering Technology (FETech),Department Chemical Engineering, Universiti Malaysia Perlis, Kampus UniCITI Alam, Sungai Chuchuh, Padang Besar, 02100 Perlis, Malaysia.
[2]School of Bioprocess Engineering, Universiti Malaysia Perlis, Kompleks Pengajian Jejawi 3,02600 Arau, Perlis, Malaysia

E-mail: zainab@unimap.edu.my

Abstract. The move towards a green bio-oil extraction technique is highlighted in this paper. The commonly practised organic solvent oil extraction technique could be replaced with a modified microwave extraction. Jatropha seeds (*Jatropha curcas*) were used to extract bio-oil. Clean samples were heated in an oven at 110°C for 24 hours to remove moisture content and ground to obtain particle size smaller than 500μm. Extraction was carried out at different extraction times 15 min, 30 min, 45 min, 60 min and 120 min to determine oil yield. The bio-oil yield obtained from microwave assisted extraction system at 90 minutes was 36% while that from soxhlet extraction for 6 hours was 42%. Bio-oil extracted using the microwave assisted extraction (MAE) system could enhance yield of bio-oil compared to soxhlet extraction. The MAE extraction system is rapid using only water as solvent which is a non-hazardous, environment-friendly technique compared to soxhlet extraction (SE) method using hexane as solvent. Thus, this is a green technique of bio-oil extraction using only water as extractant. Bio-oil extraction from the pyrolysis of empty fruit bunch (EFB), a biomass waste from oil palm crop, was enhanced using a biocatalyst derived from seashell waste. Oil yield for non-catalytic extraction was 43.8% while addition of seashell based biocatalyst was 44.6 %. Oil yield for non-catalytic extraction was 43.8% while with addition of seashell-based biocatalyst was 44.6%. The pH of bio-oil increased from 3.5 to 4.3. The viscosity of bio-oil obtained by catalytic means increased from 20.5 to 37.8 cP. A rapid and environment friendly extraction technique is preferable to enhance bio-oil yield. The microwave assisted approach is a green, rapid and environmental friendly extraction technique for the production of bio-oil bearing crops.

1. Introduction

Research towards a green, efficient and sustainable oil extraction technique and the conventional soxhlet extraction using hazardous chemicals was compared. Conventional extraction techniques are laborious, time consuming, involving large amounts of solvent, and ultimately, may cause some target molecule degradation and partial loss of volatiles. The use of a non-conventional technique such as microwave assisted extraction (MAE) using water as extractant and a green technique using seashell waste as biocatalyst was studied. The main advantages of green technology such as higher yield of bio-oil, reduction in extraction time and quality improvement of bio-oil are evident in this study. However each technique requires an optimised operating condition in order to obtain high yield and good quality of bio-oil. Efficient cost effective extraction is often not only a matter of technique and extraction conditions, in fact the type of equipment such as microwave and reactor can strongly affect oil yield [1].

MAE was used to obtain bio- oil from *Jatropha* crop and MAE is more favourable compared to Soxhlet extraction (SE). MAE does not consume organic solvent, low cost and has extremely shorter extraction time compared to SE. MAE showed shorter reaction time compared to SE with similar yield of 40 % and 48 % [2]. Also, the catalytic application for extraction using pyrolysis technique enhanced yield and quality of bio-oil. Rohim *et al.* [3] pyrolyzed empty fruit bunch (EFB) of oil palm in a fixed bed reactor with waste eggshell to determine the effect of amount of catalyst loading and pyrolysis conditions on the oil yield. The maximum bio-oil yield was 37 % at 500 °C of pyrolysis temperature and 10 % of catalyst loading.

Extraction of bio-oil from non-edible oil crop should be efficient and sustainable. Therefore, to maximize the desired product quality and quantity it is important to minimize costs and environmental concerns. This study is aimed to evaluate bio-oil extraction from *Jatropha curcas* and empty fruit bunch (EFB) using microwave assisted extraction and fixed bed reactor to obtain high yield and quality of bio-oil and to investigate the effect of the seashell waste biocatalyst on the bio-oil production.

2. Materials and Methods

2.1 Microwave assisted extraction (MAE)

2.1.1. Sample preparation. *Jatropha curcas* seeds were supplied by Agrotechnology Centre at University Malaysia Perlis situated north of Malaysia. Only fully ripened *Jatropha curcas* seeds were collected, cracked and the shells were removed. White and ripe seeds were selected while damaged seeds were discarded. The seeds were cleaned and dried at 105 °C for 30 minutes. Ground seeds were sieved through 500 μm sieves to obtain a uniform powder for the extraction process.

2.1.2. Experimental. 10 g *Jatropha* seed powder was placed in 500 mL round bottom flask and placed inside microwave cavity space. A condenser was attached to the round bottom flask. An open vessel microwave system operating at 700 watt maximum power with frequency 2450 MHz was used. The parameters studied were extraction time (60, 80, 100, 120, and 140 minutes), microwave power (200, 300, 400, 500, 600 and 700 watts) and amount of water (70, 100, 130, 160, 190, 210 mL). The extracted oil from MAE was separated from solid residue by vacuum filtration using Buchner filtration assembly using Whatman filter paper No. 1. The complete separation of oil from water was done in Rotavapor (BUCHI R-205) fitted with a controlled temperature oil bath (BUCHI B-490). Oil yield was calculated as follows:

$$\text{Oil yield (\%)} = m_o/m_s \times 100 \quad (1)$$

$m_{o(g)}$ = The mass of ground *Jatropha* seeds

$m_{s(g)}$ = The mass of oil yield

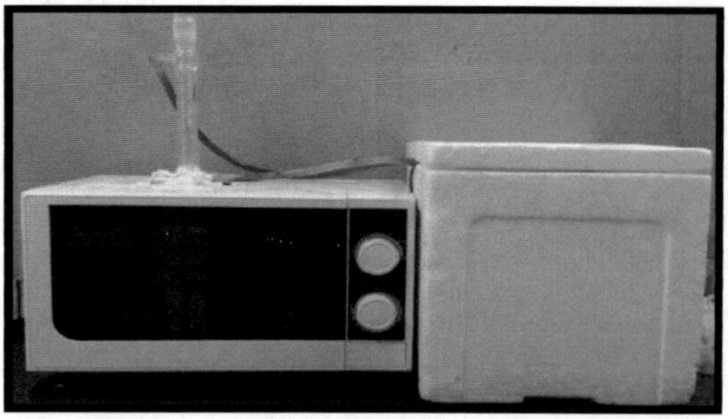

Figure 1. Microwave assisted extraction equipment

2.1.3. Physico-chemical analysis

2.1.3.1 Acid value. 2 g of oil was weighed into 250 mL conical flask. 20 mL of neutralize alcohol was mixed with the oil. The mixture was heated for 3 minutes using a water bath. The mixture was allowed to cool down and titrated with 0.1N alcoholic potassium hydroxide solution. Few drops of phenolphthalein were added as an indicator. A blank sample was titrated simultaneously.

2.1.3.2. Free fatty acid (FFA). FFA is the amount of sodium hydroxide or potassium hydroxide (in milligrams) needed to neutralize the free acids in one gram of sample. Usually free fatty acid (FFA) is expressed as oleic acid percentage according to equation below:

$$\% \text{ FFA as oleic acid (C18:1)} = \frac{28.2 \times N \times V}{W} \quad (2)$$

Where N = Normality of NaOH solution;
V = Volume of NaOH solution used in mL;
W= Weight of sample;

5 g of *Jatropha* oil sample was placed into 250 mL conical flask. 50 mL of neutralised solvent was mixed with the oil. The mixture of *Jatropha* oil sample and neutralised solvent was heated at 40 °C on the hot plate. The sample was titrated with 0.1 N standard NaOH solutions while continuously shaking the flask gently until the first permanent pink colour was observed and persisted for 30 seconds.

2.1.3.3. Iodine value. 0.2 g *Jatropha* oil was weighed into a glass vial. The vial was then transferred into 500 mL flask. 20 mL of cyclohexane was added into the glass vial. 25 mL of Wijs solution was added followed by insertion of stopper. The mixtures were gently shaken and kept in the dark for 1 hour. 100 mL of water and 20 mL of the potassium iodide solution was added into the mixture. It was titrated with the standard sodium thiosulphate solution ($Na_2S_2O_3$). Titration was considered completed when the yellow colour of the mixture disappeared. Subsequently, 2 mL of the starch indicator was added and the titration was prolonged while shaking the vial vigorously until the blue colour disappeared. The blank sample was also carried out simultaneously with the test samples.

2.1.3.4. Peroxide value. 2 g *Jatropha* oil was weighed and mixed with 30 mL acetic acid-chloroform (3:2 v/v) in conical flask. 0.5 mL of saturated potassium iodide was pipetted into flask. The flask was kept swirling for about 1 minute before addition of distilled water. The starch solution which was prepared was added to the solution drop wise.

Titration of the solution was done by adding 0.01 N sodium thiosulphate solutions ($Na_2S_2O_3$) slowly with continual and mild shaking. Vigorous shaking of the solution during titration when approaching the end-point to unfetter the remaining iodine from the chloroform layers. Carefully, $Na_2S_2O_3$ solution was added drop wise until the blue color vanished. Simultaneously, titration of the blank sample was similarly determined.

2.1.3.5. Saponification value. 2 g of Jatropha oil was weighed precisely into 250 mL round bottom flask. 25 mL of 0.5 N alcoholic potassium hydroxide solutions were added into the oil. The mixture was refluxed gently using water bath with condenser attachment for 1 hour. The mixture was allowed to cool for 30 minutes. The mixture was titrated with 0.5 N hydrochloric acid solution using phenolphthalein as indicator. Titration of blank sample was similarly determined.

2.1.3.6. Fatty acid composition. The determination of fatty acid composition was analyzed by conversion of oil to fatty acid methyl esters prepared by adding 950: l of n-hexane 50 mg of oil followed by 50: l of sodium methoxide. The mixtures were vortex for 5 s and allowed to settle for 5 min.The top layer (1: l) was injected into a gas chromatograph (Model GC-14A, Shimadzu Corporation, Kyoto, Japan) equipped with a flame-ionisation detector and a polar capillary column (BPX70 0.25), 0.32 mm internal diameter, 60 m length and 0.25: m film thickness (SGE Incorporated, USA) to obtain individual peaks of fatty acid methyl esters. The detector temperature was maintained at 240 ˚C and column temperature at 110 °C held for one minute and increased at the rate of 8 ˚C /min to 220 °C and held for one minute. The run time was 32 minutes. The fatty acid methyl ester peaks were identified by comparing their retention time with the standards. Relative content of fatty acid was calculated based on the peak area of a fatty acid species to the total peak area of all the fatty acids in the oil sample.

2.2 Fixed bed reactor

2.2.1. Sample preparation. The empty fruit bunch (EFB) was obtained from a palm oil mill located in Kedah, North Malaysia. The sample was cleaned and chopped into smaller sizes and dried at 105 ˚C for 24 hours in an oven. The sample was ground using a crusher and then sieved to obtain desired particle size.

Calcium oxide (CaO) used as catalyst in this study was collected from nearby seaside and washed with distilled water and dried in oven for 24 hours at 110 ˚C. The seashells were finely ground. The powdered sample was sieved using stainless steel test sieve. The calcium oxide was prepared by calcinations of the sea shells at a temperature of 800 ˚C in the furnace for 2 hours. The calcined sample was cooled in a desiccator, the CaO was again ground in grinder. The biocatalyst was stored in desiccators before analysis.

2.2.2. Experimental. Extraction of sample was carried out using a fixed bed reactor system. The components of the system were a bucket, furnace, nitrogen cylinder, condenser and thermocouple, to monitor reaction temperature and attached with the chiller/refrigerator bath for cooling. The sample was placed in the reactor bucket and the weight of the reactor tube with sample was recorded. The chiller/refrigerator bath temperature was set up at 0 ˚C to make sure the condenser was cooled. The reactions such as pyrolysis temperature and nitrogen flow rate were set up prior to the experiment. Nitrogen was allowed to purge for 30 minutes.

The yield obtained from the experiment was weighed. The mass of the reactor tube with biochar and glass wool were also weighed. After production of oil, the products obtained were weighed according to equations (3)-(5).

$$\text{Biochar yield (\%)} = (\text{char product (g)}) / (\text{EFB sample (g)}) \times 100\% \quad (3)$$

$$\text{Bio-oil yield (\%)} = (\text{liquid product (g)}) / (\text{EFB sample (g)}) \times 100\% \quad (4)$$

$$\text{Gas yield (\%)} = 100\% - (\text{Biochar yield (\%)} + \text{Bio oil yield (\%)}) \quad (5)$$

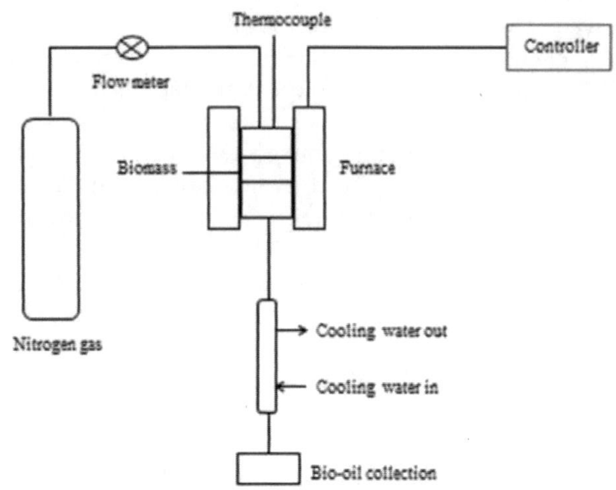

Figure 2. Schematic of reactor equipment

2.2.3. Catalytic pyrolysis. The experiment to study the effect of addition of catalyst on pyrolysis yields was conducted by mixing the EFB with different weight of the catalyst at 5, 10, 15, 20 wt. % maintained at optimum pyrolysis condition. The liquid product was collected and weighed. The bio-oil was obtained using rotary evaporator at a temperature of 58 ˚C in which the solvent was evaporated and removed. The solid biochar was collected and weighed. The syngas yield was calculated from the balance of overall material.

2.2.4. Characterization of bio-oil
2.2.4.1. pH. In order to evaluate the corrosive property of the bio-oil products, the pH of the bio-oil was measured using a pH meter. The electrode was directly dipped into bio-oil sample. The recorded value was expressed as a pH of bio-oil.

2.2.4.2. Viscosity. The viscosity of the pyrolytic oil was determined at 25˚C using rheometer (HAKEE Rheostress 1). Cone and plate type (Measuring Cup Z 43 (Series 1) and plate PP 35 Ti, D=35mm geometry was used to determine the viscosity where temperature was controlled accurately within 0.05˚C by HAAKE DC-50 temperature controller. A series of rheological data were collected and the average was recorded as viscosity of the oil.

3. Results and Discussion

3.1. Effect of extraction time on Jatropha oil yield
All the results obtained were presented in tables 1, 2 and 3 for extraction time, microwave power and amount of water used respectively. The extraction times tested were 60, 80, 100, 120, and 140 minutes

at 210:10 water to sample ratio (mL/g) and power of 700 watts. As extraction time increased from 60 minutes to 120 minutes, the oil yield also increased with the optimum yield at 39 %. From 120 minutes to 140 minutes, the yield obtained was stagnant which indicated the extraction reached equilibrium. Microwave energy work selectively and diffuse very well within polar molecules such as water. Water thus works well as an extraction solvent replacing hazardous organic solvents such as hexane and petroleum ether often used in extraction of oil. Microwave energy emitted as waves work well with water rather than hexane.

Energy emitted during extraction process help in rupturing the cell wall of the seed sample thus releasing the oil. The oil yield using microwave extraction was lower compared to yield obtained using Soxhlet apparatus which was 48.16 % [2]. However the yield obtained in this study could be considered as high when compared to the oil yields obtained using several extraction methods ranging from 20-60 % [4]. Several factors such as climate, storage and variety influence the oil yield of *Jatropha curcas*.

Table 1. Effect of extraction time on *Jatropha* oil yield.

Time (minutes)	Yield (%)
60	9.2
80	12.6
100	25.5
120	38.9
140	38.8

3.2. Effect of extraction power on Jatropha oil yield

Jatropha oil was extracted at various power (watt) ranging from 200, 300, 400, 500, 600 and 700 watts at 120 minutes extraction time and 210:10 g water to sample ratio (mL:g). The maximum power supply was 700 watts. The oil yield for each watt power used at 200, 300, 400, 500, 600 and 700 watt were 5.2, 11.3, 18.4, 28.6, 33.6, and 38.7 % respectively which showed an increased in yield as power was increased. The extraction technique using microwave technique lies in its ability to expose sample to heat caused by microwave field [5]. The heating ability by this technique is efficient and fast compared to conventional extraction method where sample was heated by conduction [6].

Table 2. Effect of extraction power on *Jatropha* oil yield.

Time (minutes)	Yield (%)
200	5.2
300	11.3
400	18.4
500	28.6
600	33.6
700	38.7

3.3 Effect of water volume on Jatropha oil yield

When extracting oil from *Jatropha* seeds, another important parameter to investigate was volume of water per sample. Weight of sample used was 10 g and extraction was carried out at a constant extraction time of 120 minutes and a power of 700 watts. The volume of water used was determined at 70, 100, 130, 160, 190 and 210 mL. There was an increase in yield from 70 mL to 160 mL and yield was stagnant from 160 to 210 mL of water insignificant differences between the values. Thus it can be concluded that volume of water of 190 mL gives highest yield of bio-oil. Water plays an important role to keep the sample moist. The dipole rotation of water when exposed to microwave energy will rotate quickly and produce heat. This process resulted in cell walls to expand and rupture thus releasing essential oil into the solvents. Sometimes, dry samples are rehydrated before extraction to encourage localized heating to take place. Moreover, moisture content is crucial in MAE because water superheats and promotes solute to be emancipated into solvents.

Table 3. Effect of water volume on *Jatropha* oil yield.

Water volume (ml)	Yield (%)
70	9.1
100	15.8
130	27.8
160	37.7
190	37.9
210	37.6

3.4. Physico-chemical characterization of bio-oil

Jatropha curcas oil was tested for its oil quality and fatty acid composition and the results obtained are as shown in table 4. The acid value, free fatty acid, iodine value, peroxide value and saponification value were 2.8 mg KOH.g^{-1}, 1.7, 102.1 mg/g, 2.23 m Eq/kg and 190.4 mg/g respectively. From this study the acid value was comparable to previous studies by several authors [7, 8, 9].

Table 4. Physicochemical properties of *Jatropha* oil yield.

Characteristic	Value
Acid value (mgKOH.g^{-1})	2.8
Free fatty acid	1.7
Iodine value(mg/g)	102.1
Peroxide value (mEq/kg)	2.2
Saponification value (mg/g)	190.4

The extracted oil showed great quality in terms of its acid value as it was lower than the permissable value which is 10. Acid value measures the existence of oxidation product that is corrosive when used in engines [10]. High unsaturation of fatty acid will give high iodine value. There are two types of oil which is drying and semidrying. Iodine value of oil higher than 125 mg/g will be categorized as drying while iodine value between 110–140 mg/g categorised as semidrying of oils. The saponification value was higher than that obtained from *Bassia longifolia* 185.5 mg KOH/g, Indonesia *Jatropha* oil 183.2 mg KOH/g and Indian *Jatropha* oil 156.2 mg KOH/g. The major composition of fatty acid was determined from gas chromatography (GC) technique and the unsaturated fatty acid content was 72.9 % followed by saturated fatty acid content of 24%. The unsaturated fatty acid comprised oleic acid 41.7 % and linoleic acid 30.8%. The saturated fatty acid was palmitic acid 17.1 % and stearic 6.9 %. The fatty acid compositions is as shown in table 5.

Table 5. Fatty acid composition of *Jatropha* oil yield.

Fatty acid	Value
Oleic acid	41.6
Linoleic acid	30.8
Palmitic acid	17.1
Stearic acid	6.9
Saturated Fatty Acid	24.0
Unsaturated Fatty Acid	72.9

3.5 Effect of calcined seashell catalyst in the pyrolysis process

In a series of pyrolysis, the experiment was determined at catalyst biomass loading of 0 wt. %, 5 wt. %, 10 wt. %, 15 wt. % and 20 wt. %, constant nitrogen flow rate of 250 cc/min, pyrolysis temperature of 638 °C, heating rate at 10 °C /min and particle size of 500-1000 μm. Figure 3 shows the effect of catalyst on oil yield. Bio-oil yield for non-catalytic extraction was 43.8 %. The optimum bio-oil yield was 44.5 % achieved at 5 % of catalyst loading. However, bio-oil yield was slightly reduced from 43.7 to 42.1 % with addition of 10 to 20 % catalyst. According to Rohim *et al.* [3] it could be due to the agglomeration of active CaO phase and the covering of basic sites by excess CaO, thus resulting in lower oil yield.

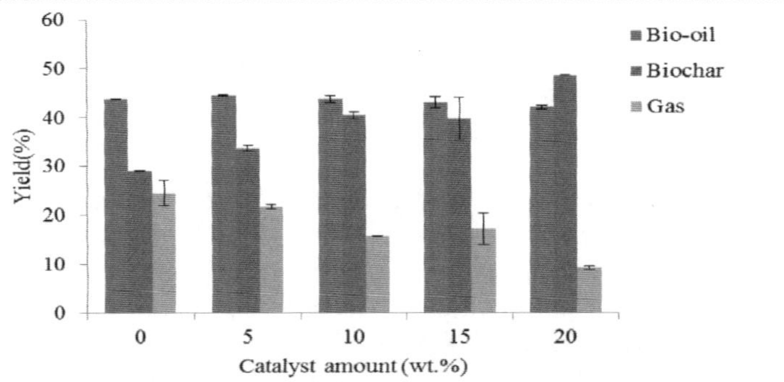

Figure 3. Effect of the amount of catalyst and bio-oil yield.

Bio-oil yield with addition of catalyst induced extraction increased from 43.8 %to 44.6 % as compared to the non-catalytic process. Therefore addition of catalyst enhanced the quality and yield of bio-oil products but production of syngas was reduced by 21.7 %.

3.6. Physicochemical characterization of bio-oil

3.6.1 Viscosity and pH. The properties of bio-oil produced from EFB are given in table 6. The optimization of bio-oil under optimized condition was used for bio-oil characterization. The pH value of bio-oil at room temperature is noticeably acidic with a pH of 3.5. This finding is in agreement with Sukiran [11] in which the presence of diluted water and volatile acids, such as acetic and formic acid, resulted in low pH of bio-oil. The pH value determined by catalytic pyrolysis of bio-oil was higher at pH 4.5 as compared to non-catalytic extraction of pH 3.5 (table 6). Bio-oil obtained by catalytic technique was less acidic therefore enhanced quality. The presence of acid in bio-oil is one of the main factors for corrosivity to materials in storage and application processes [12].

Table 6. Physical and chemical properties of bio-oil.

Properties	Non-catalytic bio-oil	Catalytic bio-oil
Viscosity(cP)	20.5	37.7
pH	3.5	4.3

Bio-oil viscosity was 20.5 cP for non-catalytic extraction of bio-oil and 37.7 cP for catalytic bio-oil extraction. Viscosity is related to the ability of the liquid to flow. The bio-oil obtained from this study was low in viscosity. This could be due to high value of water content, causing it to be less viscous. The presence of water content in bio-oil is mainly due to reaction from the lignin present in the raw material which was approximately 53.4 % [13].

4. Conclusion

The maximum oil yield was obtained with optimum extraction parameters of extraction time of 120 minutes, microwave power of 700 watts and 190ml of water used. The bio-oil yield extracted using MAE showed good quality based on its physico-chemical and fatty acid composition. Water used as an extractant solvent is clean, abundant, readily available and importantly non-hazardous to human health, living things and atmosphere. It is a fast extraction technique and enhanced quality of bio-oil. Similarly, the use of waste seashell as a biocatalyst improved bio-oil extraction from biomass waste such as EFB. This natural biocatalyst enhanced yield and quality of bio-oil. The extraction techniques and parameters used in this study showed potential to be explored and developed as a green technology to reduce environmental pollution. This process is also cost effective since water is cheaper than chemical solvent. Natural catalyst from waste seashells is also environment friendly and cost effective compared to chemically based calcium oxide.

5. References

[1] Giancarlo C, Arianna B and Laura O 2011 Green extraction techniques: For high-quality natural products *Agro Food Industry Hi Tech* **22-6** 56-59.

[2] Ahmad N, Hamzah Z and Hashim O 2015 Comparative study between microwaves assisted extraction and soxhlet extraction technique for bio-oil extraction from *Jatropha curcas J. Eng. Sci. Tech.* **8** 9-16.

[3] Rohim R, Ahmad R and Ibrahim N 2014 Effect of pyrolysis paramater and catalyst on oil production from empty fruit bunch. *8th Malaysian Technical Universities Conference on Engineering and Technology (MUCET)* 10-11 November 2014 Melaka Malaysia.

[4] Raja A S, Robinson D S and Robert C L 2011 Biodiesel production from *Jatropha*

oil and its characterization *Res. J. Chem. Sci.* **1** 81-87.

[5] Saoud A A, Yunus R M and Aziz R A 2006 Microwave assisted extraction of essential oil from eucalyptus: Study of the effects of operating condition *J. Eng. Res.* **3** 31-37.

[6] Ugarte G A C, Becerra G P J, Morales M E S and Malo A L 2013 Microwave-assisted extraction of essential oils from herbs. *J. Micro. Power Electro. Energy* **47** 63-72.

[7] Okullo A, Temu A K, Ogwok P and Ntalikwa J W 2012 Physico-chemical properties of biodiesel from *Jatropha* and castor oils. *Inter. J. Renew. Energy Res.* **2** 47-52.

[8] Nayak B S and Patel K N 2010 Physicochemical characterization of seed and seed oil of *Jatropha curcas* L. collected from Bardoli (South Gujarat) *Sains Malay.* **39** 951–955.

[9] Salimon J and Ahmed W A 2012 Physicochemical characteristics of tropical *Jatropha curcas* seed oil *Sains Malay.* **41** 313–317.

[10] Girisha S T, Ravikumar K, Mrunalini B R and Girish V 2014 Comparative study of extraction methods and properties of non-edible oils for biodiesel production *Asian J. Plant Sci. Res.* **4** 28-35.

[11] Sukiran M A 2008 Pyrolysis of empty oil palm fruit bunches using the quartz fluidized fixed bed reactor Master Science Thesis Universiti Malaya.

[12] Xu Y, Hu X, Li W and Shi Y 2011 Preparation and characterization of bio-oil from biomass Progress in Biomass and Bioenergy Production Progress in Biomass and Bioenergy Production.http://www.intechopen.com/books/progress-inbiomass-and-bioenergy-production/preparation-and-characterization-of-bio-oil-from-biomass.

[13] Abnisa F, Daud W M A W and Sahu J N 2011 Optimization and characterization studies on bio-oil production from palm shell by pyrolysis using response surface methodology. *Biom and Bioenergy* 3604-11.

IOP Conf. Series: Materials Science and Engineering **133** (2016) 012054 doi:10.1088/1757-899X/133/1/012054

Low cost method for manufacturing a data acquisition system with USB connectivity

V Niculescu[1], R A Dobre[2,3], E Popovici[2]

[1] Electronic Devices, Circuits and Architectures Department, Politehnica University of Bucharest, Bucharest, Romania

[2] Telecommunications Department, Politehnica University of Bucharest, Bucharest, Romania

E-mail: rdobre@elcom.pub.ro

Abstract. In the process of designing and manufacturing an electronic system the digital oscilloscope plays an essential role but it also represents one of the most expensive equipment present on the typical workbench. In order to make electronic design more accessible to students and hobbyists, an affordable data acquisition system was imagined. The paper extensively presents the development and testing of a low cost, medium speed, data acquisition system which can be used in a wide range of electronic measurement and debugging applications, assuring also great portability due to the small physical dimensions. Each hardware functional block and is thoroughly described, highlighting the challenges that occurred as well as the solutions to overcome them. The entire system was successfully manufactured using high quality components to assure increased reliability, and high frequency PCB materials and techniques were preferred. The measured values determined based on test signals were compared to the ones obtained using a digital oscilloscope available on the market and differences less than 1% were observed.

1. Introduction

The world is heading into an era in which everything tends to become "smart" – smart phones, smart homes, smart cars etc. – with communications representing an essential pillar for this development. Since the Internet of Things (IoT) idea, almost every device is designed in such way that it can be connected to the Internet [1], becoming accessible and controllable from anywhere. Another word that characterizes the near future is "open" as more and more open source software [2]. Also raw hardware resources (electronic components) become available to anyone interested since their prices keep dropping. On top of that electronic devices become smaller and are designed to be power saving [3].

Many engineers and hobbyists want to contribute to the IoT development and having these mentioned tools at their disposal is great, but not always sufficient. In order to design systems which contain a communication module, a digital oscilloscope or acquisition board that could be used for signal visualization represents a prerequisite, but their price make them not so easily available. Since signal bursts are to be investigated, a more affordable analog oscilloscope is not very useful in this case.

The paper proposes a solution to replace the expensive digital oscilloscope making available the development of electronic modules to the wide community of electronic enthusiasts, solution which is

[3] Iuliu Maniu blvd, no. 1-3, A building, room A301, Bucharest, Romania, 061071, rdobre@elcom.pub.ro

represented by a low cost, low manufacturing complexity, small physical size acquisition system with decent speed performances (up to 33 Msps – megasamples per second), 60 V peak-to-peak input dynamic range which sends the acquired data to a USB connected PC that runs the dedicated developed application. All the components of the proposed system are described in detail in this document as follows: in section 2 the hardware part is explained and section 3 is dedicated to the software application details. Section 4 presents the main results: the manufactured board, the user interface of the software application and some measurements (an acquired signal and a frequency measurement comparison). Finally, section 5 concludes the paper mentioning also the future improvements to be made.

2. The electronic blocks of the acquisition board

The hardware part of the acquisition system will be known as the acquisition board. It represents the largest part of the system and consists of four functional blocks: the analog input block which has the role to condition the input signal in order to bring it in the input voltage range of the analog-to-digital converter (ADC), the digital block responsible mainly for the analog-to-digital conversion and data sending, the USB communication block needed to transmit the acquired data to a PC for visualization purposes and the power supply block. Each of them will be detailed in this section. The block diagram of the whole acquisition system is illustrated in figure 1.

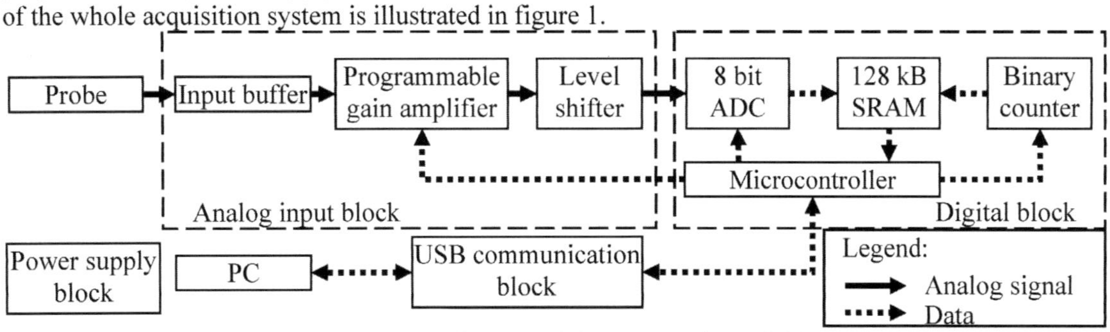

Figure 1. The block diagram of the proposed acquisition system.

2.1. *The analog input block*

The main purpose of this block is to process the input signal in order to make it fit the input voltage range of the ADC circuit. An 8 bit converter can be used with good performances only if the input signals span as much as possible over the input range of the ADC to keep good signal to noise ratio because uniform quantization is used. Also, the input signals can have positive or negative voltage values, but the input voltage range of the used ADC is between 0 V and 5 V.

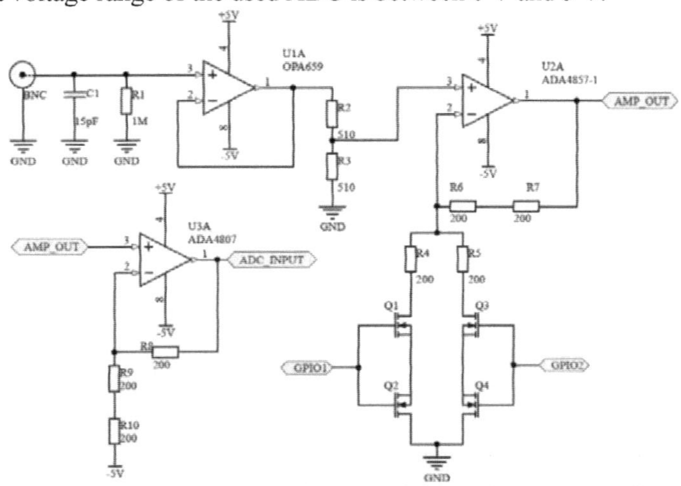

Figure 2. The schematic diagram of the analog input block.

This functional block solves these two problems and has three main parts: an input buffer, a programmable gain amplifier and a level shifting stage. The schematic diagram of the analog input block is presented in figure 2.

2.1.1. The input buffer. If a x10 probe is used, the current flowing through the attenuator represented by the impedance of the probe and the input impedance of the acquisition board (the resistor denoted R1 in parallel with the capacitor referenced as C1) is in the microamperes range, so the current drawn by its load (the next stage of the board) should be much smaller. To prevent any issues that could occur, a voltage follower implemented with an OPA659 JFET input operational amplifier was introduced between the output of the attenuator and the programmable gain amplifying stage, representing the input buffer stage.

2.1.2. The programmable gain amplifier. Low amplitude input signals must reach the converter's input amplified in order to maintain a good enough signal to noise ratio. A programmable gain amplification stage is needed for this purpose and is implemented around an ADA4857-1 operational amplifier which fits this application as it is unity gain stable, has a very large bandwidth and low noise.

The components that are part of this stage can be observed in figure 2: the operational amplifier (U2A), four resistors (R4, R5, R6, R7) and four MOSFET (metal–oxide–semiconductor field effect transistor) transistors (Q1, Q2, Q3, Q4). It is a non-inverting amplifier configuration so the gain is set by the resistors in the negative feedback network. R4 and R5 can be introduced or removed from the feedback network using the four aforementioned transistors which act as two digitally controlled switches controlled by two general purpose input-output (GPIO) pins of the microcontroller. The microcontroller was omitted from figure 2 because it is not considered part of the analog block. Still, GPIO1 and GPIO2 represent the connections between the microcontroller and the analog input block. If the microcontroller pin denoted GPIO1 is pulled "logic high", the Q1 and Q2 transistors will conduct and will connect R4 to ground, making it part of the feedback network. The other branch containing Q3, Q4 and R5 is controlled by another microcontroller pin (denoted GPIO2) and has an identical behavior. If both switches are off (all transistors are in cutoff mode), neither R4 nor R5 are in the feedback network and a simple unity gain configuration is obtained. If only one digitally controlled switch (either the one implemented with Q1 and Q2 or the one implemented with Q3 and Q4) is switched on, R4 or R5 will become part of the feedback network and the gain of the amplifier will be:

$$G = 1 + \frac{R_6 + R_7}{R_x} \tag{1}$$

In equation (1) Rx will be R4 or R5 depending on which switch is open, resulting a gain value equal to 3 according to the values presented in the schematic shown in figure 2. In the case where both switches are open, in a similar manner it can be shown that the gain value will be equal to 5 since Rx will be represented by R4 and R5 connected in parallel. There were not considered any higher gain values to avoid errors introduced by the offset voltage of the operational amplifier. A complementary pair of MOSFET transistors is needed in each switch to make the unity gain configuration to function properly. Both positive and negative parts of the signal must be blocked in this case, but if only one transistor was used in each switch, only one part, depending on the used transistor type (n channel or p channel), will be blocked while the opposite part will pass through the anti-parallel diode of each transistor resulting in a half-wave rectification behavior. In the proposed configuration, the anti-parallel diodes of the transistors contained in each switch will be connected back to back avoiding this rectification issue.

2.1.3. The level shifting stage. The maximum input range of the OPA659 operational amplifier is ±3 V. The signal at the output of the programmable gain amplifier can still have both polarities while the ADC is unipolar with the input voltage range between 0 V and 5 V. Since the maximum peak-to-peak voltage of the input signal (6 V) is larger than the input range of the ADC even when the programmable gain amplifier is set to its lowest gain value, a voltage divider implemented with R2 and R3 will be used to scale the input signal before adding the necessary offset. The attenuated input signal must then be shifted towards positive values, operation which can be done using an analog level shifter which is the last stage before the ADC and it is implemented around an ADA4807 rail-to-rail input and output operational amplifier which can be identified on the schematic diagram presented in figure 2 (U3A). The other components that are part of this stage are three resistors (R8 ÷ R10). The signal at the output of the level shifting stage is defined by:

$$\text{ADC\_INPUT} = \left(1 + \frac{R_8}{R_9 + R_{10}}\right) \cdot \text{AMP\_OUT} - \frac{R_8}{R_9 + R_{10}} \cdot (-5\,V) = 1.5 \cdot \text{AMP\_OUT} + 2.5\,V \quad (2)$$

In equation (2) with ADC_INPUT is denoted the output signal of the stage presented in this subsubsection (which will be connected to the input of the ADC) and with AMP_OUT is denoted the signal at the output of the programmable gain amplifier, in order to easily identify them on the schematic diagram presented in figure 2. It can be observed that this stage amplifies the input signal with a gain equal to 1.5 and then adds a constant voltage equal to 2.5 V to it.

To recap, the input signal is scaled down (in order to reduce its peak-to-peak value as motivated in subsubsection 2.1.3) using the voltage divider implemented with the resistors R2 and R3, bringing its extreme values to ±1.5 V. This signal is amplified by the programmable gain amplifier if it is necessary (in this limit example it is not necessary since the lowest programmable gain other that 1 is 3 which will cause the output of the next stage to clip) and then is applied to the input of the level shifter. Finally, at the input of the ADC, according to equation (2), the voltage will swing between 0.25 V and 4.75 V assuring a good span over the permitted input range.

In the circuit design process only low noise components were considered, which had a sufficient slew rate value. Because the maximum input swing of the OPA659 buffer is 6 V peak-to-peak, the input domain of measurement becomes ±30V, using an x10 probe. A low-pass RC filter with a cutoff frequency around 50 MHz was introduced between the output of the level shifting stage and the ADC in order to reduce the bandwidth of the system and consequently the system's noise.

2.2. *The digital block*

This block contains the microcontroller (PIC32MZ2048EC), the ADC, a 128 kB static random access memory (SRAM), a logic level shifter and a binary counter. In order to be able to study not only audio signals which can be analyzed even using directly the soundcard of a PC [4], a data acquisition system must contain at least a medium speed ADC. The data output rate of such converters may be overwhelming for a low cost microcontroller as it may not be able to process the samples fast enough thus a block of discrete circuits representing a fast data storage solution must be used as a data buffer between the ADC and the microcontroller.

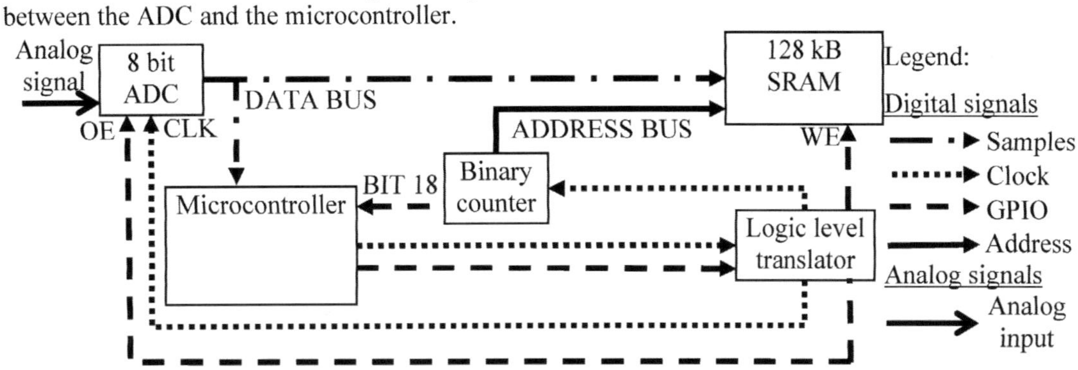

Figure 3. The detailed functional diagram of the digital block.

The functioning overview of the digital block is briefly presented further: because the ADC can work with an output data rate the microcontroller cannot handle directly (33 Msps), the samples provided by ADC are at first stored in a very fast SRAM. For easily addressing the whole memory, an 18 bit binary counter is used, the first 17 being used for addressing and the 18th representing a "memory full" indicator. When the memory gets full, the microcontroller reads it with a lower frequency and packs the data in order to send it to a USB connected PC (using the USB communication block) for visual representation. The functional diagram of this block is illustrated in figure 3.

The data bus is an 8 bit parallel bus used for writing the data into the SRAM. The same bus is used when the microcontroller reads the samples stored in SRAM. The address bus is a 17 bit parallel bus used for setting the address location in which the data sample will be stored. The clock line delivers the clock signal to the synchronous logic circuits: the ADC and the binary counter. The microcontroller works with 3.3 V logic levels and all the other circuits work using TTL (transistor-transistor logic, 5 V) levels. A logic level translator implemented using a 74HC7014 digital buffer circuit with Schmitt-triggered inputs and 5 V as supply voltage is needed to assure the communication between the microcontroller and the other digital circuits. The "logic high" threshold of its inputs is 3.1 V, so the signals from the microcontroller will be correctly interpreted. Some microcontroller pins are 5 V tolerant when used as inputs, so these were used to receive data in order to avoid another logic level conversion. A GPIO pin of the microcontroller is connected to the WE (write enable) input of the SRAM to distinguish between writing and reading operations and another one to the OE (output enable) input of the ADC to permit blocking its data output even when clock signal is received on the CLK (clock) input. This state is needed when reading the SRAM or else the value stored in the memory and the output of the ADC would conflict.

The acquisition algorithm is structured in several phases as follows:

- In the first phase, the system waits for the acquisition request sent by the PC;
- The second phase starts: the WE and OE commands are issued, the microcontroller generates the clock signals (their frequency depending on the sample rate set in the PC application) for the ADC and the binary counter which increments the memory addresses of the SRAM. On every negative edge of the clock signal the ADC outputs an 8 bit value (the current sample) and the counter increments the memory address. The sample is stored when the memory location is incremented as the memory is asynchronous and does not need a write command. The second phase runs in a loop until the most significant bit of the counter (denoted BIT 18 in figure 3) becomes "logic high" indicating that the whole memory was written, triggering a microcontroller interrupt which will start the next phase;
- In the third phase the OE and WE commands are deactivated, the SRAM is put in reading mode and the clock frequency is set to 1 MHz. The samples acquired in the previous phase are transferred into the microcontroller;
- Finally, a data packet is constructed in the PIC32 in order to be sent to the computer, using the USB communication block;
- The process is restarted as soon as the PC sends another acquisition request.

2.3. *The USB communication block*

This block is built using a MCP2200 chip. It implements a RS232 communication over USB connection. The PC recognizes the module as a COM port. Even if it does not permit very fast data rates, the serial communication is easy to be integrated in any software application because of its simplicity. The chip considered in this design supports baud rates up to 921600 symbols per second.

2.4. *The power supply block*

The main component of this block is a TPS65130RGER positive and negative output DC to DC converter. This block is able to convert the USB voltage to ±8 V. The voltages needed in this design (±5 V and 3.3 V) are then obtained using linear, low dropout voltage regulators.

3. The software application

A PC application was designed to display the acquired signal and to permit the user to easily modify the acquisition parameters. It is a standalone application and it was developed using C# programming language. The refresh rate of the screen is controlled using a timer. Whenever it overflows, a message which asks for a new data packet is sent to the microcontroller and a new acquisition cycle starts. In the GUI (graphical user interface) the user can set the sample rate between 1 ÷ 33 MHz, the trigger level and slope, the probe type that was used (x1 or x10) and the COM port used for the communication. The trigger controls are software defined: the application scans the received signal, identifies the trigger conditions set by the user and displays the signal starting from the first sample after the trigger event. In order to permit the acquisition board be connected to the PC before or after the application was launched, a COM refresh button was added. The vertical axis is graded in volts and the horizontal axis is graded in sample indexes. Knowing the sample rate and the sample indexes, time intervals can be calculated.

4. Results and discussions

The hardware part was designed using an open source CAD (computer aided design) software suite and then manufactured. In order to obtain a module as small as possible to assure portability, SMD (surface mounted device) components were preponderantly used. All the electronic components were manually soldered. As a plus, the board is entirely supplied using the USB connection, so it does not require any external power source. Two copper layers were enough for the design, keeping the manufacturing costs low. For an easier debugging process, the microcontroller and the serial converter were chosen as plug-in modules so they can be easily detached if necessary. The final manufactured version of the proposed acquisition board is presented in figure 5.

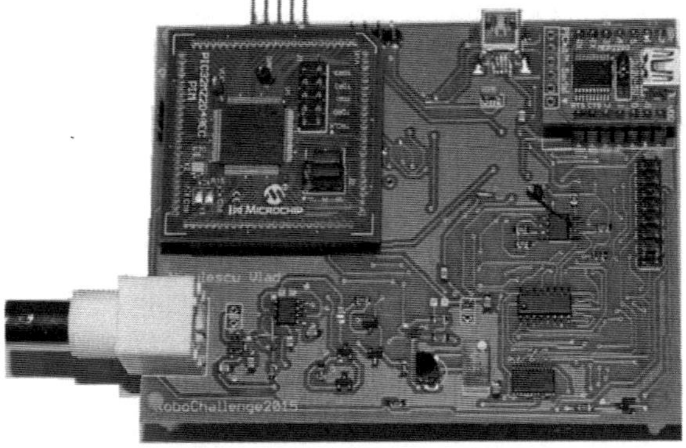

Figure 4. The manufactured acquisition board.

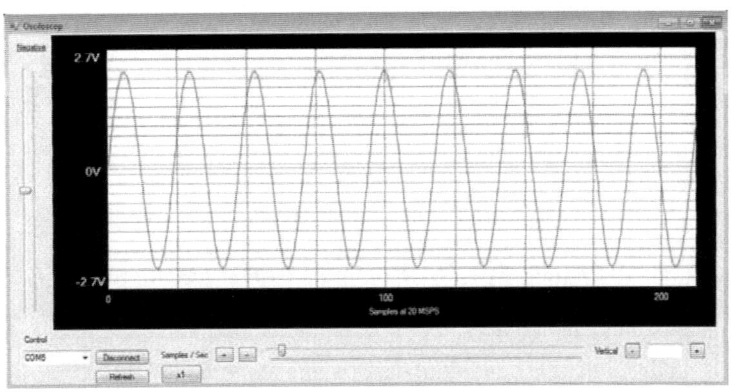

Figure 5. The graphical user interface of the PC software application.

The GUI of the developed application can be observed in figure 5 displaying an acquired sine wave validating the correct functioning of the system. All the elements described in the previous section can be easily identified. Because the waveform of the input signal is assumed to be unknown, the only possible interpolation is linear. The measured frequency for a 1 MHz sine wave acquired with the system differed just by 1% compared to the same measurement executed with a Tektronix TDS1001 oscilloscope. Any oscilloscope passive probe can be used for the measurement process.

5. Conclusions

More and more electronics enthusiasts, engineers or not, are interested in developing electronic modules which contain at least one communication block, contributing to the Internet of Things growth. The tools needed for designing and debugging such systems are not affordable in general, while the price of the electronic components involved in the communication blocks lowers as time passes. A tool which would not leave a hole in one's budget and which would help the debugging of such systems is very attractive. The present paper proposes a compact, low cost, medium speed acquisition system with USB connectivity which would represent a handy tool on anyone's workbench.

The proposed system contains two main parts: a hardware part and a PC software application. The hardware part is represented by an analog signal conditioning block, a digital block, a USB communication block and a power supply block. The analog block is represented by an input buffer, a programmable gain amplifier and a level shifting stage, all three circuits working with the main purpose to make the input signals' voltage range to match the input range of the analog-to-digital converter. Since the sampling rate can have important values (33 megasamples per second), the microcontroller may not be able to process the samples directly so a data buffer was introduced between the analog-to-digital converter and the microcontroller. It was implemented using a static random access memory which is addressed using a 17 bit binary counter. The microcontroller provides the clock and other signals needed for this digital block to work correctly. After the memory gets full, the microcontroller reads its contents at a lower speed, and sends them to the PC application using the USB communication block after they have been properly packed. The software application displays the acquired signal and is used to set the acquisition parameters.

The manufactured board is compact since most components are surface mounted devices and its cost is low because only two electrical layers were enough for the interconnection structure to be successfully designed. The board is USB powered avoiding the need of any other external power sources, this advantage contributing to the portability of the system.

The measured signals were compared to the ones acquired using a Tektronix TDS1001 oscilloscope and the differences were minor, only 1% errors.

Future improvements could include: making the board battery powered and replacing the USB communication with a wireless one in order to be able to make remote measurements. More options are to be added in the PC software application like fast Fourier transform computation, automatic measurement procedures (for peak-to-peak value, frequency etc.), cursors etc.

References

[1] A. Drumea and P. Svasta, 2011 *Designing low cost embedded systems with ethernet connectivity*, Proc. 17th Int. Symposium for Design and Technology in Electronic Packaging (Timişoara, RO, 20–23 October 2011) pp. 217–220.

[2] A. Radovici and I. Culic, 2013 *Open cloud platform for programming embedded system*, 11th RoEduNet Intl. Conf.: Networking in Education and Research (Sinaia, RO, 17–19 January 2013), pp. 1–2.

[3] C. Marghescu and A. Drumea, 2015 *Modelling and simulation of energy harvesting with solar cell*, Proc. SPIE 9258 Advanced Topics in Optoelectronics, Microelectronics, and Nanotechnologies, pp. 92582L–92582L-8.

[4] D. Yi, 2010 *Research and Design of Virtual Oscilloscope Based on Sound Card*, Intl. Conf. on Electrical and Control Engineering (Wuhan, CHN, 25–27 June 2010), pp. 1566-1569.

IOP Conf. Series: Materials Science and Engineering **133** (2016) 012055 doi:10.1088/1757-899X/133/1/012055

Antimicrobial property of zinc based nanoparticles

V. Chiriac[1,2,3], D. N. Stratulat[1,2,3], G. Calin[1,3], S. Nichitus[1,3], V. Burlui[1,3], C. Stadoleanu[1,3], M. Popa[1,2,3], I. M. Popa[2]

[1]*Apollonia* University of Iasi, Romania
[2]*Gheorghe Asachi* Technical University, Faculty of Materials Science and Engineering, Dimitrie Mangeron Street, no. 61A, 700050, Iaşi, Romania.
[3] *Academician Ioan Haulica* Institute of Researches, Apollonia University of Iasi

E-mail: m_gabriela2004@yahoo.com

Abstract. Pathogen bacteria strains with wide spectrum can cause serious infections with drastic damages on humans. There are studies reflecting antibacterial effect of nanoparticles type metal or metal oxides as an alternative or concurrent treatment to the diseases caused by infectious agents. Synthesised nanoparticles using different methods like sol-gel, hydrothermal or plant extraction were tested following well-established protocols with the regard to their antimicrobial activity. It was found that zinc based nanoparticles possess strong synergistic effect with commonly used antibiotics on infection tratment.

1. Introduction

Last years a lot of serious health issues was caused by the incresing resistence of pathogen agents towards antibiotics. Researches could reveal novel application of metals as intrinsic antimicrobial activity by combining advanced technologies of material science and material science respectively (Figure 1).

According to the reports, metal and metal oxide nanoparticles represent a group of materials which were investigated in respect to their antimicrobial effects. In present study, our attention was focused on the recent discoveries concerning antimicrobial activity of metal and metal oxide nanoparticles together with their mechanism of action. Results showed that the particle size was the essential parameter which determined the antimicrobial effectiveness of the metal nanoparticles. By combining classical therapy with metal based nanoparticles might be an important choise to overcome the current bacterial resistance to antibiotics[1].

Zinc oxide type nanoparticles proved to have bactericidal effects on Gram-positive and Gram-negative bacteria as well as the spores which are resistant to high temperature and high pressure [2].

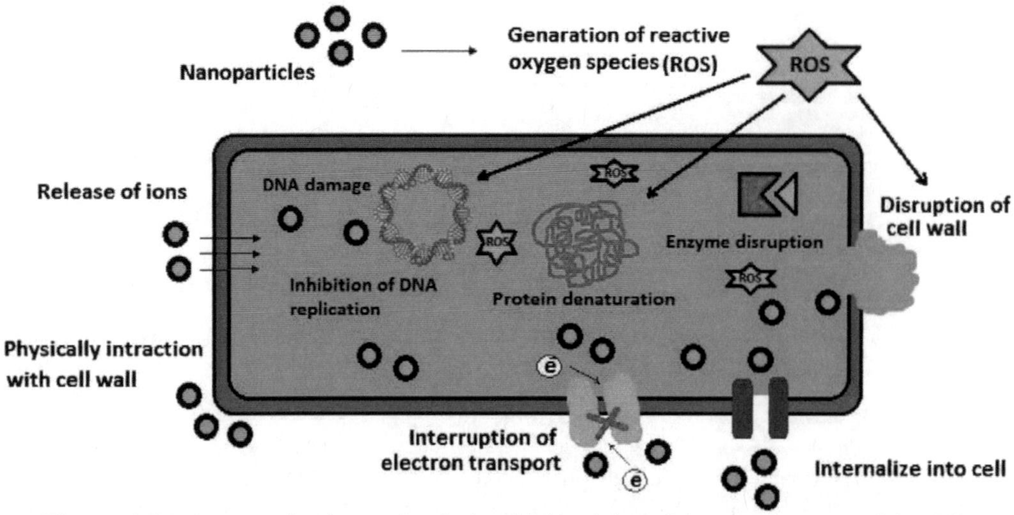

Figure 1. Various mechanisms of antimicrobial activity of the metal nanoparticles [1].

2. Materials and methods

In recent years biologically synthesized nanomaterial has become a major field of nanotechnology. A leaf aqueous extract of Azadirachta indica (L.) was used to obtain ZnO based nanoparticles and their antimicrobial activity was investigated. Results described an increase antimicrobial activities due to the increase of H2O2 concentration from the surface of ZnO which depended on concentration of ZnO nanoparticles (50, 100, 200 μg/mL). Further more, zinc oxide nanoparticles exhibited an interesting antibacterial effect on both Gram positive and Gram negative bacteria at micromolar concentration [3].

Zinc oxide nanoparticles (ZnO NPs) were also synthesized using a simple and eco-friendly method based on leaf extract of Moringa oleifera. Wurtzite hexagonal structure and presence of functional groups of both leaf extract and ZnO NPs were revealed using XRD and FT-IR patterns. FE-SEM and EDX technique used for textural characteristics highlighting showed their particles size, morphology and topography, intense and narrow width of zinc and oxygen with high purity and crystalline respectively. Studies results of antibacterial activity (Figure 2) revealed that intense area of inhibition was observed Gram positive bacteria followed by the Gram negative bacteria and fungal at concentration of 200 μg/mL of ZnO NPs [4].

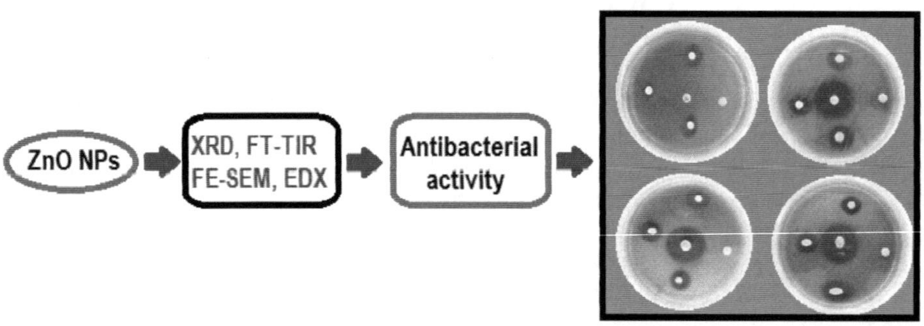

Figure 2. Antimicrobial activity of ZnO nanoparticles form *Moringa oleifera* leaf extract [4]

Another biological process using V. trifolia plant leaf extract was developed to synthesize zinc oxide nanoparticles presented in figure 3. Physical-chemical characterization by X-ray diffraction FT-IR spectroscopy, SEM- EDX assessment showed small crystalline size at higher volume of extract, functional groups present from the leaf extract such as alcohols, phenols, 10 amines, aromatic and aliphatic amines and the presence of pure aspect zinc and oxygen without a different impurity peaks.

It was shown that green synthesized zinc oxide nanoparticles are more useful in both environmental purification and clinical laboratory [5]. Antimicrobial efficacy of ZnO has been investigated against Escherichia coli in simple model systems and in typical topical products containing antioxidants, chelating agents, electrolytes, titanium dioxide pigment resulting that antibacterrial activity of zinc oxide nanoparticles was partially inhibited by NaCl and MgSO4 salts.

By adding ZnO in complex formulations, microbiological quality was significantly improved in spite of the presence of other ingredients that influenced antimicrobial activity [6].

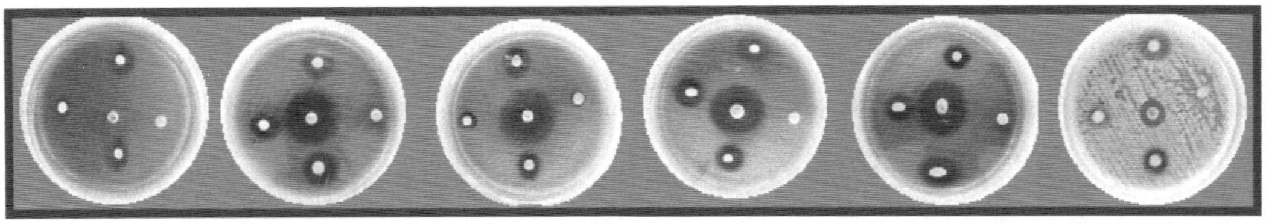

Figure 3. Zones of inhibition (mm) of various human pathogenic microorganism effects of various antimicrobial agents and its equivalent bar diagram and Gas Chromatogram of leaf sample of *Vitex trifolia.*

Other researches used zinc oxide nanoparticles with hexagonal, spherical shape with particle size smaller than 100 nm from Murraya koenigii leaf extract. Results showed that zinc based nanoparticles had intense activities on Staphylococcus aureus (14.0±0.50mm) and Bacillus subtilis (13.8±0.76mm) at concentration of 200 μg/mL [7].

3. Experimental part

Clear waters is a primary need for human health. Recycling naturally waters is not enough for population and undrinkable according to standard parameters. Presently, pathogenic bacteria significantly contaminate most of the natural waters. Globally, approximate 700 million peoples are facing huge scarcity of water, predicting nearly about 1.8 billion people face water scarcity problem in 2025 [8]. Annually, arround 12 million people die because of water borne disease (WHO 2012) [9] and majority of these diseases occurring in underdeveloped countries is due to the ingestion of contaminated waters [10]. Advanced nanotechnology and nano material applications respectively open the oportunities to improve the waste water treatment methods including decontamination of the nano- and mycro- contaminants due to their reactive surface inducing coating on pollutant decreasing the toxicity of compounds and destroying pathogen agents. Antimicrobial activity of zinc based nanomaterials makes them useful in antibiotic simultaneous treatment to significantly improve healing [11]. Several synthesis methods are available for the preparation of pure and doped ZnO nanomaterials, like hydrothermal, hydrolysis, sol–gel, vapor condensation, spray pyrolysis and organic precursor flame decomposition [12–17].

Generaly metal nanomaterials like silver, gold and zinc exhibited important antimicrobial activities [18–20]. Bacteria resistance developing against antibiotics and metals creates health problems, leading to inability of classical medicine to treat efficient several diseases [21]. For human welfare there has been a special attention for stable antimicrobial metal nanoparticles development [22]. Nanomaterial based antimicrobial medicines especially zinc oxides are more efficient in hard conditions [23].

Nowadays, researchers focussed on antibacterial effect and stability of ZnO nanoparticles also known to be safe materials for human and ecosystem [24, 25]. More over, zinc oxide nanoparticles were dopped in wallpaper of hospitals to prevent microbial load on walls and nosocomial infection respectively [26]. An enhanced antimicrobial activity of ZnO nanparticles and other metal oxides like Al2O3, SiO2 and TiO2 was due to the presence of water molecules on its surface being capable to generate free radicals of hydroxyl and oxygen species responsible for oxidative stress in treated bacterial cells of Escherichia coli, B. subtilis and Pseudomonas sp.

4. Results and discussions

Chemically synthesized cobalt doped ZnO nanoparticles were found to be crystalline with a single phase as confirmed by XRD (Figure 4) and SEM (Figure 5) features.

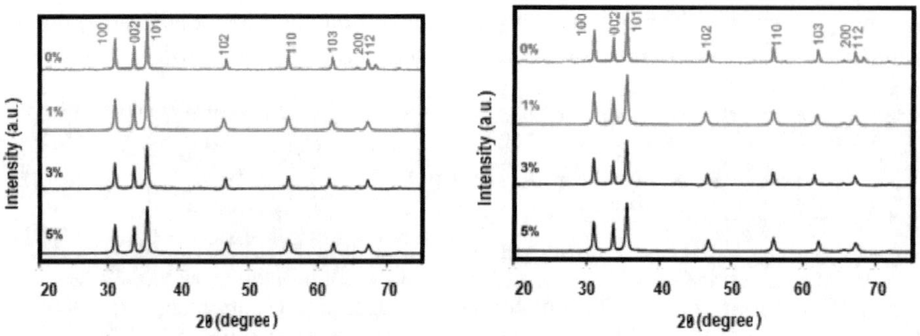

Figure 4. X-ray diffraction pattern of 0% pure ZnO, 1% of Co doped ZnO, 3% of Co doped ZnO and 5% of Co doped ZnO.

It was found that increasing the percentage of Co from 0% to 5% in ZnO, leads to an enhanced crystallite size from 20.5 to 25.7 nm.also experienced a colour change, but of a smaller value (ΔE^*_{ab} = 7.06), in comparison with that of C2 area, where ΔE^*_{ab} = 9.58.

SEM images revealed that, after treatment, Co dopped nanoparticles morphology changed from rod to spherical shaped. Cobalt dopped zinc oxide nanomaterials were tested as a bactericidal agent resulting that this treatment and exposure of sunlight enhanced the antibacterial activity against Escherichia coli and Vibrio cholerae at 50 μg concentration [27].

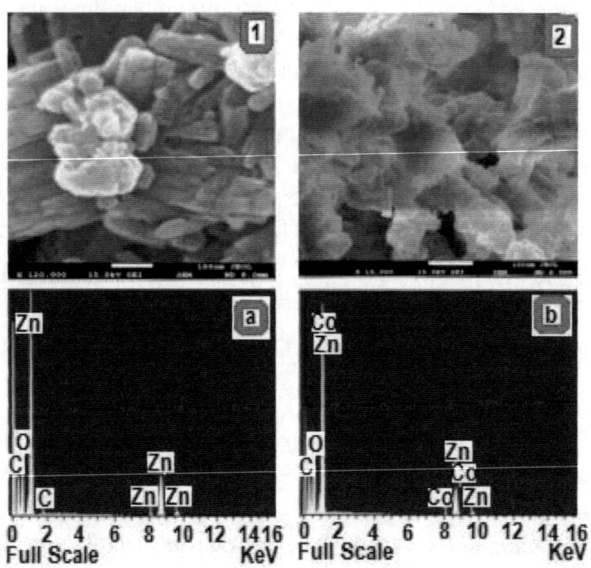

Figure 5. Doping analysis of ZnO and Co doped ZnO nanoparticles by SEM coupled with energy dispersive X-ray spectroscopy, here, pure ZnO (1 and a), 1% Co doped ZnO (2 and b).

Antimicrobial activity of zinc oxide powders was investigated on the five microbial strains of Escherichia coli; Staphylococcus aureus; Pseudomonas aeruginosa; Candida albicans; Aspergillus brasiliensis in order to assessed relationships between the structural parameters and physicochemical characteristics of zinc oxide particles and their antimicrobial activity.

Zinc oxide particles with interesting textures as small size, large specific area and high porosity exhibit higher antibacterial effect according to their action mechanisms [28]. Antimicrobial activity against various bacterial and fungal strains is repesented by zinc ions (Zn2+) since zinc oxide nanoparticles releases Zn2+ ions in aqueous solution contributing to the antimicrobial effectivness of ZnO as shown in figure 6.

Researchers studied the contribution of the soluble zinc species regarding antimicrobial activity of ZnO on microbial cultures in broth medium. The Zn2+ released in the broth significantly contributed to the overall antibacterial effect of zinc oxide nanoparticle. Soluble zinc species and ZnO powders possessing larger specific area showed specificity with respect to the microbial strains. [29].

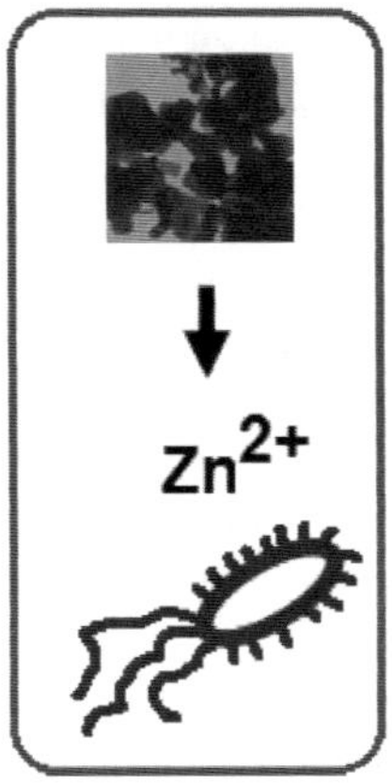

Figure 6. Action mecanism of Zn^{2+} on microbial cell

5. Conclusions

Interdisciplinary researches that include nanotechnology, pharmacology, physiology, physics, chemistry and medicine represent nowadays a key direction for nanoparticles development in order to prevent or treat serious diseases caused by dangerous infectious agents. Antibacterial activity of zinc based nanoparticles is enhanced by their morphological characteristics and physicochemical properties mostly due to the positive surface charge of metal that binds to the negatively charged surface of tested bacteria. Regardless of the used synthesis method, zinc oxide or cobalt doped ZnO nanostructures with low toxicity have shown an important antibacterial effect and can be used as a suitable alternative in pharmaceutical industries.

Acknowledgment

PECAFROM - SOP HRD/144/6.3/S/127928 co-financed by European Social Fund through the Sectoral Operational Programme Human Resources Development 2007-2013. Invest in people

References:

[1] Dizaj S M, Lotfipour F, Barzegar-Jalali M, Mohammad H Z and Khosro A 2014 Antimicrobial activity of the metals and metal oxide nanoparticles *Materials Science and Engineering C* **44** pp 278–284.

[2] Malarkodi C, Rajeshkumar S, Paulkumar K, Vanaja M, Gnanajobitha G and Annadurai G 2014 Biosynthesis and Antimicrobial Activity of Semiconductor Nanoparticles against Oral Pathogens *Bioinorg. Chem. Appl.* pp 1–10

[3] Elumalai E and Velmurugan S 2015 Green synthesis, characterization and antimicrobial activities of zincoxide nanoparticles from the leaf extract of Azadirachta indica (L.) *Applied Surface Science* **345** pp 329–336

[4] Elumalai K, Velmurugan S, Ravi S, Kathiravan V and Ashokkumar S 2015 Green synthesis of zinc oxide nanoparticles using Moringa oleifera leaf extract and evaluation of its

antimicrobial activity *Spectrochimica Acta Part A: Molecular and Biomolecular Spectroscopy* **143** pp 158–164
[5] Elumalai E, Velmurugan S, Ravi S, Kathiravan V and Adaikala Raj G 2015 Bio-approach: Plant mediated synthesis of ZnO nanoparticles and their catalytic reduction of methylene blue and antimicrobial activity *Advanced Powder Technology* **26** pp 1639–1651
[6] Pasquet J, Chevalier Y, Couval E, Bouvier D and Bolzinger M A 2015 Zinc oxide as a new antimicrobial preservative of topical products: Interactions with common formulation ingredients *International Journal of Pharmaceutics* **479** pp 88–95
[7] Elumalai K, Velmurugan S, Ravi S, Kathiravan V and Ashokkumar S 2015 Bio-fabrication of zinc oxide nanoparticles using leaf extract of curry leaf (Murraya koenigii) and its antimicrobial activities *Materials Science in Semiconductor Processing* **34** pp 365–372
[8] Water Scarcity, United Nation, International Decade for Action Water for Life (2014) 2005–2015
[9] ***www.who.int/infectious-diseasereport/pages/textonly.html (2012)
[10] Baruah S, Kitsomboonloha R and Myint M T Z 2009 Nanoparticle applications for environmental control and remediation, in: R.S. Chaughule, R.V. Ramanujan (Eds.), Nanoparticles: synthesis characterization and applications *American Scientific Publishers* pp 195–216
[11] Li Q, Mahendra S, Lyon D Y, Brunet L, Liga M V, Li D and Alvarez P J J 2008 Antimicrobial nanomaterials for water disinfection and microbial control: potential applications and implications *Water Res.* **42** pp 4591–4602
[12] Wiesner M R, Lowry G V, Alvarez P, Dionysiou D and Biswas P 2006 Assessing the risks of manufactured nanomaterials *Environ. Sci. Technol.* **40** pp 4336–4345
[13] Ansari S A, Khan M M, Lee J and Cho M H 2014 Highly visible light active Ag@ZnO nanocomposites synthesized by gel-combustion route *J. Ind. Eng. Chem.* **20** pp 1602–1607
[14] Djerdj I, Zvonko J, Denis A and Markus N 2010 Co-doped ZnO nanoparticles: minireview *Nanoscale 2* pp 1096–1104
[15] Hirano S I 1987 Hydrothermal processing of ceramics *J. Am. Ceram. Soc.* **66** pp 1342–1344
[16] Arshad M, Azam A, Ahmad A S, Molla S and Naqvi A H 2011 Effect of Co substitution on the structural and optical properties of ZnO nanoparticles synthesized by Sol Gel Route *J. Alloys Comp.* **509** pp 8378–8381
[17] Ulrich G D and Riehl J W 1982 Aggregation and growth of submicron oxide particles in flames *J. Colloid. Inter. Sci.* **87** pp 257–265
[18] Ansari S A, Khan M M, Ansari M O, Lee J and Cho M H 2013 Biogenic synthesis, photocatalytic, and photoelectrochemical performance of Ag–ZnO nanocomposite *J. Phys. Chem. C.* **117** pp 27023–27030
[19] Oves M, Khan M S, Zaidi A, Ahmed A S and Ahmed F 2013 Antibacterial and cytotoxic efficacy of extracellular silver nanoparticles biofabricated from chromium reducing novel OS4 strain of Stenotrophomonas maltophilia *PLoS ONE 8,* e59140
[20] Azam A S, Ahmed M, Oves M S and Khan S and Habib S 2012 Antimicrobial activity of metal oxide nanoparticles against Gram-positive and Gram-negative bacteria: a comparative study *Int. J. Nanomed.* pp 6003–6009
[21] Schito G C 2006 The importance of the development of antibiotic resistance in Staphylococcus aureus *Clin. Microbiol. Infect.* **12** pp 3–8
[22] Saravanan R, Khan M M, Kumar V, Mosquera E, Gracia F, Narayanang V and Stephen A 2015 ZnO/Ag/Mn2O3 nanocomposite for visible light induced industrial textile effluent degradation, uric acid and ascorbic acid sensing and antimicrobial activity *RSC Adv.* **5**, 34645
[23] Sawai J 2003 Quantitative evaluation of antibacterial activities of metallic oxide powders (ZnO, MgO and CaO) by conductimetric assay *J. Microbiol. Methods* **54** pp 177–182
[24] Stoimenov P K, Klinger R L, Marchin G L and Klabunde K J 2002 Metal oxide nanoparticles as

bactericidal agents *Langmuir* **18** pp 6679–6686

[25] Fu G, Vary P S, Lin C and Anatase T 2005 TiO2 nanocomposites for antimicrobial coatings *J. Phys. Chem. B.* **109** pp 8889–8898

[26] Richards E 2006 Antibacterial wallpaper *Chemtech.* **11**

[27] Oves M, Arshad M, Khan M S, Ahmed A S, Azam A et al. 2015 Anti-microbial activity of cobalt doped zinc oxide nanoparticles: Targeting water borne bacteria *Journal of Saudi Chemical Society* **19** pp 581–588

[28] Pasquet J, Chevalier Y, Couval E, Bouvier D, Noizet G, Morlière C and Bolzinger M A 2014 Antimicrobial activity of zinc oxide particles on five micro-organismsof the Challenge Tests related to their physicochemical properties *International Journal of Pharmaceutics* **460** pp 92– 100

[29] Pasquet J, Chevalier Y, Pelletier J, Couval E, Bouvier D and Bolzinger M A 2014 The contribution of zinc ions to the antimicrobial activity of zinc oxide *Colloids and Surfaces A: Physicochem. Eng. Aspects* **457** pp 263–274

IOP Conf. Series: Materials Science and Engineering **133** (2016) 012056 doi:10.1088/1757-899X/133/1/012056

The effect of erosion and abrasion on surface properties of composite resin

S Stoleriu[1], S Andrian[1], G Pancu[1], I Nica[1], A Munteanu[2], A Balan[3] and G Iovan[1]

[1] Odontology-Periodontology and Fixed Prosthesis Department, "Grigore T. Popa" University of Medicine and Pharmacy of Iaşi, Faculty of Dental Medicine, 16 Universitatii Str., 700115, Iaşi, Romania
[2] Department of Machine Tools, "Gheorghe Asachi" Technical University of Iaşi, Blvd. D. Mangeron, 59 A, 700050 Iaşi, Romania
[3] Surgical Department, "Grigore T. Popa" University of Medicine and Pharmacy of Iaşi, Faculty of Dental Medicine, 16 Universitatii Str., 700115, Iaşi, Romania

E-mail: stoleriu_simona@yahoo.com

Abstract. The aim of the study was to evaluate the surface roughness of two commercial composite resins submitted to erosive attack, to abrasive wear and to association of erosive and abrasive challenge. Standardized samples of G-ænial anterior (GC Company) and Essentia (GC Company) composite resins were randomly split in 6 groups. In group 1 the samples were maintained in artificial saliva until the evaluation of surface roughness. In group 2 the samples were submitted only to erosive attack, in group 3 only to abrasive challenge and in groups 4 ,5, and 6 the erosive attack was followed by abrasive challenge immediately (group 4), 30 minutes after the erosive attack (group 5) and one hour after the erosive attack (group 6). The specimens were evaluated using surface roughness measuring tester SJ-210 (Mitutoyo Corporation, Japan) and the mean surface roughness values (Ra, µm) of each specimen were registered. A significantly increase of both composite resins surface roughness was recorded after erosive attack and abrasive challenge. Toothbrushing 60 minutes after acidic contact determined no significant differences in surface roughness of composite resins.

1. Introduction

Oral cavity is a complex environment, where the teeth are exposed to acids coming from bacterial metabolism, beverages, food or stomach content or to mechanical wear or to temperature changes [1, 2]. In the latest years an increase in acidic beverages consumption was observed and related to that an increase in the prevalence of dental erosion was reported. Enamel, dentine and cement are prone to chemical dissolution in acidic environment [3]. Except the loss of dental hard tissues, acidic agents can also affect the materials for filling. The contact of erosive factors with the composite resins might lead to degradation of the resin matrix, releasing of filler due to abrasion or chewing forces or alteration of coupling agent [4].

Generally, the abrasive effects in the oral cavity are related to the abrasiveness of toothpastes and toothbrushing technique. Previous studies showed a direct correlation between the size of the filler and the amount of material loss [5, 6]. The shape of the fillers can also influence the resistance of the composite resin to abrasive challenge.

One of the clinical results of these chemical and mechanical phenomena is the increase of the surface roughness of restorative materials. This can lead to higher bacterial colonization due to their retention in the surface irregularities, which can determine periodontal problems [7]. The surface roughness can be a cause of changes in materials colour and translucency, which might affect the aesthetic aspect of final restorations [8-11].

The aim of the study was to evaluate the surface roughness of two commercial composite resins submitted only to erosive attack, only to abrasive wear and to association of erosive and abrasive challenge.

2. Materials and Methods

2.1. Samples preparation

Two composite resins were chosen for this study: G-ænial anterior (GC Company) and Essentia (GC Company). Thirty samples of each material having 25 mm in length, 7 mm in width and 2 mm in height were obtained by placing the composite resin in contact with a transparent matrix between two glass slabs in order to flatten the surface. The samples were built-up in one increment of 2mm. Every sample was light cured for 40 seconds in one step, using a LED curing light unit (LED B, Guilin Woodpecker Medical Instrument Co., Ltd, China), with an intensity of the light source of 850-1000mW/cm^2 and a wavelenght of 420-480 nm. The samples of each composite resin were equally split in 6 groups. In group 1 the samples were kept in artificial saliva (AFNOR NF S90-701) until the evaluation of surface roughness. In group 2 the samples were submitted only to erosive attack, in group 3 only to abrasive challenge and in groups 4, 5, and 6 the erosive attack was followed by abrasive challenge immediately (group 4), 30 minutes after the erosive attack (group 5) and one hour after the erosive attack (group 6).

2.2. Erosion attack simulation

Each sample of groups 2-6 was immersed in 10 mL of Coca Cola® (Coca-Cola Company) two times a day, 3 minutes each time. Fresh beverage was used for every immersion. Between the immersion periods, the samples were maintained in artificial saliva.

2.3. Abrasion simulation

To simulate the abrasive challenge due to toothbrushing, a device created by the authors was used (figure 1).

Figure 1. Toothbrushing machine.

The device is composed by:1. the weight that ensures a constant pressure on tooth brushes; 2. tooth brushes support that have the possibility to move on vertical direction on two gripper bars and allow toothbrushes lifting and descending under the pressure during tooth brushing; 3. toothbrushes; 4. tube to inject the dentifrice directly on the tooth brushes head; 5. system for fixing the samples; 6. electric motor having 30W and 50Hz; 7. support table to mentain the motor and the sliding system; 8.

mechanism to transform the rotation movement in forward and back movements; 9. sliding table which is stable in vertical direction but allowes the movements in horizontal direction; 1o. fixing and sliding bars; 11. sample fixed in correct position; 12. adhesive bands to prevent scraching of the samples when fixing and releasing the samples; 13. tray to collect the used tootpaste

The device was design to simulates the forward and back movements during toothbrushing. The amplitude of the movements was 30 mm (15 mm in each direction) and the frequency was 60 cycles/minute. Toothbrushes having medium hardness of the bristles were used in this study (Colgate Classic Deep Clean). The weight applied on every toothbrush was 250 g. Colgate Total toothpaste diluted with water (1:2 by volume) was used as dentifrice. In this study was simulated the toothbrushing for 2 minutes, twice a day, 30 days. On each sample the calculated mean number on toothbrushing cycles during a toothbrushing of 2 minutes was 20.

2.4. Surface roughness evaluation

The specimens were evaluate using surface roughness measuring tester SJ-210 (Mitutoyo Corporation, Japan) (figure 2), having the measuring force of 0.75 mN, traversing speed of 0.5 mm/s and the cutoff length of 0.8µm. The mean surface roughness values (Ra, µm) of each specimen were obtained from ten measurements from the center of each sample in different directions.

Figure 2. Surface roughness measuring tester.

3. Results

Examples of profiles for some composite resins samples in control and study groups are presented in figure 3.

	G-ænial anterior	Essentia
Group 1	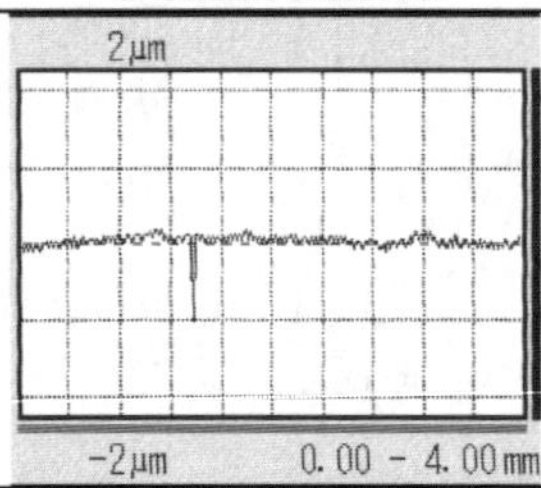	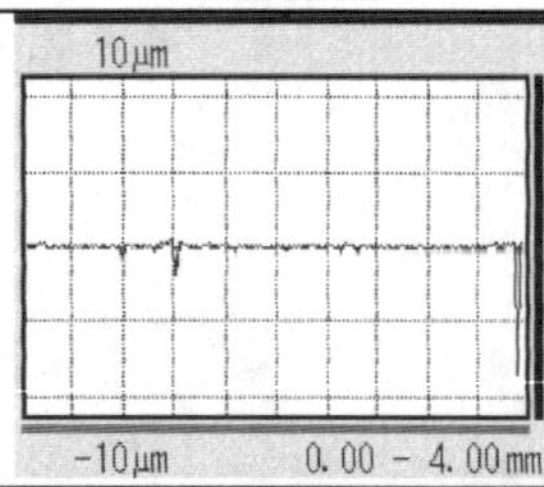
Group 2	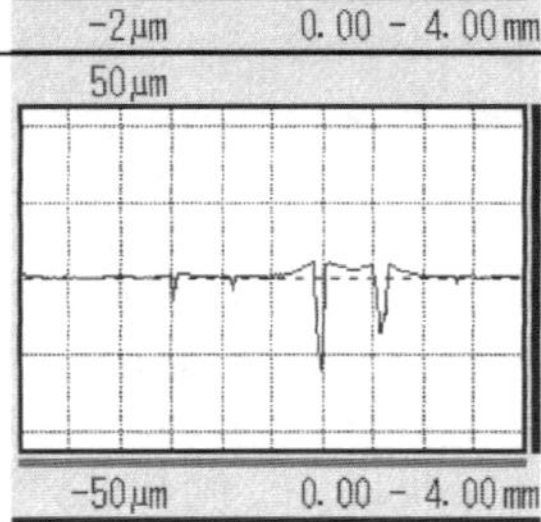	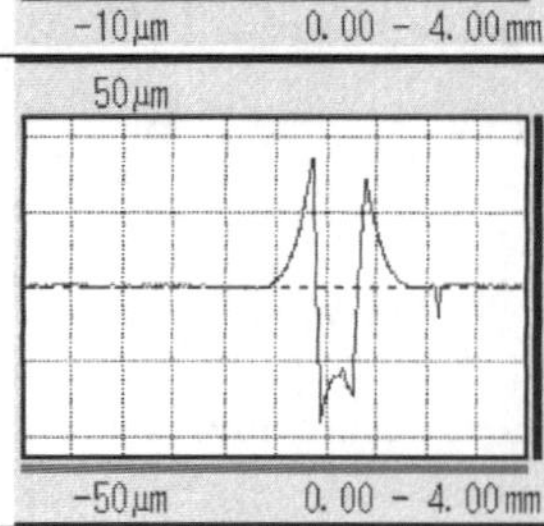

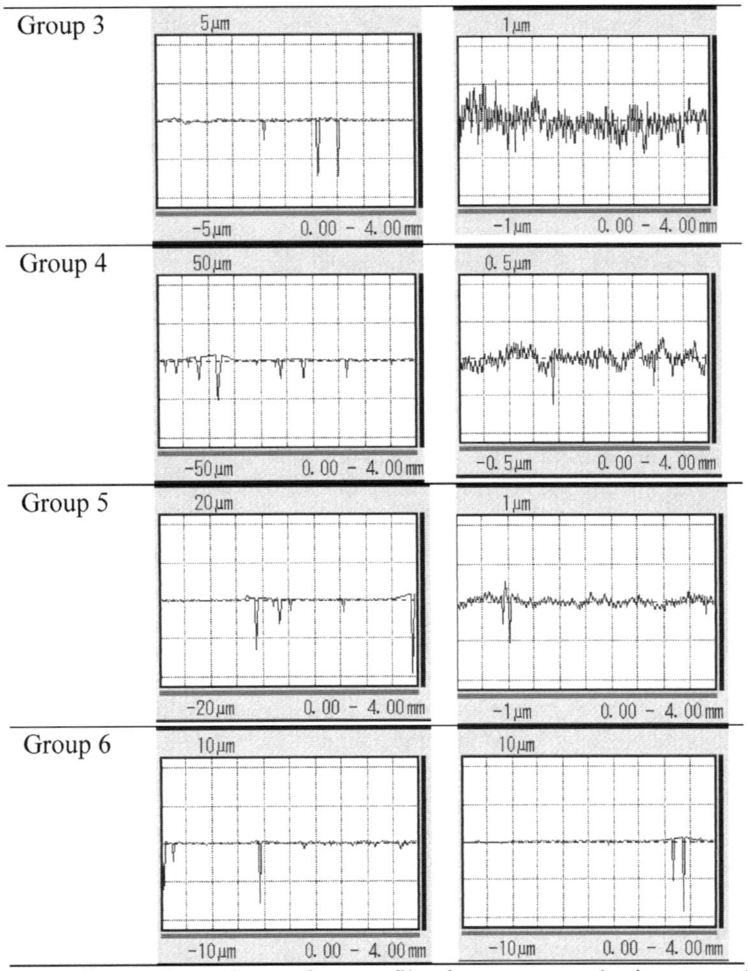

Figure 3. Composite resins surface profiles for some samples in groups 1-6.

The mean values of surface roughness (µm) and standard deviation of the samples in all six groups are presented in table 1.

A significantly increase of both composite resins surface roughness was recorded after erosive attack when compared to control group (ANOVA and post hoc Bonferroni statistical tests, $p<0.05$). Besides, after toothbrushing, the surface roughness of both composite resins increased significantly (ANOVA and post hoc Bonferroni statistical tests, $p<0.05$). When toothbrushing was simulated immediately after the erosive attack, the highest mean surface roughness was obtained. Toothbrushing at 30 minutes and at 60 minutes after the erosive attack leaded to a lower surface roughness of composite resins when compared to the group where toothbrushing was simulated immediately after the erosive attack. No significantly differences were recorded in the surface roughness values when brushing was simulated at 60 minutes after the acid attack and in the group where only the abrasive challenge was simulated.

G-ænial Anterior composite resin recorded significantly higher surface roughness values when compared to Essentia composite resin, in control group and in all six study groups (ANOVA and post hoc Bonferroni statistical tests, $p<0.05$).

Table 1. Mean values of surface roughness (μm) ± standard deviation

	Group 1	Group 2	Group 3	Group 4	Group 5	Group 6
G-ænial	0.070±0.02 Aa	0.142±0.04 Ba	0.090±0.02 Ca	0.424±0.02 Da	0.159±0.02 Ea	0.098±0.03 Fb
Essentia	0.042±0.03 Aa	0.093±0.03 Ba	0.056±0.03 Ca	0.297±0.01 Da	0.078±0.01 Ea	0.061±0.02 Fb

Similar capital letters in a column represent significant statistical difference between materials according to ANOVA and post hoc Bonferroni statistical tests ($p<0.05$). Similar small letters in a line represent significant statistical difference between groups, according to ANOVA and Bonferroni tests ($p<0.05$).

4. Discutions

In our study the two commercial composite resins presented different surface roughness. The amount, size, shape, distribution and hardness of filler particles, resin matrix conversion percentage, together with the interaction with the organic matrix, can explain the differences in composite resins roughness [12]. In control group and in study groups, G-ænial composite resin also presented higher surface roughness when compared to Essentia. G-ænial is a hybrid composite restorative with a combination of two types of pre-polymerized resin fillers (silica and strontium and lanthanoid) having the mean size of 16-17 μm, silica as inorganic filler with more than 100 nm size and fumed silica as inorganic filler with less than 100 nm size.

Degradation of composite resins is in a direct relation with the storage solution [13]. It was showed that the pH of immersion solution can accelerate the erosive wear. A lot of factors have the capability to influence the behaviour of restorative materials in acidic conditions [14-16]. In the present study both composite resins were affected by the contact with acidic beverage. Many studies showed that Coca-Cola or Coca-Cola-like beverages are one of the most aggressive erosive agents. The composite resin degradation in low pH solutions is due to hydrolytic degradation [17, 18]. First step in erosive degradation is represented by water absorption, that can further pass inside the resin matrix, at the interface with the fillers or other defects, compromising their reinforcing effects [19]. Also factors like the cross-link nature of the resin matrix, and the solvent sorption uptake have a direct effect on composite resins degradation [20]. Previous studies showed that the increase of composite resins roughness in acidic solutions is probably due to their soften surface that leaches the resin components and displaces the filler particles [21-23].

In the present study tothbrushing using a toothbrush having medium hardness of bristles and a toothpaste with low relative dentine abrasivity for thirty days leaded to higher surface roughness of both composite resins. Similar results were found in other studies [8, 24-26].

The final surface roughness of both composite resins after erosive attack, abrasion challenge or association of both processes was less then 0.2 mm, that is considered to be critical for bacterial retention [27]. For the patient surface roughness lower then 0.25-0.5 mm are impossible to be detected [28], so we can consider that the behaviour of both studied composite resins is clinically safe.

3. Conclusions

The exposure to acidic beverages like Coca-Cola increases the surface roughness of composite resins. Toothbrushing using a toothbrush having medium hardness of bristles and a toothpaste with low relative dentine abrasivity also increases the surface roughness of composite resins. Abrasive challenge immediately after and 30 minutes after an acidic attack determines the highest alteration of the surface state of composite resin. To a lesser alteration of composite resins surface roughness, toothbrushing 60 minutes after an acidic challenge is recommended.

References

[1] S. Wongkhantee, V. Patanapiradej, C. Maneenut and D. Tantbirojn, 2006 *J. Dent.*, 34, pp. 214.
[2] A. C. Valinoti, B. G. Neves, E. M. da Silva and L. C. Maia, 2008 *J. Appl. Oral Sci.*, 16, pp. 315.

[3] S. Stoleriu, G. Iovan, A. Georgescu, A. V. Sandu, M. Roşca and S. Andrian, 2012 *Rev. chim. Bucharest*, 63 (1), pp. 68.
[4] I. M. Hamouda, 2011 *J. Esthet. Restor. Dent.*, 23, pp. 315.
[5] A. B. de Paula, S. B. Fucio, G. M. Ambrosano, R. C. Alonso, J. C. Sardi and R. M. Puppin-Rontani, 2011 *Oper. Dent.*, 36, pp. 670.
[6] S. B. de Fúcio, A. B. de Paula, F. G. de Carvalho, V. P. Feitosa, G. M. Ambrosano, and R. M. Puppin-Rontani, 2012 *Am. J. Dent.*, 25, pp. 315.
[7] J. Park, C. Song, J. Jung, S. Ahn and J. Ferracane, 2012 *Oper. Dent.*, 37, pp. 532.
[8] J. da Costa, A. Adams-Belusko, K. Riley and J. L. Ferracane, 2010 *J. Dent.*, 38 (Suppl 2), pp. 123.
[9] Y. K. Lee, H. Lu, M. Oguri and J. M. Powers, 2005 *J. Prosth. Dent.*, 94, pp. 370.
[10] H. Lu, L. B. Roeder, L. Lei and J. M. Powers 2005 *J. Esthet. Restor. Dent.*, 17, pp. 108.
[11] M. Villarroel, N. Fahl, A. M. De Sousa and O. B. De Oliveira Jr., 2011 *J. Esthet. Restor. Dent.*, 23, pp. 73.
[12] S. Gladys, B. Van Meerbeek, M. Braem, P. Lambrechts and G. Vanherle, 1997 *J. Dent. Res.*, 76, pp. 883.
[13] I. Nica, G. Iovan, A. Ghiorghe, S. Stoleriu, G. Pancu and S. Andrian, 2014 *Rev. materiale plastice*, 3, pp. 282.
[14] I. Ikeda, M. Otsuki, A. Sadr, T. Nomura, R. Kishikawa and J. Tagami, 2009 *Dent. Mater. J.*, 28, pp. 679.
[15] Y. K. Lee and J. M. Powers, 2005 *J. Biomed. Mater. Res. B Appl. Biomater.*, 73, pp. 361.
[16] L. Han, A. Okamoto, M. Fukushima and T. Okiji, 2008 *Dent. Mater. J.*, 27, pp. 455.
[17] R. G. Chadwick, J. F. McCabe, A. W. Walls and R. Storer, 1990 *Dent. Mat.*, 6, pp. 123.
[18] A. U. J. Yap, C. L. Chew, L. F. K. L. Ong and S. H. Teoh, 2002 *J. Oral Rehabil.*, 29, pp. 87.
[19] N. K . Sarkar, 2000 *J. Biomed. Mater. Res.*, 53, pp. 371.
[20] A. Gopferich, 1996 *Biomaterials*, 17(2), pp. 103.
[21] A. L. Briso, L. P. Caruzo, A. P. Guedes, A. Catelan and P. H. dos Santos, 2011 *Oper. Dent.*, 36, pp. 397.
[22] V. V. Badra, J. J. Faraoni, R. P. Ramos, R. G. Palma-Dibb, 2005 *Oper. Dent.*, 30, pp. 213.
[23] A. B. Neamat, L. Han, A. Okamoto and M. Iwaku, 2000 *J. Esthet. Dent.*, 12, pp. 97.
[24] R. R. de Moraes, L de S. Goncalves, A. C. Lancellotti, S. Consani, L. Correr-Sobrinho and M. A. Sinhoreti, 2009 *Oper. Dent.*, 34, pp. 551.
[25] R. R. Moraes, S. Ribeiro Ddos, M. M. Klumb, W. C. Brandt, L. Correr Sobrinho and M. Bueno, 2008 *Braz. Oral Res.*, 22, pp. 112.
[26] R. Takahashi, J. Jin, T. Nikaido, J. Tagami, R. Hickel and K. H. Kunzelmann, 2013 *Dent. Mat. J.*, 32, pp. 75.
[27] C. M. Bollen, P. Lambrechts and M. Quirynen, 1997 *Dent. Mat.*, 13, pp. 258.
[28] C. S. Jones, R. W. Billington and G. J. Pearson, 2004 *Brit. Dent. J.*, 196, pp. 42.

IOP Conf. Series: Materials Science and Engineering **133** (2016) 012057 doi:10.1088/1757-899X/133/1/012057

Study of switching electric circuits with DC hybrid breaker, one stage

T Niculescu[1*], M Marcu[2], F G Popescu [3]

[1] Department of Control Engineering, Computers, Electrical Engineering and Power Engineering, University of Petrosani, 20 Univerisitatii str., Petrosani, Romania
[2] Department of Control Engineering, Computers, Electrical Engineering and Power Engineering, University of Petrosani, 20 Univerisitatii str., Petrosani, Romania
[3] Department of Control Engineering, Computers, Electrical Engineering and Power Engineering, University of Petrosani, 20 Univerisitatii str., Petrosani, Romania

E-mail: niculescu_titu@yahoo.com

Abstract. The paper presents a method of extinguishing the electric arc that occurs between the contacts of direct current breakers. The method consists of using an LC type extinguishing group to be optimally sized. From this point of view is presented a theoretical approach to the phenomena that occurs immediately after disconnecting the load and the specific diagrams are drawn. Using these, the elements extinguishing group we can choose. At the second part of the paper there is presented an analyses of the circuit switching process by decomposing the process in particular time sequences. For every time interval there was conceived a numerical simulation model in MATLAB-SIMULINK medium which integrates the characteristic differential equation and plots the capacitor voltage variation diagram and the circuit dumping current diagram.

1. Introduction

In the DC circuits that contain power breakers, a big problem is the disconnecting the loads, because an electric arc that can destroy its contacts occurs. In these situations we can use an additional circuits that are connected in parallel over the main breaker. These circuits generate a counter current injection which opposes the current breaker and results a current zero through the main breaker and the electric arc will be extinguished.

The basic electric circuit for switching DC systems is shown in figure 1:

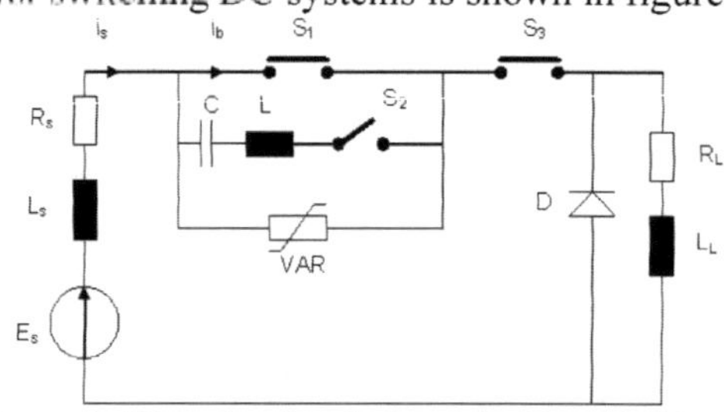

Figure 1. Electric circuit for switching DC loads.

* 20 Univerisitatii str., Petrosani, Romania, niculescu_titu@yahoo.com

The circuit from figure 1 contains a DC power E_s which has the intern parameters R_s and L_s, in series with a main breaker S_1. S_3 is an electric separator and at initial moment, a load with R_L and L_L parameters is connected. In parallel with S_1 a switching circuit is connected. This consists in the capacitor C, the inductivity L and the auxiliary switch S_2 (usually is a semiconductor).

For limiting the over voltages, the VAR varistor is applied across the main breaker, and the freewheeling diode D will take the current when its slope is negative.

The auxiliary switch S_2 is closed immediately after disconnection to extinguish the arc in the main breaker S_1. Energy stored in the capacitor C, is released and that generates a current that opposes the current through the breaker, resulting a zero current through its contacts. If the arc is not extinguished at the first current zero, the capacitor C and the inductor L form an oscillating circuit which will produce a second current zero. After arc extinguishing, the capacitor C is loaded again from the source E_s and is ready for a new circuit switching.

2. Analysis of circuit switching

We consider the continuous equivalent voltage across of the extinguishing circuit U_{lc}. In this case the equivalent resistance R, the voltage differential equation is given by (1):

$$LC\frac{d^2u_c(t)}{dt^2} + RC\frac{du_c(t)}{dt} + u_c(t) = U_{lc} \tag{1}$$

where the capacitive current is:

$$i_c(t) = C\frac{du_c(t)}{dt} \tag{2}$$

Equation (1) can be written in the equivalent form:

$$\frac{d^2u_c(t)}{dt^2} = \frac{1}{LC}\left[U_{lc} - u_c(t) - RC\frac{du_c(t)}{dt}\right] \tag{3}$$

Based on equations (2) and (3), the numerical simulation scheme in MATLAB-SIMULINK medium of arc current through the main breaker S_1 was released (Figure 2):

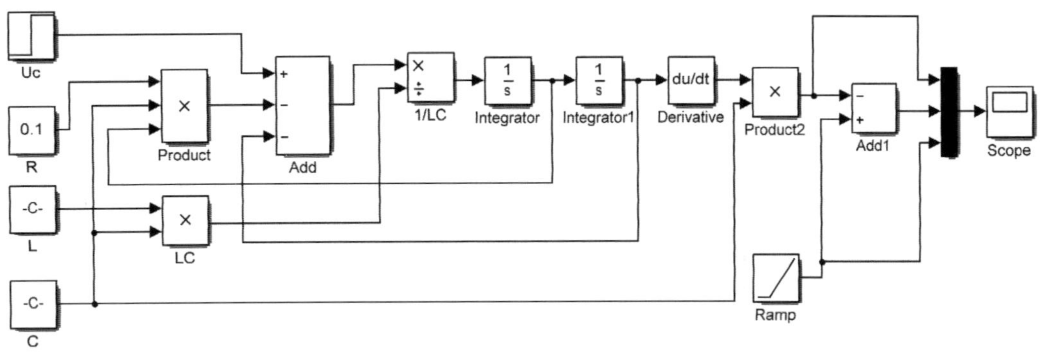

Figure 2. SIMULINK model of breaker currents.

The above scheme (figure 2) generates currents diagram from Figure 3. As is shown in figure 3, the current value i_{t1} is reached at time t_1 (reported at the time of load disconnection), At this moment voltage capacitor C is U_{C1} value. At this time t_1 a counter current injection in opposite to i_s occurs through the main breaker and the breaker current begins to decrease. At time t_2 it becomes zero and the capacitor voltage decreases to U_{C2} value.

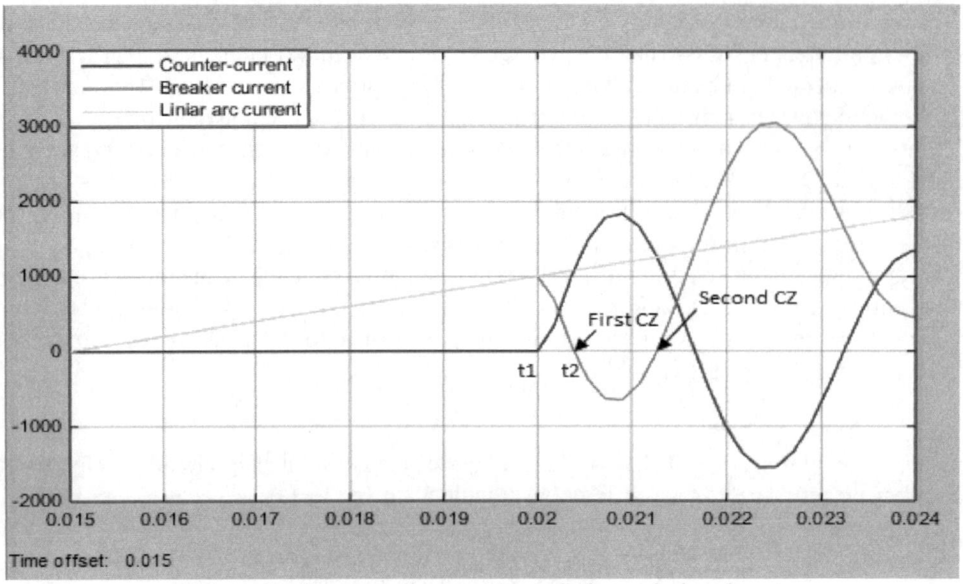

Figure 3. Currents diagram through the main breaker.

Forcing zero current through the main breaker lids to arc extinguishing. The SIMULINK models presented were made for the following values of electric parameters: U_c=2000[V]; R=0.1Ω; C=500μF; L=500μH.

3. Analyses of switching process

For the study of switching process there is used the circuit from figure 4.

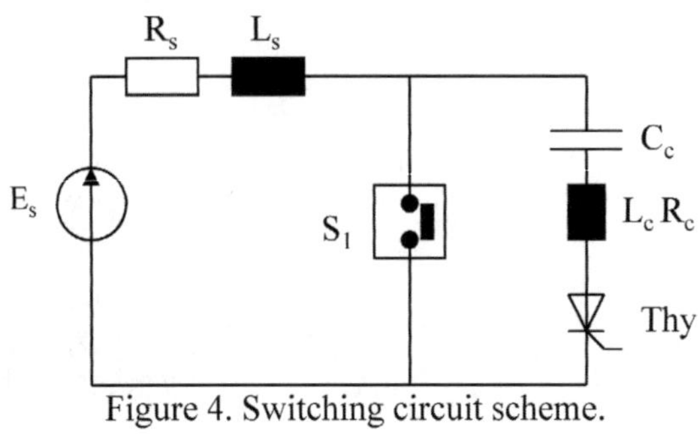

Figure 4. Switching circuit scheme.

The current interrupting process can be splitted in intervals, where every interval is described by a differential equation. These differential equations can be integrated using the MATLAB-SIMULINK medium, which allows plotting the variation curves for the current and capacitor voltage.

There are considered the following time periods:

- t_1 – time after that S_2 closed and dumping process begins;
- t_2 – time after that occurs the first current zero through the main breaker S_1;
- t_3 – time after that the source current was zero and C capacitor is fully charged at maximum.

3.1. Variation of electric parameters in the time interval [0 – t_1]

Throughout this time interval, the E_s source current passes the loop in figure 5 and is described by the differential equation (4):

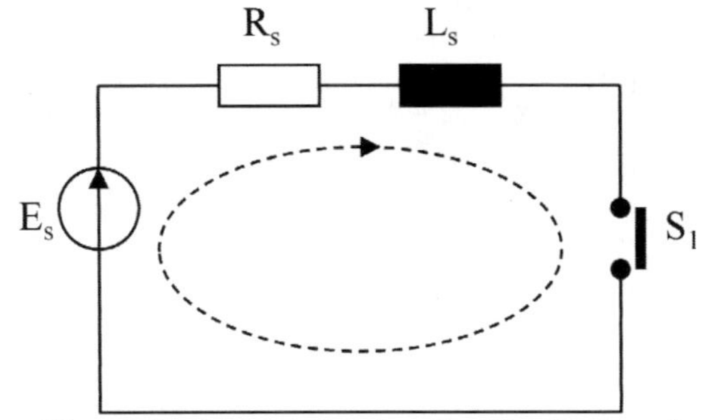

Figure 5. Current flowing for [0-t_1] interval.

$$L_s \frac{di_s}{dt} + R_s i_s = E_s \tag{4}$$

If the initial conditions for this interval are took into account, the circuit current expression is:

$$i_s(t) = \frac{E_s}{R_s}(1 - e^{-\frac{t}{T}}) \tag{5}$$

Where $T = L_s/R_s$ is circuit time constant.
Its variation is an exponential function.

3.2. Variation of electric parameters in the time interval [$t_1 - t_2$]

At the t_1 moment the oscillation in extinguishing circuit begins, and counter current through the main breaker occurs.

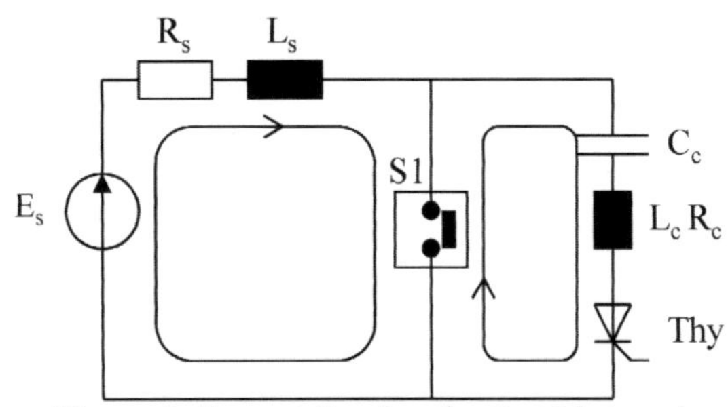

Figure 6. Current flowing for [t_1-t_2] interval.

The capacitor voltage which belongs to the extinguishing circuit is described by the second order differential equation:

$$LC\frac{d^2u_c}{dt^2} + RC\frac{du_c}{dt} + u_c = U_{c1} \tag{6}$$

Where U_{c1} is the voltage at which the C_c capacitor is charged at t_1 moment.

Differential equation (6) is integrated using MATLAB-SIMULINK medium. The numerical simulation scheme which plots the capacitor voltage variation and the current through the extinguishing circuit is presented in figure 7:

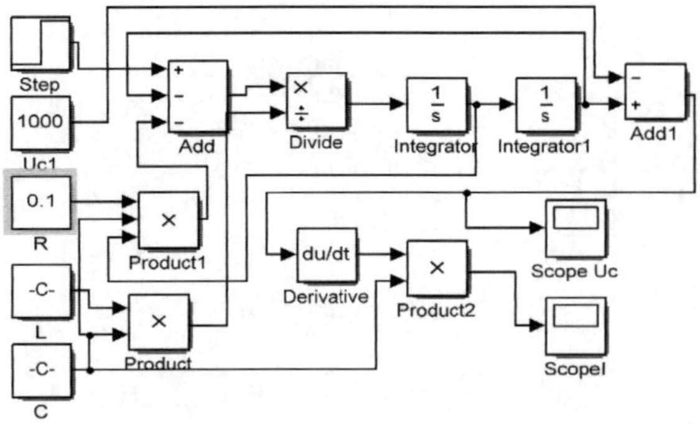

Figure 7. SIMULINK model for oscillating voltage and capacitor current for [t_1-t_2] interval.

This model generates the following diagrams (figure 8 and figure 9):

At moment t_2, the current through the main breaker S_1 became zero and the voltage across the capacitor reached U_{c2}.

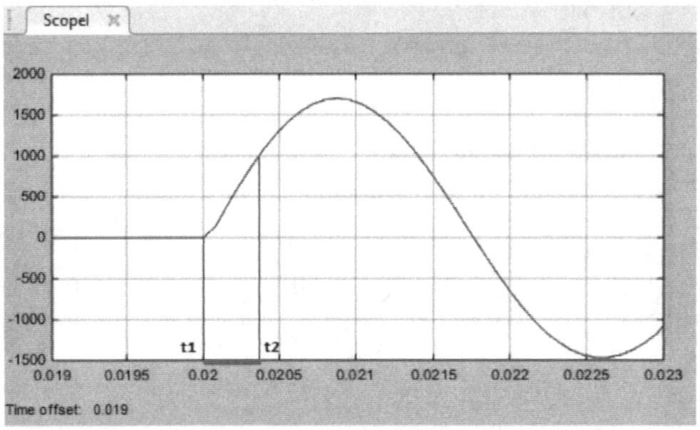

Figure 8. Current variation in [t_1-t_2] interval.

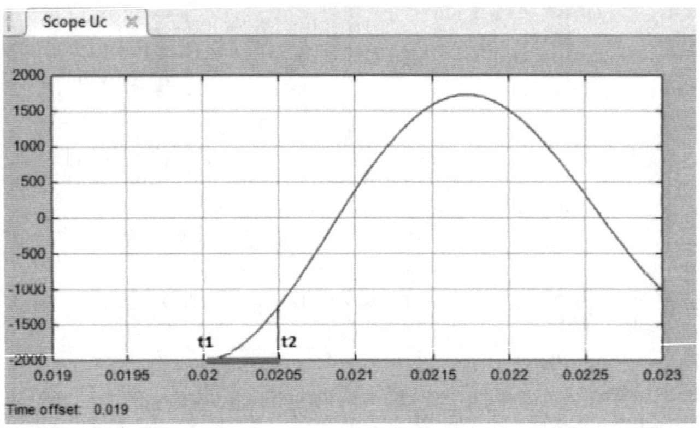

Figure 9. Capacitor voltage variation in [t_1-t_2] interval.

3.3. Variation of electric parameters in the time interval [$t_2 - t_3$]

In this interval, source current becomes equal to the capacitor current. The capacitor voltage at the initial moment is U_{c2}, and the current is I_{t2}. After the opening of main breaker, current flowing occurs according to scheme in figure 10:

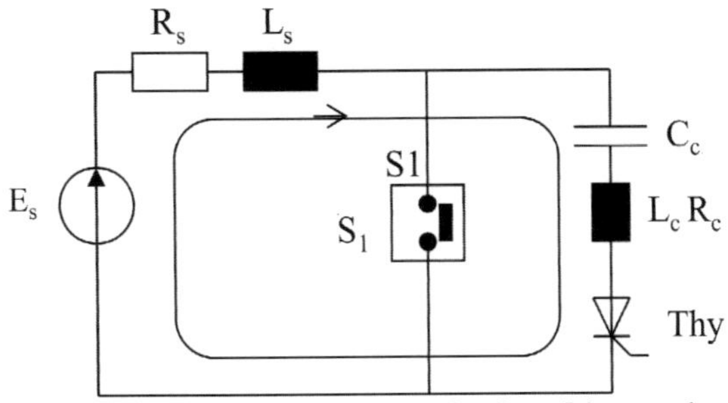

Figure 10. Current flowing for [t_2-t_3] interval.

The capacitor voltage which belongs to the extinguishing circuit, is described by the second order differential equation:

$$C_c(L_s + L_c)\frac{d^2u_c}{dt^2} + C_c(R_s + R_c)\frac{du_c}{dt} + u_c = U_{c2} \tag{7}$$

Differential equation (7) is integrated using MATLAB-SIMULINK medium. Numerical simulation model which plots the capacitor voltage variation and the current through the extinguishing circuit is presented in figure 11:

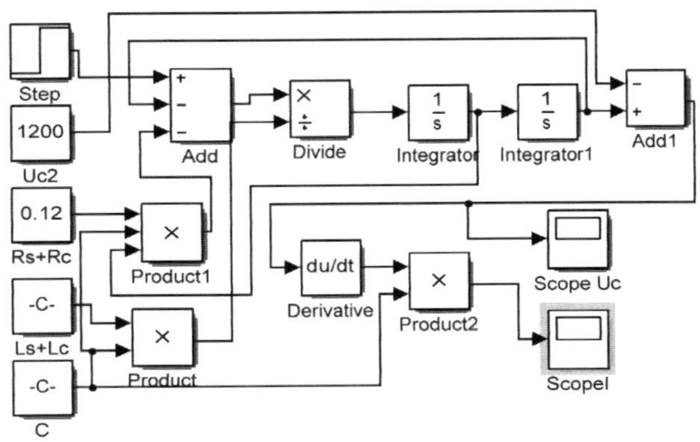

Figure 11. SIMULINK model for oscillating voltage and current through the capacitor for [t_2-t_3] interval.

The SIMULLINK model in figure 11 generates the following diagrams:

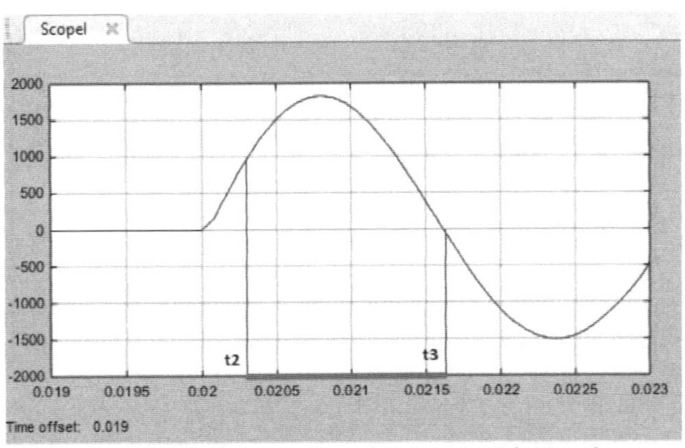

Figure 12. Current variation in [t_2-t_3] interval.

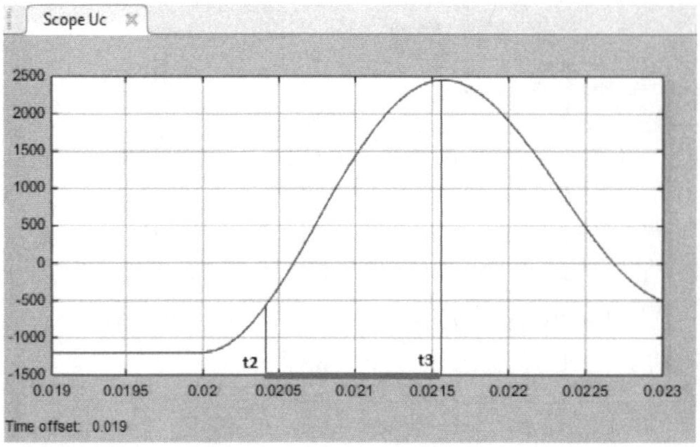

Figure 13. Capacitor voltage variation in $[t_2-t_3]$ interval.

At time t_3, the source current will be zero when the capacitor C_c was fully charged, see figure 13.

The SIMULINK models presented (Figure 7 and Figure 11) were made for the following values of electric parameters: U_c=2000[V]; R=0.1Ω; R_s=0.02 Ω; C=500μF; L=500μH; $L_s \approx 0$.

4. Conclusions

The auxiliary switch S2 was chosen unidirectional (solid state switch – thyristor) and the counter-current from a resonant LC circuit could provide two current-zeros. From this point of view exist two opportunities for circuit interruption.

Another way of interrupting the load current, involves the using of a bidirectional switch (vacuum breaker). In this case, after the source current was commutated, the current oscillation occurs for several time periods according to the circuit damping, until the final capacitor voltage becomes equal with the supply voltage. During oscillation, energy is transferred between source and LC circuit.

Depending on the commutation current value chosen, an appropriate switch might be found from the available power semiconductors. Solid-states switches were commonly vulnerable to increasing slope currents. But the switch had to be able to withstand surge counter-currents when switching on and surge voltages when switching off. The power semiconductors allow high current to be switched by lowering its frequency. When the frequencies of LC circuit are high, a good method for interrupting the load current, consists of using a vacuum breakers.

References

[1] A. Atmaji, 2000 *Direct Current Hybrid Breakers,* Data Library Technical University Eindhoven.

[2] M. Bartosik, 1997 *Progress in D.C. breaking,* Proc. 8th Int. Conf. Switching Arc Phenomena, Lodz, Poland 3-6 sept, Vol.2, pp. 25-41.

[3] Benouar, M., 1984 *Pyrotechnique circuit intrerupterfor the protection of electrical systems,* IEEE Trans. On Power Apparatus and Systems, Vol. Pas-103, No. 8.

[4] C.M. Franck, 2011 *HVDC Circuit Breakers: A Review Identifying Future Research Needs,* IEEE Trans. on Power Delivery, vol. 26, pp. 998-1007.

[5] M. Ghinea, V. Firteanu, 2000 *MATLAB calcul numeric- grafica – aplicatii,* Editura Teora, Bucuresti.

[6] W. Grieshaber, J.P. Dupraz, M. Collet, 2011 *Direct current during commutation and interruption: relation between interruption duration, insulation level and energy in the transmission means,* in print a MatPost11, 23&24, Lyon-France.

[7] M. Marcu, L. Samoilă, F.G. Popescu, 2013 *Power Active Filter Based on Synchronous Reference System Theory,* 1st WSEAS International Conference on Industrial and

Manufacturing Technologies (INMAT '13), Vouliagmeni, Athens, Greece, May 14-16.

[8] M. Merlin, T. Green, P. Mitchenson, D, Trainer, D. Critchley, R Crookes, 2010 *A new hybrid multi level Voltage-Source Converter with DC fault bloking capability*, 9th IET Intenational Conference on AC and DC Power Transmission 2010 ACDC, pp. 1-5.

[9] T. Niculescu, D. Pasculescu, 2006 *The using of Matlab program to study the transitory phenomena in the circuits with ideal elements* A-XI-A Editie A Conferintei Stiintifice a Facultatii de Inginerie, cu participare internationala, pp. 197-199.

[10] T. Niculescu, I. Uţu, M. Marcu, 2006 *The utilization of MATLAB-SIMULINK software to study the transitory phenomena in the d.c. circuits – microCAD*, International Computer Science Conference, University of Miscolk, pp. 59-62.

[11] T. Niculescu, 2006 *Analiza circuitelor electrice prin simulare in spatiul MATAB,* Editura Universitas, Petrosani, ISBN(10)973-741-025-4; ISBN(13)978-973-741-025-2.

[12] T. Niculescu, M. Niculescu, 2010 *The study of fundamental electrical circuits transitory phenomena using MATLAB software,* Simpozion international multidiscilplinar UNIVERSITARIA SIMPRO 2010, Electrical Engineering and Energetics System Control, Applied Informatics and Computer Engineering, Editura Universitas Petrosani, pp. 31-34.

[13] T. Niculescu, D. Pasculescu, L. Pana, 2010 *Journal Wseas Transactions on Circuits and Systems*, Vol. 9, pp.430-439.

[14] T. Niculescu, 2012 *Technology and Engineering Applications of Simulink*, Chapter 2: Study of Inductive-Capacitive Series Circuits Using the Simulink Software Package ISBN 978-953-51-0635-7, Hard cover, Publisher: InTech, Published: May 23, under CC BY 3.0 license, in subject Technology.

[15] T. Niculescu, M. Marcu, L. Pană, F. Popescu, 2013 *Study of switching electric circuits with DC hybrid breaker, one stage* – Proceedings of the 15th International Conference on Automatic Control, Modelling & Simulation (ACMOS '13), Brasov, Romania, June 1-3, ISSN: 1790-5117 ISBN: 978-1-61804-189-0, pp. 37-41.

[16] S. Halunga-Fratu, O. Fratu, 2004 *Simularea sistemelor de transmisie analogice si digitale folosind mediul MATLAB/SIMULINK*, Editura MatrixRom, Bucuresti.

IOP Conf. Series: Materials Science and Engineering **133** (2016) 012058 doi:10.1088/1757-899X/133/1/012058

Sensory evaluation based fuzzy AHP approach for material selection in customized garment design and development process

Y Hong[1,2,3*], A Curteza[1], X Zeng[2], P Bruniaux[2], Y Chen[3]

[1] College of Textile, Leather and Industrial Management, "Gheorghe Asachi" Technical University of Iasi, B-dul. D. Mangeron Nr. 67 700050, Iasi, Romania
[2] GEMTEX, ENSAIT, 2 allée Louise et Victor Champier, 59056 Roubaix Cedex 1, France
[3] College of Textile and Clothing Engineering, Soochow University, Suzhou 215021, China

E-mail: yannichonghk@gmail.com

Abstract. Material selection is the most difficult section in the customized garment product design and development process. This study aims to create a hierarchical framework for material selection. The analytic hierarchy process and fuzzy sets theories have been applied to mindshare the diverse requirements from the customer and inherent interaction/interdependencies among these requirements. Sensory evaluation ensures a quick and effective selection without complex laboratory test such as KES and FAST, using the professional knowledge of the designers. A real empirical application for the physically disabled people is carried out to demonstrate the proposed method. Both the theoretical and practical background of this paper have indicated the fuzzy analytical network process can capture expert's knowledge existing in the form of incomplete, ambiguous and vague information for the mutual influence on attribute and criteria of the material selection.

1. Introduction

Material selection has been considered to be the main task for the designers in the garment design and development process. Utilizing the ideal material assists enterprises to follow market position, satisfy the customers, and shorten the product development cycle, permitting to quick feedback from the market. The increasing requirement for customized design strengthened the importance of the fabric selection, especially for the design with special requirement such as disability, functionality and protection [1].

Traditionally, designers select an appropriate fabric material based on their knowledge and experience for fashion product development. When preparing a customized design collection, the designers will search a collection of fabric, referring to the current trend and the requirement of the customer. Then, based on the desired style of the garment, fabrics are generally selected by garment designers according to a number of physical and basic sensory criteria such as fabric hand and fabric appearance. The sensory evaluation based on individual sensory criteria can be fast and effective for

* Address: 10 rue Charles Quint 59100 Roubaix France, Email: yannichonghk@gmail.com

the designers, but the decision-making can be difficult. Also, as the designers have no specific training about the physical parameters of the fabric, there are test systems developed to help the designers acquire the mechanical properties such as KES. But the test is time consuming and the KES test machine is expensive. The result can be very precise but it is not so applicable because of the time and expense problem [2].

In practice, the issues for selecting the most suitable fabric have emerged as the most crucial topics for the designers. The requirements from the market and the similarity among the fabrics influentially affect the designers to make the decision. The selection of fabrics for specific market is a multi-criteria decision making problem. There are considerable numbers of decision models have been developed to analysis the factors from the goal, such as the preference ranking organization method, the analytical hierarchy process (AHP), discrete choice analysis, and data envelopment analysis. Using these models, the weight and the grade of importance can be quantized to demonstrate the preference of different factors in the material selection. If the importance of a criterion can be captured properly, the quality of the decision-making for material selection will be enhanced correspondingly. However, the available information in a multi-criteria decision process is usually uncertain, vague, or imprecise, and the criteria are not independent in the real situation. In the real world, this usually exists when the brand image selection is full of uncertainty and the imprecision of human subjective judgment. Thus, in this paper fuzzy sets theory is used to cope with these situations [3].

In this context, a sensory evaluation based fuzzy AHP approach is proposed for material selection in customized garment design and development process.

2. Methodologies and scope of study

It can be easily found that, the main difficulties of in this study are as follows:

(1) Acquisition of customer's requirements and needs;

(2) Application of textile knowledge for the analysis and comparison of customer's requirements and needs;

(3) Knowledge representation and classification. The methodologies and scope will be discussed in these aspects.

2.1. Knowledge acquisition and classification

For the acquisition of the customer's requirement and expert's experience and knowledge, fuzzy set, fuzzy numbers and fuzzy semantic representation will be applied.

Fuzzy sets and fuzzy numbers: The applications of fuzzy techniques are mainly concentrated on domestic appliances like washing machines, air conditioners, cameras and camcorders as well as public devices such as water treatment systems, port cranes, subways and ventilation systems.

Compared with other tools of artificial intelligence, fuzzy techniques are particularly efficient for processing information uncertainty and imprecision related to human factors. It is for this reason that researchers have paid much attention to fuzzy techniques when developing decision support systems in uncertain environments, such as finance and medical diagnosis [4].

Fuzzy semantic representation: Normally, when in the evaluation of certain things, people will use descriptive language to express their feelings at most of the time. In order to make the computer identify these things, a way to quantify the human thought patterns with different semantics to be expressed in the form of the number must be chosen. This study uses semantic variables method to convert fuzzy semantic value in the form of Triangular Fuzzy Numbers [5].

2.2. Multi-criteria decision-making (MCDM)

MCDM deals with a selection problem under the presence of multiple (a finite number of) decision criteria and alternatives. Many exponents of MCDM are available which have enjoyed a wide acceptance in the academic area and many real-world applications. Weighted sum model (WSM), Weighted product model (WPM), the Analytic hierarchy process (AHP), Revised analytic hierarchy process (RAHP), Technique for order preference by similarity to ideal solutions (TOPSIS), and elimination and choice translating reality (ELECTRE) are among the most popular ones. Each of these methods has their own characteristics and background logic, as well as merits and demerits. It is almost impossible to decide which one is the best decision making method. The choice of the method depends on the complexity of the decision problem [6].

3. General research scheme using fuzzy AHP method based on sensory evaluation

In this study, a sensory evaluation based fuzzy AHP approach is proposed for material selection in customized garment design and development process. A real empirical application for the physically disabled people is carried out to demonstrate the proposed method. The general research scheme is described in figure 1.

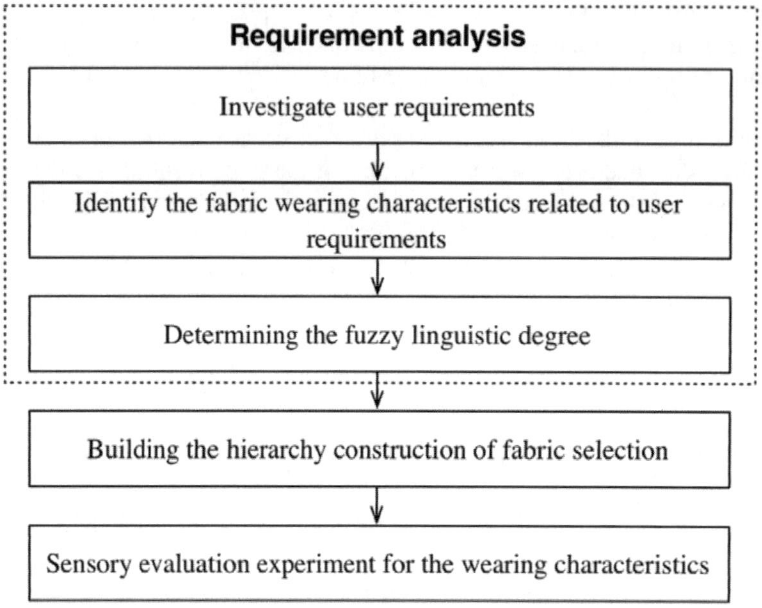

Figure 1. General research scheme using fuzzy AHP method based on sensory evaluation.

Firstly, the requirements from the market (consumer) will be analyzed. Then, the related wearing characteristics will be discussed. Fuzzy set will be applied in the following procedure to capture the importance of the requirements and wearing characteristics. Next, a collection of sensory evaluation experiment will be carried out to compare the pre-selected fabrics based on different wearing characteristics.

The final fabric will be selected based on the score ranking, determined by the sensory evaluation score and their corresponding weight value given from the analytical hierarchy process.

4. Construction of fuzzy AHP in material selection

The analytic hierarchy process (AHP) is developed to help decision makers modeling a complex problem in a hierarchical structure showing the relationships of the goal, objectives (criteria), sub-objectives, and alternatives (figure 2). All influencing factors can also be included. AHP enables decision-makers to derive ratio scale priorities or weights as opposed to arbitrarily assigning them. AHP has the following functions: (1) helping the decision-makers structure the issues into goal, attributes (criteria) and alternative, (2) permitting to obtain the weight of all alternatives that are deficient with respect to one or more objectives can compensate by their performance with respect to other objectives. The AHP procedure involves six essential steps [7].

1. Define the unstructured problem
2. Developing the AHP hierarchy
3. Pair-wise comparison

4. Estimate the relative weights

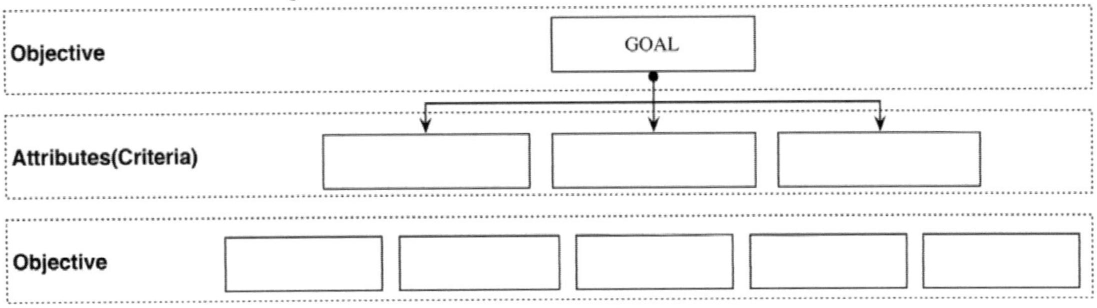

Figure 2. Hierarchy structure of decision problem.

4.1. Formation the structure of decision hierarchy and d eveloping the AHP hierarchy

4.2. The goal in the analytical hierarchy process is to select the most suitable fabric aiming at the specific target market. To get the goal, the requirements should be considered. Then, the wearing characteristics related to the requirements can be defined.

Let B be a set of n requirement criteria, denoted as $B=\{b_1, \ldots, b_n\}$.
Let C be a set of m wearing characteristics, denoted as $C=\{c_1 \ldots c_m\}$.
Then, the problem of material selection can be decomposed into a hierarchical structure.

4.3. Pair-wise comparison

In this step, the relational data for comparing the alternatives or options are generated. This requires the decision maker to evaluate the relative importance and then formulate pair-wise comparison matrices of elements at each level of the hierarchy relative to each activity at the next higher level. In AHP, if a problem involves M alternatives and N criteria, then the decision maker has to construct N judgment matrices of alternatives of $M \times M$ order and one judgment matrix of criteria of $N \times N$ order. The Triangular Fuzzy Numbers was introduced to quantify the 7-point scales of linguistic description of the intensity of importance. Table 1 shows the relation between the intensity of importance on an absolute scale and Triangular Fuzzy Numbers [8].

Table 1. Relationship Between the intensity of importance on an absolute scale and Triangular Fuzzy Numbers.

Linguistic Descriptions	**Triangular Fuzzy Numbers**
Extreme Unimportant	(0,0,1)
Very Unimportant	(0,1,3)
Rather Unimportant	(1,3,5)
Average	(3,5,7)
Rather Important	(5,7,9)
Very Important	(7,9,10)
Extreme Important	(9,10,10)

4.4. Estimate the relative weights

The relative importance of different elements in each pair-wise comparison matrix can be determined using the Triangular Fuzzy Numbers in table 1.

First, the relevant importance of the fabric requirements with respect to the goal should be discussed. In this process the experience of wearer should be acquired. Sensory evaluation method was applied in the process. *M* disabled people were interviewed to compare the importance of the fabric requirements with respect to the goal.

N experts in textile engineering were interviewed compare the importance and the wearing characteristics. In this process the experience and knowledge was applied and simulated.

If m is the wearing characteristics, m=1, 2, ……, 9, i is the fabric requirement, i =1, 2, 3, 4, n is the number of experts, n=1, 2, ……, N, the result can be represented with Triangular Fuzzy Numbers in the format as

$$\tilde{t}_{im} = (t_{imn1}, t_{imn2}, t_{imn3}) = \left(\frac{1}{k}\sum_{1}^{n} t_{imk1}, \frac{1}{k}\sum_{1}^{n} t_{imk2}, \frac{1}{k}\sum_{1}^{n} t_{imk3}\right) \quad (1)$$

For H criteria the size of the comparison matrix *(H1)* will be $H*H$ and the entry H_{ij} will devote the relative importance of criterion i with respect to the criterion j. in the matrix, $H_{ij=1,}$ when i=j and $H_{ji=i}$ $H_{ij.}$

The relative weights (W) of matrix A is obtained from following equation:

$$AW=\lambda max W, \quad (2)$$

Where λmax is the biggest eigenvalue of matrix A, I = unit matrix.

$$\lambda = \frac{1}{n}\sum_{i=1}^{n}\frac{(AW)_i}{w_i} = \frac{1}{n}\sum_{i=1}^{n}\frac{\sum_{j=1}^{n} a_{ij} w_j}{w_i} \quad . \quad (3)$$

4.5. Check the consistency and Obtain the overall rating

In this step the consistency property of matrices is checked to ensure that the judgments of decision makers are consistent. For this end some pre-parameter is needed. Consistency Index (CI) is calculated as:

$$CI = \frac{\lambda max - N}{n - 1} \quad (4)$$

The consistency index of a randomly generated reciprocal matrix shall be called to the random index (RI), with reciprocals forced. Using a sample size of 100 generated an average RI for the matrices of order 1–15. The table of random indexes of the matrices of order 1–15 can be seen in Saaty. The last ratio that has to be calculated is CR (Consistency Ratio). Generally, if CR is less than 0.1, the judgments are consistent, so the derived weights can be used. The formulation of CR is:

$$CR = CI/RI \quad (5)$$

Then the relative weights of decision elements are aggregated to obtain an overall rating for the alternatives as follows:

$$w_k = \frac{\sum_{j=1}^{n} a_{kj}}{\sum_{i=1}^{n} \square \sum_{j=1}^{n} a_{kj}} . \qquad W=(w_1, w_2, \cdots, w_n)^T \quad (6)$$

where λmax means the maximum Characteristic root of the pairwise comparison matrix A, W is the corresponding eigenvector of the maximum Characteristic root. After this, the weight values should be synthesized to get the relevant importance of the wearing characteristics.

Table 2. Relative weights of the wearing characteristics.

Wearing Characteristics	**Weight**
C_1	U_1
C_2	U_2
C_3	U_3
…	…
C_M	U_M

5. Empircal study

This study investigates a fabric selection for disabled people in the customized garment design and development process to demonstrate the proposed sensory evaluation based fuzzy AHP method.

5.1. Identifying the customer's requirements

The need for disabled people is more complicated than that of normal people, leading to the different and special requirement of fabric selection. Elements of fabric requirement for physical disabilities should be investigated and analyzed.

In order to understand the current needs and indeed thoughts of physically disabled people about fabrics, interviews and survives were conducted in this research. These requirements includes:

Requirements from the disabled body part: Terminals of the disabled body part, such as stump of the amputation, feet of polio patients, etc., often face the problems of poor blood circulation, low skin temperature and prone to skin allergies. More soft and warm fabric is needed in these areas.

Requirements from the auxiliary tools: Auxiliary tool plays a very important role in daily life of physically disabled. Usually auxiliary tools will have a direct contact with human body. In this case, fabrics with moisture management and air management can meet the requirements for wearers well.

Requirements from safety and protection: Most of persons with disabilities have the problem of the mobility so the safety and protection performance should be put forward to meet wearer requirements. For example, physically disabled are easy to tumble, so the fabric should have a good wearing continuance.

5.2. Identifying the fabric wearing characteristics related to customer's requirements

Based on the interview and survey of the fabric requirements for disabled people, we can also easily find that most of the requirements have a lot of relationship with the fabric wearing characteristics. From the view of product design, relationship should be investigated and analyzed between the requirements and fabric wearing characteristics. Wearing characteristics include: moisture management, air management, warmth retention property, flexibility, wearing continuance, antimicrobial resistance, stain resistance, flame resistance, antistatic, softness and other properties. But not all of these wearing characteristics will influence the wearer's requirements a lot, so a rough selection should be conducted. After the interview of the textile expert, the following wearing characteristics are considered to be the most important criteria related to the wearer's requirements:
Softness, flexibility, moisture management, warmth retention property and wearing continuance.

5.3. Establishing the hierarchical structure

The Investigation of user requirements and fabric wearing characteristics can offer the information for the formation of decision hierarchy and also it can easily to be found that the user requirements and fabric wearing characteristics have the relationship of Hierarchy. Based on this the decision hierarchy can be formatted as figure 3.

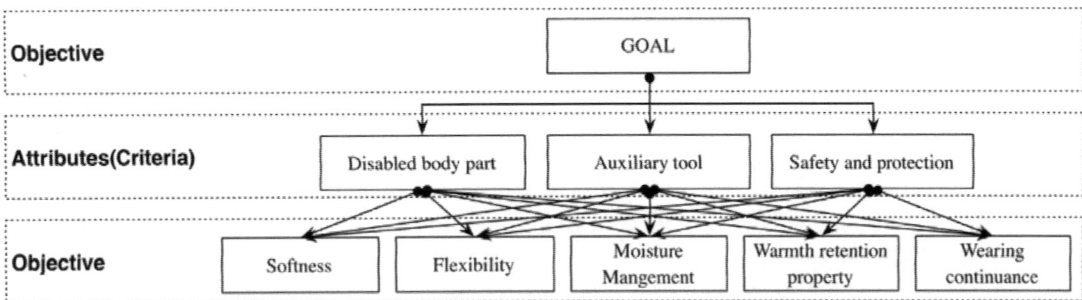

Figure 3. Hierarchical structure of fabric selection for physically disabled people.

5.4. Experiment I (Fabric selection modelling)

20 disabled people were interviewed to compare the importance of the fabric requirements with respect to the goal. *30* experts in textile engineering were interviewed compare the importance and the wearing characteristics. In this process the experience and knowledge was applied and simulated.

After the calculation the weights for the obtained wearing characteristics are in table 3.

Table 3. Relative importance among Wearing characteristics.

Wearing Characteristics	**Weight**
C_1	23.9
C_2	18.3
C_3	27.6
C_4	16.5
C_5	13.7

Then the model for the selection is established.

5.5. Experiment II (Fabric selection for the garment design aiming at physically disabled people)

Here an experimental design case discussed using the proposed model. The design case is to design a summer dress for physically disabled people. Based on the request, five ideal fabrics are preselected by the designers. Then, 10 experienced designers are invited to carry out the sensory evaluation experiment. Then the wearing characteristics are compeered to have a ranking. Different scores corresponding to different wearing characteristics are given from 1-5. At last, the weights of the characteristics are Complex calculated with the sensory evaluation scores. The best fabric is then selected.

6. Conclusions

In this study, the analytic hierarchy process and fuzzy sets theories have been applied to establish a dynamic Material selection model. The requirements of the customers can be fully considered and analyzed. A list of wearing characteristics and their corresponding weights can be obtained using the proposed model based on the customer's requirements. Then, sensory evaluation experiment of preselected fabric permits to a quick and effective fabric selection based on the wearing characteristics. The analytic hierarchy process and fuzzy sets theories have been applied to mindshare the diverse requirements from the customer and inherent interaction/interdependencies among these requirements. Sensory evaluation ensures a quick and effective selection without complex laboratory test such as KES and FAST, using the professional knowledge of the designers. A real empirical application for the physically disabled people is carried out to demonstrate the proposed method. Both the theoretical and practical background of this paper have indicated the fuzzy analytical network process can capture expert's knowledge existing in the form of incomplete, ambiguous and vague information for the mutual influence on attribute and criteria of the material selection.

References

[1] L.I. Iezzoni, et al., 2014 *General health, health conditions, and current pregnancy among U.S. women with and without chronic physical disabilities*. Disabil Health J., 7(2), pp. 181-8.

[2] Y. Zhu, et al., 2010 *A general methodology for analyzing fashion oriented textile products using sensory evaluation*. Food quality and preference, 21(8), pp. 1068-1076.

[3] X. Zeng, D. Ruan, and L. Koehl, 2008 *Mathematics and Computers in Simulation*, 77(5-6), pp. 443-452.

[4] X. Liu, et al., 2008 *Mathematics and Computers in Simulation*, 77(5-6), pp. 512-521.

[5] X. Chen, et al., 2015 *Knowledge-Based Systems*.

[6] L.-Z. Lin, and T.-H. Hsu, 2011 *Applied Soft Computing*, 11(1), pp. 561-573.

[7] A.H.I. Lee, W.-C. Chen, and C.-J. Chang, 2008 *Expert Systems with Applications*, 34(1), pp. 96-107.

[8] L. Koehl, et al., 2008 *Analysis and identification of fashion oriented industrial products using fuzzy logic techniques*. in Intelligent System and Knowledge Engineering, ISKE, 3rd International Conference on IEEE.

IOP Conf. Series: Materials Science and Engineering **133** (2016) 012059 doi:10.1088/1757-899X/133/1/012059

Quality Engineering Tools Focused on Designing Remote Temperature Measurements for Inaccessible Locations by Using Light Components Parameterization of the Heated Materials

M. Rîşteiu[1], R. Dobra[2*], D. Pasculescu[2], A. Ahmad Mohammad[3]

[1] Department of Computer Science, "1 Decembrie 1918" University Alba-Iulia, Gabriel Bethlen Str., No.5, 510009 Alba Iulia, Romania
[2] Department of Computers and Power Engineering, University of Petrosani, 20 Universitatii str, Petrosani, România
[3] Department of Physics, Faculty of Science, P. O. Box 741, University of Tabuk, Tabuk, 71491, Saudi Arabia.

E-mail: remusdobra@upet.ro

Abstract. This paper is focused on research dedicated to measure the bucket wheel bearing temperature of the bucket wheel excavator (BWE). It proposes a measurement method for heating friction materials because is difficult to detect the temperature variation in the bearing. The major issue is to detect the generated infrared light according to the material detection. The temperature is considered the major signal of the wheel reliability and a remotely temperature detection method is proposed and because the sleeve bearing is a crucial part of the excavator a predictive measurements system for buckled wheel axis system was designed.

1. Introduction

In lighting design they are two large efforts: one is dedicated by the lamps producers, researchers and users in order to realize efficient lighting systems based on semiconductor devices [1]; second development direction is dedicated to correct lighting from the eye sensitivity point of view [1']. On the other hand, power LED lighting systems are more suitable, and more convenient for automation, lighting scattering managing, with intelligent implementations. A very important demand of LED lighting applications, out of indoor or outdoor lighting is the correctly render of the colors [2, 3]. According to [2, 3], if the demands are satisfied, the tendencies of practical implementations are:

- good color rendering for material analyzing,
- good color rendering for safety,
- high reliability devices and simple commanded devices,
- smart lighting (Adaptive Environment Lighting - AEL),
- energy efficiency and fast light switching.

* remusdobra@upet.ro

2. Typical switch mode power supply (SMPS) topologies

The implementation of LED lighting has some requirements and the effort of performing these requirements driven the industry toward smaller, lighter and more efficient electronics, development of the Switch Mode Power Supply (SMPS) is an issue. A very good synthesis of topologies commonly used to implement SMPS is presented in [4] and [5]. Although certain topology may be used in many applications, there is for any application a best suited topology. The choice of the best topology must take into account [4] several factors, like: the relation between input and output voltage, number of outputs, requirement for electric isolation, values for input/output voltages and currents and the maximum duty cycle.

1.1. Transitory regime analysis of the designed boost-up module with TPS55340

For the paper purpose, in this section we are proposing two issues to be solved. One issue is related to transient analysis, and second issue is related to the components related thermal stress. There are several ways for running such designing simulation scenarios as the Spice model which is available on the TI sites can be exported as other tools, but also can be used in the web simulation tools of TI called WEBench [6-12]. Running WEBench designer with many solutions, based on our criteria, the circuit diagram is shown in figure 1. The simulation interface of the WEBench uses four designing areas (A-schematic designing, B- menu area, C- simulation tools, D- thermal evaluation), which are available on entire session simulation. For the simulation, based on the TPS55340 datasheet, the block diagram for parameters controlling is shown in figure 2.

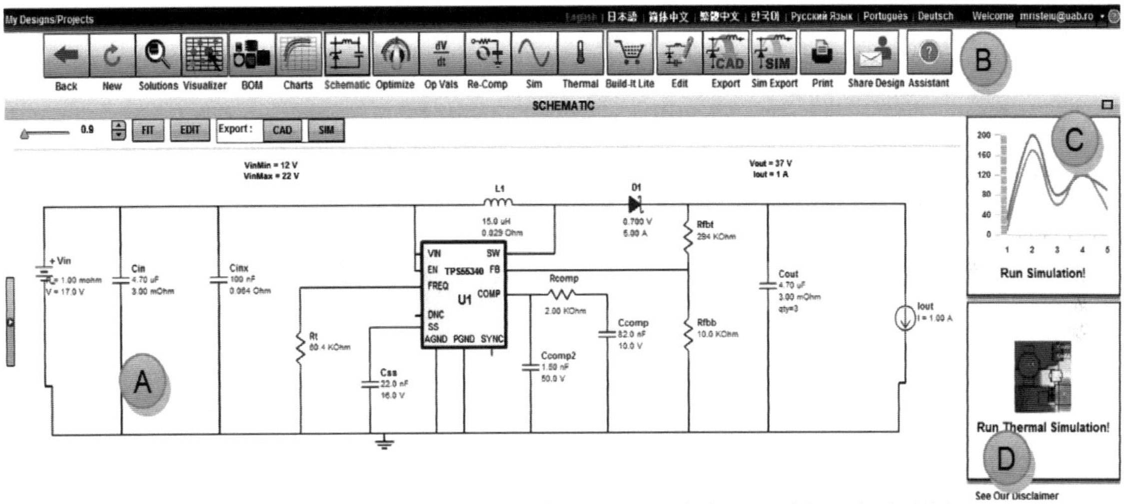

Figure 1. The graphical interface of WEBench designer with main facilities.

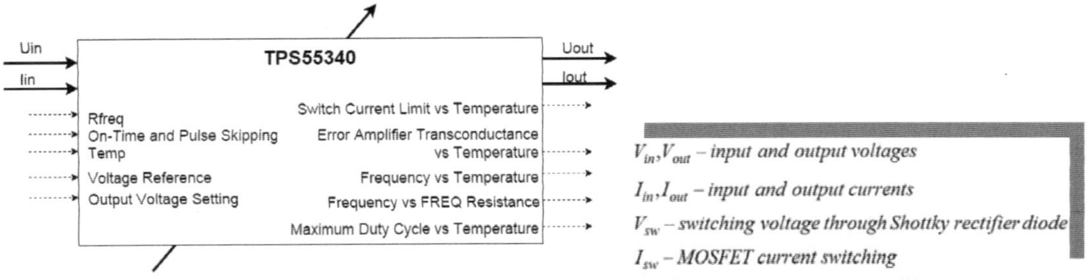

Figure 2. The block diagram of the TPS55340 for parameters controlling.

At simulation start-up, based on our input parameters (Vin = 12V, Vout =37V, Iout =1 A), the WEBench offer some recommended values for the major requirements [Uout, Iout] = f [Uin, Iin], but they are many other possibilities to control some output parameters according to some local conditions input values (operational temperature, switching frequency, duty cycle). By analyzing some simulation reports, they are some critical points inside to circuit diagram. They are shown in figure 3.

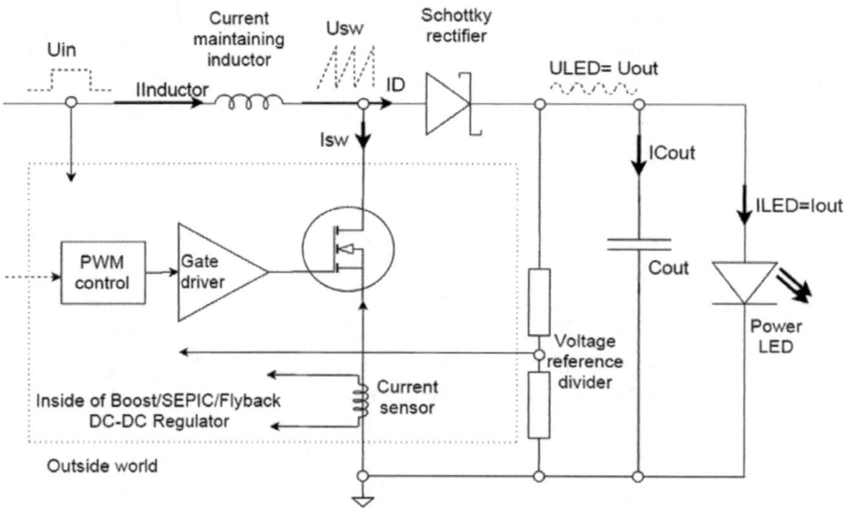

Figure 3. TPS55340 connected into a power supply boost-up circuit.

Some other stressing signal variations are inductor current (*IInductor*), and output capacitor current (*ICout*). First, as figure 3 suggests, we are simulating the transitory output voltage simulation according to the input supplying.

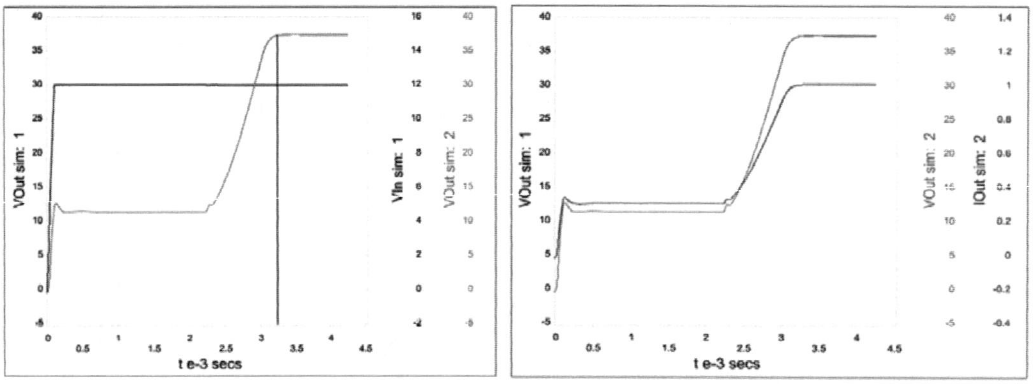

a) b)

Figure 4. a) The first simulation diagram: the output voltage according to the input voltage, with default setting up parameters; b) The output current runs similar with output voltage.

1.2. Measuring visible and infrared radiation with micro sensor TLS2561

The measuring method is based on two integrating analog-to-digital converters (ADC) that integrate currents from two photodiodes. Integration of both channels occurs simultaneously. Upon completion of the conversion cycle, the conversion result is transferred to the Channel 0 and Channel 1 data registers, respectively. The transfers are double-buffered to ensure that the integrity of the data is maintained. After the transfer, the device automatically begins the next integration cycle.

It combines one broadband photodiode (visible plus infrared) and one infrared-responding photodiode on a single CMOS integrated circuit capable of providing a near-photopic response over an effective 20-bit dynamic range (16-bit resolution).

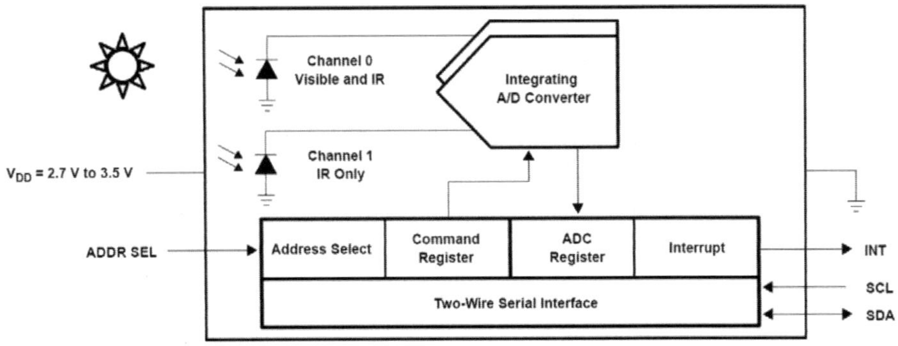

Figure 5. Measuring visible and infrared radiation with micro sensor TLS2561.

Two integrating ADCs convert the photodiode currents to a digital output that represents the irradiance measured on each channel.

1.3. Measuring visible color, with infrared filter using TCS3472

The TCS3472 device provides a digital return of red, green, blue (RGB), and clear light sensing values. An IR blocking filter, integrated on-chip and localized to the color sensing photodiodes, minimizes the IR spectral component of the incoming light and allows color measurements to be made accurately. The high sensitivity, wide dynamic range, and IR blocking filter make the TCS3472 an ideal color sensor solution for use under varying lighting conditions and through attenuating materials. The TCS3472 color sensor has a wide range of applications including RGB LED backlight control, solid-state lighting, health/fitness products, industrial process controls and medical diagnostic equipment. In addition, the IR blocking filter enables the TCS3472 to perform ambient light sensing (ALS). Ambient light sensing is widely used in display-based products such as cell phones, notebooks, and TVs to sense the lighting environment and enable automatic display brightness for optimal viewing and power savings. The TCS3472, itself, can enter a lower-power wait state between light sensing measurements to further reduce the average power consumption.

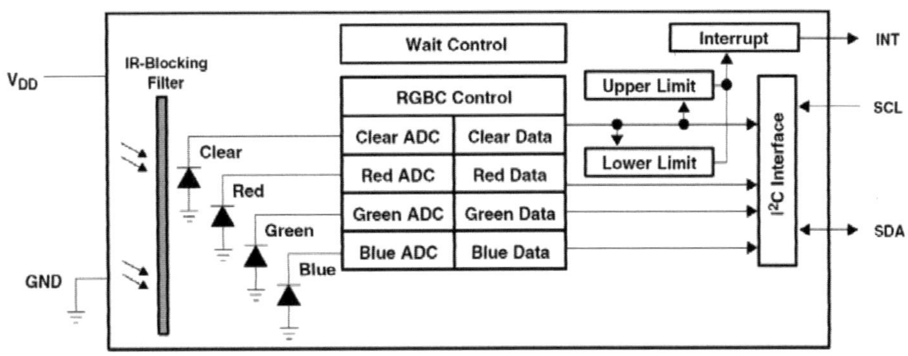

Figure 6. Measuring visible color, with infrared filter using TCS3472.

The four integrating ADCs simultaneously convert the amplified photodiode currents to a 16-bit digital value. Upon completion of a conversion cycle, the results are transferred to the data registers, which are double-buffered to ensure the integrity of the data. All of the internal timing, as well as the low-power wait state, is controlled by the state machine. Communication of the TCS3472 data is accomplished over a fast, up to 400 kHz, two-wire I2C serial bus. The industry standard I2C bus facilitates easy, direct connection to microcontrollers and embedded processors.

1.4. Measuring proximity/UV/ambient light sensor with SI 1145

The Si1145/46/47 is a low-power reflectance-based, infrared proximity, ultraviolet (UV) index, and ambient light sensor with I2C digital interface and programmable event interrupt output. This touchless sensor IC includes an analog-to-digital converter, integrated high-sensitivity visible and infrared photodiodes, digital signal processor, and one, two, or three integrated infrared LED drivers with fifteen selectable drive levels.

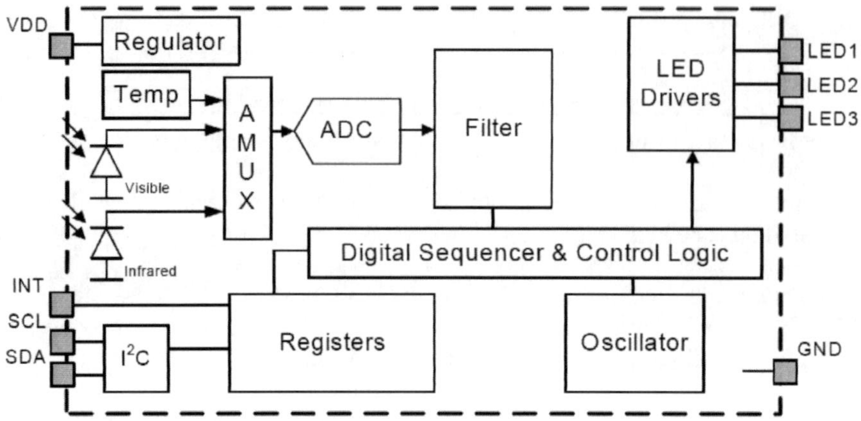

Figure 7. Measuring proximity/UV/ambient light sensor with SI 1145.

The Si1145/46/47 offers excellent performance under a wide dynamic range and a variety of light sources including direct sunlight and can also work under dark glass covers. The photodiode response and associated digital conversion circuitry provide excellent immunity to artificial light flicker noise and natural light flutter noise. With two or more LEDs, the Si1146/47 is capable of supporting multiple-axis proximity motion detection. These devices are provided in a 10-lead 2x2 mm QFN package and are capable of operation from 1.71 to 3.6 V over the –40 to +85 °C temperature range.

The UV Index is a number linearly related to the intensity of sunlight reaching the earth and is weighted according to the CIE Erythemal Action Spectrum and has been standardized by the WHO and includes a simplified consumer UV exposure level as shown in figure 8. This weighting is a standardized measure of human skin response to different wavelengths of sunlight from UVB to UVA.

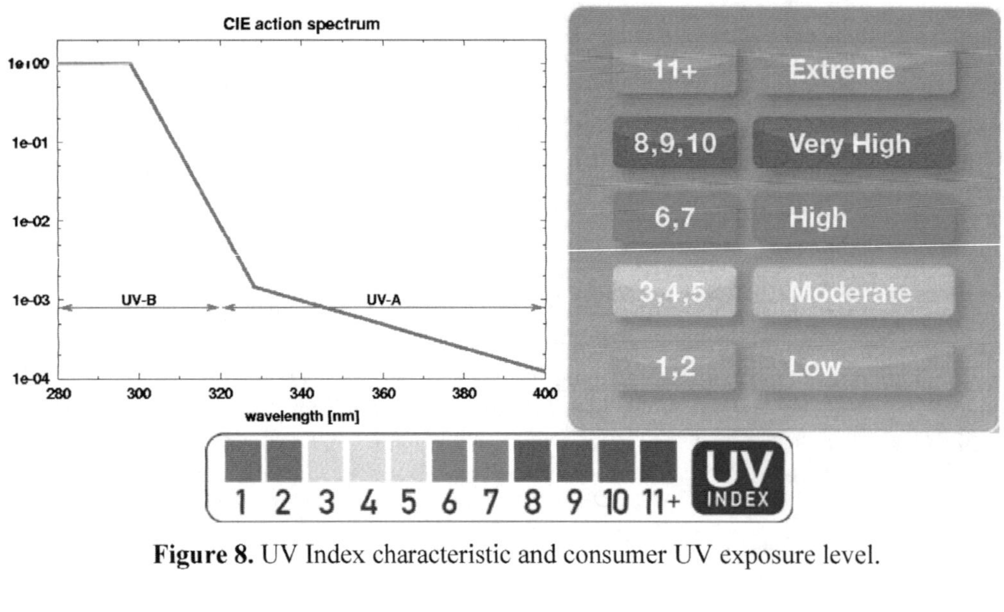

Figure 8. UV Index characteristic and consumer UV exposure level.

When the calibration parameters are recovered they show up at I2C registers 0x22 to 0x2D. These are the same registers used to report the VIS, IR, PS1, PS2, PS3, and AUX measurements. Once the calibration parameters have been recovered the routine Si114x_set_ucoef is used to modify the default values that go into the UCOEF0 to UCOEF3 UV configuration registers to remove normal part-to-part variation. The typical calibrated UV sensor response vs. calculated ideal UV Index is shown in figure 11 for a large database of sunlight spectra from cloudy to sunny days and at various angles of the sun/time of day.

3. Designing remote temperature device for inaccessible locations

One urge required action is to measure the bucket wheel bearing temperature of the bucket wheel excavator (BWE). The issue is related to the BWE maintenance. The temperature is considered the major signal of the wheel reliability [1,2], but at this moment there almost impossible to detect the temperature variation in the bearing. We propose here a remotely temperature detection method. We are starting with general approach for infrared detection [3], where the major issue is to detect the generated infrared light according to the material detection. The authors of [3] underline the major factors that influence the temperature remote measurements. In this paper we are developing our research related to the external light that influences the measurements. Next figure shows the detection principle we are taking into account.

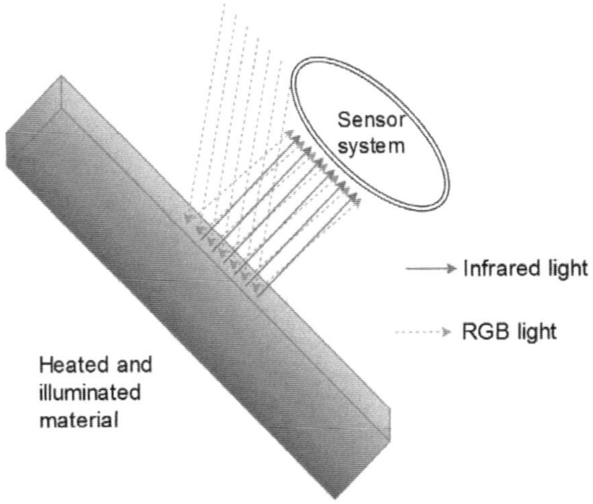

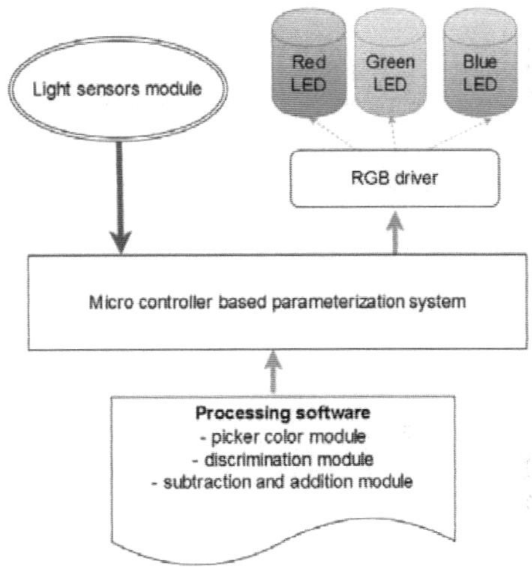

Figure 9. a) The sensors module detects the entire radiated energy; b) Hardware software architecture of the lab parameterization system.

The sensors module captures the entire radiation: infrared radiation as a material temperature estimation, and reflected temperature when the material is variable illuminated. Because the sensors detect kmol $m^{-2}s^{-1}$, (quantic model of the light) it is required to extract RGB content from the measured light signal. In order to implement this method, we switch now to the frequency model of the light that means that we are using the addition and subtraction method of the beams of lights.

According to this method, we can subtract part of the lights if they have equal intensity. It means that each primary color of light has a secondary color of light as its complement. In order to manipulate this method we have design hardware- software architecture and the block diagram of this architecture is shown is presented in figure 9. The sensors module captures the entire radiation: infrared radiation as a material temperature estimation, and reflected temperature when the material is variable illuminated.

In order to create this measuring and discrimination system we have implemented the microcontroller- based architecture around an 8 bit system, that communicate with the software on a PC for data processing. The main way of components dialog is shown next:

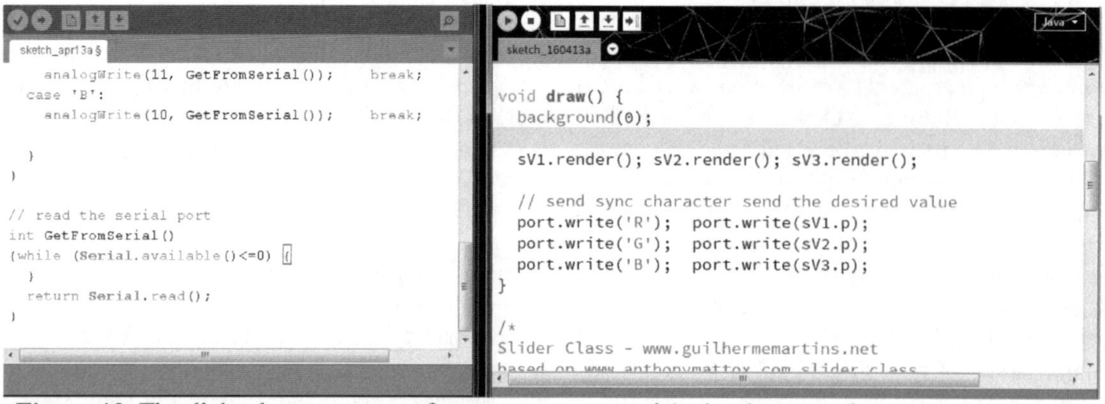

Figure 10. The dialog between two software components of the hardware- software implementation.

After the embedded software handshake (ports communication pairing and synchronization), during the execution data from a processing software is send as RGB parameters to the execution software. The upper figure shows the data exchange between two components. The interfaces for manual control during calibration process are shown next.

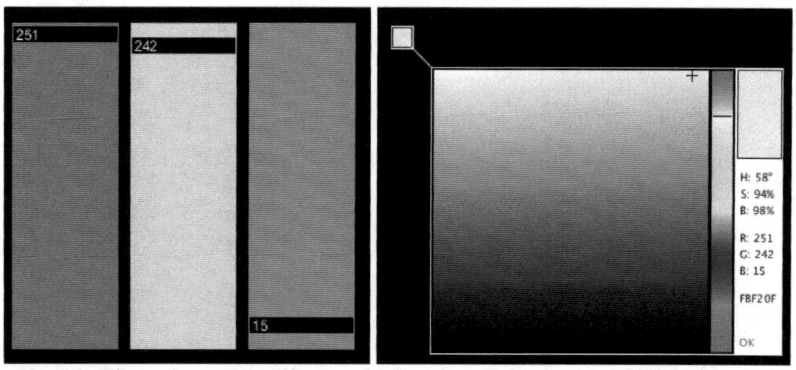

Figure 11. The color controlling tools developed for manual color discrimination.

In both figure there is a handling facility to control/ manipulate every single component of the RGB high power driver.

4. Testing and results

The laboratory testing intended to discriminate the light components. For our experiment we have used a laboratory calibration device Fluke 712B with PTH100 (385) temperature probe. Next figure shows capture screen of the Fluke device during calibration process.

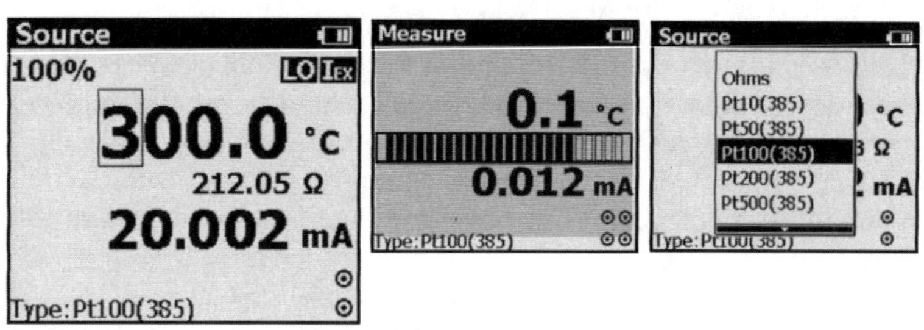

Figure 12. Calibration process done with Fluke 712B, and PT100(385) temperature probe.

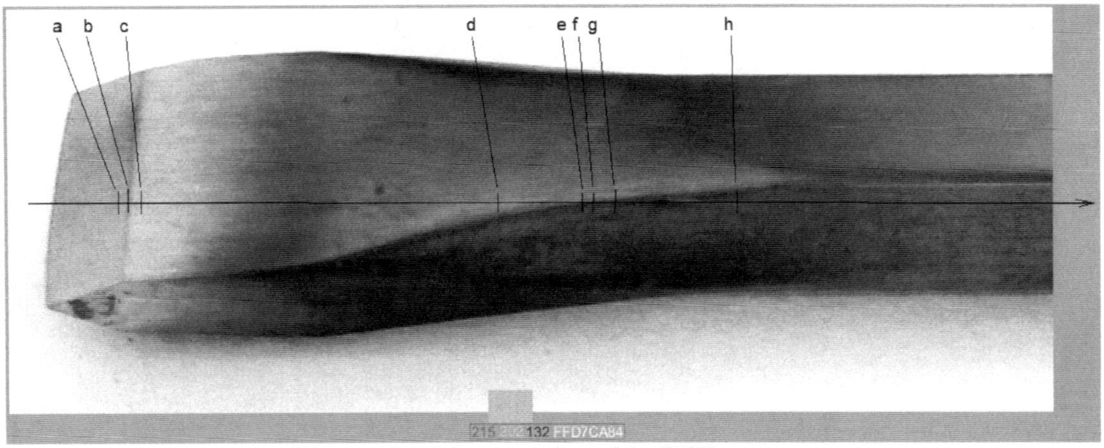

Figure 13. Light extraction of the RGB components.

The upper figure shows the processing application for light components discrimination, which is used to subtract the RGB components of the captured light by the sensors module. The results are shown in next table, where the values are in Celsius degrees.

Region	*a*	*b*	*c*	*d*	*e*	*f*	*g*	*h*
After RGB subtraction	290	310	300	170	145	135	110	110
FLUKE measured	310	310	310	170	170	170	120	120
Infrared termal camera	350	350	350	140	130	130	130	130

The comparative graph representation is shown next.

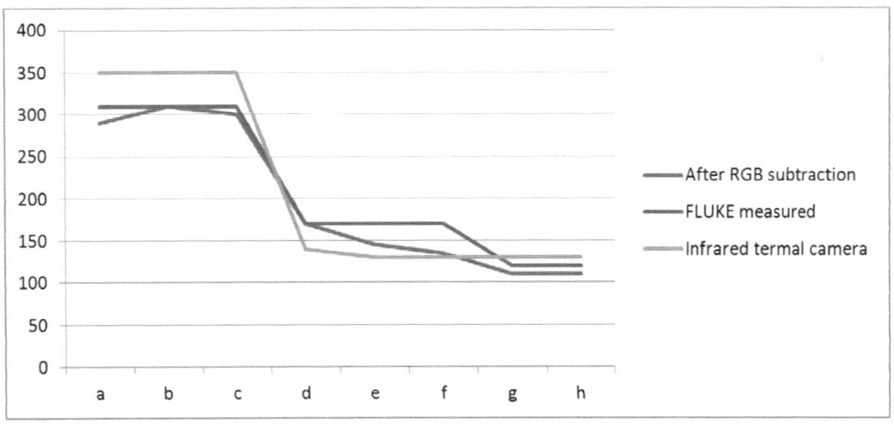

Figure 14. The comparative representation of the temperature measured values using three methods.

5. Conclusions

We have developed a hardware-software system for remote temperature measurement with RGB errors compensation. The system uses a driver for high power LED in order to ensure controlled light spectrum.

The system is tested by comparing a calibration method with our implemented method and one consecrated measuring method. If we consider the calibration method as reference, we can conclude that the infrared camera is not sensitive (the error exceed 20%). Our implemented method offers one possibility to detect border situation. Our method offers a measuring solution in 16 bits resolution, and in an error range of 5%.

The method must be tested also taking into account the errors introduced by the measuring distance.

Acknowledgements
This paper has been initiated by CEO Oltenia Romania. When they asked to design a predictive measurements system for buckled wheel axis system, where sleeve bearing is a crucial part of the excavator and where there is not possible to apply a classical method for temperature measurements. They supported our experimental setup requirements for the research.

References

[1] Shareef H, Mohamed A, Uddin S and Hannan HA 2012 LEDs as Energy Efficient Light Bulbs: a Bibliographical Survey *International Review on Modelling and Simulations* 5(1) pp 357-361
[2] Camacho O, Freescale Semiconductor, Automotive High Brightness LED Control, Application Note 2010 at http://cache.freescale.com/files/microcontrollers/ doc/app_note/AN4105.pdf
[3] Fred Schubert E 2015 Human eye sensitivity and photometric quantities
[4] Kamil M, Microchip Technology Inc., Switch Mode Power Supply (SMPS) Topologies (Part I), online 2015 at http://ww1.microchip.com /downloads/en /AppNotes/01114A.pdf
[5] Semolič B, Jovanović P, Kovačev S and Obradović V 2008 Improving Repair Management of Bucket Wheel Excavator, SRs1200 by Application of Project Management Concept, Strojniški vestnik *Journal of Mechanical Engineering* **54**(7-8) pp 565-573
[6] Bagherpour R, Technical and economical optimization of surface mining processes – Development of a data base and a program structure for the computer-based selection and dimensioning of equipment in surface mining operations - Dissertation zur Erlangung des Grades eines Doktor-Ingenieurs, Mining Industry Progresses Clausthal University of Technology, Department of Surface Mining and International Mining, Institute of Mining.
[7] Bower S M, Kou J and Saylor J R 2009 A method for the temperature calibration of an infrared camera using water as a radiative source, *REVIEW OF SCIENTIFIC INSTRUMENTS 80* 095107
[8] Bersani A, Microchip Technology Inc., Switch Mode Power Supply (SMPS) Topologies (Part II), online 2015 at http:// ww1.microchip. com/downloads/ /AppNotes/01207A.pdf.
[9] TPS55340 Integrated 5-A Wide Input Range Boost/SEPIC/Flyback DC-DC Regulator datasheet
[10] *** WEBench - http://webench.ti.com/webench5/power/webench5.cgi?lang_chosen=en_US &VinMin =14.0&VinMax=22.0&O1V=61&O1I=1.0&op_TA=30, last visited in June 2015
[11] *** Directorate C: Public Health and Risk Assessment , Scientific Committee on Emerging and Newly Identified Health Risks - SCENIHR - http://ec.europa.eu/health/ph_risk/risk_en.htm
[12] *** Initial Evaluation of a DC/DC Switch Mode Power Supply Application Note, Rohde & Schwarz, online 2015 at http://www.rohde-schwarz.de/file/1TD04_0e _RTO _DC-DC-Converter.pdf

IOP Conf. Series: Materials Science and Engineering **133** (2016) 012060 doi:10.1088/1757-899X/133/1/012060

The implications of free 3D scanning in the conservation state assessment of old wood painted icon

Marius Munteanu[1, 2*], Ion Sandu[1, 3]

[1] „Al. I. Cuza" University of Iasi, Doctoral School of Chemistry and Earth and Life Sciences, Faculty of Geography and Geology, Iaşi, România

[2] „Al. I. Cuza" University of Iasi, Arheoinvest Platform, Blvd. Carol I 11, 700506, Iasi, Romania

[3] Romanian Inventors Forum, Str Sf. Petru Movila 3, L11, III/3, 700089, Iaşi, Romania

E-mail: munteanu.maryus@gmail.com

Abstract. The present paper presents the conservation state and the making of a 3D model of a XVIII-th century orthodox icon on wood support, using free available software and cloud computing. In order to create the 3D model of the painting layer of the icon a number of 70 pictures were taken using a Nikon DSLR D3300, 24.2 MP in setup with a Hama Star 75 photo tripod, in loops 360° around the painting, at three different angles. The pictures were processed with Autodesk 123D Catch, which automatically finds and matches common features among all of the uploaded photographs in order to create the 3D scene, using the power and speed of cloud computing. The obtained 3D model was afterwards analyzed and processed in order to obtain a final version, which can now be use to better identify, to map and to prioritize the future conservation processes and finally can be shared online as an animation.

1. Introduction

During the last few years, the modern investigating and examination techniques have played an important role in the conservation and preservation of old paintings. In order to carry out initial researches of the cultural heritage object, several noninvasive examination methods can be performed (optical microscopy, scanning electron microscopy). However, none of these methods provides data on the geometric shape and metric analysis of the object surface. This gap in the initial optical examination of an artwork can be filled with optical 3D scanning [1].

In the last decade, the 3D examination technique was introduced in the examination processes of cultural heritage, which allowed restorers a more in depth and variable examination opportunity. Due to equipment limitations, the scanning technique was mostly used in the case of structures and sculptures until now [2].

New improvements in the field of 3D scanning has now allowed us to create 3D models of the painting surface, models that can be used to study the conservation state of the painting layer, to observe the planar surface changes that can take place after a consolidation processes, or (in the case of more accurate scanning) even to create three dimensional reproductions through new 3D printing techniques [3-7].

Although the technique involves a certain investment in scanning apparatus and data processing technology in order to obtain accurate and printable 3D copies of the painting surfaces, there are other

IOP Conf. Series: Materials Science and Engineering **133** (2016) 012060 doi:10.1088/1757-899X/133/1/012060

cheaper options for obtaining a 3D scanning of a painting for study and conservation purposes. One of these options involves using the free available Autodesk 123D Catch software to create your 3D models using your own 360^0 pictures of the painting.

The paper wants to present the process of obtaining a 3D copy of a XVIII-th century icon obtained through the above mentioned software, without any costs what so ever. The purpose of this paper is to encourage restorers to use new examination techniques, to share their 3D models obtained through various scanning techniques with the cultural heritage community and not only, and to prove the easiness to obtain a 3D proof of their work.

2. Icon presentation and conservation state

The painting taken into study is a XVIII-th icon, done in egg tempera on a lime wood support, part of the Arheoinvest laboratory collection. The lime wood support (L= 45cm, W= 38cm, h= 2cm) is made out of three wood panels, kept together by two back crossbeams. The wood support was covered with a canvas, in order to ensure its resistance and to restrict the dimensional modifications caused by humidity variations. A thick layer of ground was applied on the canvas, which served as an appropriate surface for painting.

The icon depicts the *Virgin Mary*, holding baby Jesus in her arms, one of the most well know orthodox iconographic representations (figure 1).

Figure 1. *Virgin Mary with baby Jesus* icon: a) Front; b) Back.

The conservation state of the icon is very poor, both the wood support and the painting layer being affected by several deterioration and degradation processes.

Approximately 20% of the painting layer has been already lost, while the rest of the painting layer suffers from longitudinal and mosaic cracks, extensive detachments, blackened areas, chromatic modifications [8] etc. The gaps are mostly located where the three panels are joined together. These severe cracks were mostly likely caused by the dimensional modifications (especially shrinkage) of the three wood panels, as a response to the environmental variations of humidity [9-13].

The painting layer also presents areas that suffer from thermal cornification and oxidative dirt gathering, deterioration processes that were favored by the fact that the icon was exposed though its liturgical purpose, to heat sources (candles, lamps etc.) (figure 2a-f).

The wood support suffers from a xilophagic attack (inactive for the moment), its effects being visible in gaps, on the back and on the sides of the panels (tunnels and fly holes). Due to this, the resistance of the support has severally decreased, leading to cracks and loss of material, especially on the sides and corners of the panel [14-15].

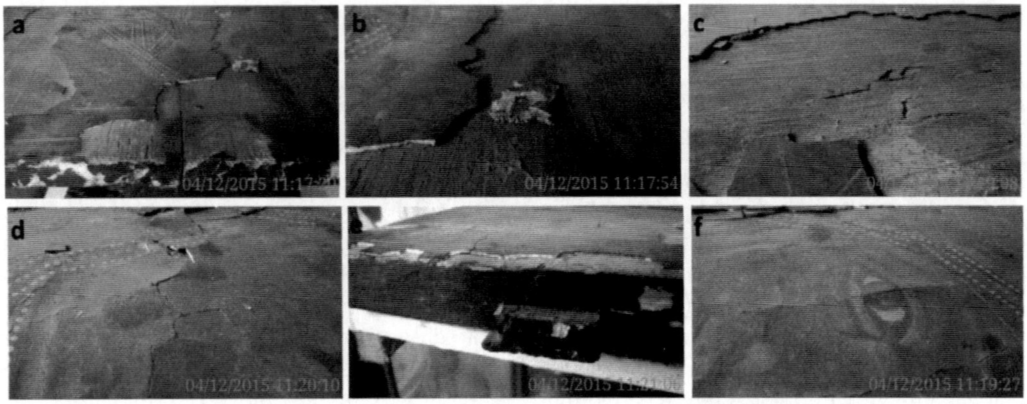

Figure 2. Deteriorations and degradations of the painting layer: a-b) dirt deposits, deep gaps; c) dirt deposits, deep gaps, xilophagic attack; d)dirt deposits, dettachaments; e) deep gaps; f) dettachaments, dirt gatherings.

A thin layer of ground, covered with color, was applied on the back of the support at a certain point in time, in order to reduce the influence of humidity (figure 3 a-d). Nowadays, this protective layer also has gaps and detachments. The fact that the icon was exposed to concentrated heat sources had also affected the support, severe burning marks can be observed in picture 3b.

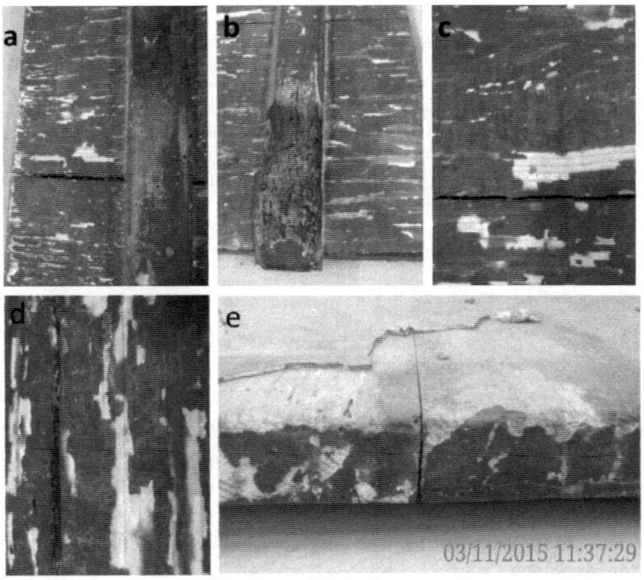

Figure 3. Deteriorations and degradations of the support: a) Dirt deposits, wood components distancing from each other; b) burn marks, dirt deposits; c) Dirt deposits, wood components distancing from each other; d) crack along the wood fiber; e) fly holes, wood components distancing from each other.

3. Experimental part

In order to obtain the required set of photos, a Nikon DSLR D3300, 24.2 MP was used in setup with a Hama Star 75 photo tripod. A number of 70 pictures were taken from all around (360°) the center of the painting layer of the icon, at approximately the same distance, without using the flash of the camera. A constant source of light was used, in order to avoid any changes in the lighting condition of the painting. The photos were then uploaded into the Autodesk 123D Catch software to be processed in order to obtain the 3D copy of the painting layer of our icon. The software automatically finds and matches common features among all of the uploaded photographs in order to create the 3D scene, using the power and speed of cloud computing [16].

4. Results and discussions

The process of uploading the pictures to the cloud and their computing through feature matching varies according to the number of picture (70 being the max number that can be used) and their properties (dimensions, quality etc). The obtained 3D model allowed us to establish first of all the point from where the photos were taken and if a certain angle or detail of the painting was not taken into examination.

In our case, the 70 pictures can be divided into three sets that were taken 360° around the center point of the painting, at three different angles, in order to ensure sufficient information on each part of the surface. The sets of photos were done in a continuous manner without interrupting the process, without changing the lighting and most importantly, without moving the object (figure 4).

Figure 4. The three loops of photos, taken at three different angles.

The fact that the 3D model can be created using cloud computing represents one the reasons that this or a similar scanning technique should be used more often by the restorers. The curators or restorers aren't required to invest in computing technique or to employ a specialist in order to obtain a workable 3D model.

The photos circle radius was diminished with each loop (first loop is the largest, while the last loop has the smallest radius), for the same reason already mentioned, but also to ensure that no shading problems would obscure details of the painting layer (figure 5).

After ensuring that the quality of the pictures and 3D model obtained is accurate, the process of examining the painting layer can begin. The viewer can now observe through simple navigation around the 3D model the deterioration and the degradation of the painting layer, without exposing the object to further manipulation or light sources (figure 6a-c). The 3D model can also be observed as a structure. This can help the exact mapping of the detachments and deep gaps of the painting layer (figure 6d-f). The mapping process can then help curators and restorers in evaluate the conservation state of the painting and to prioritize future interventions [17].

The same gaps and detachments can be observed at different angles and magnifications, in order to determine their dimensions and high/depth.

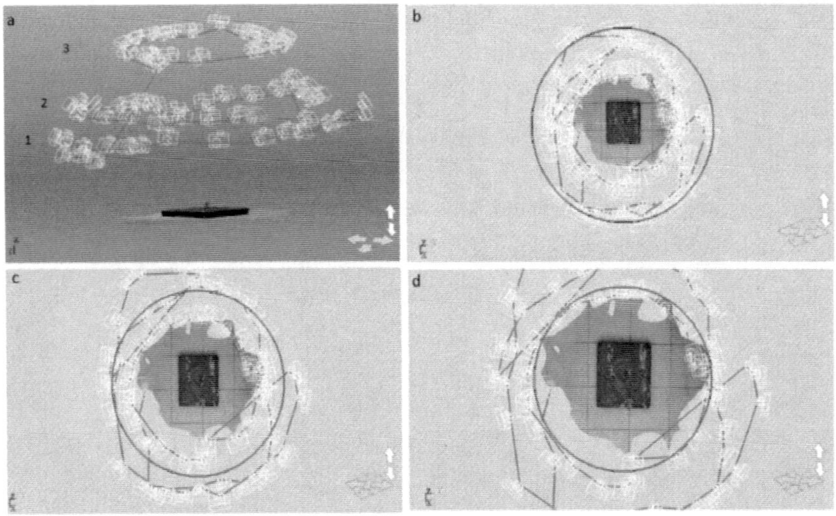

Figure 5. The radius of the three loops: a) general view of the three loops; b) radius of the first loop; c) radius of the second loop; d) radius of the third loop.

Figure 6. The obtained 3D model of the icon: a-c) front and sides; d-f) front and side structure.

In order to improve the aspect of the final 3D model, the background can be removed (figure 7a-f). The model can also serve as a digital testimony of the conservation point of the painting at a certain point and a comparison base after a period of time or after a restoration process. Using another feature of the program, the 3D model can be presented as a animation, which can be then directly shared online on the restorers YouTube account. This can be the next step in making people aware on the conservation problems of cultural heritage paintings and can also serve as a communication channel between specialists, allowing them to inspect the painting remotely and to offer their professional opinion.

The fact that some the 3D models created through Autodesk 123D Catch can be used after as examples by the software company is another way to share the conservation problems of a painting with people that are not cultural heritage specialists.

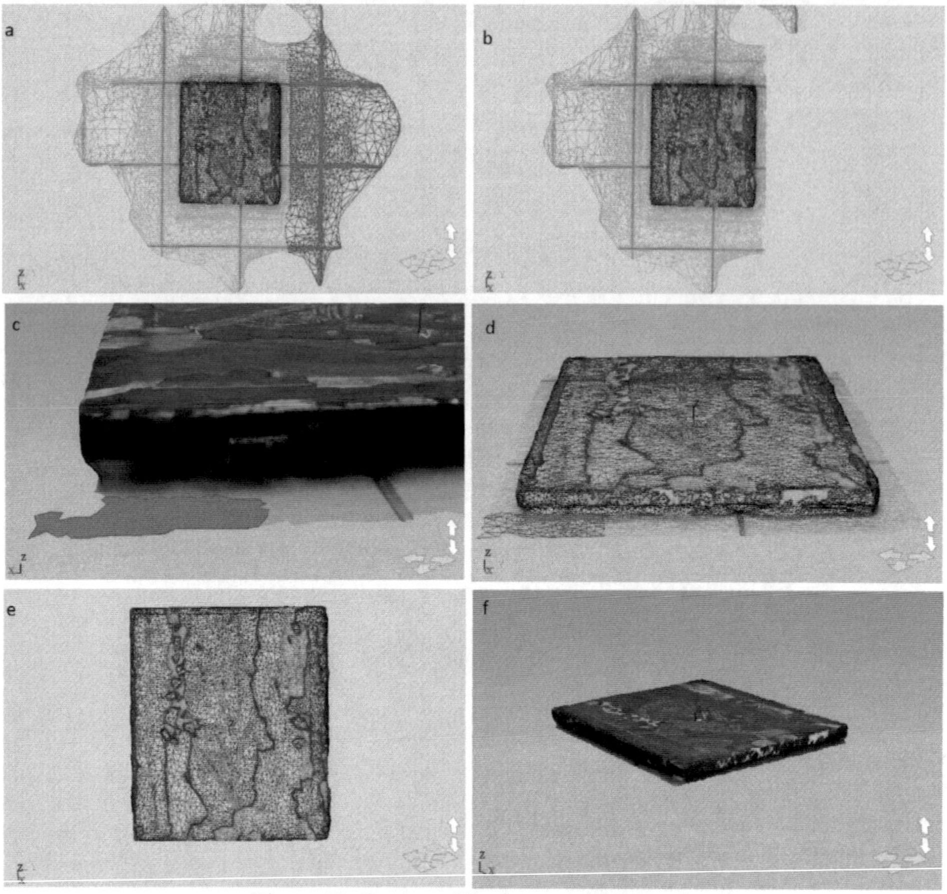

Figure 7. Stages of background removal.

The obtained and processed model can also be used to create a 3D print of the painting. In order to do this, the digital model created in Autodesk 123D Catch should be processed using free software, Meshmixer, to improve its characteristics.

5. Conclusions

The *Virgin Mary with baby Jesus* icon, part of the Arheoinvest laboratory collection, was used as part of an attempt to create a free 3D model of a painting. A number of 70 pictures were taken from all around (360°) the painting layer of the icon, at approximately the same distance, and were then uploaded into the Autodesk 123D Catch software to be processed through cloud computing, in order to create a 3D model of the icon. Navigating around the 3D model, the viewer can now observe the deterioration and the degradation of the painting layer, without exposing the object to further manipulation or light sources. This can help the exact mapping of the detachments and deep gaps of the painting layer, in order to prioritize the required conservation steps. The obtained 3D model can then be share online as an animation between specialists, allowing them to inspect the painting remotely and provided advices and change opinions.

The fact that the 3D model is easy to obtain and doesn't require significant investments should be an incentive for all restorers to apply a more technological and modern approach to the process of painting restoration.

References

[1] K. Gorecka, A. Rzeszutek, 2015 *Does restorer need a scanner? Optical methods in canvas painting diagnostic*, Proceedings of SPIE 9662, Photonics Applications in Astronomy, Communication, Industry and High Energy Physics Experiments, 11 September.

[2] * * *, 3D Icons Case Studies, Paceprint, Dublin, Ireland, 2015.

[3] T. Zaman, P. Jonker, B. Lenseigne, J. Dik, 2014 *Heritage Science*, 2, Issue 23, pp. 1-10.

[4] L. Yi, Z. Liu, K. Wang, X. Lin, M. Chen, S. Peng, K. Yang, J. Wang, 2016 *Applied Physics A-Materials Science and Processing*, 122, Issue 4, art. No. 329.

[5] Y. Volpe, R. Furferi, L. Governi, G. Tennirelli, 2014 *International Journal of Computer Aided Engineering and Technology*, 6, Issue 1, pp. 88-112.

[6] W.S. Elkhuizen, T. Zaman, W. Verhofstad, P.P. Jonker, J. Dik, J.M.P. Geraedts, 2014 *Topographic scanning and reproduction of near-planar surfaces of paintings, Proceedings of SPIE*- The International Society for Optical Engineering, 9018, art. No. 901809, San Francisco, United State, 03 February-04 February.

[7] G. Guidi, L.L. Micoli, S. Gonizzu, P.R. Navarro, M. Russo, 2013 *3D digitizing a whole museum: A metadata center workflow*, Proceedings of the 1st International Congress on Digital Heritage, 2, art. No. 6744768, Marseille, France, 28 octomber-01 november.

[8] E. Ghelardi, I. Degano, M. P. Colombini, J. Mazurek, M. Schilling, H. Khanjian, T. Learner, 2015 *Dyes and Pigments*, 123, pp. 396-403.

[9] D. Hunt, 2012 *Journal of Cultural Heritage*, 13S, pp. S10-S15.

[10] M. Łukomski, 2012 *Journal of Cultural Heritage*, 13S, pp. S90–S93.

[11] D.P. Vici, P. Mazzanti, L. Uzielli, 2006 *Journal of Cultural Heritage*, 7, pp. 37–48.

[12] B. Rachwał, Ł. Bratasz, M. Łukomski and R. Kozłowski, 2012 *Strain*, 48, pp. 366–374.

[13] M.F. Mecklenburg, C.S. Tumosa, D. Erhardt, 1998 *Structural response of painted wood surfaces to changes in ambient relative humidity*, Painted Wood: History and Conservation, Los Angeles, The Getty Conservation Institute, pp. 464–483.

[14] C.M. Popescu, M.C. Popescu, C. Vasile, 2010 *Carbohydrate Polymers*, 79, pp. 362-372.

[15] R. A. Blanchette, 1995 *A Guide to Wood Deterioration Caused by Microorganisms and Insects, in The structural conservation of panel paintings*. Procedings of a symposium at the J. Paul Getty Museum, edited by K. Dardes and A. Rothe.

[16] http://www.123dapp.com/catch, accessed on 22.02.2016.

[17] L. Granero-Montagud, C. Portalés, B. Pastor-Carbonell, E. Ribes-Gómez, A. Gutiérrez-Lucas, V. Tornari, V. Papadakis, R.M. Groves, B. Sirmacek, A. Bonazza, I. Ozga, Vermeiren, J., K. Van Der Zanden, M. Föster, P. Aswendt, A. Borreman, J.D. Ward, A. Cardoso, L. Aguiar, F. Alves, P. Ropret, J.M. Luzón-Nogué, C.Dietz, 2013 *Deterioration estimation of paintings by means of combined 3D and hyperspectral data analysis, Proceedings of SPIE* - The International Society for Optical Engineering, Optics for Arts, Architecture, and Archaeology IV, Munich, Germany, 15 May -16 May.

IOP Conf. Series: Materials Science and Engineering **133** (2016) 012061 doi:10.1088/1757-899X/133/1/012061

Colour change evaluation on UV radiation exposure for *Păun-Repedea* calcareous geomaterial

V Pelin[1], I Sandu[2], M Munteanu[1], C T Iurcovschi[1], S Gurlui[3], AV Sandu[4], V Vasilache[2], M Brânzilă[5] and I G Sandu[6,7]

[1]*Alexandru Ioan Cuza* University, Faculty of Geography – Geology, Doctoral School of Chemistry and Life Science and Earth, Carol I Blvd, no. 20A, Iași, 700505, Romania.
[2]*Alexandru Ioan Cuza* University, Interdisciplinary Platform ARHEOINVEST, Carol I Blvd, no. 11, 700506, Iași, Romania.
[3]*Alexandru Ioan Cuza* University, Faculty of Physics, Carol I Blvd, no. 11, Iași, 700506, Romania.
[4] Faculty of Technology, Universitas Ubudiyah Indonesia (UUI), Jln Alue Naga Desa Tibang, Banda Aceh, Indonesia
[5]*Alexandru Ioan Cuza* University, Faculty of Geography – Geology, Carol I Blvd, no. 20A, Iași, 700505, Romania.
[6]*Gheorghe Asachi* Technical University, Faculty of Materials Science and Engineering, Dimitrie Mangeron Street, no. 61A, 700050, Iași, Romania.
[7]Romanian Inventors Forum, Sf. PetruMovilă Street, no. 3, 700089, Iași, Romania.

E-mail: munteanu.maryus@gmail.com

Abstract. When talking about the preservation treatments that can be applied to natural stones used in different constructions, the surface hydrophobization plays an important part, especially when referring to porous surfaces like the calcareous oolithic stones specific to Repedea area, Iași County, Romania. The present paper presents a method that evaluates the hydrophobization efficiency of two types of pellicles, involving UV artificial ageing and colorimetric analysis of the treated surfaces. The evaluation was done through continuous colorimetric monitoring and by comparing the evolution of the chromatic modifications of the two treated surfaces with the original colorimetric values and with the witness area, which was exposed to UV radiations under the same conditions, but left chemical untreated. The techniques used during this experiment were: CIE $L^*a^*b^*$ colorimetry, OM, SEM-EDX, UV radiation exposure and Spectrum Irradiance Measurement.

1. Introduction

Geomaterial definition includes soils, rocks and minerals which are mostly used in the construction and structuring of buildings, in natural form or processed form but excludes the products fabricated through complex technological processes, based upon an advanced knowledge of engineering [1]. With the passing of time, any one of geomaterial can be subject to deterioration or degradation, through the influence of environmental factors, the main factor of aggression being environmental humidity, weathering and the presence of urban pollution [2-4]. The occurrence of urban/industrial aerosols type and their impact in the ambient air over the Iași area (Romania) is already proved by specialized literature by changes of radiative forcing due to the properties of aerosols to absorb and scatter differently the optical radiation [5,6]. As a result, the sunlight radiation represents another factor of alteration, because it is able to generate undesired effects [4,7], especially in the case of hydrophobic treatments, through the chromatic variation of treated surfaces, due to UV radiation exposure. Thus, is necessary to apply a waterproofing and coating treatment with various chemical solutions, current commercially available. Therefore, this paper proposes presenting a collection of data that afford for the analysis of chromatic deviation and the accelerated aging from the lithic surface of the calcareous stone sample exposed to UV radiation, after carrying out hydrophobic treatments.

2. Materials and methods

The lithic material exposed to UV radiations is a sedimentary calcareous rock, form during the Sarmatian geological age, and was sampled from Păun – Repedea village, Iași County [8].

The two chemical products used for hydrophobization and tested are commercially available and known as LTP Mattstone® and LTP Mattstone H2O®.

CIE *L*a*b colorimetry** was done using a Lovibond® RT 300 spectrophotometer (Reflectance Tintometer D65/10°) which allowed the measuring of the chromatic modifications, in the exact same spots, for the entire duration of the investigation.

The porosity and the degree of permeability of the hydrophobized surfaces were studied using a Zeiss Axio Imager® A1m optic microscope, with an attached Axio Cam and specialized software. The **optical microscopy** (OM) was also used to evaluate the changes regarding the texture, morphology and chromatic modification that took place after the treatment of the investigated surfaces.

Scanning Electron Microscopy (SEM) coupled with **energy dispersive X-ray** (EDX) allowed the recording of the micrographic analyzed area and recording the spectrum based on which to determine elemental composition for lithic materials. A scanning electron microscope model Vega II LSH – Tescan®, coupled with an EDX detector, model Quantax QX2 – Bruker - Roentec® was used.

UV radiation exposure was done with a predominant wavelength 300 - 365nm by using four high-pressure mercury vapour lamps (Osram Ultra Vitalux® 300W - 230V). The intensity of the light produces by these lamps was measured both analogically (TYPMDLx lux meter) and digitally (LX 1010B lux meter).

Comparative **spectrum irradiance measurement** have been studied by means of a high-resolution monochromator (Princeton Instruments® Acton SP2750i, 750 mm focal length) and an intensified CCD camera (Roper Scientific Princeton Instruments® PI MAX3-1003-UNIGEN2, 1024x1024 pixels, minimum gating time 2ns). The complete spectrum was obtained by using the "step-and-glue" procedure of the monochromator/camera software, which automatically increments the grating position once a given (8nm wide).

The humidity content of the sample surface was measured with a non-invasive technique, using a Gann Hydromette® (Compact B humidity meter) that works on the principle of measuring the dielectric constant, through high radio frequency.

The investigations were carried out in constant laboratory conditions, at a temperature of 21°C and relative humidity (RH) of 60%.

3. Experimental part

After probing, the sample of lithic material was cut into dimensions and divided into three areas of investigations, with equal surfaces (of around $6cm^2$). The three obtained areas were noted using the symbols C0, C1 and C2 (figure 1). After that, the stone sample was stored in desiccators for 48 h before chemical treatment. The two hydrophobic products were noted using the symbols S1 (LTP Mattstone®) and S2 (LTP Mattstone H2O®). Each product was used for one area as follows: S1 for C1 and S2 for C2. The area C0 is the witness area, left untreated, but exposed to the UV radiations, under the same conditions like the two treated areas.

IOP Conf. Series: Materials Science and Engineering **133** (2016) 012061 doi:10.1088/1757-899X/133/1/012061

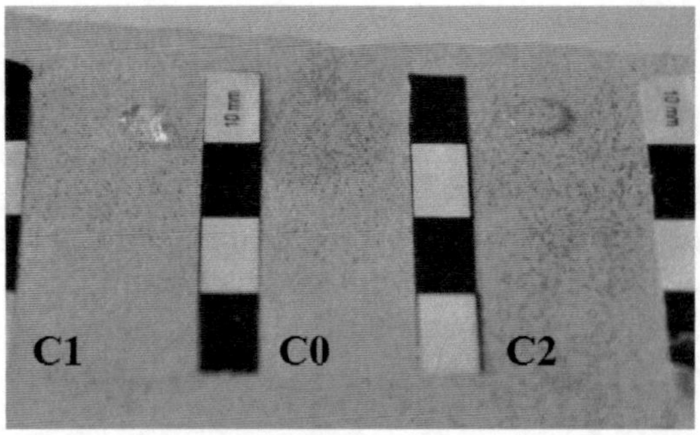

Figure 1. The dripping water test on the investigated and treated surfaces. Pictures taken at 24 h after the treatment.

After the three areas were delimitated, a first set on initial colorimetric measurements was taken, in conditions of constant humidity of the lithic surfaces. According to the technical sheets, a quantity of 0.18 mL from each chemical product was applied using a soft brush on every investigated surface, in successive stages. In the case of the S1 solution, a pause of 2 hours was taken between applying the layers, while the S2 solution was applied wet on wet.

At 24h after treatment, distilled water was applied in droplets for the macroscopic evaluation of the hydrophobic degree of the surfaces being treated (figure 1). As can be observed in figure 1, the surface corresponding with the control area (C0) absorb the water instantaneously and stain the surface compared to the treated areas, where the drops of water do not reach the pores and also do not stain the surfaces. The volume of distilled water used was 0.05 mL for every area investigated.

In order to evaluate the hydrophobic efficiency of the two products on the surfaces C1 and C2, the evaporation time for the distilled water drops used to determine the degree of permeability was calculated: t_1 /C1 = 170 minutes and t_2 /C2 = 135 minutes.

Furthermore, the hydrophobic degree was also evaluated at a microscopic level (figure 2), using distilled water drops with a volume of up to 4 μL for each investigated surface.

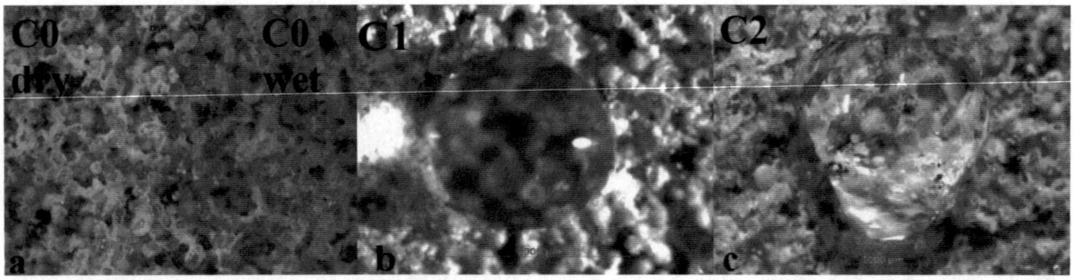

Figure 2. Optical microscopy (5X) from the porous surface of the investigated sample. Images taken during the dripping process of the distilled water over the three surfaces :(a) C0 – dry / wet; (b) C1; (c) C2.

The porosity of the investigated surfaces can be observed in figures 2(a), (b) and (c), and is influenced by the oolithic structure specific to the lithic material from Repedea area. The porosity was also analysed through SEM (figure 3), while the elemental composition of the sample was determined through EDX (table 1) thus confirming the presence of chemical elements specific to sedimentary calcareous rocks [9,10]. The presence of carbonate minerals was confirmed through a rapid drop test using HCl, the effervescent results specific to this chemical reaction being observed in figure 4.

Figure 3. SEM micrographic images for oolithic calcareous sample: a) 100X image ; (b) 500X image.

Figure 4. HCl dripping rapid test.

Table 1. The elemental composition for calcareous sample, obtained through EDX spectrum.

Element	[norm. wt.-%]	[norm. at.-%]
Calcium	47.41	31.47
Silicon	18.39	17.42
Aluminium	1.63	1.60
Magnesium	0.90	0.98
Iron	1.55	0.74
Potassium	2.40	1.63
Carbon	0.12	0.27
Oxygen	27.60	45.89
Total	100	100

After 48h of chemical treatment, the surfaces were exposed to UV radiation in laboratory conditions, total period of exposure being on 24h for each surface. In this case, the colorimetric variation was monitored every two hour, taking into consideration the reference values after the application of chemical treatment, for areas C1 and C2. At the same as the colorimetric measurements, the surfaces humidity of each area investigated was also monitored, so as to evaluate its possible influence upon colorimetric radiation.

The four UV lamps were distributed in a radial pattern, in half a circle with a radius on 40 cm, the centre being the place where the lithic sample is fixed. Two of the lamps were centrally positioned, while the other two lamps were fixed on the sides, all at an angle of 36° reporting to the planar surface of the sample.

4. Results and discussions

Concerning the colorimetric evolution, the colour change was measured for each coordinate (L^*, a^* and b^*), as compared to its initially value, on the same sample and in the same point, whereas the total change of colour (ΔE^*_{ab}) was calculated in accordance with the following equation [11-13]:

$$\Delta E^*_{ab} = \sqrt{(\Delta L^*)^2 + (\Delta a^*)^2 + (\Delta b^*)^2}\ ,$$

where L^* represent the lightness /darkness coordinate and varying from 0 (black) to 100 (white), a^* is the red/green coordinate ($+a^*$ indicating red and $-a^*$ green), and b^* means the yellow/blue coordinate ($+b^*$ indicating yellow and $-b^*$ blue).

The results of the colorimetric measurements carried out at 24h after the chemical treatments are presented in table 2. In the case of measurements carried out simultaneously with the UV radiation exposure the colorimetric variations for control and treated areas are presented in table 3 and figure 5.

Also, in table 2 and table 3 are presented the humidity from investigated areas on all experimental procedure`s phase.

Table 2. Total colour change after chemical treatment and the moisture content from the surface of the material (h%).

Investigated area	before treatment	after treatment	
	h%	ΔE^*_{ab}	h%
C0	0.22	-	-
C1	0.18	7.06	0.19
C2	0.17	9.58	0.18

Table 3. Total colour change after UV exposure and the moisture content from the surface of the material (h%).

	area C0		area C1		area C2	
hours	ΔE^*_{ab}	h%	ΔE^*_{ab}	h%	ΔE^*_{ab}	h%
0	0	0.22	0	0.19	0	0.18
2	0.57	0.18	1.05	0.18	1.19	0.17
4	0.68	0.20	1.10	0.19	1.24	0.19
6	0.62	0.18	1.17	0.16	1.27	0.16
8	0.69	0.18	1.30	0.17	1.44	0.16
10	0.75	0.17	1.47	0.16	1.58	0.15
12	0.81	0.18	1.51	0.16	1.65	0.16
14	0.84	0.18	1.57	0.16	1.80	0.16
16	0.86	0.17	1.67	0.16	1.84	0.15
18	0.87	0.18	1.75	0.16	1.91	0.16
20	0.88	0.17	1.81	0.15	1.93	0.16
22	0.92	0.17	1.81	0.15	2.04	0.16
24	0.99	0.17	1.88	0.15	2.17	0.16

It should be mentioned that, throughout all the investigations stage, the water content from each surface of the sample varied insignificantly (table 2 and table 3). In this case, the humidity of surfaces did not influence the chromatic deviation of the areas being investigated. In the case of the two treated areas of studied geomaterial, the chromatic aspect is influenced by the way in which the treatment solutions interact with the porous surfaces of samples [14]. So, it can notice the colour tendency to accentuate more on the surface C2, treated with the product S2 (figure 1). This fact is also evidenced by the data represented in table 2, where it can be clearly seen that treatment with the solution S2 determined notable value of total colour change, in accordance with the specialized literature in this domain [11,12]. The C1 area treated with S1 product, also experienced a colour change, but of a smaller value ($\Delta E^*_{ab} = 7.06$), in comparison with that of C2 area, where $\Delta E^*_{ab} = 9.58$.

The colour tendency to change simultaneously with the exposure at UV for a period of 24h (figure 5) is also greater in the case of area C2 ($\Delta E^*_{ab} = 2.17$), in comparison with the area C1 ($\Delta E^*_{ab} = 1.88$) and the witness area C0 ($\Delta E^*_{ab} = 0.99$).

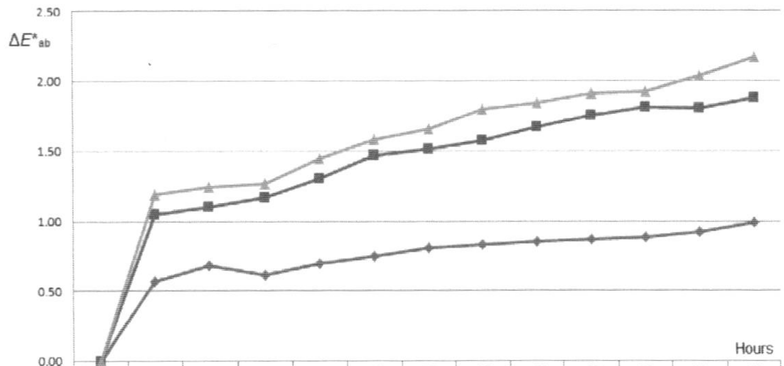

	Hours	2	4	6	8	10	12	14	16	18	20	22	24
C-0	0.00	0.57	0.68	0.62	0.69	0.75	0.81	0.84	0.86	0.87	0.88	0.92	0.99
C1	0.00	1.05	1.10	1.17	1.30	1.47	1.51	1.57	1.67	1.75	1.81	1.81	1.88
C2	0.00	1.19	1.24	1.27	1.44	1.58	1.65	1.80	1.84	1.91	1.93	2.04	2.17

Figure 5. The graphic of ΔE^*_{ab} evolution on the hydrophobised surfaces exposed 24h to the UV radiations.

Concerning the dripping process on optical microscopy, figure 2a shows how the distilled water drop is instantly absorbed in smaller surface, from untreated zone C0, and this small area becomes wet and darker. In figure 2c, we can observe the distilled water drop tendency to flatten on area C2, process observed also in macroscopic mode (figure 1 – area C2). This indicates o lower hydrophobic capacity of the S2 product [13], in comparison with the hydrophobic effect obtained in area C1, treated with S1 (figure 1 and figure 2b).

The differences between the two water drops shapes that come in contact with the treated surfaces were confirmed by the evaporation times. The distilled water drop applied on C2 has a faster evaporation time (135 minutes) due to its flatter surface, while the distilled water drop applied on C1 has a more spherical shape, which result in a longer evaporation time (170 minutes). By correlating the way the distilled water drop flattens on the treated area C2 (figure 1 and figure 2c) with the higher values of chromatic modifications (ΔE^*_{ab}) after the chemical treatment and after the UV exposure, we can certainly state that the solution S2, while is eco-friendly, has some disadvantages in comparison which the effects and results obtained by using the S1 solution (obtained by mixing organic compounds that pose a threat to the environment).

In order to evaluate the UV exposure efficiency, the Osram Ultra Vitalux® lamps wavelengths were determined and compared to those of the solar spectrum using Princeton Instruments®. Figure 6 shows solar wavelengths measured at noon, around the laboratory, which is located on this coordinate: 47°1736`N latitude and 27°5713`E longitude.

Also, figure 6 present the UV lamps spectrum which allowed us to calculate the intensity rapport between the two sources and for each wavelength. Thus, in figure 7 it is shown the ratio between both spectrum and it can observed the UV lamps radiation have an intensity of 250 to 2700 higher than solar light radiation, for the spectral domain of 290 to 380nm. By analyzing the chromatic modification values (C1 - $\Delta E^*_{ab} = 1.88$ and C2 - $\Delta E^*_{ab} = 2.17$), obtained after 24h exposure to UV lamps radiations (with approximately $30x10^3$ lux - intensity light) we can determine the fact that the two solutions, S1 and S2, have a certain UV resistance.

The ΔE^*_{ab} value (0.99) of the witness area C0, exposed to the UV radiations in the same conditions for 24h, is lower than the ΔE^*_{ab} values for the other two areas, already mentioned. When comparing the same ΔE^*_{ab} value for area C0 with the ΔE^*_{ab} value of 2.94, registered on another witness area from another old calcareous rock exposed to UV radiations for 24h [15], we can state that the artificial aging rate can vary accordingly with the age of the sample rock, with the time period between sampling and testing. By analyzing this aspect, we can deduce that a new lithic material, recently extracted, could have a reduced colour change at UV radiation exposure in laboratory conditions, than a similar geomaterial, that has already been aged through long time exposure in a natural microclimate, before being sampled and investigated in the same laboratory conditions.

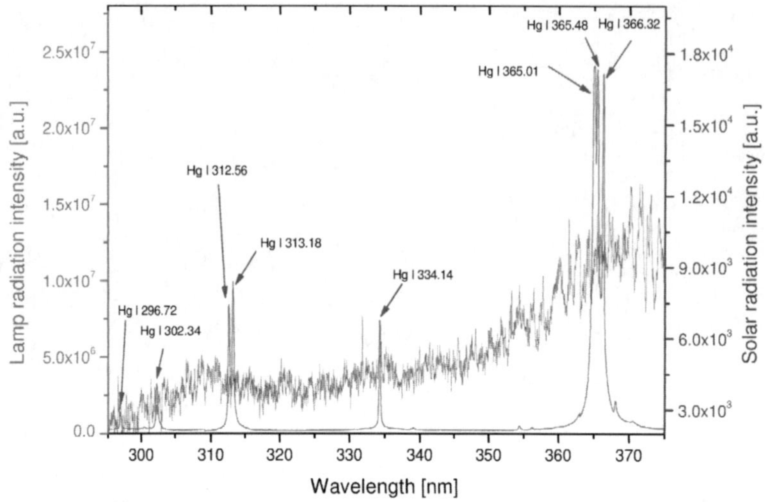

Figure 6. Lamps and solar radiation intensity.

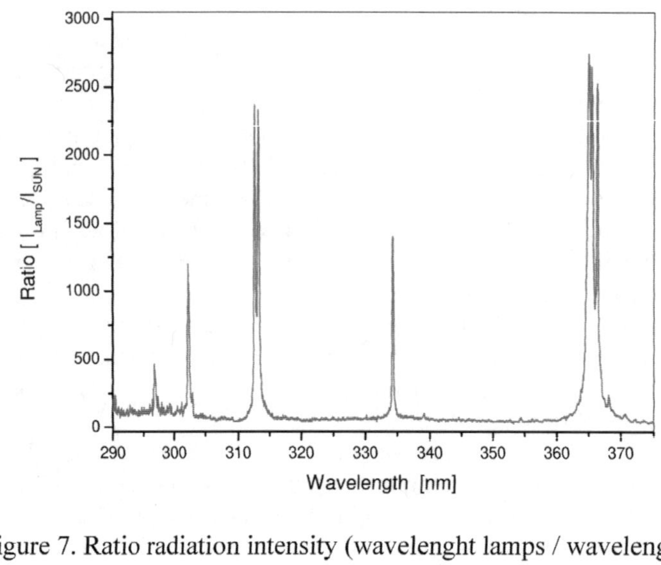

Figure 7. Ratio radiation intensity (wavelenght lamps / wavelenght Sun).

5. Conclusions

Due to the porous nature of the calcareous rocks from Păun-Repedea area, Iaşi County, any surface hydrophobization action through pellicles is both necessary and welcomed, especially when this geomaterial was used in building and/or restoration processes of constructions that have historical and cultural value. The investigations have confirmed the hydrophobic capacity of the two selected products, S1 and S2, with the notice that although they have a certain resistance to UV radiation, the samples treated changed colour in a significant manner, especially the LTP Mattstone H20® product, aspect also confirmed by the technical sheet provided by the manufacturer.

Taking into consideration the principles of scientific conservation of stone historical monuments, which require a minimal colour change during any kind of intervention [12,16], the use of S2 is not recommend in the case of constructions with historical value.

For any other maintenance process of porous calcareous rocks like those from Păun-Repedea area, the final decision, on which product to use from the two presented in the paper, belongs directly to the beneficiary and/or the involved specialists.

Acknowledgment
We would like to thank the company Rock Star Construct Ltd – Bucharest for technical support. Also, we would like to thank assoc. prof. Dragoş Cristian Achiţei and assoc. prof. Nicanor Cimpoeşu (*Gheorghe Asachi* Tehnical University - Iaşi) for samples preparation.

References
[1] P.G. Fookes, 1991 *Q J Eng Geol*, 24, Issue 1, pp. 3–15.
[2] I. Sandu, M. Brânzilă and I.G. Sandu, 2009 *The Scientific Conservation of Stone Monuments* (*Conservarea ştiinţifică a monumentelor de piatră*) "Al. I. Cuza" University Publishing Iaşi pp. 125-131.
[3] P. Brimblecombe, 2011 Environment and Architectural Stone – chapter in *Stone in Architecture* (editors: Siegesmund S. and Snethlage R.) Springer-Verlag Heidelberg Berlin pp. 317-345.
[4] D. Grossi, E. A.Del Lama, J. Garcia-Talegon, A. C.Iñigo and S. Vicente-Tavera, 2015 *Int J Conserv Sci*, 6 Issue 3, pp. 313-322.
[5] M. M. Cazacu, A. Timofte, F. Unga, B. Albina and S. Gurlui, 2015 *J Quant Spectrosc Ra*, 153, Special Issue SI, pp. 57–64.
[6] M. M. Cazacu, A. Timofte, I. Balin, D. G.Dimitriu and S. Gurlui, 2011 *Environ Eng Manag J*, 10, Issue 1, pp. 139–145.
[7] D. Camuffo, 2014 *Microclimate for Cultural Heritage. Conservation, Restoration and Maintenance of Indoor and Outdoor Monuments* (Second Edition) Elsevier San Diego CA pp. 158-163.
[8] C. Grasu, M. Brânzilă, G. C. Miclăuş and I. Boboş, 2002 *Sarmatian System of Foreland Basin of the Eastern Carpathians* (Sarmaţianul din sistemul bazinelor de foreland ale Carpaţilor Orientali) Technical Publishing Bucharest, pp. 287-291.
[9] B. Simionescu, M. Olaru, M. Aflori and F. Doroftei, 2011 *High Perform Polym,* 23, Issue 4, pp. 326-334.
[10] I. Kalmár, 1991 *New Data on Sedimentology and Geochemistry for Repedea – Iaşi Oolites* (Date noi privind sedimentologia si geochimia oolitului de Repedea-Iaşi) *"Grigore Cobălcescu" Geological Seminar* 3 "Al. I. Cuza" University Publishing IaşI, pp. 36-58.
[11] J. Schanda, 2007 *Colorimetry. Understanding the CIE System* Wiley-Interscience Hoboken & New Jersey, pp. 86-90.
[12] C. M. Belfiore, M. F. La Russa, A. Pezzino, E. Campani and A. Casoli, 2010 *Appl Phys A-Mater,* 100, Issue 3, pp. 835-844.
[13] F. M. Helmi and Y. K. Hefni, 2016 *Int J Conserv Sci*, 7, Issue 1, pp. 29-40.
[14] L. Luvidi, A. M. Mecchi, M. Ferretti and G. Sidoti, 2016 *Int J Conserv Sci*, 7, Special Issue 1, pp. 311-322.
[15] V. Pelin, I. Sandu, S. Gurlui, M. Brânzilă, V. Vasilache, E. Borş and I.G. Sandu, 2016 *Color Res Appl*, 41, Issue 3, Special Issue SI, pp. 317-320.
[16] Z. Youssef and F. Kharrat, 2015 *Int J Conserv Sci*, 6, Issue 4, pp. 587-600.

IOP Conf. Series: Materials Science and Engineering **133** (2016) 012062 doi:10.1088/1757-899X/133/1/012062

A concept design of three rudders-shaped like body in columns for low-drag USV

M N Azzeri[1,3], F A Adnan[1,2*], M Adi[1,2], M Z Md Zain[1]

[1] Faculty of Mechanical Engineering, Universiti Teknologi Malaysia, 81310 UTM Skudai, Johor, Malaysia.
[2] Marine Technology Center, Transportation Research Alliance, Universiti Teknologi Malaysia, 81310 UTM Johor Bahru, Johor, Malaysia.
[3] Faculty of Science and Defence Technology, Universiti Pertahanan Nasional Malaysia, Kem Sg. Besi, 57000, Kuala Lumpur, Malaysia.

*faizul@utm.my

Abstract. This paper presented a new design for the unmanned surface vessel (USV) platform with a self-manoeuvring system which is capable of collecting the same data as a hydrography boat. This platform was designed with three hulls that were placed in triangle position. The hulls designed were in the form of rudders-shape and were vertically placed as a slender body shape using NACA 64-0012 profile. This provides the USV with low-drag characteristic. The application of stability and resistance theories investigated the effect of the configuration position of the three hulls for this platform. The results revealed that a larger configuration distance between the three hulls will lead to a reduction in resistance and the platform will be in highly stable condition. The relationships derived from these findings should produce a stable and low-drag platform to accomplish the design concept of three rudders-shaped like body in columns for low-drag USV. This concept may help us to accomplish the design requirements that are related to low-drag and minimum power operation.

1. Introduction

In view of ocean measurement activities such as recording wave data, meteorological data, current data, sea surface data, and other oceanographic measurements, the measurements are usually done using hydrographic survey vessel which can only perform in a short period of time and it is costly. A longer period of measurement is required for certain fields such as oil and gas and marine renewable energy industries [1].

Therefore, a new design of a platform with a self-manoeuvring system and capable of collecting the same data as a hydrography boat is needed. This new ocean platform can be performed with minimal human interaction, and without any danger to users.

A previous project from [2] had tried to improve the design of a stationary buoy as a data collection platform to be an ocean measurement platform as a hydrographic survey platform. Previously, a stationary buoy is designed with a cylinder-shaped buoy and sensors are placed for specific measurement in the ocean such as recording wave data. The cylinder-shaped part at the bottom of the buoy which is immersed in water have a high drag characteristic and it will drift according to the current and waves. The primary objectives of the previous research were to design and construct a stationary buoy that will keep its position at one reference point while making specific measurements on the ocean surface.

Innovation has been made to design the platform for sea floor measurement using a multibeam sonar. The platform needs to move with minimum movement speed usually at 2-3 knots in a straight line from one point to another in order to be measured. In addition, a wave motion monitoring device was also created to measure and record wave data in certain sea areas.

This platform was designed in this shape in order to reduce the drag effect. The initial design was a large cylinder column and it was changed to three small cylinder size such as shown in Figure 1 (a) and Figure 1 (b) and it can hold the weight of the original platform on the top. As shown in Figure 1.1 (c) and Figure 1.1 (d), the design of the column was modified to a slender body shape, after it was analysed using computational fluid dynamics (CFD) in order to reduce the cross-sectional area and to reduce the drag effect as much as possible. This column will always work facing the current weather in motion or in a stationary position by changing the column angle. A rudder-shaped buoy was designed to be the steering system and an electromechanical drive train could be developed so minimum electrical power will be used. The rudder-shaped shall present the least amount of drift in any direction when it is affected by wave, wind, and currents. Therefore, it requires the least amount of power input for heading corrections.

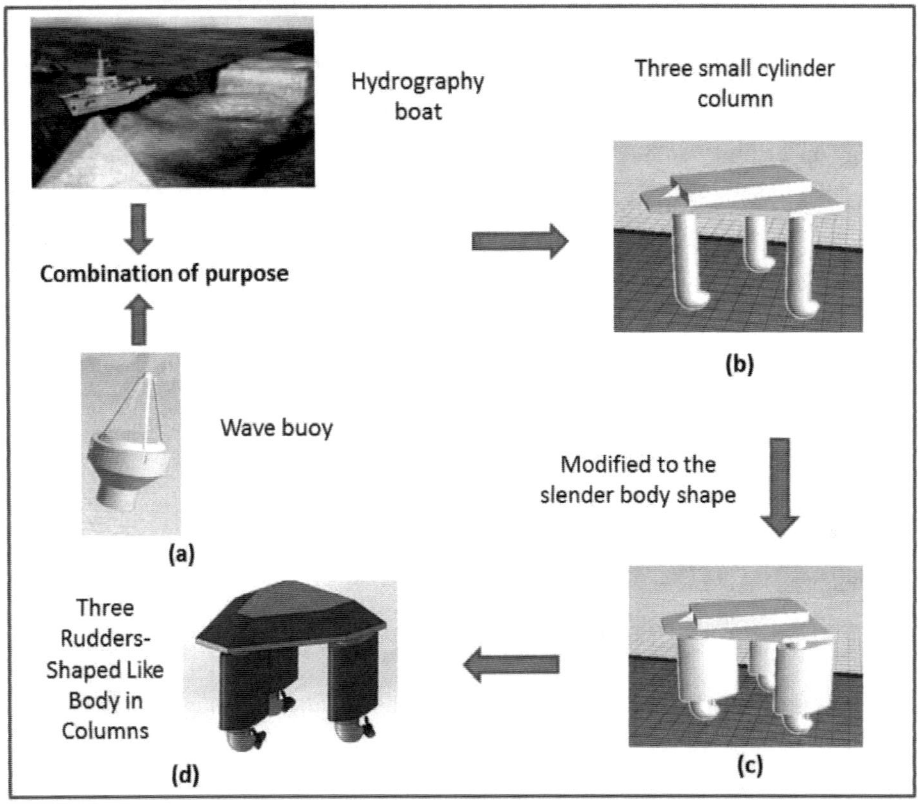

Figure 1. The concept of rudders-shaped design. (a) Combination of the purpose of hydrography boat and stationary wave buoy; (b) the original design of three large cylinder column; and (c) and (d) are the modified three columns to a rudders-shaped design.

A platform that is floating on the water should also be stable when operating at the sea. Nowadays, multihull vessels are stable platforms and it produces less friction when it moves on the water surface. For a ship that has three hulls, there are several advantages which are good stability, resistance, and seakeeping performance compared to monohull that has the same displacement. The investigations on the resistance of three hulls have proven that such hulls have lower resistance at high speed compared to catamarans (two-hull) and monohull of similar displacement [3]. Other advantages of three hulls are more deck space, better in stability, and better passengers' comfort. This study is to investigate the floating platforms' stability and resistance on the effect configuration distance between the three hulls to lead the platform in highly stable condition and reduction in resistance.

2. Design methodology

In principle, the design was made to convert into three slender bodies similar to the studies of monohull vessel which was changed to multihull vessel in increasing the speed of the ship with a reduction in the power required. This type of hull design needs to increase the slenderness ratio of the hull vessel. According to Dubrovsky (2008) [4], the term tricore refers to an identical three-hull vessel of the same shape and the term trimaran is associated with two smaller side hulls called outriggers and with a larger central main hull.

Having three separate hulls creates a higher total wetted surface area compared to similar monohull or catamaran (two-hull). A higher wetted surface area creates a relatively high resistance at low speeds thus increasing the frictional resistance. The use of a slender body create a low impact on the wave resistance at high speed. This is based on the assumption that stated that as a vessel becomes finer, the wave-making resistance decreases [3]. Wave-making resistance is also affected by the interference of separate hull wakes. The main characteristic of an identical three-hull vessel is the strong interaction between the wave systems generated by each hull. The side hulls will result in wake interfaces and the optimum placement of the side hull will reduce this resistance. By using optimum placement of the side hulls in combination of a slender hull form can result in much lower resistance at high speeds compared to both catamaran and monohull designs [3].

For this design, the NACA 64-0012 profile was chosen for these three-hull shape design. This NACA series' profile produces a high lift coefficient effectively and economical low-drag coefficient for specific feature of hull hydrodynamics [5]. The mutual interference of the hulls and the wave-making component in the total resistance of the three hulls using NACA series' profile will improve the specific feature of multihull hydrodynamics.

The non-dimensional coefficients were used to calculate the ship resistance as follows [6]:

$$C_T = C_F + C_R \tag{1}$$

From International Towing Tank Conference (ITTC) 57, the frictional resistance coefficient of trimaran was calculated using the equation below:

$$C_{FT} = C_{FMH}\left(\frac{S_{MH}}{S_T}\right) + C_{Fout}\left(\frac{2S_{out}}{S_T}\right) \tag{2}$$

Where C_{FMH} and C_{Fout} are the frictional resistance coefficients while S_{MH}, S_{out} and S_T are the wetted surfaces. The small sign of *MH* is for main hull, *out* is for side hull and *T* is the trimaran respectively.

The interference factor was defined:

$$I_F = (C_R - C_{RNI})/C_{RNI} \tag{3}$$

The "no-interference" component of the residuary resistance is C_{RNI} and is defined as:

$$C_{RNI} = C_{RMH}\left(\frac{S_{MH}}{S_T}\right) + C_{Rout}\left(\frac{2S_{out}}{S_T}\right) \tag{4}$$

Where C_{RNI} and C_{RT} are referred to

$$\left(\frac{2\nabla^{out}}{\nabla^T}\right) = \left(\frac{2\nabla^{out}}{\nabla^T}\right)_{SC} \tag{5}$$

$$\left(\frac{2\nabla^{out}}{\nabla^T}\right) = \left(\frac{2\nabla^{out}}{\nabla^T}\right)_{RC} \tag{6}$$

The application of stability theory is needed in order to investigate the effect of configuration position of the three hulls on this platform stability. Based on the study of the stability of the vessel, a *GZ* curve is usually drawn up various angles to see the values of righting lever *GZ*. In order to determine the vessel's stability at the respective angle, the righting lever must be coupled with a corresponding angle of heel [7]. The body is said to be in stable equilibrium if this couple will return the body to its original position. *GZ* may be expressed by the following equation:

$$GZ = GM \sin\varphi \tag{7}$$

GM is the metacentric height and φ is the angle of heel. For a given position of *G*, as *M* can be fixed for small inclinations, *GM* will be constant for any particular waterline. More importantly, G can be varied with the loading of the ship even for a given displacement.

The transverse metacentric height, GM_t can be calculated using the equation:

$$GM_t = KM_t - KG \tag{8}$$

The distance between the keel and metacentre both in the transverse or longitudinal direction, can be expressed as:

$$KM = KB + BM \tag{9}$$

While BM_t can be expressed as:

$$BM_t = \frac{I_x}{\nabla} \tag{10}$$

∇ is the total volume of displacement and I_x is the second moment of area or the moment inertia of a waterplane about its centreline.

And KB can be calculated using the equation:

$$KB = \frac{\int A_t y\,dz}{\int A_t dz} \tag{11}$$

From the definition of the centre of buoyancy, y is used to denote the half ordinates and A_t is the transverse section of the plane.

3. Basic design

This research designed a new concept of USV where three hulls are attached to the platform. This design is an improvement from a previous project by Adi et al. [2] in which to improve the design of the stationary buoy as a data collection platform to be an ocean measurement platform as a hydrographic survey platform. This platform was designed to have three hulls which are placed in triangle position. Each hull or the so-called rudders-shape like is vertically placed as a slender shape and it functions as a fin.

By using NACA 64-0012 profile, the slender shape provides the USV with low-drag characteristic which will need minimum power requirements in manoeuvring and station-keeping. Each hull will put aside the azipod thruster system at the bottom of the hull. Each hull is rotatable with the azipod attached. This means the azipods function as USV propulsion and steering system. The prototype design of this new USV is shown in Figure 2. The principal dimensions prototype of the USV model is shown in Table1.

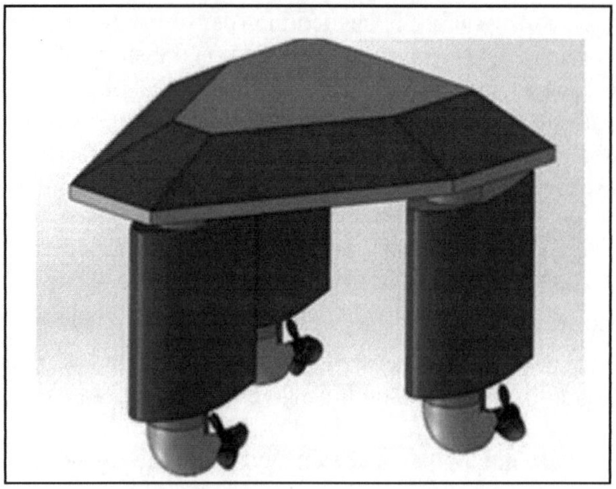

Figure 2. The prototype design of USV

Table 1. Principal particulars of the model

Platform	Length (L)	1.5 m
	Breadth (b)	1.5 m
	Draft (d_{em})	0.8 m
	Displacement (w)	80 kg
Rudder Shaped Like Body	Span (dm)	0.6 m
	Chord (L)	0.4 m
	Rudder area (A_R)	0.24 m^2
	Aspect ratio (Λ)	1.5
	Projectile Area (A_x)	0.12 m^2
Propeller	Number of blades	3
	Propeller diameter (Dp)	0.15 m

4. Analysis approach

The drag and the stability of the platform need to be defined in order to solve the problems due to the configuration position of the three hulls. The drag analysis was conducted by experimental methods in which towing tank was used for resistance test and MAXSURF software (by simulation method) was used for stability analysis.

The low drag USV Model tested at Southampton Solent University (SSU) had used the towing tank for the resistance test. The model was designed by scaling down half of the prototype size. The design only consisted of three fibreglass moulded hulls and one plywood as a platform and also four selected distance points between the three hulls which are placed in a triangle position, (the model is shown in Figure 3). The size of the model was 0.7 m long, 0.85 m wide, and 0.4 m depth with a total weight of 22 kg. The test was conducted in calm water. The model then was towed by a towing carriage at six speeds and four configuration locations of hulls. The four configuration distance locations of the hull are shown in Figure 4. In order to guide the design, manufacture, and testing of the models, ITTC [8] recommended procedures were followed as closely as possible, to improve the test results.

Figure 3. Model design of the USV

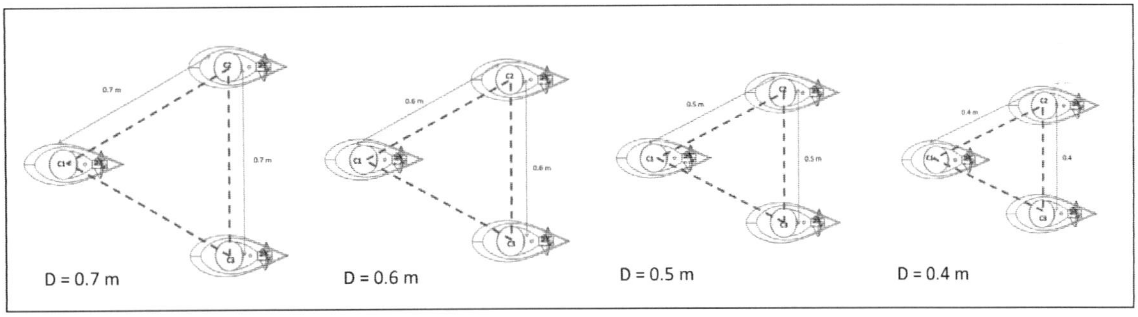

Figure 4. Four configuration location of the hull for model design of USV

5. Results and discussion

The analysis was carried out through numerous resistance test and it identified the specific findings for the configuration distance of three hulls. By referring to Figure 5, it can be stated that the resistance increases when the speed increases for all configurations of three hulls. Only three distance configurations of the test results were recorded. The smaller configuration distance of 0.4 m was in unstable condition when the test model was performed. This result can be concluded that a larger configuration distance between three hulls will lead to a reduction in resistance for identical hulls that have the same speed. A larger distance placement of the hulls will result in wake interfaces, thus reduces the resistance [3]. The combination of a slender hull using the NACA profile form and the optimal placement of hulls can result in much lower resistance for this new platform. The relationships derived from these findings should serve as a stable platform for more advanced research of multihull vessels.

The stability simulations of the three hulls for various combinations of distances were considered. The simulations used the MAXSURF software version 19 and this software has been widely used in the analysis and design of the ship [7].The analysis related to stability in this investigation focused on the effects of varying the hulls' positions in the transverse stability. Static and dynamic stability curves can be calculated using the same methods and conventional monohull [9]. The shape of the stability curve depends on the hull lines.

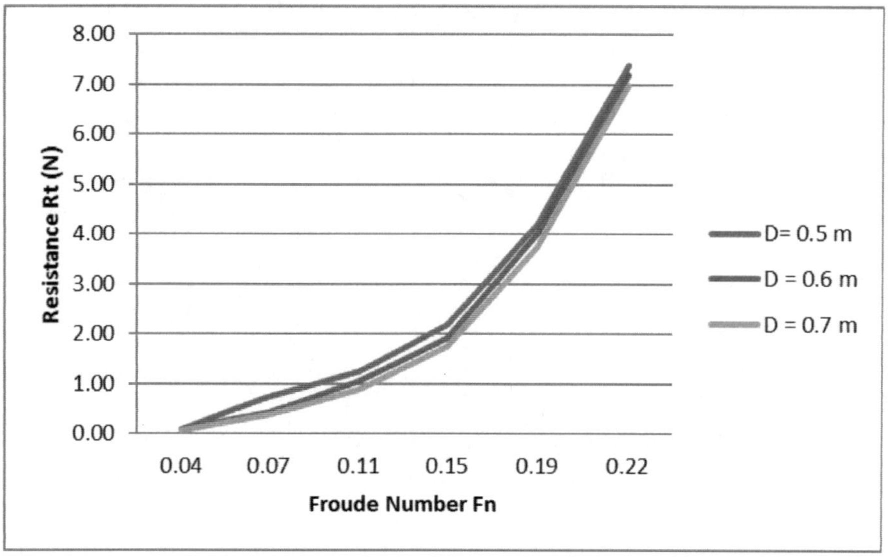

Figure 5. The resistance against Froude number for three configuration locations of hull for model design of USV

Figure 6 shows a systematic increment of the righting lever, GZ as the configuration ratio increases. This result also concludes that a smaller configuration of hull distance, the GZ values become smaller and for distance value D = 0.4 m, GZ value is negative as seen unstable platform. The three-hull vessel's GZ curves are shown to be well S-shaped with a short linear portion and a very wide range of positive stability up to 60° when the configuration distance becomes larger.

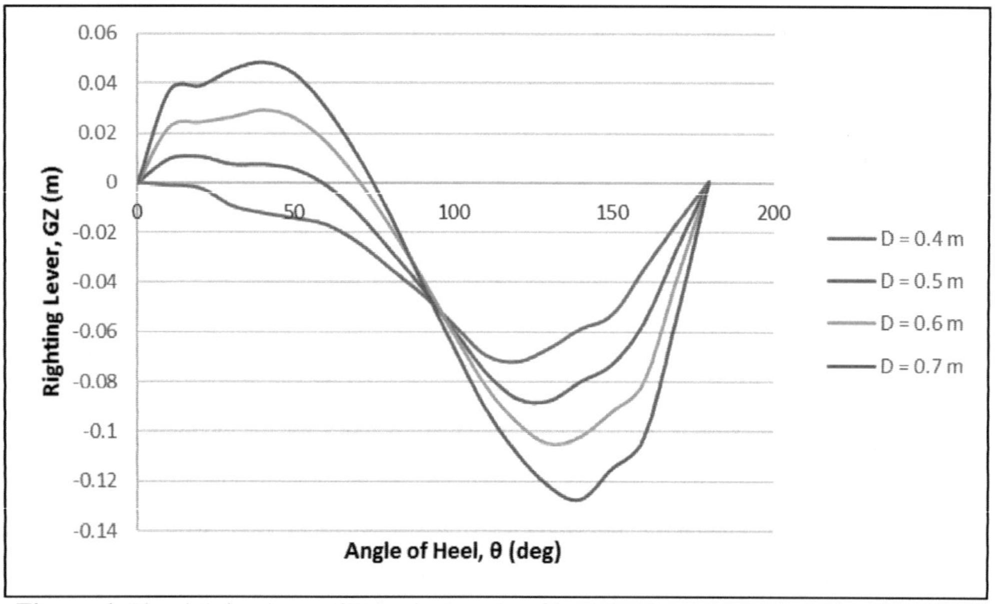

Figure 6. The righting lever, GZ against angle of heel for four configuration locations of hull for the model design of USV

The criterion is the maximum GZ value of a multihull should occur at an angle of at least 15° to satisfy this criterion in accordance with the International Maritime Organization (IMO) Code on Intact Stability [10]. The optimization of the configuration of the three hulls is important for multihulls from the transverse stability, it must directly comply with standardized code such as the IMO Code on intact stability for all types of ships.

6. Conclusion

The analysis to identify the specific differences of the optimization of configuration distance of the three hulls to be a stable and low resistance platform. The relationships derived from these findings should act as a stable and low-drag platform to accomplish the design concept of three rudders-shaped like body in columns for low-drag USV. This concept may help us to accomplish the design required which is related to low-drag and minimum power operation. A control system needs to be developed and installed in order to make it self-manoeuvring using specific controllers and sensors. An autonomous system needs to be developed using a specific programme as the core to the self-manoeuvring vehicle.

Acknowledgements

The authors wish to acknowledge the reviewer(s) for their helpful advice and comments. The authors also wish to thank the Research Management Center, UTM, for their financial support of this study through UTM Research University Grant (RUG) number Vot 01G76.

References

[1] M. N. Azzeri, F. A. Adnan, and M. Z. Md Zain, 2015, Review of Course Keeping Control System for Unmanned Surface Vehicle, Jurnal Teknologi, vol. 74, pp. 11-20,.

[2] M. Adi, O. Yaakob, M. A. Kamal, and C. W. Ng, 2006, Evaluation of Seakeeping Analysis of a Fishing Vessel using a Wave Buoy and Onboard Motions Monitoring Device, Proceedings the 1st International Conference on Natural Resources Engineering and Technology INRET.

[3] P. K. Sahoo, T. Mynard, J. Mikkelsen, and D. McGreer, 2008, Numerical and Experimental Study of Wave Resistance for Trimaran Hull Forms, 6th International Conference on High-Performance Marine Vehicles, Naples, Italy.

[4] V. Dubrovsky, 2008, Three Identical Hulls Instead of Two Ones: Ultimate Options of Longitudinal Shift, Ships and Offshore Structures vol. 3, pp. 145-148.

[5] J. Liu, F. Quadvlieg, and R. G. Hekkenberg, 2015, Impacts of Rudder Profiles on Ship Manoeuvrability, International Conference on Marine Simulation and Ship Manoeuverability (MARSIM '15), Newcastle University.

[6] C. Francesco, M. Amedeo, M. Salvatore, and P. Claudio, 2002, Experimental and Numerical Study on the Efficiency of Trimaran Configuration for High-Speed Very Large Ship, 6th Symposium on High Speed Marine Vehicles (HSMV), Naples, Italy.

[7] M. D. Claire and M. G. Andrea, 2012, Stability Analysis of a Tricore, International Journal of Mechanical, Aerospace, Industrial, Mechatronic and Manufacturing Engineering, vol. 6, pp. 597-606.

[8] ITTC, 2008, Final report and recommendations to the 25th ITTC, Proceedings of 25th international towing tank conference, Fukuoka, Japan.

[9] V. Dubrovsky and A. Lyakhovitsky, 2001, Mullti Hull Ships, Backbone Publishing Company, USA.

[10] IMO, 1993, Code on Intact Stability for All Types of Ships Covered by IMO Instruments, Resolution A749(18), London, UK.

IOP Conf. Series: Materials Science and Engineering **133** (2016) 012063 doi:10.1088/1757-899X/133/1/012063

Methodology for appreciation the manufacturing castings from perspective of circular economy

V F Soporan[1], M Crișan[1], T Lehene[1], A L Pop[1*]

[1] Tehnical University of Cluj Napoca, Bd. Muncii 103-105, Cluj-Napoca, Romania

E-mail: vfsoporan@gmail.com

Abstract. This paper presents a methodology for studying the trends toward developing circular economy, comprising the following steps: description the field analyzed statistically and technological, establishing assessment criteria, establishing the event which will be analyzed and its description and evaluation of the event from the perspective of circular economy. The case study was designed to the manufacture castings through the analysis, after the methodology proposed, by the work of the 71-th World Foundry Congress in 2014 from Bilbao. The conclusion of this approach shows us that the manufacture castings through the trends expressed fall in the principles of circular economy. At the level of resources needed for the manufacturing process of castings - specifically immaterial resources, materials and energy, are analyzed at all levels, specific phases, namely: 1. Information analysis specific for the field; 2. Analysis preparation in terms of manufacturing technology and technological infrastructure available; 3. Analysis of manufacturing castings; 4. Analysis of the results of the manufacturing process; 5. Analysis of exploitation and disposal at end of life cycle; 6. Analysis of other phases that contribute to achieving castings; 7. Analysis of the entire process in terms of respecting the environment and sustainable development. The analysis method on a field belonging to the specific requirements of circular economy is made by analyzing a global event. This comprises the following stages: Establish specific criteria for the analyzed domain in relation to the circular economy; Specifying event works falling within the criteria set, which will be analyzed, and justification for selection.

1. Considerations and motivations of thematic and methodological approach

Circular economy is considered, in the scientific community, political, economic and administrative, as a model of industrial reconstruction that its principles are taken at our sustainability management systems, in order to ensure operation in the economic and social constraints by setting circuit materials and energy in a technological process. At the same time, variant circular pattern appears as a lever for industrial competitiveness, involving, inter alia, the regulatory capacity for dialogue, communication and development of innovation processes. Despite technological improvements that contribute to a better use of material resources and energy, their consumption and waste generation increased steadily worldwide. Therefore, the circular economy can not be understood outside of this context and is a response to the phenomenon of depletion, wanting, while being an alternative model of development, combining economic growth with a vision needed on term sustainability long. In this way, linear development model is replaced with circular pattern.

IOP Conf. Series: Materials Science and Engineering **133** (2016) 012063 doi:10.1088/1757-899X/133/1/012063

Selecting this topic, in the industrial production is justified by the following generalized observations:

- Circular economy is seen as a lever for economic growth by alloying with the ecological and economic ones;
- Circular economy, acting within the perimeter of limited resources, contribute, through specific methods upper and transform them effectively;
- Regulatory framework addresses the circular economy society actors for their success in economic and social activities;
- Quantitative targets specific circular economy, varies according to the targets for recycling and reducing waste disposal;
- Regulatory framework should include the use of material flows macro indicator;
- The existence of public policy instruments that support circular economy, in particular those reported production, purchase and consumption.

Considering the specific manufacture castings to be a major consumer of energy and material resources, it is estimated that the principles of circular economy are required ensuring the development of production in the current cast.

The work presents a method to evaluate the trends towards a circular economy to manufacture castings. In a brief speech, the proposed method has the following steps: description field studied in statistical and technological establish evaluation criteria, setting event which will be analyzed in terms of economy and evaluation of the event from the perspective of circular economy, along with the summary findings of the study on membership trends of circular economy. In the case study analysis that the prospects for production of castings in terms of circular economy, especially global event is the last World Conference of Foundry- 2014 by which the assessment is proposed.

2. Description of castings in statistical production and technology

To justify the importance of addressing the production of castings from the perspective of circular economy, are presented in the following, some data on production and consumption dynamics justifying its important material and energy globally. The global production of castings, as one of the 47 - Fourth Report of world production ("47th Census of World Casting Production" - Modern Casting December 2013 - table 1) was in the year 2012 100 834 681 tons, of which 72,438,512 tons of iron castings (weighting 71.83%), 11,299,044 tons of steel castings (11.2% share) and 17,096,125 tons of castings in ferrous alloys (weighting 16.95%).

The analysis phase of the manufacturing process of castings from the point of view of economy circulation is presented in table 1, where they are divided into stages and the stages of the work.

The measures taken are presented in terms of circular economy to the quality in terms of energy and material consumption and minimize negative impacts on the environment and health.

Table 1. The analysis phase of the manufacturing process of castings from the point of view of circular economy

Nr.	Phase process	Phase step process	Phase activities
1	1. Analysis of domain-specific information;;	1.1. market analysis	1.1.1. global macroeconomic stability analysis
			1.1.2. analyze developments in raw material costs
			1.1.3. analysis of the evolution of energy costs
			1.1.4.analyze developments beneficiaries of casting pieces
		1.2. analysis of regulations affecting the production of castings	1.2.1. analysis of raw material cunsumption regalations
			1.2.2.analysis of energy resources management development
			1.2.3. analysis of issues environmental regulations(management of polluants, water, soil waste,etc)

			1.2.4. analysis of employment regulations and human resource management
			1.2.5. specific spatial regulation analysis and land development
			1.2.6. analysis of fiscal regulations and eco-taxis
			1.2.7. analysis of circuits specific financial regulations
			1.2.8. analysis of financial instruments regulations and policy enforcement
		1.3.analysis of development processes and technological infrastructure	1.3.1. making processes analysis
			1.3.2. Analysis of realization of mould processes
			1.3.3. analysis of the casting process
			1.3.4. analyze the process of cleaning and finishing
			1.3.5. heat treatment process analysis
			1.3.6. process analysis and surface coating paint castings
			1.3.7. analysis and processing activities edging castings
			1.3.8. analysis of the preparation for shipment and delivery
2	Manufacturing preparation .Analysis of point	2.1. constructive concept analysis	2.1.1. constructive database management functionalities required in conjunction with historical typology and technology of casting alloys
			2.1.2. management of traditional instruments establishing constructive forms
			2.1.3. management tools for computer aided design
			2.1.4. global information management on constructive potential coverage of functional needs
		2.2. analysis of technological design	2.2.1. general analysis of casting technology depending on the nature of the alloy, such as technology and construction features of the functionalities required (wall thickness, overall dimensions and differences dimesional to be insured)
			2.2.2 power system analysis and optimization
			2.2.3. solidification by correlation analysis and optimization feeder system, the cooling of constructive typologies designed
			2.2.4. analysis of the cooling process and design optimization from the point of view of stresses which are formed in the casting
			2.2.5. optimization of the technological constructivism alloy composition, the casting characteristics of the shape and conditions of casting
		2.3. technological preparation of production analysis	2.3.1. selection, analysis and optimization of process flow steps
			2.3.2. choice or imposition technological infrastructure needed for the process flow steps
			2.3.3. design and implementation models, forms, devices

			and process flow steps
3	Analysis of manufacturing casting pieces	3.1. direct activities	3.1.1. flow forming mixtures
			3.1.2. development alloys flow
			3.1.3. actual flow casting
			3.1.4. flow finishing of casting pieces
			3.1.5.flow of processing and protection of the casting pieces
			3.1.6. flow exhaust and dust extraction activities
			3.1.7. other streams present in the direct production process
		3.2 indirect activities	3.2.1flow transport and handling of materials entering into production
			3.2.2. use of equipment and material flow protection
			3.2.3. necessary to ensure the technological process flow utilities
			3.2.4. maintenance flow assurance activities of technological equipment
			3.2.5.flow current activities and general cleaning technology
			3.2.6.flow activities integrated and sustainable waste management and other similar productive household
			3.2.7. other indirect flows
4	Analysis of the results of the manufacturing process	4.1. step analysis useful part process manufacturing – castings delivered to beneficiaries	4.1.1. useful part is delivered directly
			4.1.2. useful part is rectified before being delivered
		4.2. part cash resulting from the manufacturing process	4.2. . the recycled part
			4.2.2. converted the by-product part
			4.2.3. converted from waste part
5	Analysis of exploitation and disposal at end of life	5.1. reused castings pieces	5.1.1. management components with the potential to be recycled
			5.1.2. developing technologies reconditioning of the casting pieces
			5.1.3. development alloys typing parts and castings to ensure interchangeability and linking overall constructive on this criterion
		5.2.recycling of the used pieces	5.2.1. establishment of collection systems based on flow assurance criteria ensuring recycling
			5.2.2. development of techniques for preparing materials recycling
			5.2.3. market promotion of recycled materials for competitive production of castings
		5.3.waste management	5.3.1. charging heredity material management

			5.3.2. application of the principles of extended producer responsibility in relation of casting pieces
			5.3.3. management of historical industrial dumps and landfills specific manufacturing processes of the casting pieces
6	Analysis of other phases contributing to the cast pieces	6.1. analyze production and distribution of the utilities	6.1.1. analysis of water resources
			6.1.2. analysis of energy resources
			6.1.3. analysis of production and distribution of compressed air
			6.1.4. analysis of production and distribution of other utilities
		6.2. analysis of other phases	6.2.1. analysis of maintenance activities
			6.2.2. analysis of transport activities
			6.2.3. analyze storage and insurance activities logistics
			6.2.4. analysis cleaning activities
			6.2.5. the analysis of other types of activities
7	Analysis of the entire process in terms of respecting the environment and sustainable development	7.1. pollution intensity	7.1.1. air pollution
			7.1.2. water pollution.
			7.1.3. soil pollution
		7.2. quantification of specific consumption per unit of product materials	7.2.1. new iron
			7.2.2. old iron
			7.2.3. recycled material
			7.2.4. ferroalloys
			7.2.5. fluxes
			7.2.6. masonry material
			7.2.7. sand casting
			7.2.8. binder
			7.2.9. timber
			7.2.10. SDV
		7.3. specific energy consumption per unit of product	7.3.1.electrical energy
			7.3.2 .methane gas
			7.3.3. recovered energy sources
			7.3.4. secondary energy sources delivered to other users to be specialized according to the specific technological process
		7.4. other consumption per unit of product	
		7.5. waste generated per unit of product	7.4.1. waste generated at the planning stage
			7.4.2. waste generated during the execution of moulds
			7.4.3. waste generated during casting

			7.4.4. waste generated during cleaning and finishing
			7.4.5. waste generated during heat treatment
			7.4.6. waste generated during dyeing and surface coating castings
			7.4.7. waste generated during the edging and processing
			7.4.8. runoff in other technological phases than those specified
		7.6. consumption of recycled material introduced into the production process of the product	7.6.1. recycled materials in the preparation phase
			7.6.2. recycled material in the stage of moulds
			7.6.3. recycled material in the casting stage
			7.6.4. recycled material during cleaning and finishing
		7.7. consumption of recycled material introduced into the production process of the product	7.7.1. recycled materials in the preparation phase
			7.7.2 recycled material in the stage of moulds
			7.7.3. . recycled material in the casting stage
			7.7.4 recycled material during cleaning and finishing
		7.8. the energy recovered into the process product production	7.8.1. recovered energy development phase
			7.8.2. the energy recovered in the stage of moulds
			7.8.3. the energy recovered in the casting stage
			7.8.4. the energy recovered in the process of cleaning and finishing
			7.8.5. energy recovered in the heat treatment step
			7.8.6. energy recovered during dyeing and surface coating castings
			7.8.7. the energy recovered during the edging and machining
			7.8.8. recovered energy other than those specified technological phases

3. Criteria for evaluation

In establishing criteria for membership on the circular economy to the analysis made in formulating J.C. Levy on the development of circular economy elements: 1. moderate use of resources as effectively as possible recycled; 2. exploitation of resources important to the upper level of their recyclability; 3. Eco-conception and cleaner production; 4. strict adherence to environmental conditions; 5. The recovery of waste and turn them into resources; 6. The generation of waste treatment without particular problems.

The analysis method on a field belonging to the specific requirements of circular economy is in terms of a global event analysis, based on established criteria. This comprises the following steps:

- Criteria for admission criteria specific to analyze the circular economy;
- Specification of works falling criteria established event, which will be supose analysis and justification of selection;
- Determination of the total weight of selected works and works in the overall structure of the event;

4. Establishing the event which will be analyzed and description and evaluation of circular economy perspective

Manufacturers castings are represented globally by the World Foundry (The World Foundry Organization) through national structures in 32 states. Since its establishment until now were held 71 world congresses, the last one being carried out in Bilbao. In 2014. this historically and high representative for the case study of the proposed procedure were used Proceedings of the 71st World Congress Foundry.

At the quantitative level, the number of papers presented in the work "71st World Foundry Congress - 2014 - Bilbao", having regard to the program event was 190, grouping them being performed in a total of 10 sections: 1 . Cast (Cast iron) with a total of 29 works; 2. Steel (Cast steel) with a total of 8 papers; 3. Ferrous and light alloys (non-ferrous and light alloys) with a total of 31 works; 4. Casting technology, equipment, manufacturing tools, robots and automation (Foundry technologies, EQUIPMENT, manufacturing, tools, robotics and automation) with a total of 30 works; 5. Implementation forms, casting cone and rapid prototyping (Moulding, core making and rapid prototyping) with a total of 22 works; 6. Advanced Engineering: Tools for design, calculation and simulation (Advanced engineering: desing, calculation and simulation tools) with a total of 18 works; 7. Energy (Energy) with a total of 5 works; 8. Environment and Sustainable Development (Enviroment and sustainability) with a number of 11 works; 9. Management, Education and Training (management, education and training) with a total of 6 papers; 10. Young Researchers (Young Researchers) with a total of 6 papers; 11. Paper poster (posters) with a total of 13 works. Reporting the number of participating countries work at the congress is made in table 2 and in table 3 shows the distribution of works on sections of the conference and their share in the criteria set.

Table 2. Number of communications relative to countries with Foundry World Congress 2014 – Bilbao

Nr.	Country	Number of works	Share works,%
1.	1 Spain	29	15,34
2.	2 Japan	25	13,22
3.	3 Poland	24	12,69
4.	4 Germany	21	11,11
5.	5 China	14	7,4
6.	6 South Korea	14	7,4
7.	7 SUA	9	4,76
8.	8 India	7	3,7
9.	9 South Africa	5	2,64
10.	10 Norway	3	1,58
11.	11 Turkey	3	1,58
12.	12 Great Britain	3	1,58
13.	13 Belarus	3	
14.	14 France, Romania,	2	1.05x3=3,15
15.	15 Ungary	1	
16	16 countries		0,53x16=8,46
Total	**33**	**189**	**100**

Table 3. Distribution of works on sections of Congress and their share of the total set

Nr.	Section	Number of works	Share works %
1.	S1:Cast iron	26	13,75
2.	Cast steel	8	4,23
3.	Non ferrous and light alloys	31	16,4
4.	Foundry technologies, echipment, manufacturing, tools, robotics and	29	15,34

	automation		
5.	Moulding, core making and rapid prototyping	22	11,64
6.	Advanced engineering: desing, calculation and simulation tools	17	8,99
7.	Energy	5	2,64
8.	Enviroment and sustainability	11	5,82
9.	Management, education and trainig	10	5,29
10.	Young researchers	17	8,99
11.	Posters	13	6,87
	Total works	**189**	**100**

5. Evaluation of the event from the perspective of circular economy

In particular, the expected analysis includes references to the keywords and the theme of the Congress said. Quantifying trends analyzed domain for its framing trend to hire "circular economy" is achieved by superimposing its characteristic elements in the priority areas addressed in the papers presented.

This estimate is based on the work carried out aiming presented in the provisional program of the 71st World Congress of Foundry - Bilbao 2014 - Advanced Sustainable Foundry (1). From this perspective, summarizes some of the characteristics of the congress:

- Keywords (emblem) of the congress: Energy (energy) Environment (environment), market trends and requirements (Market trends and Demands) and innovation (innovation);
- Sections of the Conference presentation to the share of the total work are summarized in table 3. Starting from the analysis of production process of castings made in table 1 is performed, following the methodology presented, reporting and works belonging steps that respond to circular economy. An example of selective analysis is made presented in table 4. In this step goes to the encoded specific measures and selection of articles from the conference corresponding to this analysis. For reasons of space allocated to the presentation of a work within it does not show the entire analysis made at each work.

Table 4. Distribution themes work on specific criteria for addressing circular economy

Nr.	Section	NC	NCAEC	PSCAEC	PCAECTL
1.	S1:Cast iron	26	9	34,61	4,76
2.	Cast steel	8	3	37,51	1,58
3.	Non ferrous and light alloys	31	12	38,69	6,34
4.	Foundry technologies, echipment, manufacturing, tools, robotics and automation	29	21	72,41	11,11
5.	Moulding, core making and rapid prototyping	22	13	59,09	6,87
6.	Advanced engineering: desing, calculation and simulation tools	17	15	88,23	7,93
7.	Energy	5	5	100	2,64
8.	Enviroment and sustainability	11	11	100	5,82
9.	Management, education and trainig	10	4	40	2,11
10.	Young researchers	17	11	64,71	5,82
11.	Posters	13	7	53,81	3,71
	Total Works	**189**	**111**		**58,73**

Note: NC – Number of works; NCAEC – Number of works on sections addressing to the circular economy ; PSCAEC – - Share the communications section addressing circular economy; PCAECTL – Share communications addressing reported total circular economy works.

Motivation membership trends in the manufacture of castings to the principles of circular economy is made by the following reasoning:

1. Keywords are proof castings production trends orientation to principles of circular economy, in that they are related to reducing energy consumption (energy), the environmental problems

(environment), tracking the actions regulated market to ensure sustainable development (market trends and requirements) and promoting innovative solutions in all sections of the life cycle of castings, especially in regulating specific material and energy consumption. From the enumeration is done through keyword notes that promotes symbolic but significant market economy principles.

2. The theme of the congress section gives strong arguments to show that they fall circular economy. Some of them are shown in the following:

- **Advanced Engineering** means the development of construction castings and technologies in a virtual space without consul material and energy;
- **Robotics and automation** of manufacturing processes translates into optimizing processes in the material and energy consumption;
- **Environmental protection and sustainable development** creates determinations and conditioning on the development and management processes, in the sense of not affect the state of the environment and natural resources. This translates into castings produced using fewer raw materials, using materials that generate little impact on the environment and human health compared to traditional materials with equivalent functionality through the installation of eco-efficient logistic solutions by using products and processes more flexible, modular, evolutionary and versatile, easier to maintain and repair, with a greater potential for recovery at end of life;
- **Energy remains** an important issue for users of energy-intensive, especially for manufacturing processes castings pursuing solutions to increase energy efficiency complex.
- **Rapid Prototyping** gives opportunities for achieving specific configurations of making parts with low weight and good mechanical performance.

3. The work addresses a problem analyzed directly or indirectly, which can be characterized as part of actions to promote and develop the circular economy on the whole flow generation, the existence and disposal of castings. Without going into details of each job submission, otherwise it would be difficult to achieve at this approach, the overall assessment of the membership of the circular economy is via the data presented in table 4. As shown therein, the share of work addressing, directly or indirectly, the issue of circular economy is about 59%. Reported number of works presented within the framework of the share ranges from 35% to 100% (the Energy and Environment section) Although assessments are loaded with a certain bias, it is considered that concerns research in the production of castings are geared towards promoting circular economy development activities.

The assessment presented is noted that in the circular economy for the production of castings are covered five of the six elements: moderate use and effective as possible recyclable resources, exploitation of resources important to the upper level of recyclability their strict adherence to conditions environment, waste recovery and turn them into resources and waste treatment without generating special problems.

6. Conclusions

In a statistical presentation of conclusions regarding the justification trends occurring in the manufacture of castings to the principles of circular economy, the analysis works Foundry World Congress - 2014 - Bilbao are:

Two (energy and environment) of the four (energy, environment, market trends and requirements, innovation) words emblem of the congress aimed at environmental protection and sustainable development, bearing in mind that addresses the major issues mentioned novative processes. This demonstrates that sustainable development is a major concern castings production to the new requirements of industrial processes developments. The third theme (market trends and requirements) demonstrates a reality, studies and research that take into account prevailing market trends domain and its new developments. To study and develop research outside global market trends is an effort that will not materialize in the development and maintenance of market positions.

Of the nine sections of paper conference presentation, three aimed at traditional fields (iron casting, casting light and non-ferrous alloys and steel casting), and six new areas which aims to support the issue of manufacture castings (advanced engineering, environmental protection and sustainable development, energy, rapid prototyping, management and training). This trend shows the evolution of manufacturing parts is supported at interdisciplinary many vital areas for future development, which may be contained in what is called emblematic "circular economy", developed below. The analysis of individual conference papers found that more than half, directly or indirectly targeting issues they fit into the theme concerns the development of circular economy.

The proposed methodology in the analysis work is found that a significant share of their themes fall into actions to strengthen circular economy development. This is a trend analysis tool for circular economy, proved to be useful in forecasting which are of particular developments in the field.

The case study was the assessment of trends towards a circular economy of production of castings demonstrates that World Foundry Congress - 2014 are in line with development issues analyzed domain. More than this analysis we believe that it provides a methodology by analyzing a field of circular economy development perspective.

References

[1] 71-st World Foundry Congress, Bilbao, 2014, Technical Programme, online to: www.71stwfc.com>More references.

[2] Economie circulaire: beneficier socio-economiques de l'ecoconception et de l'ecologie industrielle, 2012, *ADEME & VOUS, Strategies & etudes*, vol. 33, pp.17.

[3] J. C. Levy, 2009, L'économie circulaire: l'urgence écologique, Presses de l'École nationale des ponts et chaussées.

[4] I. Sandu, D. Aparaschivei, V. Vasilache, I. G. Sandu, O. Mircea, 2012 *REV. CHIM. (Bucharest)*, 63, No. 5, pp. 495-500.

[5] S. O. Jimoh, P.S.A. Irabor, I.U. Abhulimen, S.O. Amiebenormon, 2013 *International Journal of Scientific & Technology Research*, Vol. 2, Issue 1, pp. 257-261.

[6] K. Das Subodh and Y.Weimin, 2007 *JOM*, pp. 57-63.

[7] D. P. Kanicki, 2000 *Modern Casting*, 12, pp. 24 - 34.

[8] N. MARCU, M. SIMINICA, M. CRISTEA, D. DASCALU, 2016, *REV. CHIM. (Bucharest)*, 67, No.3, pp. 589-594.

[9] W.-Y. Dzan, 2011 *International Journal of Systems Applications, Engineering & Development*, Vol. 5, Issue 4, pp. 529-536.